TABLE OF CONTENTS

GEOLOGIC BASINS I
CLASSIFICATION, MODELING, AND PREDICTIVE STRATIGRAPHY

CLASSIFICATION

MODELING

PREDICTIVE STRATIGRAPHY

TABLE OF CONTENTS FOR

GEOLOGIC BASINS II
EVALUATION, RESOURCE APPRAISAL, AND WORLD OCCURRENCE OF OIL AND GAS

EVALUATION

RESOURCE APPRAISAL

WORLD OCCURRENCE OF OIL AND GAS

GEOLOGIC BASINS I
CLASSIFICATION, MODELING, AND PREDICTIVE STRATIGRAPHY

COMPILED BY
NORMAN H. FOSTER
AND
EDWARD A. BEAUMONT

TREATISE OF PETROLEUM GEOLOGY
REPRINT SERIES, NO. 1

PUBLISHED BY
THE AMERICAN ASSOCIATION OF PETROLEUM GEOLOGISTS
TULSA, OKLAHOMA 74101, U.S.A.

ISBN: 0-89181-400-0

INTRODUCTION

This book is part of a series of reprint volumes that in turn are part of the *Treatise of Petroleum Geology*. The *Treatise of Petroleum Geology* was born during a discussion we had at the Annual AAPG Meeting in San Antonio in 1984. When our discussion ended, we decided to write a comprehensive textbook in petroleum geology, directed not at the student, but at the practicing petroleum geologist. The project to put together one textbook gradually evolved into a series of three different publications: the Reprint Series, the Atlas of Oil and Gas Fields, and the Handbook of Petroleum Geology; collectively these publications are known as the *Treatise of Petroleum Geology*. Together with input from the Advisory Board of the Treatise of Petroleum Geology, we designed this entire effort so that the set of publications will present the state-of-the-art in petroleum exploration knowledge and application. The Reprint Series provides a collection of the most up-to-date previously published literature; the Atlas is a collection of detailed field studies chosen to illustrate the various ways oil and gas have accumulated in the world; and the Handbook is a professional explorationist's guide to the latest knowledge in the various areas of petroleum geology and related fields.

Papers in the Reprint Series volumes are meant to complement the chapters of the Handbook. Reprint papers were selected mainly on the basis of their usefulness today in petroleum exploration and development.

This book and its companion volume together contain six sections: (1) Classification, (2) Modeling, and (3) Predictive Stratigraphy; and (4) Evaluation, (5) Resource Appraisal, and (6) World Occurrence of Oil and Gas. The papers in the first section, Classification, describe a variety of ways to view a basin's architecture. Classifications are always limiting, but they provide a necessary frame of reference from which to compare and contrast the multitude of geologic features that affect the petroleum geology of a basin.

The Modeling section papers present integrated techniques to unravel the geological history of a basin. Such a history allows us to predict the maturation, migration, and accumulation of petroleum. The techniques range from empirical to analytical; all are completely dependent on the quality of geologic data available.

Predictive Stratigraphy comprises a group of papers whose main aim is to narrow the possibilities of the nature of the sedimentary column in a basin, before many wells are drilled. Most of these papers discuss global geologic processes, such as plate tectonics, sea level changes, and climatic cycles, and how they determine the stratigraphic makeup of a basin. Understanding these processes in time and space increases our ability to predict the nature of the stratigraphic column and its petroleum potential ahead of the drill bit.

Papers in the Evaluation section discuss both quantitative and qualitative methods for evaluating the petroleum potential of a basin. Most describe ways to break basins down systematically into the component parts that affect petroleum geology, and ways to analyze each part to determine its contribution to the petroleum potential of a basin. These techniques allow the geologist to assess the volume of petroleum a particular basin might have generated, and to determine where it might be located.

Resource Appraisal papers discuss quantitative methods for assessing the petroleum resources contained the world's sedimentary basins. Most are statistically based.

In the final section, World Occurrence of Oil and Gas, the papers tell us where petroleum is found today, how much is found in these places, and why it is found where it is. Most of these papers use hindsight to speculate further about where we are most likely to find petroleum in the future.

On behalf of the Advisory Board of the Treatise and the officers and members of the AAPG, we are proud to present the first of the Reprint Series volumes. We hope that these will put important information, necessary to petroleum exploration, at the fingertips of the oil finder.

Edward A. Beaumont   Norman H. Foster
   Tulsa, Oklahoma   Denver, Colorado

TREATISE OF PETROLEUM GEOLOGY
ADVISORY BOARD

CLASSIFICATION

REALMS OF SUBSIDENCE

A. W. BALLY[1] AND S. SNELSON[1]

ABSTRACT

In this review, three major basin families are differentiated:

1. *Basins on rigid lithosphere* and not associated with the formation of megasutures. These include rifts, Atlantic-type margins, and cratonic basins.

2. *Perisutural basins* on rigid lithosphere and associated with the formation of megasutures (e.g., compressional zones that encompass all deformational, igneous, and metamorphic products of a large orogenic cycle). Deep-sea trenches are perisutural basins associated with Benioff or B-subduction zones. Foredeeps are perisutural basins that are associated with and adjacent to Ampferer or A-subduction zones (e.g., zones where limited amounts of continental lithosphere are subducted).

3. *Episutural basins* within a megasuture. These include forearc basins, backarc basins, and basins that are related to the superposition of megashear systems on megasutures.

Basins on rigid lithosphere and certain *episutural basins* (such as backarc basins) are frequently initiated by thermally controlled rifting processes. Whether the thermal events lead to lithospheric uplift and attenuation, or whether they are themselves the consequence of lithospheric stretching remains to be evaluated for each basin class. Subsidence in these basins may be further modified by sediment loading, subcrustal flow, and density changes in the lower crust and mantle.

Perisutural basins of the deep-sea trench and foredeep type are probably due to lithospheric bending modified by loading with thrust sheets and folds associated with mountain-building processes, and by filling with sediments derived from the adjacent mountains.

RÉSUMÉ

Dans cet article, on a distingué trois familles de bassins importants:

1. *les bassins placés sur la lithosphère rigide* et sans rapport avec formation de mégasutures; cela inclut ''les rifts,'' les marges, type atlantiques, et les cratoniques;

2. *les bassins de perisuture* placés sur la lithosphère rigide et liés à la formation de mégasutures (les zones de compression, ce qui inclut tous les produits de déformation, ignés et métamorphiques d'un grand cycle orogénique). Les fosses océaniques sont des bassins de périsuture, associés avec des zones de ''Benioff'' ou de subduction 'B'. Les avant fosses sont des bassins de périsuture associés et adjacents aux zones d'Ampferer ou de subduction 'A' (zones où subductent des paquets limités de lithosphère continentale).

3. *les bassins d'épisuture* dans une mégasuture. Cela inclut les bassins en avant des arcs insulaires, les bassins en arrière de ces arcs, et les bassins qu'on rapporte à la superposition de systèmes du mégacisaillement sur des mégasutures.

Les processus d'extension, thermiquement controlés, sont fréquemment à l'origine des *bassins sur la lithosphère* et de certains *bassins d'épisuture* (par exemple les bassins en arrière des arcs). Pour chaque classe de bassin, reste à évaluer si les périodes thermiques sont à l'origine d'un soulèvement de la lithosphère et à son attenuation ou s'ils sont eux mêmes la conséquence d'un étirement de la lithosphère.

Dans ces bassins, l'importance de la subsidence peut être modifiée sous la charge de sédiment, le flux subcrustal, les changements de densité dans la croûte inférieur et le manteau.

---

[1]Shell Oil Company, P. O. Box 481, Houston, Texas 77001

Reprinted by permission of the Canadian Society of Petroleum Geology, from *Bulletin of Canadian Petroleum Geology*, January 1980, pp. 9–75.

La flexure lithosphérique doit être responsable des *bassins de périsuture* dans les fosses océaniques ou du type avant fosse, avec une modification de charge correspondant aux écailles de chevauchement et aux plis, associés à la genèse des montagnes, et de remplissage de sédiments provenant des montagnes alentour.

"Prudenter granoque salis columbarium legendum est."

INTRODUCTION

In years past, explorationists took the existence of sedimentary basins for granted. Their efforts concentrated on mapping and exploring the sedimentary basins of the world in ever-increasing detail. More recently, however, the genesis and evolution of sedimentary realms of subsidence have received a great deal of attention.

Academic geologists, for example, confronted with the criticism that plate tectonics did little to explain vertical crustal movements, have responded by proposing models that relate many sedimentary basins to initial thermal processes followed by cooling and loading by the sediments themselves (Sleep, 1971).

Petroleum geologists likewise have been increasingly concerned with basin genesis. This interest stems from the knowledge that thermal considerations are important both in the evolution of basins and in the generation of hydrocarbons from organic-rich source beds. For example, see Hood and Castaño (1974), Hood *et al.* (1975), Dow (1977), Tissot *et al.* (1974 and this volume), Tissot (1977), and Tissot and Welte (1978).

We here will discuss realms of subsidence in a format that expands upon a previously published basin classification by Bally (1975). This classification is based on plate tectonic concepts and should prove useful in making qualitative comparisons between basins. We are under no illusions that it — or any other classification for that matter — will contribute significantly to hydrocarbon volume forecasting. This is because we are of the opinion that the combination of critical limiting factors for hydrocarbon accumulations vary infinitely from basin to basin and from play to play. We are keenly aware of the uniqueness of each hydrocarbon play.

Our intent, then, is through classification to fit seemingly "unique" observations into an ordered hierarchy which, at the very least, provides a means of structuring our ignorance. In this spirit, Audley-Charles *et al.* (1977) have shown the usefulness of our classification for differentiating the tectonic setting of major deltas.

GLOBAL GEOLOGY

Key summaries of plate tectonics are given by Wyllie (1971), Cox (1973), and Le Pichon *et al.* (1973). For terminologic clarification, see Dennis and Atwater (1974). Short and most readable summaries have been given by Oxburgh (1974), Smith (1976), Wilson (1976), and McKenzie (1977). Particularly relevant recent symposia have been edited by Burk and Drake (1974), Bott (1976c), Talwani and Pitman (1977), and Watkins *et al.* (1979).

Many of the objections to plate tectonics have been raised by authors who started from different points of departure, e.g., oscillatory vertical tectonics and associated inversion phenomena (Beloussov, 1960, 1970, 1975), fixist concepts (Meyerhoff, 1970b,c; Meyerhoff *et al.*, 1971, 1972), convection currents rising under the continents and sinking below mid-ocean ridges (Keith, 1972) and finally, the concept of an expanding earth (Jordan and Beer, 1971; Carey, 1975, 1977; Steiner, 1977).

In accepting plate tectonics as a working hypothesis, we made the judgement that opposing views leave relatively more questions unanswered. They do, however, focus on very substantial problem areas, some of which are particularly important for a broader understanding of the genesis of sedimentary basins such as vertical epirogenetic movements and worldwide fracture patterns (rhegmatic patterns).

The basic plate tectonic jargon is illustrated on Figure 1. An outer rigid shell (the lithosphere) of the earth overlies a weak, less viscous and hotter zone (the asthenosphere). The bottom of the lithosphere is not very well defined and not easily measured (Jordan and Fyfe, 1976). Or to quote Dubois *et al.* (1977): "The lithospheric thickness depends essentially on the criterion of definition chosen for that concept" (for a thermal description, see also Pollack and Chapman, 1977). For tectonic models it is often assumed that plates remain rigid over geologic periods in excess of 10^6 to 10^8 years.

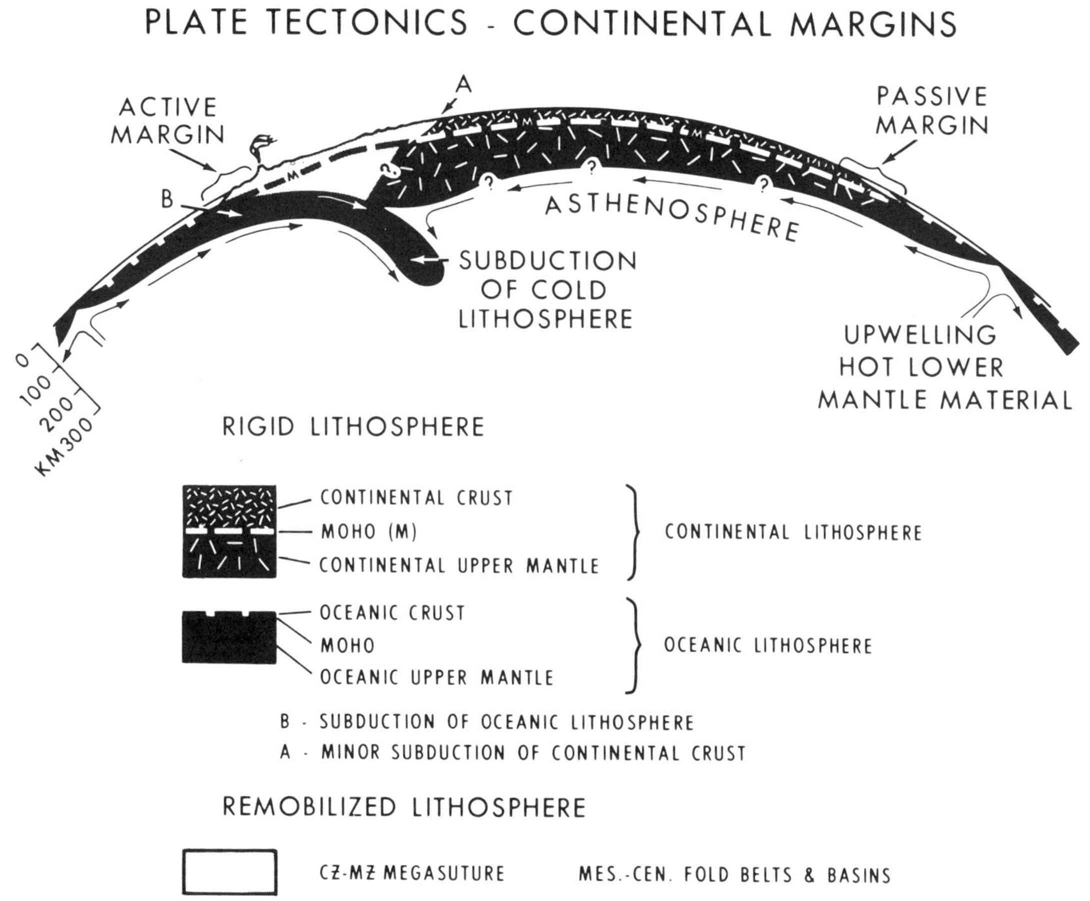

PLATE TECTONICS - CONTINENTAL MARGINS

Fig. 1. Plate tectonic terminology and continental margin types. (No horizontal scale; vertical scale is only an approximation.)

The crust-mantle boundary or the Mohorovičic discontinuity lies entirely within the present lithosphere. It is a seismic velocity discontinuity associated with a sudden increase from lower velocities to velocities greater than 8 km/sec. The significance, the geologic age, the nature of the rocks across the Moho is poorly understood and the subject of much debate. We prefer concepts that visualize the Moho to reflect a downward transition from gabbro to peridotite for oceanic crust, and granulites to gabbro and then to peridotite for continental crust. Our preference is based on observations made by others on ophiolites in mountain ranges (for a summary, *see* Clague and Straley, 1977; Coleman, 1977) and studies of the Ivrea-zone in the Southern Alps (Niggli, 1968; Fountain, 1976).

Figure 2 illustrates the thickness of the crust. Oceanic crust typically is 4-9 km thick while continental crust has thicknesses ranging from 25 km to 70 km. A number of processes may attenuate continental crustal thicknesses to transitional thicknesses (Green, 1977). As we will see later, such processes may play a significant role in the formation of basins.

THICKNESS OF CRUST

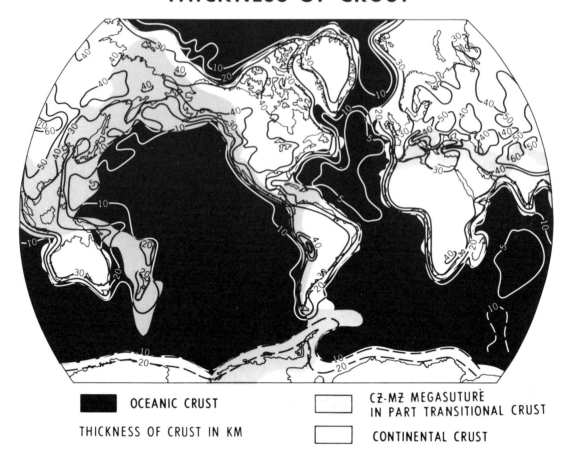

OCEANIC CRUST

THICKNESS OF CRUST IN KM

CZ-MZ MEGASUTURE
IN PART TRANSITIONAL CRUST

CONTINENTAL CRUST

Fig. 2. Thickness of crust and crustal types. Simplified and modified after Cummings and Schiller (1971).

The present lithosphere of the earth is divided into a number of rigid plates, that are circumscribed by the historic earthquake belts (for a review, *see* Sykes, 1972). A simplified earthquake distribution is shown on Figure 3.

Based on earthquake motion studies and other criteria, it is useful to differentiate three plate boundary types:

1. Extensional rifts associated with shallow focus earthquakes;

2. Transform fault boundaries also associated with shallow focus earthquakes and characterized by strike-slip first motions;

3. Predominantly compressional "subduction boundaries" with their associated shallow-intermediate-deep focus earthquakes of Benioff zones.

Note that plate boundaries are to a large extent unrelated to the distribution of oceanic or continental crust. For instance, the extensional boundary of the mid-ocean rift leaves the oceanic crustal domain to bisect continental crust in the area of the Red Sea, the East African rifts, and the Gulf of California-Great Basin area. Also the earthquake belts associated with the Circum-Pacific foldbelts continue into the intracontinental earthquake zones of the Himalayas and Central Asia and then proceed into the Alpino-Mediterranean area. Figure 4 shows the distribution of present-day lithospheric plates and their names.

CONTINENTAL MARGINS

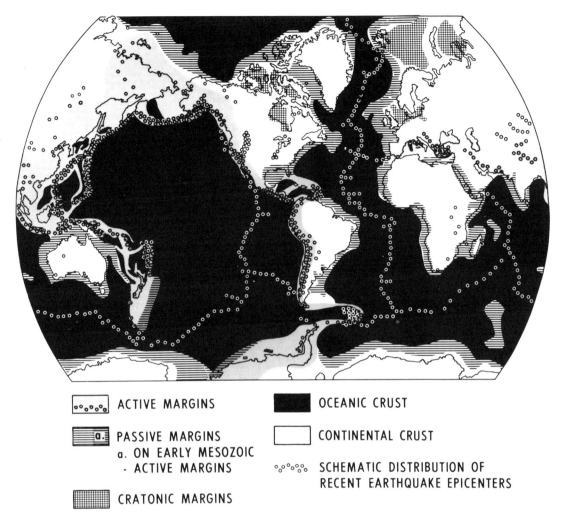

ACTIVE MARGINS

PASSIVE MARGINS
a. ON EARLY MESOZOIC
  - ACTIVE MARGINS

CRATONIC MARGINS

OCEANIC CRUST

CONTINENTAL CRUST

SCHEMATIC DISTRIBUTION OF
RECENT EARTHQUAKE EPICENTERS

Fig. 3. Schematic earthquake distribution; passive and active continental margins and submarine cratonic basins.

Figure 3 serves to show that the earthquake distribution combined with crustal characteristics form the basis for a subdivision of continental margins as follows:

1. *Passive* (or diverging margins of some authors) are continental margins with very limited or no seismic and volcanic activity. The intraplate margins straddle the boundary between continental and oceanic crust.

2. *Cratonic margins and basins* lie entirely on continental crust.

3. *Active* (or converging margins of some authors) are associated with intensive earthquake activity and spectacular volcanism. These margins form at converging plate boundaries where the cold rigid lithosphere is sinking deep into the hotter, less viscous asthenosphere. Island arcs and marginal basins on their concave side are very characteristic of active margins.

4. *Both passive and active margins* are often affected by transform fault systems, in which case the structural characteristics are dominated by strike-slip tectonics and — as in the case of active margins — the associated earthquakes.

LITHOSPHERIC PLATES

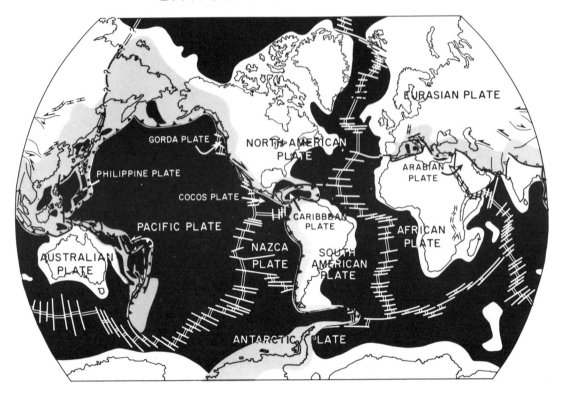

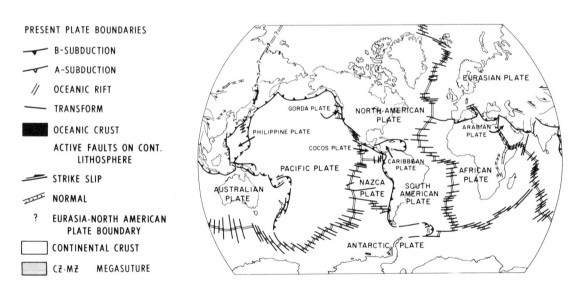

PRESENT PLATE BOUNDARIES

B-SUBDUCTION

A-SUBDUCTION

// OCEANIC RIFT

TRANSFORM

OCEANIC CRUST

ACTIVE FAULTS ON CONT. LITHOSPHERE

STRIKE SLIP

NORMAL

? EURASIA-NORTH AMERICAN PLATE BOUNDARY

CONTINENTAL CRUST

CZ-MZ MEGASUTURE

Fig. 4. Distribution of lithospheric plates. Lower inset shows plate names without differentiating crustal types.

Determining the outline, distribution and movement history of ancient plates is difficult, and various aspects have to be considered. The seafloor spreading hypothesis documents the history of today's oceanic lithosphere backward into early Jurassic time. At mid-ocean ridges new oceanic lithosphere is emplaced between older lithosphere. The imprint of magnetic reversals on the igneous rocks formed during this process allows us to follow and reconstruct the history of a spreading ocean. Spectacular confirmation of this concept was obtained by the JOIDES drilling program. The age of the sediments overlying what are believed to be the topmost major basalt flows has been predicted by the theory with remarkable accuracy. True enough, some sediments were encountered alternating with these basalts, but so far they have not been demonstrated to be substantially older. In other areas, seismic reflections underlying oceanic basalts are recognized and their calibration by drilling could lead to substantial revisions of the basic hypothesis, in case these reflections would represent layers substantially older than predicted by the theory.

Maps of the age of the sea floor have been published by Pitman *et al.* (1974) and Heezen and Fornari (1976). These maps are based upon magnetically defined stripes of the oceanic crust that record relatively orderly accretion of igneous material in a dominantly extensional regime. Many reconstructions based on "stripe matching" have been published (e.g., Pitman and Talwani, 1972; Talwani and Eldholm, 1977; Hilde *et al.*, 1977; Srivastava, 1978; Norton and Sclater, in preparation). All of them help to determine the shape, former position, and age of the extensional plate boundaries (extensional scar).

Conversely, the combined Mesozoic-Cenozoic mobile fold belts, or megasutures, record the complex accretion and deformation in a dominantly compressional regime. In other words, the pristine Mesozoic-Cenozoic oceanic crust and the associated passive margins can be viewed as the extensional counterpart of the contemporaneous compressional global fold belts with their associated active margins. This concept was developed earlier by Wilson (1968) and is emphasized again in this paper because of its usefulness as a support for a proposed basin classification.

Megasutures[1] can be best described using the worldwide Cenozoic-Mesozoic mobile fold belts as an example (Fig. 5). In this example the megasuture boundaries are so defined as to include all products of Cenozoic-Mesozoic orogenic and associated igneous activities (Cenozoic-Mesozoic megacycle). Emphasis is laid on the dominant compressional mode of deformation, but we are mindful of the observation that in these belts, extensional deformation and basin formation is widespread. The processes responsible for the extensional deformation are deemed to be subordinate and a consequence of complex subduction processes.

Consequently, in this description the marginal seas of the Western Pacific, the Western Mediterranean, the Caribbean and the Scotia Sea are all included in the megasutures. These basins should be viewed as a member of the class of episutural basins as will be explained later.

Megasutures show a record of folding, thrust faulting, and igneous activity. They are often associated with ductile flow of crustal rocks. Intrusive and metamorphic structures are extensive. Thus, megasutures — when viewed over time spans in the order of 10^7 to 10^8 years — are very mobile realms. It is unwise to treat them as part of rigid plates, when in fact they are pervasively plastically deformed. Because of the great difficulties in palinspastically reconstructing megasutures, it is most hazardous to extrapolate paleomagnetic and paleogeographic control data from them onto the adjacent, more rigid cratons.

To avoid confusion a brief comment regarding the term "suture" is needed. The term is widely used (e.g., Dewey and Burke, 1973; Ziegler, A. M. *et al.*, 1972; Burke *et al.*, 1977) to describe narrow, relatively linear zones which mark the boundary between two collided continents or else arc-continent collisions. When ophiolites are encountered in these zones, they

---

[1]In our earlier paper we referred to the compressional (C) megasuture. For ease in usage, we suggest dropping the (C).

are usually interpreted to be tectonically emplaced remnants of oceanic crust. The linearity of many sutures on a map is often due to superimposed faulting that is much later than the suturing process itself. The term "suture" thus generally implies close juxtaposition of differing continental paleogeographic realms and, consequently, is more restricted than the term "megasuture" here described.

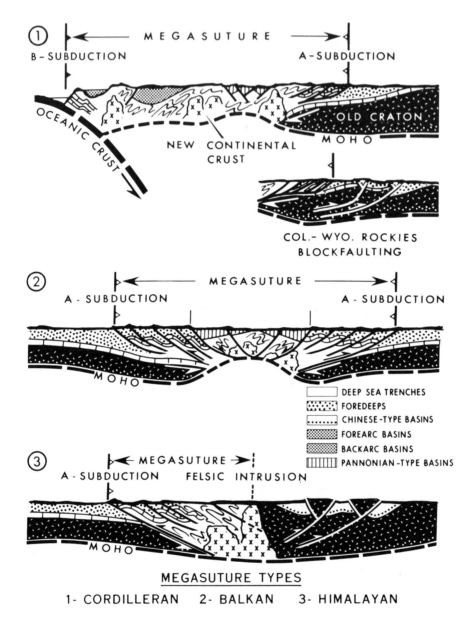

Fig. 5. Diagram illustrating the boundaries of three types of megasutures and their relation to some basins types, after Bally (1975).

Instead of "megasuture," many authors refer to mobile belts (Bucher, 1924, 1957; *see also* Dennis, 1967), a term which, however, is also used in other and different contexts and which we feel is too vaguely defined. Furthermore, our megasuture is not synonymous to "orogenic belt" and "mountain belt" because it includes the deeply depressed basins within these belts. The megasuture lumps the products of several orogenic phases into major cycles, i.e., Mesozoic-Cenozoic, Paleozoic and several Precambrian cycles. These cycles are concep-

8

tually related to Sutton's (1963, *fide* Read and Watson, 1975) chelogenic cycles, but our subdivisions differ from that author in that we envisage two Phanerozoic cycles which he preferred to lump with the Grenville.

A megasuture may be likened to a wide weld that is formed during continental and/or island arc collisions. We visualize extensive basement remobilization as suggested by the wide range of radiometric age determinations from basement rocks in folded belts. Such age determinations often reflect mobilization and/or cooling of basement rocks.

To put the boundaries of megasutures in proper perspective, it is relevant to discuss the concept of subduction. Trümpy (1975) has reviewed that concept. He pointed out that the notion was first introduced by Ampferer (1906), whose views over the years were followed by many other Alpine geologists. These geologists in essence noted that on their cross sections the width of the sedimentary section was substantially in excess of the width of available basement surface. It was therefore concluded that fairly large volumes of presumably sialic basement had to be swallowed at greater depths ("Verschluckung"). Amstutz (1951, 1955) then introduced an equivalent French term ("subduction") which was subsequently anglicized by White *et al.* (1970) to become part of the formal plate tectonic nomenclature (*see also* Dennis and Atwater, 1974). While the original Alpine concept was tied to the notion that sialic crust was being subducted, in plate tectonics, the term refers to a zone where dominantly oceanic lithosphere is being subducted, whereas subduction of buoyant continental lithosphere appears to be of only minor importance in cases of continental collision. As Trümpy (1975) points out, Alpine subduction zones visibly involve upper sialic basement, but this occurs on a relatively small scale. Although it cannot be demonstrated that continental lithosphere is being disposed at depth, in a number of cases, this can be reasonably deduced from palinspastic reconstruction and also from recent earthquakes. The inferred subduction of continental lithosphere is in contrast to the overwhelming seismologic evidence for downgoing slabs of oceanic lithosphere along Benioff (B-subduction) zones.

Because we would like to focus on a basin classification, we are only interested in the subduction boundaries of megasutures as they relate to the location of basins. The following main boundary types are differentiated referring to the Cenozoic-Mesozoic megasuture.

1. *Outer boundary of Cenozoic Benioff or B-subduction zones*, where an oceanic lithosphere slab dips under the sialic continent or an island arc deep into the mantle. Active B-subduction zones are today accompanied by present-day shallow-intermediate-deep focus earthquakes and/or substantial Cenozoic deformation. Seismologists are studying the dynamics of these zones (Sykes, 1972; Oliver *et al.*, 1973) in considerable detail. Figure 6 shows a sketch of a compilation made by von Huene *et al.* (1979) to illustrate the character of B-subduction along the Aleutian trench and its associated volcanic arc.

Figure 7(a, b, and c) shows some sketches of B-subduction zones from published seismic lines that illustrate shallow décollement or scraping of sediments off the top of the oceanic basaltic layer (Beck, 1972; Beck and Lehner, 1974, 1975; Hamilton, 1977).

Depending upon sedimentary thickness, rate of deposition versus rate of slab descent, and rheologic properties of the rocks deformed, a great variety of tectonic styles can be observed in accretionary zones along convergent or active margins. Imbrications (e.g., Java Trench — Fig. 7b), a combination of imbrications and listric normal growth faults (e.g., Colombia — Fig. 7a), and simple décollement folds (e.g., Barbados, Mascle *et al.*, 1977) have been reported. Finally, some cases are reported where the oceanic crust is clearly involved in the subduction process (Kroenke, 1972; Kulm *et al.*, 1973). Seely (1977) suggests that the vergence of folds and thrusts associated with B-subduction is not always toward the ocean but may be toward the continent. In other localities subduction ceased completely, for example, the Palawan subduction zone in the late Miocene was buried by younger sediments (Hamilton, 1977; Fig. 7c).

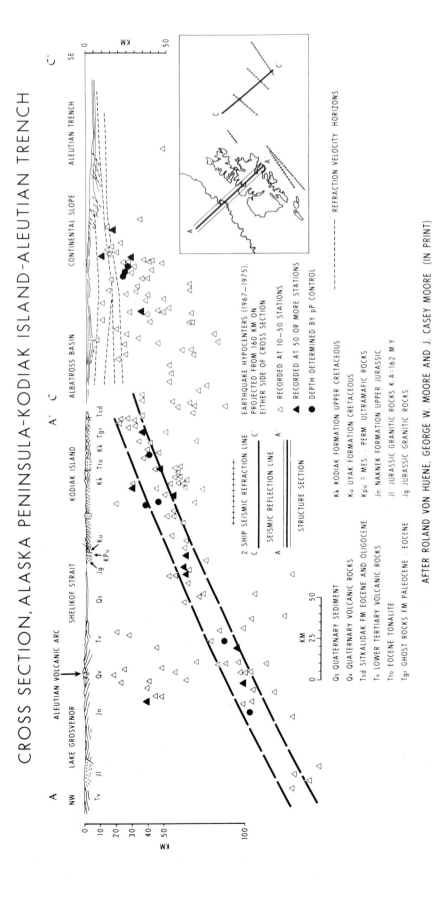

Fig. 6. Cross section across Alaska Peninsula-Kodiak Island and Aleutian trench. Simplified after and with permission of von Huene et al. (1979).

10

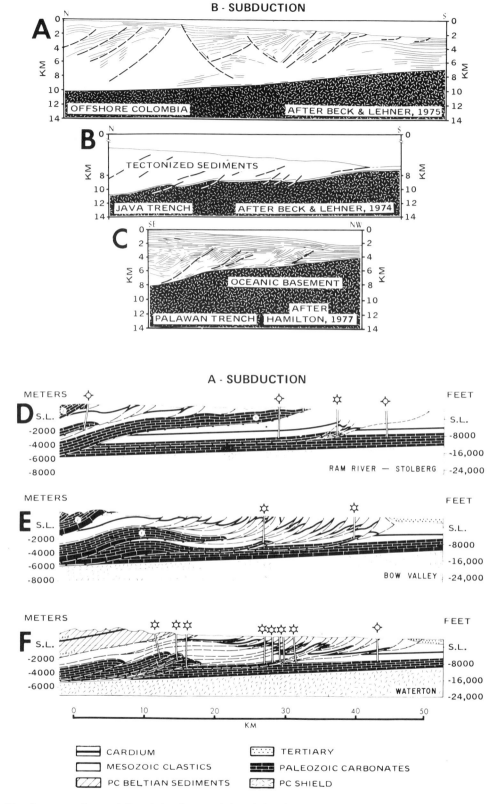

Fig. 7. A comparison of B-subduction and A-subduction at the same scale. A. Offshore Columbia (after Beck and Lehner, 1975); B. Java trench (after Beck and Lehner, 1974); C. Palawan trench (after Hamilton, 1977); D., E., F. Southern Canadian Rocky Mountains (after Bally et al., 1966).

In many cases it would appear that deformation in B-subduction zones occurs in a high pore pressure regime. Shouldice (1971) described this for the subduction zone off Vancouver Island. Oceanic sediments that reach the subduction zone maintain a substantial amount of their porosity and waters as they are only partially consolidated. This may lead to the peculiar tectonic style of scraped-off, semi-consolidated sediments often referred to as mélange. Mélanges are often only loosely differentiated from landslide-like tectonic deposits (olistostromes) that are often found on the submarine slopes of active margins.

A good case can be made for the underlying oceanic crust also containing substantial volumes of water, that are fixed during the cooling of the basalts and shallow intrusives formed during emplacement of the new oceanic crust (Fyfe, 1974, 1976).

It can therefore be inferred that the surface of the downgoing slab contains much water and may be in a high pore pressure regime. This would allow virtually frictionless "silent" subduction and explain why earthquake epicenters appear to occur mainly within the more rigid slab.

2. *Outer boundary of Cenozoic-Mesozoic Alpino-type or A-subduction zone,*[1] (Fig. 5) where some amounts of sialic crust may be disposed at intermediate depths below the megasuture. This is accompanied by large-scale décollement folding and widespread overthrusting of the overlying sedimentary cover. For instance, in the Rocky Mountains of Western Canada, reconstructions of cross sections that are based on reflection seismic data indicate that the width of the reconstructed sedimentary cover is substantially in excess of the width of the underlying continental substratum (Bally *et al.,* 1966; Gordy *et al,* 1975; Fig. 7d - f). All evidence precludes conventional gravity sliding from an uplifted area. Therefore, we infer substantial disposal at depth of possibly an earlier attenuated continental substratum underlying the Beltian-Paleozoic Cordilleran miogeosyncline.

Arguments for A-subduction can also be applied to the front of the Paleozoic Ouachita and Appalachian folded belts.

Some authors favoring plate tectonics view A-subduction as subordinate and correctly relate it to continental collisions as already envisaged by Argand (1924). Indeed, some A-type subduction zones as in the Alps and the Himalayas and the Zagros follow continental collisions and thus they are the successor of earlier B-subduction zones that were active prior to the collision. Reconstructions in the Alpine area (Laubscher, 1965; Trümpy, 1969) support subduction of a sialic basement. In a few cases the A-subduction process may be associated with recent earthquake belts (e.g., Himalayas and Zagros).

Bird *et al.* (1975), Toksöz and Bird (1977) and Bird (1978) published quantitative thermal-mechanical models of continental subduction associated with continental collisions.

Other A-type subduction zones such as those of the eastern boundary of the North American Cordillera may be viewed as the shallower conjugate (antithetic) homolog of more or less coeval B-subduction. Note, however, that in the Cordilleran case, A-subduction ceased during the Paleogene, while B-subduction on the West Coast north of the Mendocino fracture zone continued into the late Cenozoic.

The style of structural deformation related to A-subduction zones varies according to the ductility contrasts within the sedimentary packages involved in the deformation. Rates and amounts of A-subduction of sialic crust are much smaller than with B-subduction of oceanic crust.

3. *Outer margin of wrench tectonism along transform faults.* In some areas (e.g., Southern California and New Zealand) the megasuture is intersected by a ridge and transform fault system which is superimposed on an earlier compressional belt.

---

[1]In the earlier paper, "A" stood for Alpino-type, a term which is not well defined. We now would like "A" to stand for Ampferer (1906) who foresaw the importance of this process.

4. *Envelope around felsic igneous intrusions.* The boundary types proposed so far are not applicable to Mongolia and China (Terman, 1974; People's Republic of China, 1975, 1976) where the B-subduction zones of the Western Pacific do not have an equivalent A-subduction belt. In that area the western boundary of the megasuture is the westernmost limit of Mesozoic intrusives. This boundary is adjacent to an area dominated by massive Mesozoic and Cenozoic foreland blockfaulting and strike-slip faulting (Dewey and Burke, 1973; Terman, 1974; People's Republic of China, 1975, 1976; Molnar and Tapponier, 1975), suggesting that foreland deformation is peripherally related to development of megasutures.

Regions dominated by Mesozoic-Cenozoic basement controlled foreland strike-slip and blockfaulting were not included in the megasuture because the comparatively more rigid style of deformation leaves the overall cratonic character of these provinces intact. Also for reasons relating to our proposed basin classification, we prefer to include such foreland areas of this type in the perisutural domain.

A further justification for not including such blockfaulted foreland regions in the megasuture relates to depth of structural decoupling. Decoupling in the adjacent A-subduction zone tends to occur preferentially at the sediment-basement interface or higher in the sedimentary column. However, blockfaulted deformation involves the basement and suggests much deeper decoupling at the base of the sialic crust or in the upper mantle. Such a deep decoupling style is well illustrated in the Colorado-Wyoming Rocky Mountains where high-angle reverse faults, overthrusts, normal faults and transcurrent faults interact (Fig. 24; Sales, 1968; Lowell, 1974).

CONDENSED TECTONIC SYNTHESIS

The surface of the earth can be subdivided as follows:

1. The Oceans, that are products of Mesozoic-Cenozoic ocean spreading and extension

2. Their compressional contemporaneous equivalent, the Cenozoic-Mesozoic megasutures of the world

3. The combined Paleozoic foldbelts which, except for Circum-Pacific areas, represent the Paleozoic megasutures now bounded on both sides by continental crust and by A-subduction zones. Paleozoic B-subduction can only be surmised, because Paleozoic oceanic crust has not been preserved in its pristine undeformed shape, and only minor amounts of Paleozoic oceanic crust occur in the form of obducted ophiolites. Two basic alternatives may be suggested: (a) the B-subduction process has been so effective that most Paleozoic oceanic crust was disposed of, or (b) Paleozoic geosynclines did not have substantial oceanic areas. For actualistic reasons, we favor the first alternative.

4. The Precambrian fold belts of the world represent several Precambrian megasutures. Again, no pristine Precambrian oceanic crust is believed to occur.

Figure 8 illustrates our synthesis and may also serve as an approximate "economic basement" map for hydrocarbon exploration. In other words: epi-Precambrian basins overlie Precambrian megasutures, epi-Paleozoic basins overlie Paleozoic megasutures, and preserved epi-Mesozoic and epi-Cenozoic basins overlie the Cenozoic-Mesozoic megasuture. To this one may add that a realistic economic basement age of the oceanic crust is best given by the age of the oceanic crust as deduced from the interpretation of magnetic anomalies that are calibrated by JOIDES holes.

13

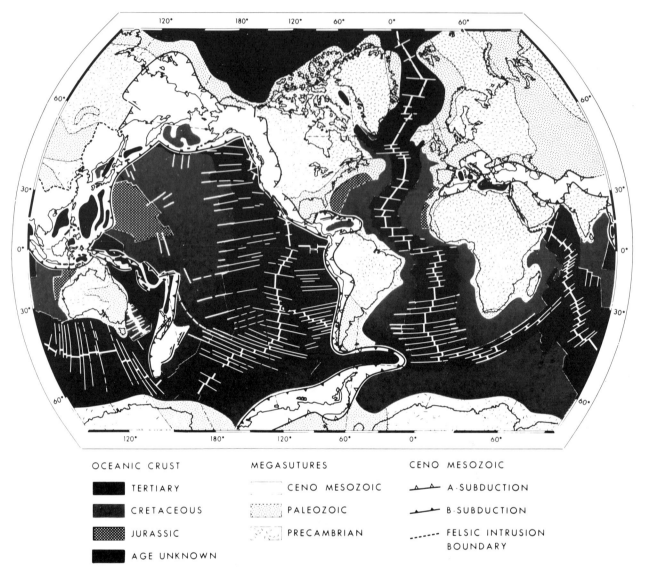

OCEANIC CRUST

- ■ TERTIARY
- ▨ CRETACEOUS
- ▨ JURASSIC
- ■ AGE UNKNOWN

MEGASUTURES

- ☐ CENO MESOZOIC
- ⠿ PALEOZOIC
- ⠿ PRECAMBRIAN

CENO MESOZOIC

- ▵▵ A-SUBDUCTION
- ▵ B-SUBDUCTION
- ------ FELSIC INTRUSION BOUNDARY

Fig. 8. Simplified tectonic map of the world, modified after Bally (1975).

BASIN CLASSIFICATION

Basins are here viewed as realms of subsidence with thicknesses of sediments commonly exceeding 1 km *that are today still preserved* in a more or less coherent form. The abyssal plains of the oceans and some of the ocean marginal seas of the Circum-Pacific contain sediment thicknesses that are often less than 1 km. They are also to be viewed as basins and are included in Figure 9. However, because of their limited interest to petroleum geologists, they will not be discussed in any detail in this paper.

Our basin definition *explicitly excludes folded belts* that involve thick and complexly deformed sediments and sometimes contain substantial hydrocarbon volumes. The basin definition also excludes other positive possibly hydrocarbon-bearing features such as intracratonic highs with a thin sediment cover.

The proposed scheme differs from earlier classifications (Klemme, 1975, 1977; Perrodon, 1971, 1977; North, 1971; McCrossan and Porter, 1973) because it is based entirely on plate tectonic concepts. It puts main emphasis on location of basins with respect to the megasuture and its A- and B-subduction boundaries. Dickinson (1976) has also classified basins with plate tectonics in mind. We have used his work to expand our earlier classification. However, the present proposal differs in a number of aspects from Dickinson's, because we attempt to structure a hierarchy of different basin types.

14

BASINS ON RIGID LITHOSPHERE

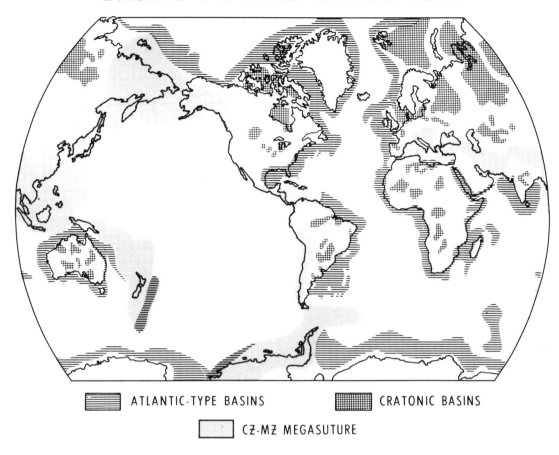

ATLANTIC-TYPE BASINS   CRATONIC BASINS

CZ-MZ MEGASUTURE

Fig. 9. Basins on rigid lithosphere.

A few words are in order to relate this classification to earlier geosynclinal nomenclature. Common American usage of "pre-plate tectonic" terminology has been discussed by King (1969a, p. 46-50). Dewey and Bird (1970) tried to reconcile the old geosynclinal concepts with plate tectonics while Dickinson (1971a) concluded that whereas "the tenets of classical geosynclinal theory have played a valuable role as a means to classify tectonic elements, they are unnecessary impediments to clear thinking in the future." For another review, see also Dott (1978). Hsü (1972a), in a most readable account, traces the evolution of geosynclinal thinking and proposes that the geosynclinal nomenclature should be restricted "to describe tectonic settings of sedimentary sequences and their associated rocks found at present continental or plate margins." With all this in mind, the following points are important:

1. The concepts of the geosyncline and classifications related to it were based largely on deductions derived from field geologic observations. These concepts evolved prior to most geophysical and oceanographic work of the past two decades. In many cases, the original concepts simply did not match the new evidence.

2. Most ortho-, mio-, and eu-geosynclines of the earlier literature have been deformed or completely destroyed by later folding and erosion processes. Consequently, today they are typically incomplete remnants of older basins. Thus the old nomenclature does not differentiate intact, pristine basins from remnants of deformed basins. From a practical point of view such a differentiation is essential. It would also appear to be conceptually desirable to make a classification independent from the hazards of reconstructing fold belt complexes. Such a procedure allows us to separate the more factual from the speculative.

15

3. Basic geosynclinal concepts such as those of Kay (1951) and Aubouin (1965) were based on paleogeographic reconstructions that are often difficult to reconcile with actualistic examples of preserved basins as shown on reflection seismic sections. For example, for some a "eugeosyncline" would encompass the entire spectrum of present-day active margin environments including its island arcs, volcanic arcs, accretionary subduction wedges, forearcs, backarcs, and marginal seas. Other authors would view the floor of the oceans, with its spilitic basement and the ubiquitous volcanic seamounts and sediment overburden as being a "eugeosyncline" or possibly a "leptogeosyncline."

4. Much has been said about the similarity of Atlantic-type shelves with miogeosynclines (Drake *et al.*, 1959). Some referred to them as miogeoclines (Dietz, 1963; Dietz and Holden, 1966). Miogeosynclines are conceptually often coupled with eugeosynclines. For Atlantic-type margins this can be arranged if one follows the view that oceanic sea floor with its sea-mounts is indeed a eugeosyncline (or leptogeosyncline when the thin sediment sequences are not overlain by thick submarine fans). Another perspective would view the continental side of a West Pacific marginal basin as a miogeosyncline. As to former miogeosynclines in folded belts, it is frequently difficult to separate them environmentally from adjacent cratonic sequences that often show comparable rates of subsidence.

It is concluded that much of the geosynclinal nomenclature has outlived its usefulness. However, familiarity with the old concepts and terms is necessary to appreciate the voluminous literature based on these concepts.

Of course, there remains the question of nomenclature of old basins presently deformed in fold belts, that have been reconstructed using various more or less hazardous palinspastic procedures. If at all necessary and to avoid introducing new terms, it may be advisable to limit the term "miogeosyncline" to essentially non-volcanic paleo-basins deformed in folded belts.

The term "eugeosyncline," however, should be abandoned. Reconstruction of former "eugeosynclinal" realms would in the future best be described in terms of the proposed basin classification (eventually to be complemented by an uplift classification) with an added prefix such as "old" or "paleo" and a statement of the time interval of subsidence.

Our basin classification differentiates three basically different families of sedimentary basins:

1. Basins located on rigid lithosphere and not associated with the formation of megasutures (Fig. 9)

2. Perisutural basins on rigid lithosphere associated with, and flanking the compressional megasutures (Fig. 5 and Fig. 18)

3. Episutural basins located upon and mostly contained within the megasuture (Fig. 5 and Fig. 24)

In the following, we will briefly sketch the main characteristics of each family. For ease of reference we will make use of a decimal classification which is shown on our Table 1.

1. Basins located on rigid lithosphere not associated with the formation of megasutures.

The distribution of this large family of basins is shown on Figure 9. Deep oceanic basins of the 112 and 113 type of our classification in general do not have sediment thickness of significant interest to petroleum geologists. They are not shown on Figure 9 and will not be discussed further, despite their vast areal extent and general importance. Rifted grabens (111) are also not shown on Figure 9.

16

TABLE 1 - BASIN CLASSIFICATION

1. **BASINS LOCATED ON THE RIGID LITHOSPHERE, NOT ASSOCIATED WITH FORMATION OF MEGASUTURES**

   11. Related to formation of oceanic crust

       111. Rifts

       112. Oceanic transform fault associated basins

       113. Oceanic abyssal plains

       114. Atlantic-type passive margins (shelf, slope & rise) which straddle continental and oceanic crust

           1141. Overlying earlier rift systems
           1142. Overlying earlier transform systems
           1143. Overlying earlier Backarc basins of (321) and (322) type

   12. Located on pre-Mesozoic continental lithosphere

       121. Cratonic basins

           1211. Located on earlier rifted grabens
           1212. Located on former backarc basins of (321) type

2. **PERISUTURAL BASINS ON RIGID LITHOSPHERE ASSOCIATED WITH FORMATION OF COMPRESSIONAL MEGASUTURE**

   21. Deep sea trench or moat on oceanic crust adjacent to B-subduction margin

   22. Foredeep and underlying platform sediments, or moat on continental crust adjacent to A-subduction margin

       221. Ramp with buried grabens, but with little or no blockfaulting
       222. Dominated by block faulting

   23. Chinese-type basins associated with distal blockfaulting related to compressional or megasuture and without associated A-subduction margin

3. **EPISUTURAL BASINS LOCATED AND MOSTLY CONTAINED IN COMPRESSIONAL MEGASUTURE**

   31. Associated with B-subduction zone

       311. Forearc basins
       312. Circum Pacific backarc basins

           3121. Backarc basins floored by oceanic crust and associated with B-subduction (marginal sea sensu stricto).
           3122. Backarc basins floored by continental or intermediate crust, associated with B-subduction

   32. Backarc basins, associated with continental collision and on concave side of A-subduction arc
       321. On continental crust or Pannonian-type basins
       322. On transitional and oceanic crust or W. Mediterranean-type basins

   33. Basins related to episutural megashear sytems

       331. Great basin-type basin
       332. California-type basins

17

EVOLUTION OF PASSIVE MARGIN

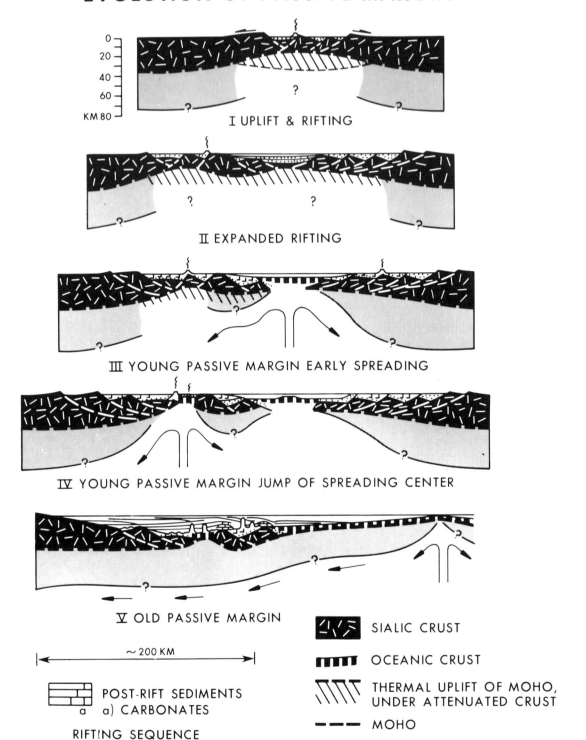

I UPLIFT & RIFTING

II EXPANDED RIFTING

III YOUNG PASSIVE MARGIN EARLY SPREADING

IV YOUNG PASSIVE MARGIN JUMP OF SPREADING CENTER

V OLD PASSIVE MARGIN

~ 200 KM

POST-RIFT SEDIMENTS
a) CARBONATES

RIFTING SEQUENCE

SALT   SALT DOME

SEDIMENTS

SIALIC CRUST

OCEANIC CRUST

THERMAL UPLIFT OF MOHO,
UNDER ATTENUATED CRUST

MOHO

UPPER MANTLE OR
LOWER LITHOSPHERE

BASE LITHOSPHERE OR
TOP ASTHENOSPHERE

Fig. 10. Evolution of passive margin.

11. Basins related to the formation of oceanic crust

In recent years a number of authors have contributed to the development of a kinematic model that explains the formation of Atlantic-type (passive or diverging) continental margins. Descriptions of the model were given by Dietz (1963), Dietz and Holden (1966), Dewey and Bird (1970), Kinsman (1975), and Falvey (1974).

The main features of this model are shown on Figure 10. In general, three main phases are visualized:

1. Rifting, commonly associated with a thermal uplift of the mantle and typified by the formation of horst and graben structures (Phases I and II of Fig. 10).

2. The onset of drifting involving the separation of continental lithosphere. Ocean crust is emplaced and accretes along mid-ocean ridges (based on the evidence of magnetic sea-floor anomalies) in the gap between continental blocks (III and IV of Fig. 10).

3. Main drifting phase, dominated by massive subsidence. Typically, the rate of sedimentation exceeds the rate of subsidence and leads to the accumulation of thick prograding sedimentary wedges (V of Fig. 10). In some cases, particularly where horst blocks are so far away and separated from the mainland that they cannot be reached by terrigenous supply, a sediment supply deficiency leads to the formation of starved margins.

A number of processes have been proposed to explain various phases of this proposed sequence of events. These we will discuss in the next paragraphs.

Some authors relate rifts to deep-seated mantle hot spots that lead to broad thermal uplifts. The crests of these uplifts break up in trilete graben systems which later became triple junctions that separate lithospheric plates. Cloos (1939) foresaw this concept which in recent years has been developed in much detail by Burke and Dewey (1973), Hoffman et al. (1974), and Burke (1977). It is also known that rifting can develop without substantial uplift; but in both cases, later in the cycle, rifting may often abort and regional subsidence may then occur as the system cools (e.g., North Sea, Fig. 11c). Continued rifting may also lead to the formation of a young proto-ocean (Red Sea-type, Fig. 11a) that with further spreading can develop into a full-fledged ocean.

Various models have been proposed to account for rifting and related extension of the continental lithosphere that is associated with high surface heat flow. The simple observation that oceanic rifts also bisect continental lithosphere as in East Africa, suggests that the asthenosphere which underlies both continents and oceans may play a significant role in allowing the lithosphere to stretch. Stretching in the upper crust may well be accommodated by listric normal faulting (Lowell and Genik, 1972; Lowell et al., 1975; McKenzie, 1978b).

As to the mechanism and style of attenuation below such a listric fault domain, clues have been provided by deep reflection and refraction surveys. Below many rifts there is an interpreted low-velocity layer overlying a sometimes elevated Moho (Fuchs, 1974; Artemyev and Artyushkov, 1971; Bott, 1976a; and others). The low-velocity layer may reflect a zone of plasticity in the lower crust which allows for extension by necking and ductile flow. Listric normal faults in the overlying brittle crust might well "bottom out" near the top of this presumably ductile, low-velocity layer (see also de Charpal et al., 1978).

The preceding paragraphs view rifts as isolated geologic objects, but it is useful to also discuss their overall global setting. The Red Sea Rift may well be viewed as the leading edge of a propagating crack as illustrated by Bowin (1974). Other rifts like the Rhine Valley Graben (Illies, 1977; Illies and Greiner, 1978) and the Baikal Rift (Zorin and Rogozhina, 1978) may be related to asthenospheric uplifts associated with the late stages of continental collisions (Molnar and Tapponier, 1975; Sengör et al., 1978). Finally, the Rio Grande Rift of New Mexico is formed during the Neogene uplift of the Western Cordillera of the United States and may be viewed within the context of formation of Basin and Range megashear tectonics (see Fig. 31).

19

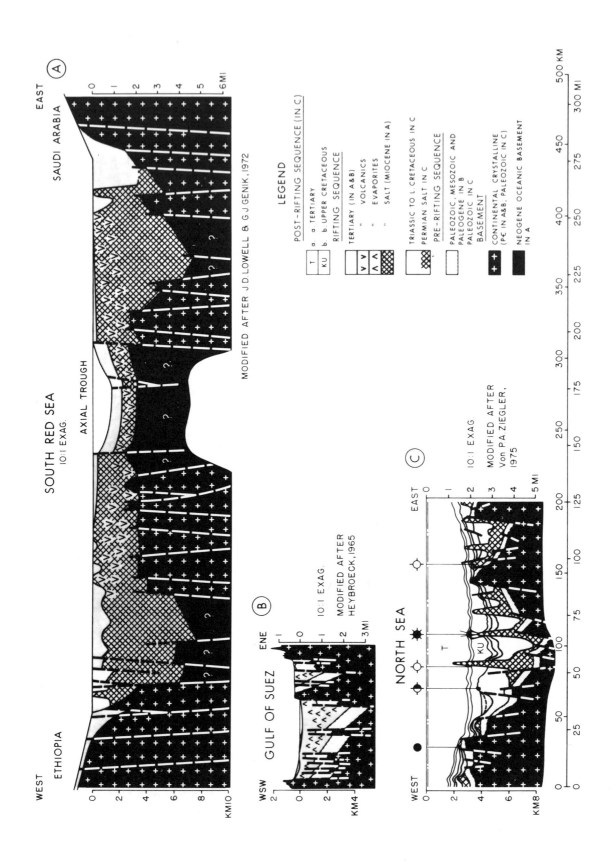

Fig. 11. A comparison of three rifted grabens. A. Red Sea — incipient ocean; B. Gulf of Suez, Tertiary rift; C. North Sea, buried graben.

20

A number of models have been proposed to explain the origin of the massive subsidence of continental margins that follows the rifting and the breakup stages. Walcott (1972) visualizes that much of the subsidence is due to flexural bending of the lithosphere under the load of sediments. Bott (1979) points out that simple Airy isostatic loading or else Walcott's model permit subsidence only amounting to about 2 or 3 times the original water depth of the basin. Because passive margin sedimentation begins with shallow water deposition, it would appear that Walcott's model may have only limited application as with the deposition of deep sea fans in a relatively deep pre-existing basin. Watts and Ryan (1976) also conclude that simple loading does not suffice to explain subsidence on passive margins. Consequently, additional ''driving forces'' for the subsidence have to be sought.

Subsidence due to cooling is an important factor in passive margin subsidence. As described by Sclater and Francheteau (1970) and Parsons and Sclater (1977), the oceanic lithosphere cools and subsides as it moves away from the generating mid-ocean ridge. The adjacent continental margin contracts thermally and rates of subsidence decay exponentially (Sleep, 1971). Additional mechanisms may also be important; for example, Bott (1971) and Bott and Dean (1972) suggest that subsidence may also be caused by gravitational outflow of lower crustal material.

111. Rifts: Examples include the great African Rifts (Baker and Wohlenberg, 1971; Baker *et al.,* 1972; Beyth, 1978; Chapman and Pollack, 1975; Degens *et al.,* 1973; Girdler, 1975; Girdler *et al.,* 1969; Pilger and Rösler, 1975, 1976), the Rhine Valley Graben (Illies and Fuchs, 1974; Illies, 1977), Lake Baikal (Artemiev and Artyushkov, 1971: Logatchev and Mohr, 1978; Logatchev and Florentsov, 1978), the Rio Grande Rift (Chapin and Seager, 1975; Bridwell, 1976; Kelley, 1977; Cordell, 1978) and others. Many rifts are buried under the sedimentary blanket of Atlantic-type margins, cratonic basins and foredeeps. Studies of rifts, particularly young rifts such as the Rhine Valley Graben and the Rio Grande Rift, allow us to observe the following:

1. As stated by Artemiev and Artyushkov (1971), the crustal extension observed is in excess of the small amount explained by simple doming. Therefore, crustal and/or lithospheric stretching has to come into play.

2. Earthquake first motion studies suggest extension. This is corroborated by observed antithetic normal faults.

3. High heat flow.

4. Low-velocity layers occur within the lower crust. (Fig. 12-1).

5. Refraction and wide angle reflection data suggest a Moho elevated with respect to the surrounding area. In young rifts, sub-Moho velocities are often low (e.g., 7.6-7.7 km/sec). Under old inactive rifts, e.g., the Dnepr-Donetz Aulacogen, the Moho is likewise elevated but mantle velocities are normal (8-8.6 km/sec), suggesting that crustal attenuation remains preserved in ancient rifts and that low-velocity layers at the bottom of the crust are converted into higher velocity mantle material (Fig. 12-3).

6. Although rift grabens are filled with sediments (in excess of 3 000 m in the Rhine Graben) the details of the filling process, i.e., progradation along rift axis versus filling from the rift flanks, are often poorly known. A number of rifts have lacustrine sourcebeds like the Rhine Valley Graben or coal deposits that may act as a hydrocarbon source. Restricted sedimentation also leads to formation of thick salt deposits and in some cases to deposition of marine oil source beds.

114. Atlantic-type passive margins: These always overlie earlier rift systems that often are (1) sub-parallel to the ocean margins, (2) near-perpendicular and oblique failed arms of triple junctions like the Benue Trough of Nigeria (Burke and Dewey, 1973) or (3) associated

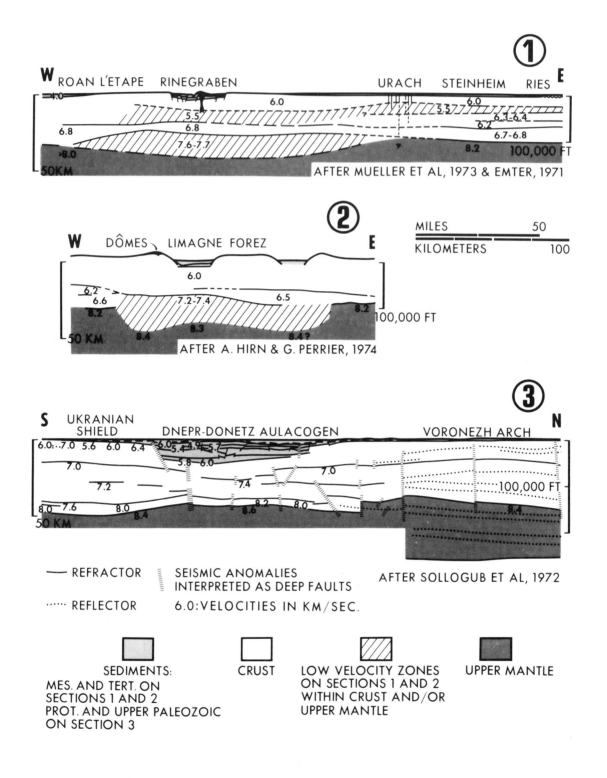

Fig. 12. Sketches of crustal profiles across graben systems. 1. Rhine Graben; 2. Limagne Graben; 3. Dnepr-Donetz Aulacogen.

22

with transform systems like the Grand Banks (Amoco and Imperial, 1973, and Fig. 14) the Gulf of Guinea (Mascle, 1976) and the margins of South Africa (Scrutton and Dingle, 1976). In all cases these margins allow us to recognize an early phase corresponding to the rifting, and a later drifting phase. Falvey (1974) postulated that both phases are often separated by an unconformity. Such unconformities are often plainly visible on seismic lines (Amoco and Imperial, 1973; Given, 1977; Schlee *et al.*, 1976, 1977). In other cases unconformities are less obvious, and the graben fill merges continuously with the overlying drifting sequence. Occasionally, more than one unconformity can be observed. All this suggests that we know little about the tectonic style and processes associated with the breakup process. Montadert *et al.* (1977), based on the results of IPOD drilling, have recognized two contrasting styles. In one case (the Rockall Bank) there was substantial subaerial relief at the end of rifting, while in another case (Bay of Biscay and Galicia) more than 2000 m of submarine relief was present at the end of the rifting phase.

For the latter example, de Charpal *et al.* (1978) have published multi-channel reflection sections that display systems of listric normal faults that merge into a visible subhorizontal sole fault. These sections therefore support the concept of crustal stretching without accompanying doming. The same authors suggest that the amount of stretching achieved by listric normal faulting may not fully account for the crustal attenuation observed on refraction sections. Therefore, crustal attenuation below rifts may be accompanied by crustal flow in the underlying lower ductile crust.

Today's passive margins and Atlantic-type basins are post-Triassic in age and evolved following the breakup of Pangea since Triassic times. Figure 13 illustrates this process, and it may be noted that the extent of passive margins increased substantially during the past 200 ± Ma. Different segments of the oceans opened up at different times and the breakup time can be determined approximately for different oceans. One can "eyeball" the separation of the rifting sequence from the drifting sequence on continental margins as follows:

Mid-Jurassic (about 180 Ma): North America-Africa (Pitman and Talwani, 1972);

Lower Cretaceous (about 130 Ma): South America-Africa (Larson and Ladd, 1973) and India-Antarctica-Australia (Markl, 1974, 1978);

Mid-Cretaceous (about 90 Ma): Newfoundland-British Isles (Srivastava, 1978);

Late Cretaceous (about 75 Ma): Greenland-North America (Srivastava, 1978); (about 80-90 Ma): India-Madagascar (Norton and Sclater, in prep.);

Lower Paleocene (about 60-65 Ma): Norwegian Sea and Baffin Bay (Talwani and Eldholm, 1977; Srivastava, 1978);

Eocene (about 50 Ma): Australia-Antarctica (Weissel and Hayes, 1972);

Upper Miocene (about 10 Ma): Red Sea and Gulf of Aden (Laughton *et al.*, 1970);

Late Miocene (about 5 Ma): Gulf of California (Larson *et al.*, 1968; Moore, 1973).

The sequence associated with the drifting phase may consist of mostly carbonates (e.g., Blake Plateau, Fig. 14; Florida-Bahamas) or dominantly clastics (Baltimore Canyon trough, Fig. 14; South Australia, Niger Delta) but most often it is a mix of the two types. Salt may be locally an important constituent near the base of the drifting sequence (e.g., Gabon: Brink, 1974). Lithologies appear primarily controlled by the climate and morphology of the hinterland.

In contrast to the rifting sequence which is characterized by block-faulting, it appears that the tectonic deformation of the drifting sequence is mostly gravity controlled (listric growth faults in soft sediments and/or salt tectonics). The complex progradation process on Atlantic-type margins is ultimately responsible for the morphology of these margins, with their flat and relatively stable shelves, unstable slopes, and the deep-sea fans of the rise.

THE BREAKUP OF PANGEA

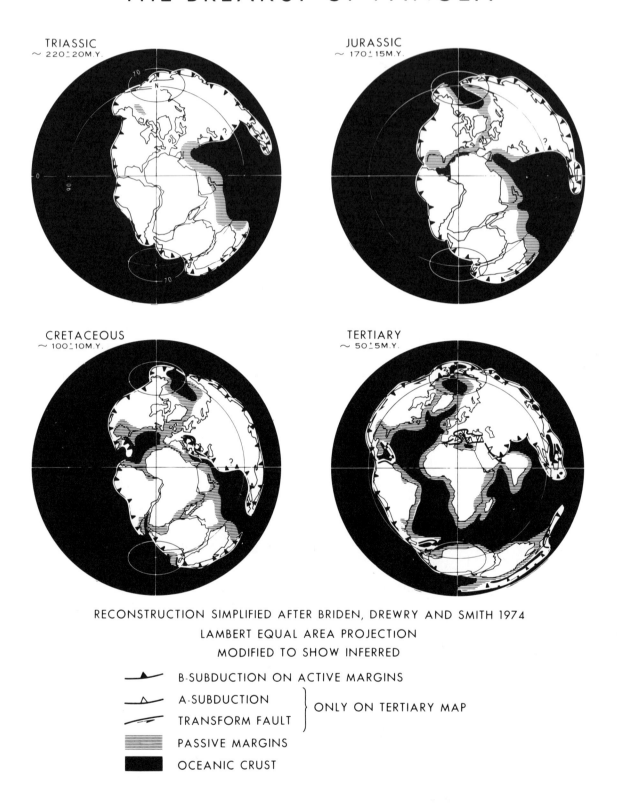

Fig. 13.   The breakup of Pangea.

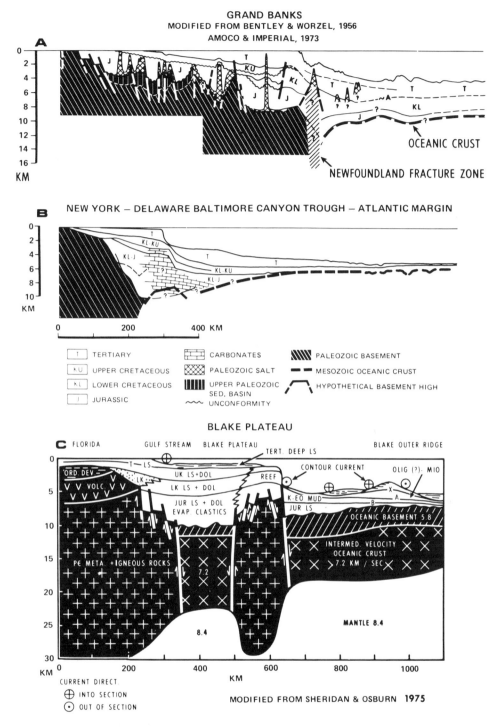

Fig. 14. Atlantic-type margins offshore eastern North America: A. Grand Banks; B. Baltimore Canyon trough; C. Blake plateau. Vertical exaggeration 20:1 for all sections.

Keen (1979) has made a detailed study of subsidence offshore Nova Scotia and Labrador. She separates the effects of sediment loading from "tectonic" subsidence due to thermal contraction. The data from wells of the Canadian Atlantic margin support in general the cooling model that predicts a linear relationship between tectonic subsidence and the square root of time since subsidence began.

The following is a selection of areas and papers that well illustrate passive margin geology as it pertains to the search for hydrocarbons:

Eastern Canada: Amoco and Imperial (1973), Pelletier (1974), Keen *et al.* (1974, 1975), Yorath *et al.* (1975), Given (1977), Srivastava (1978); *Eastern U.S.:* Sheridan (1976), Schlee *et al.* (1976, 1977), Grow and Markl (1977), Grow *et al.* (1979), Buffler *et al.* (1978); *Brazil:* Asmus and Ponte (1973), Campos *et al.* (1974), Kumar *et al.* (1976), de Almeida (1976), Ponte *et al.* (1976, 1977, and this volume); *Atlantic Coast of Africa:* Driver and Pardo (1971), Brink (1974), Rabinowitz *et al.* (1975), Rabinowitz (1976), Delteil *et al.* (1976), Scruton and Dingle (1976), Mascle (1976), Lehner and de Ruyter (1977), Rabinowitz and LaBrecque (in press), Shipley *et al.* (1978); *Australia:* Smith (1968), Martinson *et al.* (1972), Thomas and Smith (1974), Robinson (1974), Boeuf and Doust (1975), Powell (1976), Willcox and Exon (1976), Veevers and Cotterill (1978); *Western Europe:* Debyser *et al.* (1971), Woodland (1975), Talwani and Eldholm (1977), Montadert *et al.* (1977).

A special case concerns somewhat circular, passive margin-like basins that no longer face spreading oceans (Fig. 15). Such basins include the Gulf of Mexico (for seismic sections, see Watkins *et al*, 1977; Worzel and Burk, 1978) and Sverdrup Basins, and possibly the Pericaspian Basin. Their centre has an attenuated continental basement or, as in the case of the Gulf of Mexico, an oceanic basement. Both the Gulf and Sverdrup Basins are superimposed over upper Paleozoic megasutures. These megasutures were bounded by flanking A-subduction zones and contain an internal late- to post-orogenic rift basin complex that may be similar to the Late Cenozoic backarc basins (1143 of Table 1) of the Pannonian (321) or West Mediterranean (322) type.

The origin of the pre-salt or proto-Gulf of Mexico Basin is, of course, highly speculative. For a review of various hypotheses, see Woods and Addington (1973) and Nairn and Stehli (1975). Such speculations have included Early Mesozoic sea-floor spreading (Walper and Rowett, 1972), remnant of a Paleozoic proto-Atlantic (Shurbet and Cebull, 1975), Mesozoic collapse, extension, and spreading between two megashears (Beall, 1973), and Late Paleozoic-Triassic collision-related backarc basins associated with A-subduction (Bally, 1975, 1976). Any hypothesis preferably should be compatible with an assumed reconstruction for a Permian Pangea. Such reconstructions vary widely among authors (i.e., Smith *et al.*, 1973; Bullard *et al.*, 1965; van der Voo *et al.*, 1976) but they all suggest that the Yucatan Peninsula took its position in the Gulf of Mexico after Permian and before Mid-Jurassic times.

The rift basin complex which underlies at least the northern Gulf Basin was initiated in the Late Pennsylvanian. Along the northern rim of the Gulf region and south of the Ouachita Fold Belt, relatively undeformed Desmoines and younger shallow water shelf carbonates and clastics are reported in subsurface (Vernon, 1971). Other Late Paleozoic sediments are reported to the west and south in Mexico and the Yucatan Peninsula (*see* Woods and Addington, 1973; Lopez-Ramos, 1969, 1972). Also dredged siltstones with authigenic glauconite from the Sigsbee Knolls have yielded 318 Ma K-Ar dates (Pequegnat *et al.*, 1971).

Although speculative, it seems reasonable that the Late Paleozoic rift complex within the megasuture had a thermal origin and was accompanied by some crustal attenuation.

In the Late Triassic, additional graben development and further attenuation probably occurred simultaneously with the well-known Triassic rifts of the eastern U.S. (e.g., Faill, 1973). These latter grabens are generally regarded to be the precursors of Early Jurassic sea-floor spreading in the early Central Atlantic.

In the Gulf region, however, this later phase of rifting may have been associated with reported Triassic batholithic activity in eastern Mexico (see Lopez-Ramos, 1972) and "eugeosynclinal" deposition farther west (de Cserna, 1971). These observations indicate a more complex Late Triassic-Early Jurassic history than along the eastern U.S. coast. In fact, the late Triassic proto-Gulf could represent a reactivated backarc basin (as suggested to us by M. T. Roberts). When and if new sea floor actually was formed during the Late Triassic-Early Jurassic we cannot be certain; however, we suspect a certain amount of spreading did occur.

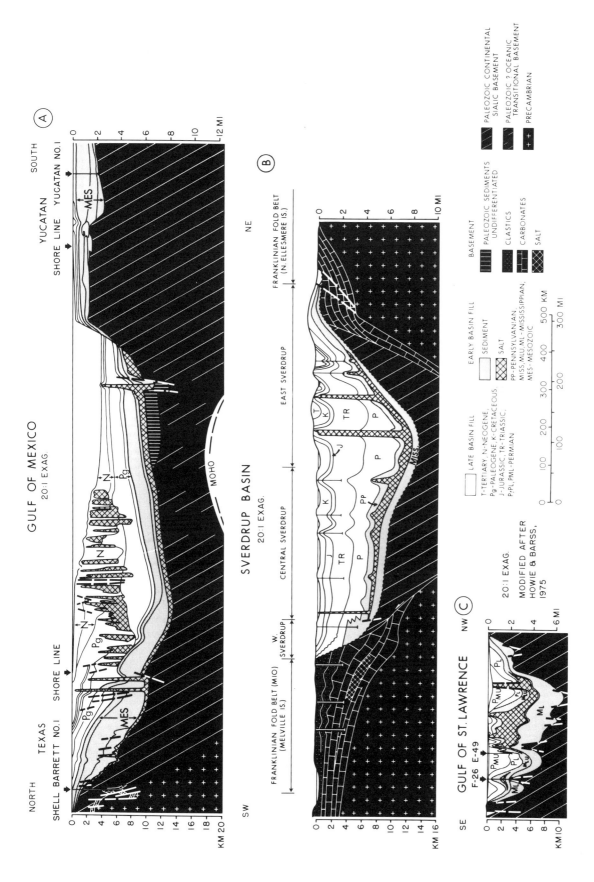

Fig. 15. A comparison of A. the Gulf of Mexico and B. the Sverdrup Basins superimposed on Paleozoic megasutures. C. The Gulf of St. Lawrence, a Paleozoic episutural basin may serve as a model for the early history of A and B. Section A based on reflection section by Watkins et al., (1977). Section B modified after a section by and with permission of J. Stout, Shell Canada Ltd.

27

The origin of the Sverdrup Basin is even less known that that of the Gulf Basin. Its history is intimately linked with the mysterious origin of the Amerasian Arctic Basin (for a review, see Bally, 1976). The Sverdrup Basin is described by Meneley *et al.* (1975), Baker *et al.* (1975), and Sweeney (1976). Meneley *et al.* document Late Pennsylvanian to Early Permian blockfaulting which to us suggests a Late Paleozoic thermal origin of the basin. The Sverdrup Basin is located entirely on a Paleozoic basement. During its Mesozoic subsidence, the basin became isolated from the Arctic Ocean to the north. Widespread diapiric folds and associated faulting have led many petroleum geologists to compare the Sverdrup with the Gulf of Mexico Basin, although the Sverdrup Basin was compressed during its late history.

Typically, the basins shown on Figure 15 contain very thick evaporite sequences suggesting that they may have been isolated from the main oceans and that they desiccated much like the Messinian Mediterranean Sea (see Hsü, 1972b; Hsü *et al.*, 1978). Thick Mesozoic and Tertiary clastic and carbonate sequences indicate massive subsidence well after the initiation of these basins. Whereas this was certainly in part due to sedimentary loading, there must have been an additional, as yet unknown, driving force.

Structural styles in the Gulf of Mexico basin class are dominated by gravity tectonics (salt and/or shale diapirism, listric normal growth faults). However, the folds of the Sverdrup Basin appear to be mostly due to relatively mild compression combined with diapirism in Late Cretaceous to Miocene times.

12. Basins located on pre-Mesozoic continental lithosphere

These are the cratonic basins shown on Figure 9. They differ from the preceding basins because they lie on continental crust and are not related to an adjacent spreading ocean. A comparison of Figure 9 with Figure 18 shows that the boundary between cratonic basins and foredeeps may often be arbitrary.

121. Cratonic basins: Cratonic basins are deceptively simple. Whereas we have plausible models for many more complex geological features, we still lack convincing explanations for the development of "simple" cratonic basins (the syneclises of the Russian literature) and their positive counterparts, the cratonic arches (anticlises). Our classification attempts to differentiate cratonic basins underlain by simple early graben systems (1211, i.e., North Sea) from those that overlie an earlier basin and range type configuration related to backarc spreading or shear tectonics (1212, i.e., West Siberia).

In the recent past a number of authors have tried to apply to the cratonic Michigan Basin the thermal contraction concepts that succeeded so well in explaining Atlantic-type basins. The Michigan Basin was selected because of its basic simplicity, the abundance of stratigraphic data and because gravity and reflection seismic data suggested that the basin was underlain by a late Precambrian graben system. A recent well and a COCORP reflection line that is to be published suggest that the fill may consist of Keweenawan red beds (about 1 000 Ma) and an igneous rock metamorphosed about 500-800 Ma ago (van der Voo and Watts, 1978).

For their models Sleep and Snell (1976) visualize a visco-elastic plate about 100 km thick and show that erosion of the flanks of the Michigan Basin may be due to rebound of the basin periphery that was dragged down during early stages of basin subsidence. Haxby *et al.* (1976) assume a thinner (30 km) elastic slab and fit the basin geometry to their model. Because the individualization of the Michigan Basin and because much of its subsidence occurred after the Ordovician, it would be desirable to be able to demonstrate the existence of an early Ordovician heating event. All evidence. however, suggests that no such event occurred, and therefore thermal contraction does not explain the subsidence in that basin. Of course, the Precambrian basin underlying the basin was formed too early to help explain the Paleozoic subsidence. Other authors (e.g., McGinnis, 1970) explain the subsidence by deep, dense intrusions in the lithosphere. Others (Haxby et al., 1976) propose that a large basic intrusion beneath the basin was transformed to eclogite. Unfortunately, convincing geophysical data for the postulated heavy masses underlying the basin are lacking. Similar basin-forming mechanisms have previously been proposed by Beloussov (1962, 1975).

Burke (1976) describes the young Chad Basin as an active cratonic basin and speculates that this basin developed in response to peripheral thermal uplifts that are associated with volcanics. It will be difficult to demonstrate the applicability of this model to older cratonic basins, because according to Burke the evidence of volcanism may be eroded from the uplifts surrounding such older basins. The outcropping arches near some of the American cratonic basins show little if any evidence of past volcanism.

Probably the best-described cratonic basin, involving an earlier rift (1211) is the North Sea Basin (Ziegler, W., 1975; Ziegler, P., 1975, 1977, and this volume). Its pre-Permian "basement" consists of part of the Precambrian Baltic Shield to the east, the Caledonides overlain by the Old Red Basin (a Devonian episutural basin) to the north, and to the south the Variscan foredeep with its underlying platform sediments. A Permian basin that overlies this complex basement and the neighboring outcropping Oslo Graben suggest early rifting.

However, in the North Sea Basin itself, the important rifting occurred during Triassic and episodic rifting continued until the end of the Cretaceous, with most of the growth occurring from Triassic to Lower Cretaceous.

The rifting sequence is overlain by a thick Tertiary clastic basin. Crustal studies indicate that crystalline crust is attenuated under the main Viking Central Graben. Other rift systems like the Oslo Graben and the Danish-Polish furrow suggest that rifting was quite widespread in northwest Europe, but never quite led to the inception of an ocean. The episutural nature of the Old Red Basin and possibly of the Permian sequence suggest Upper Paleozoic crustal attenuation preceding the attenuation by the main rifting event. For more details on the North Sea Basin, see also Woodland (1975).

The West Siberian Basin could serve as a type for a cratonic basin located on a former Pannonian-type backarc basin (1212). Detailed information on West Siberia is contained in Kontorovich et al. (1975), Rudkevich (1976) and in an English summary by Clarke et al. (1977). A basement map by Surkov (1975) and Figure 16 show the Mesozoic Siberian Basin to be mostly underlain by the Variscan Megasuture that also is outcropping in the Urals, by Late and Early Caledonian (Salairian) folded belts in the southern sector and by Late Precambrian (Baikalian) folds and much older Precambrian terrains to the east.

Superimposed on this varied basement are a number of small late Hercynian basins, as well as a series of narrow Triassic rifts that trend in a southerly direction as an extension of the vast expanse of the Triassic Tunguska traps of the Siberian Platform to the northeast (see Fig. 17). Clearly, a late to post-orogenic (post-Hercynian) thermal event is responsible for the initiation of the West Siberian Basin and the structural evolution of the Upper Paleozoic and Triassic "intermediate structural level" described by Zhavrev et al. (1975).

The subsequent subsidence of the basin involves first the deposition of continental Lower Jurassic beds followed by Upper Jurassic marine sequences containing significant source beds. These were deposited in an area virtually surrounded by elevated land masses. This provided a restricted environment which could well be compared physiographically — but not climatically and genetically — with the environment that today allows the deposition of organic-rich muds in the Baltic Sea (Schott, 1968). The Cretaceous of West Siberia is marine in the basin center and continental at the margins. It is capped by a thin veneer of continental Tertiary deposits.

It is tempting to explain the subsidence of the West Siberian Basin in terms of cooling with an exponential decrease of subsidence rates, following a Triassic rifting event. Precise subsidence rates are not currently available, but a generalized subsidence curve published by Rudkevich (1976) suggests a much more complex evolution with rates of subsidence

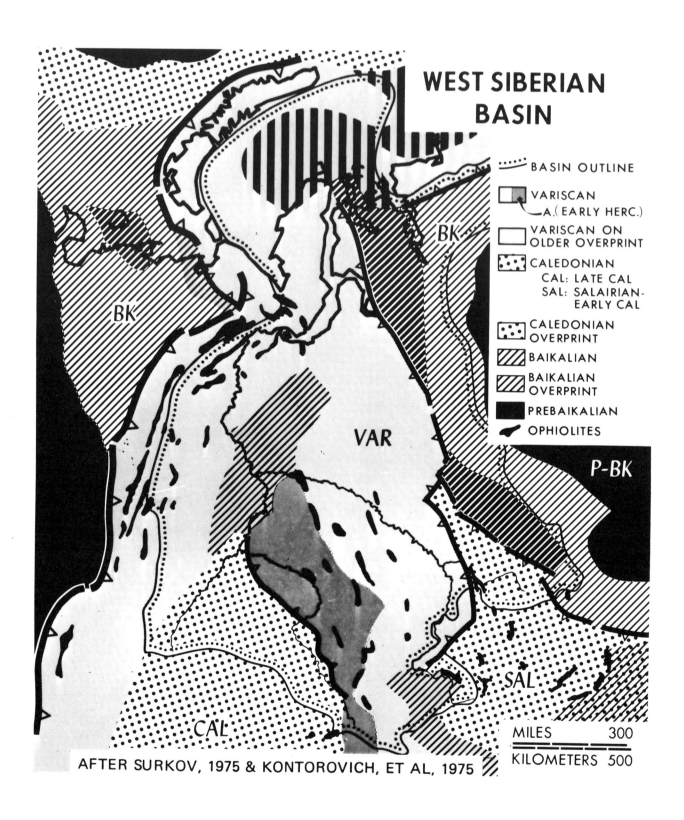

Fig. 16. The basement underlying the West Siberian Basin, after Surkov (1975, in Kontorovich *et al.*, 1975).

30

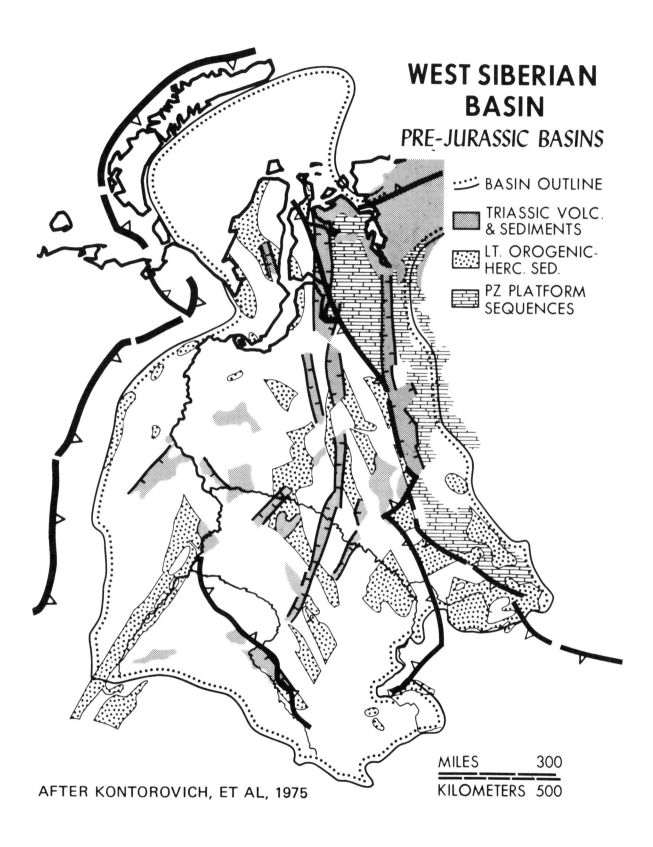

WEST SIBERIAN BASIN

PRE-JURASSIC BASINS

- ⋯ BASIN OUTLINE
- ▨ TRIASSIC VOLC. & SEDIMENTS
- ▨ LT. OROGENIC-HERC. SED.
- ▨ PZ PLATFORM SEQUENCES

MILES 300
KILOMETERS 500

AFTER KONTOROVICH, ET AL, 1975

Fig. 17. Pre-Jurassic basins of intermediate structural level of West Siberian Basin after Kontorovich *et al.,* (1975).

31

apparently decreasing from Lower to Upper Jurassic, increased from Upper Jurassic to Mid-Cretaceous and decreasing again from Mid-Cretaceous to Holocene. The situation is further complicated by the fact that the center of the West Siberian Basin is today an area of very high geothermal gradients (Tamrazyan, 1971). The Moho underlying the West Siberian Basin is at about 35-50 km depth and only slightly shallower than the present 45 km Moho depth beneath the Urals to the west and beneath the Siberian Platform to the east (Belayevsky *et al.*, 1968).

The evolution of the cratonic basins of Brazil differs substantially from the basins described so far. The evolution of the major basins has been summarized by Bigarella (1973) for the Amazon and Maranhao. Putzer (1962), Soares and Landim (1976), and de Quadros (1976) described the Parana Basin. In simple terms all three basins share a lower Paleozoic clastic glacial — continental — marine cycle (Siluro-Devonian in the Amazon and the Parana; Ordovician-Lower Carboniferous in the Maranhao). They also share another Upper Paleozoic dominantly continental clastic cycle (Upper Carboniferous and Permian for Amazon and Parana, Upper Carboniferous-Triassic with some evaporites in Maranhao).

Of great importance and typical for the Brazilian basins is the widespread igneous activity during the early Mesozoic. In the Amazon and Maranhao Basins tholeitic magmas were intruded in the forms of numerous sills and dykes, and extruded as more limited lava flows during 120-220 Ma (de Rezende, 1971; de Rezende and de Brito, 1973). It appears that these igneous events led to the individualization of the Amazon Basin, which may well be a failed rift system.

In the Maranhao and Parana Basin the age of basaltic lavas is younger, i.e., between 120-130 Ma (Bigarella, 1973). Particularly interesting is the fact that the extensive Serra Geral basalts that underlie and flank the Parana Basin find their equivalent counterpart in the Etendeka Plateau of southern Africa (Pacca and Hiddo, 1976). Soares and Landim (1976) describe how an Afro-Brazilian arch (today preserved as the Punta Grossa and Serro Do Mar Arches) formed during Triassic and Jurassic times and preceded the lower Cretaceous igneous event that led to the individualization of the Parana Basin.

Some Paleozoic growth is indicated on the published sections across all three major Brazilian basins. But the basic shape of the basins was acquired during and following the early Mesozoic igneous events. The shallow upturned rims of the basins suggest that subsequent subsidence due to lithospheric cooling was minimal.

All the examples we gave show that we are far from understanding the development of cratonic basins. The models that have been proposed do not satisfy the observations, and the knowledge we have of the initiation of cratonic basins is not sufficient to provide useful answers. There is even considerable doubt in the authors' minds whether our subdivision in two types (1211 and 1212 on Table 1) is justified. However, we have yet to see convincing seismic data that any cratonic basin is not somewhere underlain by a rifted structure.

2. Perisutural basins, on rigid lithosphere associated with the formation of compressional megasuture

Perisutural basins are adjacent to megasutures and are located on a rigid lithosphere (Fig. 18).

In our classification we differentiate deep-sea trenches that are caused by B-subduction processes, from foredeeps that relate to A-subduction boundaries. Both basin types may be viewed as homologous moats that are superimposed on ramps dipping toward and under the megasuture. In Central Asia and China, an A-subduction boundary does not exist; instead we selected the envelope around Mesozoic and Cenozoic felsic intrusives as the megasuture boundary. Basins that appear to be associated with the formation of that boundary will be referred to as Chinese-type basins.

PERI-SUTURAL BASINS

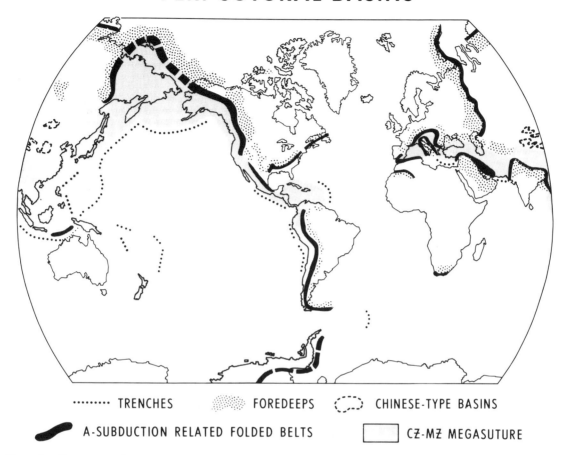

TRENCHES ⋯⋯⋯   FOREDEEPS ▒▒▒   CHINESE-TYPE BASINS ⸨⸩

A-SUBDUCTION RELATED FOLDED BELTS ━   CZ-MZ MEGASUTURE □

Fig. 18. Perisutural basins.

Foredeeps that on Figure 18 are not associated with the Cenozoic-Mesozoic boundary are all associated with Paleozoic A-subduction boundaries.

21. Deep sea trenches

The setting of deep sea trenches is shown on Figures 5, 18 and 27. On the arc side, trenches are flanked by the accretionary subduction complex or inner slope. Oceanward, the outer slope of the trench often merges into an outer rise that crests 100 m higher and lies some 100-200 km from the axis of the trench (Watts and Talwani, 1974). This outer high is explained by most authors to be due to flexure of an elastic lithosphere with a load on one end (Caldwell *et al.*, 1976). Normal faulting is commonly associated with the formation of the rise. Occasionally, oceanic crust underlying the outer slope is thrust faulted (southeast Pacific: Kroenke, 1972; Peru Trench: Kulm *et al.*, 1973).

Seismic profiles across typical trenches show pelagic sediments concordantly over rugged oceanic basement. The arcward dip of the basement extends under the trench and can commonly be followed for some distance under the accretionary wedge of the inner slope. Both basement and overlying sediments are often broken by normal faults.

Overlying the outer slope ramp is the trench fill (or trench wedge) with subhorizontal layers onlapping onto the outer slope sequence, but locally deformed on the arc-side of the trench (von Huene, 1974; Scholl *et al.*, 1974). Schweller and Kulm (1978) summarize the

33

filling modes of deep sea trenches. They differentiate four important facies: pelagic sediments of the oceanic plate, terrigenous plate sediments, the trench wedge, and deep sea fan deposits. The filling of a trench is a function of sediment supply and the convergence rate of the plates. Low sediment supply (often climatically controlled) and high to moderate convergence rates can lead to essentially empty trenches, while high sediment supply and low convergence rates will cause the trench to be buried under large submarine fans (i.e., Astoria Fan of the U.S. west coast and Bengal Fan overlying the Indo-Burma-Andaman Trench). Moderate sediment supply and moderate convergence rates lead to the formation of classical trench wedges that are often filled by turbidites that were transported along long axial channels. Fine illustrations of trench profiles can be found in papers by several authors in Talwani and Pitman (1977).

In the literature much attention has been paid to the Timor Trench. While in the Java Trench to the west, we can observe clearcut B-subduction, in the Timor Trench we witness a case of incipient A-subduction, which therefore will be discussed under the next heading.

Because of their great water depth, deep sea trenches cannot at this time be viewed as reasonable targets for petroleum geologists. Their filling history is, however, of some importance because trench sediments can end up in the deformed accretionary wedges of the B-subduction zones. For instance, when deep sea fan facies are involved, the possibility exists for potential clastic reservoirs in the accreted structures. However, it must be stated that pelagic sediments and deep sea fan deposits of the Pacific and Indian Oceans are not known for their organic richness.

22. Foredeep and underlying platform sediments

Foredeeps[1] are perisutural basins related to A-subduction boundaries. Compared to deep sea trenches, they can be viewed as homologous moats superimposed upon ramps dipping toward the megasuture. Both are in dynamic systems, and as a result, with time the depocenters of these basins commonly migrate away from the megasuture axis. Whereas trench fill deposits occur in oceanic crust domains, foredeeps develop over platform complexes overlying continental crust. In the classic literature, early foredeep deposits are commonly characterized by deeper water "flysch"-type sequences. Late foredeep deposits are characterized by shallower water "molasse"-type sediments. However, because these terms have been poorly understood and often misapplied, we suggest that they be abandoned.

We like to include the underlying platform and its complete updip extension in our definition of a foredeep. Here our classification differs from that adopted by McCrossan and Porter (1973) because these authors split the foredeep — as we define it — into Craton Centre basins with thin sequences mostly deposited during periods of maximum transgression and Craton Margin basins that correspond to the thickened wedge of sediments that occurs downdip from the Craton Centre province. The two differing approaches reflect the difficulty in drawing boundaries between foredeeps, cratonic basins, and the updip extensions of both of these basin types.

The fluid systems and migration paths for hydrocarbons have to be viewed together for the foredeep and its underlying platform. Also, in many cases, it can be shown that source beds in the underlying platform and in the basal foredeep sequence mature by virtue of the load provided by the thick overlying foredeep sequence. We therefore prefer to separate the complete foredeep ramp from individual cratonic basins. We do realize, however, that the

---

[1] Because "foredeep" is so commonly used, we would like to retain the term (see also Dennis, 1967). We would like to remind the reader that the term refers to a basement depression and has no morphologic-bathymetric connotation. We prefer "foredeep" over the synonymous "exogeosyncline" of Kay (1951) because, as mentioned earlier, we would like to abandon the old geosynclinal nomenclature.

"pigeonholing" process is particularly difficult across regional arches of differing magnitude that may separate cratonic basins from foredeep ramps.

We subdivide our foredeep class into: 1. Foredeeps on ramps with little or no contemporaneous block faulting, and 2. Foredeeps on ramps dominated by contemporaneous block faulting. In both cases, however, the platforms may have a history of aulacogen or graben development that predates the foredeep development.

221. Foredeep ramp with buried grabens, but with little or no blockfaulting: The basic sequence of a foredeep and underlying ramp includes, from top to bottom:

1. The foredeep sequence *sensu strictu,* that is often referred to as a clastic wedge (King, 1977) and has a dual sediment source: one from the adjacent A-subduction related folded belt and another from the adjacent cratonic area. The clastic influx correlates with uplifts in neighboring fold belts as suggested by correlations between K/Ar radiometric ages in orogenic core areas and sandstone depositional phases in the foredeep (Hadley, 1964; Bally *et al.,* 1966). Under certain climatic conditions, foredeep sequences may be filled by carbonates (i.e., Middle East). Shales in the lower portions of foredeep sequences often qualify as source beds. Syndepositional folding and/or faulting is observed in a number of foredeeps (Miall, 1978).

2. An unconformity that may occur between the platform complex and the foredeep *sensu strictu* or else a sedimentary wedge containing numerous unconformities (Bally *et al.,* 1966). This marks the transition to the new foredeep regime, and typically affects very large areas of the whole adjacent continent.

3. A platform complex, that is viewed as the updip extension of an ancient passive margin (shelf or miogeosyncline). Within this complex may be early, thicker sequences which relate to the presence of aulacogens or graben systems. These graben thicks often trend at high angles to the adjacent fold belt (Shatsky, 1946, 1961; Hoffman *et al.,* 1974). Overlying the rifted sequence is an essentially undisturbed sequence of sediments that represent an ancient Atlantic-type "drifting sequence".

Let us illustrate the three basic sequences listed above by a few examples and show that all stages are not always well represented. The Upper Paleozoic Volga-Ural foredeep *sensu strictu* overlies the Russian Platform and contains sediments shed from the Urals (Fig. 19; Nalivkin, 1976). Much of the updip extension of the foredeep sequence is in carbonate facies. No significant unconformity separates the foredeep sequence from the underlying platform sequence.

A simplified cross section (Fig. 19) reveals a number of underlying Proterozoic aulacogens. A map shows that much of the rifting took place during the Late Precambrian-Vendian and Riphean time (Fig. 20a). The oldest rift (Kaltasa) located in front of the Middle Urals was formed between 1 000-1 400 Ma. It and all other somewhat younger aulacogens continued to subside through late Riphean and early Vendian time (1 000-600 Ma). A sequence of late Vendian-Cambrian (Fig. 20b) sediments could well be interpreted as corresponding to the "drifting sequence" of an Early Paleozoic opening ocean that overlies ancient failed arms related to triple junctions. In the southern Russian Platform, no equivalent late Vendian-Cambrian sequence was deposited; and in fact, rifting there appears to have lasted into the upper Paleozoic. The Dnepr-Donets Aulacogen, for example, opened in Mid-Devonian (Fig. 19, Fig. 20c). Higher and relatively unfaulted Permian and younger drift sequences suggest the latest Paleozoic opening of an early Caspian ocean basin to the southwest that corresponds to the Pericaspian Basin (*see also* Burke, 1977).

RUSSIAN PLATFORM

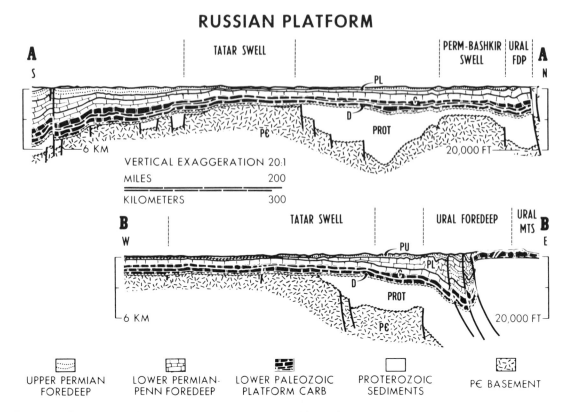

Fig. 19. Russian platform geological sections, modified after a section by and with permission of BEICIP (1976). Lines of sections indicated on Figure 20C.

The Western Canada foredeep (Fig. 21) has its continuation into Montana and the Southern Rocky Mountains. The axis of the foredeep *sensu strictu* migrated with time and with the advancing A-subduction front (Bally *et al.*, 1966; Armstrong and Oriel, 1965). Deposition began with continental clastics in the Neocomian and Aptian. It was followed by an Albian to mid-Upper Cretaceous marine interval with a number of transitional and marine sands that alternate with organic-rich shales that qualify as source beds. Finally, the uppermost Cretaceous consists of a regressive sequence leading from marine to transitional and continental clastics. For patterns of deposition, see Eisbacher *et al.* (1974), Alberta Society of Petroleum Geologists (1964), Rocky Mountain Association of Geologists (1972), and Cook and Bally (1975). The alternation of continental and marine cycles with basal source beds is characteristic for many foredeeps (East Venezuela Basin, Meta-Barinas Basin, Putumayo-Maranon of Equador and Peru, and Alpine Molasse Basin). A dual sediment source — one from the mountains, the other from the craton — is also typical.

Coastal onlap of the foredeep on the adjacent folded belt is no longer intact in Canada but in central Utah, Armstrong (1968) has illustrated the onlap of Lower and Upper Cretaceous conglomeratic sequences on the older Sevier orogenic belt.

The underlying platform sequence of Western Canada includes Proterozoic (Beltian and Windermere) sediments and Paleozoic carbonates. The latter include the prolific hydrocarbon-bearing reefs of Alberta. Interposed between the Paleozoics and the overlying foredeep *sensu strictu* is a Permo-Jurassic sedimentary wedge containing multiple unconformities (Bally *et al.*, 1966). To the east, the transitional wedge sequence merges into one unconformity below the foredeep sequence that truncates a pre-Cretaceous subcrop. The wedge sequence and its underlying pre-Cretaceous subcrop is indicative of supra-regional lithospheric tilting movements that affected the whole western craton of North America prior to initiation of a foredeep deposition. The pre-Cretaceous subcrop map is well illustrated in Cook and Bally (1975, p. 207). Other excellent references to the Paleozoic and Mesozoic evolution of the region can be found in Alberta Society of Petroleum Geologists (1964); Rocky Mountain Association of Geologists (1972); Parsons (1973), and Torrie (1973).

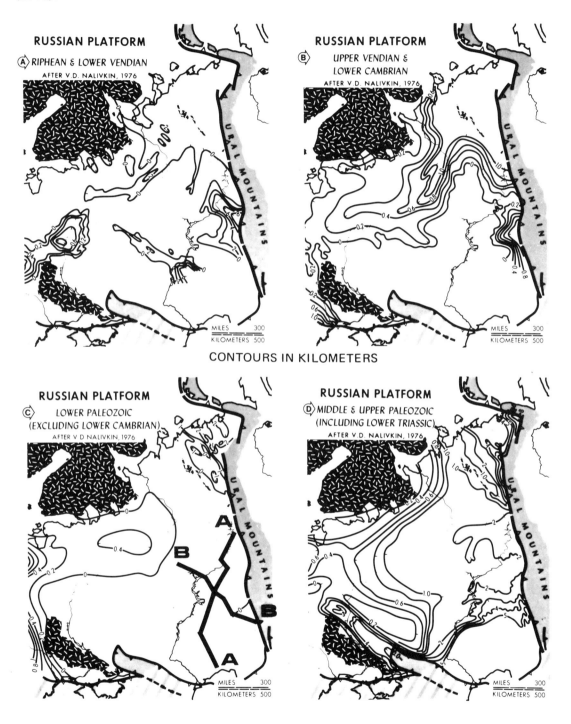

Fig. 20. Four stages of the evolution of the Russian platform, the Volga Ural foredeep and the Dnepr-Donetz Aulacogen. Simplified after Nalivkin (1976).

37

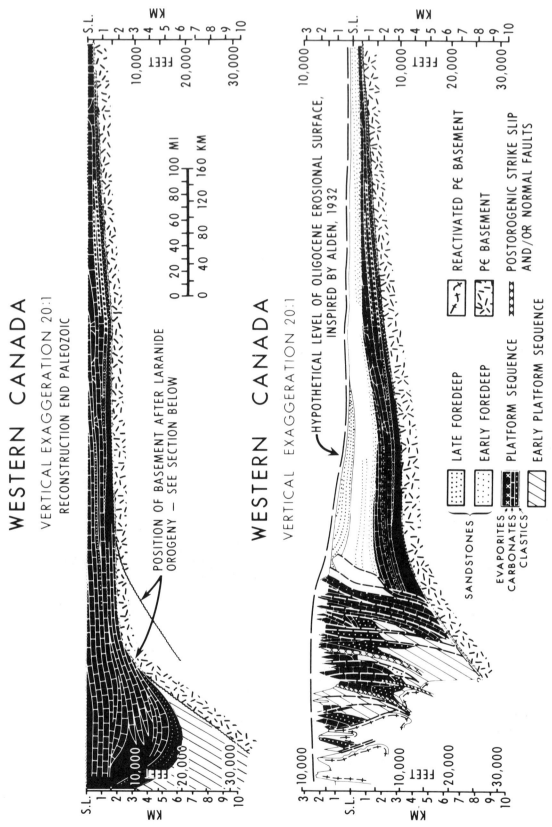

Fig. 21. Upper section: Paleozoic reconstruction of the Western Canada Basin; Lower section: cross section across Western Canada Basin (east on the right).

38

Rifting antecedents that underlie the Paleozoic platform are ill-defined. The distribution of thicker and thinner Proterozoic rifts may be associated with the initial formation of the Cordilleran geosyncline (see Kanasevich *et al.,* 1968; Stewart, 1972; Burke and Dewey, 1973; Wright *et al.,* 1974; Harrison *et al.,* 1975; Burke, 1978). The classic Late Proterozoic Athapuscow Aulacogen (Hoffman *et al.,* 1974) is known to extend in subsurface under the northeastern portion of the western Canada foredeep. As with the Russian examples, it is difficult to work out a precise time relationship between the rifting phase proper and the overlying Late Proterozoic sequence that may well represent Atlantic-type passive margin deposition of a "drifting sequence." Note that farther south some postulated Proterozoic rifts have been inverted in Laramide times, e.g., the Big Snowy and the Uinta Mountains.

The Appalachian Basin (Fig. 22) is a complex basin which includes several Paleozoic foredeeps (Colton, 1970; Meckel, 1970; McIver, 1970). Cambro-Ordovician carbonates overlie a Late Precambrian-Early Cambrian rift system, and are followed by a Mid-Ordovician-Silurian foredeep clastic wedge reflecting Taconic deformation to the east. At the base of the foredeep clastics are organic-rich shales (e.g., Utica). A brief carbonate interlude during Silurian and Mid-Devonian is succeeded by a younger Upper Devonian to Pennsylvanian foredeep clastic wedge. The base of that wedge again consists of black shales. A similar Devonian clastic wedge occurs in the foredeep of the Innuitian Fold Belt of Northern Canada. The Mid-Devonian Canol shales of the Northwest Territories are the basal source beds.

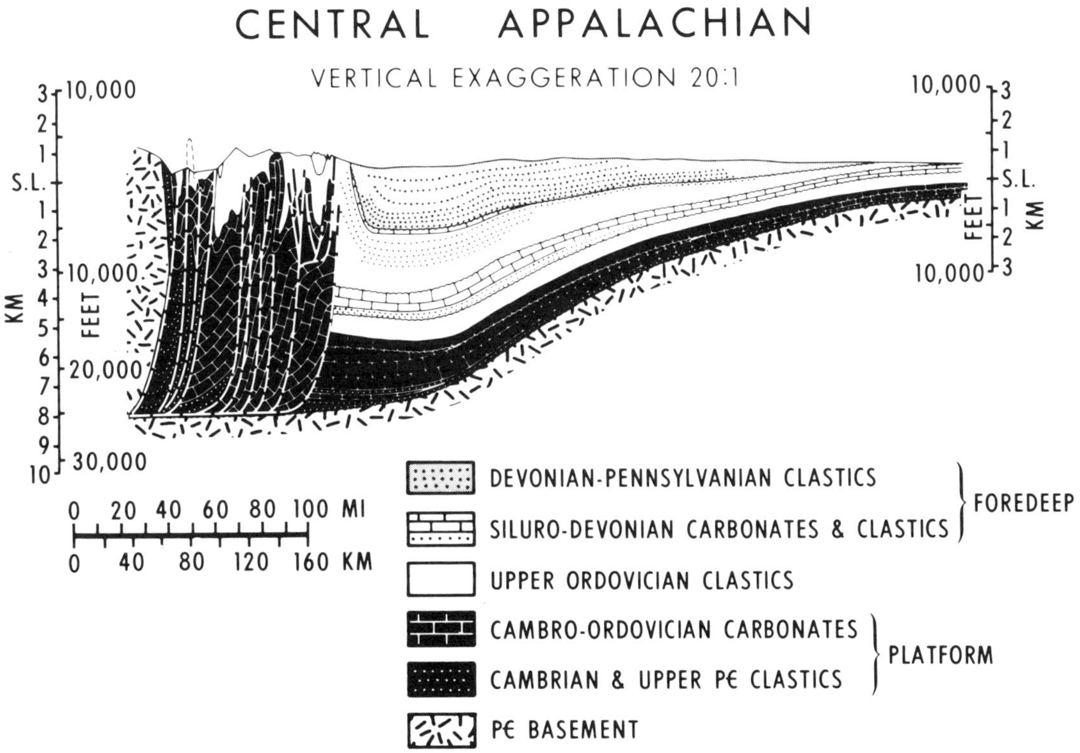

Fig. 22.   Cross section across Central Appalachian Basin (northwest on the right).

On a larger scale and immediately preceding the deposition of the Upper Paleozoic clastic wedge is a continent-wide unconformity. A pre-Upper Devonian subcrop pattern is observed that suggests broad warping of the North American lithosphere. This event formed two supra-regional arches, i.e., the Transcontinental Arch that extends from Minnesota to Arizona, and the Peace River Arch in Western Canada (Bally, 1975; Cook and Bally, 1975, p. 75). This phase is analogous to the Pre-Cretaceous tilting that developed in Western Canada prior to the foredeep sequence.

Descriptions of other foredeeps can only briefly and selectively be quoted. Galley (1958, 1971), Hills (1968), Hill (1971), and Nicholas and Rozendal (1975) discuss the setting of the foredeep that is associated with the Upper Paleozoic Ouachita-Marathon Fold Belt. The Southern Oklahoma Aulacogen (Hoffman et al., 1974) is a major, lower Paleozoic rifting feature that divides the underlying platform sequence into separate areas to the east and to the west. The foredeep as now exposed in the Ouachitas contains widespread Pennsylvanian turbidite sequences. Graham et al. (1975) postulated that these deposits swept into the basin laterally in an overall setting similar to the present Bengal Fan. One may add to this the general observation that foredeeps that are filled with thick turbidite sequences often are also characterized by synsedimentary down to the "folded belt" normal faulting. Such faulting is observed all along the Ouachita foredeep (Buchanan et al., 1968; Berry and Trumbly, 1968; Hopkins, 1968) and, in a very similar setting, in the deep structures of the St. Lawrence Lowlands of Quebec. The normal faulting suggests extension of an attenuated lithosphere in response to flexural lithospheric bending, similar to phenomena observed on the outer rise of deep sea trenches.

Summaries of the evolution of the Swiss Molasse Basin and its underlying pre-Tertiary subcrop have been given by Büchi et al. (1965, 1977); the foredeep of the Eastern Alps and the Carpathians is described by several authors in Mahel (1974b); and finally a fine description of the hydrocarbon geology and habitat of the Rumanian foredeep is offered by Paraschiv (1975). The Bureau de Recherches Géologiques et Minières et al. (1973) provided another fine illustration of foredeep evolution for the Aquitaine Basin.

We have classified the whole Middle East, exclusive of the Zagros fold belt and its continuation into Turkey, as a foredeep. The enormous economic importance of this area does warrant a detailed description, but few overviews have been published; and fortunately, an excellent up-to-date summary by Murris (this volume) is available to the reader. In simple terms, the Arabian Platform can be viewed as a faulted ramp dipping towards the Zagros Mountains. There is as yet no evidence for any pre-Paleozoic rifting phase but the extensive presence of the Infracambrian Hormuz salt formation in the southern Arabian-Persian Gulf and the Zagros Range (Stöcklin, 1968) suggests an incipient Atlantic-type margin for the Early Paleozoic sequence. The uniqueness of this Atlantic-type margin can be perceived on the Triassic reconstruction of Figure 14. It can be seen that at that time the Arabian Platform was part of a northernmost Gondwana passive margin that extended eastward into northern India (with the Infracambrian salt of the Salt Ranges) and over to the west coast of Australia. There the Fitzroy trough of the Australian Canning Basin and the southern end of the Bonaparte Gulf represent Mid-Paleozoic failed arms of triple junctions that faced the southernmost margin of the same Tethys Sea.

The subsequent history of the Middle East is one of a steadily narrowing embayment and the final closure of the Mesozoic Tethys. Such an embayment may offer a fine setting for the restricted environments that led to the source rock deposition and the enormous hydrocarbon richness of the Middle East. The structural evolution of the northern margin of the Tethys embayment is highly complex and includes the incorporation of small cratonic masses such as the Lut-block and possibly Tibet into the Alpino-Himalayan chain (Stöcklin, 1974; Stoneley, 1974; Takin, 1972). The southern Arabian Platform remained a passive Atlantic-type margin throughout the Triassic, Jurassic, and Early Cretaceous. Low-relief basement structures appear to be growing throughout this time. A major, pre-Cenomanian unconformity is due to regional warping of the Arabian Platform. Another unconformity follows at the end of the Cenomanian. In Late Cretaceous time, an arc now represented within the Zagros Range collided with the Arabian Peninsula (see Ricou, 1971; Bird, 1978), and a Late Cretaceous and Tertiary foredeep sequence loaded the Arabian platform. Much of that foredeep sequence appears to be deformed today in the Zagros Mountains.

The inception of a foredeep merits special attention and can be seen in areas where B-subduction zones pass laterally into A-subduction zones as can be seen in places where a continent reaches a B-subduction boundary. Such is the transition from the Bengal Gulf into the

Indian Platform to the north (Curray and Moore, 1974; Stoneley, 1974). Probably the most detailed description of an incipient A-subduction occurs in the Timor and Arafura Seas where the continental mass of Australia collides with the Indonesian island arc. The geology of Timor has received much attention and has recently been summarized by Audley-Charles et al. (1974, 1977), Audley-Charles (1976), Carter et al. (1976) and Curray et al. (1977). The detailed reflection lines published by Montecchi (1976) display the northerly-dipping ramp of the Atlantic-type margin of continental Australia that is broken by a number of normal faults. This ramp can be followed for some distance under an accretionary wedge that underlies the inner trench wall. The trench fill has subhorizontal layers that onlap onto the cratonic ramp. The arc side of the trench fill is folded and overthrust by the adjacent accretionary wedge.

Similar examples of submarine A-subduction have been published by Finetti and Morelli (1972) and Finetti (1976) for the transition from the Ionian Sea into the southern Adriatic Sea. Cruz et al. (1977) have provided detailed seismic illustrations of the onlap relations of a submarine foredeep sequence on the adjacent folded belt in the Veracruz Basin of Mexico. There, Miocene conglomerates were shed into a bathyal environment to provide reservoirs for important gas accumulations.

The interesting quantitative models of Bird and Toksöz (1977) and Bird (1978) have yet to be tested by detailed studies of the subsidence history and the record of thermal history of foredeeps.

A glance at Figure 21 shows some aspects that may need explanation by such models. A comparison of the end-Paleozoic reconstruction of Western Canada with a present profile indicates that subsidence of the foredeep may be accentuated by loading of thrust sheets on the downdip extension of the foredeep ramp. The dip of that ramp is determined with reasonable accuracy by reflection seismic data. Note that the basement hinge-zone has shifted with time in keeping with the concept of a migrating foredeep axis. We ask: is the loading by thrust sheets and the sediment filling of a marginal moat sufficient to explain the subsidence of a foredeep?

There is reason to suspect that additional factors may play an important role. Laubscher (1976, 1978) called attention to the paradox of contemporaneous normal faulting (sub-parallel to the strike of the folded belt) and compressional features in the foreland of the Alpine folded belt. As mentioned earlier, the same phenomena are also seen in the foreland of the Ouachitas-Marathon belt and the St. Lawrence Lowlands. Laubscher suggests that the main masses of the neighboring folded belt constitute an ''orogenic float'' that is cushioned by masses of comparatively low viscosity. Thus, in foreland areas, structurally detached masses remain afloat while the underlying lithosphere is being subducted. Decoupling by the low viscosity cushion allows compressional forces to prevail within the ''orogenic float'' while the underlying foreland lithospheric bulge shows extensional normal faulting much like the homologous lithospheric bulge associated with a deep sea trench. The existence of low viscosity-low velocity layers is suggested by interpretations of crustal seismic profiles across the Alps and Apennines (Angenheister et al., 1972; Giese et al., 1973).

Additional aspects have to be taken into account in modelling foredeep basins. The frequent occurrence of supra-regional unconformities preceding the inception of foredeeps is poorly mapped and explained. The unconformities suggest extensive flexuring of continental lithosphere, preceding the formation of the foredeep sensu strictu.

Finally, the ultimate configuration of the present-day morphology of many foredeep areas suggests widespread late uplift and tilting away from the mountain range. Bally et al. (1966) refer to a late morphorogenic phase in the Northern Rocky Mountains. Similar observations are made in the Southern Rockies (Epis and Chapin, 1975). The dashed line on Figure 21 is not supported by accurate morphologic observations, but suggests a projected position of an Oligocene terrace level over the summit level of the adjacent Rocky Mountains, following

41

reasoning that has been developed by Alden (1932). According to these observations, the subsidence of foredeeps is reversed by uplift following cessation of A-subduction.

222. Foredeep dominated by blockfaulting: The simple concept of a foredeep ramp flanking the A-subduction boundary of megasutures does not hold for a number of areas. Most notably these exceptions include that portion of Soviet Central Asia that lies north of the fold belt that extends from the Pamir, over northern Afganistan into the Elburz Mountains of Northern Iran; the Venezuelan and the Columbian Andes; and the Southern Rocky Mountains of the United States. In all these areas the foreland is disrupted by blockfaulted structures that raise the underlying crystalline basement to form elevated mountains.

Because the situation is best documented in the United States Rocky Mountains, our discussion will be limited to that area. The A-subduction boundary of the Western Cordillera and its characteristic décollement structures can be followed southward with continuity into the vicinity of Las Vegas, Nevada. Drewes (1978) has attempted to trace a somewhat controversial boundary farther southward through southwestern Arizona into the Sierra Madre Oriental of Mexico. Figure 31 illustrates the eastern A-subduction boundary, and also shows areas of the southern Rocky Mountains that are characterized by basement-involved blockfaulting. The age of the blockfaulting is essentially Laramide. This coincides only with the final stages of the thrusting and foredeep development in western Canada and Montana. However, paleogeographic maps of the western United States illustrate that the foreland Rockies had an earlier history distinct from that in western Canada (*see* Cook and Bally, 1975). These antecedents are listed as follows: 1. an Archean core forms the center of the Rocky Mountain province in Wyoming; 2. the pre-Mid Devonian Transcontinental Arch trends in a northeast-southwest direction and extends from Minnesota into Arizona; 3. Pennsylvanian isopachs reflect Late Paleozoic blockfaulting (e.g., Uncompaghre Uplift) intersecting the Transcontinental Arch at right angles. Pre-Pennsylvanian subcrops suggest left-lateral strike-slip offsets of the axis of the Transcontinental Arch along a northwesterly striking fault zone (ibid., p. 105); and 4. Permian, Triassic and Jurassic isopachs and subcrops all show an east-west-trending northern boundary to the province.

In contrast to the above, Cretaceous isopachs (except for the Late Cretaceous) and patterns of sedimentation in the U.S. Rocky Mountains are generally similar to those of the Western Canadian foredeep. The distinctive uplifts and intervening basins (Fig. 23) are Paleocene-Eocene in age and coincide roughly with the timing of the easternmost décollement thrusts of the adjacent A-subduction zone (Armstrong and Oriel, 1965; Dorr *et al.*, 1977; Fig. 23).

The origin of the blockfaulted Central Rocky Mountains has been the subject of much debate. Many authors have viewed the basement blocks as "upthrusts" that were due solely to vertical uplift (e.g., Prucha *et al.*, 1965; Stearns, 1971, 1975). Others have emphasized strike-slip tectonics (Stone, 1970) while still others emphasized an overthrust style (Berg, 1962; Berg and Romberg, 1966). Particularly ingenious was the discussion by Sales (1968) that proposed the superposition of a west-northwest left-lateral couple on an overall compressive system. His plaster models duplicated the overall pattern of deformation quite well.

New light has been shed on the problem by some deep crustal seismic reflection sections obtained by the COCORP consortium (Smithson *et al.*, 1978a, and in press; Brewer *et al.*, in press). Figure 24 incorporates the result of this work and shows that the bounding fault on the west flank of the Wind River Mountains dips at an angle of about 30-35° to a depth of at least about 24 km, and possibly to depths exceeding 30 km. The seismic reflections are from the overthrust fault zone itself.

Thus, here we have one case where the controversy is resolved in favor of an overthrust. This particular thrust can be traced deep into the lower crust and may well merge into an overall decoupling level near the upper mantle-lower crust boundary. Our cross section suggests that such a Laramide intrabasement decoupling level may well underlie the whole

CENTRAL ROCKY MOUNTAINS

WYOMING

TERTIARY — VOLCANICS ▦ · PALEOZOIC ■

SEDIMENT ▦

PRECAMBRIAN SEDIMENT ■

MESOZOIC □ CRYSTALLINE ▨

NORMAL FAULT ⊢⊣ REVERSE FAULT ▲ LINE OF SECTION ·········

THRUST FAULT △ UNDIFFERENTIATED FAULT — (SEE FIG. 24)

COCORP COVERAGE ▬

0   50   100   150 MI.

0   50 100 150 200 KM.

Fig. 23. Geologic sketch map of Central Rocky Mountains with line of section of Figure 24.

43

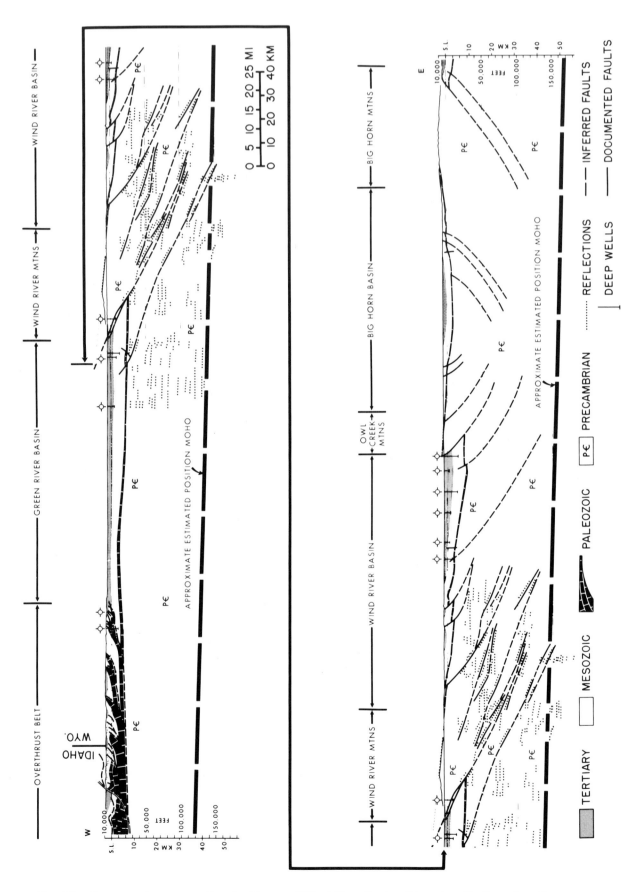

Fig. 24. Cross section across Powder River and Wind River Basins, Wyoming. For line of section, see Figure 23.

blockfaulted foreland Rocky Mountain province. The existence of such a level would permit both over-thrusting and strike-slip tectonics. The difference in tectonic style between the folded belt and the foreland would be that in the former, compressive decoupling would occur within or at the base of the sediment package, while in the foreland Rockies, such decoupling would occur at the crust-mantle boundary zone. Note that in the E-W striking Uinta Mountains, a deep-seated Proterozoic aulacogen was inverted by this process.

The breakup of the ancestral westerly dipping ramp in the Central Rocky Mountains was preceded in Wyoming by a broad regional eastward tilt and the formation of the Moxa-Labarge Arch of the western Green River Basin during Mesaverde (Campanian) time (Gill and Cobban, 1966). The unconformity documented by these authors is also associated with the growth of some of the smaller structures in that area. This suggests that lithospheric "conditioning" for the Laramide breakup occurred already during the Campanian.

The Paleocene and Eocene filling history of these Rocky Mountain basins has been described by Keefer (1965, 1970) and Bradley (1964). A summary is given in the Atlas of the Rocky Mountain Association of Geologists (1972) and Andersen and Picard (1974). The Green River, Washakie and Uinta basins are particularly well known for their lacustrine oil shales (Bradley, 1964). Production from the Green River shales has been established in the Uinta Basin (Lucas and Drexler, 1975; Fouch, 1975).

We believe that compressional deformation leading to blockfaulting such as we find in the Laramide Rockies was also common in Paleozoic time. Similar tectonic processes probably account for the structural evolution of the hydrocarbon-rich Midland, Delaware, Anadarko, and Ardmore basins. The structural growth in these blockfaulted basins coincided with the formation of the Ouachita-Marathon foredeep and adjacent thrust belt.

The inversion of former aulacogens during Paleozoic foreland blockfaulting (including strike-slip faulting) can be observed in the Wichita Mountains and also in the Donbas folds of southern Russia.

In summary, the genesis of blockfaulted basins in the foreland is due to compressional forces emanating from the adjacent megasutures. A decoupling level at about the crust-mantle boundary may be postulated, although more data are required to firm up the concept. The stratigraphic sequence in blockfaulted basins differs from the simple foredeep ramp in that they additionally contain sedimentary sequences directly derived from the adjacent uplifts.

23. Chinese-type basins

The geology of China in many respects is quite unusual. The early history of Asia, and particularly China, involved the amalgamation of several Paleozoic continental blocks and several island arcs into a major continent. Ziegler, A. M. et al. (1977) attempted to trace the migration of some of these blocks into their final assembly as the Laurasian part of Pangea. By the end of the Paleozoic, we recognize three major Precambrian platforms in China that have been successfully captured during the Paleozoic accretionary processes (the Tarim, North China, and South China Platforms — for a simple sketch map, see Dott and Batten, 1971, p. 497). As the later Mesozoic-Cenozoic megasuture formed, further accretion occurred. In this manner, the Lut block of Iran, possibly a Tibetan block, and the Indochina Platform became welded to the Paleozoic nucleus. Finally, in the Upper Cretaceous to Pliocene, the Arabian block collided with Asia, as did the Indian block in Cenozoic time (Molnar and Tapponier, 1975). The complex Mesozoic-Tertiary accretionary processes and collisions are poorly understood in detail; however, it is certain that east of the Pamir Mountains and into northern China, accretionary tectonics were not associated with any décollement fold belts that might be suggestive of A-subduction. Instead, to define the northern megasuture boundary, we have drawn an ill-defined envelope around all Mesozoic and Tertiary felsic intrusives.

Meyerhoff (1970a) and Meyerhoff and Willums (1976, 1978) have summarized the petroleum geology of the basins of China, and Terman (1974, 1976) has produced maps and sketch sections of these basins. The tectonic and geologic setting is shown by the People's Republic of China Geologic and Tectonic maps (1975). The geology is summarized by the Chinese Academy of Geological Sciences (1976a, b) and Huang (1960, 1978).

We include in our Chinese basin category the following major basins: Ordos, Pre-Nanshan, Tsaidam, Tarim, Turfan, and Dzungaria. The dominating characteristic of all these basins is that they are filled with continental Tertiary and Mesozoic sequences and that they are bounded by either wrench faults or else by what appear to be basement-controlled thrust faults similar to those we described from Wyoming. Their origin thus appears to be compressional and similar to the Rocky Mountain basins and the basins of Oklahoma and west Texas. The Chinese basins differ from the North American examples in that they do not contain an obvious A-subduction-related foredeep sequence, and also in that they are filled mostly with continental clastics (Mesozoic-Tertiary). Source beds presumably may be similar to the Green River shales.

The pre-Mesozoic history of the Chinese-type basins is known only in a very sketchy fashion. In the Tarim and Ordos Basin, a Precambrian basement and Paleozoic platform sequence underlies the basin. The Tsaidam, Turfan, and Dzungarian basins are probably underlain by folded and metamorphosed Paleozoic basement.

We exclude several major onshore Chinese basins from the class here discussed: The Szechuan basin is bounded by Tertiary thrusts to the northwest, and a late Paleozoic to Mesozoic folded belt to the southeast. Its history is complex and we have somewhat arbitrarily included this basin in the foredeep category. The North China Sungliao and the Kiangsu basins have all the characteristics of episutural basins and are inside our envelope of felsic intrusions.

3. Episutural basins located and mostly contained in a megasuture

Because these basins are small, it is difficult to show their precise outline on a world map of the size of Figure 25. The formation of most episutural basins is intimately related to subduction processes and the formation of associated marginal seas on active margins (Karig, 1971, 1974; Packham and Falvey, 1970). Active margins today are not only characterized by a high earthquake frequency, but they also are conspicuous for their volcanic activity (Fig. 26) which is presumed to be generated from hydrous, tholeitic, silica-saturated (Ringwood, 1977) magmas that are formed near the Benioff zone at depths of about 80-150 km. Fractionation of these magmas leads to the formation of basaltic to basaltic-andesitic intrusive and extrusive bodies. Erosion of such rocks then leads to the deposition of greywacke and volcaniclastic suites.

Episutural basins are shortlived because they tend to be caught up in orogenic processes that follow their formation. Therefore, the number of presently intact older episutural basins is not great. Examples include: Mesozoic: Sacramento-San Joaquin Basins, the Sungliao Basin, and probably the Gulf of Pohai of China; Paleozoic: the Gulf of St. Lawrence (Fig. 15c), and the Sidney Basin of Australia.

31. Episutural basins associated with B-subduction

Classification and genesis of some of the basin types associated with active margins has been discussed by Karig (1971, 1974). Recent summaries are given by Seely and Dickinson (1977), Toksöz and Bird (1977), and Poehls (1978). The generalized conceptual model and terminology are illustrated on Figure 27, which is based on sketches by Karig (1974), Green (1977), Seely and Dickinson (1977), and Dickinson and Seely (1979).

46

EPI-SUTURAL BASINS

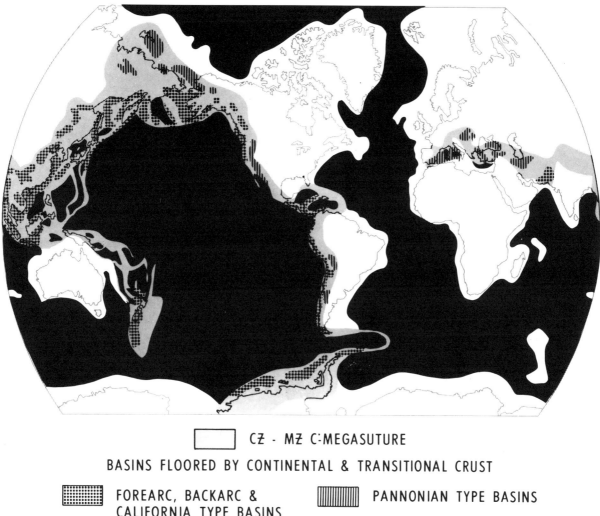

Fig. 25. Episutural basins. Preserved Paleozoic episutural basins (e.g., Gulf of St. Lawrence, Sidney Basin) are not shown.

Note that the width of forearc areas is dictated by the fact that arc volcanoes typically stand between 90 and 150 km above an inclined Benioff zone. Consequently, steep Benioff zones have narrow forearc zones and gently dipping Benioff zones have wide forearc zones.

311. Forearc basins: Seely and Dickinson (1977) in Figure 28 have clarified the nomenclature of forearc basins. The formation of these basins is related to the formation of a subduction complex that consists of terrigeneous as well as scraped-off oceanic sediments. The subduction process is often initiated within oceanic plates; from this it follows, according to Seely and Dickinson, that some oceanic crust can be expected to underlie the forearc basin.

While the arcside flank of forearc basins often involves simple onlap on an arc massif, it is often observed that the oceanward flank of forearc basins appears to be progressively rotated and steepened in harmony with the growth of the deeper imbricated structures of the accretionary subduction complex. At first glance, it would appear that subsidence of forearc basins

47

CENOZOIC - MESOZOIC MEGASUTURE OF THE WORLD AND BOUNDARIES

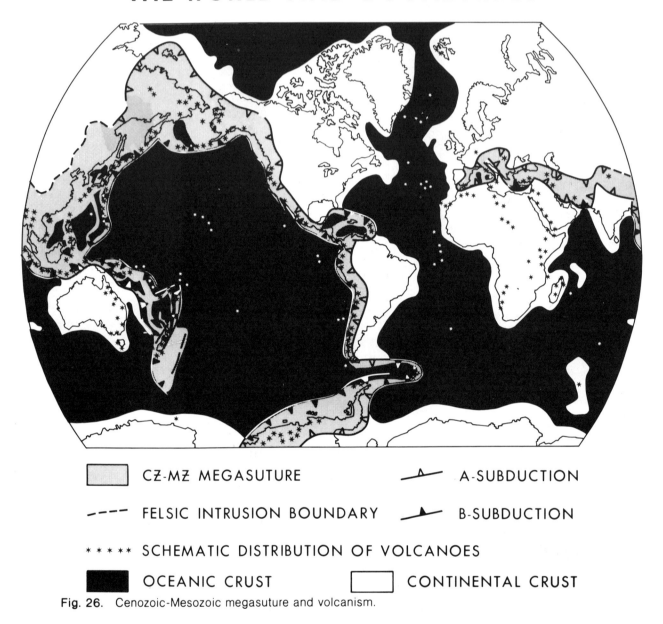

CZ-MZ MEGASUTURE — A-SUBDUCTION

---- FELSIC INTRUSION BOUNDARY — B-SUBDUCTION

* * * * * SCHEMATIC DISTRIBUTION OF VOLCANOES

OCEANIC CRUST — CONTINENTAL CRUST

Fig. 26. Cenozoic-Mesozoic megasuture and volcanism.

is mainly a tectonic response associated with the overall compressional regime of the forearc region. However, as the offshore Colombia section on Figure 7a shows, gravitational factors also come into play. Growth faulting is frequently associated with forearc regions and their unstable slopes. This and the formation of olistostromes (or submarine landslides) is but one variation on the theme of unstable slope tectonics in forearc regions. Extensional normal faulting also dominates in the forearc region of the Marianas Arc (Hussong and Uyeda, 1978).

The well-explored Cook Inlet Basin is an example of a forearc basin (Kirschner *et al.*, 1973).

312.Circum-Pacific backarc basins: Poehls (1978) sums up the various hypotheses that may lead to the creation of backarc (or in his terminology, interarc) basins. In some dynamic models, the subducted slab provides frictional heat that causes melting of mantle material, followed by the ascension of magmas and the formation of a secondary mantle current behind the island arc (*see* Hasebe *et al.,* 1970, and the discussion in Sugimura and Uyeda,

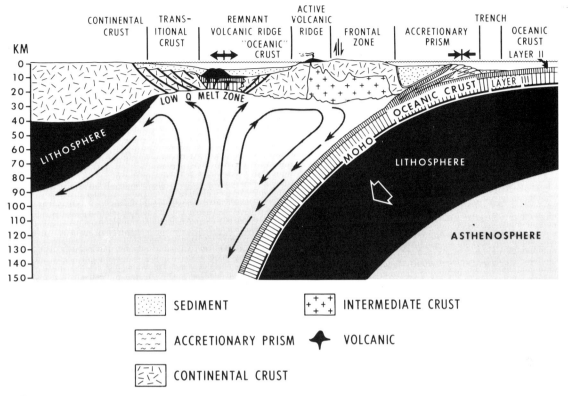

SEDIMENT

ACCRETIONARY PRISM

CONTINENTAL CRUST

INTERMEDIATE CRUST

VOLCANIC

Fig. 27. Active margin and the formation of a backarc basin, modified after Green (1977) and Toksöz and Bird (1977).

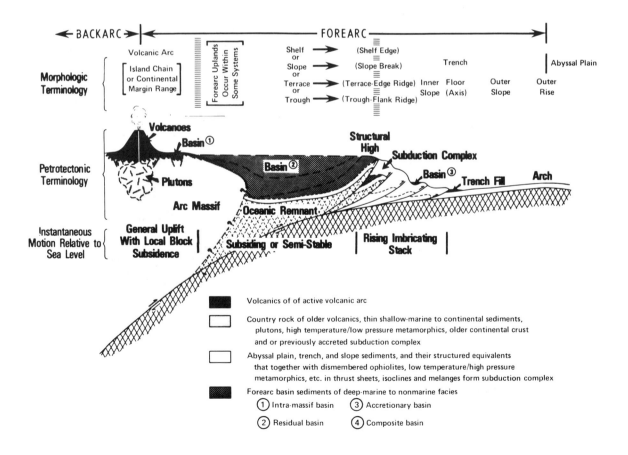

Fig. 28. Forearc basin and trench nomenclature, after Seely and Dickinson (1977).

49

1973). Other dynamic models (Toksöz and Bird, 1977) visualize that the subducted cold oceanic lithospheric slab induces convective circulation in the mantle wedge above the slab (Fig. 27). Toksöz and Bird's model and its calculation postulates a time interval of 20-40 Ma between initiation of subduction and the beginning secondary spreading behind the slab. The model is consistent with the gradual warming and weakening of the overriding lithosphere. An expected consequence of this model would be evidence for a rifting phase lasting some 20-40 Ma that precedes the actual opening of the marginal basin and the formation of its oceanic crust. Such a rifting phase appears to be roughly compatible with evidence for rifting on the continent-side margins of most backarc basins.

Kinematic variations of marginal sea formation have been interpreted as due to changes in plate geometry and spreading directions, the effect of continental collisions, and the effects of subduction of oceanic ridges, aseismic ridges, and oceanic plateaus. Many of these are briefly reviewed by Poehls (1978), who further makes the important observation that active backarc basins may occur only where the subducted plate can interact with the plate behind the arc. He notices that this condition occurs where trenches are terminated or offset by strike-slip faults and where large changes in subduction rates occur. As a consequence a tensional regime is created in the backarc basin area that is due to shear coupling across a strike-slip transform fault separating two adjacent plates. Sillitoe (1977) provides an example of this, with the offset of metallogenetic belts between Korea and Japan. He postulates that the offset is due to Mid-Miocene transform faulting associated with the opening of the Sea of Japan.

In contrast with the preceding arguments and following the earlier work of Wilson and Burke (1972) and others, Molnar and Atwater (1978) do not envisage a direct dynamic link between subduction and the formation of backarc basins. They point out that spreading does not occur behind all arcs, and that the downgoing slab may not contribute a substantial force in the opening of interarc basins. Instead, these authors note that interarc spreading is common where the ocean floor is old (i.e., older than say 50-100 Ma) while the subduction of younger ocean floor is strikingly associated with the formation of Cordilleran-type mountain ranges. The sea floor of marginal basins is created by extension and spreading because the heavy old oceanic lithosphere is pulled down faster than the adjacent plates can converge. Interarc spreading, according to Molnar and Atwater, would be preferentially initiated along pre-existing weaknesses such as volcanic arcs.

Table 2 classifies backarc basins floored by oceanic crust associated with B-subduction (3121). Because these basins rarely contain significant sediment thicknesses, we have only a few comments. Criteria for the recognition of remnants of ancient marginal basins include abundant volcaniclastics and/or deep water sediments (Karig *et al.*, 1975; Churkin, 1974b; Hussong and Uyeda, 1978).

TABLE 2 - OCEANIC BACKARC BASINS AFTER TOKSÖZ AND BIRD (1977)

Actively-spreading basins, with high heat-flow	Mariana, Lau-Havre, Scotia Sea
Inactive basins, with high heat-flow	S. Fiji, West Philippines, Sea of Japan (Fig. 28)
Inactive mature basins, with normal heat-flow	Fiji Plateau, Sea of Okhotsk, Parece-vela Basin
Undeveloped basins, involving captured oceanic basement that has not yet reached the spreading stage, following its capture by an island arc	E. Bering Sea, maybe Caribbean

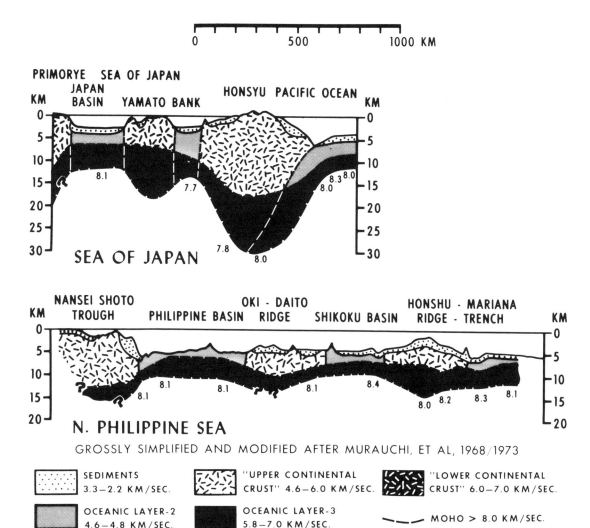

WEST PACIFIC - MARGINAL BASINS

VERTICAL EXAGGERATION 20:1

Fig. 29. West Pacific marginal backarc basins.

Backarc basins floored by continental or intermediate crust, associated with B-subduction (3122), may be viewed as a permanently or temporarily aborted precursor to a full-fledged marginal sea-type backarc basin. They are often hydrocarbon bearing and a number of them remain to be explored. They are mostly confined to shallow-water areas and typically contain thick sedimentary sequences. They commonly display a lower rifted sequence overlain by a less faulted sequence that subsided as a presumably previously heated underlying continental lithosphere cooled.

Most of the shallow-water marginal basins of the Western Pacific, including the prolific Sumatra and Java Sea Basins, are included in this class of backarc basin. Such basins are usually elongate depressions located cratonward of a volcanic arc. Their underlying basement consists of igneous and metamorphic rocks which were formed during earlier island arc settings (Katili, 1975; Ketner *et al.*, 1976). The basement dips generally towards and into the volcanic arc.

51

In the Western Pacific, the rift phase associated with this type of basin is commonly a late Paleogene-early Neogene phenomenon. The overlying "cooling" sequence includes the remainder of the Neogene. Heat flow data from these basins are spotty and range from high to normal values. Only a few detailed descriptions of these basins have been published (Todd and Pulunggono, 1971; Pulunggono, 1976; Hamilton, 1978; Ben Avraham and Emery, 1973; Wageman et al., 1970).

With continued extension, the continental crust underlying a backarc basin may open to form oceanic segments. For example, the Aegean Sea includes pre-Neogene continental islands located within an overall oceanic crust (McKenzie, 1978a). McKenzie postulates that a precursor Aegean continental crust was stretched by a factor of two, thus explaining the high heat flow in the area and the low velocities of the upper mantle. A similar origin may account for the Thyrrenian Sea (Ogniben et al., 1975).

Some backarc basins of this class show conspicuous décollement thrust and folds on their arc side. Such structures are often associated with regional strike-slip faults that coincide with or follow the folding. These features are observed in Sumatra (Posavec et al., 1973), Java, and Hokkaido. Probably the most spectacular display of this type of deformation is seen on Taiwan (Meng and Chou, 1976; Suppe, 1976; Suppe and Wittke, 1977; Ministry of Economic Affairs, Republic of China, 1974). Fold vergence toward the marginal basin suggests that this may well represent the late closing by subduction of the marginal basin itself. Such late subduction of marginal seas and associated deformation of sediments may be interrupted by periods during which subduction ceased. This in turn can lead to uplift of the folded belt and burial of its frontal portions under thick deltaic clastic wedges. This situation is well illustrated on the Palawan sketch of Figure 7c. The major hydrocarbon provinces onshore and offshore Sarawak, Sabah, and Kalimantan are in such a tectonic setting.

Not much has been published about the dynamics and kinematics leading to the elimination of marginal basins. Yet we sense that this process is of great importance if we are to understand mountain-building. Possible explanations that have been proposed include the polarity reversal of subduction zones (Murphy, 1973; Roeder, 1973). They also may include processes that involve changes in subduction rates across transform faults analogous — but reverse — from those invoked by Poehls (1978) for the formation of backarc basins. Geological evidence for the closing of marginal basins has recently been presented by Churkin (1974a) and Churkin and Eberlein (1977) for the Western Cordillera, by Scheibner (1976) for the Paleozoic Tasmanides of Australia and by Biq (1976) for the ranges of Taiwan. The Interior Jurassic-Cretaceous basins of Alaska and British Columbia also represent collapsed backarc basins (see Eisbacher, 1974, 1977).

31. Backarc basins associated with continental collision and on concave side of A-subduction arc

Another type of backarc basin is associated with continental collisions. These basins have an attenuated continental crust and sometimes an oceanic center, high heat flow values, and extensive normal faults (with listric faults predominating). They frequently occur in the wake of major intramontane strike-slip fault zones.

Because continental collisions do lead to the elimination of earlier Benioff zones and because they often terminate with the formation of A-subduction boundaries, it is tempting to relate the formation of basins of the (32) type to the A-subduction process itself.

The collision of Africa and the Arabian Peninsula with the European continent caused the formation of the Alpino-Mediterranean mountain ranges (Argand, 1924; Dewey et al., 1973; Biju-Duval et al., 1976, 1977). During and following the later phases of that collision, a number of late Paleogene-Neogene basins formed by extension behind A-subduction arcs. We differentiate basins where the extension did not advance sufficiently to lead to opening of an oceanic basin (321: Pannoni-type basins) from basins that opened and have a center underlain by oceanic crust (322: W. Mediterranean-type basins).

The Pannonian Basin of southeast Europe is superimposed on the alpine compressional megasuture that resulted from the collision of the African Promontory of Argand (1924) and Horvath and Channel (1977) with the European Plate. The Carpathians, the Dinarides, and the Alps surround the Pannonian Basin. The deformation of these mountain ranges lasted from the Cretaceous into the Late Tertiary. With the completion of the continental collision, a backarc basin formed. A number of authors have recently described and discussed this basin (Bleahu *et al.*, 1973; Boccaletti *et al.*, 1976; Horvath *et al.*, 1975; Horvath and Stegena, 1977; Kutas *et al.*, 1970, 1975; Cermák, 1974). Much geologic detail is summarized in Paraschiv (1975) and Mahel (1974). The map on Figure 30 shows Neogene thrusting and folding that is associated with A-subduction and the simultaneous subsidence of a foredeep on the outer periphery of the Carpathians. At the same time extensive backarc basins are formed; these are: the Transylvanian, Pannonian, and Vienna Basins.

CARPATHIAN FOREDEEP PANNONIAN BASIN

A   AFTER CARPATHIAN—BALKAN ASSOC. TECTONIC MAP. MAHEL, 1974

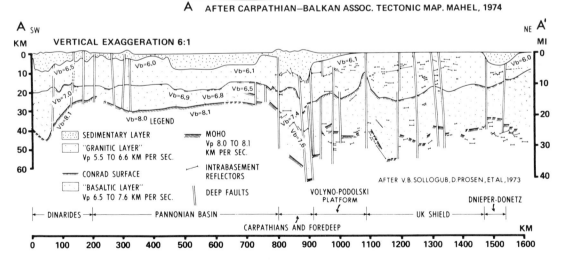

Fig. 30. Simplified map and crustal section across Pannonian Basin. The map is much simplified after the Tectonic Map of the Carpatho-Balkanian Regions (Mahel, 1974); the section after Sollogub *et al.*, (1973).

Sollogub *et al.* (1973) have shown that the Pannonian Basin is underlain by an attenuated crust (Fig. 30). Cermák (1974); Kutas and Gordiyenko (1970) and Cermák and Hurtig (1977) demonstrated high heat flow in the basin. Structural sections showing details of the bottom of the basin are scarce except for the oil-producing Vienna Basin where sections based on drilling and reflection seismic data show Late Miocene thrusting on the foreland and low-angle normal faulting in the Vienna Basin (Kapounek *et al.*, 1963, 1965; Kröll and Wiesender, 1972; Kröll and Wessely, 1973; Mahel, 1974b). Typically, the Pannonian Basin shows at its base a Late Paleogene blockfaulted sequence suggesting a thermal event. The Neogene sequence is less faulted and suggests subsidence due to cooling. However, such a model is weakened by the fact that the Pannonian Basin still exhibits high heat flow today.

How much of the attenuation of the Pannonian crust is due to stretching in a backarc region and how much is due to subcrustal erosion (Stegena *et al.*, 1975) is currently being debated.

Following the previously summarized model by Toksöz and Bird (1977), it is easy to visualize transitions that lead from basins with only continental lithospheric attenuation (i.e., Pannonian Basin) to basins involving continental lithospheric attenuation with opening of small oceanic windows. We classify the latter basins as West-Mediterranean type (322) basins. Their evolution has been discussed by Mulder (1973), Mulder *et al.* (1975), Biju-Duval and Montadert (1977a, b), Biju-Duval *et al.* (1978), Mulder and Parry (1977), Angelier (1977), and Vroman (1978). In these examples Oligocene-Early Miocene blockfaulted structures are overlain by a less faulted younger sequence that contains some of the well-known Messinian evaporites of the Mediterranean. Similar settings may account for the early evolutionary stages of the Late Paleozoic-Triassic formation of the Gulf of Mexico, the Gulf of St. Lawrence (Howie and Barss, 1975), and the Sverdrup Basin (Meneley *et al.*, 1975).

A particularly interesting type of backarc basin associated with a continental collision is the Black Sea with its oceanic crust and its great thickness of (?)Upper Cretaceous and Tertiary sediments. Following Letouzey *et al.* (1977), we are dealing with a marginal sea that was formed in Late Cretaceous time, maintained its identity until today, and thus remained sheltered from the effects of the collision of the Arabian Platform with the Eurasian continent.

33. Basins related to episutural megashear systems

Earlier we mentioned that Pacific backarc systems (Poehls, 1978) as well as Pannonian-type basins are often associated with the lateral termination of strike-slip faults. In the Western Cordillera of North America the megasuture is dominated by late and complex megashear systems (Carey, 1958). These lead to the development of a large number of small sedimentary basins. In the following, we find it useful to differentiate Great Basin-type basins from California-type basins.

331. Great Basin-type basins: The Laramide orogeny with its A-subduction related décollement fold belts and its basement-involved blockfaulting of the Western U.S. ceased during the Eocene. This coincided roughly with the progressive termination of subduction of the Farallon plate off the southern U.S. West Coast. The cessation of subduction began when the eastern Pacific Rise and west-adjacent Pacific plate reached the North American plate some 29 Ma ago (Atwater and Molnar, 1973). Due to the inherited north-northwesterly motion of the Pacific plate relative to the American plate, the entire Cordillera following the collision became subjected to extensive right-lateral shear. This wide shear domain was bounded on the southwest by the San Andreas fault system, on the northeast by the Tintina-Rocky Mountain Trench, and in the southeastern U.S. by the Rio Grande Rift. Carey originally proposed the idea of such a dextral megashear couple in 1958, although its timing was not discussed. The concept was later reintroduced and expanded by Wise (1963) and Hamilton and Myers (1966).

Basin-and-Range structural style extends far beyond the Great Basin physiographic province itself (Fig. 31). It can be found from southeast British Columbia as far south as North Central Mexico. Within the U.S., besides Nevada, Utah, and southeast California, such structures are found in the cratonal areas of southern Arizona and the Southern Rocky Mountains of New Mexico.

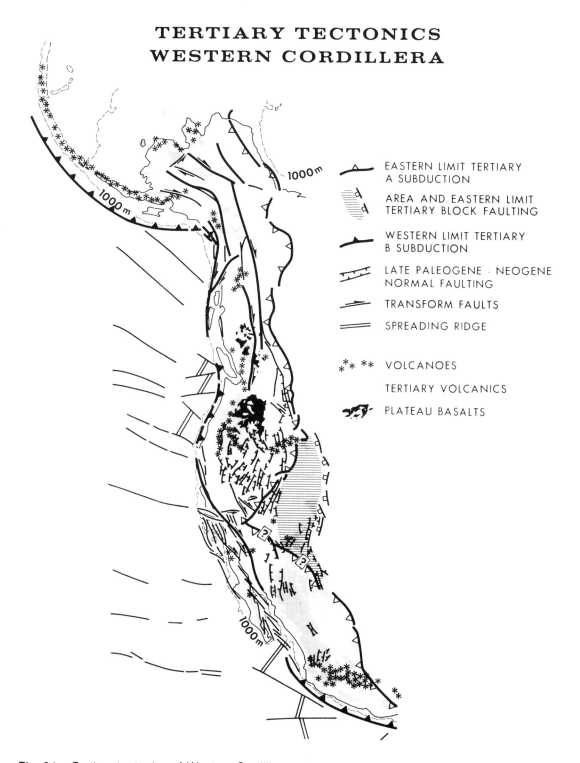

Fig. 31. Tertiary tectonics of Western Cordillera.

55

Broad arching, tilting, listric normal and/or strike-slip faulting variously characterize Basin-and-Range structure depending upon the locality.

The geophysical characteristics of the Great Basin have been well summarized by Sass *et al.* (1971), Scholz *et al.* (1971 and Fig. 32), Thompson and Burke (1974) and Smith (1977). They include high seismicity, a relatively thin crust with a low-velocity upper mantle, and high heat flow. Calc-alkaline andesitic volcanism prevailed during the mid-Cenozoic, but suddenly changed to bimodal basaltic rhyolite in the Plio-Pleistocene.

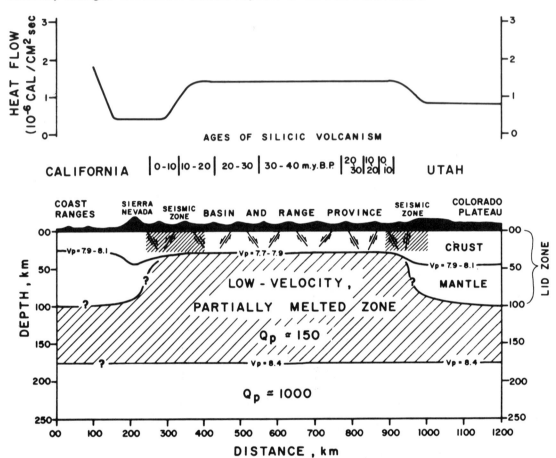

Fig. 32. Schematic geophysical section across central Nevada, after Scholz *et al.,* (1971).

Because of their importance in this basinal setting, a few words regarding the role of listric normal faults may be in order (*see* Dennis, 1967, for definition and history of term). The curved geometry of listric normal faults provides a mechanism capable of attenuating brittle crust. Conceptually, such mechanical distension should at depth be accompanied by the stretching of a "soft" ductile lower layer (*see* our discussion on rifts). Although it is not clear which comes first, lithospheric attenuation is accompanied by a rise in the asthenosphere and a concomitant increase in geothermal gradients (McKenzie, 1978b).

Longwell (1945) was one of the first to suggest the possibility that some of the normal faults in the Great Basin could flatten with depth. Bally *et al.* (1966) using seismic reflection data, demonstrated the existence of listric normal faults in the subsurface of the southern Rocky Mountains of Canada, and they inferred their existence in the southern Rocky Mountain Trench and northern Montana. Excellent additional seismic illustrations were recently published from the Basin and Range in Utah by McDonald (1976). More recently, other workers in the Great Basin explained the antithetic rotation of beds into underlying fault planes by a flattening of these fault planes at depth (Wright and Troxel, 1973; Profett, 1977; Anderson, 1971). Reflection seismic data in sections across individual valleys of the Great Basin show the rotation of beds into a listric normal fault plane in subsurface as illustrated by the example shown in Figure 33.

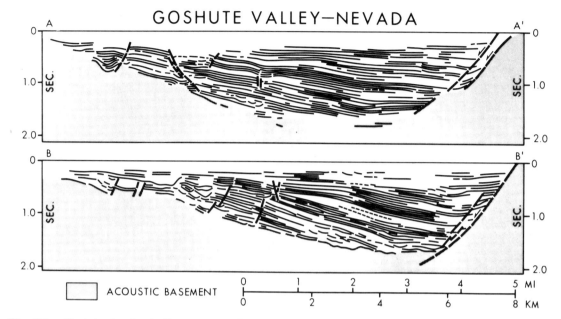

Fig. 33. Sketch of seismic lines across Goshute Valley, northeast Nevada, with permission of Shell Oil Company.

Recent work on the metamorphic core complexes of the Cordillera (Davis, G. H., 1975; Davis, G. A. *et al.,* 1977; Davis and Coney, 1979, Compton *et al.,* 1977) reveals some of the geologic details of what appear to be exhumed deep portions of listric normal fault systems. These authors suggest that the basement underlying growth faults deformed in a ductile manner forming what Davis and Coney (1979) call mega-boudins. A décollement zone associated with listric normal faults roughly coincides with the base of the sediments but is underlain by a zone of ductile flow in the metamorphic carapace of the basement complex.

In conclusion, listric normal faults may well provide the mechanism for attenuation of a brittle portion of the crust. The depth to which such listric normal faults may occur in basins must still be determined by future geophysical work. The relation of such "listric fault basins" to strike-slip faulting is illustrated on Figure 34.

How does the Great Basin-type differ from the Pannonian-type (321) basin? Both basin types are characterized by high heat flow, attenuated crusts, and widespread faulting including listric normal faults and strike-slip faults. Both can be viewed to occur in the wake of major strike-slip fault systems (of the Alps and the Canadian Rockies, respectively); volcanic manifestations are similar in both basins; however, the Great Basin is dominated by the tectonics of a complex megashear couple. High ranges separate individual basins in great contrast to the wide plains of the Pannonian Basin.

The Great Basin is, in fact, a lump term for a large number of basins and ranges, whereas the Vienna, Pannonian, and Transylvanian basins are all part of a large backarc region that lacks mountain ranges and does not appear to be as dismembered as the Basin and Range province of the U.S. Of importance is the difference in the precise time correlation of the A-subduction process in the adjacent mountain ranges and the initiation of normal faulting. In the U.S. Cordillera, normal faulting events are in general later than the Eocene cessation of

A-subduction. However, in the Pannonian and Transylvanian, thrust faulting in the folded belt, subsidence in the foredeep, and normal faulting and subsidence in the backarc area occur for all practical purposes simultaneously. In the Vienna Basin, however, backarc normal faulting occurs mostly after the thrust faulting.

There are also significant differences between Great Basin-type basins and the California-type basins that are described later. Both were initiated at about the same time, that is, when the Pacific Plate reached the North American Plate and diffused right lateral strike-slip decoupling of the two plates along the San Andreas fault system began. They differ because the basement underlying the Basin and Range province consists mostly of alpino-type folds and thrusts that deformed in A-subduction related processes and their associated basement remobilization. In some areas (e.g., Arizona and New Mexico) even cratonic shelf sequences of the foreland and their underlying "stable" basement are broken up in basin-and-range fashion. The sedimentary fill is almost exclusively continental. In contrast, California-type basins (34), as will be shown in the next paragraphs, are underlain by forearc accretionary wedges, former forearc basins, and arc massifs.

California-type basins are dominantly filled by marine sequences, while Great Basin sequences are filled with continental sediments. Compressional folding and strike-slip faulting dominate California; often these disrupt the integrity of a basin. In the Great Basin, however, extensional tectonics dominate.

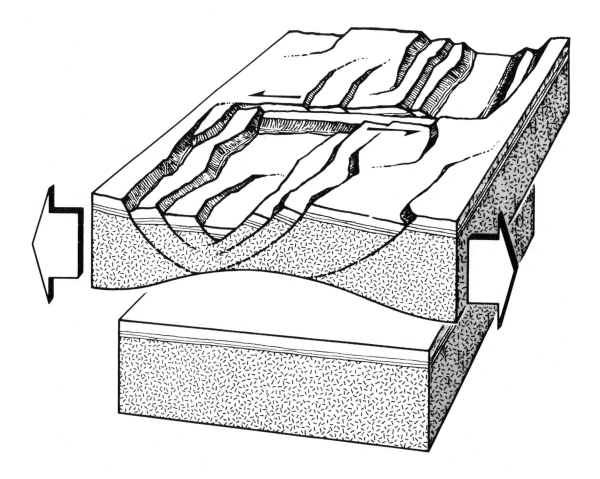

Fig. 34. Sketch showing illustration strike-slip faulting listric normal faulting and crustal attenuation. Modified after Liggett and Ehrenspeck (1974), and Liggett and Childs (1974).

332. California-type basins: An overall summary of the evolution of California-type basins has been given by Blake *et al.* (1978). These authors and many others (such as Atwater and Molnar, 1973) concluded that before the Oligocene, the Farallon plate on the east flank of the mid-Pacific rise was being subducted under the west coast of North America.

Due to this early subduction history, the basins of California are commonly underlain by three kinds of "economic basement": 1. The Franciscan formation with its chaotic (mélange) structures that may be viewed as an accretionary wedge that separated a deep sea trench to the west from the American Plate, 2. The granitic-metamorphic Klamath-Sierran basement which may be viewed as the roots of an island arc active through much of the Mesozoic but ceasing to be active during Late Cretaceous time, and 3. The Great Valley sequences of Upper Jurassic to Upper Cretaceous age that today are interpreted as dominantly deepwater forearc basin deposits (Ingersoll, 1976; Dickinson and Seely, 1979). Age equivalents of the Great Valley sequence can be found in Washington and Oregon and as far north as the Nanaimo Basin of British Columbia.

Based upon the relative quiescence of igneous activity in California and neighboring states in the latest Cretaceous and Paleogene (75-40 Ma), it has been postulated that subduction off much of the California coast either ceased temporarily or became more shallowly dipping (*see* discussions by Dickinson and Snyder, 1978; Cross and Pilger, 1978). This time interval may or may not have coincided with the formation of a proto-San Andreas fault (Suppe, 1970; Clark and Nilsen, 1973) which would account for reportedly different offsets of late Cretaceous (70± Ma) and Mid-Paleocene (60± Ma) geologic features.

The question as to the timing and location of a proto-San Andreas fault has important implications relative to paleogeographic reconstructions of Paleocene-Eocene basins in California. The main issue is whether the Salinian basement block was or was not a paleogeographic element which bifurcated an elongate Paleogene basin which developed over the late Cretaceous forearc basin. Clarke and Nilsen (1973) postulated 220-240 km of Late Cretaceous-Paleocene right lateral slip for such a fault. Johnson and Normark (1974) handle the postulated discrepant offsets without a proto-San Andreas, and most recently Graham (1978) has suggested that such slip is no greater than 185 km, and may be as little as zero. In any case, the influence of a Paleocene proto-San Andreas fault on California paleogeography now seems greatly reduced, or perhaps even eliminated.

We are inclined to view the Paleogene California margin as essentially one of oblique subduction up until the Pacific-Farallon spreading center collided with the North American margin 29 Ma ago (Atwater, 1970). In our view, the Early Tertiary was essentially characterized by an elongate forearc basin which developed east of an outer ridge and a west-adjacent subduction complex. Growing structures often resulted in unconformities along the edge of the outer high now exposed along the west flank of the Great Valley. Additional unconformities at the base of and within the Paleocene are associated with spectacular submarine canyons mapped in the Sacramento Basin (Almgren, 1978). Relics of these Paleogene basins are frequently preserved below the Neogene fill in California basins.

Only following the events that led to the formation of the San Andreas fault do the basins of California acquire their own individuality. The structural evolution has been described in some detail by Atwater and Molnar (1973). About 29-25 Ma ago the Pacific plate that was to the west of the mid-Pacific rise came into direct contact with the North American plate leading to formation of a right-lateral shear zone that separated North America from the Pacific plate. This ridge-trench collision was responsible for the inception of the Neogene basins of California. It was accompanied by a widespread regression with deposition of redbeds (Sespe facies). Transgressive marine conditions were initiated with shallow water lower Miocene (Vaqueros) sandstones. The transgression culminated with the deposition of the deep water Middle Miocene Monterey formation with its source beds. These were well described by Bramlette (1946) who related the siliceous nature of this formation to contemporaneous volcanic activity. A link between volcanism and the genesis of source beds appears to be probable. The volcanic activity proceeded from the northeast (at about 27 Ma) to the southwest (at about 14 Ma) supporting the concept of an East Pacific Rise progressively intersecting the North American continent (Yeats, 1968).

The southern sector of the San Andreas fault proper was initiated only some 10-15 Ma ago, with possibly as much as 300 km right-lateral slip since that time (Blake *et al.*, 1978), most of which occurred during the last 5 Ma. Concurrently with this, a spectacular acceleration of subsidence rates occurs. In southern California Yeats (1978) reports subsidence rates of about 700 m/Ma with an acceleration of 1 000 m/Ma after the Miocene and with maximum rates of 2 000 to 4 000 m/Ma in the Ventura Basin.

Much attention has been paid to the mechanics of formation and the subsequent deformation of California-type basins. Crowell (1974a and b, 1976) has offered and summarized most of the pertinent basin-forming concepts by differentiating the following: 1. *Basins due to crustal megarifting,* where the rifting has led to the formation of oceanic crust as with the Gulf of California; 2. A lesser amount of extension will lead to the formation of *pull-apart basins* (like those shown on Fig. 34) that are underlain by an attenuated crust and may have some volcanics in the basal sequence of the basins. The detailed studies of the Ridge Basin done by Crowell (1952, 1954) provide an illustration of the filling history of this type of basin (a particularly well-illustrated Paleozoic example of this type of basin has been shown by Steel, 1976); 3. *Fault zone rifting* associated with bends in strike-slip faults. If the bends are such that the fault blocks move apart, a pull-apart basin will form; 4. *Fault margin sagging* occurs where the adjacent fault blocks converge over a bend of the strike-slip fault. This results in compressional folding and thrust faulting. The load of the overriding block may depress the underlying block leading to the formation of an elongate basin; 5. *Fault wedge basins* occur where strike-slip faults diverge, and conversely uplifts are associated with the convergence of strike-slip faults; 6. *Fault flank depressions,* en echelon folds and intervening synclinoria are often associated with strike-slip faults. These are well-displayed on the west side of the San Joaquin valley, where presumably the more ductile Franciscan basement permits this style of deformation. Folds in that area — as do most folds in California-type basins — show growth with time (Harding, 1976). This is reflected particularly well during the previously described late Neogene acceleration of displacement and subsidence along the San Andreas fault in southern California (Yeats, 1968).

Are California-type basins unique? The combination of an antecedent forearc history with later strike-slip faulting related to the attempted subduction of a spreading ridge may be repeated in the Bellingham-Nanaimo Basin that straddles the U.S.-Canada border, the Hecate Strait Basin of British Columbia, and the Eastern Gulf Alaska Basins. Similar basins are also known from the Philippines and New Zealand. Although many of these basins have marine sections, none of them have source beds as obviously rich as the Monterey Formation.

CONCERNING OIL RICHNESS OF VARIOUS SEDIMENTARY BASINS

To illustrate the varying economic significance of particular examples of basins within the same family, we will use only oil yield ranges using the following code:

1. more than 10^8 barrels ($16 \times 10^6 m^3$) per 10^3 km^2 (ultra rich)

11. 10^7 - 10^8 barrels ($1.6 - 16 \times 10^6 m^3$) per 10^3 km^2 (very rich)

III. 10^6 - 10^7 barrels ($160 - 1 600 \times 10^3 m^3$) per 10^3 km^2 (lean)

IV. less than 10^6 barrels ($160 \times 10^3 m^3$) per 10^3 km^2 (very lean)

The reader must realize that the different basins we compare have a different exploration maturity, which in truth does not permit realistic comparison between basins. Thus, the yield class listed behind a basin name, e.g., Bighorn (II) can only give a very general idea. Our yield ranges are based on proven recoverable ultimates of oil and thus exclude estimated potential volumes or volumes from oil shales or tar sands.

60

For almost all basin types, we can find examples that have yielded insignificant or no amounts of oil, despite significant exploration efforts. Let us therefore reiterate the following qualifications from an earlier paper (Bally, 1975). Yield figures are a function of 1. exploration maturity, 2. volume of available mature source beds (three basic types with varying richnesses), 3. volume of trapped pore space, and 4. volume of oil that has escaped during geologic times and that may have been dispersed in the "plumbing system." All this makes the evaluation of yield figures difficult and puts severe limitations on their usefulness.

Table 3 lists examples of oil richness of basins on the rigid lithosphere that are related to the formation of oceanic crust. The table suggests a wide range of richnesses within rifts and Atlantic-type margins that does not allow sensible volume predictions on the basis of analogs. For unexplored rifts, not enough is known to predict whether the rift fill is dominantly marine or continental. We also do not understand the geologic reasons which in some cases permit and in other cases inhibit the formation of organic-rich source beds in the rifting sequence.

TABLE 3 - OIL RICHNESS BASINS RELATED TO FORMATION OF OCEANIC CRUST (II)

Rifts (111):

    I.   Gulf of Suez;

    II.   Tucano-Reconcavo;

    III.   Rhine Valley graben;

No Production: Oslograben, East N. American Triassic grabens, Baikal rift, East African graben.

Proto-oceanic rifts (1112):

No Production: Red Sea, Afar

Atlantic-type margins overlying earlier rifted systems and transform systems (1141,1142):

    I.   Niger Delta, Sirte Basin:

    II.   Sergipe-Alagoas, Cabinda, Perth basin, Carnavaron basin, Gabon, Cambay;

    III.   Cuanza

    IV.   Sao Luiz Barreinhas:

No Production: Despite variable exploration efforts, no significant oil volumes have as yet been found in the Nova Scotia offshore, NW Africa shelf, South Africa shelf, Amazon Delta

Atlantic-type margins overlying former backarc basins of Pannonian (321)
or Mediterranean (322) type:

    I.   Gulf of Mexico, Tampico-Tuxpan, Isthmus-Campeche;

    IV.   Sverdrup Basin.

There is no set of systematic data available that would allow us to compare the distribution of source beds within the "drifting" sequence of Atlantic-type margins. Furthermore, the deep water extension of Atlantic-type margins under the continental slope and rise remains virtually unexplored at this date. Dow (1978) has very succinctly summarized the difficulties inherent in making intelligent reserve forecasts for essentially unexplored continental margins, and his comments mitigate against excessive optimism in assessing the future potential of these provinces. Keen's fine study (1979) extolls the virtues of maturation and subsidence studies but falls short in explaining the lack of exploration success on the Nova Scotia Atlantic margin.

61

Table 4 lists oil richnesses of cratonic basins without differentiating their earlier genetic setting (type 1121 versus type 1212). Note that the rich North Sea and West Siberian Basins fall in the II Richness class, because these basins occupy a very large area. If one were to focus on oil richnesses associated only with the underlying graben systems, these basins would show a class I richness.

TABLE 4 - OIL RICHNESS OF CRATONIC BASINS

II. North Sea, West Siberia, NW German Salt Basin, North Sahara, Illinois Basin

III. Williston basin, Michigan Basin, Cooper Basin, SW Netherlands

IV. Paris Basin, Polish Basins

No significant volumes: Amazon Basin, Maranhao Basin, Parana Basin

The overall richness range among all cratonic basins is again difficult to explain because systematic quantitative source bed data are lacking. The formation of source beds in West Siberia and the North Sea is related more directly to the overall evolution of the basins. This contrasts with some of the North American basins where, for instance, the Devonian-Mississippian black shales of the Kaskaskia sequence are widespread over much of the continent, while upper Paleozoic source beds are more limited to the basins proper.

Table 5 illustrates that perisutural basins are some of the richest oil basins of the world. Note that if one were to include tar sand reserves (*see* Demaison, 1977) the eastern Venezuela basins and the Western Canada basin would be on a par or even exceed the richness of the Middle East. The richness of foredeep ramps is particularly striking. On these ramps, basement-controlled paleostructures and stratigraphic traps dominate. Allowing for the difficulty in exploring for such traps, it can readily be concluded that less-explored foredeep ramps may have a substantial potential left.

TABLE 5 - OIL RICHNESS OF PERISUTURAL BASINS

Ramps with buried grabens, but with little or no blockfaulting (221):

I. Middle East, E. Venezuela

II. Arctic Slope, Western Canada (including Sweetgrass Arch), Volga-Ural, N. Caucasus, Cuyo-Mendoza, Dnepr-Donetz

III. Putumayo-Maranon, Neuquen, Alpine Molasse Basin, Aquitaine, Pechora, Appalachian, Szechuan

IV. Pre-Rif Basin, Adriatic Basin

Ramps dominated by later blockfaulting (222):

I. Maracaibo, Ardmore, Central Basin Platform, S. Caucasus

II. Wind River, Uinta, Bighorn, Powder River, Delaware, Anadarko

III. San Juan, Arkoma, Paradox, Palo Duro

IV. Val Verde

Chinese basins (23):

II. Tsaidam

III. Dzungaria, Turfan

IV. Tarim, Ordos

Here as with many of the other basins, little systematic data are published on source bed distribution and maturation. Consequently, it is difficult to explain some of the less prolific foredeep basins of the world. It is, however, easy to visualize how foredeep basins are located between incipient folded belts that isolate the adjacent cratons from open ocean circulation. It would follow that incipient foredeeps should be characterized by restricted marine environments propitious for the generation of source beds.

We remind the reader that fold belts adjacent to foredeeps also show wide variations in richness and type of hydrocarbon accumulation. This is shown by a comparison of the oil-rich Zagros and the Carpathian foothills with the relatively oil-lean foothills of the Andes, the Appalachian, and the Rocky Mountain folded belts.

Note also that all Chinese-type basins produce exclusively from continental lacustrine sequences and that their richness may be compared with that of the Uinta Basin.

Table 6 shows again a substantial richness spread among various types of episutural basins. Forearc basins appear to be lean with two noteworthy exceptions. The richness spread among backarc basins is puzzling and may be related to climatic difference. These may account for the relatively rich source beds in the Indonesian region and the possible lack of them in the well-explored but more northerly basins of Taiwan and Japan. Pannonian-type basins are lean with the impressive exception of the Vienna Basin. The wide disparity among oil richness of episutural basins can only be partly explained by the variety of their genesis. Clearly, the rich California basins owe much of their richness to the peculiar set of circumstances that led to the deposition of the Monterey formation.

In conclusion, we reiterate what we said in the introduction: the classification of basins does little to improve our hydrocarbon volume forecasting ability.

TABLE 6 - OIL RICHNESS OF EPISUTURAL BASINS

Forearc basins (311):

	II.	Cook Inlet, Progreso-Talara
	IV.	Gulf of Alaska (an old oil field Katalla: no commerical production)
No Production:		A large number of other circum-Pacific forearc basins saw some exploration but as yet have not yielded significant amounts of oil. Some gas is reported in the Mentawai basin south of Sumatra.

Backarc basins associated with B-subduction floored by continental or intermediate crust (3122):

	I.	Sumatra
	II.	West Java-Sunda, Kutei-Tarakan, on and offshore Brunei
	III.	East Java-Madura, Barito
	IV.	Taiwan, Akita-Yamagata, Niigata
No Production:		A large number of these basins remain virtually unexplored, i.e., Bristol Bay, Yellow Sea, South China Sea.

Pannonian-Type basins associated with A-subduction (321):

	II.	Vienna Basin
	IV.	Pannonian Basin

Great Basin-type basin (331):

	IV.	Great Basin (only minor production)

California-type basins (332):

	I.	Santa Maria, San Joaquin, Ventura, Los Angeles
	III.	Taranaki (New Zealand)

Of all the proposed mechanisms of subsidence, thermal causes and effects of sediment loading have received the greatest attention. Sediment loading and other subsidence problems have been studied and reviewed by Walcott (1970, 1972), Sleep and Snell (1976), Watts and Ryan (1976), Poulet (1977), Keen (1979), Bott (1979), and Turcotte (in preparation). From these papers, we conclude that although sediment loading can account for substantial subsidence and sediment thicknesses, it does not explain initial basin-forming processes nor does it adequately account for all of the subsidence involved. Sediment loading from prograding clastic wedges (e.g., Mississippi, Ganges, and Niger deltas) is assumed to substantially increase the magnitude of crustal subsidence for these already existing deep water basins. This very reasonable assumption is often difficult to prove. The stable base underlying thick deltaic wedges is rarely mapped. Rates of subsidence, derived from reflection seismic sections only reflect the vagaries of salt and shale tectonics, and are not easily translated into figures for overall basin subsidence.

For cratonic basins as well as for Atlantic-type margins, which are initiated in shallow water and generally continue to be filled by shallow-water sediments, additional subsidence "driving forces" are required beyond the sediment load itself (Watts and Ryan, 1976).

The thermal models for subsidence were largely inspired by plate tectonic concepts. Such models explain, for example, oceanic crust subsidence as an isostatic response due to cooling and resultant thickening of lithosphere, as it moves laterally away from mid-ocean ridges where the oceanic lithosphere is created (*see* Parsons and Sclater, 1977). In addition to this example, the thermal cooling concept for subsidence has also been used to explain subsidence and thicknesses of drift-phase related sequences on Atlantic-type margins. A detailed documentation of the effect of subsidence due to sediment loading and that due to thermal contraction is given by Keen (1979). However, in this environment, sediment loading and possibly lateral ductile flow in the lower crust must be considered as important secondary factors (for discussion, see Sleep, 1971, 1972; Sleep and Snell, 1976; Bott, 1971, 1973; Bott and Dean, 1972). Another factor to be considered in this regime is whether or not the cooling continental-ocean crust boundary is rigidly coupled, or whether differential faulting or warping occurs along it during the subsidence phase (*see* Turcotte *et al.*, 1977; Bott, 1979).

Subsidence and flank uplifts related to rifting, the common precursor to Atlantic-type margins, are topics still under much debate. Whereas we have discussed questions relative to extensional mechanisms such as listric normal faulting in the brittle upper crust and probable ductile flow within an underlying low velocity layer earlier in this paper, we have not specifically reviewed subsidence in this regime.

Several authors (i.e., Poulet, 1977; McKenzie, 1978b) point out that whereas simple heating and thermal expansion of the lithosphere will create uplift, subsequent cooling alone will only restore the elevated lithosphere to its original position. It would appear that in the absence of "pulling apart" of crustal blocks (McKenzie, 1978b), actual net subsidence would only occur if subaerial or subcrustal erosion of lithosphere develops, or if original deep crustal or mantle densities were increased. The two erosional mechanisms have been discussed earlier by Hsü (1965) and Gilluly (1964), respectively.

In backarc basins, thermal models can be readily applied as long as the subsidence process is not modified by structural loading by thrust sheets or else by the elimination of the backarc basin due to subduction processes.

Cratonic basin subsidence still remains a puzzle. Simple models of thermal subsidence as attempted by Sleep and Snell (1976) cannot be reconciled with the observation that in many cratonic basins the thermal rifting phase occurs much too early to be related to the subsidence of some of the overlying sequences. Among other explanations, basalt-eclogite phase changes offer one interesting alternative (*see* O'Connell and Wasserburg, 1967; Haxby *et al.*,

1976; Mareschal and Gangi, 1977). Unfortunately, convincing evidence for an eclogitic mantle is lacking. The model of Burk (1976) for the Chad Basin emphasizing the role of peripheral uplifts lacks geophysical documentation and is not easily applied to older cratonic basins.

It may be well to point out that cratonic basin subsidence should be viewed in the context of the evolution of the whole continental platform on which they are located. Isopach maps of North America (Cook and Bally, 1975) show that many cratonic basins maintain their individuality during some periods while during other periods (mostly times of foredeep subsidence) their subsidence merges into the subsidence of much larger areas. To understand cratonic arches, understanding of continent-wide subcrop patterns is important. Since Levorsen's (1960) pioneering work, little thought has been given to the nature of the lithospheric flexuring required to generate such supra-regional subcrops.

Relative to foredeeps, we are not aware of any published quantitative models that adequately describe their subsidence. Reflection seismic evidence in several cases allows us to map the downdip continuation of the ramp that underlies the foredeep. Such ramps are loaded by thrust sheets that often evolve and advance from the inside of the folded belt to the outside. This causes a corresponding migration of the foredeep and the underlying "hinge zone" of the underlying platform. In contrast to subsidence by sediment loading in a sedimentary basin, the loading process here is not limited by water depth, because folded belts form mountain ranges. The question is whether loading by thrust sheets and filling of the adjacent moat with sediments is the only cause for subsidence of foredeeps, or whether lithospheric flow associated with the postulated A-subduction process also contributes to foredeep subduction.

Further, quantitative models are also required to explain the "rebound" uplift of foredeeps and the adjacent folded belts following completion of the compressive phase of mountain-building.

Our intention thus far has been to show differences between various basin types and to point out where useful models of subsidence apply to some basin types. In general, we note the lack of sufficient data to adequately test quantitative models. What we have not discussed yet is the question of correlatable global subsidence patterns among the various realms of subsidence. To this end a few words in regard to stratigraphic sequences may be in order.

Stratigraphic sequences that are bounded by top and basal unconformities or their correlative conformities (Mitchum, 1977) have been described by Sloss (1963) and Wheeler (1963) for North America. Sloss (1972a, b, 1976) also demonstrated the synchroneity of such sequences between the North American craton and the Russian platform. More recently, Soares et al., (1978) correlated these sequences with sequences in South America.

Vail et al. (1977) developed a method that permits us to recognize such sequences on seismic reflection lines. Vail et al. offer global correlations of sequences and explain these correlations in terms of global cycles of sea level changes. They take great care not to confuse rises and falls of sea levels with transgressions and regressions. The latter reflect only the balance between rate of sea-floor subsidence and rate of sediment supply and do not necessarily correlate with sea level changes, a concept well illustrated earlier by Curray (1964). The curves of Vail et al. are characterized by slow sea level rises and abrupt sea level falls. The basic construction of such eustatic curves also measures large-scale subsidence. However, to establish a better approximation to a first- and second-order eustatic curve, these authors calibrate their global cycle curve with the theoretical curve of Pitman (1978) in which eustatic sea level changes are mostly ascribed to changes in volume and elevation of mid-ocean ridges through time: fast-spreading, expanding ridges cause the sea level to rise; contraction with decreasing spreading rates causes sea level to lower. The effects of glaciation and other climatic factors are only responsible for smaller or third-order sea level oscillations (*see also* Keen, 1979).

65

The relation between subsidence mechanisms and eustatic sea level changes has been explored by Sleep (1976). He points out that major intercontinental platform unconformities may be due to either eustatic sea level changes or else to correlative continental uplifts. Where thermal contraction may be the cause of subsidence, basins would persist in subsiding irrespective of eustatic fluctuations. Sleep's computations show that small sea level fluctuations could produce significant unconformities even in rapidly subsiding basins. Vail *et al.* (1977) take this as confirmation of the hypothesis that interregional unconformities are primarily due to eustatic sea level changes and not to uplift. Our view differs because we cannot help being impressed by the extensive interregional subcrop patterns that underlie many of the North American cratonic sequences. To us, these reflect tectonic effects — namely, broad-scale lithospheric warping which occurs at the beginning of North American foredeep cycles. Sloss (1976) also notes that eustatic controls are not sufficient to explain sedimentation and deposition on cratons and therefore he, too, is searching for global mechanisms that control subsidence.

Sloss (1972a, b), Rona (1973), and Whitten (1976, 1977) all suggest the existence of globally synchronous events of increasing basin subsidence, with Whitten specifically correlating short intra-Cretaceous periods of rapid subsidence. Unfortunately, we lack more precise correlations of subsidence rates between many different basins. We do, however, sense that synchroneity among some global geodynamic basin-generating mechanisms may be real. The existing documentation for this is clearly inadequate, but we are impressed by some very general observations:

1. Preserved Mesozoic and Paleozoic backarc basins are very rare. A large majority of the world's preserved backarc basins were initiated by rifting during Eocene-Oligocene time (Ta and Tb global supercycle of Vail *et al.,* 1977). Their further subsidence due to cooling, or else their opening into oceanic marginal basins, occurred during the Neogene.

Rona and Richardson (1978) provide a context for this observation. They point out that during the Eocene a major reorganization of global plate motion pattern occurred. This involved a reorientation of motions with large N-S components in motions with large E-W components and was due to an increase in length of collisional plate boundaries. Such a reorientation can be expected to initiate new E-W sea-floor spreading.

2. Substantial cratonic subsidence of Tertiary age is rare and appears to be limited to the North Sea Basin, whereas we have many cratonic basins of either Mesozoic or Paleozoic age.

3. Most of the undeformed passive margins of the world today had a Mesozoic and/or Tertiary origin. This is related to the tectonic breakup of Pangea and the reason for Mesozoic-Cenozoic first-order cycle of Vail *et al.* (1977). By analogy it is tempting to relate major lower Paleozoic realms of subsidence to the breakup of Morel and Irving's (1978) Precambrian (600 Ma) supercontinent.

4. Subsidence of the Atlantic margin of the U.S. broadly correlates with the subsidence of the Rocky Mountain foredeep. In the future, it may be possible to correlate the initiation dates and subsidence history of the foredeep sequences with the spreading history of the oceans.

The above observations suggest a certain global synchroneity of the tectonic initiation and subsidence of some basin groups. Obviously, however, so far these are only apparent relationships which will require more study and understanding before meaningful conclusions can be drawn.

FUTURE STUDIES

Our classification reveals the genetic diversity among varying realms of subsidence, yet there is also some order. We need to reconcile some of the very elegant theoretical subsidence models that have been proposed with geologic and geophysical realities. To this end, integrated studies of several basin types are needed. We need to combine conventional stratigraphic and subsurface studies with detailed industry-quality reflection profiles and deep crustal refraction and reflection data, in addition to gravity, magnetic, and heat flow studies.

Also, more precise reconstructions of the subsidence history of sedimentary basins are needed. Van Hinte (1978) has quite cogently laid out the procedure for a more reliable geohistorical analysis of basins including proper evaluation of paleontologic, paleoecologic and radiometric methods. Adequate correction for the effects of compaction is needed. In addition to present heat flow measurements, the thermal history should be completed with the relevant information from clay compaction and organic maturation studies (for an example, *see* Keen, 1979).

Early in our paper and with our richness listings, we indicated that we are far from being able to predict potential reserves for less explored basins. To this end it will be useful to systematically study the geologic distribution of various source bed types and their maturation history as they relate to the evolution of different basin types.

The incentive for earth scientists in industry to continue working together with their colleagues in academic institutions is great. This is because the practical concerns for future hydrocarbon exploration will require a better understanding of the basic processes that led to the formation of the great sedimentary basins of the world.

ACKNOWLEDGMENTS

As we have written an eclectic paper, we had to rely on the work of many authors. Our list of references is indeed a most inadequate acknowledgment. We selected mainly references that would permit the reader to dig further into the subject. As a consequence, we often neglected to trace the origin of concepts we assimilated over the years. For this we apologize to the originators of these ideas.

Our manuscript was reviewed by U. Allan, K. Arbenz, K. Burke, W. Dickinson, and L. Sloss. We are most grateful for their many helpful suggestions. A large number of them were incorporated in our final version. We also thank H. Scott and D. Branson for drawing most of our illustrations and K. Ziegler for carefully typing and retyping our manuscript. Finally, we would like to thank Shell Oil Company for permission to publish this paper.

REFERENCES

Alberta Society of Petroleum Geologists, 1964, Geological history of Western Canada: Calgary, Alberta, 232 p.

Alden, W. C., 1932, Physiography and glacial geology of eastern Montana and adjacent areas: U.S. Geol. Surv. Prof. Paper 174.

Almgren, A. A., 1978, Timing of Tertiary submarine canyons and marine cycles of deposition in the southern Sacramento Valley, California: *in* D. J. Stanley and G. Kelling, *eds.*, Sedimentation in submarine canyons, fans, and trenches; Dowden, Hutchison and Ross, Inc., p. 276-290.

Amoco Canada and Imperial Oil, 1973, Regional geology of the Grand Banks: Bull. Can. Petrol. Geol., v. 21, p. 479-503.

Ampferer, O., 1906, Über das Bewegungbild von Faltengebirge, Austria: Geol. Bundesanst., Jahrb., v. 56, p. 539-622.

Amstutz, A., 1951, Sur la zone dite des racines: Arch. Sci., v. 4, p. 319.

_____, 1955, Structures Alpines, Subductions successives dans l'Ossola: Acad. Sci. Paris, C. R., Ser. D, v. 241, p. 967-969.

Andersen, D. W. and Picard, D. M., 1974, Evolution of synorogenic clastic deposits in the intermontane Uinta Basin of Utah: *in* W. R. Dickinson, *ed.*, Tectonics and sedimentation; Soc. Econ. Paleont. Mineral. Spec. Pub. 22, p. 167-189.

Anderson, R. E., 1971, Thin skin distension in Tertiary rocks of southeastern Nevada: Geol. Soc. Am. Bull., v. 82, p. 43-58.

Angelier, J., 1977, Sur l'évolution tectonique depuis le Miocène supérieur d'un arc insulaire Méditerrané: L'arc Égéen: Rev. de Géogr. Physique et de Géologie Dynamique, v. 19, Fasc. 3, p. 271-294.

Angenheister, G., Bögel, H., Gebrande, H., Giese, P., Schmidt-Thome, R. and Zeil, W., 1972, Recent investigations of surficial and deeper crustal structure of the eastern and southern Alps: Geol. Rundschau, v. 61, p. 349-395.

Argand, E., 1924, La tectonique de l'Asie, Compt. Rend. IIIe: Cong. Int. Geol. Liège, Imprimerie Vaillant-Carmanne (translated and edited by A. V. Carozzi), New York, Hafner Press, 218 p.

Armstrong, R. L., 1968, Sevier orogenic belt in Nevada and Utah: Geol. Soc. Am. Bull., v. 79, p. 429-458.

Armstrong, F. C. and Oriel, S. S., 1965, Tectonic development of Idaho-Wyoming thrust belt, Am. Assoc. Petrol. Geol. Bull., v. 49, p. 1847-1866.

Artemjev, M. E. and Artyushkov, E. V., 1971, Structure and isostasy of the Baikal rift and the mechanism of rifting: J. Geophys. Res., v. 76, p. 1197-1211.

Asmus, H. E. and Ponte, F. C., 1973, The Brazilian marginal basins: in A. E. M. Nairn and F. C. Stehli, eds., The ocean basins and margins, v. 1, The South Atlantic Ocean; Plenum Press, p. 87-134.

Atwater, T., 1970, Implications of plate tectonics for the Cenozoic Tectonic evolution of western North America: Geol Soc. Am. Bull., v. 81, p. 3513-3536.

_____, and Molnar, P., 1973, Relative motion of the Pacific and North American plates deduced from sea floor spreading in the Atlantic, Indian and South Pacific Oceans: in R. L. Kovach and A. Nur, Proc. Conf. on tectonic problems on San Andreas Fault System; Stanford University, Geol. Sciences, v. 13, p. 136-148.

Aubouin, J., 1965, Geosynclines: Developments in Geotectonics, v. 1, Amsterdam, Elsevier, 335 p.

Audley-Charles, M. G., 1976, Mesozoic evolution of the margins of Tethys in Indonesia and the Philippines: Proc. Indonesian Petr. Assoc., Jakarta, 5th Ann. Convention, p. 1-28.

_____, 1977, Interpretation of a regional seismic line from Misool to Siram: implication for regional structure and petroleum exploration: Oil and Gas J., v. 23, p. 20-23.

_____, Carter, D. J. and Barber, A. J., 1974, Stratigraphic basis for tectonic interpretations of the outer Banda arc, eastern Indonesia: Proc. Indonesian Petrol. Assoc., p. 25-44.

_____, Curray, J. R. and Evans, G., 1977, Location of major deltas: Geology, v. 5, p. 341-344.

Baker, B. H. and Wohlenberg, J., 1971, Structure and evolution of the Kenya Rift Valley: Nature, v. 229, p. 538-542.

_____, Mohr, P. A. and Williams, L. A. J., 1972, Geology of the eastern rift system of Africa: Geol. Soc. Am. Spec. Paper 136, 67 p.

Baker, D. A., Illich, H. A., Martin, S. J. and Landin, R. R., 1975, Hydrocarbon source potential of sediments in the Sverdrup Basin: in C. J. Yorath, E. R. Parker and D. J. Glass, eds., Canada's continental margins and offshore petroleum exploration; Can. Soc. Petrol. Geol., Mem. 4, p. 545-556.

Bally, A. W., 1975, A geodynamic scenario for hydrocarbon occurrences: Proc. the 9th World Petrol. Congr., Tokyo, v. 2 (Geology), Applied Sci. Pub., Ltd., Essex, England, p. 33-44.

_____, 1976, Canada's passive continental margins — a review: Mar. Geophys. Res., v. 2, p. 327-340.

_____, Gordy, P. L. and Stewart, G. A., 1966, Structure, seismic data and orogenic evolution of Southern Canadian Rockies: Bull. Can. Petrol. Geol., v. 14, p. 337-381.

Beall, R., 1973, Plate tectonics and the origin of the Gulf Coast Basin: Trans. Gulf Coast Assoc. Geol. Soc., v. 23, p. 109-114.

Beck, R. H., 1972, The oceans, the new frontier in exploration: Australian Petroleum Expl. Assoc. J., v. 12, pt. 2, p. 7-28.

_____, and Lehner, P., 1974, Oceans, new frontier is exploration: Am. Assoc. Petrol. Geol. Bull., v. 58, p. 376-395.

_____, _____, with collaboration of Diebold, P., Bakker, G. and Doust, H., 1975, New geophysical data on key problems of global tectonics: Proc. 9th World Petrol. Congr., Tokyo, v. 2, Applied Sci. Pub., Ltd., Essex, England, p. 3-17.

BEICIP (Bureau d'études industrielles et de coopération de l'Institut Français de Pétrole), 1976, The petroleum industry in the USSR, Petroleum geology of the sedimentary basins, v. 1 and v. 2.

Beloussov, V. V., 1960, Development of the earth and tectogenesis: J. Geophys. Res., v. 65, p. 4127-4146.

_____, 1962, Basic Problems in Geotectonics: New York: McGraw Hill, 809 p.

_____, 1970, Against the hypothesis of ocean-floor spreading: Tectonophysics, v. 9, p. 489-511.

_____, 1975, Foundations of Geotectonics (in Russian): Moscow, Nyedra, 260 p.

Belyayevsky, N. A., Borisov, A. A., Volvosky and Schukin, Yu. K., 1968, Transcontinental crustal sections of the USSR and adjacent areas: Can. J. Earth Sci., v. 5, p. 1067-1078.

Ben-Avraham, Z. and Emery, K. O., 1973, Structural framework of Sunda Shelf: Am. Assoc. Petrol. Geol. Bull., v. 57, p. 2323-2366.

Berg, R. R., 1962, Mountain flank thrusting in Rocky Mountain foreland, Wyoming and Colorado: Am. Assoc. Petrol. Geol. Bull., v. 48, p. 1019-1032.

_____, and Romberg, F. E., 1966, Gravity profile across the Wind River Mountains, Wyoming: Geol. Soc. Am. Bull., v. 77, p. 646-656.

Berry, R. M. and Trumbly, W. D., 1968, The Wilburton gas field, Arkoma Basin, Oklahoma: *in* L. M. Cline, *ed.*, A guidebook to the geology of the Western Arkoma Basin and Ouachita Mountains, Oklahoma; Oklahoma City Geol. Soc., p. 86-103.

Beyth, M., 1978, A comparative study of the sedimentary fills of the Danakil depression (Ethiopia) and Dead Sea rift (Israel): Tectonophysics, v. 46, p. 357-367.

Bigarella, J. J., 1973, Geology of the Amazon and Parnaiba Basins: *in* A. E. M. Nairn and F. C. Stehli, *eds.*, The Ocean basins and margins, v. 1, The South Atlantic Ocean; Plenum Press, p. 25-83.

Biju-Duval, B., Letouzey, J. and Montadert, L., 1976, Structure and evolution of the Mediterranean basins: Inst. Franc. Petr., Div. Géol., ref. 24 422, 57 p. Also *in* Initial reports of Deep Sea Drilling Project, 1978, v. 42 A, pt. 1, p. 951-984.

_____, Dercourt, J. and LePichon, X., 1977, From the Tethys Ocean to the Mediterranean Seas: A plate tectonic model of the evolution of the western Alpine system: *in* B. Biju-Duval and L. Montadert, *eds.*, Structural history of the Mediterranean Basins: Editions Technip., p. 143-164.

_____, and Montadert, L., 1977a, Introduction to the structural history of the Mediterannean Basins: *in* B. Biju-Duval and L. Montadert, *eds.*, Structural history of the Mediterranean Basins; Editions Technip. p. 1-12.

_____, _____, *eds.*, 1977b, Structural History of the Mediterranean Basins: Internat. Symposium, Split; Editions Technip., 448 p.

Bird, Peter, 1978, Initiation of intracontinental subduction in the Himalaya: J. Geophys. Res., v. 83, p. 4975-4987.

_____, Toksöz, M. N., and Sleep, N. H., 1975, Thermal and mechanical models of continent convergence zones: J. Geophys. Res., v. 80, p. 4405-4416.

Biq, Chingchang, 1976, Alpinotype Taiwan revisited: Petroleum Geol. of Taiwan, no. 13, p. 1-14.

Blake, M. C. Jr., Campbell, R. H., Dibblee, T. W. Jr., Howell, D. G., Nilsen, T. H., Normark, W. R., Vedder, J. C. and Silver, E. A., 1978, Neogene basin formation in relation to plate tectonic evolution of San Andreas fault system, California: Am. Assoc. Petrol. Geol. Bull., v. 62, p. 344-372.

Bleahu, M. D., Boccaletti, M., Manetti, P. and Peltz, S., 1973, Neogene Carpathian arc: A continental arc displaying the features of an island arc: J. Geophys. Res., v. 78, p. 5025-5032.

_____, Horváth, F., Loddo, M., Mongelli, F. and Stegena, L., 1976, The Tyrrhenian and Pannonian basins: a comparison of two Mediterranean basins: Tectonophysics, v. 35, p. 45-69.

Boeuf, M. G. and Doust, H., 1975, Structure and development of the southern margin of Australia: Australian Petroleum Explor. Assoc. J., p. 33-43.

Bott, M. H. P., 1971, Evolution of young continental margins and formation of shelf basins: Tectonophysics, v. 11, p. 329-327.

_____, 1973, Shelf subsidence in relation to the evolution of young continental margins: *in* D. H. Tarling and S. K. Runcorn, *eds.*, Implications of continental drift to the earth sciences; London and New York, Academic Press, v. 2, p. 675-683.

_____, 1976a, Formation of sedimentary basins of graben type by extension of the continental crust: Tectonophysics, v. 36, p. 77-86.

_____, 1976b, Mechanisms of basin subsidence — an introductory review: Tectonophysics, v. 36, p. 1-14.

_____, ed._, 1976c, Sedimentary basins of continental margins and cratons: Spec. Issue Tectonophysics, v. 36, p. 1-3.

_____, 1979, Subsidence mechanisms at passive continental margins: _in_ J. S. Watkins, L. Montadert, and P. W. Dickerson, _eds.,_ Geological and geophysical investigations of continental margins; Am. Assoc. Petroleum Geologists Mem. 29, p. 3-10.

_____, and Dean, D. S., 1972, Stress systems at young continental margins: Nature, Physical Sci., v. 235, p. 23-25.

Bowin, C., 1974, Migration of a pattern of plate motion: Earth and Planetary Sci. Letters, v. 21, p. 400-404.

Bradley, W. H., 1964, Geology of Green River formation and associated Eocene rocks in southwestern Wyoming, and adjacent parts of Colorado and Utah: U.S. Geol. Surv. Prof. Paper 496A, 86 p.

Bramlette, M. N., 1946, The Monterey formation of California and the origin of its siliceous rocks, U.S. Geol. Surv. Prof. Paper 212.

Brewer, J., Smithson, S. B., Kaufman, S. and Oliver, J., 1979, The Laramide orogeny: evidence from COCORP deep crustal seismic profiles in the Wind River Mountains, Wyoming: J. Geophys. Res.

Briden, J. C., Drewry, D. J. and Smith, A. G., 1974, Phanerozoic equal-area world maps: J. Geol., v. 82, p. 555-574.

Bridwell, R. J., 1976, Lithospheric Thinning and the Late Cenozoic Thermal and Tectonic Regime of the Northern Rio Grande Rift: New Mexico Geol. Soc. Guidebook, 27 Field Conf., p. 283-292.

Brink, A. H., 1974, Petroleum geology of Gabon Basin: Am. Assoc. Petrol. Geol. Bull., v. 58, p. 216-235.

Buchanan, R. S. and Johnson, F. K., 1968, Bonanza gas field — a model for Arkoma Basin growth faulting: _in_ L. M. Cline, _ed.,_ A guidebook to the geology of the Western Arkoma Basin and Ouachita Mountains, Oklahoma; Oklahoma City Geol. Soc., p. 75-85.

Bucher, W., 1924, The Pattern of the earth's mobile belts: J. Geol. v. 28, p. 707-730.

_____, 1957, The Deformation of the Earth's Crust: New York, Hafner Press, 518 p.

Büchi, U.P., Lemcke, K., Wiener, G. and Zimaars, J., 1965, Geologische Ergebnisse der Erdölexploration auf das Mesozoikum im Untergrund des Schweizerischen Molassebeckens: Bull. Ver. Schweiz. Petrol-Geol. Ing., v. 32, p. 7-38.

_____, and Schlanke, S., 1977, Zur Palaögraphie der Schweizerischen Molasse: Erdöl, Erdgas Zeitschrift, ÖGEW Sonderausgabe, p. 57-69.

Buffler, R. T., Shipley, T. H. and Watkins, J. S., 1978, Blake continental margin seismic section: Am. Assoc. Petrol. Geol., Seismic Section No. 2.

Bullard, E. C., Everett, G. E. and Smith, A. G., 1965, The Fit of the continents around the Atlantic: _in_ A symposium on continental drift; Phil. Trans. Roy. Soc. London, v. A258, p. 41-51.

Bureau de Recherches Géologiques et Minières, Elf-Re., Esso-Rep., and SNPA, 1973, Géologie du Bassin d'Aquitaine, 27 plates.

Burk, C. A. and Drake, C. L., _eds.,_ 1974, The geology of continental margins: New York, Springer-Verlag, 1009 p.

_____, 1976, The Chad Basin: An active intracontinental basin: Tectonophysics, v. 36, p. 197-206.

Burke, Kevin, 1977, Aulacogens and continental breakup: _in_ F. A. Donath, _ed.,:_ Ann. Review Earth and Planetary Sci., v. 5, p. 371-396.

_____, 1978, manuscript: Intracontinental rifts and aulacogens.

_____, and Dewey, J. F., 1973, Plume-generated triple junctions: key indicators in applying plate tectonics to old rocks: J. Geol., v. 81, p. 406-433.

_____, _____, and Kidd, W. S. F., 1977, World distribution of sutures; the sites of former oceans: Tectonophysics, v. 40, p. 69-99.

Caldwell, J. G., Haxby, W. F., Karig, D. E. and Turcotte, D. L., 1976, On the applicability of a universal elastic trench profile: Earth and Planetary Sci. Letters, v. 31, p. 239-246.

Campos, C. W. M., Ponte, F. C. and Miura, K., 1974, Geology of the Brazilian continental margin: _in_ C. A. Burke and C. L. Drake, _eds.,_ The geology of continental margins: New York, Springer-Verlag, p. 447-461.

Carey, S. W., 1958, The tectonic approach to continental drift, _in_ S. W. Carey, _ed.,_ Continental drift — a symposium; Tasmania Univ., Geol. Dept., p. 177-355.

_____, 1975, The Expanding earth — an essay review; Earth Sci. Reviews, v. 11, p. 105-143.

_____, 1977, The expanding earth: Developments in Geotectonics Series, no. 10, Amsterdam, Elsevier, 488 p.

Carter, D. J., Audley-Charles, M. G. and Barber, A. J., 1976, Stratigraphical analysis of island arc — continental margin collision in Eastern Indonesia: J. Geol. Soc. London, v. 132, p. 179-198.

Cermák, V., 1974, Temperature-depth profiles in Czechoslovakia and some adjacent areas derived from heat-flow measurements, deep seismic sounding, and other geophysical data: Tectonophysics, v. 26, p. 103-109.

_____, and Hurtig, E., 1977, Preliminary heat flow map of Europe: Geophys. Inst. Czechosl. Acad. Sci. 141-31, Praha 4, Czechoslovakia.

Chapin, C. E. and Seager, W. R., 1975, Evolution of the Rio Grande Rift in the Socorro and Las Cruces Areas: New Mexico Geol. Soc. Guidebook, 26 Field Conf., Las Cruces County, p. 297-321.

Chapman, Davis S. and Pollack, Henry N., 1975, Heat flow and incipient rifting in the Central African plateau: Nature, v. 256, p. 28-30.

Chinese Academy of Geological Sciences, 1976a, An outline of the geology of China: Peking, 22 p.

_____, 1976b, On tectonic systems: Inst. of Geomechanics: Peking, 22 p.

Churkin, M. Jr., 1974a, Paleozoic marginal ocean basin — volcanic arc systems in the Cordilleran foldbelt: in R. H. Dott and R. H. Shaver, eds., Modern and ancient geosynclinal sedimentation; Soc. Econ. Paleont. Mineral. Spec. Pub. 19, p. 174-192.

_____, 1974b, Deep-sea drilling for landlubber geologists; the Southwest Pacific, and accordion plate tectonics analog for the Cordilleran geosyncline: Geology, v. 2, p. 339-342.

_____, and Eberlein, G. D., 1977, Ancient borderland terranes of the North American Cordillera: Correlation and microplate tectonics: Geol. Soc. Am. Bull., v. 88, p. 769-786.

Clague, D. A. and Straley, P. F., 1977, Petrologic nature of the oceanic Moho: Geology, v. 5, p. 133-136.

Clarke, S. H. and Nilsen, T. H., 1973, Displacement of Eocene strata and implications for the history of offset along the San Andreas fault, central and northern California: in R. Kovach and A. Nur, Proc. conf. on tectonic problems of the San Andreas fault system: Stanford U., Geol. Sciences, v. 13, p. 358-367.

Clarke, J. W., Girard, D. W., Peterson, J. and Rachlin, J., 1977, Petroleum geology of the West Siberian Basin and a detailed description of the Samotlor oil field: U.S. Geol. Surv. Open File Report 77-871.

Cloos, H., 1939, Hebung-Spaltung-Vulkanismus: Elemente einer Geometrischen Analyse irdischer Grossformen: Geol. Rundschau, v. 30, p. 405-527.

Coleman, R., 1977, Ophiolites, Ancient Oceanic Lithosphere: Berlin, Springer-Verlag, 229 p.

Collette, B. J., 1968, On the subsidence of the North Sea area: in D. T. Donavan, ed., Geology of Shelf Seas; Oliver and Boyd, Edinburgh, p. 15-30.

Colton, G. W., 1970, The Appalachian Basin — its depositional sequences and their geologic relationships: in G. W. Fisher et al., eds., Studies of Appalachian geology, Central and Southern, Interscience Publishers, p. 5-47.

Compton, R. R., Todd, U. R., Zartman, R. E. and Naeser, C. W., 1977, Oligocene and Miocene metamorphism, folding, and low-angle faulting in northwestern Utah: Geol. Soc. Am. Bull., v. 88, p. 1237-1250.

Cook. T. D. and Bally, A. W., eds., 1975, Stratigraphic Atlas of North and Central America: Princeton University Press, 272 p.

Cordell, Lindrith, 1978, Regional geophysical setting of the Rio Grande rift: Geol. Soc. Am. Bull., v. 89, p. 1073-1090.

Cox, A., ed., 1973, Plate Tectonics and Geomagnetic Reversals: W. H. Freeman and Company, 702 p.

Cross, T. A. and Pilger, R. H., 1978, Constraints on absolute motion and plate interaction inferred from Cenozoic igneous activity in the western United States: Am. J. Sci., v. 278, p. 865-902.

Crowell, J. C., 1952, Probable large lateral displacement on the San Gabriel fault, Southern California: Am. Assoc. Petrol. Geol. Bull., v. 36, p. 2026-2035.

_____, 1954, Geology of the Ridge basin area, Los Angeles and Ventura Counties, California: California Div. of Mines, Bull. 170, Map Sheet.

———, 1974a, Origin of late Cenozoic basins in southern California: *in* W. R. Dickinson, *ed.,* Tectonics and sedimentation: Soc. Econ. Paleont. Mineral. Spec. Pub. 22, p. 190-204.

———, 1974b, Sedimentation along the San Andreas fault, California: *in* R. H. Dott, Jr., and R. H. Shaver, *eds.,* Modern and ancient geosynclinal sedimentation; Soc. Econ. Paleont. Mineral Spec. Pub. 19, ·p. 292-303.

———, 1976, Implications of crustal stretching and shortening of coastal Ventura Basin, California: *in* D. G. Howell, *ed.,* Aspects of the geologic history of the California continental borderland: Am. Assoc. Petrol. Geol., Pacif. Sec., Misc. Pub. 24, p. 365-382.

Cruz Helu, P., Verdugo, R. V. and Bárcenas, R. P., 1977, Origin and distribution of Tertiary conglomerates, Vera Cruz Basin, Mexico: Am. Assoc. Petrol. Geol. Bull., v. 61, p. 207-226.

Cummings, D. and Shiller, G. I., 1971, Isopach of the earth crust: Earth Sci. Reviews, v. 7, p. 97-125.

Curray, J. R., 1964, Transgressions and regressions: *in* Papers in marine geology, Shepard Comm. Volume, Chapter 10; MacMillan Co., New York, p. 175-203.

———, and Moore, D. G., 1974, Sedimentary processes in the Bengal deep-sea fan and geosyncline: *in* C. A. Burk and C. L. Drake, *eds.,* Geology of continental margins: Springer-Verlag, New York, p. 617-627.

———, Shor, G. C. Jr., Raitt, R. W. and Henry, M., 1977, Seismic refraction and reflection studies of crustal structure of the eastern Sunda and western Banda arcs: J. Geophys. Res., v. 82, p. 2479-2489.

Davis, G. A., Evans, K. V., Frost, E. G., Lingrey, S. H. and Shackelford, T. J., 1977, Enigmatic Miocene low-angle faulting, southeastern California and west-central Arizona — suprastructural tectonics?: Geol. Soc. Am. Prog. Abs., v. 9, p. 943-944.

Davis, G. H., 1975, Gravity induced folding of a gneiss dome complex, Rincon Mountains, Arizona: Geol. Soc. Am. Bull., v. 86, p. 979-990.

———, and Coney, P. J., 1979, Geological development of the Cordilleran metamorphic core complexes: Geology, v. 7, p. 120-124.

de Almeida, F. F. M., *ed.,* 1976, Continental margins of Atlantic type: Proc. Internat. Symposium on Continental Margins of Atlantic Type, Ann. Acad. Brasil, Cien. 48, Sao Paulo, Brazil, 1975.

Debyser, J., Le Pichon, X., and Montadert, L., 1971, Histoire structurale du Golfe de Gascogne: Editions Technip., 2 vols.

de Cserna, Z., 1971, Mesozoic sedimentation, magmatic activity and deformation in northern Mexico, *in:* The geologic framework of the Chihuahua Tectonic Belt; West Tex. Geol. Soc., Midland, p. 99-117.

de Charpal, O., Guennoc, P., Montadert, L., and Roberts, D. G., 1978, Rifting, crustal attenuation and subsidence in the Bay of Biscay: Nature, v. 275, p. 706-711.

Degens, Egon T., von Herzen, Richard P., Wong, How-Kin, Deuser, Werner G., and Jannasch, Holger W., 1973, Lake Kivu: Structure, chemistry and biology of an east African rift lake, Geol. Rundschau, Bd. 62, p. 245-277.

Delteil, J. R., Rivier, F., Montadert, L., Apostolescu, V., Didier, J., Goslin, M. and Patriat, P. H., 1976, Structure and sedimentation of the continental margin of the Gulf of Benin: *in* F. F. M., de Almeida, *ed.,* Continental margins of Atlantic type, Ann. Acad. Brasil, Cian. 48, p. 51-66.

Demaison, G. T., 1977, Tar sands and supergiant oil fields: Am. Assoc.Petrol. Geol.,v. 61, p. 1950-1961.

Dennis, John G., *comp.* and *ed.,* 1967, International tectonic dictionary English terminology: Commission for the Geological Map of the World, Am. Assoc. Phys. Geol. Mem. 7, Tulsa, Oklahoma.

———, and Atwater, T. M., 1974, Terminology of geodynamics: Am Assoc. Petrol. Geol. Bull., v. 58, p. 1030-1036.

de Quadros, L. P., 1976, Efeito das Intrusões de Diabasio em Rochas sedimentares do L'este e Sul da Bacia do Parana: Bull. Tec. Petrobras, v. 19, p. 139-155.

de Rezende, W. M., 1971, O Mechanismo de Introsões de diabasio nas bacias Paleozoicas do Amazonas e do Maranhão: XXV Cong. Brasileiro de Geol., v. 13, p. 124-127.

———, and de Brito, C. G., 1973, Avaliacão Geologica da bacia Paleozoica do Amazonas: XXVII Cong. Brasileiro de Geol., p. 228-243.

Dewey, J. F., and Bird, J. M., 1970, Plate tectonics and geosynclines: Tectonophysics, v. 10, p. 625-638.

_____, and Burke, K., 1973, Tibetan, Variscan, and Precambrian basement reactivation: products of continental collision: J. Geol., v. 81, p. 683-692.

_____, Pitman, W. C., III, Ryan, W. B. F. and Bonnin, J., 1973, Plate tectonics and the evolution of the Alpine system: Geol. Soc. Am. Bull., v. 84, p. 3137-3180.

Dickinson, W. R., 1970, Relation of andesites, granites, and derivative sandstones to arc-trench tectonics: Rev. Geophys. v. 8, p. 813-860.

_____, 1971, Plate tectonic models of geosynclines: Earth and Planetary Sci. Letters, v. 10, p. 165-174.

_____, ed., 1974, Tectonics and Sedimentation: Soc. Econ. Paleont. Mineral. Spec. Pub. 22, 204 p.

_____, 1976, Plate tectonic evolution of sedimentary basins: in Plate tectonics and hydrocarbon accumulation: Am. Assoc. Petrol. Geol., Short Course, New Orleans, p. 1-56.

_____, and Hatherton, T., 1967, Andesitic volcanism and seismicity around the Pacific: Science, v. 157, p. 801-803.

_____, and Seely, D. R., 1979, Structure and stratigraphy of forearc regions: Am. Assoc. Petrol. Geol. Bull., v. 63, p. 2-31.

_____, and Snyder W. S., 1978, Plate tectonics of the Laramide Orogeny, Geol. Soc. Am. Mem. 151, p. 355-365.

Dietz, R. S., 1963, Collapsing continental arises; an actualistic concept of geosynclines and mountain building: J. Geol., v. 71, p. 314-333.

_____, and Holden, J. C., 1966, Miogeoclines in space and time: J. Geol., v. 74, p. 566-583.

Dorr, J. A., Jr., Spearing, D. R. and Steidtmann, J. R., 1977, Deformation and deposition between a foreland uplift and an impinging thrust belt: Hoback Basin, Wyoming: Geol. Soc. Am. Spec. Paper 177, 82 p.

Dott, R. H., Jr., 1978, Tectonics and sedimentation a century later: Earth Sci. Rev., 14, p. 1-34.

_____, and Batten, R. L., 1971, Evolution of the Earth: McGraw Hill, 649 p.

_____, and Shaver, R. H., eds., 1974, Modern and ancient geosynclinal sedimentation: Soc. Econ. Paleont. Mineral. Spec. Pub. 19, 380 p.

Dow, W. G., 1978, Petroleum source beds on continental slopes and rises: Am. Assoc. Petrol. Geol. Bull., v. 62, p. 1584-1606.

Drake, C. L., Ewing, M. and Sutton, G. H., 1959, Continental margins and geosynclines; the east coast of North America north of Cape Hatteras: in Physics and Chemistry of Earth; Pergamon Press, London, v. 3, p. 110-198.

Drewes, H., 1978, The Cordilleran orogenic belt between Nevada and Chihuahua: Geol. Soc. Am. Bull., v. 89, p. 641-657.

Driver, Edgar, S. and Pardo, Georges, 1971, Seismic traverse across the Gabon continental margin: in C. A. Burk and C. L. Drake, eds., The geology of continental margins: Springer-Verlag, New York, p. 293-295.

Dubois, J., Dupont, J., Lapouille, A. and Recy, J., 1977, Lithospheric bulge and thickening of the lithosphere with age examples: in Geodynamics in the Southwest Pacific; Editions Technip., p. 371-380.

Eisbacher, G. H., 1974, Evolution of successor basins in the Canadian Cordillera: in R. H. Dott and R. H. Shaver, eds., Modern and ancient geosynclinal sedimentation: Soc. Econ. Paleont. Mineral. Spec. Pub. 19, p. 274-291.

_____, 1977, Mesozoic-Tertiary basin models for the Canadian Cordillera and their geological constraints: Can. J. Earth Sci., v. 14, p. 2414-2421.

_____, Carrigy, M. A. and Campbell, R. B., 1974, Paleodrainage pattern and late orogenic basins of the Canadian Cordillera: in W. R. Dickinson, ed., Tectonics and sedimentation: Soc. Econ. Paleont. Mineral. Spec. Pub. 22, p. 143-166.

Emter, D., 1971, Ergebnisse Seismischer Untersuchungen der Erdkruste und des Obersten Erdmantels in Südwestdeutschland: Dissertation, Univ. Stuttgart, 108 p.

Epis, R. C. and Chapin, C. E., 1975, Geomorphic and tectonic implications of the post Laramide, late Eocene erosion surface in the Southern Rocky Mountains: Geol. Soc. Am. Mem. 144, p. 45-74.

Faill, R. T., 1973, The tectonic development of the Triassic Newark-Gettysburg Basin in Pennsylvania: Geol. Soc. Am. Bull., v. 84, p. 725-740.

Falvey, D. A., 1974, The development of continental margins in plate tectonic theory: Australian Petroleum Explor. Assoc. J., p. 95-106.

Finetti, I., 1976, Mediterranean ridge: a young submerged chain associated with the Hellenic arc: Boll. di Geofisica, Teorica e Applicata, v. 19, p. 31-36.

_____, and Morelli, C., 1972, Wide scale digital seismic exploration of the Mediterranean Sea: Boll. di Geofisica, Teorica e Applicata, v. 14, p. 291-342.

Fisher, G. W., Pettijohn, F. J., Reed, J. C. Jr. and Weaver, K. N., *eds.*, 1970, Studies of Appalachian Geology Central and Southern: Interscience Publishers, 460 p.

Fouch, T. P., 1975, Lithofacies and related hydrocarbon accumulations in Tertiary strata of the western and central Uinta Basin, Utah: *in* Rocky Mountain Assoc. Geol., 1975 Symposium, p. 163-173.

Fountain, D. M., 1976, The Ivrea-Verbano and Strona-Ceneri zones, northern Italy: A cross section of the continental crust, new evidence from seismic velocities of rock samples: Tectonophysics, v. 33, p. 145-165.

Fuchs, K., 1974, Geophysical contributions to Taphrogenesis: *in* J. H. Illies and K. Fuchs, *eds.*, Approaches to taphrogenesis; Schweitzerbart, Stuttgart, p. 420-432.

Fyfe, W. S., 1974, The ocean ridge environment: heat and mass transfer (abs.): Ann. Mtg. Geol. Assoc. Canada, St. Johns, p. 30.

_____, 1976, Hydrosphere and continental crust: growing or shrinking?: Geoscience Canada, v. 3, p. 82-83.

Galley, J. E., 1958, Oil and geology in the Permian Basin of Texas and New Mexico: *in* L. G. Weeks, *ed.*, Habitat of oil: Am. Assoc. Petrol. Geol., p. 395-446.

_____, 1971, Summary of Petroleum Resources in Paleozoic Rocks of Region 5 — North, Central and West Texas and Eastern New Mexico: Am. Assoc. Petrol. Geol. Mem. 15, v. 1, p. 726-737.

Giese, P., Morelli, C. and Steinmetz, L., 1973, Main features of crustal structure in central and southern Europe based on data of explosion seismology: *in* S. Müller, *ed.*, Tectonophysics, v. 20, p. 367-379.

Gill, J. R. and Cobban, W. A., 1966, Regional unconformity in Late Cretaceous, Wyoming: U.S. Geol. Surv. Prof. Paper 550B, p. 20-27.

Gilluly, J., 1964, Atlantic sediments, erosion rates, and the evolution of the continental shelf: some speculations: Geol. Soc. Am. Bull., v. 75, p. 483-492.

Girdler, R. W., 1975, The great negative Bouguer gravity anomaly over Africa: EOS, v. 56, p. 516-519.

_____, Fairhead, J. D., Searle, R. C. and Sowerbutts, W. T. C., 1969, Evolution of rifting in Africa: Nature, v. 224, p. 1178-1182.

Given, M. M., 1977, Mesozoic and early Cenozoic geology of offshore Nova Scotia: Bull. Can. Petrol. Geol., v. 25, p. 63-91.

Gordy, P. F., Frey, F. R. and Ollerenshaw, N. C., 1975, Structural geology of the foothills between Savanna Creek and Panther River SW Alberta, Canada: Guidebook, Can. Soc. Petrol. Geol., Can. Soc. Explor. Geophys., Explor. Update 1975, Calgary.

Graham, S. A., 1978, Role of the Salinian Block in evolution of the San Andreas fault system: Am. Assoc. Petrol. Geol. Bull., v. 62, p. 2214-2231.

_____, Dickinson, W. R. and Ingersoll, R. V., 1975, Himalayan-Bengal model for flysch dispersal in Appalachian-Ouachita system: Geol. Soc. Am. Bull., v. 86, 273-286.

Green, A. R., 1977, The evolution of the earth's crust and sedimentary basin development: *in* J. Heacock, *ed.*, The Earth's crust; Am. Geophys. Union, Geophysical Monograph 10, p. 1-17.

Grow, J. A. and Markl, R. G., 1977, IPOD-USGS Multichannel Seismic Reflection Profile from Cape Hatteras to Mid-Atlantic Ridge: Geology, v. 5, p. 625-630.

_____, Mattick, R. E. and Schlee, J. S., 1979, Multichannel seismic depth sections and interval velocities over outer continental shelf and upper continental slope between Cape Hatteras and Cape Cod: *in* J. S. Watkins, L. Montadert, and P. W. Dickerson, *eds.*, Geological and geophysical investigations of continental margins; Am. Assoc. Petrol. Geol. Mem. 29.

Hadley, J., 1964, Correlation of isotopic ages crustal heating and sedimentation in the Appalachian region: *in* W. D. Lowry, *ed.*, Tectonics of the Southern Appalachians: Virginia Poly. Inst., Dept. Geol. Sci., Mem. 1, p. 33-46.

Hamilton, W., 1977, Subduction in the Indonesian region: *in* M. Talwani and W. C. Pitman, *eds.*, Island Arcs, Deep Sea Trenches and Back-arc Basins: Maurice Ewing Ser. 1, Am. Geophys. Union, p. 15-31.

_____ , 1978, Tectonic map of the Indonesian region: U.S. Geol. Surv. Map I-875-D.

_____ , 1979, Tectonics of the Indonesian region: U.S. Geol. Surv. Prof. Paper 1078.

_____ , and Myers, W. B., 1966, Cenozoic tectonics of the Western United States: Rev. Geophys., v. 4, p. 509-549.

Harding, T. P., 1976, Tectonic significance and hydrocarbon trapping consequences of sequential folding synchronous with San Andreas faulting, San Joaquin Valley, California: Am. Assoc. Petrol. Geol. Bull., v. 60, p. 356-378.

Harrison, J. E., Griggs, A. B. and Wells, J. D., 1975, Tectonic features of the Precambrian Belt basin and their influence on post-Belt structures: U.S. Geol. Surv. Prof. Paper 866.

Hasebe, K., Fujii, N. and Uyeda, S., 1970, Thermal processes under island arcs: Tectonophysics, v. 10, p. 335-355.

Haxby, W. F., Turcotte, D. L. and Bird, J. M., 1976, Thermal and mechanical evolution of the Michigan Basin: Tectonophysics, v. 36, p. 57-75.

Heezen, B. C. and Fornari, D. J., 1976, Geological Map of the Pacific Ocean, 1:35 000 000: Geol. World Atlas Sheet 20, UNESCO, Paris.

Heybroek, F., 1965, The Red Sea Miocene evaporite basin: *in* D. C. Ion, *ed.*, Salt basins around Africa; Inst. Petroleum, London, p. 17-40.

Hilde, T. W. C., Uyeda, S. and Kroenke, L., 1977, Evolution of the western Pacific and its margin: Tectonophysics, v. 38, p. 145-165.

Hill, Charles S., 1971, Future petroleum resources in pre-Pennsylvanian rocks of north, central and west Texas and eastern New Mexico: Am. Assoc. Petrol. Geol. Mem. 15, v. 1, p. 738-751.

Hills, John M., 1968, Gas in Delaware and Val Verde Basins, West Texas and Southeastern New Mexico: Am. Assoc. Petrol. Geol. Mem. 9, v. 2, p. 1394.

Hirn, A. and Perrier, G., 1974, Deep seismic sounding in the Limagne Graben: *in* J. K. Illies and K. Fuchs, *eds.*, Approaches to taphrogenesis; Stuttgart, E. Schweizerbart'sche Verlagsbuchhandlung, p. 329-340.

Hoffman, P., Dewey, J. F. and Burke, K., 1974, Aulacogens and their genetic relation to geosynclines, with a Proterozoic example from Great Slave Lake, Canada; *in* R. H. Dott and R. M. Shaver, *eds.*, Modern and ancient geosynclinal sedimentation; Soc. Econ. Paleont. Mineral. Spec. Pub. 19, p. 38-55.

Hood, A. and Castaño, J. R., 1974, Organic metamorphism: its relationship to petroleum generation and application to studies of authigenic minerals: U.N. ESCAP, CCOP Tech. Bull., v. 8, p. 85-118.

_____ , Gutjahr, C. C. M. and Heacock, R. L., 1975, Organic metamorphism and the generation of petroleum: Am. Assoc. Petrol. Geol. Bull., v. 59, p. 986-996.

Hopkins, H. R., 1968, Structural interpretations of the Ouachita Mountains: *in* L. M. Cline, *ed.*, A guidebook to the geology of Western Arkoma Basin and Ouachita Mountains, Oklahoma; Okla. City Geol. Soc., p. 104-108.

Horváth, F. and Channell, J. F. T., 1977, Further evidence relevant to the Adriatic African promontory as a paleogeographic premise for Alpine orogeny: *in* B. Biju-Duval and L. Montadert, *eds.*, Structural history of the Mediterranean Basins: Editions Technip., p. 133-142.

_____ , and Stegena, L., 1977, The Pannonian Basin: a Mediterranean interarc basin: *in* B. Biju-Duval and L. Montadert, *eds.*, Structural history of the Mediterranean Basins; Editions Technip., p. 333-340.

_____ , _____ , and Géczy, B., 1975, Ensimatic and ensialic interarc basins: comments on "Neogene Carpathian arc: a continental arc displaying the features of an 'island arc' " by M. D. Bleahu, M. Boccaletti, P. Manetti, and S. Peltz: J. Geophys. Res., v. 80, p. 281-285.

Howie, R. D. and Barss, M. S., 1975, Paleogeography and sedimentation in the Upper Paleozoic, Eastern Canada: *in* C. J. Yorath, E. R. Parker, and D. J. Glass, *eds.*, Canada's continental margins and offshore petroleum exploration; Can. Soc. Petrol. Geol. Mem. 4, p. 45-57.

Hsü, K. J., 1965, Isostasy, crustal thinning, mantle changes, and the disappearance of ancient land masses: Am. J. Sci., v. 263, p. 97-109.

_____, 1972a, The concept of the geosyncline, yesterday and today: reprint from Trans. Leicester Literary and Philosophical Soc., v. LXVI, Leicester, Ratnett & Co. Ltd.

_____, 1972b, When the Mediterranean dried up: Sci. American, v. 277, p. 27-33.

_____, and Bernoulli, D., 1978, Genesis of the Tethys and the Mediterranean: in K. Hsü, L. Montadert et al., Initial reports of the Deep Sea Drilling Project, v. 42, part 1, U.S. Govt. Printing Office, Washington, D.C., p. 943-950.

_____, Montadert, L., Bernoulli, D., Cita, M. B., Erickson, A. J., Garrison, R. E., Kidd, R. B., Mélières, E., Müller, C. and Wright, R., 1978 History of the Mediterranean salinity crisis: in K. Hsü, L. Montadert et al., Initial reports of the Deep Sea Drilling Project, v. 42, part 1; U.S. Govt. Printing Office, Washington, D. C., p. 1053-1077.

Huang, T. K., 1960, Die Geotektonischen Elemente im Aufbau Chinas. 2 parts: Geologie, no. 8, p. 841-866; no. 9, p. 715-733.

_____, 1978, An outline of the tectonic characteristics of China: Eclogae Geol. Helv., v. 71, p. 611-636.

Hussong, D. and Uyeda, S., 1978, Leg 60 ends in Guam: Geotimes, v. 23, p. 19-22.

Illies, J. H., 1977, Ancient and recent rifting in the Rhinegraben: Geol. Mijnbouw, v. 56, p. 329-350.

_____, and Fuchs, K., eds., 1974, Approaches to taphrogenesis: Inter-Union Commission on Geodynamics, Scientific Rept. No. 8, E. Schweizerbart'sche Verlagsbuchhandlung, Stuttgart.

_____, and Greiner, G., 1978, Rhinegraben and the Alpine system: Geol. Soc. Am. Bull., v. 89, p. 770-782.

Ingersoll, R., 1976, Evolution of the late Cretaceous forearc basin of northern and central California: Unpublished Ph. D. thesis, Stanford University.

Johnson, J. D. and Normark, W. R., 1974, Neogene tectonic evolution of the Salinian Block, west-central California: Geology, v. 2, p. 11-14.

Jordan, P. and Beer, A., 1971, The Expanding Earth: Internat. Series of Monographs, Pergamon Press, v. 37, 199 p.

Jordan, T. H. and Fyfe, W. S., 1976, Lithosphere-asthenosphere boundary, a Penrose Conference report: Geology, v. 4, p. 770-772.

Kanasewich, E. R., Clowes, R. M., and McCloughan, C. M., 1968, A buried Precambrian rift in western Canada: Tectonophysics, v. 8, p. 513-527.

Kapounek, J., Koelbl, L. and Weinberger, F., 1963, Results of new exploration in the basement of the Vienna Basin: Proc. 6th World Petrol. Congr., Sec. 1, p. 205-221.

_____, Kröll, A., Papp, A., and Turnovsky, K., 1965, Die Verbreitung von Oligozän, unter-und Mitteleozän in Niederdesterreich: Erdöl Zeitschrift Wien, v. 81, p. 109-116.

Karig, D. E., 1971, Origin and development of marginal basins in the western Pacific: J. Geophys. Res., v. 76, p. 2542-2561.

_____, 1974, Evolution of arc systems in the western Pacific: Ann. Rev. Earth Planetary Sci., v. 2, p. 51-75.

_____, Ingle, J. C., Bouma, A. H., Ellis, H., Haile, N., Koizumi, I., Ling, H. Yi., MacGregor, I. D., Moore, C., Ujiie, H., Watanabe, T., White, S. M. and Yasui, M., 1975, Initial reports of the Deep Sea Drilling Project, v. 31; Washington, U.S. Govt. Printing Office, 927 p.

Katili, J. A., 1975, Volcanism and plate tectonics in the Indonesian Island arcs: Tectonophysics, v. 26, p. 165-188.

Kay, M., 1951, North American Geosynclines: Geol. Soc. Am. Mem. 48, 143 p.

Keefer, W. R., 1965, Stratigraphic and geologic history of the uppermost Cretaceous, Paleocene, and lower Eocene rocks in the Wind River Basin, Wyoming: U.S. Geol. Surv. Prof. Paper 495 A.

_____, 1970, Structural geology of the Wind River Basin: U.S. Geol. Surv. Prof. Paper 495 D.

Keen, C. E., 1979, Thermal history and subsidence of rifted continental margins — evidence from wells on the Nova Scotian and Labrador shelves: Can. J. Earth Sci., v. 16, p. 505-522.

_____, and Keen, M. J., 1974, The Continental margins of eastern Canada and Baffin Bay: in C. A. Burk and C. L. Drake, eds., The geology of continental margins; Springer-Verlag, New York, p. 381-389.

_____ _____, Barrett, O. L. and Heffler, D. E., 1975, Some aspects of the ocean continent transition of the continental margin of eastern Canada: Geol. Surv. Can. Paper 74-30, p. 189-197.

Keith, M. L., 1972, Ocean floor convergence; A contrary view of global tectonics: J. Geol., v. 80, p. 249-276.

Kelley, V. C., 1977, Geology of Albuquerque Basin, New Mexico Discussion — origin and development: New Mexico Bur. Mines Min. Res. Mem. 33, p. 50-55.

Ketner, K. B., Kastowo, Modjo, S., Naeser, C. W., Obradovich, J. D., Robinson, K., Suptandar, T. and Wikarno, 1976, Pre-Eocene rocks of Java, Indonesia: J. Res. U.S. Geol. Surv., v. 4, p. 605-614.

King. P. B., 1969a, The Tectonics of North America — a discussion to accompany the Tectonic Map of North America, Scale 1: 5 000 000: U.S. Geol. Surv. Prof. Paper 628.

_____ , 1977, The Evolution of North America — revised edition: Princeton University Press, 197 p.

Kinsman, D. J. J., 1975, Rift valley basins and sedimentary history of trailing continental margin: in A. C. Fisher and S. Judson, eds., Petroleum and global tectonics, Princeton University Press, p. 83-128.

Kirschner, C. E. and Lyon, C. A., 1973, Stratigraphic and tectonic development of Cook Inlet petroleum province: in M. G. Pitcher, ed., Arctic geology: Am. Assoc. Petrol. Geol. Mem. 19, p. 396-407.

Klemme, H. D., 1975, Giant oil fields related to their geologic setting: a possible guide to exploration: Bull. Can. Petrol. Geol. v. 23, p. 30-66.

_____ , 1977, One-fifth of reserves lie offshore: Oil and Gas J., Petroleum 2000, August, p. 108-128.

Kontorovich, A. E., Nesterov, I. I., Salmanov, F. K., Surkov, V. S., Trofimuk, A. A. and Ervye, Yu. G., 1975, Geology of Oil and Gas of West Siberia (in Russian): Moscow: Nedra, 679 p.

Kroenke, L. W., 1972, Geology of the Ontong Java plateau: unpublished Ph.D. thesis, University of Hawaii, 119 p.

Kröll, A. and Wiesender, H., 1972, the origin of oil and gas deposits in the Vienna Basin (Austria): 245th Intern. Geol. Congr., Section 5, p. 153-160.

_____ , and Wessely, G., 1973, Neue Ergebnisse beim Tiefenaufschluss im Wiener Becken: Erdoel. Erdgas Zeitschrift, v. 89, p. 400-413.

Kulm, L. D., Scheidegger, K. F., Prince, R. A., Dymond, J., Modre, T. G., Jr., and Hussong, O. M., 1973, Tholeitic basalt ridge in the Peru trench: Geology, v. 1, p. 11-14.

Kumar, N., Bryan, G., Gorini, M., and Carvalho, J., 1976, Evolution of the continental margin off northern Brazil: Sediment distribution and carbon potential: in F. F. M. de Almeida, ed., Continental margins of Atlantic type, Ann. Acad. Brazil Cien., v. 48, p. 131-144.

Kutas, R. I. and Gordiyenko, V. V., 1970, Thermal fields of the eastern Carpathians: Geothermics, Spec. Issue 2, p. 1063-1066.

_____ , Lubimova, E. A. and Smirnov, Ya. B., 1975, Heat flow map for the European part of USSR, and its geologic-tectonical analysis: in E. A. Lubimova and M. N. Berdichevsky, eds., Geothermal and Moscow Geoelectrical Research in USSR; Moscow, Nauka, 220 p.

Larson, R. L., Menard, H. W. and Smith, S. M., 1968, Gulf of California: a result of ocean-floor spreading and transform faulting: Science, v. 161, p. 781-784.

_____ , and Ladd, J. W., 1973, Evidence for the opening of the south Atlantic in the early Cretaceous: Nature, v. 246, p. 227-266.

Laubscher, H. P., 1965, Ein Kinematisches Modell der Jura Faltung: Eclogae Geol. Helv., v. 58, p. 231-318.

_____ , 1976, Foreland folding: the northern foreland of the central Alps: 25th Intern. Geol. Cong. Abs., Sidney, p. 687.

_____ , 1978, Foreland folding: Tectonophysics, v. 47, p. 325-337.

Laughton, A. S.. Whitmarsh, R. B. and Jones, M. T., 1970, The evolution of the Gulf of Aden: Royal Soc. London Philos. Trans., v. 267, p. 227-266.

Lehner, P. and De Ruyter, P. A. C., 1977, Structural history of Atlantic margin of Africa: Am. Assoc. Petrol. Geol. Bull., v. 61, p. 961-981.

Le Pichon, X., Francheteau, J. and Bonnin, J., 1973, Plate Tectonics: New York, Elsevier, 300 p.

Letouzey, B., Biju-Duval, B., Dormel, A., Gonnard, R., Kristchev, K., Montadert, L. and Sungurlu, D., 1977, The Black Sea: a marginal basin geophysical and geological data: in B. Biju-Duval and L. Montadert, eds., Structural history of the Mediterranean Basins: Editions Technip., p. 363-376.

Levorsen, A. I., 1960, Paleogeologic maps: W. H. Freeman and Company, 174 p.

Liggett, M. A. and Ehrenspeck, H. E., 1974, Pahranagat shear system, Lincoln County, Nevada: Argus Explor. Co., Rept. of Inv., NASA-CR-136388, E74-10206, 10 p.

_____, and Childs, J. F., 1974, Crustal extension and transform faulting in the southern Basin Range Province: Argus Explor. Co., Rept. of Inv., NASA-CR-137256, E74-10411, 28 p.

Logatchev, N. A. and Mohr, P. A., 1978, Geodynamics of the Baikal rift zones: Tectonophysics, v. 45, p. 1-13.

_____, and Florentsov, N. A., 1978, The Baikal system of rift valleys: Tectonophysics, v. 45, p. 1-13.

Longwell, C. R., 1945, Low-angle normal faults in the basin-and-range province: Am. Geophys. Union Trans., v. 26, p. 107-118.

Lopez-Ramos, E., 1969, Marine Paleozoic Rocks of Mexico: Am. Assoc. Petrol. Geol. Bull., v. 53, p. 2399-2417.

_____, 1972, Estudio del Basamento ígneo y metamórfico de las zonas norte y Poza Rica: Bull. Mexican Geol. Petrol., v. 24, p. 267-323.

Lowell, J. D., 1974, Plate tectonics and foreland basement deformation: Geology, v. 2, p. 275-278.

_____, and Genik, G. J., 1972, Sea floor spreading and structural evolution of southern Red Sea: Am. Assoc. Petrol. Geol. Bull., v. 56, p. 247-259.

_____, _____, Nelson, T. H., and Tucker, P. M., 1975, Petroleum and plate tectonics of the southern Red Sea: in A. G. Fisher and S. Judson, eds., Petroleum and global tectonics; Princeton University Press, p. 129-158.

Lucas, P. T. and Drexler, J. M., 1975, Altamont-Bluebell: a major fractured and overpressured stratigraphic trap, Uinta Basin, Utah: in D. W. Bolyard, ed., Deep Drilling Frontiers of the Central Rocky Mountains; Denver, Rocky Mountain Assoc. Geol.

McCrossan, R. G. and Porter, J. W., 1973, The geology and petroleum potential of the Canadian sedimentary basins — a synthesis: in R. G. McCrossan, ed., The future petroleum provinces of Canada — their geology and potential; Can. Soc. Petrol. Geol. Mem. 1, p. 589-720.

McDonald, R. E., 1976, Tertiary tectonics and sedimentary rocks along the transition basin, and range province to plateau and thrust belt province, Utah: Rocky Mountain Assoc. Geol. Guidebook, p. 281-371.

McGinnis, L. D., 1970, Tectonics and the gravity field in the continental interior: J. Geophys. Res., v. 75, p. 317-331.

McIver, N., 1970, Appalachian turbidites: in G. W. Fisher et al., Studies of Appalachian Geology, Central and Southern; Interscience Publishers, p. 69-81.

McKenzie, D., 1977, Plate tectonics and its relationship to the evolution of ideas in the geological sciences: Daedalus, v. 1, p. 97-124.

_____, 1978a, Active tectonics of the Alpine-Himalayan belt: the Aegean Sea and surrounding regions: Geophys. J. Roy. Astr. Soc., v. 55, p. 217-254.

_____, 1978b, Some remarks on the development of sedimentary basins: Earth and Planetary Sci. Letters, v. 40, p. 25-32.

Mahel, M., ed., 1974a, Tectonic map of the Carpathian-Balkan regions and their foreland, 1:1 000 000: Carpathian-Balkan Assoc., Bratislava, 12 sheets.

_____, 1974b, Tectonics of the Carpathian Balkan regions — explanation to the tectonic map of the Carpathian-Balkan regions and their foreland: Geol. Inst. of Dionyz Stur. Bratislava, 454 p.

Mareschal, J. C. and Gangi, A. F., 1977, Equilibrium position of a phase boundary under horizontally varying surface loads: Geophys. J. Astro. Soc., v. 49, p. 757-772.

Markl, R., 1974, Evidence for the breakup of eastern Gondwanaland by the early Cretaceous: Nature, v. 251, p. 196-200.

_____, 1978, Further evidence for the early Cretaceous breakup of Gondwanaland off southwestern Australia: Marine Geology, v. 26, p. 41-48.

Martinson, N. W., McDonald, D. R. and Kaye, P., 1972, Exploration on the continental shelf off northwest Australia: Austral-Asia Oil Gas Review, v. 19, p. 8-13, 15-16, 18.

Mascle, A., Biju-Duval, B., Letouzey, J., Montadert, L. and Ravenne, C., 1977, Sediments and their deformations in active margins of different geological settings: in Intern. Symposium on Geodynamics in Southwest Pacific; Noumea (New Caledonia) Editions Technip., Paris, p. 327-344.

Mascle, J., 1976, Le Golfe de Guinée: example d'évolution d'une marge Atlantique en cisaillement. Mem. Soc. Geol. Fr. N. Serie 128.

Meckel, L. D., 1970, Paleozoic alluvial deposition in the central Appalachians: a summary: *in* G. W. Fisher *et al., eds.,* Studies of Appalachian geology, Central and Southern; Interscience Publishers, p. 49-67.

Meneley, R. A., Hanao, D. and Merritt, R. U., 1975, The northwest margin of the Sverdrup Basin: *in* C. J. Yorath, E. R. Parker and D. J. Glass, *eds.,* Canada's continental margins and offshore petroleum exploration; Can. Soc. Petrol. Geol. Mem. 4, p. 531-544.

Meng, C. Y. and Chou, J. T., 1976, Petroliferous Taiwan basins in framework of western Pacific Ocean: *in* M. Halbouty *et al., eds.,* Circum-Pacific energy and mineral resources; Am. Assoc. Petrol. Geol. Mem. 25, p. 256-260.

Meyerhoff, A. A., 1970a, Developments in mainland China 1949-1968: Am. Assoc. Petrol. Geol. Bull., v. 54, p. 1567-1580.

_____ , 1970b, Continental drift: implications of paleomagnetic studies, meteorology, physical oceanography, and climatology: J. Geol., v. 78, p. 1-51.

_____ , 1970c, Continental drift, II: high-latitude evaporite deposits and geologic history of Arctic and North Atlantic Oceans: J. Geol., v. 78, p. 406-444.

_____ , and Briggs, R. S., Jr., 1972, Continental drift, V: proposed hypothesis of earth tectonics: J. Geol., v. 80, p. 663-692.

_____ , and Meyerhoff, H. A., 1972a, Continental drift, IV: the Caribbean "plate": J. Geol., v. 80, p. 34-60.

_____ , _____ , 1972b, "The new global tectonics" major inconsistencies: Am. Assoc. Petrol. Geol. Bull., v. 56, p. 269-336.

_____ , and Teichert, C., 1971, Continental drift, III: late Paleozoic glacial centers, and Devonian-Eocene coal distribution: J. Geol., v. 79, p. 285-321.

_____ , and Willums, J. O., 1976, Petroleum geology and industry of the People's Republic of China: United Nations ESCAP, CCOP Technical Bulletin, v. 10, p. 103-212.

_____ , and Willums, J. O., 1978, China: an oilman's look behind the great wall: Intern. Petroleum Encyclopedia, p. 413-419.

Miall, A. D., 1978, Tectonic setting and syndepositional deformation of molasse and other nonmarine-paralic sedimentary basins: Can. J. Earth Sci., v. 15, p. 1613-1632.

Ministry of Economic Affairs, Republic of China, 1974, Geologic map of Taiwan, 1: 250 000, 4 sheets.

Mitchum, R. M., Jr., 1977, Seismic stratigraphy and global changes in sea level, Part II; Glossary of terms used in seismic stratigraphy: *in* C. E. Payton, *ed.,* Seismic stratigraphy — applications to hydrocarbon exploration; Am. Assoc. Petrol. Geol. Mem. 26, p. 205-212.

Molnar, P. and Tapponier, P., 1975, Cenozoic tectonics of Asia: effects of a continental collision: Science, v. 189, p 419-426.

_____ , and Atwater, T., 1978, Interarc spreading and Cordilleran tectonics as alternates related to the age of subducted oceanic lithosphere: Earth and Planetary Sci. Letters, v. 41, p. 330-340.

Montadert, L., Roberts, D. G., Auffret, G., Bock, W., DuPeuple, P. A., Hailwood, E. A., Harrison, W., Kagami, H., Lumsden, D. N., Müller, C., Schnitker, D., Thompson, R. W., Thompson, T. L. and Timofeev, P. P., 1977, Rifting and subsidence on passive continental margins in the North East Atlantic: Nature, v. 268, p. 305-309.

Montecchi, P. A., 1976, Some shallow tectonic consequences of subduction and their meaning to the hydrocarbon explorationist: *in* M. Halbouty *et al., eds.,* Circum-Pacific energy and mineral resources: Am. Assoc. Petrol. Geol. Mem. 25, p. 189-202.

Moore, D. G., 1973, Plate edge deformation and crustal growth, Gulf of California structural province: Geol. Soc. Am. Bull., v. 84, p. 1883-1906.

Morel, P. and Irving, E., 1978, Tentative paleocontinental maps for the early Phanerozoic and Proterozoic: J. Geol., v. 86, p. 535-561.

Mulder, C. J., 1973, Tectonic framework and distribution of Miocene evaporites in the Mediterranean: *in* Messinian events in the Mediterranean: Kon. Ned. Akad. Wetensch., Amsterdam, Geodynamic Report 7, p. 44-59.

___ , Lehner, P. and Allen, D. C. K., 1975, Structural evolution of the Neogene salt basins in the eastern Mediterranean and the Red Sea: Geol. Mijnb., v. 54, p. 208-221.

___ , and Parry, G. R., 1977, Late Tertiary evolution of the Alboran Sea in the eastern entrance of the Strait of Gibralter: in B. Biju-Duval and L. Montadert, eds., Structural history of the Mediterranean Basins; Editions Technip., p. 363-376.

Müller, S., Peterschmitt, E., Fuchs, K., Empter, D., and Ansorge, J., 1973, Crustal structure of the Rhinegraben area: Tectonophysics, v. 20, p. 381-391.

Murauchi, S. and Yasui, M., 1968 (Fide Sugimura, A., and Uyeda, S., 1973), Investigations in the seas around Japan: Kagaku, v. 38, p. 4.

Murphy, R. W., 1973, The Manila trench — west Taiwan foldbelt, a flipped subduction zone: Geol. Soc. Malaysia Bull., v. 6, p. 24-42.

Nairn, A. E. M. and Stehli, F. G., 1975, The Ocean Basins and Margins, v. 3: The Gulf of Mexico and the Caribbean: Plenum Press, New York, 705 p.

Nalivkin, V. D., 1976, Dynamics of the development of the Russia platform structures: Tectonophysics, v. 36, p. 247-262.

Nicholas, R. L. and Rozendahl, R. A., 1975, Subsurface positive elements within Ouachita Foldbelt in Texas and their relation to Paleozoic cratonic margin: Am. Assoc. Petrol. Geol. Bull., v. 59, p. 193-216.

Niggli, E., ed., 1968, Symposium "Zone Ivrea-Verbano": Schweizerische Miner. und Petrogr. Mitteilungen, Bd. 48, Heft 1, 355 p.

North, F. K., 1971, Characteristics of oil provinces — a study for students: Bull. Can. Petrol. Geol., v. 19, p. 601-658.

Norton, I. D. and Sclater, J. G., in preparation, A model for the evolution of the Indian Ocean and the breakup of Gondwanaland.

O'Connell, R. J. and Wasserburg, G. J., 1967, Dynamics of the motion of a phase change boundary to changes in pressure: Review Geophys., v. 5, p. 329-410.

Ogniben, L., Parotto, M. and Praturlon, A., 1975, Structural model of Italy, 4 maps and explanatory notes: Cons. Naz. delle Ricerche Quaderni de "La Ricerca Scientifica", Roma, 502 p.

Oliver, J., Isacks, B., Barazangi, M. and Mitronovas, W., 1973, Dynamics of the down-going lithosphere: Tectonophysics, v. 19, p. 133-147.

Oxburgh, E. R., 1974, The Plain man's guide to plate tectonics: Proc. Geol. Assoc., v. 85, p. 299-357.

Pacca, I. G. and Hiddo, F. Y., 1976, Paleomagnetic analysis of Mesozoic Serra Geral Basaltic lava flows in southern Brazil: Anais Acad. Bras. Cienc., v. 48, p. 211-213.

Packham, G. H. and Falvey, D. A., 1970, An hypothesis for the formation of marginal seas in the western Pacific: Tectonophysics, v. 11, p. 79-109.

Paraschiv, D., 1975, Geology of the Romanian hydrocarbon deposits: Inst. de. Geol. si Geof., Studi Tehnice si Econ., Ser. A., no. 10, 363 p. (Eng. summary, p. 347-363).

Parsons, B. and Sclater, J. C., 1977, An analysis of the variation of ocean floor bathymetry and heat flow with age: J. Geophys. Res., v. 82, p. 803-827.

Parsons, W. H., 1973, Alberta: in R. G. McCrossan, ed., The future petroleum provinces of Canada — their geology and potential: Can. Soc. Petrol. Geol. Mem. 1, p. 73-120.

Pelletier, B. R., ed., 1974, Offshore Geology of Eastern Canada, v. 1: Concepts and Applications of Environmental Marine Geology: Geol. Surv. Can. Paper 74-30, pt. 1.

People's Republic of China, 1975, Tectonic map of China, 1: 4 000 000, 1 sheet.

___ , 1976, Geologic map of China, 1: 4 000 000, 1 sheet.

Pequegnat, W. E., Bryant, W. R. and Harris, J. E., 1971, Carboniferous Sediments from the Sigsbee knolls, Gulf of Mexico: Am. Assoc. Petrol. Geol. Bull., v. 55, p. 116-123.

Perrodon, A., 1971, Essai de classification des bassins sédimentaires; Sciences De La Terre, v. 16 p. 197-227.

___ , 1977, Concepts, modèles et logique des bassins sédimentaires: Bull. Centres Rech., Explor-Prod. Elf-Aquitaine, v. 1, p. 111-130.

Pilger, A. and Rösler, A., eds., 1975, Afar depression of Ethiopia: Inter-Union Commission of Geodynamics Sci. Rept. no. 14, E. Schweizerbart'sche Verlagsbuchhandlung, Stuttgart, 416 p.

80

———, ———, 1976, Afar between continental and oceanic rifting: Inter-Union Commission on Geodynamics Sci. Rept. no. 16, E. Schweizerbart'sche Verlagsbuchhandlung, Stuttgart, 216 p.

Pitman, W. C., III, 1978, Relationship between eustacy and stratigraphic sequences of passive margins: Geol. Soc. Am. Bull., v. 89, p. 1389-1403.

———, and Talwani, M., 1972, Sea floor speading in the north Atlantic: Geol. Soc. Am. Bull., v. 83, p. 619-646.

———, Larson, R. L. and Herron, E. M., 1974, The age of the ocean basins, magnetic lineations of the oceans, 2 maps: Geol. Soc. Am. Map Series.

Poehls, K., 1978, Inter-arc basins: a kinematic model: Geophys. Res. Letters, v. 5, p. 325-328.

Pollack, H. N. and Chapman, D. S., 1977, On the regional variation of heat flow, geotherms, and lithospheric thickness: Tectonophysics, v. 38, p. 279-296.

Ponte, F. C. and Asmus, H. E., 1976, The Brazilian marginal basins: current state of knowledge: *in* F. F. M. de Almeida, *ed.,* Continental margins of Atlantic type; Ann. Acad. Brasil, Cien. 48, p. 215-216.

Ponte, F. C., Dos Reis Fonseca, J. and Gamarra Morales, R., 1977, Petroleum geology of eastern Brazilian continental margins: Am. Assoc. Petrol. Geol. Bull., v. 61, p. 1470-1482.

Posavec, M., Taylor, D., Van Leeuwen, Th. and Spector, A., 1973, Tectonic controls of volcanism and complex movements along Sumatran fault system: Geol. Soc. Malaysia Bull., v. 6, p. 43-66.

Poulet, M., 1977, Mécanisme de formation des bassins sédimentaires de marge stable: Bull. des Centres de Recherches Explor-Prod. Elf-Aquitaine, v. 1, p. 131-145.

Powell, D. E., 1976, The geological evolution of the continental margin off northwest Australia: Australian Petrol. Explor. Assoc. J., v. 16, p. 13-23.

Profett, J. M., Jr., 1977, Cenozoic geology of the Yerington district, Nevada, and implications for the nature and origin of basin and range faulting: Geol. Soc. Am. Bull., v. 88, p. 247-266.

Prucha, J. J., Graham, J. A. and Nichelson, R. P., 1965, Basement controlled deformation in Wyoming province of Rocky Mountain foreland: Am. Assoc. Petrol. Geol. Bull., v. 49, p. 966-992.

Pulunggono, A., 1976, Recent knowledge of hydrocarbon potentials in sedimentary basins of Indonesia: *in* M. Halbouty *et al., eds.,* Circum-Pacific energy and mineral resources; Am. Assoc. Petrol. Geol. Mem. 25, p. 239-249.

Putzer, H., 1962, Geologie von Paraguay: Berlin, Borntraeger, 183 p.

Rabinowitz, P. D., 1976, Geophysical study of the continental margin of southern Africa: Geol. Soc. Am. Bull., v. 87, p. 1643-1653.

———, Cande, S. C. and LaBreque, J. L., 1975, The Falkland escarpment and Agulhas fracture zone: the boundary between oceanic and continental basement at conjugate continental margins: *in* F. F. M. de Almeida, *ed.,* Continental margins of Atlantic type; Anais. Acad. Brasil, Cien. 48, p. 240-252.

———, and LaBrecque, J. L., in press 1979, The Mesozoic south Atlantic Ocean and evolution of its continental margins, J. Geophys. Res.

Read, H. H., and Watson, Janet, 1975, Introduction to geology, v. 2, Earth History, Part I: Early stages of earth history: John Wiley & Sons, New York, p. 11-15.

Ricou, L. E., 1971, Le croissant ophiolitique péri-arabe, une ceinture de nappes mises en place au Crétacé supérieur: Rev. de Géographie Physique et de Géologie Dynamique, v. 12, Fasc. 4, p. 327-350.

Ringwood, A. E., 1977, Petrogenesis in island arc systems: *in* M. Talwani and W. C. Pitman III, *eds.,* Island Arcs, Deep Sea Trenches, and Back-arc Basins; Maurice Ewing Ser. 1, Am. Geophys. Union. p. 311-324.

Robinson, V. A., 1974, Geologic history of the Bass Basin: Australian Petrol. Explor. Assoc. J., p. 45-49.

Rocky Mountain Association of Geologists, 1972, Geologic Atlas of the Rocky Mountain Region, United States of America: Denver, W. W. Mallory, *ed.,* 331 p.

Roeder, D. H., 1973, Subduction and orogeny: J. Geophys. Res., v. 78, p. 5005-5024.

Rona, P. A., 1973, Worldwide unconformities in marine sediments related to eustatic changes of sea level: Naturel Phys. Sci., v. 244, p. 25-26.

———, and Richardson, E. S., 1978, Early Cenozoic global plate reorganization: Earth and Planetary Sci. Letters, v. 40, p. 1-11.

Rudkevich, M. Ya., 1976, The history and the dynamics of the West Siberian platform: Tectonophysics, v. 36, p. 275-287.

Sales, J. K., 1968, Crustal mechanics of Cordilleran foreland deformation: a regional and scale model approach: Am. Assoc. Petrol. Geol. Bull., v. 52, p. 2016-2044.

Sass, J. H., Lachenbruch, A. H., Munroe, R. J., Green, G. W., and Moses, T. H., 1971, Heat flow in the western United States: J. Geophys. Res., v. 76, p. 6376-6413.

Scheibner, E., 1976, Explanatory notes on the tectonic map of New South Wales: Department of Mines, Geol. Surv. of New South Wales, Australia, 283 p.

Schlee, J. S., Behrendt, J. C., Grow, J. A., Robb, J. M., Mattick, R. E., Taylor, P. T. and Lawson, B. J., 1976, Regional geologic framework of northeastern United States: Am. Assoc. Petrol. Geol. Bull., v. 60, p. 926-951.

_____, Martin, R. G., Mattick, R. E., Dillon, W. P. and Ball, M. M., 1977, Petroleum geology on the United States Atlantic-Gulf of Mexico margins: in Proc. of Southwestern Legal Foundation, Exploration and Economics of the Petroleum Industry, v. 15; New York, Matthew Bender Co., p. 47-93.

Scholl, D. W., 1974, Sedimentary sequences in the north Pacific trenches: in C. A. Burk and C. L. Drake, eds., The geology of continental margins; Springer-Verlag, New York, p. 493-504.

_____, and Marlow, M. S., 1974, Sedimentary sequence in modern Pacific trenches and the deformed circum-Pacific eugeosyncline: in R. H. Dott, Jr., and R. H. Shaver, eds., Modern and ancient geosynclinal sedimentation, Soc. Econ. Paleont. Mineral. Spec. Pub. 19, p. 193-211.

Scholz, C. H., Barazangi, M. and Sbar, M. L., 1971, Late Cenozoic evolution of the Great Basin, western United States, as an ensialic inter-arc basin: Geol. Soc. Am. Bull., v. 82, p. 2979-2990.

Schott, W., 1968, Nordeutsches Wealdenbecken und Ostseebecken: Geol. Jb, v. 85, p. 919-940.

Schweller, W. J. and Kulm, L. V., 1978, Depositional patterns and channelized sedimentation in active eastern Pacific trenches: in D. J. Stanley and G. Helling, eds., Sedimentation in submarine canyons, fans and trenches; Dowden, Hutchinson and Ross Inc., p. 311-324.

Sclater, J. C. and Francheteau, 1970, The implications of terrestrial heat flow observations on current tectonic and geochemical models of the crust and upper mantle of the earth: Royal Astron. Soc. Geophys. J., v. 20, p. 509-537.

Scrutton, R. A. and Dingle, R. V., 1976, Observations on the processes of sedimentary basin formation at the margins of southern Africa: Tectonophysics, v. 36, p. 143-156.

Seely, D. R., 1977, The significance of landward vergence and oblique structural trends on trench inner slopes: in M. Talwani and W. E. Pitman III, eds., Island arcs, deep sea trenches and back-arc basins; Maurice Ewing Ser. 1, Am. Geophys. Union, Washington, D.C., p. 379-393.

Seely, D. R. and Dickinson, W. R., 1977, Structure and stratigraphy of forearc regions: Am. Assoc. Petrol. Geol. Contin. Educ. Notes ser. 5.

Sengör, A. M. C., Burke, K. and Dewey, J. F., 1978, Rifts at high angles to orogenic belts: tests of their origin and the upper Rhine Graben as an example: Am. J. Sci., v. 278, p. 24-40.

Shatski, N. S., 1946, Basic features of the structures and development of the east European platform comparative tectonics of ancient platforms: SSSR, Akad. Nauk 12 v., Geol. Ser. no. 1, p. 5-62.

_____, 1961, Vergleichende Tectonik alter Tafeln: Berlin, Academie-Verlag, 220 p.

Sheridan, R. E., 1976, Sedimentary basins of the Atlantic margin of North America: Tectonophysics, v. 36, p. 113-132.

_____, and Osburn, W. L., 1975, Marine geological and geophysical studies of the Florida - Blake Plateau - Bahamas area: in C. J. Yorath, E. R. Parker, and D. J. Glass, eds., Canada's continental margins and offshore petroleum exploration; Can. Soc. Petrol. Geol. Mem. 4, p. 9-32.

Shipley, T., Buffler, R. T. and Watkins, J. S., 1978, Seismic stratigraphy and geologic history of Blake Plateau and adjacent western Atlantic: Am. Assoc. Petrol. Geol. Bull.; v. 62, p. 792-812.

Shouldice, D. H., 1971, Geology of the western Canadian continental shelf: Bull. Can. Petrol. Geol., v. 19, p. 405-436.

Shurbet, D. H. and Cebull, S. E., 1975, The age of the crust beneath the Gulf of Mexico: Tectonophysics, v. 28, p. T25-T30.

Sillitoe, R. H., 1977, Metallogeny of an Andean-type continental margin in south Korea: implications for opening of the Japan Sea: *in* M. Talwani and W. C. Pitman III, *eds.*, Island arcs, deep sea trenches and back-arc basins: Maurice Ewing Ser. 1, Am. Geophys. Union, Washington, D.C., p. 303-310.

Sleep, N. M., 1971, Thermal effects of the formation of Atlantic continental margins by continental break-up: Royal Astron. Soc. Geophys. J., v. 24, p. 325-350.

_____, 1972, Reply to "Comments on thermal effects of the formation of Atlantic continental margins" as made by J. P. Foucher and X. Le Pichon; Royal Astron. Soc. Geophys. J., v. 29, p. 47-48.

_____, 1976, Platform subsidence mechanisms and "eustatic" sea-level changes: Tectonophysics, v. 36, p. 45-56.

_____, and Snell, N. H., 1976, Thermal contraction and flexure of mid continent and Atlantic marginal basins: Royal Astron. Soc. Geophys. J., v. 45, p. 125-154.

_____, and Sloss, L. L., 1978, A deep borehole in the Michigan Basin: J. Geophys. Res., v. 83, p. 5815-5819.

Sloss, L. L., 1963, Sequences in the cratonic interior of North America: Geol. Soc. Am. Bull., v. 74, p. 93-114.

_____, 1966, Orogeny and epeirogeny: the view from the craton: Trans. New York Acad. Sci., v. 28, p. 579-587.

_____, 1972a, Synchrony of Phanerozoic sedimentary-tectonic events of the North American craton and the Russian platform: Intern. Geol. Cong., 24th Session, Sec. 4, p. 24-32.

_____, 1972b, (Fide Whitten, 1976), Concurrent subsidence of widely separated cratonic basins: Geol. Soc. Am. Progr. Abs., v. 4, p. 668-669.

_____, 1976, Areas and volumes of cratonic sediments, western North America and eastern Europe: Geology, v. 4, p. 272-276.

_____, and Speed, R. C., 1974, Relationships of cratonic and continental margin tectonic episodes: *in* W. R. Dickinson, *ed.*, Tectonics and sedimentation; Soc. Econ. Paleont. Mineral. Spec. Pub. 22, p. 98-119.

Smith, A. G., 1976, Plate tectonics and orogeny: a review: Tectonophysics, v. 33, p. 215-285.

_____, Briden, J. C. and Drewry, G. E., 1973, Phanerozoic world maps: *in* N. F. Hughes, *ed.*, Organisms and continents through time; Spec. Paper in Palaeontology No. 12, London, Palaeont. Assoc., p. 1-42.

Smith, J. C., 1968, Tectonics of the Fitzroy wrench trough, Western Australia: Am. J., Sci., v. 266, p. 766-776.

Smith, R. B., 1977, Intraplate tectonics of the western North American plate: Tectonophysics, v. 37, p. 323-336.

Smithson, S. B., Brewer, J. A., Kaufman, S., Oliver, J. and Hurich, C., 1978, Nature of the Wind River thrust, Wyoming, from COCORP deep reflection and gravity data: Geology, v. 6, p. 648-652.

_____, _____, _____, _____, _____, in press 1979, Structure of the Laramide Wind River uplift, Wyoming, from COCORP deep reflection and gravity data: submitted to J. Geophys. Res.

Soares, P. C. and Landim, P. M. B., 1976, Comparison between the tectonic evolution of the intracratonic and marginal basins in south Brazil: *in* F. F. M. de Almeida, *ed.*, Continental margins of Atlantic type; Anais Acad. Bras. Cienc., v. 78, p. 313-324.

_____, _____, and Fulfard, V. J., 1978, Tectonic cycles and sedimentary sequences in the Brazilian intracratonic basins: Geol. Soc. Am. Bull., v. 89, p. 181-191.

Sollogub, V. G., Prosen, D., *et al.*, 1973, Crustal structure of central and southeastern Europe by data of explosion seismology: Tectonophysics, v. 20, p. 1-33.

Srivastava, S. P., 1978, Evolution of the Labrador Sea and its bearing on the early evolution of the North Atlantic: Royal Astron. Soc. Geophys. J., v. 52, p. 313-357.

Stearns, D. W., 1971, Mechanisms of Drapefolding in the Wyoming Province: 34th Ann. Field Conf., Wyoming Geol. Assoc. Guidebook, p. 124-243.

_____, 1975, Laramide basement deformation in the Bighorn basin, the controlling factor for structures in layered rocks: 27th Ann. Field Conf., Wyoming Geol. Assoc. Guidebook, p. 149-158.

Steel, R. J., 1976, Devonian basins of western Norway - sedimentary response to tectonism: Tectonophysics, v. 36, p. 207-244.

Stegena, L., Géczy, B. and Horváth, F., 1975, Late Cenozoic evolution of the Pannonian Basin: Tectonophysics, v. 26, p. 71-90.

Steiner, J., 1977, An expanding earth on the basis of sea-floor spreading and subduction rates: Geology, v. 5, p. 313-318.

Stewart, J. H., 1972, Initial deposits in the Cordilleran geosyncline; evidence of a late Precambrian (<840 m.y.) separation: Geol. Soc. Am. Bull., v. 83, p. 1345-1360.

Stille, H., 1924, Grundfragen der vergleichenden Tektonik: Berlin, Borntraeger, 413 p.

Stöcklin, J., 1968, Salt deposits of the Middle East: Geol. Soc. Am. Spec. Paper 88, p. 157-181.

_____, 1974, Possible ancient continental margins in Iran: in C. A. Burk and C. L. Drake, eds., The geology of continental margins; Springer, New York, p. 873-888.

Stone, D. S., 1970, Wrench faulting and Rocky Mountain tectonics: The Mountain Geologist, v. 6, p. 67-79.

Stoneley, R., 1974, Evolution of the continental margins in Iran: in C. A. Burk and C. L. Drake, eds., The geology of continental margins; Springer, New York, p. 339-906.

Sugimura, A. and Uyeda, S., 1973, Island Arcs: Japan and its Environs: Developments in geotectonics: Elsevier, 247 p.

Suppe, J., 1970, Offset of late Mesozoic basement terrains by San Andreas fault system; Geol. Soc. Am. Bull., v. 74, p. 357-362.

_____, 1976, Décollement folding in western Taiwan: Petroleum Geol. Taiwan, no. 13, p. 25-35.

_____, and Wittke, J. H., 1977, Abnormal pore fluid pressures in relation to stratigraphy and structure in the active fold and thrust belt of northwestern Taiwan: Petroleum Geol. Taiwan, no. 14, p. 11-24.

Surkov, V. S., 1975, Tectonic map of the basement of the west Siberian platform and its frame: in Kontorovich et al., Geology of Oil and Gas of West Siberia: Moscow, Nedra.

Sutton, J., 1963, Long term cycles in the evolution of the continents: Nature, v. 198, p. 731-735.

Sweeney, J. T., 1976, Evolution of the Sverdrup basin, Arctic Canada: Tectonophysics, v. 36, p. 181-196.

Sykes, L. R., 1972, Seismicity as a guide to global tectonics and earthquake prediction: in A. R. Ritsema, ed., The upper mantle; Tectonophysics, v. 31, p. 393-414.

Takin, M., 1972, Iranian geology and continental drift in the Middle East: Nature, v. 235, p. 147-150.

Talwani, M. and Eldholm, O., 1977, Evolution of the Norwegian-Greenland Sea: Geol. Soc. Am. Bull., v. 88, p. 969-999.

_____, and Pitman, W. C., III, eds., 1977, Island Arcs, Deep Sea Trenches and Back-arc Basins: Maurice Ewing Ser. 1, Am. Geophys. Union, Washington, D.C., 470 p.

Tamrazyan, G. P., 1971, Siberian continental drift: Tectonophysics, v. 11, p. 433-460.

Terman, M. J., 1974, Tectonic map of China and Mongolia: 1: 5 000 000, 2 sheets: Geol. Soc. Am. Map Series.

_____, 1976, Sedimentary basins of China and their petroleum potential: SEAPEX Proc., v. 3, p. 125-129.

Thomas, B. M. and Smith, D. N., 1974, A summary of the petroleum geology of the Carnarvon Basin: Australian Petrol. Explor. Assoc. J., v. 14, pt. 1, p. 66-76.

Thompson, G. A. and Burke, D. B., 1974, Regional geophysics of the Basin and Range province: Ann. Review Earth Planetary Sci., v. 2, p. 213-238.

Tissot, B., 1977, The application of the results of organic geochemical studies: in G. D. Hobson, ed., Developments in Petroleum Geology, v. 1; London, Applied Science Publishers, p. 53-82.

_____, Durand, B., Espitalié, J. and Combaz, A., 1974, Influence of nature and diagenesis of organic matter in formation of petroleum: Am. Assoc. Petrol. Geol. Bull., v. 58, p. 499-506.

_____, and Welte, D. H., 1978, Petroleum formation and occurrence, a new approach to oil and gas exploration: Springer-Verlag, New York, 537 p.

Todd, D. F. and Pulunggono, A., 1971, Sunda Basin — important new Indonesian oil province: Am. Assoc. Petrol. Geol. (abs.), v. 5, p. 367.

Toksöz, M. N. and Bird, P., 1977, Formation and evolution of marginal basins and continental plateaus: in M. Talwani and W. C. Pitman III, eds., Island arcs, deep sea trenches and back-arc basins; Maurice Ewing Ser. 1, Am. Geophys. Union, Washington, D.C., p. 379-393.

Torrie, J. E., 1973, Northeastern British Columbia: in R. G. McCrossan, ed., The future petroleum provinces of Canada — their geology and potential: Can. Soc. Petrol. Geol. Mem. 1, p. 151-186.

Trümpy, R., 1969, Die Helvetischen Decken der Ostschweiz: Versuch einer palinpastischen Korrelation und Ansätze su einer kinematischen Analyse: Eclogae Geol. Helv., v. 62, p. 105-138.

____, 1975, On crustal subduction in the Alps: *in* M. Mahel, *ed.,* Tectonic problems of the Alpine System: Bratislava, Slovak Academy of Sciences, p. 121-130.

Turcotte, D. L., in preparation, Models for the evolution of sedimentary basins.

____, and Ahern, J. L., 1977, On the thermal and subsidence history of sedimentary basins: J. Geophys. Res., v. 82, p. 3762-3766.

____ ____, and Bird, J. M., 1977, The state of stress at continental margins: Tectonophysics, v. 42, p. 1-28.

Vail, P. R., Mitchum, R. M., Jr., Todd, R. G., Widmier, J. M., Thompson, S. III, Sangree, J. B., Bubb, J. N. and Hatlelid, W. G., 1977, Seismic stratigraphy and global changes of sea level: *in* C. E. Payton, *ed.,* Seismic stratigraphy — applications to hydrocarbon exploration; Am. Assoc. Petrol. Geol. Mem. 26, p. 49-212.

Van der Linden, W. J. M. and Wade, J. A., *eds.,* 1974, Offshore geology of eastern Canada, v. 2: Regional geology: Geol. Surv. Can. Paper. 74-30, pt. 2.

Van der Voo., R., Mauk, F. J. and French, R. B., 1976, Permian-Triassic continental configurations and the origin of the Gulf of Mexico: Geology, v. 4, p. 177-180.

____, and Watts, D. R., 1978, Paleomagnetic results from igneous and sedimentary rocks from the Michigan basin borehole: J. Geophys. Res., v. 83, p. 5844-5848.

Van Hinte, J. E., 1978, Geohistory analysis — applications of micropaleontology in exploration geology: Am. Assoc. Petrol. Geol. Bull., v. 62, p. 201-222.

Veevers, J. J. and Cotterill, D., 1978, Western margin of Australia: Geol. Soc. Am. Bull., v. 89, p. 337-355.

Vernon, R. C., 1971, Possible future potential of pre-Jurassic western Gulf basin: *in* Future petroleum provinces of the United States: Am. Assoc. Petrol. Geol. Mem. 15, v. 2, p. 954-979.

Von Huene, R., 1974, Modern trench sediments: *in* C. A. Burk and C. L. Drake,*eds.,* The geology of continental margins: Springer-Verlag, New York, p. 207-211.

____, Moore, G. W. and Moore, J. C., 1979, Cross-section Alaska Peninsula — Kodiak Island — Aleutian Trench: Map and Chart Series MC-28A, Geol. Soc. Am.

Vroman, A. J., 1978, On the origin of the Mediterranean and other marginal seas: Tectonophysics, v. 46, p. 219-225.

Wageman, J. M., Hilde, T. W. C. and Emery, K. D., 1970, Structural framework of east China and Yellow Sea: Am. Assoc. Petrol. Geol. Bull., v. 54, p. 1611-1643.

Walcott, R. I., 1970, Flexural rigidity, thickness, and viscosity of the lithosphere: J. Geophys. Res., v. 75, p. 3941-3954.

____, 1972, Gravity, flexure and the growth of sedimentary basins at a continental edge: Geol. Soc. Am. Bull., v. 83, p. 1845.

Walper, J. L. and Rowett, C. L., 1972, Plate tectonics and the origin of the Caribbean Sea and the Gulf of Mexico: Trans. Gulf Coast Assoc. Geol. Soc., v. 22, p. 105-106.

Watkins, J. S., Ladd, J. W., Shaub, F. J., Buffler, R. T. and Worzel, J. L., 1977, Seismic Section WG-3, Tamaulipas Shelf to Campeche Scarp, Gulf of Mexico, Am. Assoc. Petrol. Geol. Seismic Section no. 1.

____, Montadert, L. and Dickerson, P. W., *eds.,* 1979, Geological and geophysical investigations of continental margins: Am. Assoc. Petrol. Geol. Mem. 29.

Watts, A. B. and Talwani, M., 1974, Gravity anomalies seaward of deep sea trenches and their tectonic implications: Geophysics, v. 36, p. 57-90.

____, and Ryan, W. B. F., 1976, Flexure of the lithosphere and continental margin basins: Tectonophysics, v. 36, p. 25-44.

Weissel, J. K. and Hayes, D. E., 1972, Magnetic anomalies in the southeast Indian Ocean: *in* D. E. Hayes, *ed.,* Antarctic Oceanology II: The Australian-New Zealand Sector: Am. Geophys. Union, Ant. Res. Ser. 19, p. 165-196.

Wheeler, H. E., 1973, Post-Sauk and pre-Absaroka Paleozoic stratigraphic patterns in North America: Am. Assoc. Petrol. Geol. Bull., v. 47, p. 1487-1526.

White, D. A., Roeder, D. H., Nelson, T. H. and Crowell, J. C., 1970, Subduction: Geol. Soc. Am. Bull., v. 81, p. 3431-3432.

Whitten, E. H. T., 1976, Cretaceous phases of rapid sediment accumulation, continental shelf, eastern U.S.A.: Geology, v. 4, p. 237-240.

____, 1977, Rapid Aptian-Albian subsidence rates in eastern United States: Am. Assoc. Petrol. Geol. Bull., v. 61, p. 1522-1524.

Willcox, J. B. and Exon, N. F., 1976, The regional geology of the Exmouth Plateau: Australian Petrol. Explor. Assoc. J., p. 1-11.

Wilson, J. T., 1968, Static or mobile earth: the current scientific revolution: Proc. American Philosophical Soc., v. 112, p. 309-320.

____, 1976, Continents adrift and continents aground: Readings from Scientific American; W. H. Freeman and Co., San Francisco, 230 p.

____, and Burke, K., 1972, Two types of mountain building: Nature, v. 239, p. 448-449.

Wise, D. V., 1963, An outrageous hypothesis for the tectonic pattern of the North American Cordillera: Geol. Soc. Am. Bull., v. 74, p. 357-362.

Woodland, A. W., ed., 1975, Petroleum and the Continental Shelf of north-west Europe, v. 1, A Halsted Press Book, John Wiley and Sons, New York, 500 p.

Woods, R. D. and Addington, J. W., 1973, Pre-Jurassic geologic framework, northern Gulf basin: Trans Gulf Coast Assoc. Geol. Soc., v. 13, p. 92-102.

Worzel, J. Lamar and Burk, C. A., 1978, Margins of Gulf of Mexico: Am. Assoc. Petrol. Geol. Bull., v. 62, p. 2290-2307.

Wright, L. A. and Troxel, B. W., 1973, Shallow-fault interpretation of basin and range structure, southwestern Great Basin: in K. A. de Jong and R. Scholten, eds., Gravity and tectonics: Wiley-Interscience, New York, p. 397-407.

____, ____, Williams, E. G., Roberts, M. T. and Diehl, P. E., 1974, Precambrian sedimentary units of the Death Valley region: in Guidebook Death Valley Region, Shoshone, California; Death Valley Publishing Co., p. 27-35.

Wyllie, P. J., 1971, The Dynamic Earth: John Wiley and Sons, New York, 416 p.

Yeats, R. S., 1968, Southern California structure, sea floor spreading, and history of the Pacific Ocean: Geol. Soc. Am. Bull., v. 79, p. 1693-1702.

____, 1978, Neogene acceleration of subsidence rates in southern California: Geology, v. 6, p. 456-460.

Yorath, C. J., Parker, E. R. and Glass, D. J., eds., 1975, Canada's continental margins and offshore petroleum exploration: Can. Soc. Petrol. Geol. Mem. 4, 808 p.

Zhavrev, I. P., Zubov, I. P., Krylov, N. A. and Semenovich, V. V., 1975, Comparative evaluation of the oil and gas potential of the Epi-Paleozoic basins of the U.S.S.R.: 9th World Petrol. Congr. Proc., v. 2, p. 83-91.

Ziegler, A. M., Hansen, D. S., Johnson, M. E., Kelly, M. A., Scotese, C. R. and Van der Voo, R., 1972, Silurian continental distribution, paleogeography, climatology, and biogeography: Tectonophysics, v. 40, p. 13-51.

____, Scotese, C. R., Johnson, M. E., McKerrow, W. S. and Bambach, R. K., 1977, Paleozoic biogeography of continents bordering on Iapetus (pre-Caledonian) and Rheic (pre-Hercynian) Oceans: in R. M. West., ed., Paleontology and plate tectonics with special reference to the history of the Atlantic Ocean; Milwaukee Public Museum, Spec. Pub. in Biology and Geology, no. 2., p. 1-23.

Ziegler, P. A., 1975, North Sea basin history in the tectonic framework of northwestern Europe: in A. W. Woodland, ed., Petroleum and the continental Shelf of north-west Europe; John Wiley and Sons, New York, p. 131-162.

____, 1977, Geology and hydrocarbon provinces of the North Sea: Geojournal, v. 1, p. 7-32.

Ziegler, W. H., 1975, Outline of the geological history of the North Sea: in A. W. Woodland, ed., Petroleum and the continental Shelf of north-west Europe; John Wiley and Sons, New York, p. 165-187.

Zorin, Y. A. and Rogozhina, V. A., 1978, Mechanisms of rifting and some features of the deep-seated structure of the Baikal rift zone: Tectonophysics, v. 45, p. 23-30.

Types of Petroliferous Basins

H. DOUGLAS KLEMME

Senior Vice President
Weeks Petroleum Limited
Westport, Conn., USA

Petroleum occurs in concentrated accumulations (fields) in depressed, sediment-filled areas (basins or provinces).[1] Worldwide, more than 600 basins and subbasins occur.[2] Of these, about a quarter by number and about half by area and volume have production in some portion to almost all of the basin (Fig. 8–1).[3] Many of the larger land basins have established production. The principal producing areas are in the more extensive landmass of the Northern Hemisphere. About half of the world's basins by area and volume and three-fourths by number are nonproductive. About one-third of these have never been test drilled and may be considered exploration frontier basins.[4]

Many basin classifications have been proposed. Petroleum industry classifications stress the basin as a container for petroleum and attempt to concentrate on defining those processes of basin evolution involved in the formation of petroleum deposits. Structural outline or form (basin architecture) and evolution or genesis are often factors on which industry classifications are based. Industry classifications have involved six to eight basin types within two to three categories. Most authors who have classified basins generally agree on the classification of about 75% of the world's individual basins and differ on about 25% of those that are either poorly known or are controversial due to a geologic evolution which often includes a change in basin type through time.[4, 5, 6, 7]

A basin classification when linked to the variability of petroleum characteristics may provide a worthwhile exercise in appraising the petroleum potential of new frontier basins or developing further production in newly developing basins. The relative usefulness of applying a look-alike or analog derived from basin classifications in appraisal of new frontier basins or estimating ultimate reserves in producing basins has been rejected by about as many specialists in petroleum geology as those that consider it useful.[8]

The dimensions and shapes of basins may be divided into both large and small sizes and linear- and circular-shaped basins. Further differences are noted in the effective basement profile or cross section of basins and the ratio of a basin's surface area to the volume of sediments contained therein. These architectural characteristics when related to the earth's crust, tectonic setting, and basin evolution (primarily in the framework of plate tectonics) result in 8 types of basins with notable subtypes (Figs. 8–2, 3).

Type 1—Interior Basins (Fig. 8—6)

Type 1 interior basins are simple, large, circular basins with a symmetrical profile. They are generally areas of Paleozoic platform deposition, and the ratio of the volume of sediments in the basin to the surface area of the basin is low. The genesis of this type of basin is poorly known. It is speculated that initial rifting or a thermal hot spot with the introduction of denser material in the subcrust preceded basin development. The denser material may have locally altered crustal buoyancy, creating a sag basin. They are located in the central portion of cratons near or upon Precambrian shield areas, They generally consist of a mixture of clastic and carbonate sediments and usually display low hydrocarbon recovery with few giant fields. Their traps for petroleum accumulation are predominantly associated with central arches or stratigraphic traps around the basin margin.

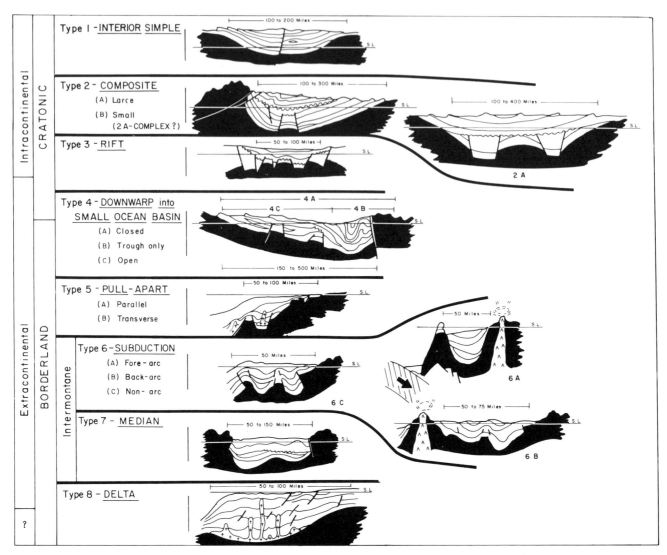

Figure 8–2. Petroleum basin types.

Type 2—Composite; 2A—Complex Basins (Fig. 8–7)

Type 2 composite basins are large, linear to elliptical intracontinental, cratonic basins with an asymmetric profile. They are generally areas that, intially, were sites of shield-derived Paleozoic platform deposits displaying characteristics similar to Type 1 basins. They became multicycle or composite in Upper Paleozoic or Mesozoic time when a second cycle of sediments derived from an orogenic uplift on the exterior margin of the craton provided sediments from an opposite source area (2-sided source), thus creating the asymmetry to the basin profile. The ratio of the volume of sediments to surface area is high. Extension during the early cycle was followed by compression during the second cycle of basin development and appears to be related to the action of the sea floor spreading on the exterior portions of cratons.

Type 2A complex basins are also multicycle basins located in exterior portions of cratons. They are large, more often elliptical basins with an irregular to asymmetrical profile. Their genesis appears to have been more complex, with multiple rifting followed by a more or less symmetrical sag resembling a Type 1 basin.

Type 2 basins generally consist of a mixture of carbonate and clastic sediments. However some are predominately clastic. Their traps are primarily large arches or block uplifts. Compression folds and stratigraphic accumulations also act as traps in most Type 2 composite basins. In many Type 2 composite basins where a large uplift or arch is present, up to half of the basin's reserves are located in one large accumula-

88

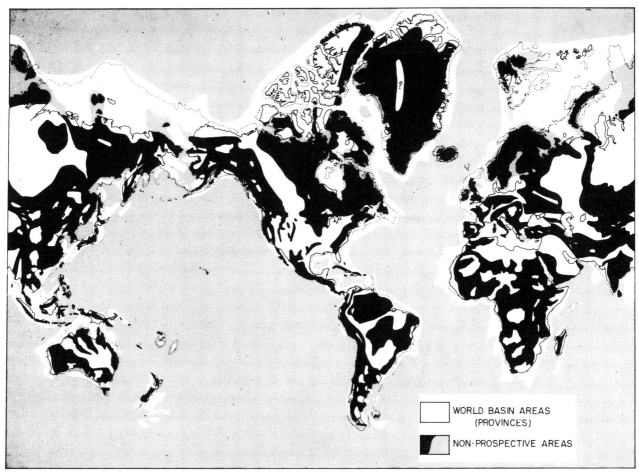

Figure 8–1.   World basin areas.

TABLE 8–1.
Basin Type in Relation to Productivity.

Basin type	% World's proven and produced reserves	% World's basin area	% (Over 4,500 ft– 18,000 ft) world's volume	% Offshore area	% Deep-water area	% Productive		% Nonproductive	
						area	volume	area	volume
Type 1	1.5	18.2	6.2	9	1	3.5	1.2	14.5	5
Type 2	25	27.3	25.4	14	5	19.6	17.7	7.4	7.7
Type 3	10	5.4	5.5	27	14	2.5	2.8	3.0	2.7
Type 4	47	17.5	26.3	38	17	12.2	19.5	5.3	6.5
Type 5	0.5	18.2	19.3	90	55	2.2	2.7	15.8	16.6
Type 6	7.5	7.1	8.8	93	50	1.8	2.3	5.2	6.5
Type 7	2.5	3.7	3.7	44	20	1.1	1.1	2.6	2.6
Type 8	6	2.6	4.8	85	50	0.6	1.1	2.0	3.7
Totals	100%	100%	100%	35+	25+	44	50	56	50

	Area
Totals	30,000,000 m² (77,000,000 km²)
	Volume
	40,000,000 m³ (165,000,000 km³)

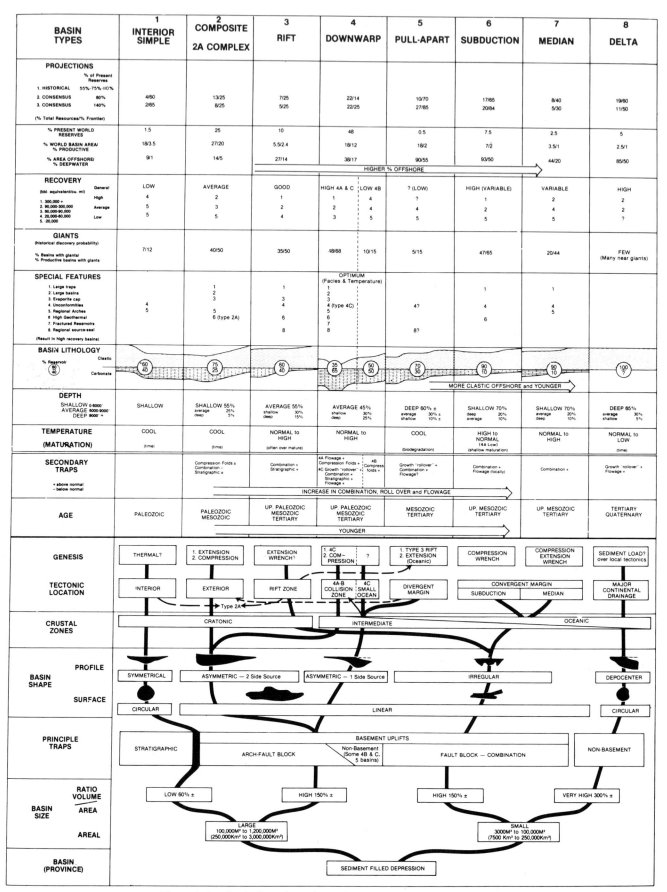

Figure 8–3. Geologic and petroleum characteristics of petroleum basins.

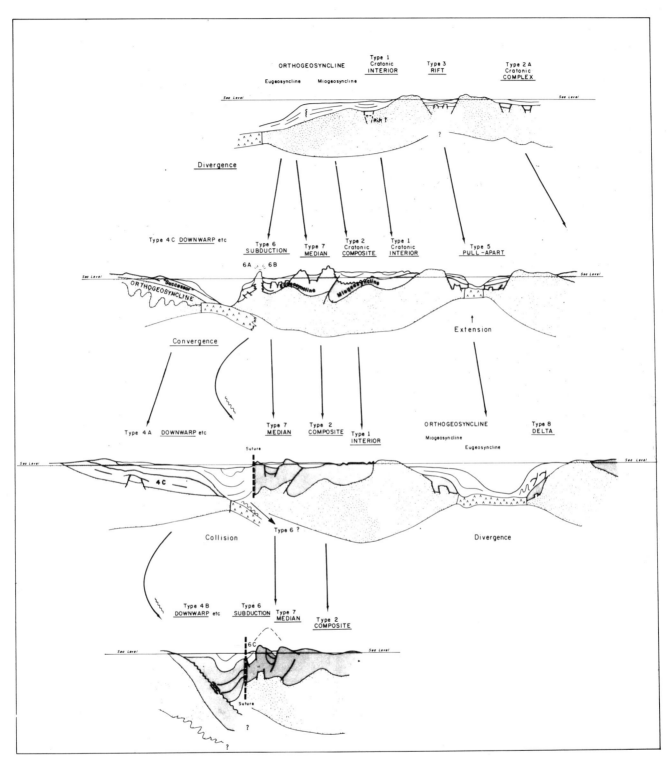

Figure 8–4. Basin evolution patterns.

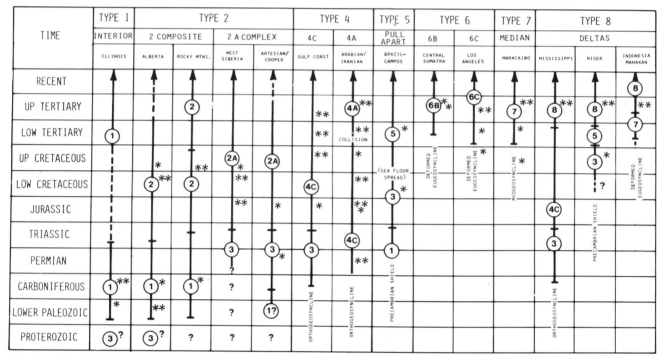

Figure 8–5. Tectonic evolution of typical basins by basin type.

tion. They generally display normal geothermal gradients and more than normal amounts of shallow production. The shallow production results from both deeper burial followed by postmaturation uplift and updip secondary migration along ramp-like carrier beds. They represent about 25% of the world's basin area and contain a similar proportion of present world petroleum reserves. About 75% of the basins are productive. In China the Ordos and Sichuan basins appear to be Type 2 composite basins.

Type 3—Rift Basins (Fig. 8–8)

Type 3 rifts are small, linear basins with an irregular profile displaying a high sediment volume to surface ratio. They appear to be a very fundamental earth structure inasmuch as they are formed at various stages during the development of almost all other types of basins. They are primarily Upper Paleozoic, Mesozoic, and Tertiary in age and are located on or near cratonic areas. About two-thirds of them are formed along the trend of older deformation belts, and one-third are developed upon Precambrian shield areas. Their sedimentary fill is most often clastic; however, where the rift opened to a warm climate sea, carbonates are often

present. Many of these linear, down-faulted grabens appear to form by (1) thermal expansion of the lithosphere during a period of high heat flow, coupled with (2) subaerial or subcrustal reduction of the lithosphere and increase in subcrustal density which results in a linear, downfaulted sag. Others appear to form by wrenching motion along older lines of weakness within cratonic plates. In some instances, these rift basins are the sites of the introduction of oceanic crustal material and therefore are the forerunners to sea floor spreading centers, while other rifts remain dormant and continue to subside with no development of spreading centers.

Because their genesis provides extension features, they form an irregular profile with considerable fault-block movement. As a result, they display more than normal combination structural-stratigraphic traps where depositional variations and unconformities are developed over differentially subsiding locks. Geothermal gradients are normal to high.

By area, these basins represent slightly over 5% of the world's basins (half are productive); however, high recovery has resulted, as they contain 10% of the world's present reserves. Where they contain marine or lacustrine sediments, they provide above-average recovery per unit volume of

92

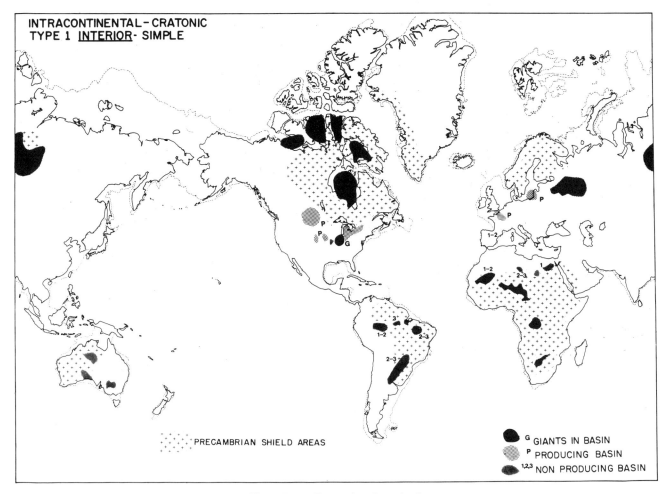

INTRACONTINENTAL – CRATONIC
TYPE 1 <u>INTERIOR</u>- SIMPLE

·:· PRECAMBRIAN SHIELD AREAS

G GIANTS IN BASIN
P PRODUCING BASIN
1,2,3 NON PRODUCING BASIN

Figure 8–6. Type 1, interior—simple.

sediments and display a normal field size distribution. In China the Songliao-North China/Bohai basins are Type 3 rift basins. This type appears to be formed by major wrench faulting in the Junggar, Turpan, Qaidam, pre-Nanshan, and northern Tarim basins.

Type 4—Downwarps (into Small Ocean Basins): A, Closed; B, Trough; C, Open (Fig. 8–9)

Type 4 downwarp (A closed and B trough) basins resemble Type 2 composite basins in size, profile and sediment volume to area ratios. Type 4 downwarp (C open) basins architecturally resemble Type 5 pull-apart basins. However, because of their unique genesis, a different temperature regime and sedimentary environment from that of Type 2 composite basins is often involved during their development. Because these basins represent a major portion of the world's reserves, they warrant a separate category.

Type 4C open downwarps are separated from the main trends of major ocean basin spreading zones and are located along the American segment of the Tethyan trend between Gondwana and Laurasia and in the Arctic. They often overlie older deformed orthogeosynclinal zones (miogeosyncline/eugeosyncline sequence) and have been labeled as successor basins. They are large, linear basins with a one-sided source and seaward asymmetry. Their genesis is related to the evolution of the small ocean basin into which they open. Considerable speculative mechanisms have been proposed for the highly complex genesis of the Arctic Ocean, Gulf of Mexico, Caribbean, eastern Mediterranean, and South China Sea where 4C basins are present.

Type 4C open basins may become Type 4A closed basins as the result of collision of continental plates (Fig. 8–4). The Middle East and South Asia Tethyan basins result from collision of the African and Indian plates with the Eurasian plate. Upon closing, a large, linear, asymmetric basin with a two-sided source (similar to Type 2 composite basins) is formed. Further plate movement appears to destroy a con-

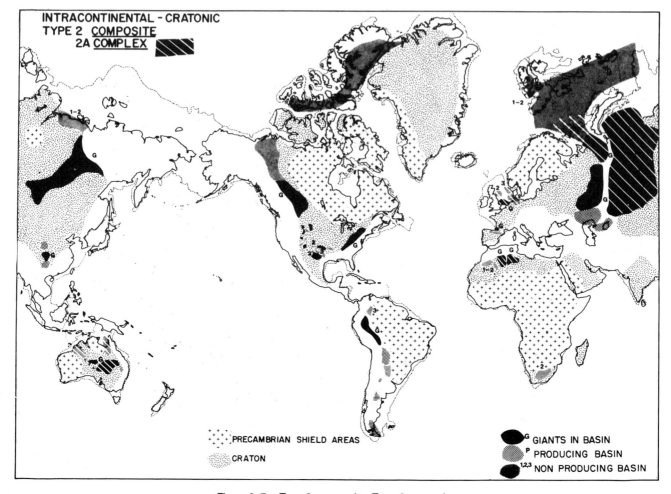

Figure 8–7. Type 2, composite; Type 2a, complex.

siderable portion of the Type 4A closed basin, leaving only a type 4B trough such as the narrow, sinuous basins south of the Himalayas on the India plate and the narrow Alpine and North African troughs.

Type 4 downwarp basins represent 18% of the world's basin area; however, they contain 48% of the world's reserves. Relatively high recovery rates in Type 4A and C may be related to the normal to often high geothermal gradients, which may provide more efficient maturation and primary and secondary migration of hydrocarbons. Often, differing from their cratonic counterparts, the sedimentary facies associated with a small, restricted ocean basin appear to provide rich source shales and considerable evaporites. In addition, along Tethys and its extension between the American continents, considerable porous carbonates were developed that provide extensive reservoir rocks. A predominance of several other special features, which enhance the fundamental petroleum character of traps, source, reservoir, and cap rocks, are present in these basin types (Fig. 8–3).

Trap types are predominately anticlinal (either drape over large arches or compressional folds) in Types 4A and 4B; more than normal stratigraphic-structural or combination traps, growth fault anticlines, and stratigraphic traps are present in 4C open basins. The presence of considerable evaporite deposits in many Type 4 basins often results in highly effective cap rocks and large flowage features. The southern Tarim basin in west-central China is reported to have some marine (Tethyan) sediments and appears to be a closed Type 4 downwarp basin, similar to Type 4 basins in the southern USSR.

Type 5—Pull-Apart Basins (Fig. 8–10)

Type 5 pull-apart basins are large and linear, displaying a high volume-of-sediment to surface ratio and a one-sided source asymmetry. They occupy the intermediate crustal zone (between thick continental crust and thin oceanic crust) and are located along the divergent margins of spreading

94

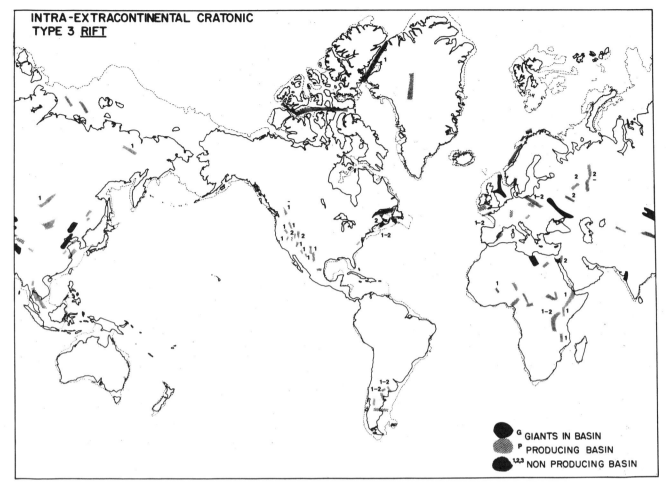

Figure 8–8.   Type 3, rift.

plates. Their genesis appears to be by rifting, which may occur either over Precambrian basement lineations (zones of weakness) or along older orogenic trends. At some later time, one or more of the rifts becomes the site of introduction of basic oceanic material and an axis of sea-floor spreading is established. As the spreading reaches oceanic distances, a post-rift series (sedimentary fan) is deposited seaward from the continent. It is speculated that subsidence is caused by both thermal cooling of the denser oceanic material (creating a buoyancy sag along the continental margin) and the weight of the sediments derived from the continental mass.

Due to an extensional genesis, most traps are tensional growth (rollover) anticlines or flowage type. Generally normal to low geothermal gradients and deeper than normal production are present where oceanic distances of spreading have occurred. The initial rifts may have had a higher heat input. Sediments are predominately clastic, although the postrift series may form as a carbonate bank as well as a clastic fan. The basins are primarily of Mesozoic and Ter-

tiary age. They represent 18% of the world's surface basin area, but due to their predominant offshore location and accessibility to petroleum industry technology (90% offshore and 55% in deep water) only 10% of these basins are productive. To date, these basins have displayed low productivity. The South China Sea (southwest of the Strait of Taiwan) appears to be a sea-floor spreading zone that has developed Type 5 pull-apart basins in coastal areas.

Two subtypes (A parallel and B transverse) are related to the axis of sea-floor spreading and the relative displacement of the coastal margin.

Type 6—Subduction Basins (Fig. 8–11)

Type 6 subduction basins are small and linear with an irregular profile. They may be subdivided on the basis of their relation to the volcanic island arc which is often present on the back side of a subduction zone. 6A fore-arc basins are located on the oceanward side of the volcanic arc; 6B back-

95

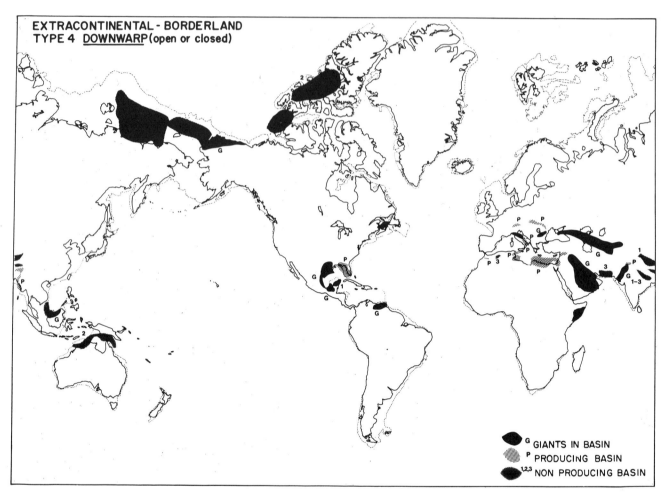

EXTRACONTINENTAL - BORDERLAND
TYPE 4 <u>DOWNWARP</u> (open or closed)

G GIANTS IN BASIN
P PRODUCING BASIN
1,2,3 NON PRODUCING BASIN

Figure 8–9.   Type 4, downward (open or closed).

arc basins are located on the back side or cratonic side of the volcanic arc; and 6C non-arc basins are located where subduction and wrench faulting have destroyed the island arc. Their genesis is related to regional compression along subducting or convergent margins. They are mainly developed over deformed eugeosynclines, which make up the basins' effective basement rocks. They are most often Tertiary in age and filled with clastic sediments. Although regional compression is present in the areas where these basins develop, considerable wrench movement takes place, creating tensional block movement over which sediments are draped. Drape over fault blocks provides the main traps, though compressional anticlines, wrench anticlines, and flowage features provide the traps in some basins.

Type 6B and 6C basins, because of the high heat flow in back of or on the craton side of subduction zones, display high geothermal gradients often providing for a highly efficient maturation and migration of petroleum. Reservoirs are predominately sandstones, ranging from shallow marine deposits to turbidites, frequently in the form of multiple pay zones. These basins represent 7% of the world's basin area and only about a quarter are productive. However, though the productive of this type comprise only 2% of the world's basinal areas, they represent over 7% of the world's reserves. Some of these basins (6B and 6C) have the highest recovery per volume of sediments of all basin types. The field size distribution for these basins commonly has one field in each basin, which contains from 25% to 45% of the basin's reserves. The outer shelf portion of the East China Sea, in back of the volcanic arc of the Ryukyus, has developed large Type 6 subduction (back-arc) basins. Some are in a second-stage process of extension.

Type 7—Median Basins (Fig. 8–12)

Type 7 median basins are small and linear with irregular profiles. They occupy the mountainous fold median zone (or

96

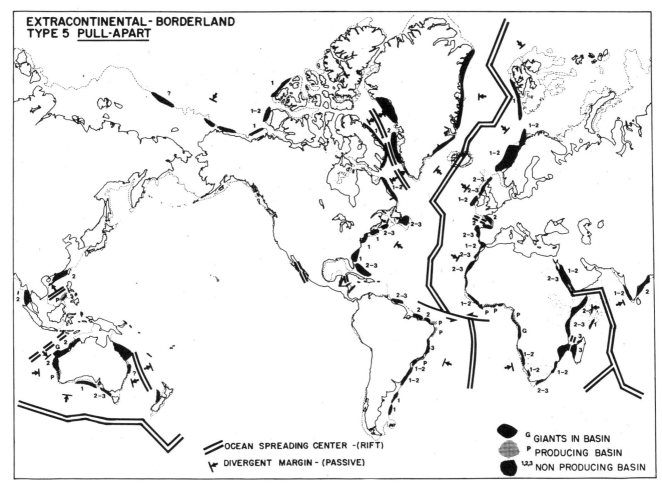

OCEAN SPREADING CENTER -(RIFT)

DIVERGENT MARGIN - (PASSIVE)

G GIANTS IN BASIN
P PRODUCING BASIN
1,2,3 NON PRODUCING BASIN

Figure 8–10.   Type 5, pull-apart.

interior portions of Cenozoic/Mesozoic megasutures), which has been developed either between an oceanic subduction zone and the basins of the craton or in the collision zone between two cratonic plates.[9] They are essentially the rifts of the median zone formed by wrench movements and foundering, creating local tension within the compressed and uplifted mountain belts surrounding the convergent margins of some continents.

Geothermal gradients appear to be normal to high, sediments are dominantly clastic, and trap types are block uplifts over which structural-stratigraphic accumulations occur—characteristics similar to Type 3 cratonic rifts. Their genesis seems to be a complex mix of subcrustal erosion and foundering together with local introduction of oceanic material and extension; in addition, considerable stress from wrench faulting has occurred.

They represent about 3.5% of the world's basin area (about one-quarter are productive) and 2.5% of the world's reserves.

Type 8—Deltas (Fig. 8–13)

Type 8 deltas are generally small, circular depocenters with an extremely high ratio of sedimentary volume to their surface area. They are most often the site of present-day "bird's foot" deltas which are prograding seaward. Their sedimentary fill is derived from major continental drainage areas. They appear to develop in any tectonic setting, with more than one-third developed over Type 4 downwarp basins, 17% along Type 5 pull-apart basins, 16% over Type 6 subduction basins, 12% in Type 3 rift basins, 12% in Type 7 median basins, and 7% over the submerged portions of Type 2 composite basins. Their location is about equally divided between divergent and convergent margins, either along open or confined coastal areas.

They are predominantly Upper Tertiary in age with an entirely clastic fill. Sediment load in a prograding depocenter has been most often assigned as the causitive factor of deltas, although the depocenter usually coincides with the regional

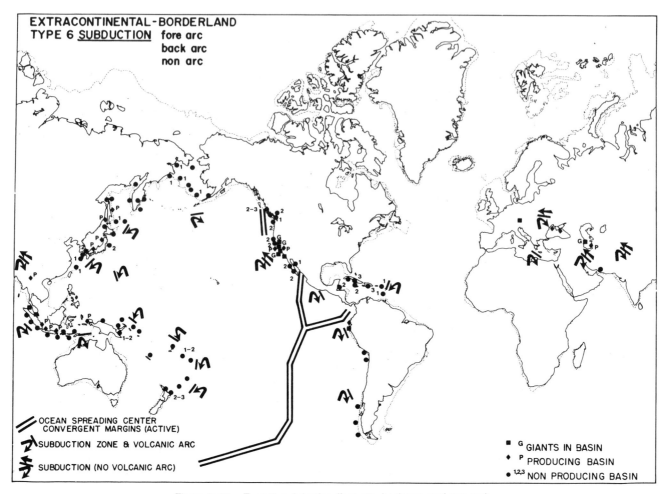

EXTRACONTINENTAL-BORDERLAND
TYPE 6 SUBDUCTION fore arc
                    back arc
                    non arc

OCEAN SPREADING CENTER
CONVERGENT MARGINS (ACTIVE)

SUBDUCTION ZONE & VOLCANIC ARC

SUBDUCTION (NO VOLCANIC ARC)

■ G GIANTS IN BASIN
♦ P PRODUCING BASIN
• 1,2,3 NON PRODUCING BASIN

Figure 8–11. Type 6, subduction (fore arc, back arc, and non arc).

tectonic zone of weakness (the triple junctions of plate tectonics). The tensional regime established by the depocenter results in nonbasement or sedimentary structural development with traps primarily provided by tensional growth (rollover) anticlines and flowage. Low geothermal gradients, perhaps due to the dampening effect of rapid deposition, result in lower-than-average depths of most accumulations. A unique field size distribution (in those deltas that have production to date) involves few giant fields, with the largest field having less than 7% of the basin's reserves. A predominance of land-derived (humic) organic matter leads to higher-than-normal gas content.

These petroleum provinces are relatively rich in hydrocarbon recovery, as they represent 2.5% of the world's basin area (40% productive) and 5% of the world's reserves.

As the crust of the earth has now been determined to have been mobile through time, it follows that this crustal mobility produces changes in the tectonics and fundamental character in basin type as individual basins develop (Fig. 8–5). Some

of these changes are recurrent in pattern and appear to be part of the specific basin classes. Thus, Type 1 interior platform basins are overlain by second-cycle sediments to form Type 2 composite basins; Type 3 rift basins become the axis of sea-floor spreading to form Type 5 pull-apart basins; Type 3 rift basins provide a more dense subcrust that sags to form Type 1 interior or Type 2A complex basins. Type 6 subduction and Type 7 median basins develop over deformed geosynclines and Type 4C open downwarp basins develop over the orthogeosynclines of older craton margins. Other changes seem to be more random and perhaps are associated with the variations in plate motions and their resulting juxtapositions. Thus, Type 4C open downwarp basins closing along continental plate collision zones become Type 4A closed downwarp basins, and eventually Type 4B trough downwarp basins, while Type 8 deltas appear to occur over any tectonic setting wherever major continental drainage reaches an ocean or sea.

Recent basin classifications have lumped basins into 3

98

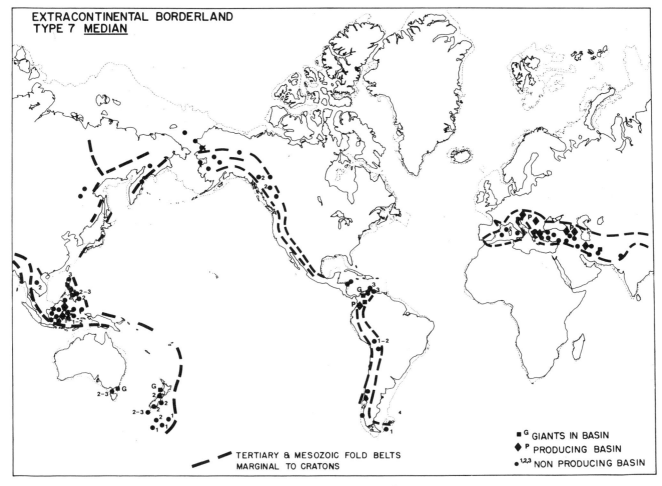

TERTIARY & MESOZOIC FOLD BELTS
MARGINAL TO CRATONS

■ G GIANTS IN BASIN
♦ P PRODUCING BASIN
● 1,2,3 NON PRODUCING BASIN

Figure 8–12.   Type 7, median.

categories: *cratonic* basins (Types 1, 2, and 4), *divergent* regions (Types 3, 5, and 8), and *convergent* regions (Types 6 and 7).[10, 11]

From left to right on Fig. 8–3, basin types, with few exceptions, show some generally consistent variations: namely, they are younger, contain less basement-controlled and more sedimentary-controlled structures, contain a greater amount of clastic rocks, are relatively smaller, involve a shift in the manner of secondary migration, occupy a higher percentage of offshore and deep water areas, and (when remote Type 1 interior basins are excluded) contain a greater percentage of high-rank frontier basins.

The thermal gradient and temperature regime in any given basin influences the depth at which petroleum formation and primary migration occur. Moreover, temperature affects petroleum viscosity, volume, pressure, and solubility. Temperature, depth, and the timing of hydrocarbon accumulation also affect the secondary changes in reservoir rock petrography. Thermal regimes of the various basin types appear to be related to tectonic location and basin evolution.[12] More heat input occurs on the backside of subduction zones (Type 6B and 6C basins) along convergent margins within many Type 3 cratonic rifts and during the initial rifting of divergent margins. Higher-than-normal heat input seems to be present in portions of Type 4 downwarps into small ocean basins. Higher geothermal gradients are often present in Type 2A basins whose evolution includes initial rifting. Type 1 interior and Type 2 composite basins generally have low heat flows. Lower gradients are present in Type 8 deltas and Type 5 pull-apart basins where rapidly deposited, thick prisms of sediments are present.

The accumulation of petroleum in any basin is related to the various trapping mechanisms available. These traps are directly related to the genesis of the basins. Structurally (visible) traps have provided the major petroleum accumulations of the world.[13] These traps were formed in either compression-dominated or tension-dominated basins. From left to right in Fig. 8–3 (with the exception of Type 6 and 7

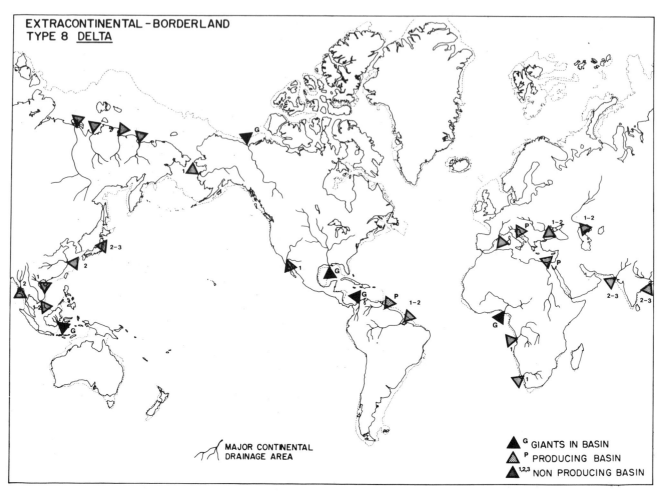

MAJOR CONTINENTAL
DRAINAGE AREA

▲ G GIANTS IN BASIN
▲ P PRODUCING BASIN
▲ 1,2,3 NON PRODUCING BASIN

Figure 8–13. Type 8, delta.

convergent margin basins), a greater proportion of sedimentary and flowage structures is present. It is also possible that a greater potential for purely stratigraphic accumulations may occur in Type 4C and Type 5 basins.

At present, the least measurable or distinguishable phase of the petroleum cycle of generation, migration, and accumulation appears to be migration. Movement of hydrocarbons from source rock to trap involves both primary migration, involving the release and expulsion of petroleum from source rock and to a considerable extent depending upon temperature and depth, and secondary migration, which involves the movement of petroleum to traps, principally by buoyancy where the tectonic and hydrodynamic character of the basin became equally important.[14] The variation in basin characteristics from left to right in Fig. 8–3 is also noted in the hydrodynamic character of basins. Type 1 to 4B basins are more mature on both an evolutionary basis and in hydrodynamic setting with considerable invasion of meteoric waters where centripetal movement along exten-

sive structural-stratigraphic ramps and a tilted potentiometric surface provide the framework for secondary migration. Type 4C and Type 8 and some Type 3 basins are younger and less mature and present more tensional or vertical fractures (often underlain by geopressured zones). Much of the sediment in these basin types is still compacting, and secondary migration may be influenced by both centrifugal and more-than-normal vertical water movement. The evolution of a basin through time often includes different tectonic environments with different hydrodynamic regimes. Variation in permeability of both the basin fill and the vertical fracturing or faults within the fill may change as a basin evolves.[15]

The various basin types have a general commonality in their dimensional and tectonic configuration. When the sedimentary fill of the eight basin types is considered, there is greater variation. Basin fill and primary reservoir quality are critical modifiers of the hydrocarbon potential. Variations in basin lithology may be linked to basin size, the time they take to be filled, and their age and tectonic setting.

100

Geologically older, larger, long-lived basin Types 1, 2, and 4 generally contain mixed clastics and carbonates, whereas small basin Types 6, 7, and 8 with a shorter fill time contain more clastics. Younger basins also often have more clastics and frequently display rapid lateral facies changes. Types of crude oil and gas-to-oil ratios appear to be initially related to lithologic types and more specifically to the influx of humic terrestrial and sapropelic marine and lacustrine organic matter.[16, 17]

High-recovery or high petroleum-yield basins in any class are often attributable to special features that enhance the fundamental petroleum parameters of source, reservoir, cap rock, and trap.[16] Special features appear to reach an optimum in many Type 4 downwarp basins.

Giant fields (over 500 million bbl or BTU gas equivalent) represent slightly more than 1% of the world's fields, yet they contribute 70% of all oil and 60% of all gas found to date.[18] The historical risk of finding giants by basin type indicates that many Type 4 downwarp (A closed and C open, Type 6 subduction (primarily 6B and 6C), and Type 3 rifts have a high chance of finding giants. Type 2 composite and 2A complex basins have many giant fields; however, by productive volume their number is average. Type 1 interior and Type 8 deltas are generally poor in giants.

In conclusion, the author believes that basin analysis and classification permit the discernment of many helpful analogies that may be applied to new or partially explored basins.

REFERENCES

1. Coury, A.B., et al., "Map of Prospective Hydrocarbon Provinces of the World," USGS, Misc. Field Studies, MR-1044 A,B,C, 1978.

2. Huff, K.F., "Frontiers of World Exploration," *Oil and Gas Journal*, 76, no. 40, 1979, p. 214.

3. Fitzgerald, T.A., et al., "Exploration in Developing Countries," Southwestern Legal Foundation, Exploration, and Economics of Petroleum Industry *Proc.*, 17, (New York: Mathew Bender & Co., 1979).

4. Huff, K.F., op cit.

5. Bally, A.W., "A Geodynamic Scenario for Hydrocarbon Occurrences," Ninth World Petroleum Congress *Proc.*, 2, 1975.

6. Klemme, H.D., "Giant Fields Related to Their Geologic Setting—A Possible Guide to Exploration," Can. Petroleum Geol. *Bull.*, 23, no. 1, 1975.

7. Wood, P.W.J., "There's a Trillion Barrels of Oil Awaiting Discovery," *World Oil*, June 1979, p. 141.

8. Klemme, H.D., "World's Oil and Gas Reserves From an Analysis of Giant Fields and Basins (provinces)," *Future Supply of Nature-Made Petroleum and Gas*, ed. R.F. Meyer (New York: Pergamon Press, 1977).

9. Bally, A.W., op cit.

10. Bally, A.W., and S.S. Nelson, "Realism of Subsidence," *Facts and Principles of the World's Petroleum Occurrences*, ed. Andrew Miall, 1980.

11. Wood, op cit.

12. Klemme, H.D., "Geothermal Gradients, Heat Flow, and Hydrocarbon Recovery," *Petroleum and Global Tectonics*, ed. A.G. Fischer and S. Judson (Princeton: Princeton University Press, 1975).

13. ———, op cit., 1977.

14. Coustau, H., et al., "Classification Hydrodynamique des Bassins Sedimentaires Utilisation Combinee avec d'Autres Methodes pour Rationaliser l'Exploration dans des Bassins Nonproductifs," Ninth World Petroleum Congress *Proc.*, 2, 1979.

15. Hunt, John M., *Petroleum Geochemistry and Geology*, (San Francisco: W.H. Freeman & Co., 1979).

16. Tissot, B.P., and D.H. Welte, *Petroleum Formation and Occurrence*, (Berlin: Springer-Verlag, 1978).

17. Klemme, H.D., op cit. 1975.

18. Halbouty, M.T., et al., "Geology of Giant Petroleum Fields," AAPG *Mem.* 14, 1970, p. 1970.

Petroleum: the Sedimentary Basin

R. STONELEY

3.1 The classification of basins

A sedimentary basin may be defined, for present purposes, as an element of the Earth's surface where sediments have accumulated to a significant thickness: lateral dimensions are measured generally in the order of hundreds of kilometres and thicknesses in the order of kilometres. Most basins have involved some degree of synsedimentary tectonic subsidence, but others are developed where sediments prograde into pre-existing deep-water areas, for example, beyond the edge of the continental shelf. It is to the sedimentary basins that the search for petroleum turns, and each basin presents its own particular problems. Attempts to classify them are useful, since they provide a basis for organising knowledge and thinking, as well as for comparison. In the estimation of the possible reserves of an unexplored basin, for example, it is logical to compare the geological conditions prevailing there with those of known basins and, with caution, to use this comparison in a preliminary evaluation.

Many classifications of sedimentary basins have been proposed (e.g. Bally, 1975; Klemme, 1975). There is no doubt of the value of these attempts to reach an overall understanding, although the variability of nature is such that if the process is taken too far, then each individual basin would fall into a category of its own. Furthermore, the tectonic setting of a basin may change during the course of its history, so that the deposits of one regime may succeed those of a completely different regime: this is a problem that was recognised many years ago (e.g. Kay, 1951).

For these reasons a very simple and broad classification is employed based on plate tectonic processes. It highlights a problem, however, that is inherent in all applications of plate tectonics to the geological past. Plate tectonics is defined through present events and any interpretation of the past record must involve extrapolation and comparisons that are, to a certain extent, subjective. Thus, we may classify a particular basin by comparison with modern analogues, but would we be correct? The problem is compounded where the basin has been through different plate tectonic environments: for example, the western Canadian basin, extending northwards from Alberta, may be considered as an original inactive continental margin, that has been involved in collision with another continental element and subsequently evolved in relation to an active margin. It is, therefore, sometimes necessary to assign a basin to different plate tectonic classifications at different stages of its evolution! This also is a subjective process, but the attempt to classify a basin should be made in terms of the interpreted plate tectonic setting *at the time the beds in question were deposited.*

Basins can be classified according to their association with (Fig. 3.1):

1 crustal spreading
2 crustal destruction (subduction)
3 transform faulting
4 intraplate environments.

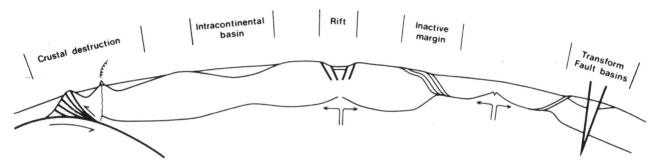

Fig. 3.1. The plate tectonic environments of the primary classes of sedimentary basins.

The last of these, the intraplate basins, evolve and, in some cases, were formed through processes that are still poorly understood and probably not directly attributable to plate tectonics (e.g. Stoneley, 1969; Beaumont, 1978). Each of these categories carries certain geological associations that are of fundamental importance to petroleum. Some generalisations are permissible, although they are certainly not exclusive and, if taken too far, can be misleading. The discussions in this section are necessarily very superficial and references are quoted as a guide to further reading: those that have been selected generally give a list of further references and are from recent, readily accessible sources.

3.2 Basins associated with crustal spreading

The basins in this category fall within the well-known 'rift-drift', or divergent continental margin, sequence of events (e.g. Thompson, 1976) (Fig. 3.2). The concern is essentially with spreading centres that develop within a continental block, since a new one within an oceanic realm is unlikely to lead to the development of a sedimentary basin of interest.

An initial uplift may herald the formation of a new ocean, but the evidence suggests that this is not always the case. Tensional rifting then leads to the emplacement of mantle material within the rift as an incipient

STAGE 1- Uplift & fracturing

STAGE 2- Rifting

STAGE 3- Further rifting; marine invasion

STAGE 4- Initiation of new ocean

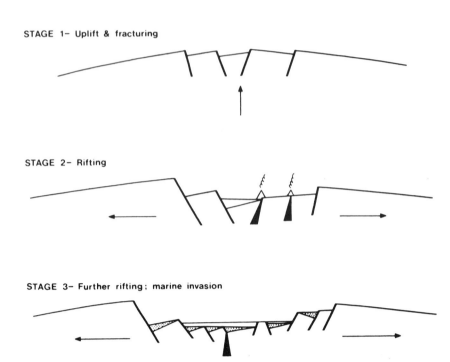

STAGE 5- Ocean formation; development of inactive margins

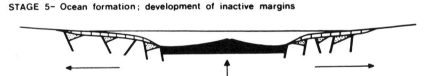

Fig. 3.2. The 'rift-drift' sequence of basin development, from the initial uplift and rifting to the formation of inactive continental margins.

ocean, which subsequently widens until a change in the global distribution of spreading centres takes place. The formation of sedimentary basins accompanies the rifting stage, but the subsequent evolution of the new inactive continental margins thus formed falls within the intraplate category 4.

The classical examples of stages of the 'rift-drift' sequence are the East African rift valleys, the Gulf of Suez, the Red Sea and the Atlantic Ocean. Upwarping is associated with the African rift system (e.g. Baker

Fig. 3.3. North Sea cross-sections, approximately east-west. A. Central North Sea; B. Northern North Sea. Phases of rifting date from the Permian to Mid Cretaceous, following which broad subsidence took place. DP, Devonian and Permian; Z, Zechstein (mainly evaporites); T, Triassic; J, Jurassic; K, Cretaceous; P-E, Palaeocene-Eocene; O-Pl, Oligocene-Pleistocene. (After Ziegler, 1977.)

et al., 1972) and the deposits are terrestrial and lacustrine, together with basaltic and alkaline volcanics. Marine invasion is a somewhat later stage, and restricted environments often lead rapidly to the precipitation of evaporites, well exemplified in the Gulf of Suez (e.g. Robson, 1971; Hassan & El-Dashlouty, 1970). Rifting may become complex, as in the southern part of the Red Sea, before the initial emplacement of mantle-derived material in the axial zones (Lowell *et al.*, 1975). Thereafter, the continental margins will continue to separate indefinitely, gradually reaching the stability of inactive shelves, well illustrated by such west African basins as Angola and Gabon (see below, p. 66).

In some cases it seems that the sequence has been interrupted and did not reach the stage of new ocean formation: such a basin is termed an *aulacogen*. The North Sea is generally regarded as an aulacogen that was aborted after the marine invasion but prior to the formation of new oceanic crust (e.g. Ziegler, 1977; Hay, 1978; Pegrum & Mounteney, 1978): it is an instructive example (Fig. 3.3). There is no evidence of initial upwarp: rather a broad depression was filled with Permian desert

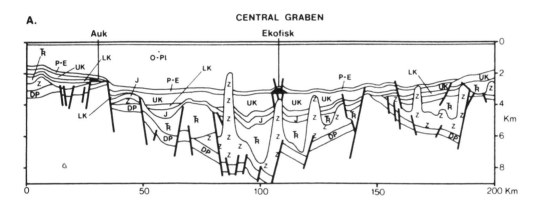

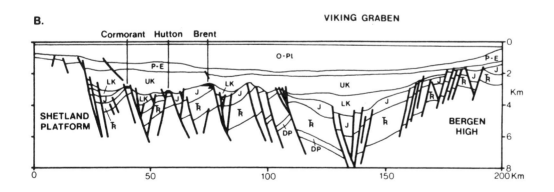

deposits (e.g. Glennie, 1972) and then carbonates and evaporites. Rifting started in the Permian but was intermittent, becoming especially pronounced in the Late Jurassic and Early Cretaceous: after this, however, as spreading was transferred to the North Atlantic, only broad subsidence continued (Kent, 1977). The actual mechanisms involved in the North Sea development are still uncertain (e.g. Ziegler, 1975; Osmaston, 1977).

PETROLEUM

Although minor quantities of gas may be encountered, the early stage of rifting is unlikely to be associated with significant petroleum potential, since the generally narrow troughs contain little or no source rock. However, once marine invasion has taken place a regime is established that is very important. Restricted seas can become enriched in organic matter and heat-flow is high, so that the source and generation potential are good; clastic detritus, and sometimes reefal growths, provide good reservoir rocks; evaporites form cap-rocks and also, perhaps, diapiric structures which, together with tilted fault blocks and drape structures, help to ensure abundant traps. One possible disadvantage may be that the geothermal gradient can be so high that sometimes the zone of gas generation is entered at relatively shallow depth: this is apparently the case in the southern Red Sea (Lowell et al., 1975).

The North Sea basin (Fig. 3.4), extending southwards into northern Germany and adjacent countries, again provides an excellent illustration (e.g. Selley, 1976; Ziegler, 1975; papers in Woodland, 1975). Gas, largely derived from the underlying Carboniferous Coal Measures of the preceding tectonic cycle, is widespread in the south in the Permian basal deposits, both in the Rotliegendes desert sands and in the early marine limestones of the Zechstein: the traps are in block-faulted and warped structures. Oil occurs in these reservoirs in the Auk and Argyll Fields, but it is believed to have been derived from a younger source (Brennand & van Veen, 1975; Pennington, 1975). Jurassic (especially Kimmeridgian) and Lower Cretaceous source rocks have fed penecontemporaneous sandstone reservoirs, of both deltaic and deep-water facies, in tilted blocks produced by intermittent rift-faulting in central and northern parts of the basin: in the south, however, traps in these beds are predominantly caused by salt motions (halokinetic), as also are those such as Ekofisk in which oil has accumulated in the Chalk of the southern Norwegian and Danish sectors. Gentle drape and compaction structures over deeper fault blocks provide the traps for oil in Tertiary reservoirs, as in the Forties Field (Walmsley, 1975).

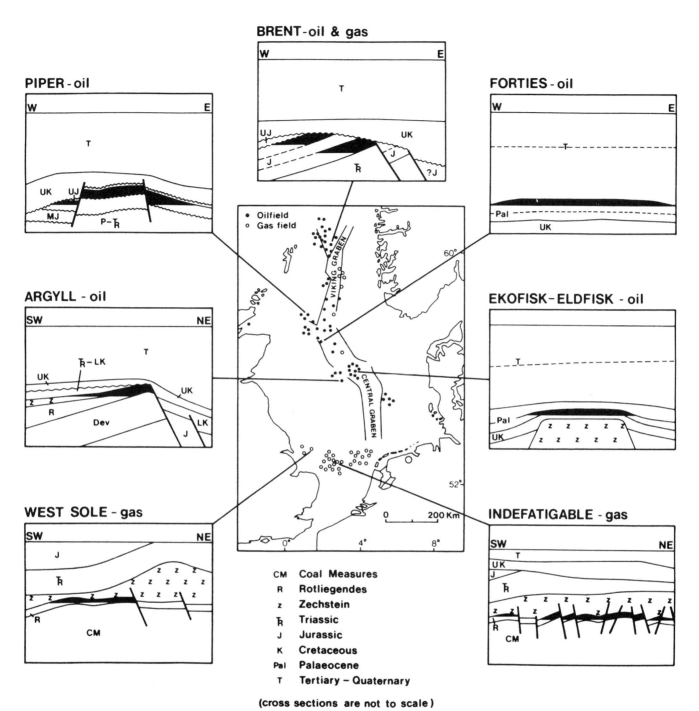

Fig. 3.4. Diagrams showing types of petroleum accumulations in the North Sea. (Based on papers in Woodland, 1975.)

107

3.3 Basins associated with crustal destruction

The basins that are developed in association with belts of plate convergence are very varied. Too close comparisons or generalisations are dangerous, especially since some basins falling under this heading may have evolved from a former inactive continental margin. Thus, for example, the Zagros basin of south-west Iran represented part of an inactive shelf from the Permian into the Neogene, and only then was involved in continental collision and subduction. Because in any classification distinctions have to be made, shelves that are tectonised as a result of collision will be considered in the context of the intraplate basins.

A convenient subdivision is provided by the island-arc environments, as follows: fore-arc/outer-arc, inter-arc basin, volcanic-arc, back-arc basin (Fig. 3.5). These environments can be recognised, with caution, both

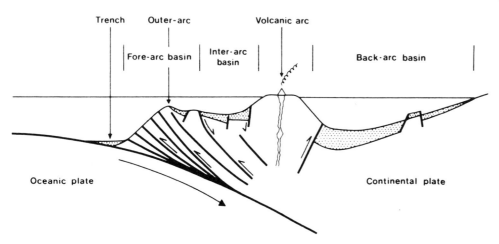

Fig. 3.5. Generalised classification of tectonic elements and basins in a zone of crustal destruction. In an oceanic realm, the back-arc basin will be one-sided and may be expanded through secondary crustal spreading.

in entirely oceanic settings and also where oceanic crust underthrusts a continental margin: indeed, a lateral passage along the strike from the one to the other can sometimes be observed. Thus, for example, the Aleutian Arc passes from an oceanic realm eastwards into continental Alaska. The various basins wil be considered following this geotectonic subdivision.

FORE-ARC BASINS

Fore-arc regions have received a lot of attention during the past few years, and an excellent review has been published recently by Dickinson & Seely (1979). This section deals largely with the deposits of an oceanic trench at the outcrop of a subduction zone. These, together sometimes with slivers of material stripped off the down-going oceanic plate, are piled up above the plane of subduction in a series of imbricate slices on the inner side of the trench. This stack may lead upwards more or less directly to the volcanic arc, as seen commonly in the more oceanic cases; or it may lead to the growth of a non-volcanic, outer island arc; or, at an active continental margin, to a series of 'coastal mountain ranges'.

In general, the sediments of such regions are immature, rapidly deposited and thick: when imbricated, the total pile may reach thicknesses of several kilometres. Turbidites, derived both from uplifted earlier sediments and from volcanic arc rocks, predominate and slump deposits are common. Of course, grain flow sands may be present and, if the rate of sedimentation outstrips subsidence, then neritic or even littoral or terrestrial beds can accumulate. Rather short-lived reefs are known. Lateral and vertical facies changes are often rapid, so that reservoir bodies are commonly of limited extent.

Structurally, the fore-arc regions are frequently complex, as a result of the compressional imbrication and also of instability-induced normal faulting. In addition, heat-flow tends to be low, so that any potential source rock is often thermally immature even at considerable depths.

Petroleum

Fore-arc regions are clearly not among the most attractive for petroleum exploration and few commercially viable fields have been discovered. Not only are most geological factors rather unfavourable, but they combine to make exploration difficult; reservoirs of limited extent and complex structure may also be associated with rugged topography, the Gulf of Alaska Cenozoic basin being an extreme example (Fig. 3.6) (Stoneley, 1967). Best known perhaps amongst the oilfields in fore-arc regions are the coastal fields of Ecuador and Peru (Fig. 3.7). They have accumulated in regressive Tertiary sediments of deep to shallow water facies, partly at least in small complexly normal-faulted 'successor' basins also subject to low-angle gliding, above the older rocks of the Coastal Ranges (e.g. Lonsdale, 1978; Travis *et al.*, 1976).

Fig. 3.6. Sketch of the inferred tectonics in the Gulf of Alaska region at an active continental margin. (Based on Stoneley, 1967.)

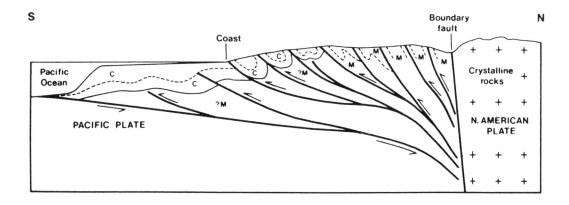

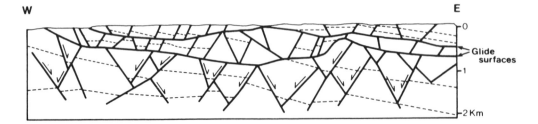

W   E

Glide
surfaces

Fig. 3.7. Talara, coastal Peru. Natural scale cross-section through the oilfields area. The block-faulting and glide structures illustrated are in the Eocene and younger paralic deposits of a regressive successor basin, overlying deep-water clays of a fore-arc environment. (After Travis *et al.*, 1976.)

INTER-ARC BASINS

These basins lie between the inner volcanic and the outer non-volcanic arcs: they are often, in the literature, still included under the heading of 'fore-arc basins' (e.g. Seely, 1979). In a regime of continental margin growth as a result of subduction, these basins will generally overlie marginal or trench deposits. They are generally narrow and filled primarily with clastic deposits, derived from both arcs and, sometimes, from the continent itself. Structural styles can be either gently compressive or tensional, depending on the expression of the underlying subduction zone.

Petroleum

The Cook Inlet of southern Alaska (Fig. 3.8), may be cited as one of the few examples of petroliferous inter-arc basins (Kirschner & Lyon, 1973; Fisher & Magoon, 1978). It has partly faulted margins, both normal and strike-slip, and contains more than 9000 m of non-marine Tertiary sediments. Oil and gas are found in the lower part of this Tertiary sequence, in sharp compressive anticlines. There is evidence, however, that the hydrocarbons originated in the underlying Mesozoic beds which, at least in part, appear to have accumulated on a temporarily stable shelf, during a period of interruption of subduction (Fisher & Magoon, 1978).

BACK-ARC BASINS

The third environment of thick sedimentation associated with active plate margins lies behind the volcanic arc. Its development can be very varied, depending on whether it is continental or oceanic, and whether or not it is associated with secondary back-arc crustal spreading. The basins are typically asymmetrical with a steep, often thrust-faulted, outer margin against the volcanic arc and a non-tectonic inner margin which, however, may be affected by basement block-faulting.

The deposits include clastics derived from both margins: from the volcanic arc and, in a continental environment, from the craton behind. The coarser detritus, however, may not reach the centre of the basin where, as for example in Sumatra, argillaceous and carbonate rocks can accumulate

110

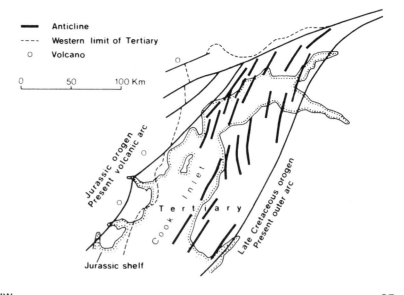

Legend:
— Anticline
---- Western limit of Tertiary
○ Volcano

0   50   100 Km

Jurassic orogen
Present volcanic arc

Late Cretaceous orogen
Present outer arc

Cook Inlet

Tertiary

Jurassic shelf

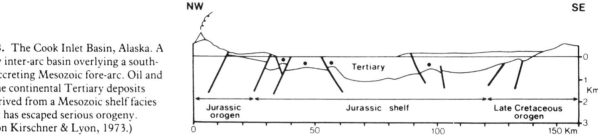

NW   SE

Tertiary

Jurassic orogen   Jurassic shelf   Late Cretaceous orogen

0   50   100   150 Km
0 1 2 3 Km

Fig. 3.8. The Cook Inlet Basin, Alaska. A Tertiary inter-arc basin overlying a southwards accreting Mesozoic fore-arc. Oil and gas in the continental Tertiary deposits were derived from a Mesozoic shelf facies belt that has escaped serious orogeny. (Based on Kirschner & Lyon, 1973.)

(de Coster, 1975). In a continental environment the filling of back-arc basins is commonly molassic, derived largely from an orogen along the volcanic belt. The deposits can be coarsely clastic and of either continental or marine facies: the tendency is for them to become generally less marine upwards, in which case petroleum can be expected to be confined to the lower parts of the sequence. This distribution may be emphasised if the basin, and indeed the whole continental margin, evolved from an earlier inactive shelf. Such is the situation in the Oriente basin, lying east of the Andes in Ecuador and Peru (e.g. Feininger, 1975; Zuñiga y Rivero et al., 1976), where Middle Cretaceous and older shelf sediments underlie thick Upper Cretaceous and Cainozoic molassic deposits (Fig. 3.9).

Lateral transitions from one crustal setting to another are known. Thus, for example, the tectonic strike passes from the continental back-arc basin of Burma southwards into oceanic Andaman Sea, where back-arc spreading is taking place (Curray et al., 1979); farther south again, it resumes a continental environment in Sumatra (Fig. 3.10). Although the situation

111

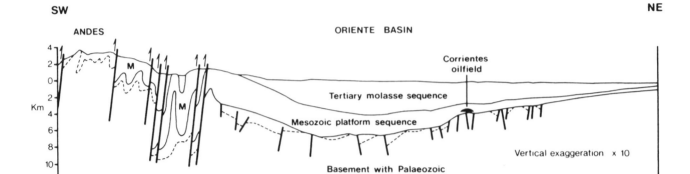

SW   NE

Fig. 3.9. The Oriente Basin of Peru has evolved as a back-arc basin since the Late Cretaceous from an earlier inactive continental margin. (Based on Zuñiga y Rivero *et al.*, 1976.)

is complicated by strike-slip faulting, the development of the volcanic arc is suppressed in the vicinity of the Andaman Sea and nearly all of the sediment is derived from the Asian mainland.

Petroleum

The wide variety in crustal environment and sedimentary filling of back-arc basins produces a corresponding variation in the occurrence of oil and gas. Reservoir rocks could, as a rule, be virtually assured, either as clastics or as carbonates, to the extent even that adequate cap-rock might not always be present. The development of source rocks will mainly be dependent upon the oceanographical environments, but they are especially likely to accumulate in basins of limited areal extent where organic matter may be concentrated. Heat-flow generally is high, particularly in the outer parts of the basin adjacent to the volcanic arc, although it decreases towards the continent: the possibility exists, for example, that the rather heavy oil found in the Oriente of Peru has migrated through considerable distances, up-dip eastwards towards the Brazilian Shield (Feininger, 1975). Structural traps, once away from the basin margin adjacent to the volcanic arc and sometimes associated orogen, are predominantly related to block movements in the underlying continental basement. The examples of Sumatra, Burma and the basins east of the Andes have already been referred to: they are indeed the most important of the back-arc producing regions.

3.4 Basins associated with transform faulting

Where there is transpression across a major strike-slip fault, or where there are deviations from its course, then the resultant stresses can lead to differential crustal warping. This is most likely to happen where transform faults pass through continental regions due to the presence of crustal inhomogeneities. Small basins are believed to have been formed in this manner, and they are commonly subject to second-order folding,

112

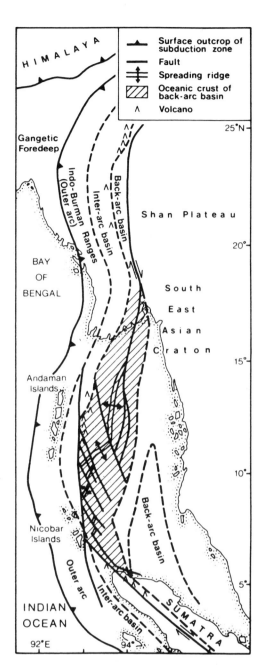

Fig. 3.10. Lateral crustal transitions of a back-arc basin, from continental in Burma, through oceanic in the Andaman Sea (oblique back-arc spreading) to continental in Sumatra. (After Curray *et al.*, 1979.)

thrusting and normal faulting. They are developed, to a greater or lesser extent, in association with most of the world's great strike-slip faults within continental blocks. Such basins may be formed, filled with sediment and uplifted very rapidly. The sediments can range over short

113

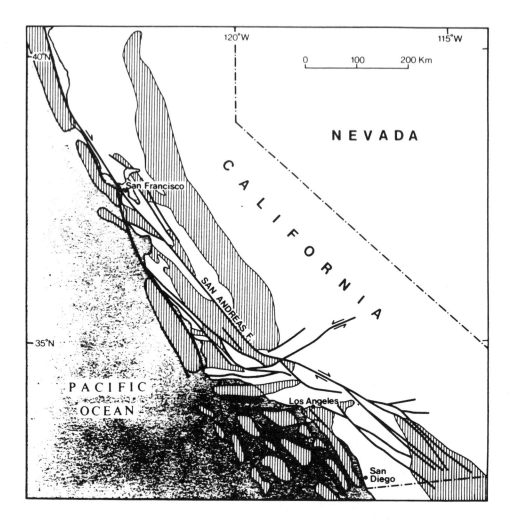

Fig. 3.11. Generalised map showing the distribution of late Cenozoic basins in California, in relation to the San Andreas transform fault system. (Adapted from Blake *et al.*, 1978 and others.)

distances from near-shore, or even continental, deposits to deep-sea turbidites and, because of localised synsedimentary tectonic movements, thicknesses will be subject to rapid variation. Geothermal gradients, similarly, will vary widely.

The supreme examples of such basins, especially in relation to their petroleum interest, must be those of central and southern California (Fig. 3.11) (e.g. Crowell, 1974; Blake *et al.*, 1978; Nardin & Henyey, 1978). Indeed, the Los Angeles Basin is the richest in the world in terms of petroleum yield per unit volume of sediment. As many as 30 small, more or less separated onshore and offshore basins of Tertiary to Recent age can be identified, all of them in some way related to the San Andreas Fault system. Their sediments were largely derived from intervening uplifts, and include almost the entire range of clastic deposits together with some carbonates. California today is situated adjacent to an area of upwelling of mineral-rich oceanic waters, leading to high organic productivity, and this factor may have contributed to the high organic content of the Cenozoic sediments. Petroleum is obtained from reservoirs ranging from coarse clastics to turbidites, and in anticlinal, fault-closed and stratigraphic traps in a great variety of local tectonic environments. There can be no doubt that the whole process from sedimentation, through trap formation to petroleum generation, migration and accumulation, is continuing up to the present time.

3.5 Intraplate basins

Sedimentary basins occur within crustal plates, entirely within the continental shields (intracratonic), superimposed upon or within fold-belts, at the margins between continents and oceans and within the oceanic realm (Section 4.4). It has already been pointed out that, strictly speaking, intraplate basins evolve through causes not directly attributable to plate tectonics: however, a consideration of them shows that some of the environments were created initially through such processes, and that plate tectonic influences may still be apparent. It is therefore appropriate that we should review them briefly.

INTRACRATONIC BASINS

The basins developed within shield areas are often surprisingly circular in outline. They may be long-lived and sedimentation can keep pace with subsidence (e.g. Sloss & Speed, 1974), so that not only do shallow-water deposits predominate but interval isopachs tend to reflect the shape of the basin. The nature of the sediments, and hence petroleum potential, will depend upon marine connections but, where they are well established, organic matter may be concentrated. Geothermal gradients tend to be approximately normal for shield areas, and traps are either related to basement swells or ridges, or are stratigraphic. Some intracratonic basins, such as those in North Africa, extend to the continental margin and, in such an environment, may later be partially involved in continental collision and orogenesis.

Outstanding examples are the Moscow Basin, where the sediments are largely continental and petroleum interest is low, and the Williston Basin in the north central United States and Canada (Fig. 3.12). The latter persisted from the early Palaeozoic to the Tertiary and the sedimentary thickness reaches 4 km. Marine connections were good and the petroleum generated within the basin has accumulated primarily over gentle basement ridges.

115

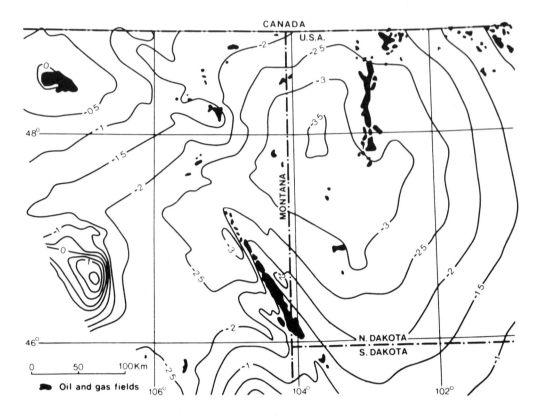

Fig. 3.12. The Williston Basin. Generalised structure contours (in thousands of feet from sea-level) on the Cretaceous, indicating the broad relationship of oil and gas fields to positive relative upwarps. (Simplified after Harding & Lowell, 1979.)

FOLD-BELT BASINS

Included here are both those basins which have developed from a formerly stable 'block' within the fold-belt and have later subsided, and those which have been superimposed on it, possibly over a tectonic sag. Examples of the former are the central European Pannonian Basin and the present day Black Sea (Degens & Ross, 1974) whilst the latter include the Vienna and Maracaibo Basins. Both categories are treated together because, as far as petroleum is concerned, they appear to have somewhat similar characteristics. They are rather restricted in area, and appear to have subsided and evolved rapidly. If an abundant supply of sediment is available, then the sediments may be coarsely clastic, even molassic, and shallow-water environments will predominate (Vienna Basin). If, on the other hand, they are relatively starved, then considerable water depths will remain (Black Sea). Marine connections may be rather restricted, leading to organic enrichment, in which case the basin may be very petroliferous. Structures may be compressional, being inherited from preceding orogeny, or purely tensional: the latter are perhaps the more common.

116

(These continue the 'rift-drift' sequence that was considered earlier.) Once equilibrium has been reached following the splitting of a continental mass, then stable continental margins will persist until they are involved in a continental collision. This may be very much later, so that the continent-ocean margin can remain in this stable state for a long period of time. Basin development is through the prograding of sediments beyond the continental edge and, generally but not always, broad subsidence of a shallow marine shelf within the margin. The continental shelf may also be subject to important gentle differential vertical movements within the basement, and, as for example in Gabon and Angola (Brink, 1974; Brognon & Verrier, 1966) (Fig. 3.13), the shelf sediments commonly

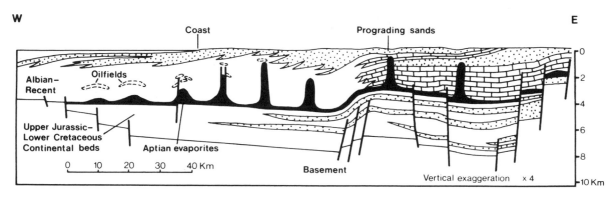

Fig. 3.13. Cross-section of the Gabon Basin, a typical inactive margin inherited from a former rift basin. Note the basal continental clastics with early evaporites and shallow marine deposits of the rift phase, overlain by oceanwards prograding beds of the inactive margin phase. (After Brink, 1974.)

overlie block-faulted, continental and early marine deposits inherited from the rifting stages of development. If evaporites are present, then diapiric structures may be expected. Inactive, or formerly inactive, margins are extremely important from the point of view of petroleum and a high proportion of the world's reserves are found in them: one has only to note that the Middle East Basin, extending from Saudi Arabia into Iran, evolved at an inactive margin until it was subjected to continental collision in the Late Tertiary (Figs 3.14 and 3.15).

The wide continental shelves reflect an aspect of plate tectonics not yet mentioned. They are, of course, very susceptible to marine transgressions and regressions, one cause being variations in the rate of sea-floor spreading: the argument is as follows, that rapid spreading gives rise to more elevated mid-ocean ridges, the displaced water resulting in a higher sea-level and transgression over the continental margins (Fig. 3.16). During slower spreading, the converse applies (e.g. Rona, 1973; Pitman, 1978). Although such sea-level variations are worldwide, their effects are particularly noticeable on the stable continental shelves, and they are extremely important from the point of view of petroleum. Transgression

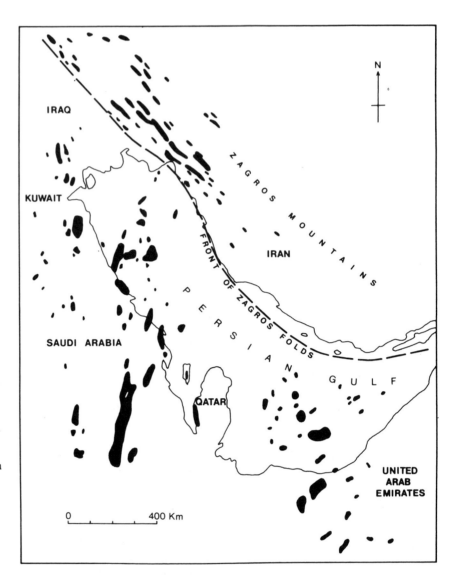

Fig. 3.14. Oilfields of the Middle East Basin. Elongations of the fields suggest their origin: those trending approximately north-south in Arabia are predominantly in structures related to basement block movements; rounded outlines of fields in the Persian Gulf are related to salt structures; north-west—south-east elongated fields in Iran represent Mesozoic oil distributed into anticlines of the Neogene Zagros foldbelt.

results in deeper water sedimentation and the accumulation of potential source rocks: shelf seas are often rich in organic matter, particularly if the continental margin is associated with oceanic upwelling. Regression leads to the accumulation of widespread sands or carbonates, perhaps reefal, and may culminate in lagoonal or sabkha evaporite precipitation. The interleaving of source rock, reservoir and cap-rock can thus be established: these favourable circumstances may be enhanced by a lateral juxtaposition of deeper water intrashelf depressions and shallow water 'highs', which can also form very broad traps. It was indeed this combination of

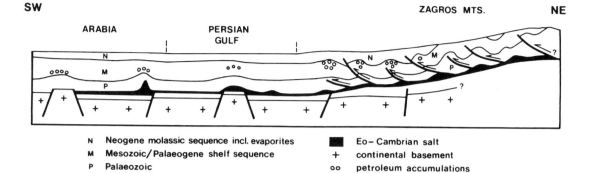

N Neogene molassic sequence incl. evaporites   ▣ Eo–Cambrian salt
M Mesozoic/Palaeogene shelf sequence   + continental basement
P Palaeozoic   oo petroleum accumulations

Fig. 3.15. Diagrammatic sketch section of the regional structure and oilfields in the Middle East Basin.

factors that is believed to have been the reason for the abundance of petroleum in the Middle East.

Sediment progradation over continental margins has, in recent years, become more widely understood, in no small measure due to the applications of seismic techniques to stratigraphy (e.g. Payton, 1977). The continental slopes have, as yet, been little explored with the drill, and they may perhaps hold out the greatest hope for significant future discoveries, largely in stratigraphically controlled traps. However, one particular such environment has already proved to be very important with respect to petroleum.

Fig. 3.16. Diagrams showing the possible worldwide effect of crustal spreading rate on sea-level.

Deltas prograding over the continental margin off the mouths of large rivers contain large reserves (Fig. 3.17). They result, in effect, in large-scale regressive sedimentary sequences, which display a fascinating association of interrelated phenomena (e.g. Chapman, 1973). Continuous deep-water shales at the base of the prograding sequence may retain much

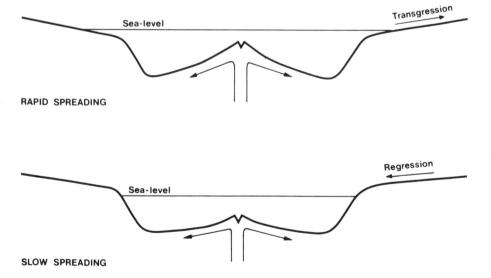

119

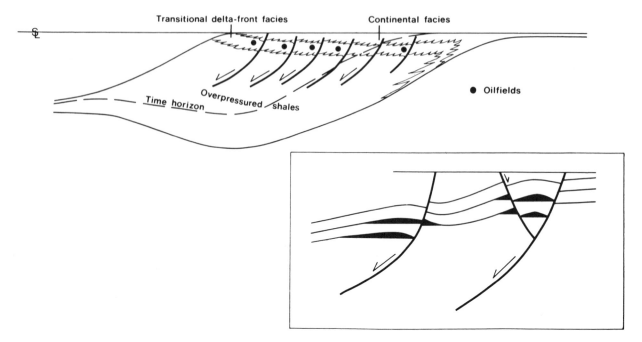

Fig. 3.17. The habitat of oilfields in major deltas. Inset: the cross-sectional relationship of growth-faults, roll-over anticlines and petroleum traps.

of their original interstitial water and are commonly under-compacted and overpressured (as compared with a hydrostatic head). As such they not only act as geothermal heat sinks but also cause instability in the overburden: this results in down-to-the-ocean slumping, expressed as synsedimentary normal faulting, and the development of associated 'roll-over' anticlines. We thus have the conditions for the generation of petroleum and its accumulation, generally in the prograded delta-front sands sealed by minor transgressive shales, in adequate but often complex structural traps. The Mississippi and Niger Deltas are the best known examples (e.g. Harding & Lowell, 1979; Weber & Daukoru, 1975; Evamy *et al.*, 1978).

A rather special situation is created when a formerly inactive shelf is involved in an interplate continental collision. A fore-deep, or molasse, basin may be developed in front of the resulting orogenic fold-belt, and may itself, at least in part, be involved in the later stages of this folding. Well-known examples, on either side of the European Alps, are the Neogene Po Valley to the south and the Molasse Basin to the north (e.g. Van Houten, 1974): the Siwalik trough on the southern margin of the Himalaya is another. Potential reservoirs and traps are commonly well developed, but the presence or absence of petroleum will depend primarily on whether or not the sea invaded the basin and, if so, whether or not a sufficient thickness of source rocks (and cap-rocks) could accumulate. It is important not to confuse petroleum endemic to the underlying, pre-orogenic shelf sequence, as is well exemplified in south-west Iran (Fig.

120

3.13), with any that may be related *sensu stricto* to the synorogenic basin, which can be barren. No indigenous petroleum is present in Iran in the thick Upper Miocene-Pliocene molassic deposits related to the fore-deep of the Zagros orogene: the oil and gas in the Zagros anticlines are derived from the Mesozoic shelf beds, and the former situation is preserved in Arabia, which has not been affected by the compressional folding.

3.6 Conclusions

This rapid and sketchy review of the plate tectonic development of sedimentary basins, and of the major habitats of petroleum, should serve to emphasise that aspects of the Earth's crust as diverse as plate tectonic processes and petroleum, in fact may be merely different manifestations of a whole spectrum of processes in the outer layers of our planet. With this in mind, the next chapter will examine in more detail the controls on petroleum accumulations, and see what impact plate tectonics may have on our understanding of them.

References

Baker, B.H., Mohr, P.A. & Williams, L.A.J. (1972) *Geology of the Eastern Rift System of Africa.* Geol. Soc. Am. Spec. Pap. 136, 67 pp.

Bally, A.W. (1975) A geodynamic Scenario for hydrocarbon occurrences. *Proc. 9th World Petrol. Congr. Tokyo,* Vol. 2, pp. 33-34.

Beaumont, C. (1978) The evolution of sedimentary basins on a viscoelastic lithosphere: theory and examples. *Geophys. J. R. astr. Soc.,* **55,** 471-497.

Blake, M.C., Campbell, R.H., Dibblee, T.W., Howell, D.G., Nilsen, T.H., Normark, W.R., Vedder, J.C. & Silver, E.A. (1978) Neogene basin formation in relation to plate-tectonic evolution of San Andreas Fault system, California. *Bull. Am. Assoc. Petrol. Geol.,* **62,** 344-372.

Brennand, T.P. & van Veen, F.R. (1975) The Auk Oil-Field. In *Petroleum and the Continental Shelf of North West Europe, Vol. 1 Geology* (Ed. A.W. Woodland) Applied Science Publishers, London, pp. 275-284.

Brink, A.H. (1974) Petroleum Geology of Gabon Basin. *Bull. Am. Assoc. Petrol. Geol.,* **58,** 216-235.

Brognon, G.P. & Verrier, G.R. (1966) Oil and geology in Cuanza Basin of Angola. *Bull. Am. Assoc. Petrol. Geol.,* **50,** 108-158.

Chapman, R.E. (1973) *Petroleum Geology: A Concise Study* Elsevier, Amsterdam, London and New York, 304 pp.

Crowell, J.C. (1974) Origin of late Cainozoic basins in southern California. In *Tectonics and Sedimentation* (Ed. W.R. Dickinson) Soc. Econ. Pal. & Min. Spec. Publ. 22, pp. 190-204.

Curray, J.R., Moore, D.G., Lawyer, L.A., Emmel, F.J., Raitt, R.W., Henry, M. & Kieckhefer, R. (1979) Tectonics of the Andaman Sea and Burma. In *Geological and Geophysical Investigations of Continental Margins.* (Eds J.S. Watkins, L. Montadert & P.W. Dickerson). Am. Assoc. Petrol. Geol., Mem. 29, pp. 189-198.

de Coster, G.L. (1975) The geology of the central and south Sumatran basins. *Proc. 3rd Ann. Convention (1974). Indonesian Petrol. Ass.* Jakarta, pp. 77-1Ħ1.

Degens, E.T. & Ross, D.A. (Eds) (1974) *The Black Sea—Geology, Chemistry and Biology.* Am. Assoc. Petrol. Geol., Mem. 20, 633 pp.

Dickinson, W.R. & Seely, D.R. (1979) Structure and stratigraphy of forearc regions. *Bull. Am. Assoc. Petrol. Geol.*, **63**, 2-31.

Evamy, D.D., Haremboure, J., Kamerling, P., Knaap, W.A., Molloy, F.A. & Rowlands, P.H. (1978) Hydrocarbon habitat of Tertiary Niger Delta. *Bull. Am. Assoc. Petrol. Geol.*, **62**, 1-39.

Feininger, T. (1975) Origin of oil in the Oriente of Ecuador. *Bull. Am. Assoc. Petrol. Geol.*, **59**, 1166-75.

Fisher, M.A. & Magoon, L.B. (1978) Geologic Framework of Lower Cook Inlet, Alaska. *Bull. Am. Assoc. Petrol. Geol.*, **62**, 373-402.

Glennis, K.W. (1972) Permian Rotliegendes of north-west Europe interpreted in light of modern desert sedimentation studies. *Bull. Am. Assoc. Petrol. Geol.*, **56**, 1048-71.

Harding, T.P. & Lowell, J.D. (1979) Structural styles, their plate-tectonic habitats, and hydrocarbon traps in petroleum provinces. *Bull. Am. Assoc. Petrol. Geol.*, **63**, 1016-1058.

Hassan, F. & El-Dashlouty, S. (1970) Miocene evaporites of Gulf of Suez region and their significance. *Bull. Am. Assoc. Petrol. Geol.*, **54**, 1686-96.

Hay, J.T.C. (1978) Structural development in the northern North Sea. *Jl. Petrol. Geol.*, **1**, 65-77.

Kay, M. (1951) *North American Geosynclines.* Mem. Geol. Soc. Am., p. 48.

Kent, P.E. (1977) The Mesozoic developments of aseismic continental shelves. *J. geol. Soc. Lond.*, **134**, 1-18.

Kirschner, C.E. & Lyon, C.A. (1973) Stratigraphic and tectonic development of Cook Inlet petroleum province. In *Arctic Geology* (Ed. M.G. Pitcher) Am. Assoc. Petrol. Geol., Mem. 19, pp. 396-407.

Klemme, H.D. (1975) Giant oilfields related to their geologic setting—a possible guide to exploration. *Bull. Can. Petrol. Geol.*, **23**, 30-66.

Lonsdale, P. (1978) Ecuadorian subduction system. *Bull. Am. Assoc. Petrol. Geol.*, **62**, 2454-77.

Lowell, J.D., Genik, G.J., Nelson, T.H. & Tucker, P.M. (1975) Petroleum and Plate Tectonics of the southern Red Sea. In *Petroleum and Global Tectonics* (Eds A.G. Fischer & S. Judson) Princeton University Press, Princeton, pp. 129-153.

Nardin, T.R. & Henyey, T.L. (1978) Pliocene-Pleistocene diastrophism of Santa Monica and San Pedro Shelves, California Continental Borderland. *Bull. Am. Assoc. Petrol. Geol.*, **62**, 247-272.

Osmaston, M.F. (1977) Some fundamental aspects of plate tectonics bearing on hydrocarbon location. In *Developments in Petroleum Geology—1* (Ed. G.D. Hobson) Applied Science Publishers, London, pp. 1-52.

Payton, C.E. (Ed.) (1977) *Seismic Stratigraphy—Applications to Hydrocarbon Exploration.* Amer. Assoc. Petrol. Geol., Mem. 26, 516 pp.

Pegrum, R.M. & Mounteney, N. (1978) Rift basins flanking North Atlantic Ocean and their relation to North Sea area. *Bull. Am. Assoc. Petrol. Geol.*, **62**, 419-441.

Pennington, J.J. (1975) The geology of the Argyll Field. In *Petroleum and the Continental Shelf of North West Europe Vol. 1 Geology* (Ed. A.W. Woodland) Applied Science Publishers, London, pp. 285-294.

Pitman, W.C. (1978) Relationship between eustacy and stratigraphic sequences of passive margins. *Bull. geol. Soc. Am.*, **89**, 1389-1403.

Robson, D.A. (1971) The structure of the Gulf of Suez (Clysmic) rift, with special reference to the eastern side. *J. geol. Soc., Lond.*, **127**, 247-276.

Rona, P.A. (1973) Relations between rates of sediment accumulation on continental shelves, sea-floor spreading and eustacy inferred from the central North Atlantic. *Bull. Geol. Soc. Am.*, **84**, 2851-72.

Seely, D.R. (1979) The evolution of structural highs bordering major forearc basins. In *Geological and Geophysical Investigations of Continental Martins.* (Eds J.S. Watkins, L. Montadert & P.W. Dickinson) Am. Assoc. Petrol. Geol., Mem. 29, pp. 245-260.

Selley, R.C. (1976) The habitat of North Sea Oil. *Proc. Geol. Assoc.*, **87**, 359-88.

The American Association of Petroleum Geologists Bulletin
V. 67, No. 12 (December 1983), P. 2175-2193, 15 Figs.

Global Basin Classification System[1]

D. R. KINGSTON,[2] C. P. DISHROON,[2] and P. A. WILLIAMS[4]

ABSTRACT

A proposed system classifies sedimentary basins, worldwide, into specific as well as general categories. The geologic history of each basin may be subdivided into cycles using three parameters: basin-forming tectonics, depositional sequences, and basin-modifying tectonics. Sedimentary basins may be simple, with one or two tectonic/sedimentary cycles, or they may be complex polyhistory basins with many different cycles and events. There are eight simple cycle types in this classification, which cover continental, continental-margin, and oceanic areas. The eight basic cycle types, their depositional fills, and tectonic modifiers have been given letter and number symbols so that the specific geologic history of each basin may be written as a formula. These formulas may then be compared and similarities or differences between basins noted.

INTRODUCTION

The main purpose of sedimentary-basin classification is to create a system whereby basins may be compared with each other and similarities or differences noted. Various basin classification systems have been proposed in recent years, e.g., Weeks (1952), Knebel and Rodriguez-Eraso (1956), Uspenskaya (1967), Halbouty et al (1970a, b), Perrodon (1971), Klemme (1971a, b, 1975), Bally (1975), Huff (1978), Bally and Snelson (1980), and Bois et al (1982). The geologic history of two continental-margin basins or two cratonic basins may be similar in general aspects, but will show important differences in detail; consequently, in the past, two basins could be compared only in very general terms. An alternative system proposed herein compares basins in both general and specific terms.

This basin classification system is based primarily on the principles of plate tectonics that have been developed by numerous authors over the past 20 years. Morgan (1968), Le Pichon (1968), Isacks et al (1968), and others provided ideas for many of the basic elements of plate tectonics, such as divergence, convergence, and transform movements. Documentation for continental breakup and divergence has been supplied by Francheteau and Le Pichon (1972), Norton and Sclater (1979), and Rabinowitz and

LaBreque (1979). Models for divergent continental margins and basins have been described by Sclater and Christie (1980) and Sawyer et al (1982). Convergent movements of plates have been investigated by many workers. Subduction and orogeny were described by Roeder (1973). Dewey and Bird (1970) and Atwater (1970) investigated convergence and mountain building. Dickinson (1973) and others described volcanism and convergence.

Esso Exploration began plate tectonic restorations on a global scale in the late 1960s. The purpose was to reconstruct movements in the earth's continental plates through geologic time, and to determine the effects of these movements on the structure, stratigraphy, and hydrocarbon occurrence of present-day basins both onshore and offshore. It was believed that by making these plate reconstructions we could recognize and predict oil potential by the development of analytical techniques for plate and continental margins. Finally, the writers wished to established valid comparisons of tectonic style and therefore of basins: their sedimentary fills, oil plays, and potentials. The result of these global plate restoration studies was the development of a basin classification system, whereby all sedimentary basins, worldwide, could be classified according to their structural genesis and evolutionary history. Contributions from Exxon domestic and overseas affiliates provided data on approximately 600 identifiable sedimentary basins worldwide. We were able to classify all of these basins within the system, the accuracy depending on the quality of the data and our collective knowledge of the regional geology.

The basic unit in this classification is the cycle, which consists of the sediments deposited during one tectonic episode. Some basins have only one sedimentary or tectonic cycle. These are called simple basins. Most basins, however, contain more than one tectonic/sedimentary cycle, and are called polyhistory basins. Figure 1 is a chart for the classification of simple basins and the identification of cycles for polyhistory basins. It should be noted that the terms "basin" and "cycle" may be used interchangeably in this system as a unit formed by one structural mode of basin formation. Basins, both simple and complex, may be classified by analyzing their geologic history in the context of plate tectonics. The major elements of this history are (1) depositional cycles or sequences, (2) basin-forming tectonics, and (3) basin-modifying tectonics.

DEPOSITIONAL SEQUENCES

The first major elements used in the basin classification system are depositional cycles and stages. A cycle is defined as the sediments deposited during one tectonic period. The minimum stratigraphic unit that can be called

---

[1]Manuscript received, February 22, 1983; accepted, May 13, 1983.
[2]Esso Exploration Inc., Houston, Texas 77024.
[3]Esso Exploration Inc. (retired), 203 Pompano, Surfside, Texas 77541.

The writers thank the management of Esso Exploration Inc. for permission to publish this paper; T. A. Fitzgerald, H. M. Gehman, Jr., D. S. McPherron, T. H. Nelson, and J. B. Sangree, for comments and assistance in revisions; and B. E. St. John, J. D. Frick, J. R. Gealy, and D. H. Roeder, who along with the writers were members of the original global studies team.

123

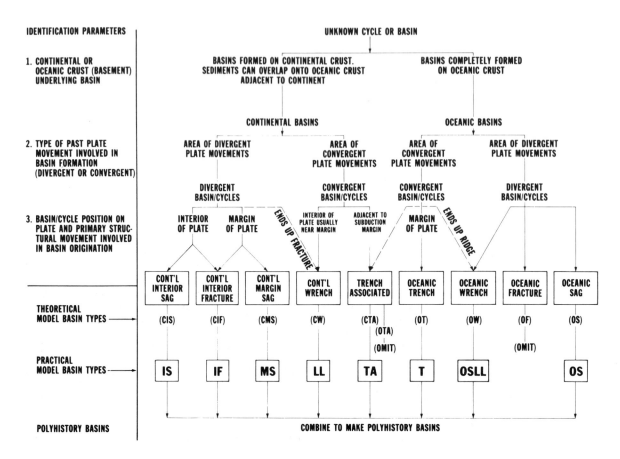

FIG. 1—Basin/cycle identification key for basin-forming tectonics. Two theoretical basin types (OTA and OF) are omitted from practical basin types.

a cycle must have significance in the development of a basin, either in thickness or span of geologic time. This allows us to lump thin units of high-shelf or wedge-edge deposits, which may form over long periods of time, into just a few cycles and to split thick prograding deposits into identifiable units.

Figure 2 shows the relation of the depositional stages to the tectonic cycle. One can think of the concept of wedge base, wedge middle, and wedge top as one sedimentary cycle with the three stages representing the three elements of one major transgressive-regressive wedge (White, 1980). Stage 1 of the cycle corresponds to a nonmarine wedge base. This includes primarily nonmarine flood-plain, lagoonal, and beach deposits, if they can be distinguished. Sedimentary types normally present are nonmarine conglomerates, sandstones, and shales. Other lithologic characteristics less commonly found are red beds, coals, volcanics, and fresh-water limestones. If the basal wedge of clastics in question is thick and over half nonmarine, it is classed as stage 1. Stage division is shown in Figure 2 at the 50% marine dashed line cutting the transition zone between wedge-edge sands and wedge-middle shales.

Stage 2 is the marine wedge middle. Lithologic types most commonly found here are marine shales, limestones, and sandstones. All massive salt is included here on the theory that thick evaporites generally indicate a marine connection or at least the drying of a marine-connected tongue. Also, massive evaporite deposition indicates enclosed depositional conditions and is generally found only in interior basins. Other less common lithologic characteristics found in stage 2 are volcanics, marine coals, flysch and other turbidites, and deep-water marls and pelagic deposits. Stage 2 may also contain the distal ends of nonmarine tongues provided these do not exceed 50% of the total.

Stage 3 is the nonmarine wedge top and the associated regional unconformity. Lithologically it typically resembles the wedge base with more than 50% nonmarine conglomerates, sandstones, shales, red beds, coals, fresh-water limestones, and minor evaporites. Post-wedge-top unconformities are included in stage 3.

The sedimentary stages should be described from the center of the depositional cycle in enclosed basins or from the thickest part of the wedge in a margin basin open to the sea on one side. Referring to a cross section of the wedge concept shown in Figure 2A, it is evident that if the portion of the basin studied is too far updip, past the pinch-out of

124

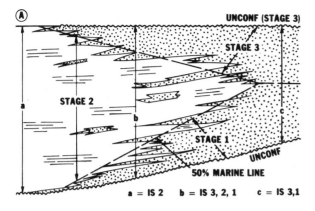

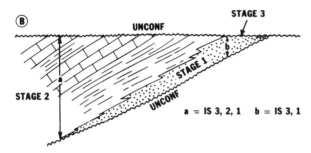

FIG. 2—Relation of stages to sedimentary wedges. (A) Sedimentary wedge showing transgressive wedge base (stage 1), wedge middle (stage 2), and regressive wedge top plus unconformity (stage 3). Dashed median line separates stages. (B) Sedimentary wedge where unconformity has cut out stage 3 and part of stage 2.

the marine wedge middle at c, only the nonmarine rocks of stages 3 and 1 will be described. Conversely, section a of Figure 2A shows that by going too far downdip, especially in continental-margin cycles, only the stage 2 marine wedge may be described.

BASIN-FORMING TECTONICS

Basin-forming tectonic style is the second major element used in classification. Figure 1 shows the identification key used in distinguishing the various basin- or cycle-forming tectonic styles. The three parameters used in the general identification of these cycle types are shown on the left in Figure 1. The composition of the crust underlying the basin is the first parameter necessary for identification. There should be no difficulty in determining crustal composition if dealing with true continental craton or true oceanic crust. Intermediate crustal composition may present a problem that can generally be resolved (Green, 1977). For the purpose of cycle identification, it need only be known whether the basin lies on continental or oceanic crust.

The type of past plate movement involved during the cycle or basin formation is the second parameter used in

Figure 1. Fundamentally, there are two types of plate movements affecting cycle or basin formation: (1) divergence and (2) convergence. It has been argued that transform movements are a third type, but these rarely show perfect side-by-side motion and generally exhibit some divergence or convergence. Small angles of convergence show up as wrenching or foldbelts, and small angles of divergence appear as normal faulting or sagging. Consequently, transform movements are not specified in the basin classification.

Convergence normally affects the margins of actively colliding plates, particularly of overriding plates. Strong convergence may be transmitted into the interior of cratonic plates along major shear zones (episodic wrenches), deforming basins well into the interior and away from the convergent margins. Convergent and divergent margins are found on both continental and oceanic crustal plates.

The third parameter used in Figure 1 is basin position on the plate (continental interior or margin) and primary structural movement (sagging, normal faulting, or wrench faulting). These combinations give rise to the theoretical model of 10 simple basin or cycle types (shown in parentheses in Fig. 1). Two of these (OTA and OF) are not considered at this time to be prospective for hydrocarbons and are omitted from the practical model, which has eight simple basin types. These eight cycle or basin types are the ones used in this basin classification. A cycle, or simple basin, is normally referred to by its abbreviated letter, thus margin sag cycles are MS, wrench or shear are LL (lateral), oceanic sags are OS, and so forth.

From the standpoint of petroleum exploration, the greatest number of cycles fall into four major categories and the rest into four minor categories. The major categories are interior sag (IS), interior fracture (IF), margin sag (MS), and wrench (LL). The minor categories are trench (T), trench associated (TA), oceanic sag (OS), and oceanic wrench (OSLL).

Globally, most of the hydrocarbons discovered to date have been found in the four major cycle categories which are associated with continental crust. All the basins within this classification consist of one or more of these cycles, plus modifying structural events. For example, a foldbelt, which is complete folding of a basin or part of one, is not considered to be a basin or cycle type, but is treated as a tectonic event that affects preexisting basins. Some of these cycle types may be split into variations for play analysis and evaluation.

Divergent Cycle Types

Divergent plate movements are the underlying cause of several important basin types. These may be single cycle (simple) basins or multicycle polyhistory basins. Models for various types of divergent fracture basins have been discussed by Shatskii (1946a, b), Klemme (1971), Lowell and Genik (1972). Margin and interior sag basin types have been discussed by numerous writers including Shatskii and Bogdanoff (1960) and Klemme (1975).

Interior sag cycles/basins (IS).—These cycles or basins are defined as being located entirely on continental crust in areas of divergence. They are found in the interior of con-

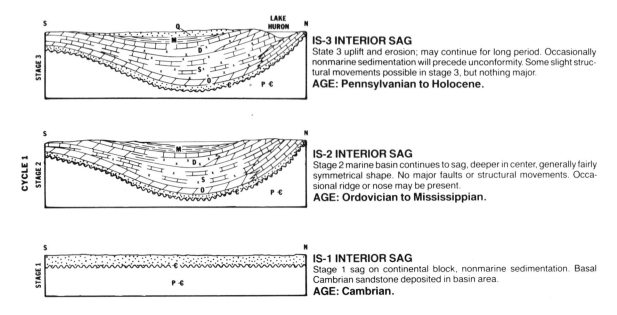

IS-3 INTERIOR SAG

State 3 uplift and erosion; may continue for long period. Occasionally nonmarine sedimentation will precede unconformity. Some slight structural movements possible in stage 3, but nothing major. **AGE: Pennsylvanian to Holocene.**

IS-2 INTERIOR SAG

Stage 2 marine basin continues to sag, deeper in center, generally fairly symmetrical shape. No major faults or structural movements. Occasional ridge or nose may be present. **AGE: Ordovician to Mississippian.**

IS-1 INTERIOR SAG

Stage 1 sag on continental block, nonmarine sedimentation. Basal Cambrian sandstone deposited in basin area. **AGE: Cambrian.**

FIG. 3—Evolution of interior sag basin illustrated by restored cross sections in three stages. Example from Michigan basin.

tinental masses, not at the plate margin, and if near the edge their axes are generally at a significant angle to the margin. Interior sag basins (IS) are normally more or less circular or oval in shape and generally do not accumulate as great a thickness of sediments as continental margin basins. They are formed by simple sagging of the continental crust as shown in Figure 3. Many interior sag basins originated in the Paleozoic. Some are simple or single cycle basins, whereas others contain several repeated sag cycles. Figure 3 shows the evolution of a typical interior sag (IS) basin. The basin simply sags with minor or no faulting. Interior sag cycles are commonly found in polyhistory or multiple cycle basins.

Interior fracture cycles/basins (IF).—This basin/cycle type is defined as being found on continental crust, either in the interior of present plates or at the crustal margins of old continental plates. Interior fracture basins are caused by divergence and tension within the continental block. Vertical horst-and-graben faulting and subsidence are the dominant features.

Figure 4 is a series of restored cross sections showing the Gulf of Suez, a typical interior fracture basin. Cycle 1 shows basin formation in the Early Cretaceous, with tension block faulting, and subsidence. Stage 1 nonmarine sandstone and shales are deposited. During stage 2, marine rocks filled the Suez graben as block faulting, and subsidence continues. Reservoir sands and carbonates are deposited over the highs while shales fill the lows. Continued subsidence and compaction cause drape of beds over high-standing fault blocks. During the third or final stage of interior fracture the basin fills or is uplifted, allowing nonmarine wedge-top sediments to be deposited; unconformities may truncate the top of the section. If this cycle is repeated with additional faulting and deposition, the succeeding cycle still would be classified as an interior

fracture basin. Should the structural style change to something else, it becomes a polyhistory or multiple-cycle basin.

Margin sag cycles/basins (MS).—Margin sag cycles/ basins are located on the outer edges of continental crust blocks in areas of divergence. The basin axes lie parallel with the continental/oceanic crust boundary, and the sediments may overlap onto oceanic crust. Such basins are referred to as being located on "Atlantic-type" margins. All margin sag basins have at least two basin-forming tectonic origins and are polyhistory basins.

The evolution of a typical margin sag basin is shown in Figure 5. The basin begins with the cracking of a cratonic mass by divergence. This first phase (stage 1 of cycle 1) as previously outlined is called interior fracture, and may resemble the present-day rift valleys of Africa. These grabens generally are filled with nonmarine sediments. During graben formation, basal block fault structures are formed and buried by sediments. Probably no continental separation occurs at this time, though deeper fracture zones may have extended through the crust. There is generally no stage 2 in this cycle, inasmuch as the continents have not separated enough to allow a marine incursion.

The next phase of basin formation, stage 3 of cycle 1, is the end of interior fracture basin development. Continued continental divergence and graben subsidence by block faulting are typically accompanied by nonmarine deposition. One or more deep fracture zones may form as potential separation centers. These fracture zones may fill with basalt intrusions and dike swarms. Stage 3 of cycle 1 normally ends with a major unconformity. Basement block faults cease differential movement near the end of the interior fracture cycle, and overlying beds generally do not show any rejuvenated movement or structure. It may be that continental rifting freezes the basement block faults

126

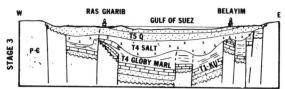

IF-3 INTERIOR FRACTURE
Stage 3, continued block faulting and subsidence. Nonmarine deposition as basin fills (deposition overcomes subsidence). Faulting not generally observed in stage 3, just sag or subsidence.
AGE: Pliocene (T5) to Holocene.

IF-2 INTERIOR FRACTURE
Stage 2, tension and block faulting continue, graben system deepens to form basin. Marine waters invade basin. Marine beds deposited as structures form; sandstone is deposited and limestone reefs form over highs; shale (marl) accumulates in lows. Marine stage 2 ends with salt deposit.
AGE: Middle Cretaceous - T4 (Miocene).

IF-1 INTERIOR FRACTURE
Stage 1, tension and rifting within continental block. Interior graben system develops by block faulting. Depressions filled with nonmarine clastics.
AGE: Early Cretaceous.

FIG. 4—Evolution of interior fracture basin illustrated by restored cross sections in three stages from Early Cretaceous to Holocene. Example from Gulf of Suez.

on either side; but whatever the cause, at this time the type of structural origin changes from interior fracture to margin sag.

Margin sag is initiated as the spreading center in the interior fracture grabens is activated and begins to grow. The continents separate and begin to move apart. Basement faults are no longer independently active, and basement begins to subside as one block. The entire edge of the continent sinks. Simultaneously, stage 1 of margin sag cycle 2 commonly begins with deposition of nonmarine beds and

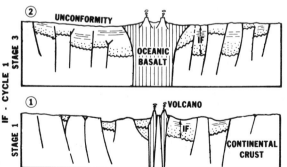

MS-2 MARGIN SAG (Tertiary)
Marine deposition, more tilting and subsidence of coastal basin seaward. Clinoform deposition of clastic wedges into deepening Tertiary ocean.

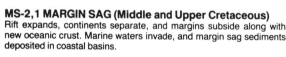

MS-2,1 MARGIN SAG (Middle and Upper Cretaceous)
Rift expands, continents separate, and margins subside along with new oceanic crust. Marine waters invade, and margin sag sediments deposited in coastal basins.

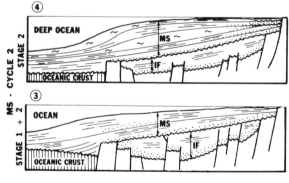

IF-3 INTERIOR FRACTURE (Lower Cretaceous)
Continued graben subsidence and fill, rift dike area expands, block faulting still active. Cycle 1 ends with major unconformity.

IF-1 INTERIOR FRACTURE (Jurassic)
Divergence causes rifting of continental block. Grabens fill with nonmarine sediments.

FIG. 5—Evolution of continental margin sag basin.

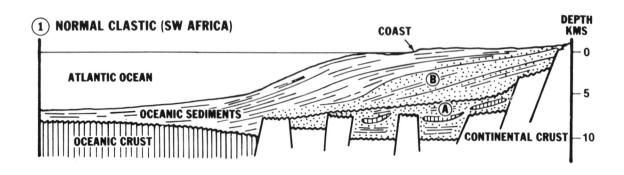

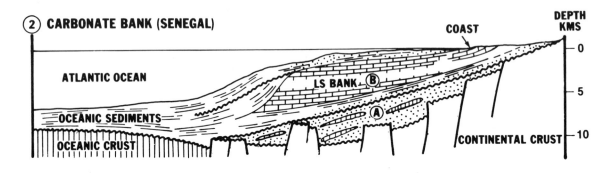

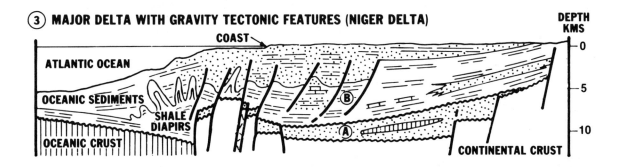

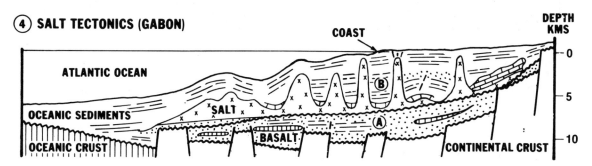

FIG. 6—Examples of four main types of continental margin sag basins classified according to cycle 2 marine fill: (1) normal clastic, (2) carbonate bank, (3) major delta, (4) salt tectonics. Most divergent basins are in one of these four categories or in combinations of two or more. For example, Tarfaya (Morocco) basin contains combination of Triassic salt (type 4), Jurassic carbonate bank (2), and Lower Cretaceous delta clastics (3). A = lower cycle 1 nonmarine series; B = upper cycle 2 marine rocks. Adapted from Beck and Lehner (1974).

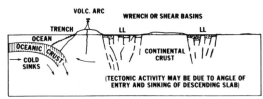

2. MIOCENE TIME
  A. Arching of upper slab. Basin formation.
  B. Trough collects deep-water sediments (some volcanic).
  C. Volcanoes on inside arc.
  D. Wrench or shear basins initiated by block faulting and differential plate movement.

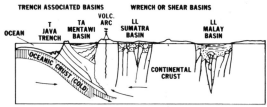

4. PRESENT DAY
  A. Trench sediments deformed. Thrusting produces new non-volcanic arc and associated basin.
  B. Volcanoes inside arc.
  C. Strike-slip basins continue to fill and deform by wrench-fault couplet.

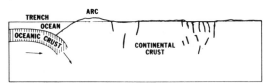

1. MIDDLE TERTIARY
  A. Convergence of 2 plates (oceanic + continental) subduction begins.
  B. Trench and arc are formed (downbending of oceanic plate causes tension in overriding plate; or cold oceanic plate simply sinks under upper plate, and no compression results).

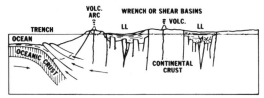

3. PLIOCENE TIME
  A. Trough continues to collect sediments which are continuously folded and thrust faulted. This is only area of compression.
  B. Volcanoes inside arc.
  C. Strike-slip tension basins fill first with nonmarine clastics, later with marine sediments. Basins sporadically wrenched (structured) as they fill.

FIG. 7—Evolution of convergent basins (shear and trench). Example from Java-Sumatra.

occasional minor evaporites.

Stage 2 of margin sag includes continued subsidence and continental separation, and is identified by marine deposition. Marine waters invade the infilled graben system for the first time, depositing clastics, carbonates, or massive salt at the base of the series. As the oceanic spreading center expands, the continents separate. New layers of oceanic crust form, and the older ones cool.

On many present-day divergent continental margins, there is evidence of very rapid subsidence and oceanward tilting of the basins, beginning in the Early Cretaceous and accelerating in the Tertiary. Marine sediments on the open shelf mark periods of highstands of sea, and nonmarine sediments and paleoslope unconformities mark the lowstands. As the ocean basins become deeper, clinoform deposition into the deep water dominates all previous patterns. The continental crust margin may appear to have subsided more than the adjacent oceanic crust, probably because of sediment loading. If salt is present, salt diapirs may intrude during the later stages of the basin formation. Delta deposits may accumulate at the mouths of rivers, with attendant gravity features, shale diapirs, and growth faults.

Margin sag basins have been divided into various general types by Beck and Lehner (1974). These are normal clastic, carbonate bank, major delta, and salt tectonics; they are illustrated in Figure 6. Most of the more than 100 margin sag basins we have identified can be placed in one of these four groups. The four margin sag basin types, however, are not subdivided from the margin sag category; all are

referred to as margin sag (MS) basins or cycles. The advantage of these groupings is in their oil-play association (see Kingston et al, 1983).

One variation of the margin sag cycle should be noted separately—it comprises "old" continental margin basins that, as suggested by depositional evidence, have been converted into interior basins by the subsequent formation of a foldbelt on the seaward side. They were margin sag cycles transformed into interior sag cycles by orogeny.

In many situations the evidence for margin sag origin is only partial, but we assume that these basins once were margins sags because any evidence to the contrary was destroyed by the formation of the foldbelt. The main reason for the separation and identification of margin sag–interior sag (MSIS) cycles is that they are the classic "asymmetrical basins" and, worldwide, many are prolific producers of hydrocarbons.

Margin sag–interior sag (MSIS) cycles were generally deposited on broad, gently dipping continental platforms with sheet sands and carbonate deposits more closely comparable to those of interior basins than those of present-day narrow continental margins. Sediments deposited in these old margin sag–interior sag (MSIS) cycles are marginal, i.e., they generally grade from coarse to fine in the paleosea direction, and show no structural or stratigraphic evidence for the existence of the other side of the basin. No clastics were being introduced from the seaward side; this indicates the existence of an "old ocean." After the foldbelt has transformed the cycle into an asymmetric interior sag basin, it is designated MSIS to differentiate it from cycles deposited as interior sags and also from modern margin sag basins.

Convergent Cycle Types

Basins classified in this category are those formed on margins or nearby interiors of two or more plates converging toward one another. Most basins on converging plates mainly exhibit tensional features. Figure 7 shows a simplified version of the relative positions and development of basins on convergent plate margins. Models for basin formation on convergent margins are well known. Seely et al (1974) have discussed trench, forearc, and backarc basins. Carey (1958), Freund (1965), Harding (1973), and Crowell (1974) have discussed various aspects of shear or wrench basins. Carey (1958) described small ocean basins.

Wrench or shear cycles/basins (LL).—These strike-slip, shear, or wrench basins are referred to here as double "L" cycles (LL), for lateral movement. These cycles are found on continental or intermediate crust. For this classification, hydrocarbon-prospective wrench basins are restricted to those found on or directly adjacent to continental crust. Wrench basins/cycles are found in areas of two or more converging plates. They are formed by a divergent wrench couplet with strike-slip faults along two or more sides, as shown in Figure 8.

Most wrench couplet or shear basins are found in the areas of present-day, or Tertiary, plate convergence. Typical areas where these basins occur are: (1) the periphery of the Pacific Ocean, including Antarctica; (2) southeast Asia; (3) the Himalayan-Alpine chain from the Solomon Islands to Spain; and (4) around the Caribbean (Antilles arc).

For most of these shear basins, stage 1 (basin initiation by divergent wrenching) appears similar to an interior fracture (IF) basin, with tensional block faulting and little or no evidence of wrenching. Nonmarine wedge-base sediments are deposited, unless the basin is initiated under water.

Sometime after the basin is initiated in stage 1 or 2, wrench deformation of the basin begins. Wrench structures may form along the flanks or within the basin. Major wrenches outline the basin, and minor wrenches form structures within the basin. If the basin is near enough to the ocean, marine sediments may be deposited; many wrench basins have no marine fill, being too far from the sea to have had invasions of marine waters.

In stage 3 of the wrench (LL) cycle, the basin is uplifted and eroded, and the continued shearing may begin to destroy the structures and parts of the basin. The term "L3FB" is used for the final stage of a basin completely wrenched to a foldbelt. Continued plate convergence may result in orogeny.

If the wrench-faulting process continues over a long enough period, it will eventually destroy the LL basin. As a result, few examples are found of preserved wrench basins older than Tertiary. Typically, pre-Tertiary wrench (LL) basins occur as wrenched foldbelts, with only fragments of the destroyed basins being recognized. However, if convergence between the two plates and, therefore, the engine driving the mechanism, has stopped at some inter-

mediate interval, the basins existing at that time may be preserved. Some portions of the wrench basins of Oklahoma (Ardmore) are believed to represent different stages of arrested wrenching. Sedimentary fill in wrench (LL) basins is extremely varied. Marine clastics, carbonates and evaporites, nonmarine clastics, volcaniclastics, and flysch-chert-ophiolites are all found, depending on depositional conditions.

Trenches (T) and trench-associated (TA) basins.—These categories of the classification system are relatively minor from a hydrocarbon-prospect standpoint. Results from past exploration have been poor, and it appears these basins have very little future prospect of containing commercial hydrocarbons.

Trench-associated cycles/basins (TA), as shown in Figure 7, are located on convergent continental plate margins, landward of the trench or nonvolcanic arc, if one is present. The basins generally are built on folded trench sediments, not continental crust, and are formed by a simple sag, often deformed by contemporaneous wrenching. Trench-associated (TA) basins are likely to be filled with a high percentage of volcaniclastic sediments, though quartz or arkosic sands may be found, given the proper nearby sediment source area. Sediments in these basins generally are marine, though some have been found to contain nonmarine materials in the lower stage (Abukuma basin of Japan). Other trench-associated basins appear to have subsided rapidly and are filled with deep-water sediments in the lower stage. Thus, two basins of diverse origins are included in the trench-associated (TA) category because of their location. Trench-associated (TA) basins easily may be confused with wrench (LL) basins. Forearc basins (Seely et al, 1974) are included in this trench-associated category. In our global basin classification system, we find many "forearc basins" with no arc association (Sabah, Palawan, Philippines, etc). As a consequence, we prefer to use the term "trench-associated basins."

Trenches (T) are located (1) on oceanic crust, and (2) at the margins of two or more converging plates. A subduction zone is formed with the trench being the "bent" portion of the lower plate, as shown in Figure 7. Present-day trenches are relatively narrow downwarps located over subduction zones. Active trenches are areas of low heat flow and are Tertiary in age; older trenches have been converted into foldbelts.

We recognize two types of trenches. The first involves one oceanic plate overriding another, forming a midocean trench such as the Mariana, Aleutian, and Philippine trenches. These features normally have little sediment fill, and the amount they do have primarily is volcanogenic and deep-water pelagic. The second type involves an oceanic plate overridden by a continental plate. This trench can receive oceanic pelagic sediments and volcanics, as well as land-derived fine clastics. These ocean-margin trenches accumulate very thick deep-water deposits. Continued convergence with its attendant compression and shearing, eventually results in a foldbelt. Most trenches and trench-associated basins end up as foldbelts, and such folded trench sediments are given the designation "FB3T."

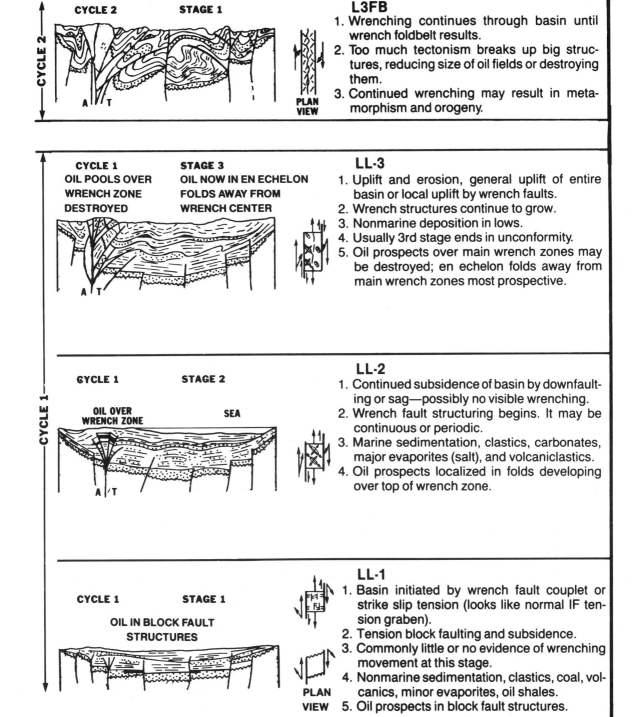

L3FB

CYCLE 2   STAGE 1

1. Wrenching continues through basin until wrench foldbelt results.
2. Too much tectonism breaks up big structures, reducing size of oil fields or destroying them.
3. Continued wrenching may result in metamorphism and orogeny.

PLAN VIEW

LL-3

CYCLE 1
OIL POOLS OVER
WRENCH ZONE
DESTROYED

STAGE 3
OIL NOW IN EN ECHELON
FOLDS AWAY FROM
WRENCH CENTER

1. Uplift and erosion, general uplift of entire basin or local uplift by wrench faults.
2. Wrench structures continue to grow.
3. Nonmarine deposition in lows.
4. Usually 3rd stage ends in unconformity.
5. Oil prospects over main wrench zones may be destroyed; en echelon folds away from main wrench zones most prospective.

LL-2

CYCLE 1   STAGE 2

OIL OVER WRENCH ZONE   SEA

1. Continued subsidence of basin by downfaulting or sag—possibly no visible wrenching.
2. Wrench fault structuring begins. It may be continuous or periodic.
3. Marine sedimentation, clastics, carbonates, major evaporites (salt), and volcaniclastics.
4. Oil prospects localized in folds developing over top of wrench zone.

LL-1

CYCLE 1   STAGE 1

OIL IN BLOCK FAULT STRUCTURES

1. Basin initiated by wrench fault couplet or strike slip tension (looks like normal IF tension graben).
2. Tension block faulting and subsidence.
3. Commonly little or no evidence of wrenching movement at this stage.
4. Nonmarine sedimentation, clastics, coal, volcanics, minor evaporites, oil shales.
5. Oil prospects in block fault structures.

PLAN VIEW

FIG. 8—Evolution of wrench or shear basins. LL = lateral. Development of convergent plate wrench in three stages (LL-3, 2, 1). L3FB shows final step in many wrench basins, that of foldbelt caused by wrenching. This is final step in process that creates basin and finally destroys it. After initiation of basin (LL-1), movement may cease in any succeeding stage. Basin may also stop strike-slip mode and change to polyhistory basin.

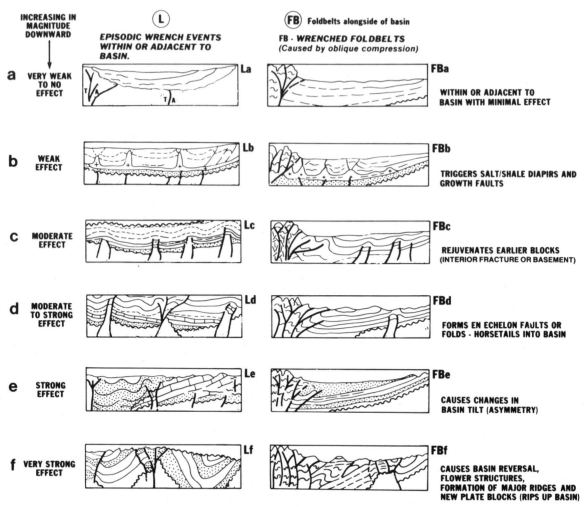

(L) *EPISODIC WRENCH EVENTS WITHIN OR ADJACENT TO BASIN.*

(FB) Foldbelts alongside of basin
FB - WRENCHED FOLDBELTS
(Caused by oblique compression)

a VERY WEAK TO NO EFFECT

La

FBa

WITHIN OR ADJACENT TO
BASIN WITH MINIMAL EFFECT

b WEAK EFFECT

Lb

FBb

TRIGGERS SALT/SHALE DIAPIRS AND
GROWTH FAULTS

c MODERATE EFFECT

Lc

FBc

REJUVENATES EARLIER BLOCKS
(INTERIOR FRACTURE OR BASEMENT)

d MODERATE TO STRONG EFFECT

Ld

FBd

FORMS EN ECHELON FAULTS OR
FOLDS - HORSETAILS INTO BASIN

e STRONG EFFECT

Le

FBe

CAUSES CHANGES IN
BASIN TILT (ASYMMETRY)

f VERY STRONG EFFECT

Lf

FBf

CAUSES BASIN REVERSAL,
FLOWER STRUCTURES,
FORMATION OF MAJOR RIDGES AND
NEW PLATE BLOCKS (RIPS UP BASIN)

FIG. 9—Tectonic modifiers of primary basin types (L and FB).

Oceanic sags (OS).—Oceanic sags are areas where oceanic crust has been formed by continental separation and the formation of a spreading center. Subsequent cooling and subsidence of the oceanic crust have caused it to sink. The sag areas that have accumulated a significant thickness of sediments are called oceanic sag (OS) basins. Most of these thick sediment areas are near continental blocks or island arcs and may be associated with either convergence or divergence. The sediments consist of pelagic material, volcaniclastics, and distal turbidites. Like the trench and trench-associated basins, oceanic sags appear to have very limited hydrocarbon potential.

Oceanic sags–wrench couplet cycles/basins (OSLL).—These are the mini-oceans or small ocean basins, floored by oceanic or transitional crust. Such basins are believed to have been formed by a large divergent wrench couplet (rather than mantle upwelling and sea-floor spreading) that opened a mini-ocean through separation of the sialic layer. These basins may be filled with pelagic materials, volcaniclastics, or distal turbidites, as in oceanic sags (OS) basins.

BASIN-MODIFYING TECTONICS

The third major element used to classify basins is basin-modifying tectonics. Basins or cycles formed by one type of tectonic movement may be changed during their history by other structural events. There are three types of basin-modifying tectonics: episodic wrenches (L), adjacent foldbelts (FB), and complete folding of a basin area (FB3) which is foldbelt formation.

Definition of Episodic Wrenches and Foldbelts

Episodic wrenches are designated by the single letter "L," and represent a wide variety of lateral movements not connected with basin/cycle origin. Episodic wrenches modify basins formed by other means and are found in basins with all ages of basement rocks. It is believed that old zones of weakness in the basement, such as old sutures, interior fracture zones, plate boundaries, etc, move periodically or episodically in response to plate movements. These plate movements are manifested at the surface by plate collisions, rotations, fragmentation, and by subduction zones. Wrench faulting or rotational movements are prevalent in basin histories on all continents and react in different ways at different times. Crustal blocks floating on the asthenosphere may be similar to the jostling of ice flows on the polar seas.

Generally, the origin of an episodic wrenching or lateral L movement is fairly easy to ascertain, given good plate-tectonic reconstructions. In some places, however, one cannot directly trace the originating event for an L movement. The event which triggers isolated wrench movements may be on some other side of the plate in question, or even on some other part of the earth. The effect of varying intensities of L movements is shown in Figure 9.

Foldbelts are caused by convergence of two or more plates. Basin areas caught in this convergence may be completely or only partly folded. Basins not completely folded are not considered to be foldbelts and are said to have been episodically wrenched. Basins completely folded are called foldbelts (FB3). Foldbelt formulas are derived from the basin that was folded. If the edge of a large basin is folded, the formula for the foldbelt is the same as that of the basin except that the youngest event will be some type of folding. Completely folded rocks, or foldbelts (FB3) are commonly found adjacent to relatively unfolded basins and, in fact, may grade into basins. The expression used in the basin formula to denote an adjacent foldbelt is "FB." Foldbelts have varying effects on adjacent basins, as shown in Figure 9.

Tectonic Modifiers

Wrenches (L) and Adjacent Foldbelts (FB)

The tectonic modifiers of primary basin types are listed in Figure 9 in order of increasing magnitude, downward from "a" to "f." Each of these effects is found associated with both episodic wrenches (L) and adjacent foldbelts (FB). The very weak "a" effect is known to occur within or adjacent to a basin with minimal structural effect. La would mean that a wrench passed through or adjacent to a basin but caused no faulting or folding visible at the surface or on seismic reflection. Porosity and permeability, however, may be affected. FBa would mean that a foldbelt was formed on the side of a basin but had no effect, of faulting or folding, on the basin itself.

Lb and FBb illustrate the "b" effect—still very weak on the scale. The "b" effect triggers salt or shale diapirs and growth faults within the basin and can cause open folds in basins adjacent to FBb. We believe that without a tectonic event of "b" intensity or stronger, salt and shale may not be triggered to flow. The movement could be described best as "jiggling." There are numerous examples, worldwide, of basins with thick salt layers and plenty of load that have never flowed to produce domes (e.g., Yakutsk, Touggourt, Permian basins). Nor do they exhibit any other evidence of post-salt deposition structuring. It can be concluded that slight plate motions or jiggling are required to initiate or trigger salt and shale diapirs and growth-fault movements. Salt and shale diapirs, once triggered by an Lb event, could continue to grow by static load

without further jiggling to keep them moving.

Lc and FBc illustrate the "c" effect which is rejuvenation of preexisting blocks, either interior fractures or basement. This jostling of older blocks can cause structural growth which may or may not reach the surface. Many of the world's giant fields owe their structural growth to "c" effects. It is important to note that the "a," "b," and "c" effects of domes or arches may not reach the top of the structured cycle, which was also the old ground surface. The first modifier to reach the old ground surface as wrench-generated faults and folds is the "d" effect, which is rated as a moderate to strong event. It is convenient to divide the list of modifiers into the ones causing "weak" effects (a, b, c) and the ones causing "strong" effects (d, e, and f).

Ld and FBd are examples of classic wrenching. Relative plate movement is enough to cause en echelon faults or folds to be well developed. Here are found the first flower structures recognizable on seismic records or visible on the surface. Horsetails (a series of en echelon faults or folds) may be seen as fanning out of foldbelts or wrench zones into basins. The Ld flower structures are fairly modest ones and do not bring basement to the surface, as they do in stronger wrench events. Le and FBe are strong plate tectonic effects which significantly alter basin tilt, causing marked basin asymmetry. Tilt in one direction, or change of tilt direction, is an "e" effect.

Lf and FBf effects are the strongest episodic wrench events we record in a basin. Basins are turned inside out or "reversed" with synclines becoming anticlines. Flower structures bring basement or very old rocks to the surface. These "f" effects also see the formation of major ridges or arches in a basin, the breakup of plates under basins, and consequent formation of new smaller basins out of the old megabasin. The basin can be ripped up extensively and still be called an "f" effect. However, if the basin is completely folded, destroyed, or altered by faulting, we would go one step further than "f" and call it a foldbelt (FB3).

It should be noted that any of the characteristic tectonics of the weaker modifiers may be found in any stronger one. For example, salt and shale diapirs and growth faults (Lb) may be found along with rejuvenated block movements (Lc). All of the tectonic parameters, salt domes, jostled blocks, flower structures, basin tilts, etc, may be found in Lf.

Several important points need to be emphasized concerning L and FB events. First, the tectonic modifiers affect, in varying degrees, basins already formed by other processes. Second, an episodic wrenching zone (L) can turn into a wrenched foldbelt (FB) along its length, as a matter of degree of wrenching. We believe that most foldbelts are caused by wrenching movements or convergence at some oblique angle (other than 180°). Third, the modifiers are described as to what effect they have on the basin, not on the wrench or foldbelt. For example, a wrench alongside a basin may be very disruptive locally but have little or no effect on the basin itself. Similarly, a foldbelt alongside a basin may be vaulted or highly deformed mountains but have little structural effect on the adjacent basin. It is the structural effect on the basin that is described in Figure 9. If the foldbelt is hydrocarbon-prospective, it will have its own description.

1. FB3U

HIGH OR UPLIFTED MOUNTAINS, IMBRICATE STRUCTURE

2. FB3B

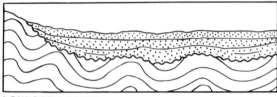

**LOW OPEN FOLDS
(MAY BE BURIED)**

3. FB3F

**LOW COMPLEX FOLDS,
IMBRICATE STRUCTURE
(MAY BE BURIED)**

4. FB3T

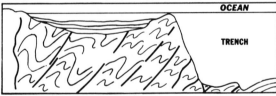

FOLDED TRENCH SEDIMENTS

5. L3FB

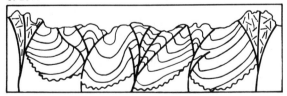

**WRENCHED TO A FOLDBELT WITH
HORSETAILS USUALLY AFTER LL-321**

6. FB3 - BASIN COMPLETELY FOLDED, TYPE UNKNOWN

FIG. 10—Foldbelt types (FB3).

Foldbelts represent sutures where past plates have converged or are still converging. This convergence results in compression and shearing motions that cause the rocks to be wrenched and folded. If ultrabasic rocks, serpentine, chert, volcanic flysch, and other oceanic sediments are found in foldbelts, it is assumed that an old area of oceanic crust was destroyed by subduction or plate collision, and the foldbelt suture is all that remains of the vanished oceanic plate.

The six different types of foldbelts (FB3) are shown in Figure 10. They are: (1) FB3U, the uplifted foldbelts or high-mountain ranges, generally having imbricate structure, such as the Canadian Rockies; (2) FB3B, the topographically low, open folds, which may be partly or almost completely buried by younger sediments, as in part of the Iranian foldbelt; (3) FB3F, the topographically low but complex folds with imbricate structure, which may be partly or almost completely buried by younger sediments, as in the Vienna basin; (4) FB3T, folded trench sediments, found on convergent plate margins; (5) L3FB, wrenched to a foldbelt (can have horsetails), invariably a former LL basin; and (6) FB3, basin completely folded, specific type unknown. Of the six foldbelt types listed, only three—FB3B, FB3F, and L3FB—presently produce commercial hydrocarbons.

BASIN CLASSIFICATION

All basins, worldwide, may be classified by using the structural and stratigraphic elements previously discussed. It is possible to combine these elements to make a formula for each of the basins within the system. The formula is a general expression of the basin's structural and stratigraphic history. It does not describe unit thickness, rock color, source, reservoir, grain sizes, and many other factors necessary for basin analysis. Therefore, there is no magic formula which can separate sedimentary basins into oil- and gas-prone versus barren. Formulas are simply useful means for summarizing the important points in a basin's structural and depositional history. In classifying a basin and writing its formula, it is important to outline only the main events in the basin history, and not to attempt to describe all detail. Too much detail results in long and needlessly complicated formulas, which are difficult to use.

Basins may be classified by comparing basin parameters, as shown in Figure 11. The data needed are regional maps and cross sections, well logs or surface sections, and regional seismic lines, if available. From the maps and cross sections, the geologic and plate tectonic history of the basin may be deduced. Major unconformity breaks within the cross sections should be restored to the old paleosurface, and a series of historically restored cross sections should be made. From these, the basin history can be broken down into cycles, stages, and tectonic events. These may be compiled as shown in Figure 11 to derive the proper basin classification and formula.

Figure 11 shows how the (1) cycles, (2) stages, and (3) tectonic events of a polyhistory basin are combined to create

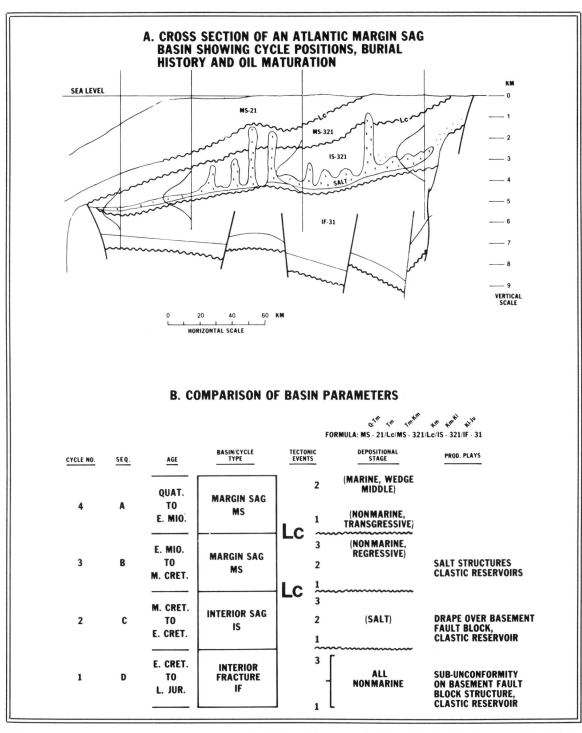

FIG. 11—Comparison of basin parameters using example (A) from Atlantic margin. Formula for basin is given at top right of B. In left column, cycles are numbered in ascending order, oldest to youngest; this gives a system of identifying geologic events as they developed during history of basin. Second column, labeled "sequence," is inversion of column 1, using letters; this enables identification of cycles or events from youngest to oldest as they would be encountered in drilling a well. Third column ties cycles to geologic age. Fourth column (boxes) gives gross cycle type and abbreviation. Next column shows tectonic events affecting basins, their strength, and when they occurred in relation to other events. Sixth column shows depositional stages of wedge top (3), middle (2), and base (1), with or without unconformity. Right column shows hydrocarbon-producing plays.

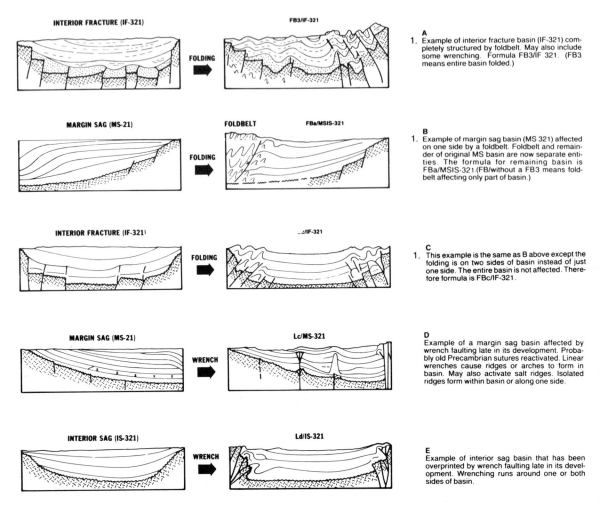

FIG. 12—Examples of polyhistory basins, showing simple divergent basins overprinted by foldbelt and wrenching tectonics.

the basin formula. Figure 11A is a cross section of the basin to be formulated. Figure 11B translates the cross section events into the formula. It should be noted that cycles are ended by any of four events: (1) a change in cycle type (basin-forming tectonics); (2) the occurrence of a significant wrench or folding event, such as L, FB, or FB3 (tectonics affecting basins); (3) a major sedimentary transgression and regression; or (4) a regional unconformity which may be caused by L, FB, or FB3 events but commonly occurs without them.

Viewing the events in the example shown on Figure 11, as they would become apparent in drilling a well, the youngest cycle is a margin sag (MS) which is still in stage 2 because the continental margin is under water. The marine wedge overlies a wedge base of early Miocene age and an unconformity; this unconformity corresponds to an episodic wrench which may have, in part, caused the unconformity. This Lc event reactivated the salt domes. The next older event is another margin sag (MS) cycle with wedge top, middle, and base (3,2,1) followed by another Lc event in the middle Cretaceous. This event initiated the salt

dome movement in the basin and probably affected the unconformity. The next cycle down is an interior sag, IS-321, with thick salt in the stage 2 portion. The bottom cycle, an interior fracture, IF-31, has an unconformity at the top which is not related to L or FB events. The oldest cycle is entirely nonmarine. Combination of these cycles, stages, and events results in the formula being written youngest to oldest as follows: MS-21/Lc/MS-321/Lc/IS-321/IF-31. The abbreviated ages of each cycle may be written above the formula, as shown in Figure 11B.

Figure 12 shows examples of how single-cycle basins are affected by foldbelts and wrenching, and how the formulas are written.

Figure 13 is a series of cross sections showing the step by step development of the Persian Gulf basin, and its formula. The basin starts out in the early Paleozoic as an interior sag, as evidenced by the Hormuz Salt. The next cycle appears to be a margin sag, existing from the Permian through the Jurassic. An Lc event at the end of Jurassic was caused probably by convergence in the Tethyan zone and jostling collision of Turkish and Iranian microplates

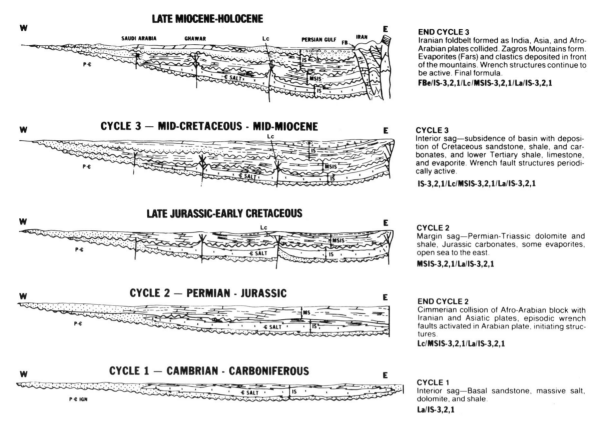

LATE MIOCENE-HOLOCENE

END CYCLE 3
Iranian foldbelt formed as India, Asia, and Afro-Arabian plates collided. Zagros Mountains form. Evaporites (Fars) and clastics deposited in front of the mountains. Wrench structures continue to be active. Final formula.

FBe/IS-3,2,1/Lc/MSIS-3,2,1/La/IS-3,2,1

CYCLE 3 — MID-CRETACEOUS - MID-MIOCENE

CYCLE 3
Interior sag—subsidence of basin with deposition of Cretaceous sandstone, shale, and carbonates, and lower Tertiary shale, limestone, and evaporite. Wrench fault structures periodically active.

IS-3,2,1/Lc/MSIS-3,2,1/La/IS-3,2,1

LATE JURASSIC-EARLY CRETACEOUS

CYCLE 2
Margin sag—Permian-Triassic dolomite and shale, Jurassic carbonates, some evaporites, open sea to the east.

MSIS-3,2,1/La/IS-3,2,1

CYCLE 2 — PERMIAN - JURASSIC

END CYCLE 2
Cimmerian collision of Afro-Arabian block with Iranian and Asiatic plates, episodic wrench faults activated in Arabian plate, initiating structures.

Lc/MSIS-3,2,1/La/IS-3,2,1

CYCLE 1 — CAMBRIAN - CARBONIFEROUS

CYCLE 1
Interior sag—Basal sandstone, massive salt, dolomite, and shale.

La/IS-3,2,1

FIG. 13—Evolution of polyhistory basin showing successive stages and formula development. Example from Persian Gulf.

against the Afro-Arabian shield. This Lc event also closed off the eastern sea margin, and changed the previous margin sag cycle to an MSIS cycle. The Persian Gulf basin remained closed from that time to the present. The final events show late Mesozoic and Tertiary interior sags, here shown as a single cycle, and the final folding of the Zagros Mountains along the eastern margin of the basin.

Figure 14 shows a form we have found useful in summarizing key characteristics of basins for classification, analysis, and assessment. The form has been completed using the Persian Gulf basin as an example. The various parameters of basin classification and assessment are listed vertically on the left of the form, whereas the geologic ages are shown horizontally, from youngest to oldest, across the top. This permits us to locate, in time, the various key parameters within a basin such as cycles and stages, basin-forming or modifying tectonics, type of sediment fill, trap types, and hydrocarbon reserve information. This form may be used to describe either an individual basin or an oil field, for comparison with others.

St. John (1980) has published a map showing the location of world sedimentary basins. We have included approximately 600 of these basins in the global classification system. It is obviously too complicated to refer to each one by its specific formula; so another description must be available for more general use. The writers have found it convenient to use the following categories. On a global basis, about two-thirds of the basins in the system may be called simple or single-cycle basins. These have only one basin-forming tectonic cycle or, if they have other cycles and tectonic events, are dominated by one type. These basins are grouped under the name of this dominant cycle such as interior fracture basins, interior sags, margin sags, wrench or shear basins, oceanic sags, oceanic wrenches, trenches, and trench-associated basins. The more complex polyhistory basins that cannot be categorized with the eight basic cycle types, make up the remaining one-third of the basins classified globally. These are referred to simply as complex polyhistory basins, and further subdivision is not proposed at this time.

Regional Cross Sections

Six regional sections, drawn across a variety of polyhistory basins, are presented in Figure 15 to illustrate examples of how the more complex basins are classified. The formula for each basin is constructed using the information available on the section. In all cases, the cross sections will be described in their order of deposition—from the bottom up. The construction of these regional cross sections is both an aid to basin classification and a useful tool in rapidly explaining the tectonic history of a basin to others.

Cross Section AA'.—This is a section of a series of sags

137

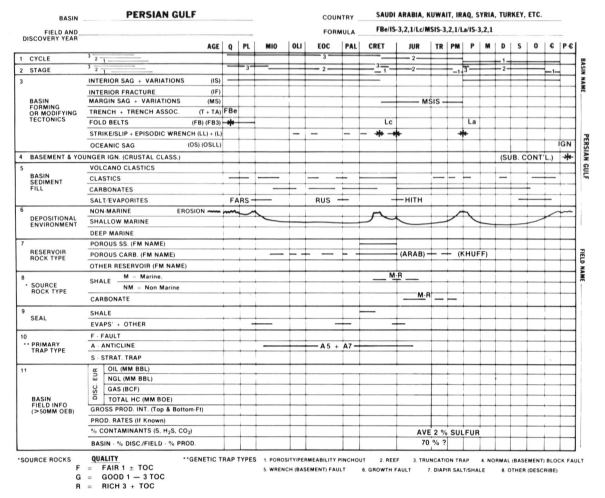

FIG. 14—Basin classification summary form.

with wrenching along one side. Three major unconformities in the section include the post-Precambrian, but exclude the present surface. The major tectonic events are associated with these unconformities so that cycles correspond to the intervals in between. The bottom cycle (1) lies on the Precambrian basement and shows no structure corresponding to the cycle. Sandstone appears to be coming into the basin from both sides with marine shale in the center. Cycle 1 appears to be an interior sag and would be written IS-321. The second cycle, which is late Paleozoic in age, also has sandstone coming in from the sides and limestone and shale in the center. This cycle appears to be another interior sag and should be written IS-321. Corresponding to the unconformity at the top of cycle 2 is an L event which affects both older cycles but not the Permian. This wrench appears to be an Ld event. The third or final cycle, of Permian age, is another interior sag, apparently with no marine stage 2, which would be written IS-31. The final formula for the cross section, written left to right, youngest to oldest, would be IS-31/Ld/IS-321/IS-321.

Cross Section BB'.—In section BB', there are three major unconformities, excluding the present surface, and the structuring is coincident with cycles or unconformities. Cycle 1 appears to be an interior fracture of Late Cretaceous age, containing sandstones for stages 1 and 3, marine shale for stage 2, and ending with an unconformity. Cycle 2, of Paleogene age, appears to be an interior sag with marine shale and limestone in stage 2 and sandstones being introduced from both sides. The cycle ends with an unconformity which coincides with wrenching activity along the right side of the cross section. The wrenching appears to be an Ld event. Cycle 3, the Neogene, is clearly another interior sag with sandstones coming in from the sides and limestone in the middle. The final formula can be written IS-321/Ld/IS-321/IF-321.

Cross Section CC'.—This cross section exhibits a reversal of tilt in the middle of its depositional history of three cycles. The bottom cycle is Late Cretaceous in age. It has a wedge base, carbonate platform with shales for stage 2, and an unconformity cutting off the wedge top for stage 3. There is no way of knowing from the cross section if this is an IS or MSIS cycle. Either would be correct. From rela-

138

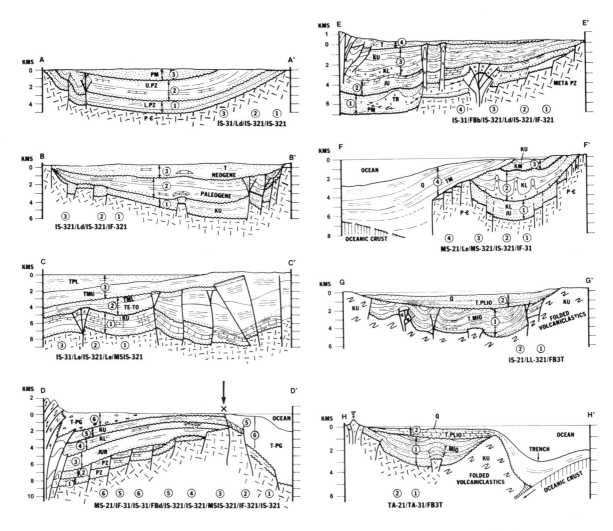

FIG. 15—Regional cross sections of complex basins, showing formulas developed from classification.

tively flat Late Cretaceous cycle, the basin tilted to the right, beginning in the early Eocene. The Paleocene may also be missing. This basin tilt corresponds to an Le event. The lower Eocene clastic wedge, thickening to the right, appears to be an IS cycle with three stages: wedge base, middle, and top plus unconformity. At the end of the early Miocene, there was another basin tilt reversing the direction to the left. This is considered to be another Le event. The last cycle of late Miocene-Pliocene would be nonmarine interior sag. The final formula for the basin would be IS-31/Le/IS-321/Le/MSIS-321.

Cross Section DD'.—Cross section DD' illustrates the complexity that may be caused when too many basin areas are included in the same formula. First, from the bottom up, there are two fairly thin Paleozoic cycles. The bottom cycle appears to be an interior sag cut by interior fractures of cycle 2. The Jurassic rocks appear to be directed into a deeper basin to the left, with no sign of the other side. We can designate this cycle MSIS-321. Both the Early and

Late Cretaceous cycles appear to be interior sags with sandstones coming in from both sides. At the end of the Cretaceous, a tectonic event affected the left side of the basin, forming a foldbelt. The effect of this foldbelt on the basin has been to change the degree of tilt but not radically enough to call it an "e"-strength event. Therefore, we would probably designate it FBd. During Late Cretaceous to Paleogene, the mountains formed, the basin subsided, and clastics were shed off the rising mountains into the basin. This cycle is another IS cycle. The formula now reads: IS-31/FBd/IS-321/IS-321/MSIS-321/IF-321/IS-321. This should have been the end of the formula, but in this extended basin area a new ocean margin formed in the Tertiary at the right side of the cross section. If these geologic events were added, the formula would read MS-21/IF-31/IS-31/FBd/IS-321/IS-321/MSIS-321/IF-321/IS-321. This cumbersome geologic-history formula illustrates one of the pitfalls of including too large an area within the basin boundaries. The solution would have

139

been to separate the area into two basins at point "X" on the section, and to give each basin a separate formula.

Cross Section EE'.—In this section, there are four major unconformities separating the cycles. At the bottom, is a Permian-Triassic interior fracture with a thick salt layer as a stage 2. No tectonic event is visible at the end of cycle 1. Cycle 2 looks like an interior sag with sandstone coming in from both sides of the basin and carbonates in the middle. At the end of cycle 2 (Jurassic), the large flower structure at the right of center was activated by wrenching (Ld) and then truncated. Possibly some salt movement was triggered at this time also, but the diapirs have younger movements which obscure the older evidence; therefore any old movements are unconfirmed. The Lower and Upper Cretaceous are lumped together into one cycle which appears to be an interior sag owing to the presence of Lower Cretaceous salt, because no evaporites are found in margin sag (MS) cycles. At the close of the third cycle is a tectonic event. A foldbelt formed on the left side that caused the salt to flow; but it does not appear to have jostled basement blocks. It probably should rate about an FBb. The final cycle is another interior sag, probably nonmarine. The final formula for the basin would be IS-31/FBb/IS-321/Ld/IS-321/IF-321.

Cross Section FF'.—In this section, a divergent continental-margin basin, four major unconformities outline basin cycles. The oldest cycle is an interior fracture with nonmarine sediments only. The second cycle is Lower Cretaceous and includes a thick salt bed. This means it must be an interior basin inasmuch as there is no salt deposition on open-ocean margins. Cycle 2 is called an interior sag, and may have ended with some L-event tectonic activity during middle Cretaceous time; however, there is no conclusive evidence on this cross section. Cycle 3, middle and Late Cretaceous, has been cut off on the left side by an unconformity. It is therefore impossible to determine, from the evidence on the cross section, if this cycle was an interior sag or a margin sag. However, because the basin is an opening continental margin, this cycle should be more marginal than the previous cycle. The best guess is that it is MS-321. The end of cycle 3 was accompanied by a strong uplift of the ridge in the basin center which changed the tilt of the basin, truncated the previous cycles, and caused salt domes to rise. This activity corresponds to an episodic wrench of Le intensity. The final cycle (4) is clearly a margin sag (MS) which occurred as the continental margin sank in the Tertiary. The final formula for this basin would read: MS-21/Le/MS-321/IS-321/IF-31.

Cross Section GG'.—The type of basin found on this cross section is commonly found around the margins of the Pacific Ocean. It is a wrench or shear LL basin overlain by an interior sag. The formula of this cross section would be IS-21/LL-321. The FB3T basement of folded trench sediments included in the formula is considered to be a key indicator of a convergent margin basin.

Cross Section HH'.—This cross section shows a convergent continental margin with volcanic arc, basin, and trench. The basin was initiated on folded volcaniclastic trench sediments in the Miocene. From the cross section, it is not possible to tell whether or not the trench-associated (TA) cycle contains a marine stage 2; because most TA cycles have marine rocks, we assume this one does also.

The basin formula would be written TA-21/TA-321/FB3T.

CONCLUSIONS

This global basin classification system identifies and compares basins in specific as well as general terms. The system is based on the genesis and evolution of basins in the context of their geologic history. The main elements used to classify basins are basin-forming tectonics, depositional cycles, and basin-modifying tectonics. Basin-forming tectonics are deduced by knowledge of the type of underlying crust, past plate tectonic history, basin location on the plate, and type of primary structural movement involved in the basin formation—such as sagging or faulting. The result is eight single tectonic-cycle or simple basin types that are termed interior sag, margin sag, interior fracture, wrench, trench, trench associated, oceanic sag, and oceanic wrench.

Basin-modifying tectonics include episodic wrenches, basin-adjacent foldbelts, and completely folded basins. These have been identified and placed on a scale of increasing magnitude, from movements of slight to major structural effects. More complex basins may contain several different tectonic cycles, plus basin-modifying tectonic events. These are called polyhistory basins. The eight simple basin types, their depositional fills, and tectonic modifiers have been given letter and number symbols so that the specific geologic history of each basin may be written as a formula. These formulas may then be compared between basins, and similarities or differences noted.

The primary uses of the global basin classification are summarized as follows: (1) to locate and identify all basins of the world in one framework (the system can expand the explorationist's viewpoint to include all possible basin types, and not just those with which he or she has had personal experience—it is an aid to exploration thinking); (2) to permit the separation of complex basins into their simple component parts, for analysis as simple units; (3) to compare plays within one or two basins of the same type, or two or more basins classified as different types; (4) to provide a system for evaluating favorable plays and risks for each basin type, because risks should be understood before venture decision; (5) to predict what geologic events must be found in a basin to improve oil prospectiveness; (6) to enhance the prediction of oil potential in unknown or little known basins by referring to known basins of the same classified type; (7) to provide a system where the paleontology, seismic stratigraphy, geochemistry, and sedimentary history of like (and different) basins can be compared and evaluated; (8) to permit location and assessment of the best specific-play areas in a basin, not just total sediment volume; and (9) to act as a vehicle for comparative assessment of hydrocarbon basins, worldwide.

SELECTED REFERENCES

Atwater, T., 1970, Implications of plate tectonics for the Cenozoic tectonic evolution of western North America: GSA Bulletin, v. 81, p. 3513-3536.
Bally, A. W., 1975, A geodynamic scenario for hydrocarbon occurrences: 9th World Petroleum Congress Proceedings, v. 2, p. 33-44.
——— and S. Snelson, 1980, Realms of subsidence, *in* A. D. Miall, ed.,

Facts and principles of world petroleum occurrence: Canadian Society of Petroleum Geologists Memoir 6, p. 9-75.

Beck, R. H., and P. Lehner, 1974, Oceans, new frontier in exploration: AAPG Bulletin, v. 58, p. 376-395.

Bois, C., P. Bouche, and R. Pelet, 1982, Global geologic history and distribution of hydrocarbon reserves: AAPG Bulletin, v. 66, p. 1248-1270.

Carey, S. W., 1958, The tectonic approach to continental drift, in S. W. Carey, ed., Continental drift: University of Tasmania, Geology Department Symposium, No. 2, p. 177-355.

Crowell, J. C., 1974, Origin of late Cenozoic basins in southern California, in Tectonics and sedimentation: SEPM Special Publication 22, p. 190-204.

Dewey, J. F., and J. M. Bird, 1970, Mountains belts and the new global tectonics: Journal of Geophysical Research, v. 75, p. 2625-2647.

Dickinson, W. R., 1973, Widths of modern arc-trench gaps proportional to past duration of igneous activity in associated magmatic arcs: Journal of Geophysical Research, v. 78, p. 3376-3389.

———— 1974, Plate tectonics and sedimentation, in Tectonics and sedimentation: SEPM Special Publication 22, p. 1-27.

Franchteau, J., and X. Le Pichon, 1972, Marginal fracture zones as structural framework of continental margins in South Atlantic Ocean: AAPG Bulletin, v. 56, p. 991-1007.

Freund, R., 1965, A model of the structural development of Israel and adjacent areas since upper Cretaceous times: Geological Magazine, v. 102, p. 189-205.

Green, A. R., 1977, The evolution of the earth's crust and sedimentary basin development: Offshore Technology Conference 2885, p. 67-72.

Halbouty, M. T., A. A. Meyerhoff, R. E. King, R. H. Dott, Sr., H. D. Klemme, and T. Shabad, 1970a, World's giant oil and gas fields, geologic factors affecting their formation, and basin classification, part I, giant oil and gas fields, in Geology of giant petroleum fields: AAPG Memoir 14, p. 502-528.

———— R. E. King, H. D. Klemme, R. H. Dott, Sr., and A. A. Meyerhoff, 1970b, World's giant oil and gas fields, geologic factors affecting their formation, and basin classification, part II, Factors affecting formation of giant oil and gas fields, and basin classification, in Geology of giant petroleum fields: AAPG Memoir 14, p. 528-555.

Harding, T. P., 1973, Newport-Inglewood trend, California—an example of wrenching style of deformation: AAPG Bulletin, v. 57, p. 97-116.

Huff, K. F., 1978, Frontiers of world oil exploration: Oil and Gas Journal, v. 76, n. 40, p. 214-220.

Isacks, B., J. Oliver, and L. R. Sykes, 1968, Seismology and the new global tectonics: Journal of Geophysical Research, v. 73, p. 5855-5899.

Kingston, D. R., C. P. Dishroon, and P. A. Williams, 1983, Hydrocarbon plays and global basin classification: AAPG Bulletin, v. 67 (follows this paper).

Klemme, H. D., 1971a, What giants and their basins have in common: Oil and Gas Journal, v. 69, n. 9, p. 85-90.

———— 1971b, To find a giant, find the right basin: Oil and Gas Journal, v. 69, n. 10, p. 103-110.

———— 1971c, Look in Permian reservoirs or younger to find supergiants: Oil and Gas Journal, v. 69, n. 11, p. 96-100.

———— 1971d, Trends in basin development: possible economic implications: World Petroleum, v. 42, n. 10, p. 47-51, 54-56.

———— 1975, Giant oil fields related to their geologic setting—a possible guide to exploration: Bulletin of Canadian Petroleum Geology, v. 23, p. 30-66.

Knebel, G. M., and G. Rodriguez-Eraso, 1956, Habitat of some oil: AAPG Bulletin, v. 40, p. 547-561.

Le Pichon, X., 1968, Sea-floor spreading and continental drift: Journal of Geophysical Research, v. 73, p. 3661-3705.

Lowell, J. D., and G. J. Genik, 1972, Sea-floor spreading and structural evolution of southern Red Sea: AAPG Bulletin, v. 56, p. 247-259.

Morgan, W. J., 1968, Rises, trenches, great faults and crustal blocks: Journal of Geophysical Research, v. 73, p. 1959-1982.

Nehring, R., 1978, Giant oil fields and world oil resources (prepared for the Central Intelligence Agency): Rand Corporation, R-2284-CIA, 162 p.

Norton, I. O., and J. G. Sclater, 1979, A model for the evolution of the Indian Ocean and the breakup of Gondwanaland: Journal of Geophysical Research, v. 84, p. 6803-6830.

Perrodon, A., 1971, Essai de classification des bassins sedementaires: Nancy, France, Sciences de la Terre, v. 16, p. 195-227.

Rabinowitz, P. D., and J. L. LaBreque, 1979, The Mesozoic South Atlantic Ocean and evolution of its continental margins: Journal of Geophysical Research, v. 84, p. 5973-6002.

Roeder, D. H., 1973, Subduction and orogeny: Journal of Geophysical Research, v. 78, p. 5005-5024.

Sawyer, D. S., B. A. Swift, and J. G. Sclater, 1982, Extensional model for the subsidence of the northern United States Atlantic continental margin: Geology, v. 10, p. 134-140.

Sclater, J. G., and P. A. F. Christie, 1980, Continental stretching—an explanation of the post-mid-Cretaceous subsidence of the central North Sea basin: Journal of Geophysical Research, v. 85, p. 3711-3739.

Seely, D. R., P. R. Vail, and G. G. Walton, 1974, Trench slope model, in C. A. Burk and C. L. Drake, eds., The geology of continental margins: New York, Springer-Verlag, p. 249-260.

Sengor, A. M. C., K. Burke, and J. F. Dewey, 1978, Rifts at high angles to orogenic belts: tests for their origin and the Upper Rhine graben as an example: American Journal of Science, v. 278, p. 24-40.

Shatskii, N., 1946a, Basic features of the structures and development of the east European platform: Akademiya Nauk SSSR Izvestiya, Seriya Geologicheskaya, n. 1, p. 5-62.

———— 1946b, The Great Donets basin and the Wichita system: comparative tectonics of ancient platforms: Akademiya Nauk SSSR Izvestiya, Seriya Geologicheskaya, n. 6, p. 57-90.

———— 1947, Structural correlations of platforms and geosynclinal folded regions: Akademiya Nauk SSSR Izvestiya, Seriya Geologicheskaya, n. 5, p. 37-56.

———— 1955, On the origin of the Pachelma Trough: Byulleten Moskovskogo Obshchestva Ispytapel'noi Prirody, Seriya Geologicheskaya, v. 5, p. 5-26.

———— and A. A. Bogdanoff, 1960, La carte tectonique internationale de l'Europe au 2,500,000°: Akademiya Nauk SSSR Izvestiya, Seriya Geologicheskaya, n. 4.

St. John, B., 1980, Sedimentary basins of the world and giant hydrocarbon accumulations: AAPG Special Publication, 23 p., 1 map.

Trümpy, R., 1960, Paleotectonic evolution of the central and western Alps: GSA Bulletin, v. 71, p. 843-908.

———— 1975, Penninic-Austroalpine boundary in the Swiss Alps—a presumed former continental margin and its problems: American Journal of Science, v. 275-A, p. 209-238.

Umbgrove, J. H. F., 1947, The pulse of the earth: The Hague, Martinus Nijhoff, 358 p.

Uspenskaya, N. Yu., 1967, Principles of oil and gas territories, subdivisions, and the classification of oil and gas accumulations: Amsterdam, Elsevier Publishing Co., 7th World Petroleum Congress, Proceedings, v. 2, p. 961-969.

Weber, M., 1921, Zum Problem der Grabenbildung: Deutsche Geologische Gesellschaft Zeitschrift, v. 73, p. 238-291.

———— 1923, Bermerkungen zur Brunchtektonik: Deutsche Geologische Gesellschaft Zeitschrift, v. 75, p. 184-192.

———— 1927, Faltengebirge und Vorlandbruche: Centralblatt Mineralogie, Geologie, Palaontologie, Abt. 5, p. 235-245.

Weeks, L. G., 1952, Factors of sedimentary basin development that control oil occurrence: AAPG Bulletin, v. 36, p. 2071-2124.

White, D. A., 1980, Assessing oil and gas plays in facies-cycle wedges: AAPG Bulletin, v. 64, p. 1158-1178.

Whiteman, A. J., G. Rees, D. Naylor, and R. M. Pegrum, 1975, North Sea troughs and plate tectonics: Norges Geologishe Undersökelse, n. 316, p. 137-161.

Wilson, J. T., 1966a, Some rules for continental drift, in G. D. Gardland, ed., Continental drift: Royal Society of Canada Special Publication 9, p. 3-17.

———— 1966b, Did the Atlantic close and then re-open?: Nature, v. 211, n. 5050, p. 676-681.

Wilson, R. C. L., 1975, Atlantic opening and Mesozoic continental margin basins of Iberia: Earth and Planetary Science Letters, v. 25, p. 33-43.

Ziegler, P. A., 1975, North Sea basin history in the tectonic framework of north-western Europe, in A. W. Woodland, ed., Petroleum and the continental shelf of north-west Europe: New York, John Wiley and Sons, p. 131-149.

Ziegler, W., 1975, Outline of the geological history of the North Sea, in A. W. Woodland, ed., Petroleum and the continental shelf of north-west Europe: New York, John Wiley and Sons, p. 165-190.

The American Association of Petroleum Geologists Bulletin
V. 67, No. 12 (December 1983), P. 2194-2198

Hydrocarbon Plays and Global Basin Classification[1]

D. R. KINGSTON,[2] C. P. DISHROON,[2] and P. A. WILLIAMS[3]

ABSTRACT

Hydrocarbon plays and the global basin classification system have important connections. After a basin has been classified, its similarities or differences with other basins may be compared. Prolific hydrocarbon basins are categorized, and the plays are compared to specific tectonic and depositional events. Some plays are controlled by basin-initiating tectonic events, such as interior fracture or interior sag, and the type of sedimentation inherent in each basin type. In more complex polyhistory basins, oil and gas plays are commonly associated with combinations of various cycles, or by basin-modifying tectonics such as episodic wrenching, subsidence, foldbelts, and basin tilt.

INTRODUCTION

This article was written in conjunction with the preceding paper in this issue, "A Global Basin Classification System," by the same authors. Cross reference to the previous paper is made, especially to various basin types, basin formulas, and figures. The principal purpose of this paper is to identify the main hydrocarbon associations and parameters coupled with each cycle type. Also, we describe some examples of complex polyhistory basins and how they may be analyzed to identify their play-location cycles and events.

Various authors, Weeks (1952), Knebel and Rodriguez-Araso (1956), Halbouty et al (1970a, b), and Huff (1978), have discussed hydrocarbon reserves in relation to various basin types. The habitat of giant oil and gas fields has been the subject of numerous papers: Klemme (1971a, b, c; 1975), Nehring (1978), and St. John (1980). Various aspects of hydrocarbon accumulations, including tectonic style and classification, have been reported by Uspenskaya (1967), Harding (1973), Ziegler (1975), Bally (1975), and Huff (1978). The age and distribution of petroleum reserves have been discussed by Bois et al (1982).

OIL-PLAY PREDICTION

In the preceding paper, the methods of basin classification are explained. After a basin is classified, it is possible to identify the productive cycle or cycles within it. Discov- ered hydrocarbon reserves for each of these productive cycles, worldwide, may then be tabulated, and evaluations of the various cycle types can be made. Productive cycles in explored basins may be compared to similar cycles and tectonic events in unexplored areas. Assessment procedures for new basin areas are not discussed here. However, we point out general oil-play controls for each of the cycle types that we have found to be of value in basin-play predictions.

Interior Sag (IS) Cycles/Basins

Interior sags are stable cycles, commonly remnants of larger Paleozoic basins. Oil may have migrated up from older preexisting larger basins into these sags when the former were broken up by orogeny. These cycles have an interior, restricted, depositional environment, normally good for source rock development. Shallow marine and nonmarine conditions also promote the widespread extent of blanket-type reservoirs and evaporite seals. Poor structural focus is a common problem in interior sag cycles. Stratigraphic traps are commonly found in interior sag (IS) cycles. The oil plays in these cycles generally are associated with basement arches, noses, and other topographic features within the basin.

Larger interior sag basins generally are more prospective; small ones commonly are too shallow to generate hydrocarbons, with too little burial for hydrocarbon maturation, especially for Mesozoic and Tertiary rocks. Interior sag cycles/basins, though commonly large in size, may be very shallow. Those larger and deeper basins of Paleozoic age generally are the most prospective for hydrocarbons. Most interior sag (IS) basins are very shallow, have few structures, and in places have poor source rocks. Most simple interior sag (IS) basins contain poor prospects or few prospects in small fields. Polyhistory basins that contain interior sag (IS) cycles affected by later structuring may become very prospective. Forty-nine percent of all interior sag (IS) cycles studied worldwide were found to produce commercial quantities of hydrocarbons.

Interior Fracture (IF) Cycles/Basins

Interior fracture basins include some of the most prolific oil basins in the world, such as Sirte in Libya, west Siberia in the USSR, and the Gulf of Suez in Egypt. Normally two main plays are in an interior fracture cycle: (1) stage 1 nonmarine sediments on basement block faults as are found in the North Sea at Brent, and at Waha and Sarir in the Sirte basin, Libya; and (2) younger stage 2 marine reservoirs draped over block fault structures as are found in the West Siberia and Sirte basins (see Kingston et al, 1983, Figure 4).

[1]Manuscript received, February 22, 1983; accepted, May 13, 1983.
[2]Esso Exploration Inc., Houston, Texas 77024.
[3]Esso Exploration Inc. (retired), 203 Pompano, Surfside, Freeport, Texas 77541.

The writers thank the management of Esso Exploration Inc., for permission to publish this paper. We also acknowledge the helpful comments of T. A. Fitzgerald, H. M. Gehman, Jr., D. S. McPherron, T. H. Nelson, and J. B. Sangree, who advised us on content and revisions.

Active block faulting normally continues throughout the history of the basin, thereby constantly rejuvenating structural growth. Episodic wrench movements may also occur. Better reservoirs are associated with structural highs, and source rocks with the lows throughout the development of the cycle—an ideal condition for local migration and entrapment. The cycle 1 nonmarine play, with its built-in source, reservoir, and structure, attains its best development in interior fracture (IF) cycles.

A semirestricted depositional environment, commonly found in basal interior fracture (IF) cycles, may produce widespread occurrence of evaporites, or blanket sand, and rich oil source rocks such as the Socna shale in Libya. No major regional tilting of interior fracture cycles/basins is likely to occur as in margin sag basins, so there is much less chance of pooled oil spilling out of structures. A few small interior fracture basins have only one active side, however, and in these a half-graben is formed with strong tilt. Size is an important factor in interior fracture basins. Generally, the bigger basins have a better chance for exploration success, although a few smaller basins, like Vienna in Austria, and the Gulf of Suez, have good production. Several smaller interior fracture basins are too shallow, and the source rocks are immature. The burial history of these basins must be studied carefully before maturation and migration of hydrocarbons can be assumed. Many smaller interior fracture basins contain red beds and volcanics that do not appear to be good oil-generating rocks.

Interior fracture basins/cycles, whose development was arrested during the early initial rifting stage, are very high risk with minimal known production in other basins. These basins normally contain a thin sedimentary section of coarse clastics or volcanics. Some of the rift valleys of Africa are examples. The interior fracture basins, whose development reached a more mature stage, are the most favored for hydrocarbon occurrence and accumulation in either early rifting or middle graben stages. Interior fracture cycles are found commonly as simple basins. They may also be found as cycles within polyhistory basins, particularly the initiating cycle. Interior fracture basins, which have been separated by continental "pull-apart" and now comprise the bottom cycle in margin sag basins, are largely unexplored at present. To date, production has been established in these continental margin interior fracture cycles in Gabon, Angola, and Brazil.

The major plays in interior fracture basins are all related to the horst and graben features that dominate the structure and stratigraphy. The largest plays are found in reservoirs directly overlying high basement blocks; examples are sandstones (Samotlor in west Siberia) and carbonates (Romashkino in the Volga Urals), both in the USSR. Of all interior fracture basins studied worldwide, 35% of them produce commercial quantities of hydrocarbons.

Margin Sag (MS) Cycles/Basins

The divergent continental margin basins are a complex group. They have a variety of potential plays, starting with block faults and nonmarine deposition of the basal interior fracture (IF) cycle. This combination of basement block fault structure, good reservoir sand, and source shale may be the best oil play in margin basins. Unfortunately, it is generally not at optimum drilling depth. In the basin center, it is probably buried too deeply (20,000 to 30,000 ft, 6,000 to 9,000 m) to be drilled economically, or it will be gas prone with low porosities at these depths. Near the margins of the basin the play is normally too shallow (< 6,000 ft, 2,000 m) and it is immature. This play must be found at the proper depth to be considered feasible. To date, not much drilling has been conducted for basal cycle interior fracture (IF) plays in margin sag (MS) basins. Most cycle 1 fault blocks, the best potential structures, freeze or cease differential movements after continental separation. Because there is no structural growth, there is normally no drape or closure of cycle 2 younger beds over the old blocks. This is different from other main cycle types, and the play must be downgraded as a result.

Above the basal interior fracture (IF) cycle, younger units may consist of interior sag or margin sag cycles. Simple margin deposition and sag with no other tectonic movements appear to be detrimental to preservation of hydrocarbons in these basins. Contemporaneous or younger salt dome uplift, growth faults, and wrench faulting appear to increase the incidence of oil discovery in these basins.

The size of margin sag basins/cycles is an important factor. If the continental shelf is narrow and small, the cycle will be highly tilted. Most young Tertiary margin sag cycles are deposited as thick sediment pods or clinoform wedges just off the outer shelf. They have a fairly low sediment volume that does not prograde like a delta or keep pace with subsidence, and consists predominantly of deep-water clinoform clastics. The subsidence and seaward tilt are perhaps the most important negative factors. In margin sag cycles, marine shales are located seaward, down-dip, and with porosity (sand) increasing updip to the outcrop. In the Late Cretaceous and Tertiary, rapid subsidence of the outer basin margin tilted many gentle to moderate structural closures, and the oil has been lost updip. Margin sag basins with a strong oceanward tilt may have lost oil from both the older and younger structures. The updip flanks of some margin sag basins contain extensive tar deposits at the outcrop, evidence of oil lost by basin tilting. Examples of this are found in the Upper Cretaceous margin sag outcrops of Nigeria, Gabon, Cameroon, Congo, and Angola.

Some margin sag cycles have normal or gentle oceanic tilt and subsidence. These cycles are commonly found in the middle stages of margin sag development. Sediments deposited in shallow-water margin sag cycles generally are thin-bedded platform deposits of sandstone, limestone, and shale. These thin cyclic units are normally not conducive to the generation and preservation of large volumes of oil.

Some margin sag cycles are major river deltas having a high sediment volume which commonly fills the subsiding basin (see Kingston et al, 1983, Figure 6). This results in extensive shallow-water deposits in the delta area, which prograde into deep water. Most explorationists believe that modern deltas are good places to look for hydrocarbons. The writers have analyzed the 60 present-day major river delta systems around the world, and only four of them are able to produce more than marginally commercial quantities of hydrocarbons. The four major deltas are the Mississippi, Niger, Mahakam, and Mackenzie. The other 56 deltas (most have been drilled at least once), have produced only very small amounts. Older margin sag delta cycles, perhaps buried or structured by younger cycles or events, generally are prolific producers. Overpressured shale sections are common features of margin sag river delta cycles.

Extensive carbonate banks with reefs are found in many margin sag cycles; however, these carbonate bank edges, which appear to be so prospective in divergent continental margin basins, have had minimal success when drilled. Several broad carbonate platforms that have been structured by younger tectonic events, occasionally may provide reservoirs for substantial oil plays, as in the Tabasco-Salinas basin of Mexico. Cretaceous and Tertiary paleoslope unconformities often are found that have cut into sediment margins, exposing reservoirs to erosion, oil loss, or water contamination, as in Tarfaya, Morocco. The percent of productive divergent margin sag cycles worldwide, compared to those nonproductive, is approximately 20%. Most margin sag basins should be considered high risk exploration ventures.

Margin Sag–Interior Sag (MSIS) Cycles/Basins

Margin sag–interior sag (MSIS) cycles/basins have many attributes of interior sag cycles, being found in the interior on continental plates and not associated with present-day continental margins. Most of the MSIS cycles were deposited on older continental plate margins, which appear to have been broad, flat, and only slightly tilted. These stable shelves accumulated extensive reservoirs, such as blanket sands and carbonates. Thick source rocks and seals commonly were deposited as well. Subsequent structuring produced the world's most prolific basin/cycle type in the Persian Gulf, the MSIS cycle with wrench anticlines, blanket carbonate reservoirs, and evaporite seals. Approximately 50% of the MSIS cycles produce commercial hydrocarbons, one of the highest ratios of cycles we have studied. The key parameters for giant field potential in these MSIS cycles are large structures and extensive reservoirs with effective seals. In the MSIS cycles of the Permian basin of Texas, the blanket reservoirs are missing, but the source rocks are prolific. Good evaporite seals are present, and although the MSIS cycle has almost 6 billion bbl of oil equivalent, most of the fields are small. In western Canada, the Alberta basin has extensive reservoirs and source rocks, but the basin is primarily unstructured. Carbonate reefs, which are really stratigraphic traps, are the main trap available in the cycle and are small. However, MSIS cycles should be examined carefully for they have an enviable record of productivity.

Wrench or Shear (LL) Cycles/Basins

Probably the most important factor affecting wrench (LL) basins is timing. Many LL basins are formed, filled, structured, and destroyed in a relatively short geologic time. The optimum combination of these short-term factors can result in highly productive basins, such as the Los Angeles basin in California. However, optimum factors are not always present, and basins may be too young and immature or too old and overstructured. In these situations, the basins must be considered to be very high risk.

A critical factor in shear basins is the presence or absence of quartz sand for reservoir rocks. Mesozoic and older age basement usually contains quartz sand sources. Tertiary crust, however, is normally composed of oceanic basalt and volcanogenic sediments, which substantially degrade reservoir quality. Wrench basins must be situated on or adjacent to good quartz-sand yielding areas to insure the existence of predictable quartz sand reservoirs; this generally means Mesozoic or older crust. Carbonate rocks and chert, as in California, are also found occasionally as reservoirs in this type of basin.

Normally a high heat flow exists in active wrench (LL) basins. Some cycles may be too hot, and therefore gas prone or overmature in the deeper parts. Because most wrench basins are very young, the source shales generally are immature, even at intermediate depths (8,000 to 10,000 ft, 2,500 to 3,000 m). An exploration pitfall of this type basin is the potential immaturity of the upper source shales, whereas the deep portion of the basin may be overmature, with shales cooked and sandstone porosity lowered by diagenesis. This results in a very thin optimum zone between rocks that are too shallow or too deep. Unless reservoir sands and structural closures are found in this optimum zone, dry holes or gas may be the result.

The composition of the two converging plates has a profound effect upon the sediment fill in the resulting wrench basins. Two converging oceanic plates will result in volcanogenic and deep-marine sediments such as chert and pelagic material, with some carbonate. This type of basin has had very little exploration success. A continental block overriding oceanic crust can be good if the upper block is a good source of sand; both marine and nonmarine rocks may be deposited. Two converging continental blocks may provide primarily nonmarine depositional conditions, in places with salt. Basins of this type are found within foldbelts and are sometimes called "foldbelt microplates."

Wrench or shear basins presently in stage 1 may not be very prospective, because the sediments and structure are too immature (see Kingston et al, 1983, Figure 8). Stage 2 can be better. Sands may have been deposited on the fault block highs and source shales in the lows. Wrench-generated structures may have formed. At this stage, oil fields may be associated with the tops of en echelon folds, the tops of flower structures, and stage 1 block fault closures. At stage 3, some of the more active wrench zones or flower structures may have been uplifted and eroded, losing the oil on top. New fields may then be found on the flanks of such features rather than on the wrench axes, on noses or the plunge of old en echelon fold sets, or on younger folds forming basinward of the wrench welt. Stratigraphic traps and tear fault or trap-door fault structures also may be productive. When the basin evolves beyond stage 3, oil prospects generally are progressively destroyed by folding.

In prolific shear basins, oil migrates from older to younger structures as the wrenching progresses. Older structures are deformed and destroyed, and the oil migrates or is lost in the process. If convergence between the two plates stops, the wrench faulting will cease, and the basins and potential plays could remain frozen in their last evolutionary stage. Examples of this are present in the Pennsylvanian wrench system in Oklahoma in the Ardmore, Anadarko, and Arkoma basins. Size is not a critical factor in LL shear basins, as most of them are relatively small.

LL shear basins are moderate risk prospects. They have many constructive and destructive features to consider, and they are short lived. The main plays in LL shear basins are sandstone reservoirs and wrench-generated block faults or anticlines. Most fields are moderate in size; the largest we have listed is Minas in central Sumatra, with almost 5 billion bbl of oil equivalent. Some LL shear basins have prolific source rocks, such as those in California and Sumatra; producible stratigraphic traps are found, as well as structures, as the oil seeks to fill all available porosity. Approximately 47% of all wrench (LL) cycles studied worldwide were found to produce commercial hydrocarbons.

Trench Associated (TA) Cycles/Basins

Trench associated (TA) cycles/basins are found along ancient or modern subduction zones at convergent plate boundaries. Trench associated cycles are commonly developed on the continental, or high side of the oceanic trench, and oceanward of the first volcanic or magmatic arc. The source of the sediment is mainly the landward side with shallow-water clastics or carbonates, and some deep-marine deposition offshore. Volcanogenic sediments comprise the bulk of the clastic rock, and porosity/permeability is likely to be destroyed by diagenesis at depths above the hydrocarbon maturation zone. Only a few of the trench associated cycles, worldwide, contained any quartz-prone reservoirs (e.g., Sabah, Malaysia). Carbonates and chert are also found on occasion. Structuring, if present, is related to subduction, and wrench fault features are commonly present. The crust under these basins is composed of tectonized or folded trench sediments (FB_3T). Trench associated cycles will eventually self-destruct and become FB_3T folded trench foldbelts.

To date, only a few commercial fields have been found in trench associated cycles. Most of these fields are in sandstone or chert reservoirs and wrench anticlines. The fields are small and have poor producing characteristics. Trench associated basins are numerous and are found on most convergent margins. Exploration drilling in basins such as Mentawi and south Java in Idonesia, Abukuma in Japan, the Gulf of Alaska, and others has produced a poor record of discoveries to date. As potential hydrocarbon prospects the basins must be considered minor and very high risk areas.

Oceanic Sags (OS) and Oceanic Wrench Basins (OSLL)

The chances are probably small that oil plays will be found in oceanic basins filled with volcanogenic/deep-water pelagic sediments. Questionable source rocks of deep-marine pelagic sediments, thermal histories of low temperatures, poor volcanogenic reservoir rocks, and probable lack of structure combine with the presence of deep water to make these basins extremely high risk. Some of the oceanic sags contain deep-water fans from river deltas that have built out over oceanic crust, as are found in the south Caspian and Black Seas in the USSR. One oil prospective area, possibly underlain by oceanic crust and which has these features, is offshore from the Niger delta. Its seismic signature, indicating oceanic crust, is one of flat, featureless overlying sediments in deep water. Unless these sediments are activated by structural movement, they do not show gravity tectonic features. Because of this lack of structure, and thermal histories of low temperatures, most oceanic basin areas associated with continental blocks must be considered high risk. Most sediment wedges filling oceanic sags are classified with margin sags (basin flank) and not with the basin center.

Polyhistory Basins

Most large hydrocarbon-producing basins are the polyhistory type, containing numerous and perhaps varied cycles and tectonic events. The oil-play controls of single cycles apply in a general way to polyhistory basins, but they are commonly modified by tectonics. The evaluation of polyhistory basins is a complex subject, and only a few salient points are discussed here.

Classification and the writing of formulas are very useful in the evaluation of complex polyhistory basins. The formulas provide a shorthand history of each basin, permitting identification of the geologic events that form the basis for the oil plays. Once the plays are identified, they may be compared with those in other similar polyhistory basins. Change in structural genesis appears to enhance oil prospects. In Kingston et al (1983, Figure 13), a series of restored cross sections shows the evolution of the Persian Gulf. In this diagram the main events are: (1) basin formation, (2) subsidence and fill, (3) structuring, and (4) renewed subsidence and "loading" of the structures. This sequence is common to many productive polyhistory basins. The loading of the structures by more subsidence and fill, commonly appears to be critical. Polyhistory basins ending with interior sag (IS) appear to be the best situated for the generation and preservation of oil and gas. The final interior sag generally does not cause tilting of the structures, but commonly provides the sediment loading which triggers maturation and migration of hydrocarbons. Too much structuring at the end of the polyhistory basin formation is generally more destructive than constructive to the pooling and preserving of hydrocarbons.

REFERENCES CITED

Bally, A. W., 1975, A geodynamic scenario for hydrocarbon occurrences: 9th World Petroleum Congress Proceedings, v. 2, p. 33-44.
———— and S. Snelson, 1980, Realms of subsidence, in A. D. Miall, ed., Facts and principles of world petroleum occurrences: Canadian Society of Petroleum Geologists Memoir 6, p. 9-75.
Beck, R. H., and P. Lehner, 1974, Oceans, new frontier in exploration: AAPG Bulletin, v. 58, p. 376-395.
Bois, C., P. Bouche, and R. Pelet, 1982, Global geologic history and distribution of hydrocarbon reserves: AAPG Bulletin, v. 66, p. 1248-1270.

Halbouty, M. T., A. A. Meyerhoff, R. E. King, R. H. Dott, Sr., H. D. Klemme, and T. Shabad, 1970a, World's giant oil and gas fields, geologic factors affecting their formation, and basin classification, Part I, giant oil and gas fields, *in* Geology of giant petroleum fields: AAPG Memoir 14, p. 502-528.

———— R. E. King, H. D. Klemme, R. H. Dott, Sr., and A. A. Meyerhoff, 1970b, World's giant oil and gas fields, geologic factors affecting their formation, and basin classification, Part II, factors affecting formation of giant oil and gas fields, and basin classification, *in* Geology of giant petroleum fields: AAPG Memoir 14, p. 528-555.

Harding, T. P., 1973, Newport-Inglewood trend, California—an example of wrenching style of deformation: AAPG Bulletin, v. 57, p. 97-116.

Huff, K. F., 1978, Frontiers of world oil exploration: Oil and Gas Journal, v. 76, n. 40, p. 214-220.

Kingston, D. R., C. P. Dishroon, and P. A. Williams, 1983, Global basin classification system: AAPG Bulletin, v. 64 (precedes this paper).

Klemme, H. D., 1971a, What giants and their basins have in common: Oil and Gas Journal, v. 69, n. 9, p. 85-90.

———— 1971b, To find a giant, find the right basin: Oil and Gas Journal, v. 69, n. 10, p. 103-110.

———— 1971c, Look in Permian reservoirs or younger to find super-giants: Oil and Gas Journal, v. 69, n. 11, p. 96-100.

———— 1975, Giant oil fields related to their geologic setting—a possible guide to exploration: Bulletin of Canadian Petroleum Geology, v. 23, p. 30-66.

Knebel, G. M., and G. Rodriguez-Eraso, 1956, Habitat of some oil: AAPG Bulletin, v. 40, p. 547-561.

Nehring, R., 1978, Giant oil fields and world oil resources (prepared for the Central Intelligence Agency): Rand Corporation, R-2284-CIA, 162 p.

St. John, B., 1980, Sedimentary basins of the world and giant hydrocarbon accumulations: AAPG Special Publication, 23 p., 1 map.

Uspenskaya, N. Yu., 1967, Principles of oil and gas territories, subdivisions, and the classification of oil and gas accumulations: Amsterdam, Elsevier Publishing Co., 7th World Petroleum Congress Proceedings, v. 2, p. 961-969.

Weeks, L. G., 1952, Factors of sedimentary basin development that control oil occurrence: AAPG Bulletin, v. 36, p. 2071-2124.

Ziegler, P. A., 1975, North Sea basin history in the tectonic framework of north-western Europe, *in* A. W. Woodland, ed., Petroleum and the continental shelf of north-west Europe: New York, John Wiley and Sons, v. 1, p. 131-150.

Sedimentary Provinces of the World — Hydrocarbon Productive and Nonproductive

Bill St. John
Primary Fuels, Inc.
Houston, Texas

A.W. Bally
Department of Geology
Rice University
Houston, Texas

H. Douglas Klemme
Geo Basins Limited
Norwalk, Connecticut

Published by
The American Association of Petroleum Geologists
Tulsa, Oklahoma 74101, U.S.A.

Introduction

The accompanying map, *Sedimentary Provinces of the World,* locates known sedimentary accumulations greater than 1000 meters (3000 feet) thick. These include basins, foldbelts, carbonate banks on oceanic crust, etc. The compilation is an attempt to inventory, classify, and rate any sedimentary deposit which might conceivably contain recoverable hydrocarbons.

The term "provinces" was deemed preferable tŏ "basins" because many of the areas are not basins in the strict sense. Oil and gas are found in overthrust or fold belts, cratonic arches, and fan deposits. Deep sea fans and cones are included as are carbonate bank build-ups on elevated oceanic crust. These latter provinces are rated as having "poor" potential but technology for recovering oil and gas continues to yield surprises.

The provinces are listed alphabetically by continental or geographical area. Various groups identify provinces by different names and known alternatives are included. Several of the provinces are divided by political boundaries rather than by geological parameters, as it was considered best to apply the names and boundaries most commonly used.

Much effort has been made over the years to classify sedimentary basins. Two of the most frequently quoted are Bally and Snelson (1980) and Klemme (1971), Tables 1 and 2. Both Bally and Klemme have classified each of the provinces presented herein. Figures 1 to 8 provide additional background for Bally's classification, and Figure 9 shows the distribution of plates as it relates to Klemme's classification.

The hydrocarbon productivity, potential productivity, or nonproductivity of each province is indicated based on the most recent available data. A host of technical publications have been the sources, chief among which have been various issues of the *AAPG Bulletin,* the *International Petroleum Encyclopedia,* and the *Oil and Gas Journal.*

Provinces having production, or potential production, from giant oil or gas fields are denoted by "G". A giant oil field is one containing 500 million barrels or more of recoverable oil; a giant gas field has at least 3 trillion cubic feet of recoverable gas. Those provinces with fields of less than giant-size (sub-giant) are classified "S" and nonproductive provinces "N". These categories are color-coded on the map.

Of the 799 provinces, 82 produce or have discoveries of giant-size oil and/or gas fields, 177 produce from sub-giant fields and 540 are non-productive. There are 208 nonproductive provinces with "fair" or better chance of eventually yielding commercial hydrocarbons and 332 rated as having "poor" or "poor-to-fair" potential.

Numerous recent and some not-so-recent announced discoveries are not yet developed. Those provinces having undeveloped but probably commercial discoveries are marked with an asterisk (*) on the attached list and are stippled and appropriately color-coded on the map. For example, the Grand Banks province offshore eastern Canada with the undeveloped giant Hibernia Field oil discovery.

The provinces are rated subjectively as to hydrocarbon potential as follows:

P   poor
F   fair
G   good
E   excellent

While numerous explorationists have shared their judgement of the ratings, the inevitable errors are the responsibility of the authors.

Numerous people have contributed to the map and provinces tabulation. Some prefer to remain unnamed but the generous assistance of the following is gratefully acknowledged: Arnold Bouma and Neal E. Barnes (deep sea fans and cones); Alan P. Bennison (Somalia); Carlos Walter Marinho Compos and Edson Machado Ribeiro (Brazil); Arthur Grantz (Alaskan offshore); Hans R. Grunau and Sam Carmalt (Europe and Far East); and Myron K. Horn, Robert E. King, W.R. Muehlberger and Fred H. Wessman (general).

Selected References

Bally, A.W., 1975, A geodynamic scenario for hydrocarbon occurrences, *in* Proceedings of the 9th World Petroleum Congress, vol. 2 (geology): Essex, England, Applied Sci. Pub., Ltd., p. 33-44.

Bally, A.W., and S. Snelson, 1980, Realms of subsidence: Mem. 6 Can. Soc. Petrol. Geols., p. 9-94.

Coury, A.B., et al, 1978, Map of prospective hydrocarbon provinces of the world: U.S.G.S., Misc. Field Studies, MR-1044 A, B, C.

Cummings, D., and G.I. Shiller, 1971, Isopach of the earth crust: Earth Sci. Reviews, v. 7, p. 97-125.

Guangming, Z., Z. Wenzhao, and H. Chaoyuan, 1982, Oil and gas accumulations in China's continental basins: Oil and Gas Jour., v. 80, no. 50, p. 129-136.

King, R.E., 1982, Recent marine exploration yields mixed results: Ocean Industry, October, p. 64-73.

Klemme, H.D., 1971a, What giants and their basins have in common; Oil and Gas Jour., v. 69, no. 9, 10, 11; part 1, p. 85-90; part II, p. 103-110, part III, p. 96-100.

—— , 1971b, Trends in basin development: World Petroleum, v. 42, p. 47-51, 54-56.

—— , 1975, Giant oil fields related to their geological setting——a possible guide to exploration: Can. Soc. Petrol. Geols. Bull., v. 23, p. 30-66.

—— , 1983, Field size distribution related to basin characteristics: Oil and Gas Jour., v. 81, n. 52, (December 26) p. 168-176.

McCaslin, J.C., ed., 1983, International Petroleum Encyclopedia: Tulsa, Pennwell Books, 418 p.

St. John, B., 1980, Sedimentary basins of the world: AAPG, map.

TABLE 1. Bally and Snelson (1980) classification, modified to differentiate oceanic crust and to include folded belts and platform basalts.

1. **BASINS LOCATED ON THE RIGID LITHOSPHERE, NOT ASSOCIATED WITH FORMATION OF MEGASUTURES**
    - 11. Related to formation of oceanic crust
        - 111. *Rifts*
        - 112. *Oceanic transform fault associated basins*
        - 113-OC. *Oceanic abyssal plains*
        - 114. *Atlantic-type passive margins (shelf, slope & rise) which straddle continental and oceanic crust*
            - 1141. Overlying earlier rift systems
            - 1142. Overlying earlier transform systems
            - 1143. Overlying earlier Backarc basins of (321) and (322) type
    - 12. Located on pre-Mesozoic continental lithosphere
        - 121. *Cratonic basins*
            - 1211. Located on earlier rifted grabens
            - 1212. Located on former backarc basins of the (321) type
2. **PERISUTURAL BASINS ON RIGID LITHOSPHERE ASSOCIATED WITH FORMATION OF COMPRESSIONAL MEGASUTURE**
    - 21-OC. Deep sea trench or moat on oceanic crust adjacent to B-subduction margin
    - 22. *Foredeep and underlying platform sediments*, or moat on continental crust adjacent to A-subduction margin
        - 221. Ramp with buried grabens, but with little or no blockfaulting
        - 222. Dominated by block faulting
    - 23. *Chinese-type basins* associated with distal blockfaulting related to compressional or megasuture and without associated A-subduction margin
3. **EPISUTURAL BASINS LOCATED AND MOSTLY CONTAINED IN COMPRESSIONAL MEGASUTURE**
    - 31. Associated with B-subduction zone
        - 311. *Forearc basins*
        - 312. *Circum Pacific backarc basins*
            - 3121-OC. Backarc basins floored by oceanic crust and associated with B-subduction (marginal sea sensu stricto)
            - 3122. Backarc basins floored by continental or intermediate crust, associated with B-subduction
    - 32. Backarc basins, associated with continental collision and on concave side of A-subduction arc
        - 321. On continental crust of *Pannonian-type* basins
        - 322. On transitional and oceanic crust or *W. Mediterranean-type* basins
    - 33. Basins related to episutural megashear systems
        - 331. *Great-basin-type* basins
        - 332. *California-type* basins
4. **FOLDED BELT**
    - 41. Related to A-subduction
    - 42. Related to B-subduction
5. **PLATEAU BASALTS**

TABLE 2. Klemme basin classification.

CRATON / ACCRETED ZONE / Continental Crust / OCEANIC CRUST	TECTONIC LOCATION	REGIONAL STRESS	BASIN AREAL SHAPE (Areal Size)	BASEMENT PROFILE	RATIO Basin Volume M^3 / Basin Area M^2 65 Prod. Basins	SEQUENTIAL BASIN ARCHITECTURAL FORM	PETROLEUM BASIN TYPES
	CONTINENTAL	Extensional Sag	Circular to Elongate (Large)	Symmetrical	95%	Sag	I CRATON INTERIOR BASINS
	CONTINENTAL	2. Compression / 1. Extension ?	Elongate to Circular (Large to Moderate)	Asymmetrical	195%	2. Fordeep / 1. Platform or Sag	II CONTINENTAL MULTICYCLE BASINS A. CRATON MARGIN – Composite
	CONTINENTAL	2. Sag / 1. Extension	Random (Large to Moderate)	Symmetrically Irregular	160%	2. Sag / 1. Rift	B. CRATON / ACCRETED MARGIN – Complex
	CONVERGENT MARGIN "COLLISION"	2. Compression / 1. Extension ?	Elongate (Large)	Asymmetrical	a) 250% b) High c) 250%	2. Fordeep / 1. Platform or Sag	C. CRUSTAL COLLISION ZONE – Convergent Plate Margin – downwarp into small ocean basin a) closed b) trough c) open
	CONTINENTAL	1. Extension (Local wrench compression)	Elongate (Narrow Fault controlled) (Small to Moderate)	Irregular	235%	1. Rift / Sag	III CONTINENTAL RIFTED BASINS A. CRATON and ACCRETED ZONE RIFT
	CONVERGENT MARGIN "CONSUMPTION"	1. Extension plus 2. Wrench compression	Elongate (Small)	Irregular	180% (Variable)	1. "Rift / Wrench" Rift / Sag	B. RIFTED CONVERGENT MARGIN—Oceanic Consumption a) Back Arc c) Median b) Transform
	DIVERGENT MARGIN "PULL – APART"	1. Extension	Elongate (Small to Moderate)	Asymmetrically Irregular	200%	1. "Rift / Drift" Rift / Sag	C. RIFTED PASSIVE MARGIN – Divergence a) parallel b) transform
	CONTINENTAL and ALL MARGINS	A. Extensional Sag B. ?	Circular to Elongate (Moderate)	Depocenter	350%	1. "Modified Sag" ?	IV DELTA BASINS – Tertiary to Recent A. Synsedimentary B. Structural
	CONVERGENT MARGIN "CONSUMPTION"	Compression and Extension	Elongate (?)	Asymmetrical	?	Subduction	V FOREARC BASINS

THICKNESS OF CRUST

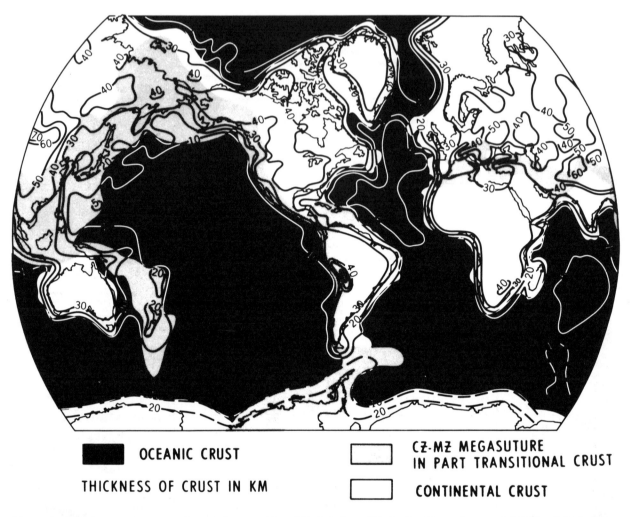

Figure 1. Thickness of crust and crustal types. Simplified and modified after Cummings and Schiller (1971). The gray band (or Cz-Mz Megasuture on this figure and all other figures) marks a zone that involves Cenozoic and Mesozoic folded belts, as well as all the sedimentary basins that occur within these folded belts.

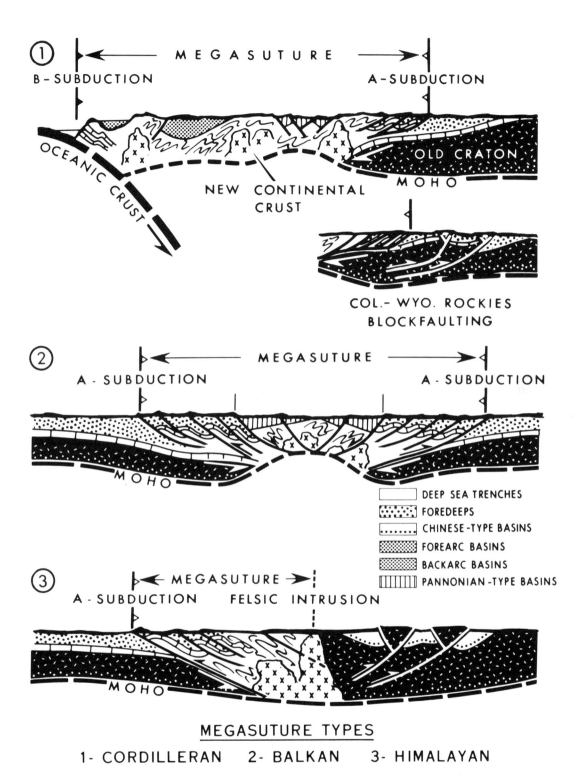

Figure 2. Sketches of three important types of megasutures and basins that are related to them (modified after Bally, 1975). Note that B- subduction involves the subduction of oceanic lithosphere and its crust, while A-subduction involves the subduction of continental lithosphere and parts of its crust. Additional megasuture boundaries are transform faults (for example, San Andreas) and an envelope around Cenozoic-Mesozoic felsic intrusives (Chinese boundary).

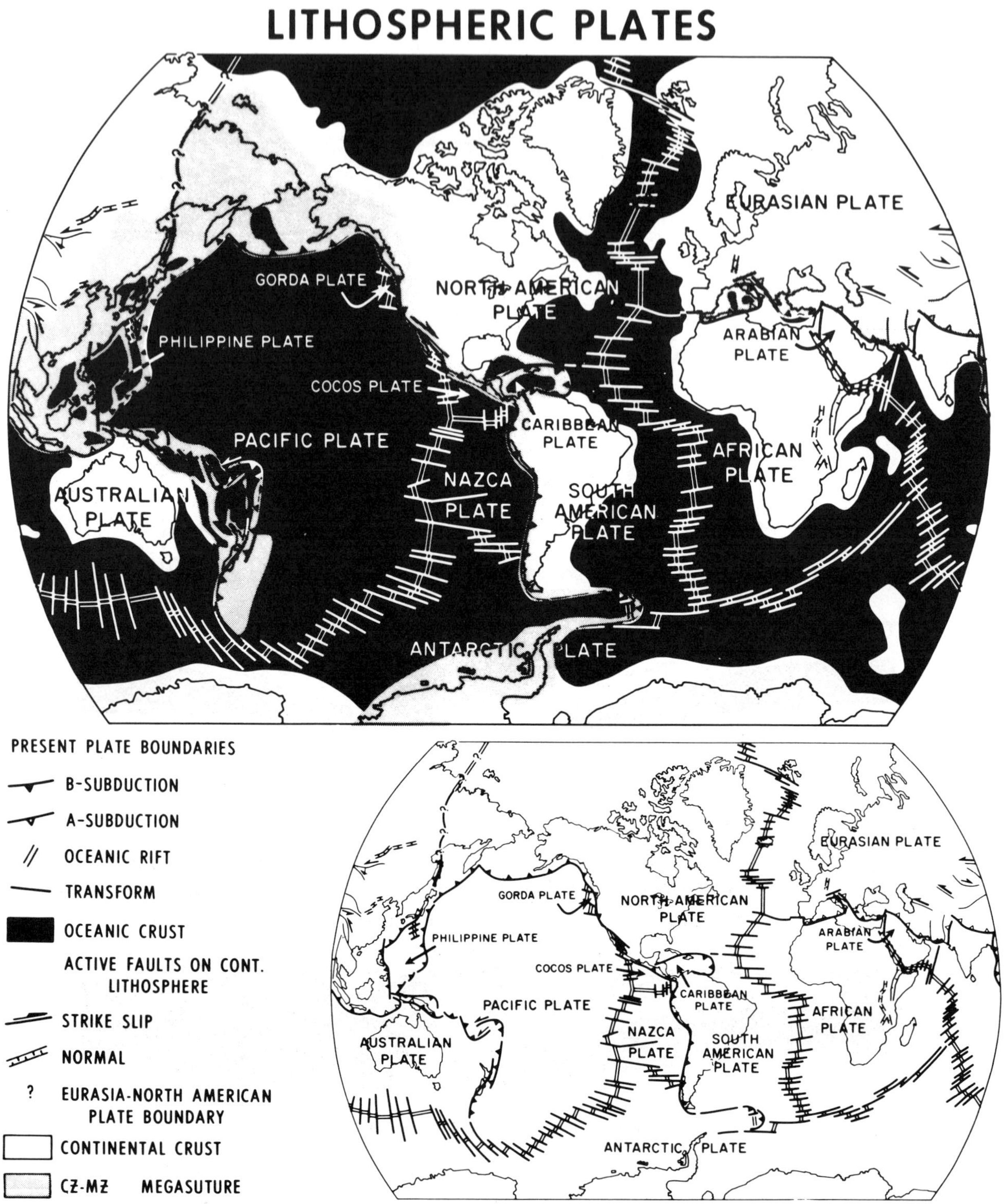

LITHOSPHERIC PLATES

Figure 3. Distribution of lithospheric plates (after Bally and Snelson, 1980). Lower inset shows plate names without differentiating crustal types. The megasuture (in gray) is the integrated product of all Mesozoic and Cenozoic subduction processes that occurred while the ocean crust formed by ridge-spreading processes during about the same time interval (i.e., Jurassic — Recent).

CONTINENTAL MARGINS

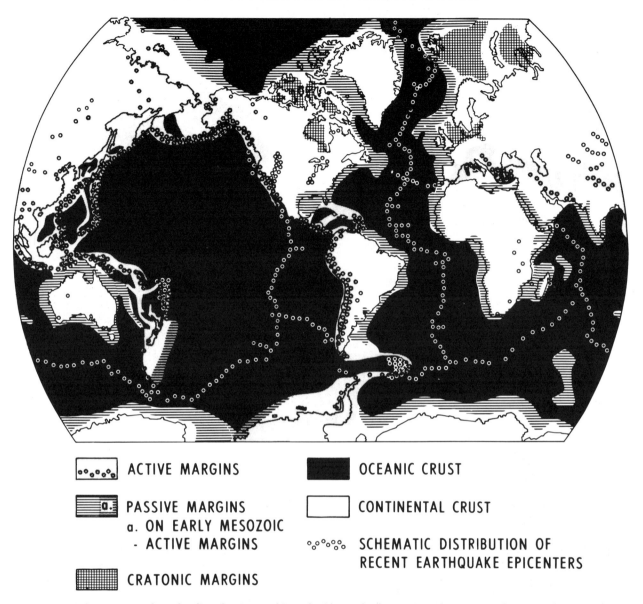

ACTIVE MARGINS

OCEANIC CRUST

PASSIVE MARGINS
a. ON EARLY MESOZOIC
- ACTIVE MARGINS

CONTINENTAL CRUST

SCHEMATIC DISTRIBUTION OF
RECENT EARTHQUAKE EPICENTERS

CRATONIC MARGINS

Figure 4. Schematic earthquake distribution and its relation to the formation of passive and active continental margins (after Bally and Snelson, 1980). Note that there are also a number of cratonic continental margins (that is, extensive offshore areas that are underlain by cratonic basins).

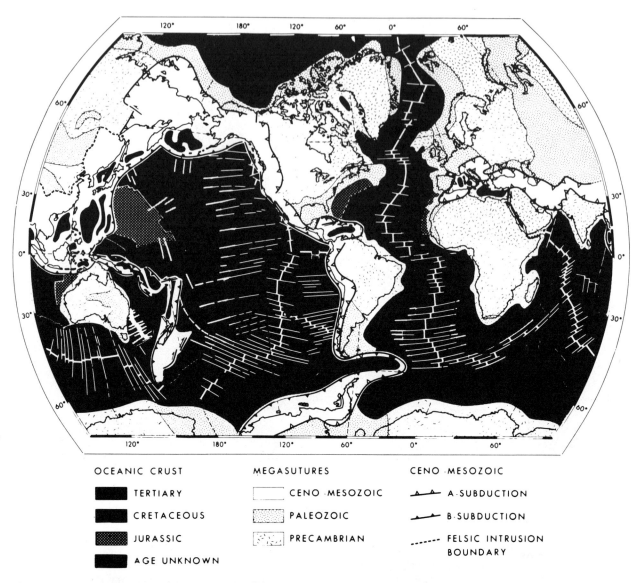

Figure 5. Simplified tectonic map of the world (modified after Bally and Snelson, 1980). This map shows that the oceans of the world were formed in Cenozoic-Mesozoic times; the age of the oceanic crust corresponds to the age of economic basement. The Cenozoic-Mesozoic megasuture (white) is the integrated product of subduction-related processes (folded belts, island arcs and sedimentary basins contained within them). Economic basement in that zone is typically Mesozoic in age. The Paleozoic megasuture is the product of Paleozoic subduction processes, and forms the economic basement of all overlying sedimentary basins. Finally the Precambrian consists of a number of Precambrian megasutures, surrounding Archean nuclei. These form the economic basement of all overlying sedimentary basins. In a nutshell this map is used as indicator of the age of economic basement underlying various types of sedimentary basins.

BASINS ON RIGID LITHOSPHERE

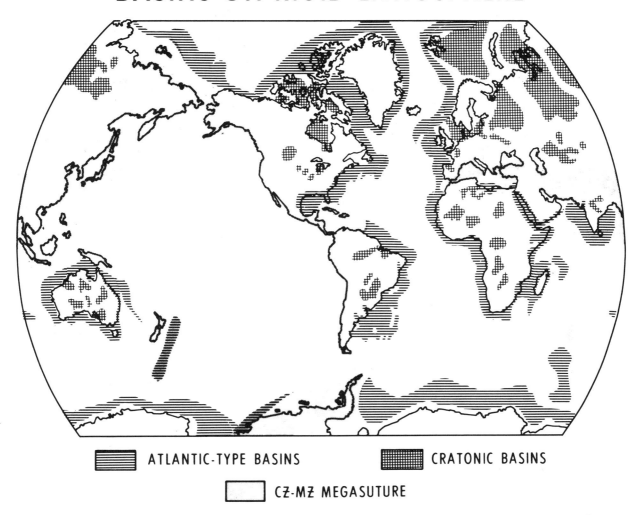

ATLANTIC-TYPE BASINS   CRATONIC BASINS

CZ-MZ MEGASUTURE

Figure 6. Basins on the rigid lithosphere not associated with the formation of megasutures (after Bally and Snelson, 1980). Basically the Atlantic-type basins correspond to the passive margins of the world and with few exceptions (for example, Gulf of Mexico) face a spreading ocean basin. Cratonic basins overlie mostly a Precambrian basement, and occasionally (for example, W. Siberian basin) a Paleozoic basement. Most basins in this class are initiated by one or more rifting events (lithospheric extension). Because of their small areal extent, rift systems that are included in this class are not shown on the map. Also, for provinces within the oceanic realm the reader should consult the "OC" provinces on our list.

PERISUTURAL BASINS

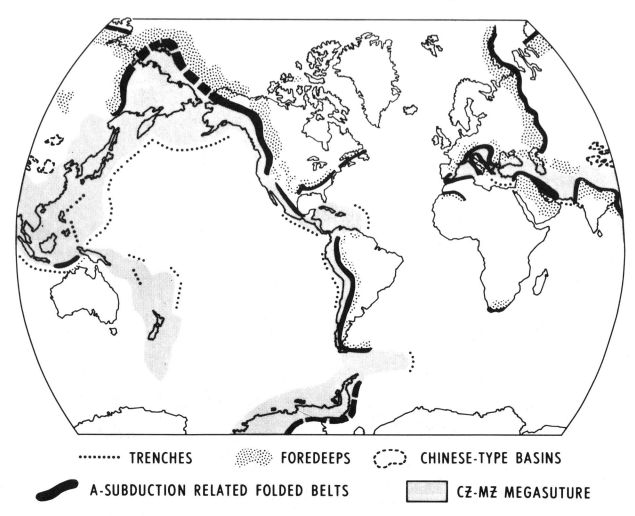

······· TRENCHES   ⬚ FOREDEEPS   ⟨⟩ CHINESE-TYPE BASINS

▬ A-SUBDUCTION RELATED FOLDED BELTS   ☐ CZ-MZ MEGASUTURE

Figure 7. Perisutural basins (after Bally and Snelson, 1980). These basins flank the Cz-Mz Megasuture (in gray). The ocean trenches mark the place where the oceanic lithospheric plate sinks under an adjacent and overriding plate. The deep sea trenches are unattractive for exploration. On the other hand, foredeep basins that are — by definition — associated with adjacent A- subduction can be prolific petroleum producers. Foredeeps are clastic-filled moats that are caused by the load of thrust sheets on a mountainward dipping monocline. The differentiation between foredeeps without (for example, Alberta) or with (for example, Wyoming) associated blockfaulting cannot be shown at the scale of the figure and the reader is referred to the specific classifications of our list. In China and Central Asia the A-subduction boundary of the megasuture is replaced by the Chinese-type boundary (that is, envelope around felsic intrusives). That boundary is also associated with the Chinese-type basins of our classification.

EPISUTURAL BASINS

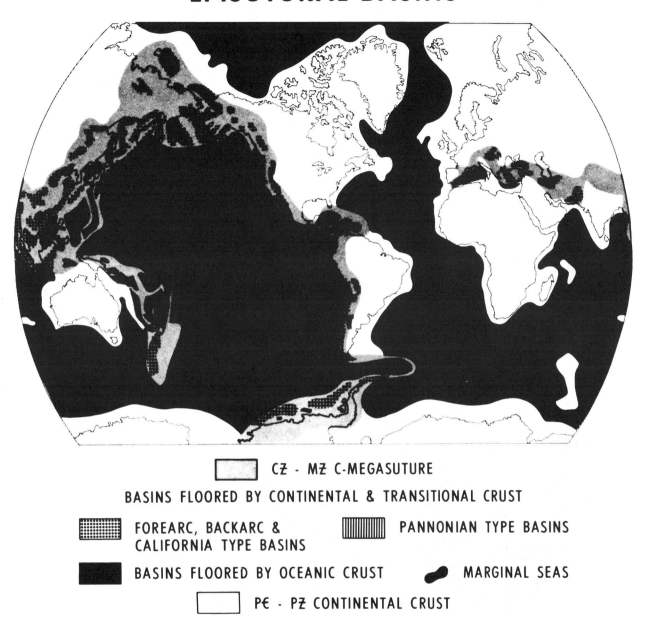

CZ - MZ C-MEGASUTURE
BASINS FLOORED BY CONTINENTAL & TRANSITIONAL CRUST

FOREARC, BACKARC &
CALIFORNIA TYPE BASINS

PANNONIAN TYPE BASINS

BASINS FLOORED BY OCEANIC CRUST

MARGINAL SEAS

PE - PZ CONTINENTAL CRUST

Figure 8. Episutural basins (after Bally and Snelson, 1980). These basins are confined within the boundaries of the megasuture. Episutural basins are associated with B-subduction processes, particularly the formation of marginal seas and in other instances with plate collisions (for example, Pannonian Basin). These basins are short-lived because they tend to get involved in orogenic processes. Most episutural basins were formed during the Tertiary. Excepting forearc basins, most episutural basins are initiated by a rifting event. Note also that these basins occupy nearly 50% of the area of the megasuture.

Figure 9. Plate boundaries relative to Klemme's basin classification.

North America and Greenland

#	Province Name	Alternative Province Name	Country	Bally	Klemme	Production	Hydrocarbon Potential
1.	Alaska, Gulf of		USA (Alaska)	311	V	N	F
2.	Alaska Fan		USA (Alaska)	113-OC		N	P
3.	Alaska Trench		USA (Alaska)	21-OC		N	P
4.	Alberta	West Canada	Canada	221	II A	G	E
5.	Alexander Fan		USA	113-OC		N	P
6.	Altar		Mexico	332	III A	S	F/G
7.	Amlia		USA (Alaska)	3122	V	N	P
8.	Amukta		USA (Alaska)	331	V	N	P
9.	Anadarko		USA	221	II A	G	E
10.	Anderson	West Canada	Canada	221	II A	N	F/G
11.	Angmagssalik	Southeast Greenland	Greenland	1141	III C	N	F/G
12.	Appalachian		USA	221	II A	G	E
13.	Arctic Shelf		Canada	1142	III C	N	F/G
14.	Ardmore		USA	222	II A	G	F/G
15.	Arguello Fan		USA	113-OC	II A	N	P
16.	Arkoma		USA	221		S	F/G
17.	Astoria Fan		USA	113-OC		N	P
18.	Baffin East	Baffin Bay & West Greenland	Greenland	1141	III C	N	F
19.	Baffin West	Baffin Bay	Canada	1141	III C	N	F
20.	Baltimore Canyon	Incl. Part of Atlantic Coastal Plain	USA	1141	III C	N	F
21.	Bay of Fundy		Canada	111	III A	N	P
22.	Bear Lake		Canada	221	II A	N	F
23.	Bellingham		USA	311	V	N	P/F
24.	Bethel		USA (Alaska)	312	III B	N	F
25.	Big Horn		USA	222	II A	S	F/G
26.	Black Mesa		USA	221	II B	N	F/G
27.	Black Warrior		USA	221	II A	S	F/G
28.	Blake Plateau & Outer Ridge	Incl. Georgia Embayment & Part of Atlantic Coastal Plain	USA	1141	III C	N	F
29.	Bodega		USA	332	III B b	N	F
30.	Bowser		Canada	321	III B c	N	F
31.	Bristol Bay		USA (Alaska)	3122	III B a	N	F/G
32.	Campeche	Incl. Yucatan Borderland & Platform & Tobasco	Mexico	114	II C c (III C Southeast)	G	E
33.	Canada		Canada	113-OC	III B c	N	P
34.	Catlow		USA	331	III B b	N	P/F
35.	Central Coastal		USA	332	III B b	N	F
36.	Channel Islands	South California Borderlands	USA	332		N	G/E
37.	Chicagof Fan		USA (Alaska)	133-OC		N	P
38.	Chihuahua		Mexico	41	II A	N	P/F
39.	Chukchi, North		USA (Alaska/USSR)	133-OC		N	P
40.	Cincinnati Arch		USA	Cratonic Arch	I - Arch	G	F/G

161

#	Name	Notes	Country	Code			
41.	Cook Inlet	Incl. Susitna Lowlands	USA (Alaska)	311	III B b	G	G
42.	Copper River		USA (Alaska)	311	III B	N	F
43.	Corrientes Trench		Mexico	21-OC		S	P
44.	Crazy Mountains		USA	221	II A	N	F/G
45.	Cumberland		Canada	1141	III C	S	F/G
46.	Dalhart		USA	222	II A	N	F
47.	Davis Strait	Southwest Greenland	Greenland	1141	III C	S	P
48.	Delgada Fan		USA	113-OC		N	F
49.	Denver	Denver-Julesberg	USA	221	I	S	F
50.	Eagle		Canada	221	II A	N	F
51.	East Texas Salt Dome		USA	1143	II C c	G	G/E
52.	Eel River		USA	332	III B b	S	F
53.	Ensenada		Mexico	332	III B b	N	F
54.	Florida/Bahama		Bahama/Cuba/USA	1141	III C	S	F
55.	Florida Deep		USA	113-OC	II C c	N	P
56.	Fort Worth		USA	221	II A	S	F
57.	Foxe		Canada	121	I	N	P
58.	Gaspe		Canada	41	II A	N	P
59.	Georges Bank		USA	1141	III C	N	F/G
60.	Georgia		Canada	311	V	N	F
61.	Gorda Fan		USA	113-OC		N	P
62.	Grand Banks	Avalon	Canada	114	III C/III A	G*	G/E
63.	Greenland East Central	East Greenland	Greenland	1141	III C	N	F/G
64.	Green River		USA	222	II A	S	F/G
65.	Guaymas	Incl. Imperial Valley, Salton Trough, Mazatlan, & Topolobampa Sub-Basins	Mexico/USA	111	III A	S	F/G
66.	Gulf Coast		Mexico/USA	1143	II C c/IV	G	E
67.	Harney		USA	331	III B c	N	P/F
68.	Hecate Fan		Canada	113-OC		N	P
69.	Henry Mountains		USA	222	II A	N	F
70.	Holitna		USA (Alaska)	321	III B	N	F
71.	Hope		USA (Alaska)/USSR	321	III B c	N	F/G
72.	Hudson Bay		Canada	1211	I	G	F/G
73.	Hudson Fan		USA	113-OC		N	P
74.	Illinois		USA	121	I	G	F/G
75.	Jones-Lancaster		Canada	121	I + III A	N	P/F
76.	Kotzebue	Selawik	USA (Alaska)	331	III B c	N	F
77.	Kodiak		USA (Alaska)	311	V	S	P/F
78.	Kaiparowits		USA	22	II A	N	F
79.	Knud Rasmussen		Greenland	121	I or II B	N	P
80.	Kronprins Christian	East Greenland	Greenland	1141	III C	N	F/G
81.	La Jolla Fan		USA	332	III B b	N	P
82.	Labrador	Incl. Newfoundland Shelf & Basin	Canada	114	III C	S*	G/E
83.	Lake Superior		Canada/USA	121	I	N	P/F
84.	Lakeview		USA	331	III B c	N	P/F
85.	Laramie		USA	222	II A	S	F/G

#	Basin	Country	Detail	Code	Class		
86.	Laurentian Fan	Canada		113-OC	II A	N	P
87.	Liard	Canada		41		S*	F
88.	Lomonosov	Canada/USSR		1141	III B b	N	P
89.	Los Angeles	USA		332	II C c	G	G/E
90.	Louisiana Salt Dome	USA		1143	IV	G	G/E
91.	MacKenzie	Canada	MacKenzie Delta, Incl. Kaktovik, Beaufort Sea	1142		G*	E
92.	MacKenzie Cone	Canada		113-OC	II A	N	P
93.	MacKenzie Plains	Canada		41		N	F/G
94.	Magdalena Fan (Mexico)	Mexico		113-OC	II A	S	P
95.	Marathon/Ouachita/Eastern Overthrust	USA		41	II A	N	F/G
96.	McKinley	Greenland	North Greenland	1142	III C	N	P/F
97.	Melville	Canada		311	II A	N	P/F
98.	Mexican Overthrust	Mexico		41	II C c	N	F/E
99.	Michigan	USA		1211	I	S	F/G
100.	Mid-Greenland	Greenland		121	II B or III A	N	P
101.	Minchumina	USA (Alaska)		331	III B	N	G
102.	Mississippi Fan	USA		113-OC	II C c	N	G
103.	Mississippi Salt Dome	USA		1143	II C c	S	P
104.	Monterrey Fan	USA		113-OC		N	F
105.	Navarin	USA (Alaska)/USSR		3122	III B b	N	P
106.	Navy Fan	Mexico/USA		332	III B b	N	P
107.	Nitinat Fan	Canada/USA		1143		N	G
108.	Northeast Gulf Salt Dome	USA		221	II C c	G	E
109.	North Slope	USA (Alaska)	Colville	331	II C c	G	F
110.	Norton	USA (Alaska)		114	III B c	N	F/G
111.	Nuwuk	USA (Alaska)	Beaufort Shelf	121	III C	N	P
112.	Ole	Greenland		21-OC	II B or III A	N	P
113.	Pacific Northeast Trench	USA		222		N	F
114.	Palo Duro	USA		222	II A	S	F/G
115.	Paradox	USA		222	II B	G	F
116.	Park, North	USA		222	II A	S	F
117.	Park, South	USA		222	II A	N	P/F
118.	Pasco	USA	Washington Tertiary	5	III B c	N	G
119.	Peel	Canada		221	II A	N	G/E
120.	Permian	USA	Incl. Delaware & Midland Sub-Basins	222	II A	G	G
121.	Peten-Chiapas	Mexico/Guatemala	Incl. Chiapas Depression & Chapayal Basin	221	II C c	s	F
122.	Piceance	USA		222	II B	s	P
123.	Portlock	USA (Alaska)		311	V	N	G
124.	Powder River	USA		222	II A	G	P
125.	Pratt	USA (Alaska)		311	V	N	P/F
126.	Purisima	Mexico		332	III B b	N	P
127.	Queen Charlotte	Canada		332	III B b	N	P

*Contains Undeveloped Discovery

128.	Quesnel		Canada	321	III B c	N	P/F
129.	Quincy		USA	5	III B c	N	P/F
130.	Raton		USA	222	II B	S	F/G
131.	Rio Grande		USA	111	III A	N	F/G
132.	Rocky Mountains	Alberta-British Columbia Foothills	Canada	41	II A	S	F/G
133.	Sabinas		Mexico	41	II C c	S	G
134.	Sacremento/San Joaquin		USA	332	III B b	G	G
135.	Salina/Forest City		USA	121	I	S	F
136.	Salinas (Mexico)	Isthmus Saline	Mexico	22	II C c	G	E
137.	Salinas (California)		USA	332	III B b	S	F
138.	Sand Wash		USA	222	II A	S	F
139.	San Juan		USA	222	II B	G	G
140.	San Lucas Fan		Mexico	113-OC		N	P
141.	San Luis		USA	222	II B	S	F
142.	Santa Maria		USA	332	III B b	s	G/E
143.	Scotia Shelf		Canada	1141	III C	s*	F/G
144.	Shumagin		USA (Alaska)	331	V	N	P/F
145.	Sigsbee Deep		Mexico/USA	113-OC	II C c	N	F/G
146.	Snake River		USA	5	III A	s	P/F
147.	South Texas Salt Dome		USA	1143	II C c	N	F/G
148.	St. George	Incl. Zhemchug, Pribilof, Otter, Dalroi, Garden, Walrun & Umnak Sub-Basins	USA (Alaska)	322	III B b	N	F/G
149.	St. Lawrence, Gulf of	Incl. Anticosti & New Brunswick-Magdalen Sub-Basins	Canada	322	III A	N	P/F
150.	St. Matthew	Incl. Hall	USA (Alaska)	3122	III B	N	F/G
151.	Sverdup		Canada	1143	II C c	G*	G/E
152.	Sydney		Canada	321	III A	N	P/F
153.	Tanana, Lower		USA (Alaska)	331	III B	N	P
154.	Tanana, Middle		USA (Alaska)	331	III B	N	P
155.	Tanana, Upper		USA (Alaska)	331	III B	N	P
156.	Tampico	Incl. Chicontepec & Tuxpan	Mexico	1134	II C c	G	G/E
157.	Tofino		Canada/USA	311	V	S	P/F
158.	Tugidak		USA	311	V	N	P
159.	Tyee	Western Oregon-Washington	USA	311	V	N	P/F
160.	Uinta		USA	222	II A	G	G/E
161.	Umatilla		USA	331	III B c	N	P/F
162.	Ungava Bay		Canada	121	I	N	P
163.	Vancouver Fan		Canada	113-OC		N	P
164.	Ventura	Incl. Santa Barbara Channel	USA	332	III B b	G	G
165.	Veracruz		Mexico	221	III C c	S	G
166.	Victoria Strait		Canada	311	I + III A	N	P/F
167.	Viscaino	Sebastian	Mexico	332	III B b	s*	F
168.	Washakie		USA	222	II A	S	F
169.	Western Overthrust		USA	41	II A	G	G
170.	Whitehorse		Canada	321	III B c	N	P

#	Province Name	Alternative Province Name	Country	Bally	Klemme	Production	Hydrocarbon Potential
171.	Williston		Canada/USA	121	I	G	G
172.	Wind River		USA	222	II A	S	G
173.	Winona		Canada	311	V	N	P/F
174.	Wollaston		Canada	121	I	N	P/F
175.	Yakima	Washington Tertiary	USA	5	III B c	N	P/F
176.	Yukon/Kandik	Kandik-Porcupine/Yukon Flats	USA (Alaska)	331	III B	N	P/F
177.	Zodiac Fan		USA (Alaska)	113-OC		N	P

*Contains Undeveloped Discovery

Central and South America

#	Province Name	Alternative Province Name	Country	Classification		Production	Hydrocarbon Potential
				Bally	Klemme		
178.	Abrolhos		Brazil	113-OC	II A	N	P
179.	Acre		Brazil	222		N	G
180.	Altiplano		Bolivia/Chile	331	III B c	N	F
181.	Amatique		Belize/Guatemala	1142	III C	N	F
182.	Amazon Cone		Brazil	113-OC	I	N	P
183.	Amazon, Lower		Brazil	1211	I	N	F
184.	Amazon, Middle		Brazil	1211	I	N	F
185.	Amazon, Mouth of the		Brazil	1141	III C/IV	S	G
186.	Amazon, Upper		Brazil	1211	I	S	F/G
187.	Ana Maria		Cuba	322-OC	III B b	S	P/F
188.	Ancud	Ancud-Osorno	Chile	311	V	N	P/F
189.	Argentina, Northwest	Incl. Carandaity	Argentina/Bolivia	221	II A	S	G
190.	Argentine		Argentina	113-OC		N	P
191.	Atacama		Chile	311	V	N	P
192.	Atrato	Pacific Coastal	Columbia	311	V	N	F/G
193.	Bahia Sul	Jequitinhonha	Brazil	1141	III C	N	F/G
194.	Barbados	Lesser Antilles	Barbados/Lesser Antilles	42	V	S	F
195.	Barreirinhas		Brazil	1142	III C	N	F/G
196.	Bato Bano	Palacios	Cuba	332	III B b	S	F
197.	Beni		Bolivia	221	II A	N	F
198.	Bonaire		Venezuela/Netherlands Antilles	332	III B b	N	F
199.	Burdwood Bank		Disputed UK/Argentina	114	III B or II C	N	F/G
200.	Campos		Brazil	1141	III C	S	G
201.	Cariaco		Venezuela	331	III B b	N	F
202.	Cauca		Columbia	322	III B c	N	P/F
203.	Cauto		Cuba	322	III B b	N	P
204.	Cayman		Cuba	322	III B b	N	P
205.	Ceara		Brazil	1142	III C	S	F/G
206.	Cesar, Central		Columbia	33	III B c	N	P/F

No.	Basin	Notes	Country	Code	Class		
207.	Chaco	Incl. Pirity	Argentina/Paraguay	121	I/II A	N	F
208.	Chile Trench, North		Chile	21-OC	III Bb	N	P
209.	Chile Trench, South		Chile	21-OC	III Bb	N	P
210.	Chiriqui	Incl. Valle Del General	Costa Rica/Panama	321	III Bb	N	P
211.	Cibao		Dominican Republic	332	III A/III C	S	P
212.	Cochinos		Cuba	322	III Bb	N	P/F
213.	Colorado	Rio Colorado	Argentina	1141	III Bc	N	F
214.	Cuba, Central		Cuba	322	II B	S	P
215.	Cuenca		Ecuador	332	V	N	F/G
216.	Cuyo	Cuyo-Atuel	Argentina	222	III C	S	P
217.	Darien	Chacunaque	Panama	311	III Bb	S	F/G
218.	Espirito Santo		Brazil	114	III C	N	F/G
219.	Falcon		Venezuela	332	III Bb	N	F/G
220.	Falkland		Disputed: UK/Argentina	1142	III C	S	F/G
221.	Grenada		Lesser Antilles	3121-OC	III Ba	S	P
222.	Guajira		Columbia/Venezuela	332	III Bb	N	F
223.	Guayaquil, Gulf of	Santa Elena	Ecuador/Peru	332	III Bb	N	F/G
224.	Guiana		French Guiana/Guyana/Surinam	114	III C	N	F
225.	Haitian/Dominican	Enriquillo Cul-de-sac	Haiti/Dominican Republic	332	III Bb	N	P/F
226.	Honduras Northcoast		Honduras	1142	III C	N	P/F
227.	Jamaica East		Jamaica	322	III B	S	P
228.	Jamaica West		Jamaica	114	III C	N	P
229.	Jatoba		Brazil	111	III A	N	P/F
230.	Jipijapa	Incl. Borbon & Doule Sub-Basins	Ecuador	311	V	N	P
231.	Limon-Boca Del Toro		Costa Rico/Panama	311	V	S	F
232.	Linares	Central Valley	Chile	311	V	N	P
233.	Llanos De Casanare	Incl. Apure & Barinas Sub-Basins	Columbia/Venezuela	221	II A	N	F/E
234.	Los Roques	La Vela Embayment	Netherlands Antilles	331	III Bb	N	F
235.	Madre De Dios		Bolivia/Peru	221	II A	S	G
236.	Magallanes		Argentina/Chile	221	II A	S	G/E
237.	Magdalena Delta		Columbia	42	IV	N	F/G
238.	Magdalena Fan (Columbia)		Columbia	3121-OC	III Bc	N	P
239.	Magdalena, Lower		Columbia	332	III Bc	S	F/G
240.	Magdalena, Middle		Columbia	332	III Bc	G	G
241.	Magdalena, Upper		Columbia	332	III Bc/II A	S	F/G
242.	Malvinas		Argentina	1142	III Bc	G	G/E
243.	Maracaibo	Incl. Chichibacoa	Venezuela	222	II Ca	G	G/E
244.	Maturin	Incl. Orinoco Tarbelt/Guarico & Tuy Sub-Basins	Venezuela	221		G	G
245.	Mid-America Trench		Central Americas	21-OC		N	P
246.	Muertos		Dominican Republic/Puerto Rico	42	V	N	P

#	Name	Description	Country				
247.	Neuquen		Argentina	221	II B	S	F/G
248.	Nicaragua Rise		Honduras/Nicaragua	1143	III C	N	P/F
249.	Nicoya		Costa Rica	311	V	N	P
250.	Nirihuau		Argentina	221	III B c	N	F
251.	Oriente	Maranon	Peru	221	II A	S	G/E
252.	Orinoco Delta	Offshore Maturin	Trinidad/Venezuela	114	IV	S	G/E
253.	Pacifico	Pacific Coastal	Columbia	311	V	N	P
254.	Panama Central		Panama	311	V	N	P
255.	Panama Gulf		Panama	311	V	N	P
256.	Para-Maranhao		Brazil	1141	III C	N	F
257.	Parana		Brazil/Paraguay/Argentina	121	I	N	P/F
258.	Paria		Trinidad/Tobago/Venezuela	332	II C a	S	F
259.	Parnaiba	Maranhao	Brazil	1141	I	N	P/F
260.	Pelotas		Brazil/Uraguay	114	III C	N	F
261.	Peru Coastal	Peruvian Shelf	Peru	311	V	N	F/G
262.	Piedrabuena	Patagonian	Argentina	1141	II B	N	F
263.	Potiguar		Brazil	1142	III C	S	G
264.	Puerto Rico		Puerto Rico	42	III B	N	P
265.	Putumayo	North Coast	Columbia/Peru/Ecuador	221	II A	G	G/E
266.	Reconcavo		Brazil	111	III A	S	F/G
267.	Rio Grande Rise		Brazil	113-OC	III A/III C	N	P
268.	Salado	Rio Salado, Santa Lucia	Argentina/Uruguay	1141	V	N	P/F
269.	Sambu		Panama	311	V	N	P
270.	San Jorge	Incl. Rio Mayo	Argentina	1141	III A	G	G
271.	San Julian	Patagonian	Argentina	1141	II B	N	F
272.	Santiago		Panama	311	V	N	P
273.	Santos		Brazil	1141	III C	N	F/G
274.	Sao Francisco	Bambui	Brazil	121	I	S	P/F
275.	Sergipe-Alagoas		Brazil	1141	III C	N	F/G
276.	Sinu-Uraba		Columbia/Panama	311	V	N	F
277.	South Georgia		Disputed: UK/Argentina	1142	III B or II C	N	P/F
278.	Tacatu		Brazil/Guyana	111	III A	S*	F
279.	Talara		Peru	111	III B b	G	G
280.	Tehuantepec	Incl. Guatemalan, Nicaraguan, & El Salvador Shelves	Central Americas	311	V	N	P
281.	Trinidad-Tobago	Tobago	Trindad/Tobago	42	III B b	N	F/G
282.	Tucano		Brazil	111	III A	S	F
283.	Ucayali		Peru	221	II A	S	F/G
284.	Ulua		Honduras	331	III B b	N	P
285.	Valdez		Argentina	1141	II B	N	P/F
286.	Venezuela, Gulf of		Venezuela	331	III B b	N	F/G
287.	Yari		Columbia	331	II A	N	P

*Contains Undeveloped Discovery

Africa and Madagascar

#	Province Name	Alternative Province Name	Country	Classification Bally	Classification Klemme	Production	Hydrocarbon Potential
288.	Aaiun		Mauritania/Spanish Sahara	1143	III C	N	F/G
289.	Alboran		Algeria/Morocco	322	III B	N	P
290.	Algerian Offshore	Incl. Chelif	Algeria	332	III Bc	N	P/F
291.	Algoa		South Africa	1142	III C	N	F
292.	Atlas	Incl. Hodna-Constantine	Algeria/Morocco	41	II Cb	S	F/G
293.	Avon Fan		Benin/Nigeria	113-OC		N	P
294.	Benue		Nigeria	111	III A	N	G/E
295.	Berbera		Somalia	1141	III C	N	P
296.	Cabinda		Congo/Gabon/Zaire/Angola (Cabinda)	1141	III C	G	G
297.	Calabar Fan		Equatorial Guinea/Cameroon/Nigeria	113-OC		N	P
298.	Cameroon		Equatorial Guinea/Cameroon	114	III C	S	G
299.	Chad		Chad/Niger/Nigeria	121	I	S	G
300.	Colomb-Bechar		Algeria		II B	N	F
301.	Congo Fan		Cabinda/Congo	113-OC	III C	N	P
302.	Coriole	Lamu	Kenya/Somalia	1141	III C	N	F/G
303.	Cuanza		Angola	1141	III C	S	G/E
304.	Cyrenaica		Libya	1141	II Ca/c	S	F/G
305.	Doba		Central African Republic/Chad	111	III A	N	F
306.	Essaouira		Morocco	1141	III C	S	F/G
307.	Etosha		Angola/Namibia	121	I	N	F
308.	Farafra		Egypt	111	III A	N	P/F
309.	Gabon	Dakhla	Gabon	1141	III C	G	G/E
310.	Gamtoos		South Africa	1142	III C	N	P/F
311.	Gao		Mali/Niger	111	III A	N	P/F
312.	Ghadames		Algeria/Libya	121	III B	G	G/E
313.	Ghana	Dahomey Coastal	Benin/Ghana/Togo	1142	III C	S	F/G
314.	Giuba	Lamu,Ogaden, Juba	Ethiopia/Kenya/Somalia	1141	II B	N	F
315.	Hafun	Somalia	Somalia	1141	II B/III C	N	F/G
316.	Illizi	Polignac	Algeria	121	I	G	G
317.	Ivory Coast		Ivory Coast	1142	III C	S	G/E
318.	Kalahari	Incl. Passarge, Nosop, & Ncojane Sub-Basins	Botswana/Namibia/South Africa	121	II B	N	P/F
319.	Kaouar		Niger	111	III A	N	F
320.	Karoo		South Africa	221	II A	N	F
321.	Kenya	Lamu	Kenya	111	II B	N	P/F
322.	Kufra		Chad/Libya/Sudan	121	I	N	P/F
323.	Lake Albert		Uganda/Zaire	111	III A	N	P
324.	Lake Edward		Zaire	111	III A	N	P

No.	Basin	Notes	Location				
325.	Lake Kivu		Zaire	111	III A	N	P
326.	Lake Nyasa		Tanzania	111	III A	N	P
327.	Lake Rudolph		Kenya	111	III A	N	P
328.	Lake Tanganyika		Tanzania/Zaire	111	III A	N	P
329.	Liberian	Incl. Sierra Leone Coastal	Sierra Leone/Liberia	1141	III C	N	F
330.	Limpopo	Mozambique	Mozambique	1141	III C	N	F/G
331.	Majunga		Madagascar	1141	III C	N	F
332.	Mauritius-Seychelles		Mauritius/Seychelles	1141		N	P/F
333.	Mbarangandu		Tanzania	121	II B	N	P/F
334.	Mediterranean, East		Egypt/Greece/Libya	1141	II Ca/c	N	P/F
335.	Mocamedes		Angola/Namibia	1141	III C	N	F/G
336.	Morondava		Madagascar	1141	III C	N	F
337.	Mozambique Fan		Madagascar/Mozambique	113-OC	III C	N	P/F
338.	Mudugh	Ogaden, Somalia	Ethiopia/Somalia	1141	II B/III C	N	F
339.	Murzuk		Libya	121	I	N	F
340.	Niger Delta		Nigeria	1141	IV	G	E
341.	Niger Fan		Nigeria	113-OC		N	P
342.	Nile		Egypt	1141	II Cc/a	N	F/G
343.	Nile Delta		Egypt	114	IV	S	G/E
344.	Nile Fan		Egypt	113-OC	IV	N	P
345.	Okavango Delta		Botswana/Zambia/Zimbabwe	121	I	N	P
346.	Orange River		Namibia/South Africa	1141	III C/IV	N	F
347.	Pelagian	Cape Bon	Libya/Tunisia	1141	II Ca	S	G/E
348.	Red Sea, West		Egypt/Ethiopia/Saudi Arabia/Sudan	111	III A/III C	N	F/G
349.	Reggane		Algeria	222	II A	N	F
350.	Rharb		Morocco	221	II Cb	S	P/F
351.	Sahara		Algeria	121	II B	G	G
352.	Senegal	Incl. Bove	Mauritania/Gambia/Guinea/Senegal	1141	III C	N	F
353.	Seychelles, West		Seychelles	1141	III C	N	P
354.	Sirte		Libya	1211	III A	G	E
355.	Sirte, Gulf of	Gulf of Sidra	Libya	1141	III A	G	F/G
356.	South African		South Africa	1142	III C	S*	F
357.	Sudan	Sudd	Sudan	111	III A	S*	E
358.	Suez, Gulf of		Egypt	111	III A	G	G
359.	Tafassasset		Niger	111	III A	N	P/F
360.	Talak		Niger	111	III A	N	P/F
361.	Tamatave		Madagascar	113-OC	III A	N	P
362.	Tanzania		Mozambique/Tanzania	1141	III C	S*	G

*Contains Undeveloped Discovery

#	Province Name	Alternative Province Name	Country	Bally	Klemme	Production	Hydrocarbon Potential
363.	Taoudenni	Incl. Tanzerouf	Mali/Mauritania	121	I	N	P
364.	Tarfaya	Incl. Rio Del Oro	Morocco/Spanish Sahara	114	III C	N	F
365.	Timimoun	Macmahon, Saoura	Algeria	222	II B	S	G
366.	Tindouf		Algeria/Morocco Spanish Sahara	221	II A	N	F
367.	Volta		Ghana	221	I	N	P
368.	Walvis		Namibia	114	III C	N	P/F
369.	Walvis Ridge		Namibia	113-OC		N	P
370.	Western Desert	Northern	Egypt	1141	II C a/c	S	F/G
371.	Zaire	Belgian Congo	Zaire	221	II B	N	F

Europe and Turkey (Excluding USSR)

#	Province Name	Alternative Province Name	Country	Classification		Production	Hydrocarbon Potential
				Bally	Klemme		
372.	Adana		Cyprus/Turkey	322	III B c	S	F/G
373.	Adriatic, North		Italy/Yugoslavia	21	II C b	S	G
374.	Adriatic, South		Albania/Italy/ Yugoslavia	221	II C b	S	G
375.	Aegean	Incl. Kavala	Greece/Turkey	3121	III B c	S	F
376.	Antalya		Turkey	221	III B c	N	P/F
377.	Aquitaine		France	221	III A	G	G
378.	Balearic	Western Mediterranean	Spain	322	III A	N	F
379.	Barents West		Norway	1211	I	N	G/E
380.	Barents Fan		Norway	113-OC		N	P/F
381.	Bay of Biscay		France/Spain	1141	III C	N	F/G
382.	Bresse	Jura	France	111	III A	S	F
383.	Cadiz	Algarve	Portugal/Spain	321	II C b	N	F
384.	Caltanisetta	Sicilian Depression	Sicily	41	II C b	S	G
385.	Cankiri		Turkey	331	III B c	N	P/F
386.	Cantabrian		Spain	332	III A	S	F
387.	Cap Ferret Fan		France	113-OC		N	P
388.	Carpathian		Czechoslovakia/ Poland/Rumania/ USSR	41	II C b	S	G
389.	Celtic		Ireland/UK	1141	III A	S	F/G
390.	Crete		Greece	3122	III B c	N	P
391.	Cuenca	La Mancha	Spain	222	II B	N	P/F
392.	Diyarbakir		Syria/Turkey	221	II C a	S	F/G
393.	Duero	Castillan	Spain	121	II B	N	F
394.	Ebro		Spain	221	II C b	N	F/G
395.	Ebro Fan		Spain	221	III A	S	F/G
396.	Faeroes		Faeroe Islands	113-OC (Seamount)		N	P
397.	Gaziantep		Syria/Turkey	221	II C a	S	F/G
398.	German, Northwest		Netherlands/ West Germany	121	II B	G	G

170

No. Name	Note	Country	Code	Stratigraphy	S	F/G
399. Graz		Austria/Czechoslovakia	321	III Bc	S	F/G
400. Guadalquivir		Spain	321	II Cb	N	F
401. Hammerfest		Norway	1141	III A	N	F/G
402. Harstad		Norway	1141	III C	N	F/G
403. Hatton		UK	1141	III C	N	P/F
404. Hebrides		UK	1141	III A	N	P/F
405. Helgeland	Western Shelf	Norway	1141	III C	N	G
406. Ionian	Incl. Epirus & Peloponesus	Greece/Italy	41	II Cb	S	G/E
407. Irish	Incl. Part of North Sea	Ireland/UK		II B	S	G/E
408. Kattegat		Denmark/Sweden	121	II B	N	F
409. Jan Mayen		Jan Mayen Island	113-OC (Seamount)		N	P
410. Ligurian	Western Mediterranean	France	322		N	G
411. Lusitania	Lisbon, Incl. Galicia	Portugal	1141	III C	N	F
412. Malatya		Turkey	331	III Bc	N	P
413. Midlands		UK	111	III A	S	F
414. Minch		UK	1141	III A	N	F
415. Moesian		Bulgaria/Rumania	221	II Cb	G	G
416. Molasse		Austria/Germany/Switzerland	221	II Cb	S	F
417. Møre	Western Shelf	Norway	1141	III C	N	F/G
418. Morella	Castillan	Spain	222	II B	N	F
419. Mus		Turkey	331	III Bc	G	G
420. North Sea, Northern	Incl. Viking Graben	Netherlands/UK/	1211	III A	G	E
421. North Sea, Southern	Incl. Central Graben	Norway/Denmark/West Germany	1211	II B/III A	G	G/E
422. Pannonian		Hungary/Rumania	321	III Bc	S	F
423. Paris		France	1211	II B	S	F
424. Po		Italy	221	II Cb/IV	G	G/E
425. Polish	Incl. NE German Lowlands, Lodz Depression, & Warsaw-Lublin-L'vov Depression	East Germany/Poland	1211	III A	S	F/G
426. Porcupine		Ireland	1141	III A	N	P/F
427. Prins Karl		Norway	113-OC	III A	N	F
428. Rhine		France/W. Germany	111	III A/IV	S	F/G
429. Rhone Delta		France	111		S	P/F
430. Rhone Fan		France	322-OC		N	P/F
431. Rockall		UK	1141	III C	N	P/F
432. Salonika		Greece	322	III Bc	N	P
433. Sardinia	Western Mediterranean	Sardinia	322		N	P/F
434. Shetlands, West		UK	1141	III C/III A	N	F
435. Sivas		Turkey	331	III Bc	N	P
436. Skaggerak	Part of North Sea	Denmark/Norway/Sweden	1211	II B	N	P/F
437. Siret		Poland/USSR	121	II B	S	F/G
438. Taranto	Incl. Calabrian	Italy	221	II Cb	S	G

*Contains Undeveloped Discovery

#	Province Name	Alternative Province Name	Country	Classification Bally	Klemme	Production	Hydrocarbon Potential
439.	Terre Noir		France	221	II C b	N	P
440.	Thrace		Turkey	331	III B c	N	F
441.	Transylvania		Romania	321	III B c	S	F/G
442.	Tromsø		Norway	1141	III A	G*	G/E
443.	Tuz Golu		Turkey	331	III B c	N	P/F
444.	Tyrrhenian		Corsica/Italy	322		S	P/F
445.	Upper Silesian		Czechoslovakia/Poland	221	II C b		F
446.	Vienna		Austria	321	III B c	G	F
447.	Vøring		Norway	1141		N	F
448.	Wessex		UK	1211	II B	S	F/G
449.	Western Approaches		France/UK	1141	III A	N	F/G

Middle East

#	Province Name	Alternative Province Name	Country	Classification		Production	Hydrocarbon Potential
				Bally	Klemme		
450.	Arabian	Persian Gulf	Bahrain/Kuwait/Neutral Zone/Iraq/Oman/Qatar/Saudi Arabia/UAE/Syria/Turkey	221	II C a	G	E
451.	Dasht-I-Kavir	Grand Kavir	Iran	331	III B c	S	F
452.	Dasht-I-Lut		Iran	331	III B	N	F
453.	Dead Sea		Israel/Jordan	111	III A	N	F
454.	Djaz Murian		Iran	331	III B	N	P/F
455.	Fartak		Oman/Yemen	111	III A	N	F
456.	Isfahan		Iran	331	III B c	N	P/F
457.	Levantine	Lebanese	Israel/Syria/Lebanon	1141	II C a/b	N	F
458.	Masira		Oman/UAE	1141	III C	N	P/F
459.	Mukalla		Yemen	1141	III C	N	P/F
460.	Oman Gulf	North Oman	Oman	21-OC	II C a	N	P
461.	Red Sea East		Egypt/Ethiopia/Saudi Arabia/Sudan	111	III A/III C	G*	G
462.	Sinai		Egypt/Israel	1141	II C a/c	S	F
463.	Tabriz		Iran	331	III B	N	F
464.	Zagros	Iranian Foldbelt	Iran/Iraq/Syria/Turkey	41	II C a	G	E

Asia and Oceania

#	Province Name	Alternative Province Name	Country	Classification Bally	Classification Klemme	Production	Hydrocarbon Potential
465.	Adavale		Australia	1143	III A Local/II B Regional	N	F
466.	Agusan		Philippines	332	II B b	N	P
467.	Akita		Japan	3122	III B a	S	F
468.	Amadeus		Australia	41	II A	S	F/G
469.	Anambas		Malaysia	3122	III A	N	F/G
470.	Andaman		Andaman Island Archipelago	42	V	N	P
471.	Andaman Sea		Burma/Indonesia/India/Malaysia	3121-OC	III C	N	F
472.	Arafura		Australia/Indonesia	1141/221	II C a	N	F/G
473.	Arckaringa		Australia	121	III A Local/II B Regional	N	P/F
474.	Assam		India	221	II C b	S	F
475.	Auckland		New Zealand	3121-OC		N	P
476.	Bali		Indonesia	3121-OC	III B a	N	P
477.	Baluchistan		Iran/Pakistan	41	II C b	N	P/F
478.	Banda North		Indonesia	3121-OC		N	P/F
479.	Banda South		Indonesia	3121-OC		S	F
480.	Bangkok		Thailand	3122	III A	N	F
481.	Banka East		Indonesia	3122	III B a	N	F
482.	Banka West		Indonesia	3122	III B a	N	F
483.	Barito		Indonesia	3122	III B a	S	F
484.	Bass		Australia	1141	III A	S*	G
485.	Bawean		Indonesia	3122	III B a	N	P/F
486.	Beagle		Australia	1142	III C	N	F
487.	Beibu Gulf	Gulf of Tonkin	China/Vietnam	3122	III B/IV	S*	G
488.	Bengal		Bangladesh/Burma/India	221/41/42	II Ca/II C b/IV	S	G
489.	Bengal Fan		Bangladesh/Burma/India	113-OC		S	F/G
490.	Benkulu		Indonesia	311	V	N	P
491.	Bicol		Philippines	332	III B b	N	P
492.	Biliton		Indonesia	3122	III B a	N	F/G
493.	Bintoeni		Indonesia	221	II C a	S	F/G
494.	Bismarck		Bismarck Archipelago	3121-OC		N	P
495.	Bohai Gulf	Offshore Huabei, Pohai	China	3122	III A	S	G/E
496.	Bombay		India	1141	III C	G	G/E
497.	Bonaparte Gulf	Incl. Sahul Platform	Australia	1141	III C/II B	S*	F/G
498.	Bone Gulf		Indonesia	3121-OC		N	P/F
499.	Bougainville		Bougainville Is.	42		N	P
500.	Bowen		Australia	121	III A Local/II B Regional	S	F/G

*Contains Undeveloped Discovery

173

No.	Basin	Notes	Region	Code	Class.		
501.	Broken Ridge		Indian Ocean	1141/113-OC	III C	N	P
502.	Browse		Australia	1141	II C c	S*	F/G
503.	Brunei-Sabah	Sarawak	Malaysia	3122	III B b	G	G
504.	Burma, North	Chindwin	Burma	3122	III B b	N	F
505.	Burma, South	Irrawaddy Embayment, Minbu	Burma	3122	III B	S	F
506.	Cagayan		Philippines	332	III A	N	P
507.	Cambay		India	1141		S	F/G
508.	Campbell		New Zealand	1141	III A or III B b	N	F
509.	Canning		Australia	1211	II B	S	F/G
510.	Canning Offshore		Australia	1141	III C	N	F/G
511.	Canterbury		New Zealand	1141	III A or III B b	N	F
512.	Capricorn		Australia	113-OC		N	P
513.	Carnarvon	Incl. Gascoyne, Onslow, & Merlinleigh Sub-Basins	Australia	1142	III A	S	F/G
514.	Carnarvon Offshore		Australia	1141	III C	N	F/G
515.	Carpentaria, Gulf of		Australia	121	I	N	P/F
516.	Cauvery	Incl. Ramnad	India	1141	III C	N	F
517.	Celebes		Indonesia/Philippines	3121-OC		N	P
518.	Central Valley		Philippines	311	V	N	P
519.	Ceram		Indonesia	41	II C a	S	F
520.	Chagos		Indian Ocean	113-OC (Seamount)		N	P
521.	Chaidamu	Tsaidam, Qaidam	China	23	III A	G	G/E
522.	Challenger		Australia/New Zealand	1141	III C	N	P/F
523.	Changdu		China	23	III B	N	F
524.	Chao Phraya		Thailand	3122	III A	S	F/G
525.	Chatham		New Zealand	1141		N	F
526.	Chuxiong		China	33	III B	N	P/F
527.	Clarence-Moreton		Australia	321		N	P
528.	Cook		Cook Islands	113-OC (Seamount)		N	G
529.	Cooper		Australia	1211	III A Local / II B Regional	G	G
530.	Coral Sea		Australia	113-OC	III B b	N	F
531.	Cota Bato		Philippines	332	I	N	P
532.	Daly River		Australia	121	III C	N	P/F
533.	Dampier	Incl. Rankin, Barrow, & Exmouth Sub-Basins	Australia	1141	III A	G*	G
534.	Dandaragon	Incl. Coolcalalaya & Byro Sub-Basins	Australia	1142	III B	S	F/G
535.	Danjo	Kyushu	Japan	3122	III B b	N	P
536.	Davao		Philippines	332	III C	N	P
537.	Drummond		Australia	121	III B a	N	F
538.	Duntroon		Australia	1141		N	F
539.	East China	Wenchow	China/Japan/South Korea	3122	III B a	S*	G/E

174

No.	Name	Notes	Country	Code	Classification		
540.	Enshunada-Kumanonada	Incl. Sagara	Japan	311	V	S	F
541.	Erlian		China	23	III A	N	P
542.	Eucla	Incl. Polda Trough	Australia	114	III C	N	F
543.	Fang		Thailand	3122	III A	S	F
544.	Fiji		Fiji Islands	42		N	P
545.	Fitzroy		Australia	111	III A	N	G
546.	Flores		Indonesia	3121-OC		N	P
547.	Galilee		Australia	121	III A Local/ II B Regional	N	F
548.	Ganges	Ganga	India	221	II C b	N	F
549.	Georgina		Australia	121	I	N	P/F
550.	Gippsland		Australia	111	III A	G	G
551.	Godavari		India	111	III A	N	F/G
552.	Gorontalo		Indonesia	311	III B c	N	P/F
553.	Great South		New Zealand	1141	III A or III B b	N	F/G
554.	Guangxi-Guizou	Kwangsi-Kweichow, Kweicho-Hupei, Tien-Chien-Kuei Platform	China	121	II B	S	F/G
555.	Hailar	Hu-lun, Ch-ih	China	1141	III A	N	P
556.	Halifax		Australia	311	III C	N	P
557.	Halmahera		Indonesia	311	III B	N	P/F
558.	Hawkes Bay	East Coast, Gisborn	New Zealand	23	III B b	S	F
559.	Heihe		China	3122	III B	N	P/F
560.	Huabei	North China, Fei-Hsi	China	3121-OC	III A	S	F/G
561.	Huon	Cape Vogel	New Guinea	41	III B	N	P
562.	Hymalaya		China	332	III B b	N	P
563.	Iloiloi		Philippines	221	II C b	N	G
564.	Indus		Pakistan	1141	IV	G	F
565.	Indus Delta		Pakistan	113-OC		N	P
566.	Indus Fan	Arabian Fan	India/Pakistan	3122	IV	N	G/E
567.	Irrawaddy Delta		Burma	311	IV	N	F
568.	Ishikari-Hidaka	Hokkaido	Japan	3122	III B bV	S	F/G
569.	Java East		Indonesia	3122	III B a	S	F/G
570.	Java West		Indonesia	32	III B a	S	F
571.	Jianghan	Tung-T'ing Hu	China		III A	N	F/G
572.	Jiuquan	Pre-Nan Shan, Chiu-Ch'uan, & Minle Sub-Basins	China		III A	G	
573.	Joban		Japan	311	V	S	F
574.	Kanto		Japan	311	V	S	F
575.	Kapuas		Indonesia		III B b	N	P/F
576.	Kerala	Incl. Karnataka	India	1141	III C	N	F
577.	Kermadec		Pacific Ocean	3121-OC		N	P
578.	Khorat		Cambodia/Thailand/Laos/Philippines	121	II C a	S*	F/G
579.	Kitakami		Japan	311	V	N	P/F
580.	Korea Bay		China/Korea	3122	III A	N	F/G

*Contains Undeveloped Discovery

175

No.	Name	Note	Region	Code	Classification		
581.	Korea, East	Incl. Kyongsang	Korea	3121-OC	III C	N	F
582.	Krishna		India	1141	III C	S	F/G
583.	Kutch		India	1141	III A	N	P/F
584.	Laura		Australia	1141		N	P
585.	Line		Line Islands	113-OC (Seamounts)		N	P
586.	Lord Howe		Australia/New Caledonia	1141	III C		P/F
587.	Luzon		Philippines	322	III B	N	P
588.	Madan	Karkar	New Guinea	3122	III B	N	P/F
589.	Mae Sot		Thailand	3122	III B	S	F
590.	Mahakam		Indonesia	3122	III Ba/IV	G	G
591.	Mahanadi		India	1141	III C	N	F
592.	Makassar, South		Indonesia	113-OC		N	P/F
593.	Makran		Iran/Pakistan	42	III Cb	G	P/F
594.	Malay		Malaysia	3122	III A	N	F/G
595.	Maldives		Maldives Islands	113-OC (Seamounts)		N	P
596.	Marianas		Marianas Islands	113-OC (Island Arc)		N	P
597.	Marquesas		Marquesas Islands	113-OC (Seamounts)		N	P
598.	Martaban, Gulf of		Burma	3122	III C	N	F
599.	Maryborough		Australia	1141	III C	N	P
600.	Meervlakte		Indonesia	322	III Bb	N	F/G
601.	Mekong		Vietnam	3122	III Cc/IV	N	F
602.	Melawi		Indonesia	322	III Bb	N	P/F
603.	Meulaboh		Indonesia	311	V	N	P
604.	Mindoro East		Philippines	311	V	N	P
605.	Mindoro Southwest		Philippines	311	V	S	P/F
606.	Miyazaki-Hyuga		Japan	311	V	N	P/F
607.	Morehead		Indonesia/Papua	211	II Ca	S	P
608.	Morotai		Indonesia	322	III B	N	P/F
609.	Murray		Australia	121	II B	N	P/F
610.	Nanyang		China		III A	N	P
611.	Naturaliste		Australia	1141	III C	S	P/F
612.	Natuna		Indonesia/Malaysia	3122	III A	S	P/F
613.	New Caledonia		New Caledonia Is.	41	II Ca	N	P/F
614.	New Guinea		Indonesia/Papua	221	II A	N	F
615.	Ngalia		Australia	222	V	N	P/F
616.	Nias		Indonesia	311	V	N	P/F
617.	Nicobar		Nicobar Islands	42		N	P
618.	Nicobar Fan		Indonesia	113-OC		N	P
619.	Nicobar Trench		Burma/India/Indonesia	21-OC			
620.	Nigata		Japan	3122	III B a	S	F
621.	Ninetyeast Ridge		Indian Ocean	113-OC (Seamounts)		N	P
622.	Norfolk, North		New Caledonia	42	III B/V	N	P/F
623.	Norfolk, South		New Zealand	42	III B/V	N	P/F
624.	Northland		New Zealand	332	III Bb	N	P/F
625.	Officer		Australia	121	II A East/II B West	N	F

No.	Name	Alt. Basin	Country	Code	Classification		
626.	Okinawa		Japan/Okinawa	42	III B a	N	P/F
627.	Ord		Australia	121	I	N	P/F
628.	Otway		Australia	1141	III C	N	F
629.	Palar		India	1141	III C	N	P/F
630.	Palau		Palau Islands	113-OC (Island Arc)		N	P
631.	Palawan North		Philippines	311	III C	S	F/E
632.	Palawan South		Philippines	3122	III B a	N	P/F
633.	Papuan	Aure	Papua	221	II C a	S	F/G
634.	Pearl River Fan		China	3122	III C	S	G
635.	Pedirka		Australia	121	III A Local/ II B Regional	N	P/F
636.	Penyu		Malaysia	3122	III A	N	P/F
637.	Perth		Australia	1142	III A	N	F
638.	Phoenix		Phoenix Islands	113-OC (Seamounts)		N	P
639.	Potwar		Pakistan	41	II C b	S	F/G
640.	Pukaki		New Zealand	1141	III A or III B b	N	F
641.	Pukaki Embayment		New Zealand	1141	III A or III B b	N	F
642.	Queensland		Australia	1141	III C	N	F
643.	Ramu		New Guinea	322	III B b	N	P/F
644.	Rason		Australia	121	III C	N	P
645.	Recherche	Bremer	Australia	1141	III C	N	F/G
646.	Saigon		Vietnam	3122	II C c	N	G/E
647.	Salawati		Indonesia	221	II C a	S	G
648.	Samoa		Samoa Islands	113-OC (Seamounts)		S	G
649.	Sarawak		Malaysia	3122	II C c	N	P
650.	Savu		Indonesia	311	V	S	G
651.	Saurashtra		India	1141	III C	N	P/F
652.	Sepik		New Guinea	322	III B b	N	P
653.	Shandu Simao	Central Yunan	China	33	III B c	N	P
654.	Shanganning	Ordos	China	23	II A	N	G
655.	Shanghai	Yellow Sea	China/Japan/Korea	3122	III A	S	G
656.	Shatskiy		Pacific Ocean	113-OC (Platform)		N	P
657.	Siargao		Philippines	332	III B b	N	P
658.	Sichuan	Szechwan	China	22	II A	G	G/E
659.	Sistan	Helmand Depression	Afghanistan/Iran	331	III B c	N	F
660.	Sokang		Malaysia	3122	II C c	N	F
661.	Solander		New Zealand	32	III B	S	F
662.	Solomon		Solomon Islands	42	III A	N	P
663.	Songliao	Sung-Liao	China	3122	III A	G	G/E
664.	South China		China/Taiwan	3121-OC		N	P
665.	South Java Trench		Indonesia	21-OC		N	P
666.	South Yellow Sea	Subei	China/Korea	3122	III A	G	G
667.	Spratly		Malaysia	3121-OC		N	P
668.	Styx		Australia	1141	III A	N	P
669.	Sulu		Indonesia/ Philippines	3121-OC		N	P

*Contains Undeveloped Discovery

177

No.	Name	Notes	Country	Code	Classification		
670.	Sumatra, Central		Indonesia	3122	III B a	G	G/E
671.	Sumatra, North		Indonesia	3122	III B a	G	G/E
672.	Sumatra, South		Indonesia	3122	III B a	S	G/E
673.	Sunda		Indonesia	3122	III B a	S	G/E
674.	Surat		Australia	121	III A Local/ II B Regional	S	F/G
675.	Sydney		Australia	321	III A	N	P
676.	Taiwan		Taiwan	41	III B a	S	F/G
677.	Taliabu		Indonesia	3121-OC		N	P/F
678.	Tarakan		Indonesia	3122	III B b/III C	S	F/G
679.	Taranaki	Incl. Greymouth	New Zealand	1141	III A North/ II C a South	G	G
680.	Tarim	Incl. Charchan, Kucho, Kashgar-Yarkano, & Pre-Kunlun Shan Sub-Basins	China	23		S	G/E
681.	Tasman		Tasmania	1141	III C	N	P/F
682.	Tasmania, East		Tasmania	1141	III C	N	P/F
683.	Tasmania, West	Incl. King Islands	Tasmania	1141	III C	N	P/F
684.	Teshio	Northeast Hokkaido, Soya	Japan	322	III B b	S	F
685.	Thai		Cambodia/Thailand/ Malaysia	3122	III A	G	G
686.	Tibet, North		China	41	III B b	N	P
687.	Tokachi	Kushiro	Japan	332		N	P/F
688.	Tokelau		Tokelau Islands	113-OC (Seamounts)	V	N	P
689.	Toki-Tosa		Japan	311		N	P/F
690.	Tonga		Pacific Ocean	3121-OC		N	P
691.	Toyama		Japan	3122	III B a	N	F
692.	Tsushima		Japan, Korea	3122	III B a	N	G
693.	Tuamotu		Tuamotu Islands	113-OC (Seamounts)		N	P
694.	Tulufan	Turfan	China	23	III A	G	G/E
695.	Vanuatu		New Hebrides Is.	42		N	P/F
696.	Vietnam		Vietnam	3122	III C	N	F
697.	Visayan		Philippines	332	III B b	N	P
698.	Waikato		New Zealand	332	III B b	N	P
699.	Wanganui		New Zealand	332	III B b	N	F
700.	Westland		New Zealand	1141	III B b	N	F
701.	Wewak	Jayapura	New Guinea	322	III B b	N	P/F
702.	Wiso		Australia	121	I	N	P/F
703.	Yamato		Japan	3121-OC		N	F
704.	Yinggehai		China	3122	III C	N	F/E
705.	Zhungeer	Dzungaria	China	23	III A	G	G/E

*Contains Undeveloped Discovery

178

USSR and Mongolia

#	Province Name	Alternative Province Name	Country	Classification Bally	Classification Klemme	Production	Hydrocarbon Potential
706.	Aginskiy		Mongolia/USSR	3121-OC	III A	N	P/F
707.	Aleutian		USA/USSR	21-OC		N	P
708.	Aleutian Trench		USA/USSR			N	P
709.	Anabar-Lena		USSR	221	II A	N	F/G
710.	Amur	Eastern Black Sea, Rioni & Georgian	USSR	221	III B c	S	F/G
711.	Anadyr		USSR	331	III B b	S	F
712.	Angara-Lena	Nera-Botuobin	USSR	221	II A	S	G
713.	Balkhash		USSR	23	III A	N	F/G
714.	Baltic		Poland/Sweden/USSR	121	I	S	F/G
715.	Barents East		USSR	121	I	N	G
716.	Black Sea		Rumania/USSR/Turkey	322-OC		S	G
717.	Bowers		USA/USSR	3121-OC		N	P
718.	Bureya, Upper		USSR	331	III A	N	P
719.	Caspian, North		USSR	1211	II C a	G	G/E
720.	Caspian, South		USSR	222	III B c	G	G/E
721.	Caucasus, North	Incl. Sivash Depression & Indolo-Kuban-Azov-Terek-Kuma Sub-Basins	USSR	222	II C a	S	G/E
722.	Chuyi	Chu-Sarysu	USSR	22	III A	N	G
723.	Dnepr-Donets		USSR	222	III A	G	G
724.	East Siberia	Incl. Oloy-Stolbovoy & Rauchan Sub-Basins	USSR	331	III B + II C c	N	P/F
725.	Fergana		USSR	222	III B c	S	G
726.	Gobi		Mongolia	23	III A	N	P
727.	Il-Yi	Lliyskaya	China/USSR	23	III A	N	F/G
728.	Indigirka-Zyryanka	Moma-Zyryanka	USSR	331	III A	N	F
729.	Irkutsk		USSR	41	II A	N	F
730.	Kamchatka, Central		USSR	332	III B b	N	P/F
731.	Kamchatka, East		USSR	311	V	N	P
732.	Kamchatka, West	Incl. Shele Kova Trough	USSR	3122	III B b	N	P
733.	Kansk		USSR	221	II A	N	F
734.	Kara Sea		USSR	1143	II B	N	G/E
735.	Kara-Kum		USSR	222	II C a	G	F/G
736.	Khan Kay		China/USSR	3122	III A	N	P
737.	Khatyrka		USSR	331	III B	N	P
738.	Komandarsky		USSR	3122	III B	N	P
739.	Kopet-Dag		USSR	222	II C a	S	F/G
740.	Kura		USSR	222	III B c	S	G
741.	Kuril-Kamchatka Trench		Japan/USSR	21-OC		N	P
742.	Kuznets		USSR	1212	II B	N	F
743.	Kyzyl-Kum		USSR	222	III A	N	F
744.	Laptev		USSR	1212	II B or III A	N	F

179

No.	Name	Note	Country	Code	Class		
745.	Lake Baikal		USSR	111	III A	N	P/F
746.	Mangyshlak		USSR	222	II C a	G	G/E
747.	Mezen		USSR	222	I	N	P/F
748.	Minisinsk		USSR	1212	II B	N	P/F
749.	Mochigmen		USSR	331	III B	N	P
750.	Moscow	Incl. Lacha & Orsha Sub-Basins	USSR	121	I	N	P/F
751.	Okhotsk, North		USSR	3121-OC		N	P
752.	Okhotsk, South		USSR	3121-OC		N	P/F
753.	Olenek		USSR	121	I	N	F/G
754.	Parapol		USSR	332	III B b	N	P
755.	Pechora	Timan-Pechora	USSR	221	II B	G	G/E
756.	Penzhina		USSR	332	III B b	N	P
757.	Pripyat		USSR	222	III A	S	F/G
758.	Sakhalin North		USSR	332	III B b	S	F/G
759.	Shanjiang	Deryugin	China/USSR	3122	III A	N	P/F
760.	Suyfun	Middle Amur	USSR	3122	III B c	N	P
761.	Tadzhik		Afghanistan/USSR	222	II C a	G	G/E
762.	Tashkent		USSR	222	III A	N	P/F
763.	Tatarsky	Tatar	Japan/USSR	3122	III B b	S	F/G
764.	Tengiz		USSR	1212	II B	N	P/F
765.	Terney		USSR	3121-OC		N	F
766.	Tunguska		USSR	121/5	I	N	G/E
767.	Turgay		USSR	222	II B	N	P/F
768.	Turkmen		Afghanistan/USSR	222	II C a	G	F/G
769.	Uda	Verkhne-Zeya	USSR	3122	III A	N	P/F
770.	Ust Urt		USSR	121	II B	S	F/G
771.	Verkhoyansk		USSR	221	II A	G	F
772.	Vilyuy		USSR	221	II A	G	F/G
773.	Volga-Ural		USSR	221	II A	G	G
774.	West Siberia	Tyumen	USSR	1212	II B	G	E
775.	Yenisey-Khatanga		USSR	221	II A	S	F/G
776.	Yermark		Greenland/USSR/Norway	113-OC		N	P
777.	Zaysan	T'a-Ch'eng	China/USSR	23	III A	N	F/G
778.	Zeya-Bureya		USSR	3122	III A	N	F

Antarctica

#	Province Name	Alternative Province Name	Country	Classification Bally	Classification Klemme	Production	Hydrocarbon Potential
779.	Amundsen		Antarctica	114		N	F/G
780.	Bellingshausen		Antarctica	114		N	F/G
781.	Bransfield		Antarctica	3121-OC		N	F
782.	Dufek		Antarctica	121	I or II B	N	P
783.	Enderby		Antarctica	1141/113-OC		N	G
784.	Graham		Antarctica	3121-OC		N	F
785.	Kerguelen East		Kerguelen	113-OC		N	P
786.	Kerguelen South		Kerguelen	113-OC		N	P
787.	Kerguelen West		Kerguelen	113-OC		N	P
788.	Marie Byrd		Antarctica	121	I or II B	N	P
789.	Orkney South		Antarctica	311		N	P/F
790.	Polar		Antarctica	121		N	P
791.	Queen Fabiola		Antarctica	121		N	P
792.	Queen Maud		Antarctica	1141/113-OC		N	P
793.	Ross East		Antarctica	1143	III A Local/ II B Regional	N	G
794.	Ross Ice Shelf		Antarctica	121		N	P/F
795.	Ross West		Antarctica	1143	III A Local/ II B Regional	N	G
796.	Scotia		Antarctica	3121-OC		N	P
797.	Scott	Adelie	Antarctica	1141/113-OC		N	G
798.	Weddell		Antarctica	221/1141	II B/III C	N	F/G
799.	Wilkes		Antarctica	121	I or II B	N	P

OTC 4843

Origin and Classification of Sedimentary Basins

by J.A. Helwig, *ARCO E&P Research*

ABSTRACT

A sedimentary basin is a domain of regional subsidence that can be characterized in space, time, and sedimentary fill. Tectonic subsidence of basins is caused by rifting, flexure, or cooling of the lithosphere. It is important to understand basin subsidence mechanisms and to explicitly incorporate them into basin classification and exploration thinking. Previous classifications tend to overemphasize the plate-tectonic setting or other geologic characteristics of basins. A classification of basins based upon both the mechanism of subsidence and the plate-tectonic setting is proposed.

INTRODUCTION

The origin and the classification of sedimentary basins are closely connected. Both are important for understanding petroleum occurrence in the context of basin history.

Over the past decade, substantial progress has been made on understanding the physics of the processes of subsidence. In addition, a wealth of information has accumulated on the nature of plate tectonics as expressed by the deformation of the earth's lithosphere. These developments make it timely to formulate a new classification of sedimentary basins which incorporates both the plate-tectonic setting and the mechanism of subsidence of a basin throughout its history.

CLASSIFICATION OF BASINS

What is a sedimentary basin? It is a physical depression at the earth's surface which accumulates and preserves sediments. The

---

References and illustrations at end of paper.

scale of basins varies greatly, from ocean basins to small ponds. Earth scientists generally restrict the term "sedimentary basin" to sediment-filled depressions of a regional (100 + kilometers) scale that have subsided deeply, requiring significant crustal deformation for their origin. Basins are further restricted to include depressions that have definable margins, as opposed to broad platforms that accumulate 1-3 kilometers of sediment over a breadth of hundreds of kilometers.

What is the proper basis for classification of sedimentary basins? Initially, basins were recognized and defined simply as thick sedimentary accumulations compared to surrounding regions. Classifications subsequently were formulated largely on the basis of the "form and origin of the contained rocks" (Kay, 1951), and elaborated in the theory of geosynclines. Classification became complex, based on additional parameters thought to be of genetic significance: the position of the basin with respect to the stable continental craton; the position and timing related to orogenic belts; the presence of volcanic rocks; and the sources of sediments (Table 1). It is clear that the form of a basin is its essential characteristic, but that the ideal basis of classification would relate to the causes of that form, as well as to the origin of the contained rocks - in short, a tectonic classification including both basin-forming processes and tectonic setting.

Tectonic classification of basins is fraught with conceptual pitfalls, burdened with semantics, and crippled by the variety and complexity of basins. With the advent of plate tectonics, the tectonic setting of basins with respect to the three types of plate boundaries, and associated features, became the evident basis for classification (Dewey and Bird, 1970; Dickinson, 1975; Klemme, 1980; Table 2) and seemed to offer an objective and comprehensive framework for classification. However, to ascribe an origin to a basin strictly in terms of its tectonic setting could lead to errors - need all back-arc basins be the same in form and origin?

Bally and Snelson (1980) reemphasized the broader tectonic setting of basins, particularly in terms of the control of basin form and fill by the thermal, mechanical and lithological properties of the lithosphere (Table 3). Their emphasis on actualistic examples make the classification most useful for Mesozoic-Cenozoic basins, formed in continental margin and orogenic settings. Kingston et al.'s (1983) classification (Table 4) correctly emphasizes basin form and complex history, but contains many subdivisions based upon either non-standard terminology or non-essential geologic parameters.

An attempt to put a variety of basins into these classifications illustrates the weaknesses of nomenclature (Table 6). The names of the classes differ, the priorities of criteria for classification differ, and confusion abounds. To properly classify basins, we need to: (1) identify essential, as opposed to modifying characteristics; (2) assign priority to these characteristics.

The form and fill of basins are their essential characteristics; all others are secondary. The ideal classification should name basins according to the essential causes of form and fill, in order of priority. Compressive and extensional stresses associated with plate motions are the principal causes of deformation of the lithosphere (subsidence, uplift), and the response characteristics of the particular region of lithosphere control the development of individual sedimentary basins. Therefore, the plate-tectonic setting of a basin is essential, but valuable only insofar as the setting provides a unique deformation and a unique lithosphere.

Unfortunately, a variety of deformational and lithospheric configurations can occur (in the same or different basins) in the same plate tectonic setting. It is true that a given tectonic setting has a tendency to have a certain recurrent configuration, but this should not be the sole basis for classification. Hence the confusion of naming a basin a "rift" on the basis of a divergent plate boundary versus a zone of lithospheric extension. Only by carefully distinguishing plate-tectonic setting and mode of lithospheric deformation can a satisfactory classification of sedimentary basins be made.

Recent advances in understanding lithospheric subsidence (Price, 1973; McKenzie, 1973) have permitted classification of basins by subsidence mechanisms (Beaumont et al., 1982; Dewey and Pitman, 1982). We need to briefly review basin physics, therefore, prior to basin classification.

BASIN PHYSICS: TWO GENETIC TYPES

In terms of physics, we would like to consider that the shape of a basin can be defined geometrically and geometry is directly related to its origin. Although complex in three dimensions, profiles of basins tend to fall into just two categories: (1) symmetrical, concave upward, normal faulted; (2) asymmetrical, wedge shape, thrust faulted (Fig. 1). These two fundamental shapes are now considered to be the two fundamental genetic types of basins, here termed rift and flexural basins (see Beaumont, et al., 1982).

To understand the origin and complex varieties of basins deriving from these fundamental two types, we must consider the thermo-mechanical properties of the lithosphere, and how extension or compression of lithosphere can cause subsidence in rift and flexural basins, respectively. Here we only summarize briefly major characteristics of form, fill and setting of these two genetic types of basins.

Rift basins subside due to extensional thinning of the lithosphere, producing: (1) initial faulting and isostatic subsidence; and (2) thermal subsidence of the cooling asthenosphere that rises to replace the lithosphere (Fig. 1b), resulting in a broad, concave basin (McKenzie, 1978). In this paper, the fault-related subsidence is called rift subsidence, and the later thermal subsidence is termed passive subsidence. Rift basins may form in a variety of plate-tectonic settings, but major rift basins form where mid-ocean ridge systems propagate into old continental lithosphere. Rifts may evolve into passive margins between continental and oceanic crust accumulating up to 15 km of sediment. The subsidence history and stratigraphy of a rift basin (or passive margin) reflect the two stages of subsidence (Fig. 1b). The typical sedimentary fill of rift basins consist of basal coarse clastic sediments associated with uplifted fault blocks, and later basin fill of fine clastic or carbonate and evaporite sediments. On passive margins, shelf facies subsequently are deposited and prograded over deep-water oceanic sediments.

The tectonic subsidence of flexural basins is due to loading and downwarping of the lithosphere driven by compressional tectonics (Price, 1973). Subsidence is controlled by tectonic thickening and overthrusting of the adjacent uplifted block, and by the distribution of loads in and on the subsiding block (Karner and Watts, 1983). Major flexural basins form at the periphery of subduction systems, are bordered by fold-thrust belts or accretionary complexes, and evolve into complex sub-basins during continental collisions. The subsidence history of a flexural basin tends to be linear in time, directly related to episodes of emplacement of thrust loads (Fig. 2b); the subsidence is terminated either by deformation and uplift as the deformed belt sweeps through the basin, or by simple uplift upon removal of load and thermal expansion. Flexural basins typically show a clastic regressive facies development marked by basal deep-water starved basin facies, followed by turbidites, and terminal filling by shelf to non-marine facies.

The forms of basins developed by rifting and flexure are markedly different (Figs. 1-4). The basement of rift basins is highly modified by block faulting, tilting of fault blocks, and volcanism, producing a bounding surface for the basin that is geologically complex with sharp structural and paleotopographic relief. The flanks of a rift ideally are symmetrical normal faults, but asymmetrical faulting can produce a tilted-wedge geometry (Fig. 3b). The floor of flexural basins is smoothly down-bowed toward the faulted flank of the basin, producing an asymmetrical wedge in cross section (Fig. 4). Both types of basins are coupled to uplifts, but foreland basins usually have only one active side and are thus characterized by asymmetric sediment supply. If sea level is high relative to lithosphere base level, rift and flexural basins might develop with a complete absence of coupled uplifts and sediment sources.

The total maximum sedimentary accumulation in a rift approaches 15 kilometers, because rifting of lithosphere creates oceanic crust that can subside to this depth under isostatic sediment loading (McKenzie, 1978). The maximum accumulation in a flexural basin may range from several kilometers to tens of kilometers, depending on the composition and strength of the lithosphere (Beaumont et al., 1982).

The thermal regime and thermal history of rift and flexural basins contrast sharply. The process of rifting causes lithospheric extension and thinning, and consequent upward rise of hot asthenosphere. In general, heat flux increases in rift basins in direct proportion to the amount of extension (McKenzie, 1978). On the other hand, no significant change in thermal regime is associated with formation of flexural basins. (Although the thermal aspect of basins cannot be directly considered for classification, it is important for maturation of hydrocarbons.)

Certain basins, particularly intraplate "sags" or cratonic basins, appear to subside due to thermal-metamorphic changes in the lower

lithosphere with no apparent extension or compression. These basins are here referred to as intraplate passive basins (Table 5).

SIMPLE BASINS

Simple basins are those having one tectonic mechanism of origin and a corresponding harmonious tectonic setting during the period of subsidence and sediment accumulation. (Simple basins here correspond to the single cycle basins of Kingston et al., 1983.) Examples are in Figures 3 and 4.

Simple rift basins develop at or near: (1) divergent plate boundaries; (2) divergent bends in transform systems; (3) divergent zones behind subduction systems. The extensional tectonic system of simple rifts remains self-similar, and consistent with major plate motions, through time. Continental rifts, transform pull-aparts, ocean basins, and passive margins are all simple rifts. The simplest type of rift basin is a symmetrical graben (Fig. 3a). In areas of pre-existing thrust faulting, asymmetrical low-angle normal faults may develop along older thrusts, producing asymmetrical successor rift basins like those in the Great Basin of the western United States (Fig. 3b). The Norton Basin of Alaska (Fig. 3c) is an example of a simple transform rift basin broken into numerous asymmetrical tilted sub-basins. Two periods of faulting (a, b; Fig. 3c) and associated sedimentary cycles are separated by unconformities. Transform rifts like those in the Gulf of California (Fig. 3d) are characterized by transform faults along two sides and rhomboidal geometry.

Simple flexural basins develop at or near: (1) convergent plate boundaries; (2) convergent bends in transform systems; (3) convergent zones behind convergent subduction systems (Fig. 4). The compressional tectonic system of simple flexural basins remains self-similar, and consistent with major plate motions, through time. Oceanic trenches and foreland basins are simple flexural basins. The Alberta Basin is an example of a simple flexure basin formed on an old continental margin by loading of a collided and deformed mountain belt (Fig. 4a). Subduction zones typically show a trench basin, deformed sediments in a tectonic dam, and dammed forearc basin all overlying the flexure of the descending oceanic plate (Fig. 4b).

CLASSIFICATION

Many basins are not simple and require classification according to age, mechanics, and tectonics. The proposed classification (Table 5) attempts to avoid the inconsistency of other classifications by carefully distinguishing plate-tectonic position and mechanics, deemed essential for classification, and pre-basin lithospheric character, etc., deemed to be of secondary importance. Modifiers specify significant sedimentary and deformational controls, geometry, and post-basin events. Terms not essential to definition of a basin should not be considered, e.g. sedimentary environments, presence of petroleum. Continental platforms are distinct tectonic categories of uncertain or speculative mechanical origin and not considered here. Based upon all the preceding, a proposed classification scheme is shown in Table 5; its applicability is particularly important for complex or "polyhistory" (Kingston et al., 1983) basins.

COMPLEX BASINS

Complex basins are those having any or all of the following during the history of subsidence and sediment accumulation: (1) multiple subsidence mechanisms; (2) a disharmonious tectonic setting; (3) changes in tectonic setting. Complex basins develop during complex history; or, they develop at or near plate boundaries of all types, but especially in collisional zones, where compressional and extensional tectonic systems do not bear a simple relationship to plate displacements in space and time.

Complex basins may show: (1) transverse geometry; (2) reactivation of compressional structures by extension, or vice versa; (3) significant displacement and multiple movement along boundaries; (4) inversion or subversion. For example, in the Java Sea back-arc region, rifting occurs and is transverse to the trend of the island arc (Fig. 5). Progressive deformation in collision zones like India-Himalaya causes uplift on displaced boundaries and isolation of sub-basins from the main Indus Plain basin (Fig. 6).

Many complex basins may be rendered simple by "unstacking" their constituent parts into a geological time series of simple basins (Green, 1983). For example, rifted passive margin basins are overlain by foreland basins in zones of continental collision (A overlying B in Fig. 4a) and should be named separately. Unstacking basins often is not easy because earlier basin history is difficult to observe and overprinted by younger structure.

Crustal collision, compressive deformation and uplift can extrude and invert a basin (Fig. 7a; Bally, 1984), leading to their eventual destruction by erosion. In thrust belts, basins may become detached from their basement, rendering paleotectonic classification difficult. Alternatively, basins may be largely concealed in the footwalls of major thrusts; such basins are here called subverted basins (Fig. 7b). Subverted basins may be detected and analyzed by geophysical techniques, balanced cross sections, and paleogeographic reconstruction. Long distance transport of basins along transform systems can lead to complex basin history, particularly in orogenic settings (Helwig, 1974; Fig. 8).

CONCLUSION

Classification of basins strictly by plate tectonic setting or tectonic process is inadequate, except where setting and process coincide in simple basins. Recent research permits a genetic (process) characterization of sedimentary basins as either rift basins, with a component of passive thermal subsidence, or flexural basins. The principal respective driving mechanisms of subsidence are thinning of continental crust, the lateral loading of lithosphere by subduction, and thermal contraction of the lithosphere. The geometry and subsidence history of a basin is further controlled by the pre-existing thermomechanical constitution of the lithosphere and by structural and sedimentary processes. The form and fill of a sedimentary basin, and the character of its subsidence history curve, ordinarily provide reasonable criteria to establish both the genetic mechanism and the plate tectonic setting of the basin. Correct classification is fundamental for understanding basin history, constructing predictive models of basin development, and evaluating hydrocarbon potential.

ACKNOWLEDGMENT

I thank the management of Atlantic Richfield Company for permission to publish this paper.

REFERENCES

1. Kay, M.: "North American Geosynclines," Geol. Soc. America Mem. 48 (1951), 143.
2. Dewey, J. F. and Bird, J. M.: "Mountain belts and the new global tectonics," J. Geophys. Res.(1970) Vol. 75, 2625-2647.
3. Dickinson, W. R.: "Plate Tectonic Evolution of Sedimentary Basins, Plate Tectonics and Hydrocarbon Accumulation," Amer. Assoc. Petrol. Geol. Short Course, (1976).
4. Klemme, H. D.: "Petroleum basins - classifications and characteristics," J. Petrol. Geol., (1980) V.3, No. 2, 187-207.
5. Bally, A. W. and Snelson, S. S.: "Realms of Subsidence," in Maill, A. D. (Ed.), Can. Soc. Petrol. Geol. Mem. 6 (1980) 1-94.
6. St. John, B., Bally, A. W., and Klemme, H. D.: "Sedimentary Provinces of the world - hydrocarbon productive and nonproductive," Am. Assoc. Petrol. Geol. (1984) map, 35.
7. Kingston, D. R., Dishroon, C. P., and Williams, P. A.: "Global basin classification system," Am. Assoc. Petrol. Geol. Bull. (1983) V. 67, No. 12, 2175-2193.

8. Price, R. A.: "Large-scale gravitational flow of supracrustal rocks, southern Canadian Rockies," in Gravity and Tectonics, edited by K. A. deJong and R. Scholten, John Wiley, New York (1973) 491-502.

9. McKenzie, D.: "Some remarks on the development of sedimentary basins, "Earth Planet. Sci. Lett. (1978) V. 40, 25-32.

10. Beaumont, C., Keen, C., and Boutilier, R.: "A comparison of foreland and rift margin sedimentary basins," Phil. Trans. Roy. Soc. London (1982) V. A305, 295-317.

11. Dewey, J. F. and Pitman, W. C.: "The Origin and Evolution of Sedimentary Basins," Tectan, Piermont, New York (1982) course notes.

12. Karner, G. D. and Watts, A. B.: "Gravity anomalies and flexure of the lithosphere at mountain ranges," J. Geophys. Res. (1983) V.88, No.B12, 10,449-10,477.

13. Mann, P. and Burke, K.: "Cenozoic rift formation in the northern Caribbean," Geology, (1984) V.12, 732-736.

14. McCarthy, J., Stevenson, A. J., Scholl, D. W. and Vallier, T. L.: "Speculations on the petroleum geology of the accretionary body: an example from the central Aleutians," Marine and Petroleum Geology, (1984) V. 1, 151-173.

15. Burbank, D. W. and Reynolds, R. G. H.: "Sequential late Cenozoic structural disruption of the northern Himalayan foredeep," Nature, (Sept. 12, 1984), 114-118.

16. McCaffrey, R., Silver, E. A., and Raitt, R. W.: "Crustal structure of the Molucca Sea collision zone, Indonesia," in The Tectonic and Geological Evolution of Southeast Asian Seas and Islands, Am. Geophys. Union, Monog. 23 (1980), 161-177.

17. Yeats, R. S.: "Large-scale Quaternary detachments in Ventura Basin, southern California," J. Geophys. Res., (1983) V.88, No.B1, 569-583.

18. Green, A. R.: "Future petroleum province exploration," Proc. World Petroleum Congr. (London) (1983) PD3(5), 167-175.

19. Helwig, J.: "Eugeosynclinal basement and a collage concept of orogenic belts," Soc. Econ. Paleont. Min. Spec. Publ. 19, (1974) 359-376.

20. Bruns, T. R.: "A model for the origin of the Yakutat block, an accretionary terrane in the Gulf of Alaska," Geology (1983) V.11, 718-721.

21. Bally, A. W.: "Tectogenese et sismique reflexion," Bull. Soc. Geol. France (1984) V.26, No.2, 279-285.

Intracratonal: (on old stable basement)	Autogeosyncline	- elliptical basin no highlands
	Zeugogeosyncline	- adjoining faulted highlands
	Exogeosyncline	- clastic wedge derived from orogenic belt
Bordering the craton: (long, narrow basins)	Miogeosyncline Eugeosyncline	- non-volcanic - volcanic
Post-orogenic:	Epieugeosyncline Taphrogeosyncline Paraliageosyncline	- adjoining complex uplifts - rift - coastal basins

TABLE 2: Klemme's basin classification

Basin Name	*Symmetry/Form (stress)	Tectonic Position
I. Craton Interior	S/ Sag (extensional)	Continental

II. Continental Multicycle
   A. Craton Margin-Composite
	A/ 2. Foredeep (compressive)
	   1. Platform/sag (extensional)
		Craton Margin

B. Craton/Accreted Margin-
       Complex
	S/ 2. Sag -
	   1. Rift (extensional)
		Continental

C. Crustal Collision Zone	A/ 2. Foredeep
	   1. Platform/sag
		Convergent Margin
		  (collision)

III. Continental Rift A. Craton & Accreted Zone Rift	I/ Rift/Sag	Continental

B. Rifted Convergent Margin
     a) back arc
     b) transform
     c) median
	I/ Rift/Wrench/Sag
	  (extensional plus wrench
	  compressive)
		Above subducting
		oceanic crust

C. Rifted Passive Margin
     a) parallel
     b) transform
	AI/ Rift/Half Sag
	  (extensional)
		Divergent Margin

IV. Delta
   A. Synsedimentary
   B. Structural
	Depocenter/Sag (extensional)
	Depocenter/?	

V. Forearc	A/ Subduction (compressive and extensional)	Convergent consuming margin

* Symmetry: S = symmetrical
              A = asymmetrical
              I = irregular

1. **BASINS LOCATED ON THE RIGID LITHOSPHERE, NOT ASSOCIATED WITH FORMATION OF MEGASUTURES**
    - 11. Related to formation of oceanic crust
        - 111. *Rifts*
        - 112. *Oceanic transform fault associated basins*
        - 113-OC. *Oceanic abyssal plains*
        - 114. *Atlantic-type passive margins (shelf, slope & rise) which straddle continental and oceanic crust*
            - 1141. Overlying earlier rift systems
            - 1142. Overlying earlier transform systems
            - 1143. Overlying earlier Backarc basins of (321) and (322) type
    - 12. Located on pre-Mesozoic continental lithosphere
        - 121. *Cratonic basins*
            - 1211. Located on earlier rifted grabens
            - 1212. Located on former backarc basins of the (321) type
2. **PERISUTURAL BASINS ON RIGID LITHOSPHERE ASSOCIATED WITH FORMATION OF COMPRESSIONAL MEGASUTURE**
    - 21-OC. Deep sea trench or moat on oceanic crust adjacent to B-subduction margin
    - 22. *Foredeep and underlying platform sediments*, or moat on continental crust adjacent to A-subduction margin
        - 221. Ramp with buried grabens, but with little or no blockfaulting
        - 222. Dominated by block faulting
    - 23. *Chinese-type basins* associated with distal blockfaulting related to compressional or megasuture and without associated A-subduction margin
3. **EPISUTURAL BASINS LOCATED AND MOSTLY CONTAINED IN COMPRESSIONAL MEGASUTURE**
    - 31 Associated with B-subduction zone
        - 311. *Forearc basins*
        - 312. *Circum Pacific backarc basins*
            - 3121-OC. Backarc basins floored by oceanic crust and associated with B-subduction (marginal sea sensu stricto)
            - 3122. Backarc basins floored by continental or intermediate crust, associated with B-subduction
    - 32. Backarc basins, associated with continental collision and on concave side of A-subduction arc
        - 321. On continental crust of *Pannonian-type* basins
        - 322. On transitional and oceanic crust or *W. Mediterranean-type* basins
    - 33. Basins related to episutural megashear systems
        - 331. *Great-basin-type* basins
        - 332. *California-type* basins
4. **FOLDED BELT**
    - 41. Related to A-subduction
    - 42. Related to B-subduction
5. **PLATEAU BASALTS**

IDENTIFICATION CRITERIA	Basins formed on continental crust (can overlap onto adjacent oceanic crust)				Basins completely formed on oceanic crust			
1. Continental vs. Oceanic Crust under basin	Continental Basins				Oceanic Basins			
2. Plate movement involved in Basin/cycle formation (Divergent or Convergent)	Divergent		Convergent		Convergent		Divergent	
3. Basin/cycle position on plate and primary structural movement involved in basin origination	Interior of Plate	Margin of Plate	Usually near Margin of Plate	Subduction margin	Margin of Plate			
Theoretical Model Basin Types	Cont'l Interior Sag	Cont'l Interior Fracture	Cont'l Margin Sag	Cont'l Wrench	Trench Associated	Oceanic Trench	Oceanic Wrench	Oceanic Fracture / Oceanic Sag

TABLE 5: Basin Classification

Essential Criteria

Age	Tectonic Position	Basin Mechanics and Form
	1. Intraplate 2. Collisional a. foreland b. intramontane 3. Convergent (subduction) a. trench b. forearc c. arc d. back arc 4. Divergent (sea-floor spreading) a. continental margin b. plateau c. back arc 5. Transform	A. Tectonic subsidence mechanism: 1. rift 2. flexural 3. passive B. Non-subsident topographic mechanism: 1. dammed (tectonically, depositionally) 2. volcanic pile 3. sediment fan 4. erosional C. Special mechanisms: 1. caldera collapse 2. salt solution/withdrawal 3. cosmic impact

Modifying Criteria

A. Pre-basin lithosphere
   1. cratonic
   2. oceanic
   3. attenuated
   4. accreted (successor)
   5. volcanic

B. Geometry
   1. symmetry in profile
   2. elongation
   3. orientation (longitudinal,
     transverse, oblique)

C. Post-basin events
   1. inversion
   2. subversion

TABLE 6: Comparison of Basin Nomenclature

Table-Reference	Michigan Basin (Upper Silurian)	Appalachian Basin (Mid-Upper Devonian)
1 – Kay	autogeosyncline	exogeosyncline
2 – Klemme	craton interior basin	craton margin foredeep
3 – Bally and Snelson	cratonic basin	perisutural foredeep
4 – Kingston et al	continental interior sag basin	continental wrench
5 – This paper	intraplate rift(?)/passive	collisional (foredeep) flexural

	Ventura Basin (Tertiary)	Aleutian Terrace (Tertiary)
1 –	epieugeosyncline	eugeosyncline
2 –	rifted convergent margin	forearc
3 –	episutural megashear	forearc
4 –	continental wrench/polyhistory?	trench associated
5 –	(1) transform rift (pre-Pliocene) (2) transform flexural (Pliocene-Rec.)	forearc dammed

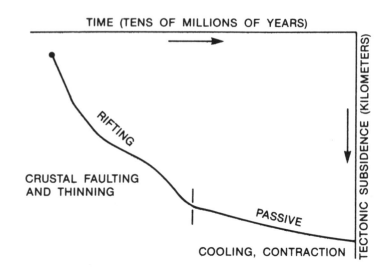

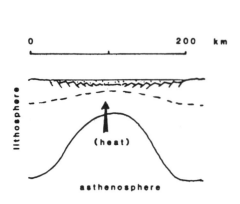

Figure 1a. Sketch of cross-section of a rift basin.

1b. Tectonic subsidence behavior of a rift basin.

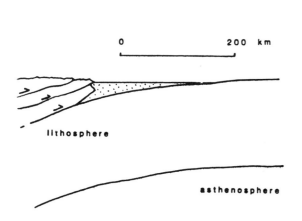

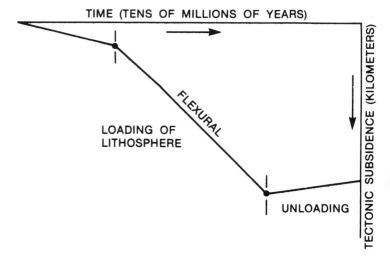

2a. Sketch of cross-section of a flexural basin.

2b. Tectonic subsidence behavior of a flexural basin.

190

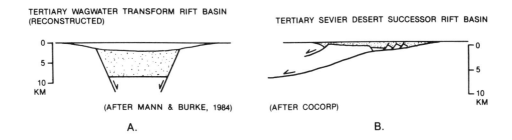

TERTIARY WAGWATER TRANSFORM RIFT BASIN
(RECONSTRUCTED)

TERTIARY SEVIER DESERT SUCCESSOR RIFT BASIN

(AFTER MANN & BURKE, 1984)

(AFTER COCORP)

A.

B.

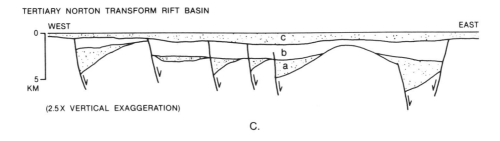

TERTIARY NORTON TRANSFORM RIFT BASIN

WEST   EAST

(2.5X VERTICAL EXAGGERATION)

C.

Figure 3. Simple rift basins: A. Simple graben. B. Asymmetrical rift. C. Multiple half-grabens.

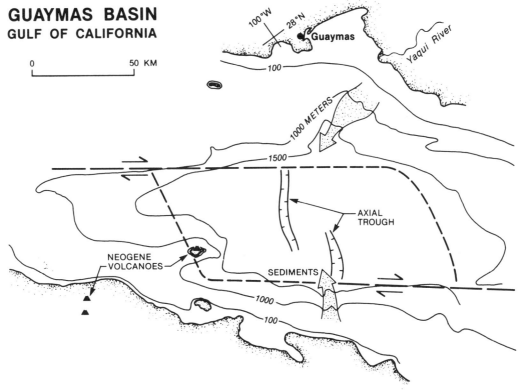

GUAYMAS BASIN
GULF OF CALIFORNIA

0   50 KM

Figure 3d. Simple transform rift basin.

191

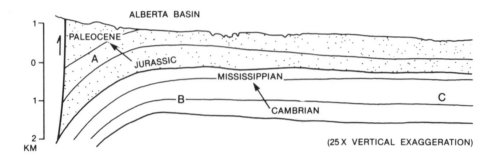

Figure 4a. Alberta collisional flexure basin (A).

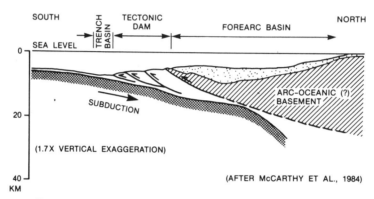

Figure 4b: Aleutian Trench flexural basin and dammed forearc basin.

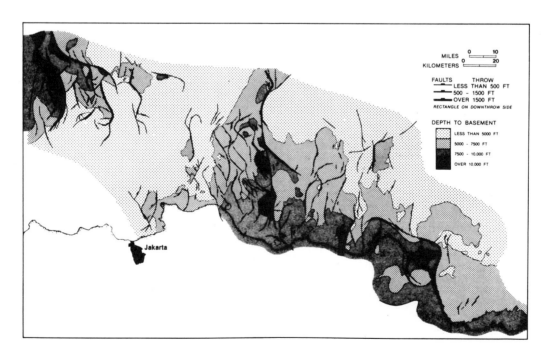

Figure 5. Northwest Java Basin, a Tertiary back-arc transverse rift basin.

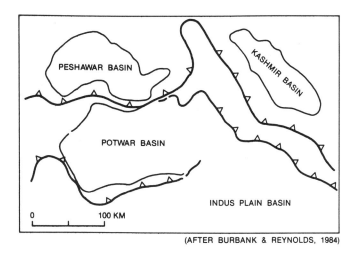

(AFTER BURBANK & REYNOLDS, 1984)

Figure 6. Indus collisional flexure basin and tectonically dammed basins.

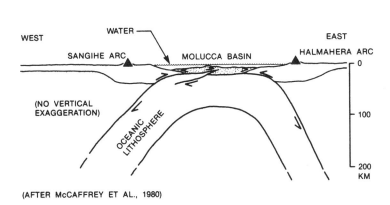

(AFTER McCAFFREY ET AL., 1980)

Figure 7a. Inverted basin, Molucca Sea.

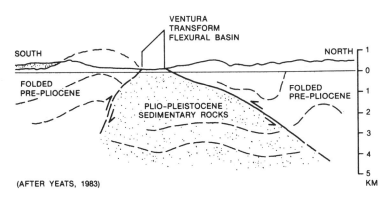

(AFTER YEATS, 1983)

Figure 7b. Subverted basin, Southern California.

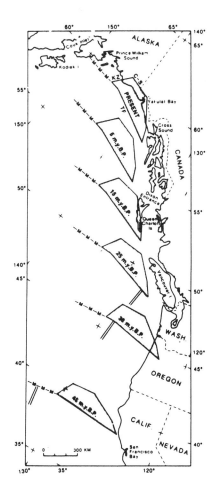

Figure 8. Travel path of Yakutat block, a transform passive basin (after Bruns, 1983).

193

MODELING

Reproduced by permission of the Geological Society, from
Hydrocarbon generation and migration from Jurassic
source rocks in the E Shetland and Viking Graben of the
northern North Sea, by J.C. Goff, in *Journal of the
Geological Society*, vol. 140, part 3, pp. 445-474, 1983.

Hydrocarbon Generation and Migration from Jurassic Source Rocks in the East Shetland Basin and Viking Graben of the Northern North Sea

J.C. Goff
Department of Energy
London, England

In the East Shetland Basin oil generation began 65 Ma ago; peak oil generation maturity occurs today at 3,250 m (0.7 percent R_o) and was first reached 40 to 50 Ma ago; the oil generation threshold is at 2,500 m. Highest oil saturations in the Kimmeridge Clay occur at 0.8 percent R_o; oil expulsion efficiencies are > 20 to 30 percent. Oil phase migration has probably occurred through oil wet kerogen laminae, and through interconnected large pores aided by low oil/water interfacial tensions. Oil migrated along strong lateral fluid pressure gradients, from overpressured source rocks in half grabens to Jurassic reservoirs in tilted fault blocks.

In the Viking Graben the Kimmeridge Clay is at oil floor maturity below 4,500 m; oil and peak oil generation began 70 to 80 and 55 to 65 Ma ago respectively; 40 Ma ago the Kimmeridge Clay passed through peak generation, and gas generation by cracking of oil had begun. Peak dry gas generation from Brent coals occurs today below 5,000 m, and began 40 Ma ago. The Frigg Field gas, probably generated from late Jurassic source rocks, migrated through microfractures in overpressured mudstones below 3,500 m; above 3,500 m methane probably migrated in aqueous solution and was exsolved in the early Tertiary aquifer.

The East Shetland Basin and the Viking Graben (Figure 1) are located in the northern North Sea Basin between the Shetland Islands and Norway (Figure 2). The Jurassic sandstones in the East Shetland Basin contain 10 billion barrels of recoverable light oil; 3 fields each have recoverable reserves greater than 1 billion barrels (Statfjord, Brent and Ninian). The Viking Graben contains major gas reserves: dry gas and associated heavy oil are trapped in early Tertiary sandstones; the Frigg Field has in place reserves of 270 billion cu m of gas and 790 million barrels of heavy oil (Heritier et al, 1979). Gas condensate has been discovered in deep, high pressure, Jurassic sandstone.

The aims of this study were to determine when these hydrocarbons were generated from their source rocks, and how they migrated from these source rocks to the traps. Both these objectives required integration of geochemical data with knowledge of the stratigraphy, geological structure and history of the study area. The geological framework of the study area, and its development, are reviewed below.

The northern North Sea Basin formed during Permo-Triassic rifting; thick Triassic red beds consisting of alluvial fan, fluvial and lacustrine clastics were deposited unconformably on Caledonian basement. Up to 1 km of early to middle Jurassic shallow-water sediments were then deposited. These comprise Hettangian-Sinemurian fluvial and marginal marine sandstones (Statfjord Formation), Sinemurian to Toarcian shallow marine shelf mudstones, siltstones and thin sandstones (Dunlin Formation), and Bajocian to early Bathonian deltaic/shallow marine sandstones (Brent Formation). Growth faulting occurred during deposition of these rocks along some fault trends, probably at least partly due to differential compaction of Triassic rocks.

The Brent Formation sandstones form a fluvial/wave dominated delta complex, up to 300 m thick, which prograded across a relatively stable shelf at least 12,500 sq km in extent. Major (down to the east) faulting, which occurred during Bathonian to Oxfordian times, created the Viking Graben and a series of westerly dipping fault blocks and half grabens in the eastern East Shetland Basin (Figure 3). Thick Bathonian to early Oxfordian marine mudstones (Heather Formation) conformably overlie the Brent Formation in half grabens; on the crests of the fault blocks, thin Heather Formation is up to 1 km thick.

Further periods of rifting occurred in Oxfordian to early Cretaceous time: Kimmeridgian faulting occurred in the western East Shetland Basin. Late Oxfordian to Portlandian, bituminous, sapropelic, mudstones (Kimmeridge Clay Formation) were deposited throughout

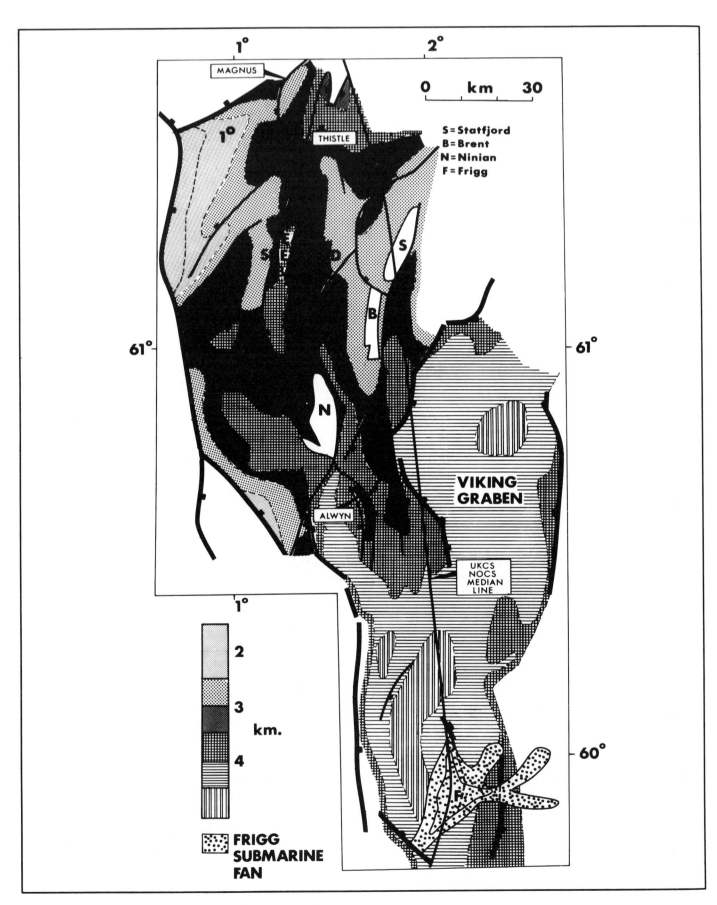

Figure 1. Depth to Base Cretaceous in the study area.

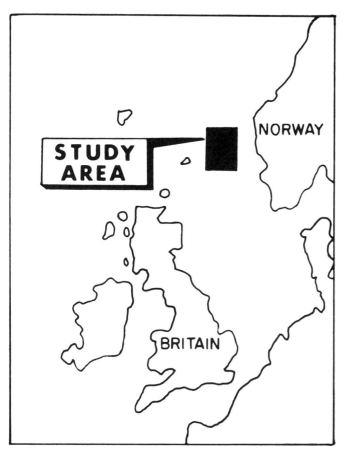

Figure 2. Location of study area.

the area, reaching thicknesses of up to 500 m in the East Shetland Basin. Submarine fan sandstones were deposited in the north-western East Shetland Basin during the Kimmeridgian. Major early Cretaceous faulting occurred in the northwest flank of the Viking Graben. Thick early Cretaceous mudstones (Cromer Knoll Group) are confined to the Viking Graben and some half grabens of the East Shetland Basin. Faulting ceased by mid-Cretaceous times except along the western fault controlled edge of the East Shetland Basin; the geometry of the Jurassic fault blocks was thus established by 100 Ma ago.

Regional subsidence occurred across the East Shetland Basin and Viking Graben during late Cretaceous time, and up to 2,500 m of deep water mudstones and thin limestones (Shetland Group) were deposited. Maximum Tertiary subsidence occurred over the Viking Graben where 2,000 to 2,500 m of sediment accumulated. Palaeocene to early Eocene submarine fan sandstones and mudstones, mid-Eocene to Oligocene mudstones and marine sandstones, the Miocene to Recent lignitic sandstones and mudstones were deposited in the Tertiary basin.

The approach used in this study was first to define the hydrocarbon source rocks and their present day maturity. The present day thermal regime was then determined by calculating heat flow in 4 wells located in contrasting structural positions (Figure 3). Well A is located on a shallow fault block, and Well B in a half graben in the East Shetland Basin. Well C was drilled on a downfaulted block on the western flank of the Viking Graben; Well D tested a deep fault block in the axial Viking Graben. A range for heat flow history at these 4 locations was deduced from their subsidence histories.

The maturation history of the source rocks in the 4 wells was calculated from their burial history, and the thermal properties of their overburden, for both constant and variable heat flow models. Timing of hydrocarbon generation was deduced from the maturation history using a correlation between calculated maturity and vitrinite reflectance. Timing of generation within the study area as a whole was then estimated from its overall thermal and subsidence history.

The efficiency of oil migration and entrapment in the eastern East Shetland Basin has been estimated by comparing the volume of oil generated with the volume of oil trapped. The mechanism of oil migration in the East Shetland Basin has been deduced by comparing the volumes of oil and compaction water which moved through the source rock during migration, and by considering the physical conditions within the source rock.

Finally the mechanism of oil and gas migration into the Frigg Field, which lies 2 km stratigraphically above Jurassic source rocks in the Viking Graben (Figure 4), is discussed. The variation of pore and fracture pressure, and of methane solubility, with depth in the Viking Graben has been studied to determine the relative importance of microfractures and water movement during migration.

SOURCE ROCKS

The source rock potential of the Cretaceous and Jurassic mudstones in the study area has been reviewed using Total Organic Carbon (TOC) analyses and determinations of Organic Matter Type. Shetland Group mudstones are lean, containing vitrinite and inertinite. Cromer Knoll Group mudstones contain 1 to 2 wt percent TOC which is predominantly inertinite (Barnard and Cooper, 1981).

"Jurassic source rocks" in the northern North Sea have weighted average TOC contents of 5.6 percent wt from 2,600 to 3,200 m, and 4.9 percent from 3,250 to 3,650 m; their non-soluble organic matter contains 80 percent sapropel and 20 percent humic/coaly material (Brooks and Thusu, 1977). Jurassic source rocks with these characteristics have only been reported from the Kimmeridge Clay Formation (Barnard and Cooper, 1981; Fuller, 1980). The Jurassic source rocks analyzed by Brooks and Thusu are thus probably from the Kimmeridge Clay. Immature Kimmeridge Clay organic matter consists predominantly of Type II kerogen (Williams and Douglas, 1980). The Kimmeridge Clay is rated as an excellent oil source rock, generating gas at high maturity levels.

The Heather Formation mudstones contain 1 to 2 percent TOC which consists dominantly of vitrinite and inertinite (Barnard and Cooper, 1981); they are rated as lean dry gas source rocks. The Brent Formation coals and vitrinite rich mudstones are excellent dry gas source rocks; the delta plain facies contains up to 10 m net of coal. The Dunlin Formation is organically lean; mudstones of its lower and middle members contain only 1 percent TOC which is dominantly inertinite (Barnard and Cooper, 1981). The

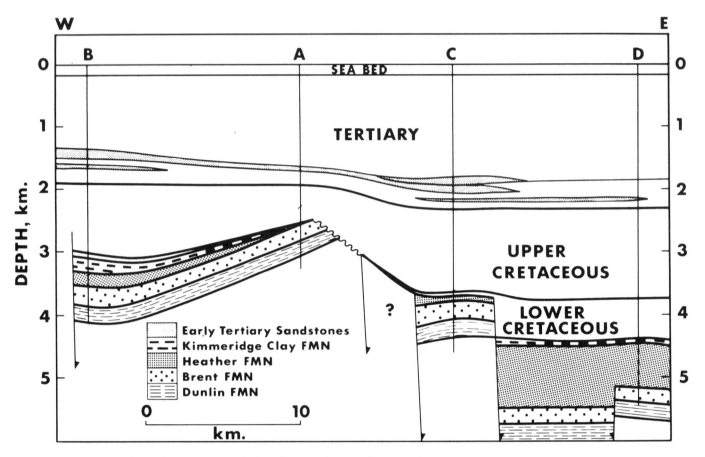

Figure 3. Cross section from the eastern East Shetland Basin to the axial Viking Graben.

upper Drake member, of Toarcian age, contains 2 percent TOC but less than 30 percent of the organic matter consists of sapropel; it thus has only limited oil potential.

The richest source rocks in the study area are thus the oil prone Kimmeridge Clay and the gas prone Brent Formation coals and coaly mudstones. Shetland Group and Heather Formation mudstones, although organically lean, reach thicknesses of 1 and 2 km respectively. They are thus capable of generating large volumes of gas.

The lithology and TOC content of the Kimmeridge Clay are very variable. In its type section in southern England it consists of carbonaceous illitic clays ($\leq$ 10 percent TOC), bituminous shales ($\leq$ 30 percent TOC), oil shales ($\leq$ 70 percent TOC) and coccolithic limestones; the bituminous shale units are up to 2 m thick (Tyson et al, 1979). The limestones are also bituminous and usually interlaminated with the oil shales, which are generally less than 10 cm thick (Gallois, 1976). In the East Shetland Basin the Kimmeridge Clay also contains thin beds of siltstone, fine-grained sandstone and dolomite.

Fuller (1980) reported an average Organic Matter Content for the Kimmeridge Clay in the North Sea Basin of 3.25 percent (equivalent to a TOC of 2.7 percent); the East Shetland Basin Kimmeridge Clay is twice as rich as this. Organic matter type varies laterally and vertically within the Kimmeridge Clay. At shallow burial depths (1,500 to 2,400 m) close to clastic sediment sources, and over stable platform areas, the Kimmeridge Clay contains

predominantly inertinite and vitrinite (Barnard and Cooper, 1981).

The high TOC content and sapropel contents of the Kimmeridge Clay in the East Shetland Basin are probably due partly to deposition in restricted fault bounded half grabens. Oil shale horizons are best developed in the middle part of the Kimmeridge Clay onshore in the United Kingdom (Gallois, 1976). A highly radioactive unit occurs at this stratigraphic level in the East Shetland Basin and Viking Graben (Figure 5); it frequently has a higher than normal electrical resistivity, suggesting it is the organically richest unit in the Kimmeridge Clay. On the crests of the fault blocks in the East Shetland Basin this unit is the only part of the Kimmeridge Clay present, indicating it was deposited during the peak Kimmeridgian transgression.

MATURITY

The present day maturity depth gradient in the East Shetland Basin and Viking Graben has been determined from vitrinite reflectance measurements on Brent Formation coals and early Jurassic to mid-Cretaceous mudstones in 12 wells. The wells were drilled in the Norwegian sector of the eastern East Shetland Basin, the axial Viking Graben, and on shallow fault blocks on the eastern flank of the Viking Graben east of the study area. The data (Figure 6) indicate a uniform present day maturity gradient across the eastern half of the study area. For a type II kerogen (Tissot and

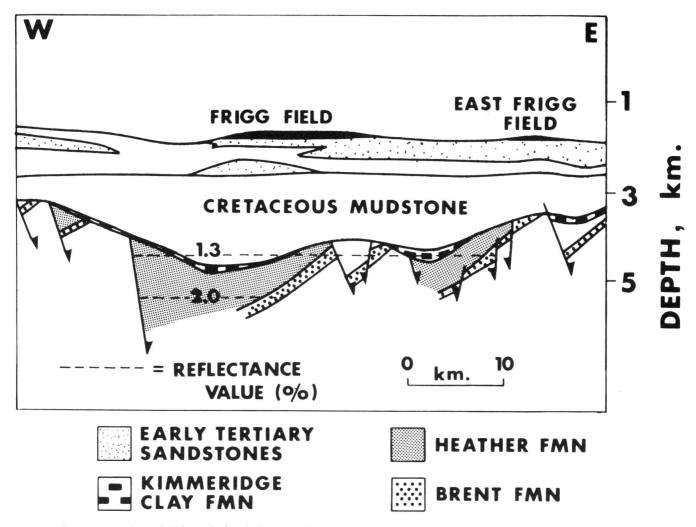

Figure 4. Cross section through Viking Graben below Frigg Field (after Heritier et al, 1979).

Welte, 1978) this gradient indicates the oil window extends from 2,550 to 4,500 m, the wet gas zone from 4,500 and 5,400 m, and that the dry gas zone occurs at depths greater than 5,400 m. Peak gas generation from coal corresponds to the boundary of the medium and lower volatile bituminous coal ranks (Hunt, 1979; Figure 5.7), corresponding to a vitrinite reflectance of 1.5 percent. This is equivalent to a burial depth of 5,000 m in the Viking Graben.

Hydrocarbon/TOC ratio data for the Kimmeridge Clay have been correlated with the vitrinite reflectance gradient to determine the vitrinite reflectance level corresponding to peak hydrocarbon generation (Figure 7). The maturity corresponding to peak hydrocarbon generation is here defined as the maturity when 50 percent of the potential hydrocarbon yield has been generated. For a Type II kerogen, 70 percent of the organic matter is capable of conversion to hydrocarbons (Tissot and Welte, 1978). When peak hydrocarbon generation is reached, 35 percent of the organic matter will thus have been converted to hydrocarbons. This corresponds to a hydrocarbon/TOC ratio in the source rock at peak generation of 0.42, if no expulsion has occurred (assuming 1 gm of organic matter is equivalent to 0.83 gm of TOC).

However, because some of the generated hydrocarbons are expelled from the source rock, the hydrocarbon/TOC ratio corresponding to peak generation will be less than 0.42. The expulsion efficiency can be estimated from the observed decrease in TOC with depth for the Kimmeridge Clay of 0.7 percent from about 2,900 to 3,400 m (assuming the TOC decrease is due to expulsion of generated hydrocarbons rather than early diagenetic processes). If ≤ 50 percent of the potential hydrocarbon yield is generated over this depth interval, which corresponds to a narrow vitrinite reflectance range of 0.6 to 0.8 percent, an expulsion efficiency ≥ 0.25 is indicated. The hydrocarbon/TOC ratio in the Kimmeridge Clay at peak generation would thus be ≥ 0.35 which corresponds to a vitrinite reflectance of 0.7 percent by extrapolation of the graph on Figure 7.

Organic matter colouration (Staplin, 1969) and source rock electrical resistivity (Meissner, 1978) can also be used to estimate source rock maturity. From 2,600 to 3,200 m plant material in the Kimmeridge Clay is light to medium brown, indicating it is moderately mature; between 3,200 and 3,650 m it is dark brown indicating it has achieved peak generation (Brooks and Thusu, 1977). Electrical resistivity increases from 2 to 3 Ωm at 2,500 to 2,600 m to a maximum

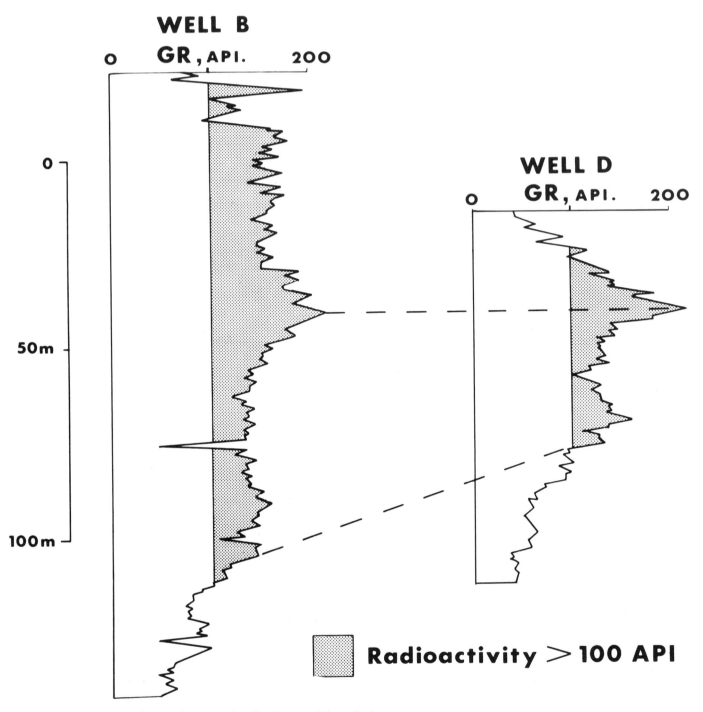

Figure 5. Gamma ray correlation of Kimmeridge Clay Basin to Viking Graben.

of 10 to 25 Ωm at 3,500 to 3,600 m (Figure 8). These data indicate peak generation has occurred between 3,200 and 3,500 m at a reflectance level of 0.7 to 0.8 percent. At a reflectance level of 1.2 to 1.3 percent (base oil window maturity) in Well D (Figure 8), the Kimmeridge Clay is highly overpressured, and has a very high transit time (115 to 135 µsec/ft) and low resistivity (<3 µm). These log responses suggest it has a high water content, and that oil, formerly present in the rock, has cracked to gas, most of which has been expelled. Low resistivities thus do not necessarily indicate immaturity.

PRESENT DAY THERMAL GRADIENTS AND HEAT FLOW

Bottom hole temperatures have been calculated in Wells B, C, and D from electric logging run temperature measurements, drilling mud circulation history, and borehole diameter, using the method of Oxburgh et al (1972). The temperature for Well A was measured during a drill stem test. Heat flows were derived from the bottom hole temperatures and thermal conductivity estimates for the rocks penetrated by each well. The thermal conductivity

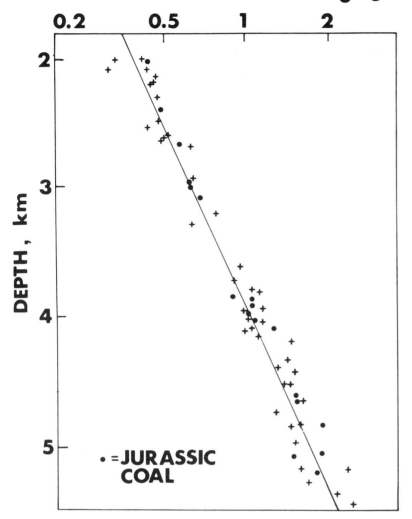

VITRINITE REFLECTANCE, R_o %

R$_o$ %	DEPTH, m
.5	2550
.7	3250
.8	3500
1.3	4500
2.0	5400

Figure 6. Vitrinite reflectance versus depth for Jurassic to early Cretaceous coals and mudstones.

of a rock (K) is a function of lithology and porosity; it varies with porosity according to the following equation (Oxburgh and Andrews-Speed, 1981);

$$K = K_w^{\phi} \cdot K_m^{1-\phi}$$

where K_w is the conductivity of water, and K_m is the conductivity of the rock matrix. Rock matrix conductivities have been determined by BP's Geophysical Research Division. The standard equation for steady state conductive heat flow (q) across a uniform rock layer of thickness, L, which has no heat production within it is given by:

$$q = \frac{k \, dT}{dL}$$

where dT/dL is the thermal gradient across the layer in the direction of heat flow.

Because of the variation of conductivity with lithology and porosity it is convenient to calculate the heat flow using the following equation (Oxburgh and Andrews-Speed, 1981):

$$q_z = \frac{T_z - T_o}{R_z}$$

where q_z, T_z, and R_z are the heat flow, temperature and thermal resistance at the depth of temperature measurement, z. T_o is the surface temperature.

The thermal resistance at depth z is defined as:

$$R_z = \frac{L_1}{K_1} + \frac{L_2}{K_2} + \frac{L_3}{K_3} + \dots$$

where $L_{1...}$ are the thicknesses of the rock layers in the overburden, and $K_{1...}$ are their thermal conductivities (Oxburgh and Andrews-Speed, 1981).

Thermal resistance was calculated as a function of depth in each well using the observed variation of lithology with

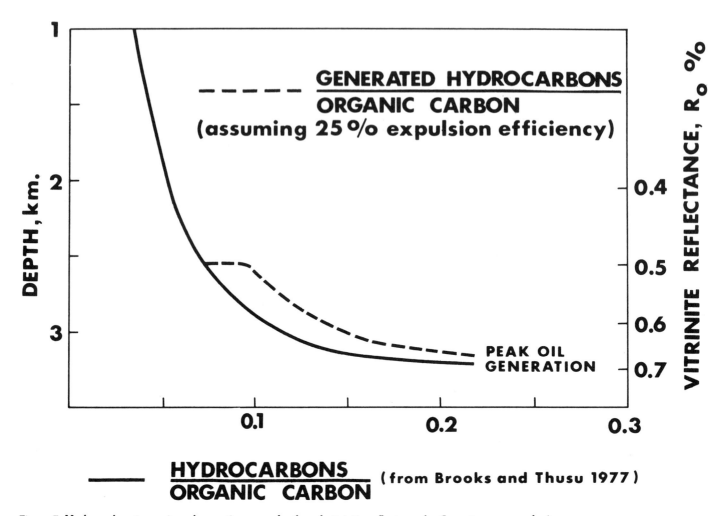

Figure 7. Hydrocarbon/organic carbon ratio versus depth and vitrinite reflectance for 'Jurassic source rocks.'

depth, and porosity/depth gradients derived from density logs and shale density measurements.

Porosity (ϕ) was calculated from the density measurements using the following equation:

$$\phi = \frac{P_m - P_r}{P_m - P_w}$$

where P_m, P_r, and P_w are the rock matrix, whole-rock and water densities respectively, using a matrix density of 2.7 gm/cc for mudstone and 2.65 gm/cc for sandstone.

Porosity/depth relationships have been modelled for each lithology assuming an exponential decrease in porosity with depth defined by:

$$\phi = \phi_o e^{-Z/m}$$

where ϕ_o is the surface porosity and m is the compaction gradient. Surface porosities of 65 percent for mud and 43 percent for sand were selected (Sclater and Christie, 1980). Calculated porosity gradients are shown in Table 2; the variation in porosity with depth for mudstones in wells C and D is shown in Figure 9.

Calculated heat flows in the 4 wells range from 0.051 to 0.061 Wm^{-2}. Heat flow varies with depth because of heat generation from decay of radioactive isotopes of potassium, uranium, and thorium in the sedimentary rocks. Surface heat flow (q_o) has been calculated for each well, using average values of heat production for mudstone and sandstone, from the following equation:

$$q_o = q_z + AZ$$

where A is the average heat generation/unit volume of rock above depth Z. The mean surface heat flow in the 4 wells is 0.060 Wm^{-2} (Table 1). This is the same as the mainland UK average heat flow (Richardson and Oxburgh, 1979).

Interval thermal gradients for Wells A-D were calculated from the thermal resistance profile and calculated bottom hole temperatures in each well. The interval gradients decrease with depth from 37 to 40°C/km for the Tertiary, 31 to 33°C/km for the Cretaceous, to 26 to 27°C/km for the Jurassic. This decrease in interval gradient with depth is attributed to the decrease in porosity, and thus increase in thermal conductivity, with depth, and to the distribution of heat sources.

Carstens and Finstad (1981) also observed a decrease in interval gradients with depth down to the base Cretaceous in the East Shetland Basin and Viking Graben. They also

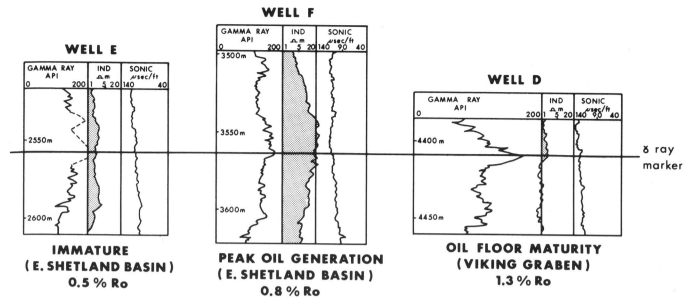

WELL F

GAMMA RAY
API
0   200
IND
Ω m
1 5 20
SONIC
μsec/ft
140 90 40

WELL E

GAMMA RAY
API
0   200
IND
Ω m
1 5 20
SONIC
μsec/ft
140   40

WELL D

GAMMA RAY
API
0   200
IND
Ω m
1 5 20
SONIC
μsec/ft
140 90 40

δ ray marker

IMMATURE
(E. SHETLAND BASIN)
0.5 % Ro

PEAK OIL GENERATION
(E. SHETLAND BASIN)
0.8 % Ro

OIL FLOOR MATURITY
(VIKING GRABEN)
1.3 % Ro

Figure 8. Kimmeridge Clay log response as a function of maturity.

calculated very high Jurassic interval gradients > 35°C/km. These high apparent Jurassic gradients probably result from the lack of correction for borehole diameter in the Horner Plot method, used by Carstens and Finstad, to calculate true bottom hole temperatures. The Horner Plot method can underestimate true formation temperature by as much as 20 percent in a 12¼ in diameter borehole, and by about 5 percent in a 8 in hole (S.W. Richardson, personal communication).

Consider a 500 m Jurassic section at 3,000 to 3,500 m depth with a true interval gradient of 26°C/km, and true bottom hole temperatures of 113°C at 3,000 m, and 126°C at 3,500 m. A 20 percent error in the Horner Plot derived temperature in 12¼ in hole at 3,000 m, and a 5 percent error in 8 in hole at 3,500 m, will give an apparent Jurassic interval gradient of 59°C/km (which is more than 100 percent too high!).

Average thermal gradients in Wells A, B, and C in the eastern East Shetland Basin are 35 to 36.5°C/km. Well D in the Viking Graben has an interval gradient of 35°C/km. Carstens and Finstad (1981) reported average gradients of 30 to 35°C/km in the central and western East Shetland Basin. These data indicate a fairly uniform present day thermal

Table 2. Compaction data for gross lithologies.

Well	Lithology	Surface Porosity	Compaction Gradient (meters)
A, B, C, D	Sandstone	0.43	5,000
A, B	Mudstone	0.65	2,000
C	Mudstone	0.65	2,350
D	Mudstone	0.65	2,500

regime in the study area.

SUBSIDENCE AND HEAT FLOW HISTORY

McKenzie (1978) proposed that the North Sea Basin formed as a result of stretching, and consequent thinning, of the continental lithosphere. He predicted that the thinning of the crust was caused by listric normal faulting. In his model an initial *fault controlled* subsidence occurred which maintained isostatic compensation; it increased with the amount of stretching, β, where

$$\beta = \frac{\text{initial lithospheric thickness}}{\text{lithospheric thickness immediately after stretching}}$$

Table 1. Thermal data for wells.

Well	Depth (mBRT)	Temperature (°C)	Time Since Circulation (hrs)	Calculated Bottom Hole Temperature (°C)	Thermal Resistance (Wm^{-2}T^{-1})	Measured Heat Flow (Wm^{-2})	Surface Heat Flow (Wm^{-2})
D	5,180	150	15				
C	4,775	153.3	21	161.5	3,055	0.051	0.057
		154.4	7.75				
B	3,940	157.2	12.5	161.7	2,554	0.061	0.065
		115.6	5				
A	2,725	121.1	9	128	2,214	0.055	0.058
		94.2	Drill stem test temperature		1,554	0.056	0.058

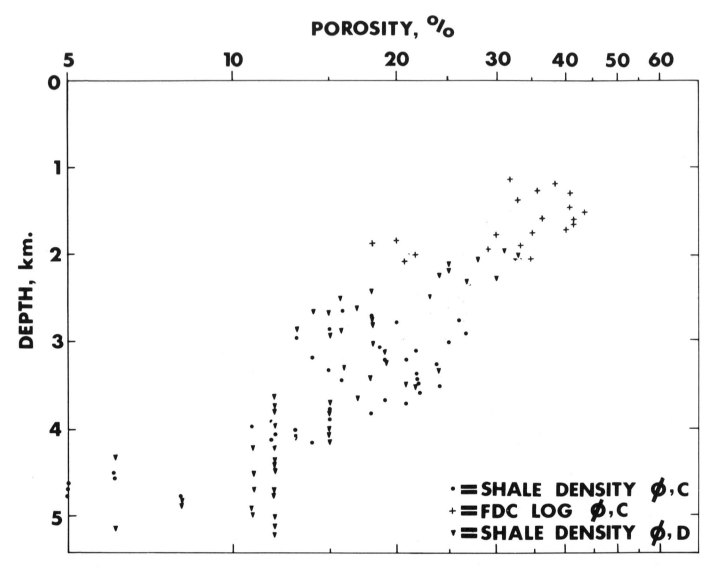

Figure 9. Mudstone porosity versus depth in Wells C and D.

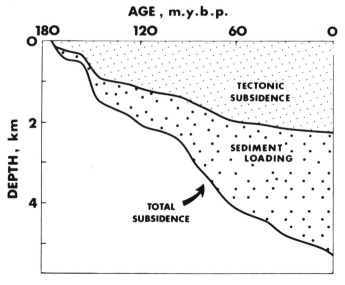

Figure 10. Subsidence history at Well D since late early Jurassic.

The initial fault controlled subsidence varies across rifted basins because of local uplift of tilted blocks. The fault controlled subsidence in the axial grabens is most likely to be representative of the amount of stretching.

Cooling of buoyant hot asthenosphere below the thinned lithosphere creates a thermal anomaly reflected by a high surface heat flow. The heat flow declines with time as the asthenosphere cools. This decline in heat flow with time can be calculated if the age and amount of stretching is known (McKenzie, 1978; Figure 2). The *age* of stretching is defined by periods of normal faulting; the *amount* of stretching, β, can be calculated from the amount of tectonic subsidence that has occurred since faulting ceased.

Deposition of sediments in a basin is an additional important subsidence mechanism (Turcotte, 1980). The subsidence of sediment loaded crust is two to three times greater than the water loaded tectonic subsidence, depending on the sediment density. The total subsidence is related to the water loaded tectonic subsidence by the following equation:

$$d_{Total} = \frac{(pm - pw)}{(pm - ps)} d_w$$

where d_{Total} is the total subsidence, d_w is the water loaded tectonic subsidence, pw is the water density, ps is the sediment density, pm is the density of the mantle.

Geophysical evidence indicates thinned crust below the northern North Sea as predicted by McKenzie's model. Seismic refraction studies across the Viking Graben indicate a value of β of about 2 (Sclater and Christie, 1980); gravity profiles across the East Shetland Basin and Viking Graben suggest the crust has been thinned from 30 km to 20 km, i.e. $\beta = 1.5$ (Donato and Tully, 1981). The initial fault controlled subsidence predicted by McKenzie's model for $\beta = 1.5$ is 800 m (Sclater and Christie, 1980). The thickness of Bathonian to early Cretaceous mudstones in the axial Viking Graben is about 2 km (Figure 3). This thickness is consistent with an initial fault controlled subsidence of about 800 m (after removing the effects of sediment loading).

There are two major problems in applying McKenzie's model to determine the heat flow history of the East Shetland Basin and Viking Graben. Firstly, several phases of faulting occurred between late Bathonian to early Cretaceous time; secondly, water depth during deposition in late Jurassic to early Tertiary time is not accurately known.

Major phases of rifting occurred in Bathonian to early Oxfordian time along the western flank of Viking Graben (Figure 3) and in late Oxfordian to Kimmeridgian time along the western edge of the East Shetland Basin; both phases of faulting occurred within the East Shetland Basin. Early Cretaceous rifting occurred in the northern part of the East Shetland Basin and possibly also along the western flank of the Viking Graben (Figure 3). Fault controlled subsidence ceased at the end of the early Cretaceous (100 Ma ago) in the East Shetland Basin and Viking Graben except along the eastern bounding fault of the East Shetland Platform. Similar diachronous rifting occurred in the East Greenland Jurassic rift (Surlyk, 1968), with main phases in the Bathonian, late Oxfordian, early Kimmeridgian and at the Jurassic/Cretaceous boundary. The faulting history suggests that several phases of crustal thinning occurred between 165 and 100 Ma ago in the study area.

The water loaded tectonic subsidence at Wells A-D has been calculated from the subsidence history (corrected for compaction) by removing the loading effects of the sediments, assuming local isostatic compensation. The tectonic subsidence and the subsidence due to sediment loading for Well D are shown in Figure 10.

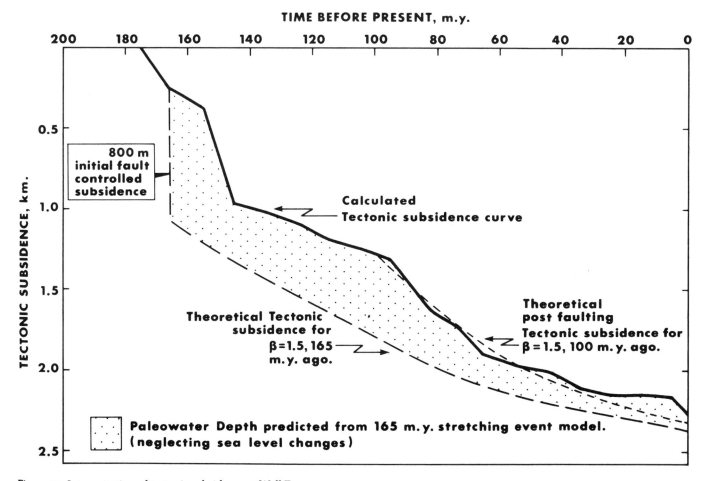

Figure 11. Interpretation of tectonic subsidence at Well D.

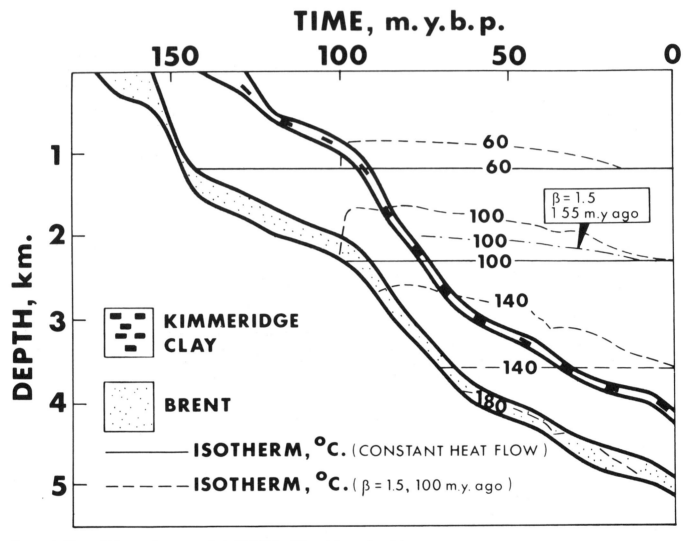

Figure 12. Thermal history of source rocks in Well D for different thermal models.

Modelling of the post-faulting tectonic subsidence in Well D gives a good fit to a stretching event 100 Ma ago with $\beta = 1.5$ (Figure 11) assuming constant water depth in the late Cretaceous and Tertiary. However, sedimentation rates were slow during the Kimmeridgian to early Cretaceous, and water depths were probably much greater than at the present day. A more realistic model for the tectonic subsidence at Well D is obtained with a stretching event 165 Ma ago (corresponding to the main Viking Graben rifting), with a stretching factor, (β), of 1.5 (Figure 11). This model predicts Kimmeridgian to early Cretaceous water depths of 500 to 700 m in the Viking Graben, and progressive shallowing during the late Cretaceous and Tertiary to the present day value of 150 m.

The variation in the age of crustal stretching (and consequent increase in heat flow) within the study area indicates that heat flow was probably very variable between 165 and 100 Ma ago. However, Jurassic source rocks were not deeply buried enough for hydrocarbon generation to occur prior to 100 Ma ago. Consequently hydrocarbon generation in the study area is insensitive to early variation in heat flow.

Two extreme heat flow models can be used to determine the timing of hydrocarbon generation in the study area. A *constant heat flow model* will underestimate maturity at a given time because it fails to account for higher heat flows during the Cretaceous and Tertiary than at the present day. The model with a *heating event 100 Ma ago* will overestimate maturity because it fails to account for early heat loss from the crust during late Jurassic and early Cretaceous rifting. The true maturity of the source rocks will lie *between* the values predicted by these two models.

THERMAL AND MATURATION HISTORY OF SOURCE ROCKS

The thermal history of the source rocks is related to the variation in heat flow, and the increase in the thermal resistance of their overburden, with time. The thermal resistance in Wells A-D was calculated as a function of time using the depositional history, and the compaction gradients for mudstone and sandstone. The maximum range for the thermal history of the Jurassic source rocks is bounded by the constant heat flow and 100 Ma heating event models.

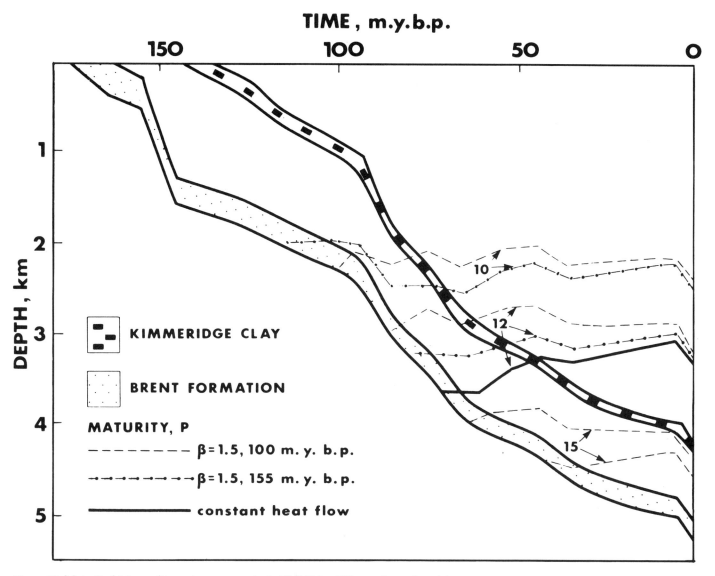

Figure 13. Maturity history of Jurassic source rocks in Well D for different thermal models.

The thermal history of the source rocks in Well D in the Viking Graben is shown in Figure 12. Source rock temperatures rose rapidly during late Cretaceous burial. During the Tertiary the constant heat flow model predicts a moderate increase in source rock temperature; the 100 Ma heating event model predicts that source rock temperatures remained relatively constant, because of a decrease in heat flow with time during the Tertiary. A model with a heating event in the late Jurassic 155 Ma ago predicts source rock temperatures intermediate between the constant heat flow and 100 Ma heating event models (see the 100°C isotherm on Figure 12).

The maturation history of the source rocks in Wells A-D was calculated by integrating time and temperature using the following equation (Royden et al, 1980):

$$P = \mathrm{Ln} \int_{o}^{t} 2^{T/10} dt$$

where P is the maturity parameter, t = time (Ma) and T = temperature (°C). This equation allows for the

approximate doubling of the reaction rate of hydrocarbon-forming reactions for every 10°C increase in temperature, in a similar way to 'Lopatin's Time Temperature Index' used by Waples (1980) to determine maturity.

The maturity parameter, P, was calculated as a function of time for each source rock, and as a function of depth at the present day in each well, using the constant heat flow and 100 Ma heating event models. The maturation history of the source rocks in Well D as predicted by the different thermal models is shown in Figure 13. The 100 Ma heating event model predicts earlier maturation of the source rocks than the constant heat flow model. The 155 Ma heating event model predicts intermediate maturation levels, except for the Brent Formation maturity prior to 90 Ma ago. The maturities predicted by the different models converge with time so that *present day* maturities predicted by the different models are similar.

The maturity/depth gradient calculated using the 100 Ma heating event model is shown in Figure 14. Note that the

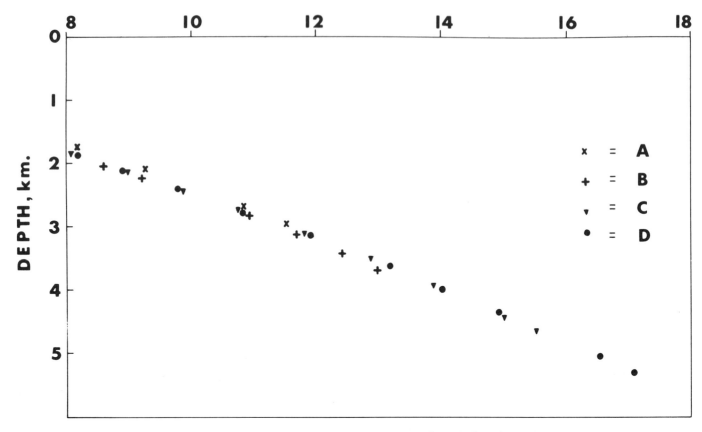

Figure 14. Calculated maturity, for 100 Ma heating event model, versus depth in Wells A, B, C, and D.

gradient is almost the same for each well, despite their different structural locations. This is consistent with the uniform maturity gradient indicated by the vitrinite reflectance data (Figure 6).

In order to determine the timing of generation from the calculated maturity, vitrinite reflectance has been correlated with P. Vitrinite reflectance measurements in Wells C and D have been correlated with P value calculated at the same depths. Vitrinite reflectance measurements from other wells have been correlated with P values obtained from the calculated regional maturity/depth gradient. In this way P has been correlated with vitrinite reflectance, and thus generation stage, for both *the constant heat flow* and *100 Ma heating event models* (Table 3). The correlation of vitrinite reflectance with P, calculated using the 100 Ma heating event model, is shown in Figure 15.

TIMING OF GENERATION

The timing of oil generation from the Kimmeridge Clay in Wells A-D has been determined from its maturation history using the correlations of vitrinite reflectance with P (Table 4). The constant heat flow model predicts *later* generation than the 100 Ma heating event model. The onset of oil generation (0.5 R_o) predicted by the constant heat flow model is up to 10 Ma later than predicted by the 100 Ma heating event model. For Well D, the top of the Kimmeridge Clay began to generate oil between 69 and 77 Ma ago and reached peak generation between 53 and 65 Ma ago.

The maturity distribution of the Kimmeridge Clay at the end of the Cretaceous (65 Ma ago) has been determined by correlating its maturity 65 Ma ago in Wells B, C, and D, with the present day Cretaceous thickness in these wells (Figure 16). This correlation shows that the Kimmeridge Clay had become mature by the end of the Cretaceous, where the present Cretaceous thickness is >1,700 m. The area of mature Kimmeridgian at the end of the Cretaceous was then deduced from a Cretaceous isopach map of the study area. The maturity distribution of the Kimmeridge Clay 40 and 20 Ma ago was estimated using the present day base Cretaceous structure contour map and correlations of maturities at these times in Wells A-D with present day depth to the Kimmeridge Clay (Figure 17, Table 5).

Table 3. Correlation of vitrinite reflectance with calculated maturity (P), and oil and gas generation from source rocks.

Vitrinite Reflectance R_o %	P for 100 Ma Heating Event Model	P for Constant Heat Flow Model	Generation Stage
0.5	10.1	9.7	Oil threshold
0.7	11.9	11.5	Peak oil generation for Kimmeridge Clay and gas generation threshold for coal
1.3	15.0	14.4	Oil floor
1.5	15.8	15.0	Peak gas generation for coal
2.0	17.2	16.3	Wet gas floor

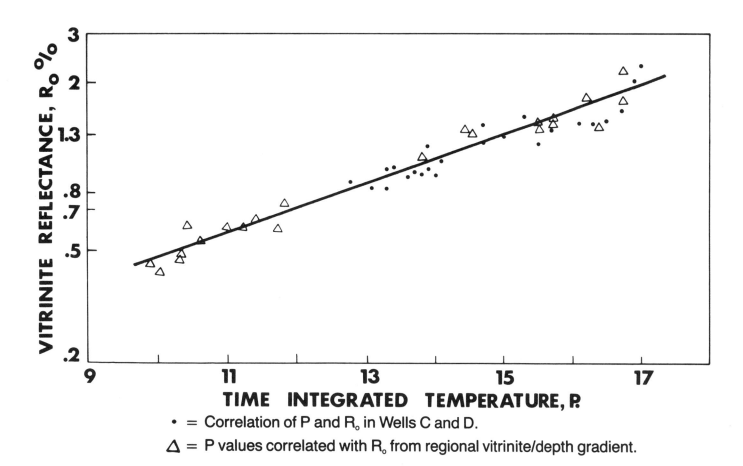

= Correlation of P and R_o in Wells C and D.

△ = P values correlated with R_o from regional vitrinite/depth gradient.

Figure 15. Correlation of vitrinite reflectance with maturity calculated using 100 Ma heating event model.

The generation history of the Kimmeridge Clay during the Tertiary is summarized in Figure 18. At the end of the Cretaceous the Kimmeridge Clay was mature throughout the Viking Graben, and in the deepest troughs and half grabens of the East Shetland Basin. In the Viking Graben peak generation was reached during the Palaeocene. By the end of the Eocene, peak generation was occurring in the trough west of the Ninian Field in the East Shetland Basin. During the Oligocene, peak generation was reached in the half grabens of the East Shetland Basin, and oil floor maturity was attained in the deepest parts of the Viking Graben. By the present day, peak generation had been established throughout the axial region of the East Shetland

Basin, and a maturity just above or greater than the oil floor was attained in the Viking Graben. A generation front has thus moved progressively updip with time within the Kimmeridge Clay from the deepest parts of the Viking Graben during Campanian time, to the structurally highest fault blocks in the East Shetland Basin at the present day. The zone of intense oil generation (0.65 to 0.9 percent R_o) has moved from the Viking Graben to the axial East Shetland Basin since the end of the Eocene.

The generation history deduced above is consistent with diagenetic studies in the Brent Formation sandstone reservoirs in the East Shetland Basin (see Figure 1 for the locations of fields discussed below). Hancock and Taylor

Table 4. Timing of maturation for Kimmeridge Clay in Wells A-D for different thermal models. a, Heating event 100 Ma ago. b, Constant heat flow.

a Maturity, P	Time Reached, Ma Before Present				b Maturity, P	Time Reached, Ma Before Present				
	A	B	C	D		A	B	C	D	
10.1	–	47	63	77	9.7	–	37	53	69	To Kimmeridge Clay
	3	57		79		3	47		71	Base Kimmeridge Clay
11.9	–	–	34	65	11.5	–	–	27	53	Top Kimmeridge Clay
	–	–		67		–	–		56	Base Kimmeridge Clay
12.5	–	–	25	59	12.1	–	–	15	44	Top Kimmeridge Clay
	–	–		62		–	–		47	Base Kimmeridge Clay
15.0	–	–	–	–	14.4	–	–	–	–	Top Kimmeridge Clay
	–	–		–		–	–		–	Base Kimmeridge Clay

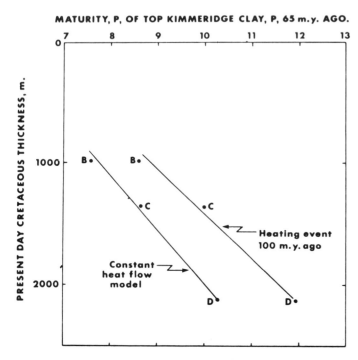

Figure 16. Maturity of Kimmeridge Clay 65 Ma ago versus present day burial depth.

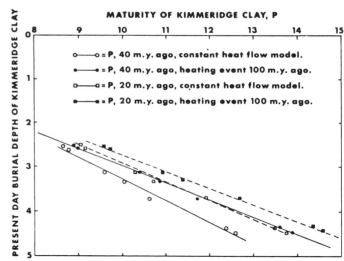

Figure 17. Maturity of Kimmeridge Clay 40 and 20 Ma ago versus present day Cretaceous thickness.

(1978) observed an upward transition from authigenic illite to authigenic kaolinite within a thick oil column in an East Shetland Basin oil field. They concluded that oil migration occurred synchronously with illite diagenesis. Sommer (1978) observed a more abrupt change from illitic cement in the oil/water transition zone to predominantly kaolinite in the oil zone in an East Shetland Basin oil well drilled by Total (presumably in the Alwyn area). The youngest radiometric dates obtained for pure illite cements in this well were 45 to 55 Ma, indicating that this field had filled up by mid-Eocene times. This age of migration is consistent with the calculated age of generation in the trough west of the Alwyn field (Figure 18).

In the Thistle Field (Hay, 1977), porosity/depth gradients within individual oil saturated sandstone units range from 1.5 to 5 percent porosity loss/100 m. These very rapid reductions in porosity with depth probably indicate progressive cessation of diagenesis in the sandstones as they filled with oil, over a long period of time. This suggests that oil migration occurred over an extended period of time in the Thistle area, which is consistent with the calculated progressive updip movement of the oil generation front of this part of the northern East Shetland Basin (Figure 18). A probable palaeo-OWC in the Magnus Field (De'ath et al, 1981) suggests that this field had filled up prior to late Tertiary tilting, which is consistent with the early onset of generation in the half graben downdip to the west (Figure 18).

The timing of gas generation from Brent Formation coals in the Viking Graben can be inferred from the maturation history of this formation in Wells C and D. Maturation analysis in Well D, using the constant heat flow and 100 Ma heating event models and the appropriate correlations of P

and vitrinite reflectance (Table 3), indicates gas generation began 70 to 85 Ma ago. Peak dry gas generation from the coals in Well D began 30 to 40 Ma ago. In Well C, on the flank of the Viking Graben, gas generation began 27 to 34 Ma ago. In the deepest part of the Viking Graben, where the Brent Formation is buried to 6 km, gas generation probably began 100 Ma ago.

SOURCE AND TIMING OF GENERATION OF FRIGG FIELD HYDROCARBONS

The Frigg Field oil is a relatively heavy (24 API), napthenic oil consisting dominantly of hydrocarbons with a carbon number $> C_{17}$; condensate dissolved in the overlying dry gas column comprises 87 percent $C_{11}+$ (Heritier et al, 1979). Heritier et al (1979) suggested, using pristane/phytane ratio data for the East Frigg Field oil, that the Frigg Field oil was sourced from Lower-Middle Jurassic rocks. However, good oil source rocks have not been described from the Dunlin and Brent formations in the Viking Graben. An alternative hypothesis is that the Frigg

Table 5. a. Correlation of palaeomaturity of Kimmeridge Clay from 100 Ma heating event model with present burial depth.

Maturity 40 Ma Ago	Present Depth	Maturity 20 Ma Ago	Present Depth, m
10.1	3,050	10.1	2,800
11.9	3,700	11.9	3,500
12.5	4,000	12.5	3,700
15.0	5,000	15.0	4,600

b. Correlation of palaeomaturity of Kimmeridge Clay from constant heat flow model with present burial depth.

Maturity 40 Ma Ago	Present Depth	Maturity 20 Ma Ago	Present Depth, m
9.7	3,100	9.7	2,800
11.5	4,000	11.5	3,600
12.1	4,300	12.1	3,800
14.4	5,400	14.4	4,800

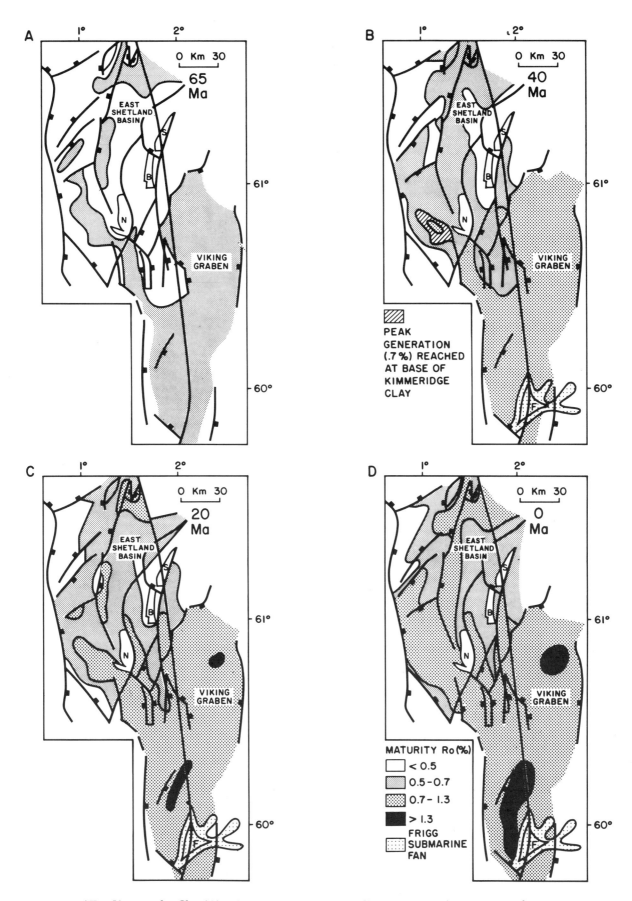

Figure 18. Maturity of Top Kimmeridge Clay (A) 65 Ma ago, (B) 40 Ma ago, (C) 20 Ma ago, and (D) at present day.

oil was generated from the Kimmeridge Clay.

The drainage area of the Frigg Field is estimated to be at least 1,500 sq km from the contour map of the top of the Frigg Sand published by Heritier et al (1979). The Kimmeridge Clay within this drainage area is buried below 4 km (Figure 4). The maturation history of the Kimmeridge Clay within the drainage area of the Frigg Field can be inferred from the maturation history of this formation in Well D (Table 4), since they have had similar burial histories. The Cretaceous to early Eocene thickness is 2,500 m in Well D, compared to 2,500 to 2,700 m in the main drainage area of the Frigg Field. The Cretaceous thickness in Well D is 2,100 m, which corresponds to the Cretaceous thickness in the deepest part of the Frigg drainage area.

The Kimmeridge Clay within the drainage area of the Frigg Field thus began to generate oil 70 to 80 Ma ago. Peak generation was reached 55 to 65 Ma ago. At the time the trap was sealed, 45 Ma ago, it had reached a maturity of 0.8 to 0.9 R_o percent equivalent, corresponding to a transformation ratio of 70 to 80 percent. The Kimmeridgian had thus been expelling oil at peak generation 10 to 20 Ma ago before the trap was sealed.

The heavy, napthenic, nature of the oil may thus result from migration of a normal light oil into a leaky trap in which sea water was circulating, with consequent water washing and biodegradation removing the light hydrocarbons and n-alkanes.

The Frigg Field methane has a δC^{13} value of 43.3 percent (Heritier et al, 1979), which is characteristic of the carbon isotope composition of methane derived from sapropelic organic matter in the maturity range 0.5 to 1.3 percent (Hunt, 1979, p. 376). This suggests the gas was sourced from the Kimmeridge Clay and not from coaly material in the Brent Formation. Unfortunately there is little analytical data on the formation of methane from sapropel at maturities less than 1.0 percent R_o (Tissot and Welte, 1978, p. 218). However, much of the methane derived generation from kerogen must be formed by cracking of previously formed oil remaining in the rock in the lower part of the oil window (0.9 to 1.3 percent R_o). Hunt (1979, p. 1964) stated that the gas yield from sapropelic kerogen is 1.5 to 2 times that of humic kerogen because of this cracking of the previously formed oil. The Kimmeridge Clay passed through the 0.9 to 1.3 percent R_o maturation interval between 50 Ma ago and the present day, largely after the trap was sealed.

VOLUMETRIC MODELLING OF OIL GENERATION AND ENTRAPMENT

The ratio of oil generated from the Kimmeridge Clay to oil trapped in Jurassic sandstones has been studied in the central and eastern area of the East Shetland Basin (Figure 19). The amount of oil generated within this area was calculated using a Geochemical Mass Balance Equation. The Geochemical Mass Balance method of reserves calculation has been discussed by White and Gehman (1979). Waples (1979) pointed out that a major problem in quantitative evaluation of oil source rocks is lack of knowledge of oil expulsion efficiencies. In this section I shall use the Geochemical Mass Balance method to determine expulsion efficiency.

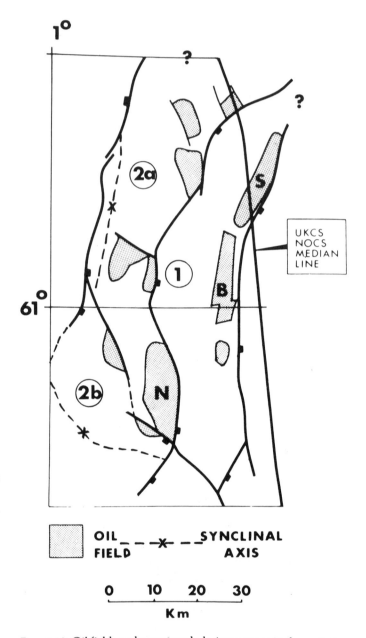

Figure 19. Oil fields and source rock drainage areas in the eastern and central East Shetland Basin.

The volume of oil generated from a unit volume of source rock is related to the amount, type, and maturity of its kerogen. The volume of generated oil that is ultimately trapped is related to the expulsion efficiency of oil from the source rock, the migration efficiency, and the sealing efficiency of the traps.

The average TOC content of the Kimmeridge Clay is 5.6 percent wt at 2,600 to 3,200 m at pre-peak generation maturity. The average TOC content at the onset of oil generation (2,550 m) was thus probably about 6 percent wt, which is equivalent to an organic matter content of 7 percent wt. The Kimmeridge Clay source rock is estimated to have a porosity of 15 percent at the onset of oil generation, using the compaction gradient for mudstone in Well B. The volume percentage of organic matter in the

Kimmeridge Clay calculated using a wt percent organic matter content of 7 percent, a porosity of 15 percent, and rock matrix and water densities of 2.7 and 1 gm/cc respectively, is 15 percent.

The amount of organic matter capable of conversion to hydrocarbons is given by its 'genetic potential.' The theoretical genetic potential of a Type II kerogen is 70 percent (Tissot and Welte, 1978). The relative proportions of sapropel and humic/coaly material in the Kimmeridge Clay kerogen indicate that the oil content of the hydrocarbons generated from the kerogen during maturation is about 80 percent.

The maturity of a kerogen can be expressed by its 'transformation ratio,' which is defined as the ratio of the amount of hydrocarbons generated to the total amount of hydrocarbons that the kerogen is capable of generating. The transformation ratio corresponding to peak generation is thus 0.5. The transformation ratio for the Kimmeridge Clay kerogen as a function of vitrinite reflectance is given in Figure 20 based on the correlation of peak oil generation with a reflectance of 0.7 percent.

Generation of a typical East Shetland Basin light oil (specific gravity = 0.84) from kerogen involves a volume increase of about 20 percent. The volume of oil generated from a unit volume of source rock is given by the following equation:

$$\text{Oil generated} = \frac{\text{bulk rock volume}}{\text{of source rock}} \times \frac{\text{organic matter}}{\text{content by volume}}$$

$$\times \frac{\text{genetic}}{\text{potential}} \times \frac{\text{fraction of oil in}}{\text{hydrocarbon yield}}$$

$$\times \frac{\text{transformation}}{\text{ratio}} \times \frac{\text{volume increase on}}{\text{oil generation}}$$

For the Kimmeridge Clay, the volume of oil generated (V_o) becomes:

$$V_o = \frac{\text{bulk rock}}{\text{volume}} \times 0.15 \times 0.7 \times 0.8$$

$$\times \frac{\text{transformation}}{\text{ratio}} \times 1.2 = 0.1 \binom{\text{bulk rock volume} \times}{\text{transformation ratio}}$$

The eastern East Shetland Basin has been divided into two drainage areas (Figure 19). The eastern boundary of drainage area 1 is defined by the Brent/Statfjord fault trend, the boundary of drainage areas 1 and 2 by the Ninian/Hutton fault trend. These fault trends converge to the north. The northern boundary of drainage area 2 is not well defined, but is probably defined by a series of faults and dip reversals. The overpressure in the Brent Formation in the eastern East Shetland Basin is 1,700 to 1,900 psi. However, the author has not studied enough pressure data to determine whether there is fluid communication within the Brent across the Ninian/Hutton fault trend.

The bulk rock volume of Kimmeridge Clay in each drainage area has been calculated from an isopach map based on well control and geological structure maps. Weighted average thicknesses for the Kimmeridge Clay are

70 m in Areas 1 and 2a, and 200 m in Area 2b.

The oil generated and trapped within each drainage area is shown in Table 6. The average entrapment/generation ratio is 25 percent, which indicates a very high oil expulsion efficiency from the Kimmeridge Clay (>25 percent) and extremely high migration and sealing efficiencies.

The apparent entrapment/generation ratio for drainage area 1 is 54 percent. However, the sonic log response of mudstones in the upper 150 m of the Heather Formation in Well B (Figure 21) in drainage area 1 suggests that this interval is a source rock. It might be placed in the Kimmeridge Clay on the basis of its sonic velocity. Recalculating the amount of oil generated in drainage area 1, including this extra interval, gives an entrapment/generation ratio of 25 percent; the average entrapment/generation ratio becomes 20 percent.

The average entrapment/generation ratio has also been calculated using average values of organic richness and hydrocarbon yield determined for immature Kimmeridge Clay samples from onshore UK and the North Sea by the BP's Geochemical Research Division. Average organic matter content is 8 percent by wt; average hydrocarbon yield during pyrolysis is 30 kg/ton rock, equivalent to a genetic potential of 37 percent. The average entrapment/generation ratio obtained using these data, and the additional source rock interval in drainage area 1, is 30 percent.

The oil entrapment/generation ratio in the eastern East Shetland Basin is 20 to 30 percent. Oil expulsion efficiency is ≥20 to 30 percent, since some oil must be lost as residual oil along migration paths, and oil generated in the upper part of the source rock may have been expelled into the Cretaceous mudstones.

OIL SATURATION IN COMPACTION FLUIDS EXPELLED FROM KIMMERIDGE CLAY

The oil saturation of the compaction fluids expelled from the Kimmeridge Clay in the drainage areas (Figure 19) during migration has been estimated by calculating the volumes of oil and water expelled. By the end of the Cretaceous the Kimmeridge Clay was mature in the deepest parts of the East Shetland Basin; migration is thus considered to have begun 65 Ma ago. The volume of oil generated is (15.4 to 20.2) × 10^9 cu m (Table 6); the volume of oil expelled using an expulsion efficiency of 30 percent is thus (4.6 to 6.1) × 10^9 cu m. The water filled porosity loss from the Kimmeridge Clay has been estimated assuming that the mudstone compaction gradient during the Tertiary was the same as at the present day. The average porosity loss for the Kimmeridge Clay is estimated to be 10 percent since the end of the Cretaceous. This porosity loss is a maximum value, since the Kimmeridge Clay is over-pressured (under-compacted) in the East Shetland Basin. The bulk rock volume of the Kimmeridge Clay is 280 × 10^9 cu m (Table 6); the maximum amount of water expelled from the Kimmeridge Clay during oil migration is thus about 30 × 10^9 cu m. The minimum oil saturation in the compaction fluids expelled during migration is thus about 13 to 17 percent.

However, 50 percent of the oil is generated over a limited

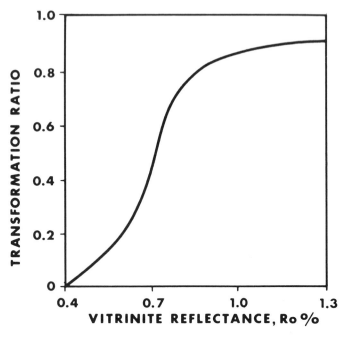

Figure 20. Transformation ratio as a function of vitrinite reflectance for Kimmeridge Clay.

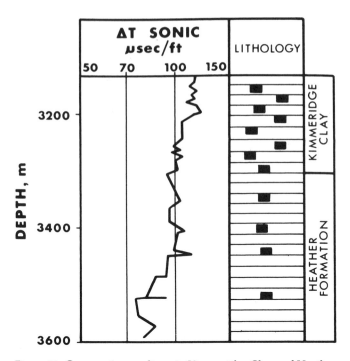

Figure 21. Compaction gradients in Kimmeridge Clay and Heather Formation in Well B.

maturation range (0.65 to 0.8 R_o percent maturity) during peak generation (Figure 20). The maturation history of the Kimmeridge Clay in Well C is representative of deeply buried Kimmeridge Clay in the East Shetland Basin. The Kimmeridge Clay in Well C passed through this maturity range 25 to 5 MA ago; its maximum water filled porosity loss is estimated at 1 to 2 percent during this time. The minimum oil saturation of compaction fluids expelled during peak generation is thus about 30 to 50 percent. These very high oil saturations indicate that oil phase primary migration occurred.

MECHANISM OF OIL MIGRATION

Several mechanisms of oil phase primary migration have been proposed: McAuliffe (1979) suggested that oil migrates through a 3 dimensional, oil wet, kerogen network. The high average organic matter content of the Kimmeridge Clay (15 to 18 vol percent) suggests that much of the rock may be oil wet. Momper (1978) and Meissner (1978) suggested that oil expulsion is a direct consequence of maturation, and that oil migrates through microfractures created by abnormal pore pressures resulting from generation. They pointed out that generated oil and residual

kerogen occupies a greater volume than the immature kerogen. Dickey (1976) suggested that oil is able to flow through source rocks at relatively low oil saturations (1 to 10 percent of the total porosity) because much of the pore water is absorbed on clay mineral surfaces and is thus not part of the *effective* porosity of the rock. Du Rouchet (1981) considered that generated oil is expelled from kerogen by compaction and that it migrates through laminae with relatively large pore throats, or through microfractures. Momper (1978) noted that about 90 to 95 percent of the pore volume of mudstones comprises small subcapillary pores (1 to 3 nm diameter). Abnormally large mudstone pores are associated with lenses of silt, microvugs, and leached zones.

Two other factors that may be important in primary migration are *creation of porosity* by conversion of kerogen to oil, and the reduction in oil/water interfacial tension with increasing temperature. The change in porosity during maturation of the Kimmeridge Clay source rock, with 30 percent oil expulsion, is summarized in Table 7. The total porosity of the source rock at first decreases as water is expelled from the rock, but then increases slightly near peak oil generation. Oil saturation in the effective porosity of the rock (comprising the abnormally large rock matrix pores

Table 6. Volume of oil generated and trapped in the central and eastern East Shetland Basin.

Drainage Area	Bulk Rock Volume ($\times 10^9$ cu m)	Weighted Average Maturity = Transformation Ratio	Oil Generated $\times 10^9$ Barrels	Trapped Oil $\times 10^9$ Barrels	Entrapment/ Generation Ratio
1	98 (180)	0.43 (0.5)	27 (57)	14.5	0.54 (0.25)
2a	102	0.51	33	9.5	0.14
2b	79	0.75	37		
Total	279 (361)	0.55 (0.56)	97 (127)	24	0.25 (0.19)

Figures in parentheses calculated using additional source rock in upper part of Heather Formation in drainage area 1.

214

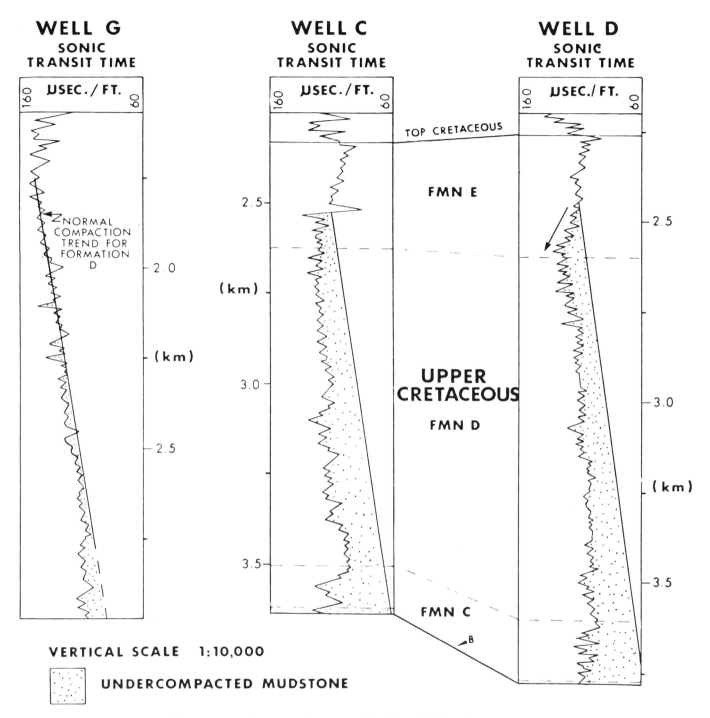

Figure 22. Sonic transit time of Cretaceous mudstones as a function of depth for Well G in the Sogn Graben, and Wells C and D in the Viking Graben.

Table 7. Estimated porosity and oil saturation of mature Kimmeridge Clay source rock.

Transformation Ratio of Kerogen	Water Filled Clay Porosity (%)	Oil Content of Rock (Vol %)	Oil Saturation in Effective Porosity† (Vol %)	Total Porosity
0.1	18	1	35	19
0.25	14	1.7*	55*	15.7
0.5	13	3.5*	70*	16.5
0.75	12	5*	80*	17

*Calculated assuming 30 percent oil expulsion efficiency.
†Effective porosity = oil-filled porosity + 0.1 (water-filled clay porosity).

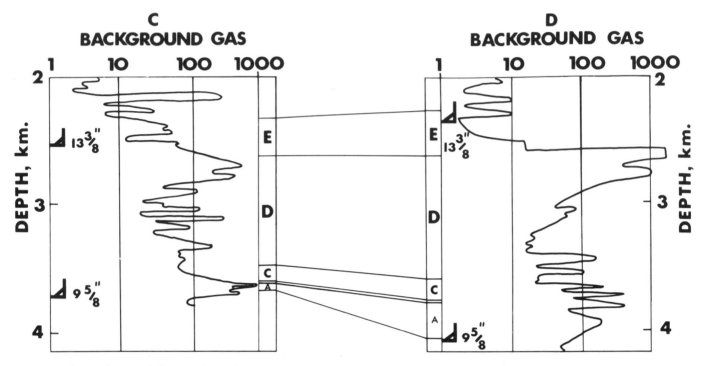

Figure 23. Gas readings in drilling mud in early Tertiary to late Cretaceous mudstones in Wells C and D.

and porosity in the kerogen laminae) increases from 35 percent at a transformation ratio of 0.1, to 55 percent at 'quarter' generation. These high oil saturations suggest that oil may be able to migrate out of the rock at relatively low maturities prior to peak generation.

Oil/water interfacial tension decreases with temperature by 0.2 to 0.4 dyne/cm°C from 25 to 70°C (Schowalter, 1979). At temperatures >70°C the oil/water interfacial tension for light oils is <5 dyne/cm. No data is available for oil/water interfacial tension at temperatures of oil generation from the Kimmeridge Clay (95 to 140°C). Clearly very low oil/water interfacial tensions <5 dyne/cm may exist in source rocks in this temperature range. Capillary pressure (the resistant force to oil phase migration through water wet rock) is directly proportional to interfacial tension:

$$P_c = \frac{2\gamma \cos \theta}{R}$$

where P_c is the capillary pressure, γ is the interfacial tension, θ is the contact angle between oil and water, R is the radius of the pore throat.

The pressure required to inject oil through a pore throat is termed the displacement pressure. Mercury/air displacement pressures for mudstones are >5,000 psi and 1,700 to 5,000 psi for siltstone. For an interfacial tension of 5 dyne/cm these correspond to oil/water displacement pressures of about 20 to 50 psi for silty laminae and >50 psi for mudstone (from Schowalter, 1979, Figure 23).

These low displacement pressures estimated for the larger interconnected pores of the source rock suggest that microfractures, created around isolated kerogen laminae, would tend to inject their oil into the more coarsely porous

laminae of the source rock. The main oil migration routes within the Kimmeridge Clay and underlying Heather Formation are thus probably the continuous oil wet kerogen laminae, and water wet, silty and sandy laminae.

The mature Kimmeridge Clay source rocks in the East Shetland Basin are overpressured. Drilling exponent data in Well F indicates approximately normal pore pressures in the late Cretaceous mudstones above 3,100 m. The estimated pore pressure gradient increases with depth to 0.63 psi/ft in the Kimmeridge clay source rock at 3,500 m.

The Brent reservoirs in the East Shetland Basin were probably normally pressured, or slightly overpressured, prior to oil migration and Tertiary loading. Oil has probably migrated along strong *lateral* fluid pressure gradients set up from actively generating, overpressured source rocks in the half grabens, towards the lower pressured Brent reservoirs.

MECHANISM OF HYDROCARBON MIGRATION INTO THE FRIGG FIELD

Primary migration into the early Tertiary aquifer of the Frigg Field requires vertical migration through 2,000 m of Cretaceous and Palaeocene mudstones (Figure 4). Possible migration mechanisms include: (1) buoyant hydrocarbon phase flow through the interconnected largest pores of the mudstone, (2) migration in aqueous solution with exsolution of hydrocarbons in the early Tertiary aquifer, (3) migration through microfractures created by abnormal pore pressures. In order to assess the relative importance of these possible migration mechanisms it is first necessary to determine the physical conditions in the Jurassic and Cretaceous mudstones, and the hydrocarbon distribution within them.

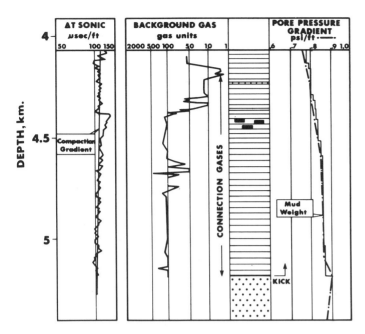

Figure 24. Pore pressure gradient, sonic transit time and drilling mud background gas in the early Cretaceous and Jurassic section of Well D.

Pore Pressure

The pore pressure distribution in the Tertiary to Jurassic mudstones below the Frigg Field can be inferred from a study of pore pressures in Well D drilled in the axial Viking Graben. Pore pressures were estimated from gas readings, mud weights required to balance minor flows of formation fluid during drilling, and from mudstone sonic transit times.

Low sonic transit times and gas readings indicate the late Cretaceous mudstones are normally pressured above 2,350 m (Figures 22 and 23). Between 2,350 m and 2,500 m transit times are less than those of normally compacted late Cretaceous mudstones in the Sogn Graben to the NNE of the study area (Figure 22). Between 2,400 and 2,600 m sonic transit time increases from 100 to 120 μsec/ft, which suggests an increase in porosity with depth. Gas readings increase gradually from 2,350 to 2,550 m and dramatically below 2,550 m. These two observations indicate that an abnormal pore pressure transition zone occurs from 2,350 to 2,550 m.

Between 2,550 m and 3,900 m mudstone transit time decreases with depth; the transit time diverges from the normal compaction trend (Figure 22), indicating a progressive increase in overpressure with depth. Between 3,950 to 4,000 m the borehole is badly caved and the sonic readings are thus unreliable. High gas readings at the base of the 12¼ in hole (Figure 23) indicate a pore pressure gradient of 0.76 psi/ft at 4,050 m at the base of the late Cretaceous mudstones.

The increase in mudstone transit time between 3,950 to 4,070 m near the base of the late Cretaceous mudstones reflects an increasing pore pressure gradient but may also be caused by a change in mudstone mineralogy. The transit time of the early Cretaceous and late Jurassic mudstones decreases with depth at a very slow rate (Figure 24), indicating a further progressive downward increase in

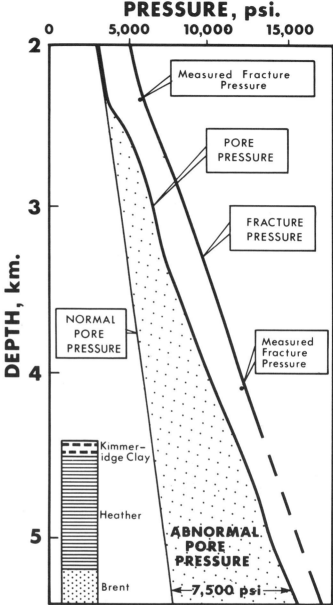

Figure 25. Pore and fracture pressures in Well D.

overpressure to 0.83 psi/ft at the base of the Cretaceous. The Kimmeridge Clay transit time is abnormally high relative to the overlying and underlying formations because of its high organic matter content. Mud weights required to control the pore pressure in the Jurassic sandstone at 5,200 m indicate a pore pressure gradient of 0.89 psi/ft at the base of the late Jurassic mudstones.

The pore pressure profile in the Cretaceous and Jurassic mudstones of Well D (Figure 25) thus comprises a compacted 'caprock' mudstone (2,250 to 2,350 m), a pore pressure transition zone (2,350 to 2,550 m) and a progressive slow increase in overpressure (2,550 to 5,200 m). The general form of this profile is characteristic of thick laterally extensive mudstones buried below normally pressured permeable overburden (Bishop, 1979). Chiarelli and Duffaud (1980) concluded that overpressures in the

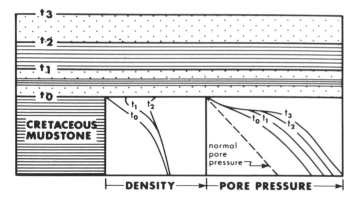

Figure 26. Change in density and pore pressure of Cretaceous mudstones in Viking Graben during successive increments of loading by Tertiary sediments (t_0-t_3) (based on Bishop, 1979).

Frigg area are associated with undercompaction. Bishop numerically modelled the evolution of the pore pressure of abnormally pressured, compacting, shale masses. The qualitative evolution of the pore pressure of the late Cretaceous mudstones as predicted by Bishop's model is shown in Figure 26.

Fracture Pressure

The Fracture Pressure Gradient is given by the sum of the least principal stress gradient and the pore pressure gradient (Eaton, 1969):

$$\frac{Pf}{Z} = \frac{(S - P_p)}{(Z)}\frac{M}{(1 - M)} + \frac{P_p}{Z}$$

where Z = depth
Pf = fracture pressure
P_p = pore pressure
S = overburden pressure
M = Poisson's Ratio

This equation was used to estimate fracture pressure gradient during drilling of Well D. The variation in Poisson's ratio with depth was assumed to be the same as that reported in the Gulf Coast by Eaton (1969). The calculated fracture pressure as a function of depth below 2,000 m is shown in Figure 25.

The fracture pressure gradient was also measured directly during formation leak off tests at 2,345 and 4,065 m. Poisson's ratio was calculated at these depths using the measured fracture gradient and estimated pore and overburden pressure gradients (Table 8).

For comparison Poisson's Ratio at 2,000 m and 4,000 m in the Gulf Coast are 0.42 and 0.47 respectively (from Eaton's plot of Poisson's ratio versus depth).

Microfracture Distribution

The 'fracture pressure' calculated using Eaton's Method, and measured during formation leak off tests, is the pressure required to propagate large scale fractures. Palciauskas and Domenico (1980) reviewed work on experimental microfracturing of sandstones, siltstones, and carbonates. They concluded that these rock types begin to microfracture

at fluid pressure about 20 percent less than the 'fracture pressure.' They suggested that this microfracture criterion probably also applies to mudstones.

The difference in pore and fracture pressure in Well D has been calculated to determine whether microfractures exist in the overpressured Cretaceous and Jurassic mudstones, using this microfracture criterion. The pore pressure (3,400 psi) at the top of abnormal pore pressure transition zone (2,350 m) is 42 percent less than the fracture pressure (5,900 psi). At the base of the transition zone (2,550 m) the pore pressure (5,000 psi) is 26 percent less than the fracture pressure (6,800 psi). These data indicate the transition zone is not microfractured.

Between the base of the transition zone and 3,600 m the pore pressure is 20 to 25 percent less than the fracture pressure. This suggests that this interval may be microfractured. Below 3,600 m the pore pressure is within 20% of the fracture pressure which indicates that active microfracturing is probably occurring.

Within the Kimmeridge Clay and Heather Formation source rocks the pore pressure is within 10 to 13 percent of the fracture pressure. The pore pressure at 5,300 m within the Brent Formation (15,200 psi) is only 10 percent less than the fracture pressure (16,800 psi).

Pore pressure within the high pressure Brent reservoirs of the Viking Graben converges updip with the calculated seal fracture pressures. Figure 27 shows the extrapolated Brent pore pressure updip of Well D, and the corresponding seal fracture pressure. At 4,500 m the extrapolated Brent pore pressure equals the fracture pressure. A Brent pore pressure measurement, reported by Lindberg et al (1980) in the Frigg area is within 5 percent of the calculated seal fracture pressure at that depth (Figure 27). These data suggest that either the microfracture criterion of Palciauskas and Domenico is incorrect, or that the calculated fracture pressures are too low, or that microfracturing does not prevent further increases in pore pressure gradient.

Well C penetrated a high pressure gas condensate column in the Brent Formation; the pore pressure is within 10 percent of the calculated fracture pressure. Retention of this gas column in the reservoir suggests that either its seal is not microfractured, or that the rate of migration into the trap at least balances leakage of gas through microfractures.

Methane Solubility

The aqueous solubility of methane in the early Tertiary to Jurassic section of Well D has been estimated from the calculated variation in pore pressure and temperature with depth using methane solubility data given by Magara (1980), and Jones (1980) (Figure 28a). Methane solubility increases from 3 standard cu m/cu m water at the top of the abnormal pore pressure transition zone to 13 cu m/cu m water at 5,400 m.

Table 8. Calculated values of Poisson's Ratio in Well D.

Depth (m)	Pore Pressure Gradient (psi/ft)	Fracture Pressure Gradient (psi/ft)	Overburden Pressure Gradient (psi/ft)	M
2,345	0.45	0.76	0.89	0.41
4,065	0.76	0.91	0.95	0.44

218

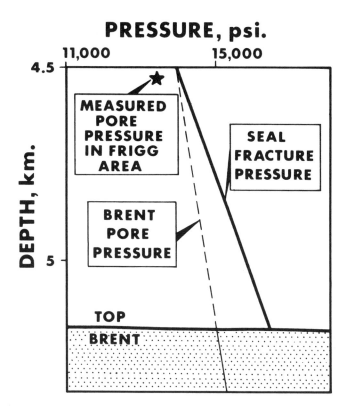

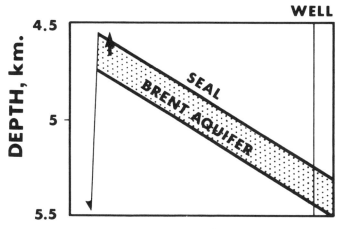

Figure 27. Convergence of reservoir pore pressure and seal fracture pressure in deep high pressure fault blocks.

Cumulative dissolved methane within a 1,500 sq km area around Well D (equivalent to the drainage area of the Frigg Field) is shown in Figure 28b. Cumulative dissolved methane was calculated from the solubility/depth plot and the mudstone compaction gradient, assuming all mudstone pores contained fresh water saturated with methane. The actual solubility of methane in pore water with a salinity of 35,000 ppm total dissolved solids is 10 to 20 percent less than the fresh water solubility (Hunt, 1979).

Volume of Gas Generated in Drainage Area of Frigg Field

The amount of gas generated within the drainage area of the Frigg Field has been estimated assuming that the average thickness and maturity of Kimmeridge Clay and Heather formations within the area are the same as in Well D. The amount of gas generated from the Heather Formation has been calculated assuming it contains dispersed coaly material, using the methane generation/maturity plot for coal (Hunt, 1979). The amount of gas generated from the Kimmeridge Clay has been estimated assuming it is derived from cracking of oil remaining in the source rock after oil generation ceased. Estimated gas yields from the Heather Formation and Kimmeridge Clay are 100×10^{10} cu m and $(100$ to $200) \times 10^{10}$ cu m respectively. Ten m of Brent coal within a drainage area equivalent to the Frigg Field would have generated 200×10^{10} cu m of methane assuming the same Brent maturity as in Well D. However the carbon isotope data suggests the Frigg gas was not sourced from Brent coals.

Variation in Light Hydrocarbon Composition and Concentration with Depth (Figure 29)

In the normally pressured early Tertiary and late Cretaceous (2,000 to 2,350 m), hydrocarbon concentrations are low; gas wetness increases from 10 percent at the top of the interval to 30 percent at the base. In the abnormal pore pressure transition zone (2,350 to 2,550 m) total gas, gasoline range hydrocarbon (C_5-C_7) concentration, and gas wetness increase dramatically with depth (Figure 29).

Very high background gas was measured in the 250 m interval below the abnormal pore pressure transition zone (2,550 to 2,800 m). Oil shows were recorded in limestone cuttings in this interval. A similar zone of high background gas occurs below the abnormal pore pressure transition zone in Well C. The much higher background gas values over this interval in Well D are due to underbalanced drilling. Analysis of canned cuttings from this interval indicate the gas is wet (40 to 80 percent C_2-C_4); C_5-C_7 hydrocarbon concentration exceeds 5,000 vol ppm of rock between 2,550 to 2,650 m.

Within the late Cretaceous mudstones below 3,300 m total gas and gas wetness increases with depth. Mud background gas increases markedly over a 50 m interval above the top of the Kimmeridge Clay (Figure 24). Total gas in the late Jurassic mudstones is nearly half an order of magnitude greater than in the Cretaceous mudstones (3,000 to 4,387 m). This reflects higher organic richness and maturity of the late Jurassic mudstones.

Evaluation of Hydrocarbon Migration Mechanisms

Methane/water interfacial tension is about 25 dyne/cm in the temperature range 100 to 300°F and pressure range 5,000 to 15,000 psi (Schowalter, 1979). Gas-water displacement pressures are > 350 psi for mudstones which have mercury/air displacement pressures > 5,000 psi for this interfacial tension. These data indicate that vertical gas columns > 300 m in height are required to inject gas through mudstone pores in the late Cretaceous section of the Viking Graben. There is no evidence for the existence of such columns; gas phase migration through mudstone pores is unlikely.

Comparison of the pore and fracture pressures strongly suggests that gas is migrating through microfractures below 3,600 m. From 3,600 to 5,150 m there is a positive correlation between total gas and gas wetness which suggests migration is occurring in gas phase. The lack of evidence of microfractures above 3,500 m, and especially

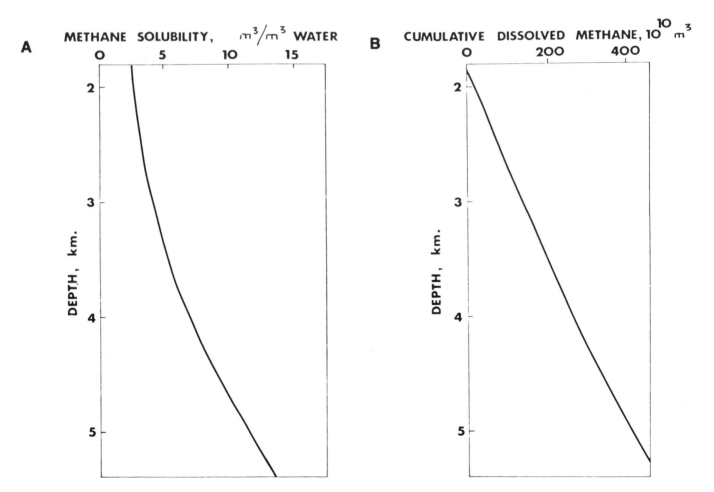

Figure 28. (A) Methane solubility as a function of depth in Early Eocene to Jurassic rocks in Well D. (B) Cumulative dissolved methane in Early Eocene to Jurassic rocks within the Frigg Field drainage area (1,500 sq km).

within the abnormal pressure transition zone and overlying normally pressured mudstones, suggests another migration mechanism occurs in the upper part of the late Cretaceous section.

Any hypothesis for the origin of the Frigg Field gas must account for its very low wet gas content (3.5 percent ethane, 0.05 percent propane + butane). This composition can be explained by migration in aqueous solution within the upper part of the late Cretaceous mudstones, and subsequent exsolution in the early Tertiary aquifer. This hypothesis requires that methane/saturated pore water exsolves sufficient methane, when it is expelled into the early Tertiary aquifer, to account for the in-place reserves of the field.

The amount of gas generated from Jurassic source rocks within the drainage area of the Frigg Field (200 to 300 × 10^{10} cu m) exceeds that required to saturate the water in the late Cretaceous mudstones between 2,250 m and 3,500 m (150 × 10^{10} cu m) (Figure 28). This suggests that pore waters in this interval are saturated with respect to methane.

The aqueous solution hypothesis can be tested by calculating the amount of pore water expelled into the aquifer from the Frigg Field drainage area since the trap was sealed, and the volume of methane exsolved (Table 9). The water loss from the Jurassic and Cretaceous mudstones in

the last 40 Ma has been calculated from the mudstone compaction gradient in Well D. The pressure drop across the early Tertiary aquifer has been estimated form Magara's solubility data for the temperature and pressure conditions prevailing in the Frig aquifer in the last 40 Ma. The calculated volumes of exsolved methane is more than enough to account for the in-place reserves of the field (27 × 10^{10} cu m).

To conclude, the migration mechanism of the Frigg gas probably involved three processes: 1, Migration as gas phase in microfractures in the over-pressured late Jurassic

Table 9. Evaluation of aqueous solubility mechanism for migration of Frigg Field methane.

*Drainage area of field	1,500 sq km
*Decrease in thickness of late Cretaceous mudstones since field was sealed	450 m
*volume of water expelled into early Tertiary aquifer from below within field drainage area	(1,500 × 10^6 × 450 cu m = 70 × 10^{10} cu m
*Pressure drop between top and bottom of aquifer	900 psi
*Methane exsolved in aquifer due to this pressure drop	0.5 m^3/m^3 water = (70 × 10^{10} × 0.5) cu m = 35 × 10^{10} cu m.

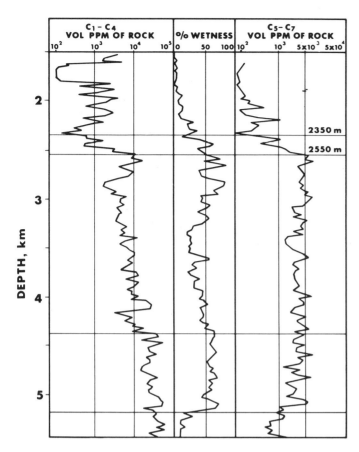

Figure 29. Total gas, gas wetness and C_5-C_7 hydrocarbons in canned cuttings samples as a function of depth in Well D.

mudstones and overlying late Cretaceous mudstones at depths greater than about 3,500 m. 2, Migration in aqueous solution through the late Cretaceous mudstones at depths shallower than about 3,500 m. 3, Exsolution of methane in the Frigg Field aquifer and subsequent buoyant flow of gas to the trap.

An aqueous solubility mechanism for the Frigg Field oil has also been evaluated. About 5×10^9 cu m of oil is estimated to have been expelled from the Kimmeridge Clay source rock within the drainage area of the Frigg Field assuming a source rock thickness of 100 m. The in-place reserves of the field are $\cong 0.1 \times 10^9$ cu m. The water loss from the Cretaceous and Jurassic mudstones within the field's drainage area is 70×10^{10} cu m. About 150 ppm of oil would thus have to be exsolved from these compaction waters in the trap to account for the in-place reserves of the field. This amount of exsolution requires an oil concentration of the order of 1,000 ppm in the compaction water. Whole oil solubilities at 100 to 150°C, equivalent to the temperatures prevailing in the Cretaceous mudstones, are < 100 ppm (Prie, 1976). An aqueous solubility mechanism thus cannot account for the in-place reserves of the field. However, the heavy polar nitrogen and sulphur bearing compounds in the oil could have migrated by this mechanism.

The relatively high and uniform gasoline range hydrocarbon concentrations in the overpressured Cretaceous mudstones suggest that large scale vertical

migration of hydrocarbons generated from the Kimmeridge Clay has occurred. The most likely migration mechanisms are oil migration through microfractured overpressured mudstones below 3,500 m, and migration through siltstones and the interconnected larger mudstone pores in response to low oil/water interfacial tensions.

CONCLUSIONS

1. The richest source rocks in the East Shetland Basin and the Viking Graben are the Kimmeridge Clay (oil and gas) and the Brent Formation coals (dry gas). Bathonian and Toarcian mudstones may also have generated oil. Thick Heather Formation mudstones are gas source rocks in the Viking Graben.

2. A uniform present day maturity gradient indicates the oil window (0.5 to 1.3 percent R_o) extends from 2,550 to 4,500 m, and the wet gas zone (1.3 to 2 percent R_o) from 4,500 to 5,400 m. Peak oil generation from the Kimmeridge Clay occurs at a reflectance level of 0.7 percent R_o (3,250 m). Highest oil saturations in the Kimmeridge Clay, inferred from resistivity logs, occur at 0.8 percent R_o (3,500 to 3,600 m). Intense gas generation from oil trapped in the Kimmeridge Clay has probably occurred in the reflectance range 1 to 1.3 percent R_o (4,000 to 4,500 m). Peak dry gas generation from Brent Formation coals occurs at a reflectance level of 1.5 percent R_o (4,000 to 4,500 m). Peak dry gas generation from Brent Formation coals occurs at a reflectance level of 1.5 percent R_o (5,000 m). At the present day the Kimmeridge Clay is mature over most of the East Shetland Basin and has reached peak generation throughout the axial region of the basin. In the Viking Graben the Kimmeridge Clay is at a maturity close to the oil floor (1.3 percent R_o).

3. Oil generation from the Kimmeridge Clay began 70 to 80 Ma ago in the Viking Graben; 65 Ma ago the Kimmeridge Clay was generating oil throughout the Viking Graben and in the deepest troughs of the East Shetland Basin. Peak oil generation was reached 55 to 65 Ma ago in the Viking Graben; 40 Ma ago peak generation had occurred in the deepest troughs of the East Shetland Basin, and throughout the Viking Graben. Twenty to 40 Ma ago the Kimmeridge Clay entered the wet gas zone in the deepest syncline. Generation of gas by cracking of oil in the Kimmeridge Clay of the Viking Graben occurred during the last 50 Ma. Gas generation from Brent Formation coals began 100 Ma ago in the Viking Graben; peak dry gas generation occurred during the last 40 Ma.

4. Oil phase primary migration from the Kimmeridge Clay has occurred in the East Shetland Basin. The oil expulsion efficiency from this source rock is < 20 to 30 percent. Migration and trapping efficiencies are very high in the East Shetland Basin: 20 to 30 percent of the oil generated from the Kimmeridge Clay is now reservoired in Jurassic sandstones. Oil saturation in compaction fluids expelled from actively generating Kimmeridge Clay was at least 30 percent. Oil migration has occurred along high fluid potential gradients set up from actively generating overpressured source rocks in the half grabens to normally pressured or slightly overpressured Jurassic reservoirs, stratigraphically below, but structurally updip of, the

source rocks. During primary migration oil probably moved through porosity created in the kerogen laminae of the source rock by oil generation, and through the interconnected larger pores of the Kimmeridge Clay and underlying Heather Formation aided by low oil/water interfacial tensions.

5. Migration of the Frigg Field methane probably involves 3 processes: 1, Migration of gas in gas phase in microfractures in the overpressured late Jurassic mudstones and overlying the late Cretaceous mudstones (at depths >3,500 m). 2, Migration in aqueous solution through late Cretaceous mudstones at depth shallower than about 3,500 m. 3, Exsolution of methane in the Frigg Field aquifer, and subsequent buoyant flow of gas to the trap. Most of the Frigg Field oil probably did not migrate in aqueous solution. Migration of oil probably occurred through microfractures, and through siltstone and the larger mudstone pores in response to low oil/water interfacial tensions.

ACKNOWLEDGMENTS

The author acknowledges many valuable discussions with BP geologists, geochemists, and geophysicists, but would emphasize that the views expressed are his own. In particular he wishes to thank Dr. A.M. Spencer for encouragement to publish this paper, Dr. J.R. Bloomer and Dr. S.W. Richardson for invaluable help with the geothermal modelling, and G.C. Speers and Dr. G. Dungworth for discussion of geochemical principles. Anita McKearney kindly typed the manuscript. The author wishes to thank the management of British Petroleum Development Ltd. for permission to publish this paper.

REFERENCES

Barnard, P.C. and B.S Cooper, 1981, Oils and source rocks of the North Sea area, in L.V. Illing and G.D. Hobson, eds., Petroleum geology of the Continental Shelf of North-West Europe: Heyden and Son, p. 169-175.

Bishop, R.S., 1979, Calculated compaction states of thick abnormally pressured shales: AAPG Bulletin, v. 63, p. 918-933.

Brooks, J., and B. Thusu, 1977, Oil-source identification and characterisation of the Jurassic sediments in the northern North Sea: Chemical Geology, v. 20, p. 283-294.

Carstens, H., and K.G. Finstad, 1981, Geothermal gradients of the northern North Sea Basin 59 to 62°N, in L.V. Illing and G.D. Hobson, eds., Petroleum geology of the Continental Shelf of North-West Europe: Heyden and Son, p. 152-161.

Chiarelli, A., and F. Duffaud, 1980, Pressure origin and distribution in Jurassic of Viking Basin (United Kingdom-Norway): AAPG Bulletin, v. 64, p. 1245-1266.

De'ath, N.G. and S.F. Schuyleman, 1981, The geology of the Magnus Oil Field, in L.V. Illing and H.D. Hobson, eds., Petroleum geology of the Continental Shelf of North-West Europe: Heyden and Son, p. 342-351.

Dickey, P.A., 1975, Possible primary migration of oil from source rock in oil phase: AAPG Bulletin, v. 59, 337-347.

Donato, J.A. and M.C. Tully, 1981, A regional interpretation of North Sea gravity data, , in L.V. Illing and H.D. Hobson, eds., Petroleum geology of the Continental Shelf of North-West Europe: Heyden and Son, p. 65-75.

Du Rouchet, J.D., 1981, Stress fields, a key to oil migration: AAPG Bulletin, v. 65, p. 74-85.

Eaton, B.A., 1969, Fracture gradient prediction and its application in oil field operations: Journal of Petroleum Technology, v. 21, p. 1353-1360.

Fuller, J.G.C.M., 1980, in J.M. Jones and P.W. Scott, 1980, Progress report on fossil fuels—exploration and exploitation: Proceedings of the Yorkshire Geological Society, v. 33, p. 581-593.

Gallois, R.W., 1976, Coccolith blooms in the Kimmeridge Clay and origin of North Sea Oil: London, Nature, v. 259, p. 473-475.

Hancock, N.J. and A.M. Taylor, 1978, Clay mineral diagenesis and oil migration in the Middle Jurassic Brent Sand Formation: Journal of the Geological Society of London, v. 135, p. 69-72.

Hay, J.T.C., 1977, The thistle field: Paper given at Bergen North Sea Conference.

Heritier, F.E., P. Lossel, and E. Wathne, 1979, Frigg Field—large submarine-fan trap in Lower Eocene rocks of North Sea: AAPG Bulletin, v. 63, p. 1999-2020.

Hunt, J.M., 1979, Petroleum Geochemistry and Geology: San Francisco, W.H. Freeman and Co., 617 p.

Jones, P.H., 1980, Role of geopressure in the hydrocarbon and water system, in Problems of petroleum migration: AAPG Studies in Geology, n. 10, p. 207-216.

Lindberg, P., R. Riise, and W.H. Fertl, 1980, Occurrence and distribution of overpressures in the Northern North Sea area: Society of Petroleum Engineers Journal, preprint paper 9339.

McAuliffe, C.D., 1979, Oil and gas migration: chemical and physical constraints: AAPG Bulletin, v. 63, p. 761-781.

McKenzie, D., 1978, Some remarks on the development of sedimentary basins: Earth and Planetary Science Letters, v. 40, p. 25-32.

Magara, K., 1980, Primary migration of oil and gas, in A.D. Míall, eds., Facts and principles of world petroleum occurrence: Canadian Society of Petroleum Geologists Memoir, n. 6, p. 173-191.

Meissner, F.F., 1978, Petroleum geology of the Bakken Formation, Williston Basin, North Dakota and Montana: Williston Basin Symposium, Montana Geological Society, p. 207-227.

Momper, J.A., 1978, Oil migration limitations suggested by geological and geochemical considerations: AAPG Course Note Series, v. 8, B1-B60.

Oxburgh, E.R., and C.P. Andrews-Speed, 1981, Temperature, thermal gradients and heat flow in the southwestern North Sea, in L.V. Illing and H.D. Hobson, eds., Petroleum geology of the Continental Shelf of North-West Europe: Heyden and Son, p. 141-151.

——, et al, 1972, Equilibrium borehole temperatures from observation of thermal transients during drilling: Earth and Planetary Science Letters, v. 14, p. 47-49.

Palciauskas, V.V. and P.A. Domenico, 1980, Microfracture development in compacting sediments: relation to hydrocarbon-maturation kinetics: AAPG Bulletin, v. 64,

p. 927-937.

Price, L.C., 1976, Aqueous solubility of petroleum as applied to its origin and primary migration: AAPG Bulletin, v. 60, p. 213-244.

Richardson, S.W. and E.R. Oxburgh, 1979, The heat flow field in mainland U.K.: London, Nature, v. 282, p. 565-567.

Royden, L., J.R. Sclater, and R.P. Van Herzen, 1980, Continental margin subsidence and heat flow: important parameters in formation of petroleum hydrocarbons: AAPG Bulletin, v. 64, p. 173-187.

Schowalter, T.T., 1979, Mechanics of secondary hydrocarbon migration and entrapment: AAPG Bulletin, v. 63, p. 723-760.

Sclater, J.G., and P.A.F. Christie, 1980, Continental stretching: an explanation of the post- Mid-Cretaceous subsidence of the Central North Sea Basin: Journal of Geophysical Research, v. 85, p. 3711-3739.

Sommer, F., 1978, Diagenesis of Jurassic sandstones in the Viking Graben: Journal of the Geological Society of London, v. 135, p. 63-68.

Staplin, F.L., 1969, Sedimentary organic matter, organic metamorphism and oil and gas occurrence: Bulletin of Canadian Petroleum Geology, v. 17, p. 47-66.

Surlyk, F., 1968, Jurassic basin evolution of East Greenland: London, Nature, v. 273, p. 130-133.

Turcotte, D.L., 1980, Models for the evolution of sedimentary basins, in A.W. Bally, ed., Dynamics of Plate Interiors: American Geophysical Union and Geological Society of America.

Tissot, B.P., and D.H. Welte, 1978, Petroleum formation and occurrence: Springer Verlag, 538 p.

Tyson, R.V., R.C.L. Wilson, and C. Downie, 1979, A stratified water column environmental model for the type Kimmeridge Clay: London, Nature, v. 277, p. 377-380.

Waples, D.W., 1979, Simple method for oil source bed evaluation: AAPG Bulletin, v. 63, p. 239-245.

——— , 1980, Time and temperature in petroleum formation: application of Lopatin's Method to petroleum exploration: AAPG Bulletin, v. 64, p. 916-926.

White, D.A., and H.M. Gehman, 1979, Methods of estimating oil and gas resources: AAPG Bulletin, v. 63, p. 2183-2203.

Williams, P.F.V., and A.G. Douglas, 1980, A preliminary organic geochemical investigation of the Kimmeridgian Oil Shales, in Advances in organic geochemistry, 1979: Oxford, Pergamon Press, p. 631-645.

American Association of Petroleum Geologists Memoir
35, *Petroleum Geochemistry and Basin Evaluation*, edited
by Gerard Demaison and Roelof J. Murris, copyright
1984, pp. 53–77.

Geological and Geochemical Models in Oil Exploration; Principles and Practical Examples

P. Ungerer
F. Bessis
P.Y. Chenet
B. Durand
E. Nogaret
Institut Francais du Pétrole
Rueil-Malmaison, France

A. Chiarelli
SNEAP
Pau, France

J.L. Oudin
Compagnie Francaise des Pétrole
Paris, France

J.F. Perrin
Université de Bordeaux, France

It is now possible to develop mathematical models that make quantitative predictions of the geological phenomena leading to oil accumulations. The models presented in this study are deterministic, that is, based on physical or chemical laws and not on statistical analysis.

A first set of models may be used during the initial stage of exploration of a sedimentary basin, when little geological data are available. Their purpose is to determine the general geological evolution and the overall petroleum potential. A backstripping model makes an automatic reconstruction in time of sedimentary basins, taking into account the progressive compaction of the sediments. It also computes the sediment load on the basement together with the subsidence variations in time, enabling a geodynamic model to be used for some basins to simulate the history of heat flow. It is then possible to reconstruct the temperature history of each sedimentary unit. A kinetic model of organic matter maturation is used to compute the possible area and timing of petroleum formation.

A second set of models may be applied when exploration is more advanced, to estimate the importance of various traps in a given petroleum province. These models, which need more extensive data than the first set, take into account the influence of migration processes on oil accumulation. The migration model gives a quantitative description of the formation of hydrocarbons and computes the pressure regime of the fluids. Thus it determines the amount of petroleum expelled from the source rocks by compaction and its possible accumulation in traps. A thermodynamic model computes hydrocarbon migration in the gas phase and its consequences for the composition of oils.

The Gulf of Lion passive margin in France, the Viking Graben in the North Sea, and the Mahakam Delta in Indonesia are used as examples.

INTRODUCTION

The increasing cost of petroleum exploration in frontier areas requires optimization of the opportunities to find commercial discoveries. On the other hand, finding more hydrocarbons in already producing zones requires the discovery of subtle traps.

The purpose of the mathematical models presented is to help in exploration for this double objective. In a given basin, these models can provide an interpretation of the processes leading to oil accumulation. Thus they enable the selection of the most favorable zones on the basis of the available geological data.

Up to now, the petroleum potential of a given area was estimated mostly in a qualitative way, looking at the possible structures, reservoirs, cap rocks, and source rocks. The processes of hydrocarbon formation, migration and accumulation were only roughly understood and the geologists would assess their chances of success from experience. This kind of approach led to the development of statistical models, based on correlations between geological parameters and oil accumulation. However, this approach needs a very large amount of data that can only be supplied by larger oil companies.

It is now widely accepted that the accumulation of oil results through a long chain of geological processes: the sedimentary organic matter (OM) is progressively buried during basin subsidence; this burial causes a temperature increase, under which influence the OM progressively generates hydrocarbons; once sufficient maturation is reached, the hydrocarbons are expelled from the source rocks, and they may migrate in carrier beds and accumulate in traps, provided a caprock is present.

During the past decade, much progress has been made in the understanding of these processes. They have been described by physical or chemical laws that enable us to set up mathematical models. These models may be applied to a given area and provide quantitative results.

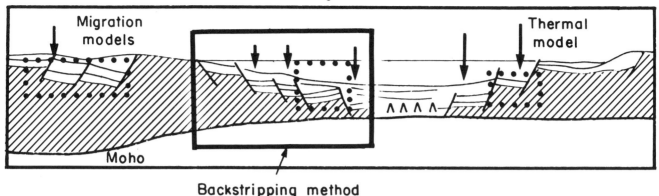

Geodynamic model

Figure 1a. General scheme showing the scale on which the various models are applied. The arrows represent the possible applications of the thermal model, while the dotted frames are related to the migration models.

Each of the models presented here (Figure 1) deals with a specific geological process:

- A backstripping model makes an automatic reconstruction of the sedimentary basin in time through successive cross sections.
- A geodynamic model is used for the basins formed by extension. It reconstructs the heat flow history.
- A thermal model computes the temperatures in the basin. It is coupled to a kinetic model of OM maturation.
- A migration model computes the amounts of petroleum generated in the source rocks and the amounts accumulated in traps through time.
- A thermodynamic model determines the amount and composition of the hydrocarbons that may migrate in gaseous solution.

VALIDITY CONDITIONS OF THE MODELS

Before discussing the various models, it is useful to review the conditions that have to be respected for their validity in practical cases of basin exploration.

1) The processes that are described by the model must be the actual processes occurring in the basin evolution and they have to be represented by correct equations. For instance, it is important for migration purposes to assume that the migration of oil occurs mainly in a separate phase from water, and not in an aqueous solution. Generally speaking, the processes are selected on the basis of theoretical considerations, typical basin examples and/or laboratory experiments.

2) The computer program has to give a good approximation of the exact mathematical solution. This numerical difficulty is easily solved in one dimensional problems. However, many uncertainties may occur in the case of two- or three-dimensional problems, and they generally lead to some error in the quantitative results.

3) The natural complexity of the basin has to be considered without excessive simplification. In each case, a conceptual system of the basin is elaborated using the available geological data (Welte and Yukler, 1981). This system should give a consistent reproduction of the stratigraphical, lithological, and geochemical data at the scale considered, and must account for small size details if necessary. For instance, thin permeable beds may have a considerable importance for the migration of hydrocarbons and therefore should not be neglected. Generally, the validity of the conceptual system is limited by the lack of data at great depths (poor resolution of seismic profiles, scarcity of deep wells).

4) The amount and quality of the input data must be sufficient to allow reasonably accurate results. Some of the models require data that are frequently obtained with low accuracy. For instance, the temperature and source rock data that are needed by the migration model are often subject to great uncertainties, resulting in important possible errors in the calculated amounts of accumulated oil.

It may be argued that these conditions are never fully satisfied in practice. They constitute, however, a goal that we should strive to reach.

THE USE OF DETERMINISTIC MODELS IN OIL EXPLORATION

In a poorly known sedimentary basin, such as an undrilled continental margin, a first set of models may be used to describe the general pattern of basin evolution. This allows in some cases an estimate of the possible maturation of the source rocks. The models are used in the following sequence: first, the burial and compaction history is reconstructed and basin subsidence through time is determined by the backstripping method; second, the temperature history of each sedimentary unit is computed from the heat flow variation with time (in some basins, for example passive continental margins, the heat flow may be derived from the results of subsidence studies based on a geodynamic model); finally, the temperature results are used as input data for the maturation model enabling the reconstruction of the oil window through time.

This first set is particularly applicable to sedimentary basins that have undergone a simple tectonic history, limited to extensional tectonics followed by general subsidence.

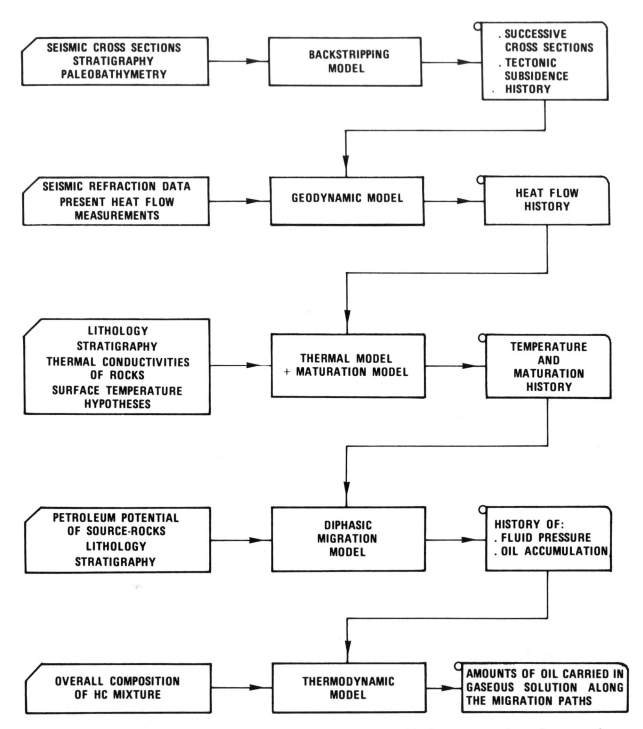

INPUT DATA
(Poor accuracy data may be
reassessed if necessary for the
consistency of the results).

OUTPUT
RESULTS

SEISMIC CROSS SECTIONS
STRATIGRAPHY
PALEOBATHYMETRY

BACKSTRIPPING
MODEL

. SUCCESSIVE
  CROSS SECTIONS
. TECTONIC
  SUBSIDENCE
. HISTORY

SEISMIC REFRACTION DATA
PRESENT HEAT FLOW
MEASUREMENTS

GEODYNAMIC MODEL

HEAT FLOW
HISTORY

LITHOLOGY
STRATIGRAPHY
THERMAL CONDUCTIVITIES
OF ROCKS
SURFACE TEMPERATURE
HYPOTHESES

THERMAL MODEL
+ MATURATION MODEL

TEMPERATURE
AND
MATURATION
HISTORY

PETROLEUM POTENTIAL
OF SOURCE-ROCKS
LITHOLOGY
STRATIGRAPHY

DIPHASIC
MIGRATION
MODEL

HISTORY OF:
. FLUID PRESSURE
. OIL ACCUMULATION

OVERALL COMPOSITION
OF HC MIXTURE

THERMODYNAMIC
MODEL

AMOUNTS OF OIL CARRIED IN
GASEOUS SOLUTION ALONG
THE MIGRATION PATHS

Figure 1b. Flow diagram indicating the ideal sequence of operation of the various models, their main input data and output results.

At more advanced stages of exploration the well data enable us to have a fairly reliable idea of source rock distribution and richness and thus to estimate quantitatively the amounts of hydrocarbons generated in the basin. It is then possible to use the second set of models, that study the migration, accumulation and dysmigration of hydrocarbons. The migration model that represents these processes for a cross section may use the paleotemperature reconstruction produced by the previous models. Its results show pressure and hydrocarbon saturation within the sediments through time. Thus it allows us to quantitatively determine the volume of hydrocarbon accumulations. In addition, the pressure reconstruction allows the prediction of probable high pressure zones. The pressure results are also used as input for the thermodynamic model, together with temperature data. This model determines the amount and composition of the hydrocarbons that migrate dissolved in the gas phase and that are left behind by exsolution along the migration path of the gas. The thermodynamic model may also be used to predict the maximum condensate content of a wet gas as function of temperature and pressure.

BURIAL OF SEDIMENTS AND SUBSIDENCE: THE BACKSTRIPPING METHOD

The reconstruction of the progressive burial of sediments and the corresponding basin subsidence is usually carried out in a qualitative way through the construction of palinspastic cross sections. These are used to predict the most favorable environments for the occurrence of source rocks and permeable units. In order to determine more precisely the thickness of the sedimentary column through time, compaction may be taken into account, as achieved by Perrier and Quiblier (1974) or Magara (1978). The backstripping method, as proposed first by Watts and Ryan (1976) is still more elaborate. It consists of the computation and automatic drawing of the successive shapes of the basin since the beginning of sedimentation. The model elaborated by the present authors accounts for the thickness variation of the sedimentary units through burial, and for paleobathymetry. The weight of the sediments on basement is computed by the model. It is thus possible to estimate what part of the subsidence that is due to the loading effect of sediments and water. The residual subsidence, not due to the sedimentary load, is computed by the model and is called tectonic subsidence (Watts and Steckler, 1981). Thus the vertical movements caused by deep seated tectonic processes may be estimated (Figure 2).

Compaction

The thickness evolution of the sedimentary column during compaction depends on the type of sediment, rate of sedimentation and loading, possibilities of drainage of expelled fluids, and mineral transformation, cementation and dissolution processes during diagenesis.

For sands, clay, and marls, compaction is mainly due to the expulsion of interstitial fluids and the volume of solid matrix remains relatively constant. For carbonates, cementation may be important but is generally compensated by dissolution. Diagenetic studies show that the dissolution

fluids enriched with calcium precipitate in the vicinity of the dissolution zone (De Charpal and Devaux, 1981). Thus, the hypothesis of solid matrix volume conservation during compaction may be considered valid for all lithologies.

Compaction is computed from average relationships between porosity and depth of burial. The porosity data may be obtained either from well logs or from core analysis. Following Magara (1978), it is assumed that the final state of compaction (or porosity) is obtained at maximum depth of burial, that is, no appreciable rebound occurs during uplift and erosion.

Paleobathymetry

Once the changes in thickness of the sedimentary column are determined, the reconstruction of the total subsidence requires a reference level for the top of the sediments. For this purpose, the paleobathymetric evolution and eustatic sea level changes have to be considered.

The paleobathymetry is estimated from biostratigraphic studies, based on the paleontological record from well data. The precision is good in shallow-water sedimentary facies but much less accurate for pelagic sediments. Through use of seismic stratigraphy techniques (Vail et al, 1977a, b), the paleocontinental shelves, slopes, and deep basins can be reconstructed, giving a first estimate of the paleobathymetry at a given time. Coupled to adequate hypotheses about eustatic sea level changes, as will be discussed later, this allows computation of total subsidence.

Additional hypotheses on the mechanical behavior of the basement have to be made for the subsequent determination of the tectonic subsidence.

Mechanical Behavior of Basement

The main effect of the sedimentary loading is to amplify the subsidence caused by deep seated thermomechanical processes. In plate tectonic theory, the lithosphere consists of rocks in a solid state overlying a molten asthenosphere. In this paper, the term lithosphere corresponds to that part of this solid layer which has an elastic behavior and is generally thinner than in the plate tectonic model.

In order to remove the separate loading effect of sediments and water on the lithosphere (backstripping), it is necessary to consider a model describing the behavior of the lithosphere under loads of large lateral extension. The first of these "isostatic models" was proposed by Airy in 1855, who assumed that the lithosphere had no strength at all. Consequently the crust behaves as a set of columns moving independently of each other (Figure 4a).

Although this very simple model was found accurate enough to account for the observed gravity anomalies in some basins, it is now believed that the lithosphere behaves rather like a rigid plate overlying a fluid asthenosphere. The main arguments supporting this "regional isostatic" behavior come from studies on oceanic features such as seamounts and ridges (see for instance Watts et al, 1975). These studies indicate that the oceanic lithosphere deforms as an elastic plate, of which the flexural rigidity (or the equivalent elastic thickness), which controls the amount and lateral distribution of the deformations, is mainly dependent on the thermal structure of the lithosphere at the time of loading. The elastic thickness of oceanic lithospheres has

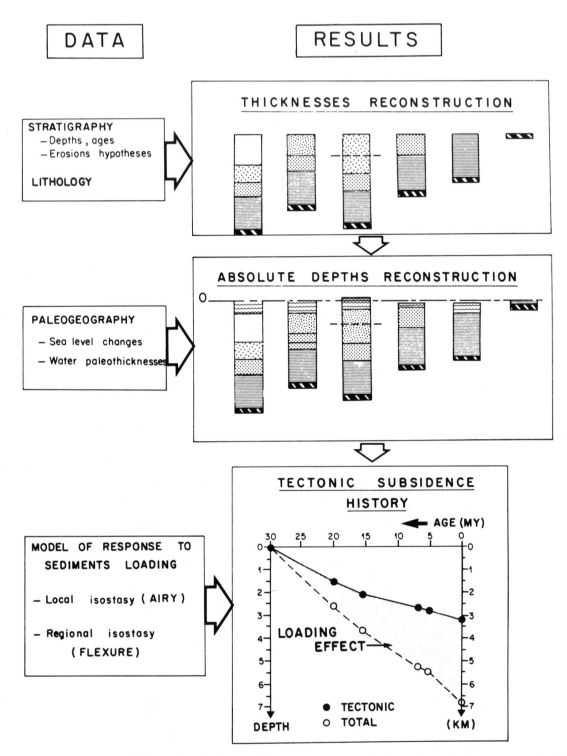

THE BACKSTRIPPING METHOD

DATA

RESULTS

THICKNESSES RECONSTRUCTION

STRATIGRAPHY
- Depths, ages
- Erosions hypotheses

LITHOLOGY

ABSOLUTE DEPTHS RECONSTRUCTION

PALEOGEOGRAPHY
- Sea level changes
- Water paleothicknesses

TECTONIC SUBSIDENCE HISTORY

MODEL OF RESPONSE TO SEDIMENTS LOADING

- Local isostasy (AIRY)

- Regional isostasy (FLEXURE)

AGE (MY)

LOADING EFFECT

DEPTH

● TECTONIC
○ TOTAL

(KM)

Figure 2. Principles of the backstripping method. Note that the reconstruction of the history of sedimentation only takes compaction into account and does not assume any geodynamic processes.

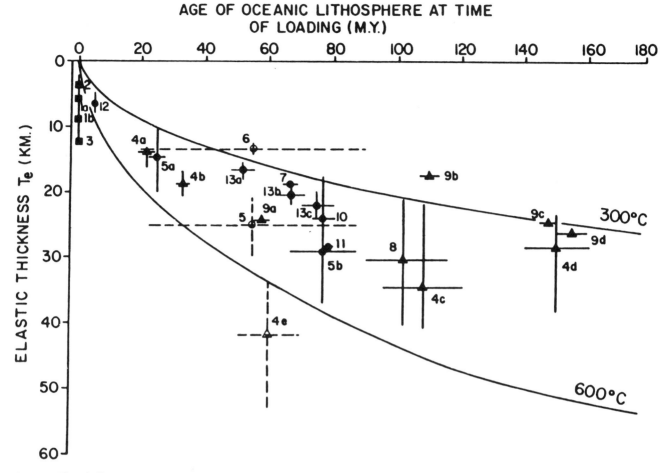

Figure 3. Plot of effective elastic thickness T_e of oceanic lithosphere against age of the lithosphere at the time of loading (from Watts et al, 1980). The estimates of T_e are based on oceanic crustal topography (solid squares), seamounts and oceanic islands (solid circles) and deep sea trench-outer rise systems (solid triangles). The 300 and 600°C isotherms are based on the cooling plate model.

been found to increase with age of loading according to the time-depth relation of an isotherm given by a simple cooling plate model (450°C ± 150°C, see Watts and Steckler, 1981; and Figure 3). Recently, Karner et al (1983) have shown that the continental lithosphere behaves similarly.

For our purpose, we use such estimates of elastic thickness evolution through time (see Beaumont et al, 1982; Watts et al, 1982) to compute the tectonic subsidence, and so estimate the crustal heat-flow with a higher accuracy. Since the lithosphere deflection at a given location in the basin depends on the whole sedimentary load, it is necessary to consider both vertical and lateral distribution of the sediments through time for the subsidence computations (Figure 4).

Eustatic Sea Level Changes

The estimate of sea level changes during basin evolution permits us to compute all crustal vertical movements with respect to a reference level common to its whole history. Many authors have proposed "eustatic curves" on the basis of worldwide correlations of stratigraphic features for various margins (Vail et al, 1977a, b). Watts et al (1982)

have shown that Vail's curves include tectonic effects and so are not exclusively related to eustatic processes. Computing subsidence for several margins in the northern Atlantic, and assuming exponential decay related to thermal contraction of the lithosphere, Watts and Steckler (1981) found that eustatic sea level changes are not so irregular as suggested by Vail. The general trend is a high sea level during the Cretaceous (100 to 300 m; 328 to 984 ft) and a slight decrease up to the present. These variations appear to be mainly controlled by the variation in accretion rate of oceanic ridges.

TECTONIC SUBSIDENCE AND THERMAL STATE OF BASINS RESULTING FROM EXTENSIONAL TECTONICS

Numerous sedimentary basins, such as rift basins and Atlantic-type continental margins, have undergone a rather simple tectonic history. Their structural evolution is characterized by an extensional tectonic phase (rifting), followed by a general subsidence period without active faulting.

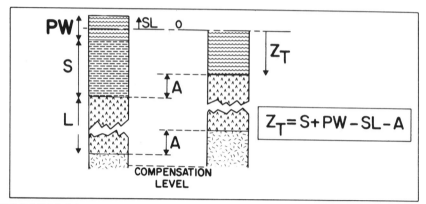

TECTONIC SUBSIDENCE COMPUTATION

1— LOCAL ISOSTASY

$$Z_T = S + PW - SL - A$$

2 — REGIONAL ISOSTASY

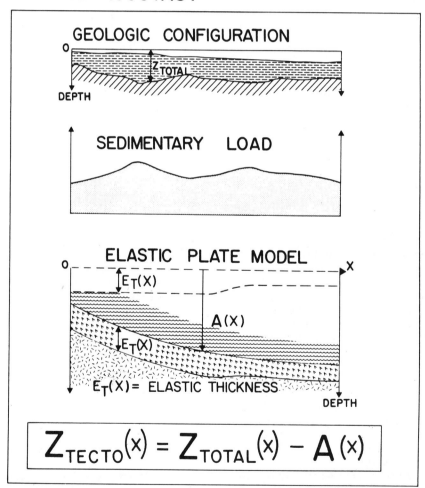

GEOLOGIC CONFIGURATION

SEDIMENTARY LOAD

ELASTIC PLATE MODEL

$E_T(x)$ = ELASTIC THICKNESS

$$Z_{TECTO}(x) = Z_{TOTAL}(x) - A(x)$$

Figure 4. Models used for tectonic subsidence computations: 1) The local isostasy model (AIRY) assumes no evolution of the mechanical properties of the lithosphere; 2) The regional isostasy model assumes a possible increase of flexural rigidity through time. Thus, at a given step, the actual deflection of the lithosphere (A[x]) is inherited from the past history of sedimentary loading. Note that the flexure rigidity is allowed to vary through space.

Passive margins studied by Keen and Barett (1981, Eastern Canada), and Montadert et al (1979, Ireland) exhibit a tectonic subsidence decaying exponentially with time. Similarly, Sclater et al (1980) have shown that the heat flow decreases with time after the rifting phase. Quantitative predictions of subsidence and heat flow patterns have been made based on thermal models of the lithosphere (Sleep, 1971; McKenzie, 1978; Falvey and Middleton, 1981; and Royden and Keen, 1980).

A practical example is the margin offshore Ireland, where for the deepest part of the basin the subsidence history has been studied by Chenet et al (1983). They have shown that subsidence is explained satisfactorily by a 50 percent stretching of the continental crust and a strong thinning of the lithosphere, including the lower part of the crust. These model predictions seem to be in good agreement with the seismic reflection and refraction data.

In order to get a better estimate of the heat flow than obtained by the model of Royden and Keen (1980), our model considers also the heat production by radioactive decay in the continental crust. Basic parameters are the stretching ratio of the continental crust, the thinning ratio of the lithosphere, and the rate of radioactive heat production within the crust.

These parameters are estimated by comparing the tectonic subsidence predicted by the model with the results obtained by the backstripping method. Available data on present heat flow and on the thinning of the continental crust from seismic refraction experiments are also taken into account. These heat flow measurements must be corrected for the effect of sedimentation. Attention must be paid to the fact that the geodynamic model described here deals only with vertical heat transfer within the lithosphere. For this reason, its predictions should only be used where tectonic subsidence does not display sudden lateral changes.

THERMAL HISTORY OF A SEDIMENTARY BASIN

Heat flow evolution, as predicted by geodynamic models or by empirical studies, may be used to determine the temperature history of a basin.

Heat Transfer

The temperature evolution of a sedimentary slice is controlled by its burial, by the heat flow across the sedimentary pile, and by the mode of heat transfer. In most sedimentary basins, conductive heat transfer predominates. Thus the temperature distribution depends on heat flow, burial rate, and thermal conductivity of the sediments.

However, when fluid circulation is important, convective heat transfer may influence the thermal gradient within the sediments. Since the geological data are generally not abundant when the model is used at an early stage of basin exploration, these processes are difficult to quantify. Therefore the application of the thermal model is limited to basins that have not been subjected to tectonic phases leading to important fracturation. Additionally, it is assumed that there is no important convective heat transfer by lateral drainage.

Thermal Model

For the basins meeting these requirements, we built a

thermal model that takes vertical heat transfer by conduction and by expulsion of fluids during compaction into account. The variations of the thermal properties of the sediments with burial are also taken into account. The model provides the temperature history of any level of the sedimentary column and is coupled to a kinetic model of OM maturation (Tissot and Espitalie, 1975). The application range is restricted to sedimentary basins of simple structure and large dimensions, and beds that are not strongly dipping. These restrictions are caused by the monodimensional character of the model. A bidimensional model using the same basic principles will be developed which will extend the application possibilities.

Possible Effects of Convection and Rapid Sedimentation Rates

The input of cold sediments at the top of the sedimentary column causes a decrease of the surface heat flow with respect to the bottom heat flow, a transient process which is important in case of rapid sedimentation (1,000 m/million years). Even at lower sedimentation rates the compaction fluids could locally modify the temperature distribution if they are expelled through preferential drainage zones.

Preliminary studies have also been carried out by Perrin (1983a, b) on the possible impact of natural convection. In this process, fluid circulation is caused by temperature related density variations of water. The two-dimensional model that has been developed shows that convection cells are likely to occur in porous and permeable media. The fluid flow greatly disturbs the geothermal gradient and the temperature could remain almost constant, vertically, over hundreds of meters.

APPLICATION TO THE CENOZOIC WESTERN MEDITERRANEAN BASIN

The first set of models—backstripping, geodynamic and thermal—has been applied to the Cenozoic Western Mediterranean basin. This basin includes the onshore Camargue basin, the Gulf of Lion continental margin and the opposite margins of Sardinia and Corsica (Figure 5a). Exploration is more advanced onshore than offshore, where few wells have been drilled. The basin opening results from an initial rifting phase during the Oligocene-lower Miocene (Biju-Duval, Montadert, 1977).

Subsidence History of the Camargue

The tectonic subsidence in the Camargue basin (up to 5,000 m, or 16,404 ft, of Cenozoic sediments, see Figure 5b) is reconstructed by the backstripping model from data on 20 exploratory wells. The subsidence pattern appears to be very similar for the different wells (Figure 5c).

The main tectonic phases affecting the basin may be assessed from the interpretation of the tectonic subsidence. Thus, the Oligocene-lower Miocene phase is characterized by a rather high subsidence rate and active normal faulting. The subsidence rate is much lower during the middle-late Miocene when the whole basin displays a general subsidence without active faulting. The complex pattern from the Pliocene to the present corresponds to the reactivation of faulting in this area, where tilting occurred due to overthrusting in the Alpine chain.

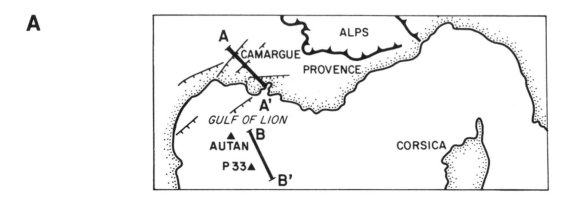

A

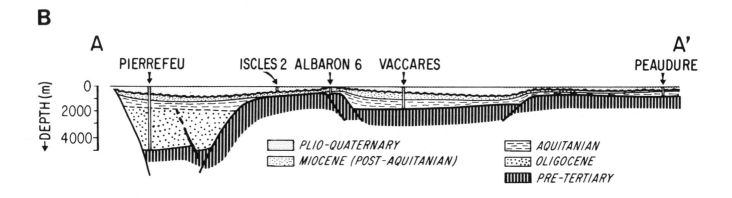

B

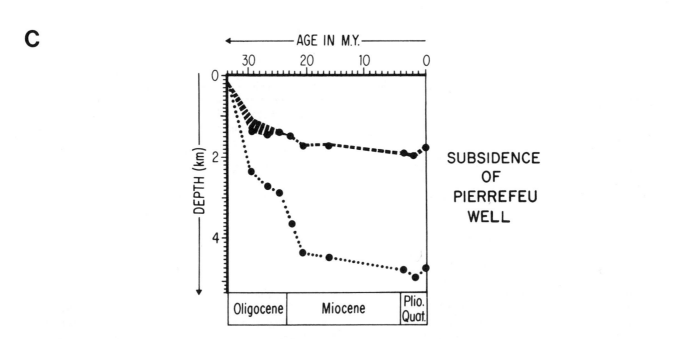

C

Figure 5a. Location map of the Cenozoic intracratonic Camargue basin and Gulf of Lion margins. Main normal faults in the Camargue and main overthrusts in the Provence are shown.
Figure 5b. Schematic cross section A-A' of the Camargue basin. Note that many normal faults die out in the Aquitanian.
Figure 5c. Subsidence curves of the Pierrefeu well and reconstitution of the progressive burial of the various sedimentary units. Heavy dashed lines—tectonic subsidence; dotted lines—total subsidence of pre-Cenozoic substratum.

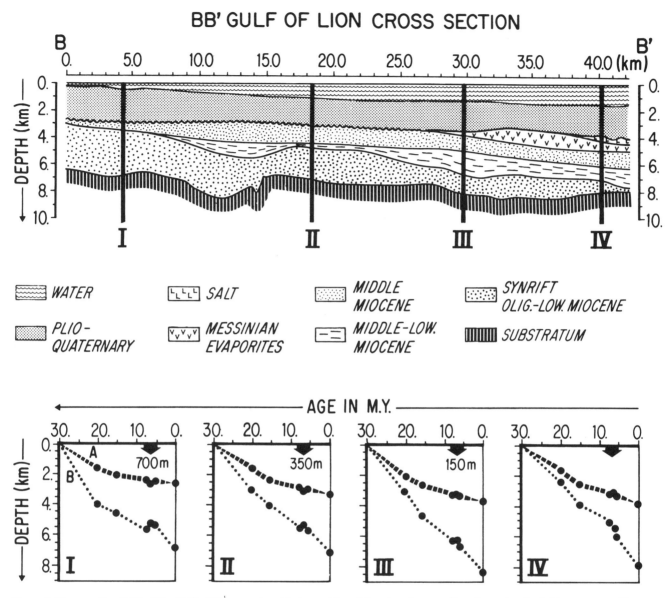

Figure 6. Cross section B-B' of the Gulf of Lion margin with the position of four imaginary wells and associated subsidence curves. The amount of erosion during the regression at the end of the Miocene is indicated below the arrow.

Subsidence History of the Gulf of Lion Margin

The evolution of the Gulf of Lion passive continental margin is reconstructed at the hand of cross sections evolving through time. The geological cross section of Figure 6 is based on a seismic profile from the continental shelf to the deep basin. The pre-Tertiary basement shows a typical rift structure with horsts, grabens, and tilted blocks. Normal faults die out in the overlying layers. The rifting is completed at the end of the Aquitanian and followed by a sedimentary prism of prograding Miocene deposits. This unit is truncated at the top by an unconformity which corresponds to a major regression at the end of the Miocene, caused by the closure of the Mediterranean basin (Cita, 1982). At this time, evaporitic sediments are deposited in the deepest part of the basin. The thick uppermost unit of

Pliocene to Quaternary age corresponds to the rapid infilling of the basin after the Pliocene transgression. Apart from local salt or clay movements, neither faults nor folds of tectonic origin affecting post-Aquitanian deposits have been detected on the seismic profiles.

The successive cross sections are based on seismostratigraphic interpretation, assuming that basement geometry is conserved from one step to the other during the postrift subsidence period. We also assumed that the tectonic subsidence did not undergo any sudden change in rate between Aquitanian and early Pliocene times, since no active tectonics have been recorded during this period. This hypothesis is consistent with the hypothesis of thermal control of postrift subsidence (McKenzie, 1978). We thus computed the amount of erosion and regression at upper

SECTION B B

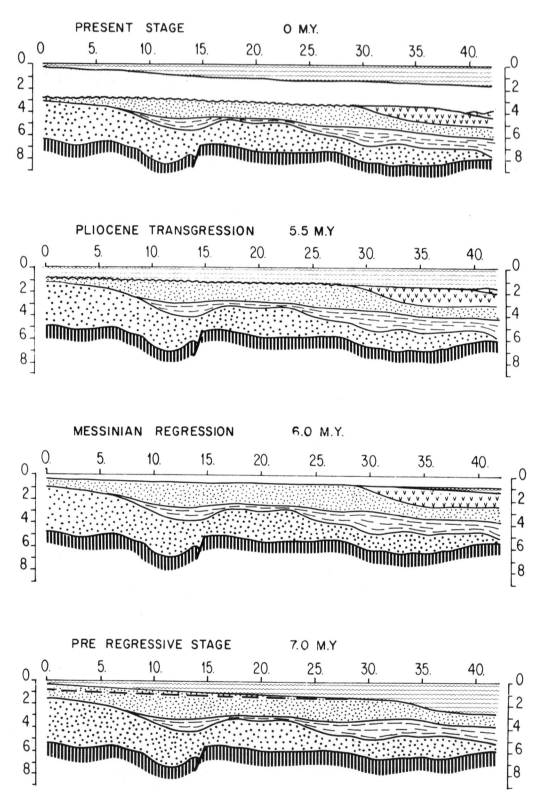

Figure 7. Reconstruction of the geometry of the Gulf of Lion from the Miocene (7 m.y.) to present. The position of the future erosion surface at 6 m.y. is indicated with a dotted line.

Miocene times (Figure 6, below). The successive shapes of the basin—before, during and after the regression event—are the final result (Figure 7).

Heat Flow and Temperature History

The geodynamic model was applied to the Gulf of Lion margin, calibrating its parameters from the tectonic subsidence reconstruction and from surface heat flow measurements. The resulting heat flow variations with time have been used as input data in the thermal model. We computed the current geothermal gradient and the temperature history of the sediments for two locations: the Antan well in the upper part of the margin, and the imaginary well P33 in the lower part (see Figures 8 and 6 for location).

As shown by Figure 8, the computed present-day geothermal gradient matches satisfactorily the measurements in the Autan well. The thermal model also indicates that the effects of the upper Miocene erosion on the present geothermal gradient have completely disappeared. However, the underlying series was cooled during the erosional events (Figure 8c) and the maturation of the organic matter was consequently slowed.

In the case of the P33 "imaginary" well, the computed thermal gradient at the top of the sediments is similar to the measured one. The theoretical gradient at depth is lower than near the surface, due to the high sedimentation rate during the Plio-Pleistocene period and the change with depth of the thermal properties (Figure 8a). High temperatures, around 250°C, are expected at the base of the sedimentary column. This suggests that if any hydrocarbons were formed here, they are presently cracked into gas.

These applications of the backstripping, geodynamic, and thermal models indicate their potential use in oil exploration. First, they enable us to compute the subsidence history of the basin, to estimate the magnitude of the main regressive events, and to determine the corresponding thickness of eroded sediments. Second, they allow us to assess the temperature history with greater confidence, which is important for maturation studies.

This kind of study has already been applied to various sedimentary basins formed by extension. Royden (1982), using similar models, has shown that the study of the subsidence of the Pannonian basin was useful for the reconstruction of the tectonic history and that the maturation predictions satisfactorily matched the geochemical data. Encouraging results have also been obtained by Royden and Keen (1980) on the eastern U.S. Atlantic margin. Of course the temperature results provided by such models may be used in further studies about the migration and accumulation processes, once more geochemical data have become available.

OIL GENERATION AND MIGRATION

As shown by Durand et al (1980) the ultimate recoverable reserves of oil amount to only 1 percent of the oil formed in all the sedimentary rocks presently found on earth. Besides, it is well known that a substantial number of apparently valid traps are not filled with hydrocarbons, even in rich zones. This suggests that the migration process

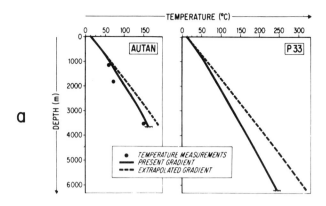

a

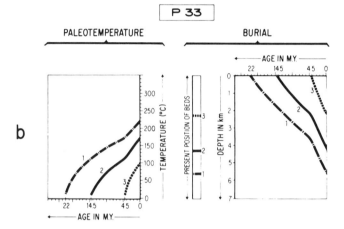

b

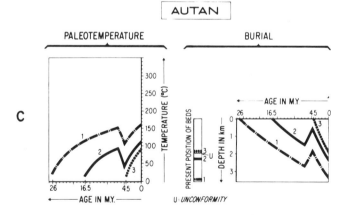

c

Figure 8a. Computed present-day geothermal gradient for the Autan well, and the P33 imaginary well (location Figure 4). The dotted line represents the temperatures extrapolated from surface measurements.

Figure 8b, c. Reconstruction of the burial history and paleotemperature for three horizons in each well. The main unconformity in Autan corresponds to the strong erosion at the end of the Miocene. The amount of erosion has been computed with the backstripping model. Notice the decrease of temperatures for beds 2 and 3 at the time of erosion, concomitant with a decrease in the overburden.

235

plays a very selective and dominant role in the occurrence of oil pools.

Physical and Chemical Processes Involved

It is generally accepted that the generation of petroleum results from maturation of the sedimentary organic matter, as controlled by temperature and time. There is no evidence that pressure plays an important role (Tissot and Welte, 1978, p. 183). The modelling of maturation has been described by several authors: Huck and Karweil (1955), Lopatin and Bostick (1973), and Tissot and Espitalie (1975).

On the other hand, migration of oil and especially expulsion from the source rocks is still not well known, although a wide range of mechanisms have been proposed: 1) migration in aqueous solution; 2) migration by diffusion; and 3) migration of hydrocarbons and water in separate phases.

Though the first two monophasic processes have been proposed by various authors in the past, current evidence and data strongly favor the third, biphasic process, which has consequently been adopted for our migration model.

The main driving force for expulsion is thought to be the compaction of the source rock under the sediment load. In the first stage of maturation, the oil saturation within the pore system of the source rock is low, so that the oil or gas phase is discontinuous, and migration is prevented by capillary forces. Later on, provided the petroleum potential is high enough, the hydrocarbon phase becomes continuous, and expulsion can start. Once the hydrocarbons have reached a permeable bed, buoyancy becomes the most important driving force.

This scheme is valid if the hydrocarbon phase is unique, which is probably the case for most source rocks within the oil generating zone (Ungerer, Behar and Discamps, 1981). However, it may not be the case at shallower depth, where the decrease in temperature and pressure can cause separation into a gas phase and an oil phase. Once individualized the gas phase moves much more easily than the oil (Nogaret, 1983). The experimentally measured solubilities of oil in gas are such that important amounts of oil may be carried in solution (Zhuze, 1974). During the upward migration, the oil moved in solution undergoes retrograde condensation when temperature and pressure decrease. At first, the least soluble components are left behind, whereas the light hydrocarbons (more soluble) condense at shallower depths.

Generation and Migration Modelling

Based on the considerations above, the modelling of migration should include the computation of: oil and gas generation, compaction, triphasic fluid flow, and phase behavior of the hydrocarbon mixtures. These four processes are interrelated, and as a consequence the migration model should intergrate all four into a unique system. It should also be at least two-dimensional, since lateral migration of petroleum is generally needed for its accumulation in traps.

This problem is even more complex than the triphasic and three-dimensional computer models that have been developed in reservoir engineering. Effectively, the porous medium is not static, but changing as a result of burial and compaction, and the process takes place on a million-year

scale. This high complexity coupled to the desire for separating the different processes involved, made us develop two models instead of one:

- A biphasic migration model (oil/water) accounts for processes 1, 2, and 3, under the assumption that the hydrocarbon mixture is monophasic.
- A thermodynamic model computes the composition of the hydrocarbon phase(s), assuming that they are in equilibrium (process 4).

We have developed basic equations for the oil generation and migration. The generation of oil is computed from the kinetic model published by Tissot and Espitalie (1975), with the difference that our model makes no provision for secondary degradation of hydrocarbons (dismutation into gas and carbonaceous residue). In one of the examples we simulated gas formation, modifying the initial oil potential, multiplying them by a coefficient which is assumed to account for the secondary cracking yield.

Hydrocarbon expulsion is influenced by pressure in the compaction process. Thus it is not possible to simply use porosity versus depth relationships as done by backstripping, as this method does not allow for possible undercompaction. Following Smith (1971), compaction is represented in our model by the law of effective stress, whereby effective stress is considered as a function of porosity. The function differs from one lithology to another, reflecting the fact that different rocks compact in different ways.

Darcy's law, extended to polyphasic flows through the relative permeability concept, has been adopted for our model. According to Matheron (1967), application of Darcy's law to very slow monophasic flows is perfectly justified. Its extension to polyphasic flows by the relative permeability concept has, on the other hand, no theoretical justification and we have to consider it as a pragmatic solution only.

Pressures in the hydrocarbon and water phases are assumed to be linked by the capillarity equation, the capillary pressure sign depending on the direction of flow. In our case it can be taken as positive, for we generally find a growing oil saturation, whether in source rocks where hydrocarbons are generated, or carrier beds. The intrinsic permeability of the porous medium is computed from the porosity and the average grain size of the rock. Source rocks or permeable beds through which fluids pass in the course of secondary migration are treated similarly, taking anisotropy into account. Furthermore, in the source rock kerogen is converted into oil, ultimately leading to an increased hydrocarbon pressure within its pores (Du Rouchet, 1981).

When pressure (P) increases in a porous rock, rupture occurs when one of the effective stress components becomes negative. This is evidenced either by creep (diapirism), or by the opening of fissures. These fissures, through which fluids may flow, close as soon as pressure drops. To account for this process, we assumed that if pressure exceeded the minimum stress S_3, rock permeability increased in proportion to the square of the difference $(P-S_3)$.

The data required by the model are mainly the geometry of the cross section, the age and lithology of the stratigraphic levels, the subsidence and temperature history, and finally the initial petroleum potential of the various

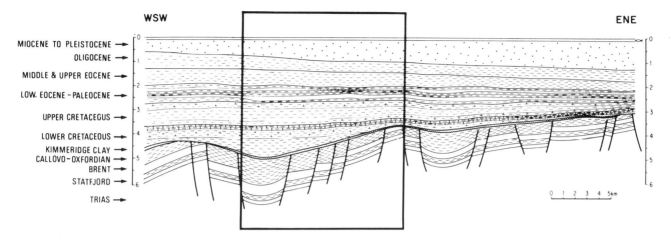

Figure 9. Schematic section of part of the Viking graben in the North Sea. The framed zone was selected for application of the migration model. It is bounded by the two largest faults which define a tilted block below the major unconformity.

sediments. The results consist of a reconstruction of compaction (porosity, pressure) of oil generation (maturation level) and of migration (oil saturations in source rocks and reservoirs).

APPLICATION EXAMPLES

Fault Block in Viking Graben, North Sea

Figure 9 represents a typical situation in the southern part of the Viking Graben (Frigg). Source rock units correspond to the Kimmeridge Clay, Heather and Dunlin formations, as well as the Brent coals, while three permeable units correspond to the Brent, Statfjord, and Paleocene sands. The section is classified by a finite element grid, using six types of lithology: sand, sandy silt, silty clay, clay, marl, and coal, with lithological and geochemical characteristics shown in Table 1.

A relatively high oil potential is assigned to the Brent coals, for analysis indicates that they are richer in hydrogen than is type III of Tissot et al (1974). A geothermal gradient of 30°C/km and a surface temperature of 5°C have been taken. It is also assumed that the faults bounding the block form impermeable barriers, but that the faults inside the block do not interrupt the continuity of the Jurassic sands.

The model plots through time the depth, pressure, and oil saturation for each element of the grid, thus presenting a scenario for oil accumulation over geological time. Figure 10 compares the state of oil saturation (Figure 10a) and pressure (Figure 10b) in the Maastrichtian, when according to the model, hydrocarbons were starting to accumulate in the upper part of the Brent sands, with the present-day situation.

Four classes of oil saturation are represented: 0-5 percent, 5-10 percent, 10-20 percent, 20 percent and over. The model indicates for the present time the existence of hydrocarbon accumulations in the upper part of the Middle Jurassic sandstone reservoirs, and also high oil saturations in source rocks in synclinal positions. The source rocks in which saturation is highest are the Brent coals, because these have the highest organic carbon content. These are also the first to expel oil, which occurs mainly where oil saturation

Table 1. Lithological and geochemical characteristics for Viking Graben, North Sea, formations.

Formation		Lithology	Mean Organic Carbon Content (% wt)	Mean Initial Oil Potential (mg/g OC)
Eocene-Paleocene		Sand	0	0
		Silty sand	0.5	210
		Silty clay	0.7	210
		Clay	1	
Senonian		Marl	0.7	70
		Clay	0.7	70
Turonian		Marl	1.5	70
Cenomanian		Marl	1.5	70
Lower Cretaceous		Clay	1.5	70
Upper Jurassic	Kimmeridge	Clay	7	310
	Heather	Clay	6.5	210
Middle Jur.	Brent	Sand	0	0
		Coal	30	220
Lower Jurassic	Dunlin	Clay	3.5	210
	Statfjord	Sand	0	0

exceeds 20 percent (Figure 15). Subsequently, the Kimmeridge Clay, Heather and Dunlin source rocks in turn supply the reservoirs. In sands with high intrinsic permeability, a low relative permeability will suffice to cause oil displacement. Secondary migration, under the

Figure 10a, b. Representation of the main results of the model applied to the Viking graben: a) oil saturation; b) excess pressure relative to hydrostatic. The first stage (at the top) corresponds to the beginning of oil expulsion (Maastrichtian). The top of the oil generation zone corresponds to a 10 percent conversion ratio of the potential. At the present-day stage, a pool has formed in the permeable levels of the Jurassic.

237

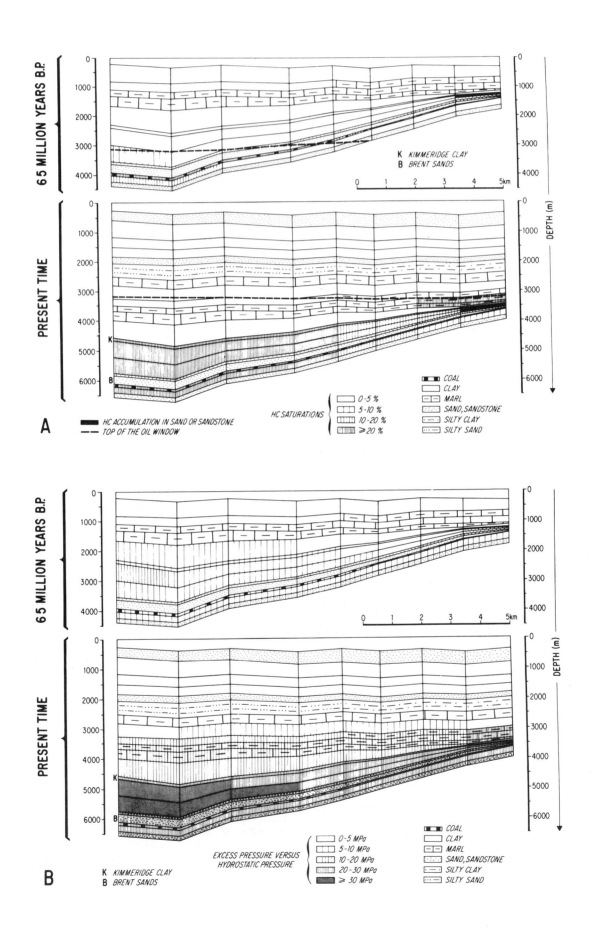

65 MILLION YEARS B.P.

K KIMMERIDGE CLAY
B BRENT SANDS

0   1   2   3   4   5km

PRESENT TIME

DEPTH (m)

K

B

■■ HC ACCUMULATION IN SAND OR SANDSTONE
--- TOP OF THE OIL WINDOW

HC SATURATIONS

0-5 %
5-10 %
10-20 %
≥20 %

COAL
CLAY
MARL
SAND, SANDSTONE
SILTY CLAY
SILTY SAND

A

65 MILLION YEARS B.P.

0   1   2   3   4   5km

PRESENT TIME

DEPTH (m)

K

B

K KIMMERIDGE CLAY
B BRENT SANDS

EXCESS PRESSURE VERSUS
HYDROSTATIC PRESSURE

0-5 MPa
5-10 MPa
10-20 MPa
20-30 MPa
≥ 30 MPa

COAL
CLAY
MARL
SAND, SANDSTONE
SILTY CLAY
SILTY SAND

B

238

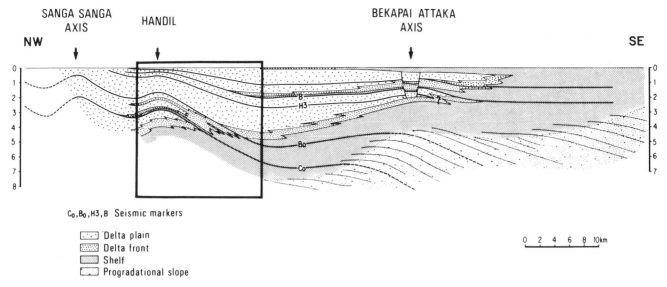

Delta plain
Delta front
Shelf
Progradational slope

0 2 4 6 8 10km

Figure 11. Schematic east-west section of the Mahakam Delta (Indonesia). The framed zone was selected for application of the migration model.

effect of buoyancy, therefore occurs with low residual saturation in the beds passed through (1 to 2 percent). The model also forecasts a dysmigration through the clay overburden of the upper part of the Brent reservoirs, which is only several tens of meters thick.

Five classes of excess pressure relative to hydrostatic are calculated: 0-5 MPa, 5-10 MPa, 10-20 MPa, 30 MPa and over. Imposed boundary conditions are: zero flow at base and through the faults bounding the block, lateral flow possible by gradual reduction of hydraulic resistance above the Lower Cretaceous, and a hydrostatic system at the upper boundary of the grid. The model calculates a hydrostatic system in the Paleocene sands and slight excess pressure in their overburden, as well as considerable excess pressure from 3,000 m (9,843 ft) down. These forecasts are in agreement with observational data. Due to the transmission, toward the top of the structure, of excess pressure treated by source-rock compaction and hydrocarbon generation, dysmigration is focused there.

Absolute excess pressure is highest in the synclinal part of the Callovo-Oxfordian Heather formation. The model thus indicates that, with the geological assumptions made at the outset, expulsion from this formation should be downward into the Brent sands. However, if there would be a lightly more permeable bed within the Kimmeridge Clay-Heather unit, this would be enough to cause lateral migration toward the structural top of the source-rock unit. This sensitivity stresses the impact of slight lithological or geometric variations on the direction of flow. Such minor variations are unfortunately impossible to predict if no wells have been drilled in the synclinal zones.

In this example, we also varied the hydrocarbon density and viscosity values to simulate a gas phase. Our objective was to understand what consequences these variations could have on the amount of possible dysmigration through the Brent reservoir seals. At the same time, we reduced the oil generating potential of the various source rocks to produce approximately the same amounts of hydrocarbons

as would have been generated if gas had been produced. Without modifying the other parameters, the model calculates a migration into the Paleocene sands of quantities of "gas" of the same order of magnitude as found in Frigg. This method, though open to criticism from the theoretical viewpoint, does reveal that the possibilities of gas displacement are much higher than for oil, all other things being equal.

The Mahakam Delta, Indonesia, a Tertiary Delta

The geological section of Figure 11 represents a situation described by Durand and Oudin (1979). The middle Miocene to Pliocene sequence is composed of interbedded sands, silty clays, and coals. The facies pattern shows a continuous progradation of the delta from west to east, with the exception of a small transgressive episode. Sand continuity is high and conditions for secondary migration are therefore favorable. On the Handil axis and in the surrounding synclinal areas, sufficient depths are reached to enable large quantities of hydrocarbons to be generated. Since the number of elements in the grid is restricted (25 vertically and 8 horizontally) we could not represent the small (meter) scale lithological features, which are common in such deltaic formations. In generalizing, we assumed that the delta plain sands were interconnected up to a present

Figure 12a, b. Representation of the main results of the model in the case of the Handil structure. a) Oil saturation. The coals, too thin to be properly represented, have oil saturations over 20 percent in the oil generation zone. The top of this zone corresponds to a 10 percent conversion ratio of the oil potential. At the present-day stage, pools have been formed in the sands and limestones at the top of the Handil structure between 1,500 and 3,000 m (4,921 to 9,843 ft), as well as in a stratigraphic trap. b) Excess pressure relative to hydrostatic.

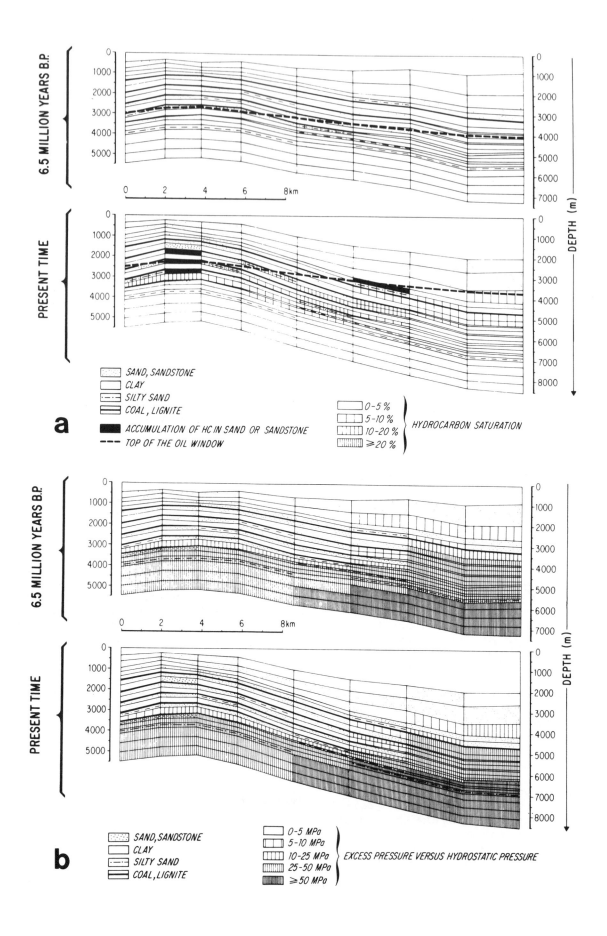

SAND, SANDSTONE
CLAY
SILTY SAND
COAL, LIGNITE
ACCUMULATION OF HC IN SAND OR SANDSTONE
TOP OF THE OIL WINDOW

0-5 %
5-10 %
10-20 %
≥20 %
} HYDROCARBON SATURATION

a

SAND, SANDSTONE
CLAY
SILTY SAND
COAL, LIGNITE

0-5 MPa
5-10 MPa
10-25 MPa
25-50 MPa
≥50 MPa
} EXCESS PRESSURE VERSUS HYDROSTATIC PRESSURE

b

Table 2. Lithological and geochemical characteristics of Handil axis and surrounding synclinal areas, Mahakam Delta, Indonesia.

Facies	Lithology	Mean Organic Carbon Content (% wt)	Mean Initial Oil Potential (mg/g OC)
Prodelta	Clays	0.0	0
Delta front	Clays	1.5	310
Delta plain	Silty sands	0.0	
	clays	2.5	310
	Clays (+ coals)	5.5	310
	Coals	35.0	310

depth of 2,800 m (9,186 ft) across the Handil structure and deeper on its flanks. The proportion and distribution of coals has been plotted to the limits of the model. Four lithologies have been defined: clay, silty sand, sand, and coal, with the lithological and geochemical characteristics shown in Table 2.

Based on available data, the geothermal gradients decline from the top of the Handil structure (31°C/km) shown to the synclinal axis (23°C/km). A surface temperature of 27°C is assumed. Applied boundary conditions are as follows: zero flow at the base of the grid and at its right hand vertical limit (syncline between the Handil and Bekapai structures), hydraulic resistance low at the left hand vertical limit down to about 2,500 m (8,202 ft). It proved much more difficult to put this section together than the previous one, because the vertical dimensions of the lithological elements are small compared to the grid dimensions.

Figure 8a shows oil saturations at −6.5 million years, the time at which the model forecasts the beginning of hydrocarbon expulsion from the source rocks, as well as present time. As in the Viking Graben case, expulsion starts first in the coals. The model suggests that the accumulated oil comes mainly from the coal beds (which again are richer in hydrogen than type III of Tissot et al, 1974), a logical consequence not only of their quantitative importance, but also because of a basic assumption on which our model is based—that expulsion is more rapid and efficient from an organic-rich sediment than from a lean one.

Repeated runs of the model calculated the present-day existence of several superimposed accumulations at the top of the structure or in flank positions, at depths corresponding to those of the real accumulations. The calculated quantities of accumulated hydrocarbons, extrapolated from the two-dimensional model to the three-dimensional real world, are of the same order of magnitude as those observed.

The calculated excess pressure (Figure 12b) shows two distinct zones: 1) the delta plain where sand continuity ensures a fully hydrostatic system; and 2) the delta front and prodelta where excess pressure is forecast.

The pressure contrast is very large at the top of the Handil structure at the transition from delta plain to delta front, where pressures reach values causing hydraulic fracturation. Figure 13 shows that the computed pressures are in agreement with the observations at Handil, with geostatic pressures below 3,000 m (9,842 ft). This matching

has been obtained by assuming that the sand beds were interconnected in the sand-rich facies, and allows us to predict the depth at which the geostatic pressures can be found in undrilled areas, such as in the syncline. The fluids move within the undercompacted zone from the syncline toward the top of the structure, because permeability is higher parallel to stratification. Vertical expulsion of these fluids from the high pressure zone occurs mainly across the anticline, where hydraulic fracturation is focused.

A large number of model runs have been made on this example in order to assess the role of the main parameters. We were thus able to quantify the impact of modification, such as permeability restrictions in migration channels, or changes in source rock distribution, or the values assigned to hydrocarbon viscosity. We were also able to determine the sensitivity of the thermal system: a 10 percent increase in the geothermal gradient results in a doubling of oil accumulation.

Possible Applications of the Migration Model

In these two modelled cases it has proved possible to account for observed data such as: location of accumulations, quantities of hydrocarbons in place, and current pressure systems, without the need for substantial modification of the parameters when switching from one case to the other. This seems to indicate that the physical and numerical approximations used are relatively satisfactory. However, in a model like ours, which involves a large number of parameters, it is always possible to use it to explain a variety of situations, as only a slight modification in lithology, geothermal gradient, or viscosity results in markedly different outcomes. We must therefore take care not to use it in its present state to describe situations where little geological knowledge and no data to check results are available, especially if quantitative results are desired. In particular, geochemical data have to be sufficient for a reasonably accurate assessment of the petroleum potential of the source rocks. For the time being, this is also necessary for any model claiming to describe migration. Furthermore, its resolving power is limited and it is illogical to try to process examples in which lithological features are much smaller in size than the possible grid units considered. In this respect, the Mahakam Delta is a borderline case.

The value of a model like ours is that in the case of well understood examples, it enables the researcher to synthesize the various types of data and to deduce their consequences with respect to generation and migration of oil. It is in this way that the examples processed have revealed the considerable importance of the thermal system. The interest in describing the thermal history of sedimentary basins better than is usually done is clearly apparent.

THERMODYNAMIC MODEL

The purpose of the thermodynamic model is to determine the phase behavior of the HC mixtures as a function of pressure and temperature along the migration paths: it determines the composition of the gas and liquid phase, provided the pressure, temperature, and bulk composition of the mixture are known. The pressure in the subsurface is

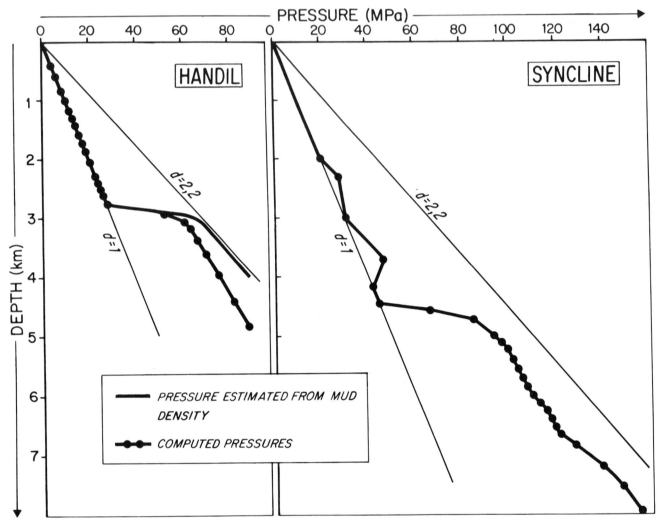

Figure 13. Comparison of pressure gradients as measured and computed by the migration model. To the left is Handil structure. The abnormal pressure zone occurs at 2,800 m (9,186 ft). To the right is synclinal location. The high pressure zone is predicted at 4,600 m (15,092 ft).

generally so high that the gas phase has liquid-like properties. For instance, at P = 30 MPa, and T = 70°C, the density of methane is about 0.18 g/cu cm, which is higher than the density of liquid methane below boiling point (0.15 g/cu cm). This justifies the use of the Regular Solution Theory, which was developed by Hildebrand (1923) to describe the solubility of liquids. The resulting model is mostly applicable to nonpolar hydrocarbons for pressures in excess of 1,000 to 2,000 m (3,281 to 6,562 ft).

Application of the Thermodynamic Model to Handil Structure, Mahakam Delta

The thermodynamic model is used to determine at what depth a typical oil/gas mixture is monophasic. At Handil, it appears that for an oil/gas ratio of 1:3 in weight, the mixture is monophasic for depths in excess of 2,800 m (9,186 ft). This is due to the pressure increase, which favors miscibility. For other locations, the boundary of the monophasic and biphasic state matches roughly with the

upper limit of the high pressure zone, which occurs between 2,800 m (9,186 ft) at the top and 4,500 m (14,764 ft) in the syncline (Figure 13).

Another use of the model is to compute the composition of the oil and gas phases during vertical migration. As mentioned earlier, the condensed oil is thought to remain mostly trapped while the gas undergoes dysmigration. The results (Figure 14) show that the initial mixture which is monophasic at 3,000 m (9,842 ft) in the high pressure zone, separates out into a heavy oil phase and a gas phase when the temperature decreases; upward, successively lighter oils are condensing. Their amount is far from negligible, since 12 percent of the initial oil is condensed between 1,000 and 2,000 m (3,281 to 6,562 ft). A consequence is that the migration of oil in gaseous solution is quantitatively important, at least in some cases. These results are also considered to explain: 1) the origin of some very light oils at low depths; 2) the increase with depth of condensate content in natural gases (Durand and Oudin, 1979); and, 3) the

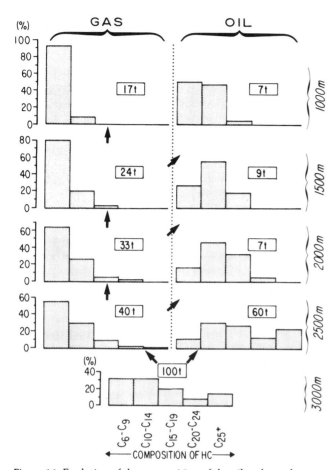

Figure 14. Evolution of the composition of the oil and gas phases during vertical migration, as computed by the thermodynamic model. The initial composition of the mixture is represented by the deep condensates that are found in the Mahakam Delta in the high pressure zone. Important amounts of successively lighter oils are condensed during the upward migration of the gas phase.

amount of the distillate fraction in reservoired oils, which shows a slight but unusual decrease with increasing depths (Oudin and Picard, 1982).

Finally it should not be forgotten that this kind of migration is subordinate to the generation and migration of important amounts of gas from the deepest parts of the basin. The reliability of the results depends presently on the estimation of these amounts rather than on the accuracy of the thermodynamic model itself.

CONCLUSIONS

The reduction of exploration risks implies better prediction of the zones, structures, or stratigraphic horizons favorable for oil and gas accumulation, so that the location of expensive geophysical surveys or exploratory wells can be optimized.

During early stages of exploration, the available geological information is scanty and incomplete. There are almost no well data and few seismic cross sections where some reflectors may be traced from not too distant explored zones. Seismostratigraphic interpretation may give an idea of the lithologies, and knowledge of worldwide anoxic

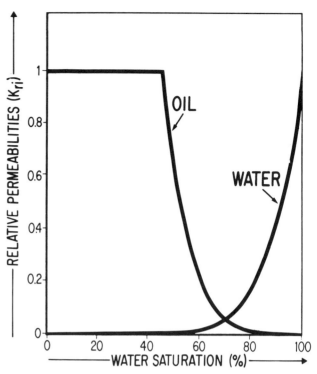

Figure 15. Shape of relative permeability curves used for modeling. For low saturation values, relative permeability is very low but not nil.

events may help in locating possible source rocks. At this stage, the models can operate in two ways:

• In the first place, the backstripping model helps to restore the evolution of the geological cross sections with time. The shape of the basin is determined at each step and the position as well as the time of formation of the traps is better understood.

• In the second place, our models allow us to depict the geodynamic and thermal evolution of the basin. The maturation model then gives an estimation of the oil generating zone on the geological cross sections, at each step of basin evolution.

These early models allow us to answer some important questions: 1) Which are the zones of the basin where the generation of oil is probable?; and, 2) If determined, does the generation occur later than trap formation?

At this stage, a selection can be made of the zones which have been charged by the possible source rocks. However, more information is needed, mainly through exploratory wells, in order to predict the most favorable structures. Once temperature data are available and the reservoirs, source rocks and the type of organic matter are better known, the migration model may be applied and helps in answering the following questions: 1) What is the time span between oil generation and its possible accumulation in the traps?; 2) What are the key parameters that control the volume of accumulated HC?; 3) Considering the time of expulsion of HC from the source rocks, are the accumulations oil or gas?; and, 4) Where and at what depth is there a risk for high pressures? In an ultimate stage, the thermodynamic model helps in determining the condensate

243

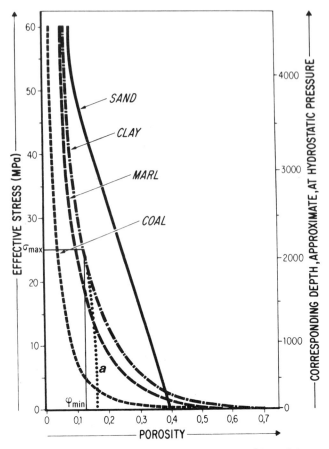

Figure 16. Effective stress versus porosity curves used in studying a North Sea block. If effective stress declines, the curve followed (a) is different from the initial one and joins it at the point of maximum compaction reached ($\emptyset_{min}$, σ_{max}).

content of the gas and the zones where retrograde condensation of oil is most probable.

However, referring to the four validity conditions mentioned in the introduction, it should be emphasized that the predictions given by the various models strongly depend on the input data and on the assumptions made when some geological parameters are not available. One of the main advantages of the mathematical models is their ability to test the influence of uncertainties in the input. They allow us to predict the structures which are in any cases favorable, whatever the uncertainties of the geological data may be. On the other hand, data acquisition can be directed toward those key parameters which control the prediction of the more favorable zones.

REFERENCES

Beaumont, C., C. Keen, and R. Boutilier, 1982, On the evolution of rifted continental margins; comparison of models and observations for the Nova Scotia Margin: Geophysical Journal of the Royal Astronomical Society, v. 70, p. 667-715.

Biju-Duval, B., and L. Montadert, eds., 1977, Structural history of the Mediterranean Basin, *in* 25th Congress of the International Commission for the Scientific Exploration of the Mediterranean Sea (Split, Yugoslavia): Paris, Ed. Technip, 448 p.

Chenet, P.Y., et al, 1983, Extension ratio measurements on the Galicia Portugal and N. Biscay continental margins; implications for some evolution models of passive continental margins, *in* J.S. Watkins and C.L. Drake, eds., Studies in continental margin geology: AAPG Memoir 34, p. 703-716.

Cita, M.B., 1982, The Messinian salinity crisis in the Mediterranean, a review, *in* H. Buckhemer, ed., Alpine Mediterranean geodynamics: Washington, D.C., Geodynamic Series, American Geophysical Union, v. 7, p. 113-140.

De Charpal, D., and Devaux, 1981, Diagenèse des calcaires granulaires de Bassin Parisien et du Nord de l'Aquitaine; conséquences sur les propriétés du réservoir: l'Institut Francais du Pétrole Internal Report.

Durand, B., and J.L. Oudin, 1979, Exemple de migration des hydrocarbures dans use série deltaique; le delta de la Mahakam, Indonésie: Bucarest, Communication au 10eme Congress Mondial du Pétrole.

—— et al, 1980, Kerogen, insoluble matter from sedimentary rocks: Paris, Technip, 519 p.

Du Rouchet, J., 1981, Stress fields, a key to oil migration: AAPG Bulletin, v. 65, p. 74-85.

Falvey, D., and M.F. Middleton, 1981, Passive continental margins; evidence for a prebreakup deep crustal metamorphic subsidence mechanism: Oceanologica Acta, n° sp., Colloque C3, 26ème C.G.I., p. 103-104.

Hildebrand, J.M., 1929, Regular solution theory: Journal of the American Chemical Society, p. 51-66.

Huck, G., and J. Karweil, 1955, Physikalische problem der inkohlung: Brennst.-Chem., v. 36, p. 1-11.

Karner, G.D., M.S. Steckler, and J.A. Thorne, 1983, Long-term thermomechanical properties of the continental lithosphere: Nature, v. 304, p. 250-253.

—— and A.B. Watts, 1982, On isostasy at Atlantic-type continental margins: Journal of Geophysical Research, v 87, p. 2923-2948.

Keen, C.E., and D.L. Barett, 1981, Thinned and subsided continental crust on the rifted margin of eastern Canada; crustal structure, thermal evolution and subsidence history: Geophysical Journal of the Royal Astronomical Society, v. 65, p. 443-465.

Lopatin, N.V., and N.H. Bostick, 1973, Geology in coal catagenesis, *in* Nature of organic matter in recent and ancient sediments: Moscow, Symp. Nauka, p. 79-90 (in Russian).

Magara, K., 1978, Compaction and fluid migration, *in* Practical petroleum geology developments in petroleum science 9: Amsterdam, Elsevier.

Matheron, G., 1967, Eléments pour une théorie des milieux poreux: Masson, 164 p.

McKenzie, D., 1978, Some remarks on the development of sedimentary basins: Earth and Planetary Science Letters, v. 40, p. 25-37.

Montadert, L., et al, 1979, North east Atlantic passive margins; rifting and subsidence processes, *in* M. Talwani, ed., Deep drilling results in the Atlantic ocean; continental margins and paleo environment: Washington, D.C., American Geophysical Union, Series 3.

Nogaret, E., 1983, Solubilité des hydrocarbures dans les gaz comprimés; application à la migration de pétrole dans les bassins sédimentaires: Paris, Thesis, May, Ecole des Mines.

Oudin, J.L., and P.F. Picard, 1982, Genesis of hydrocarbons in the Mahakam Delta and the relationship between their distribution and the over pressured zones: Jakarta, Indonesia, Presented at the XIth Annual Indonesian Petroleum Association Convention.

Perrier R., and S. Quiblier, 1974, Thickness changes in sedimentary layers during compaction history: AAPG Bulletin, v. 58, p. 507-520.

Perrin, J.F., 1983a, Trasferts thermiques dans les bassins sédimentaires: Revue de l'Institut Francais du Petrole, v. 38.

——, 1983b, Modélisation du champ thermique dans les bassins sédimentaires; application au bassin de la Mahakam, Indonésie: University of Bordeaux, Doctoral Thesis.

Royden, C., 1982, Subsidence and heat-flow in the Pannonian Basin: Boston, MA, Massachusetts Institute of Technology, Ph.D. Thesis.

—— and C.E. Keen, 1980, Rifting processes and thermal evolution of the continental margin of eastern Canada determined from subsidence curves: Earth and Planetary Science Letters, v. 51, p. 343-361.

——, J. Sclater, and R.P. Von Herzen, 1980, Continental margin subsidence and heat-flow; important parameters in formation of petroleum hydrocarbons: AAPG Bulletin, v. 64, p. 173-187.

Sclater, J.G., C. Jaupart, and D. Galson, 1980, The heat-flow through oceanic and continental crust and the heat loss of the earth: Review of Geophysics and Space Physics, v. 18, p. 269-311.

Sleep, N.H., 1971, Thermal effect of the formation of Atlantic continental margins by continental break-up: Geophysical Journal of the Royal Astronomical Society, v. 24, p. 325-350.

Smith, J.E., 1971, The dynamics of shale compaction and evolution of pore fluid pressure: Mathematical Geology, v. 3, p. 239-269.

Tissot, B., et al, 1974, Influence of nature and diagenesis of organic matter in formation of petroleum: AAPG Bulletin, v. 58, p. 499-506.

—— and J. Espitalie, 1975, L'évolution thermique de la matière organique des sédiments; applications d'une simulation mathématique: Revue de l'Institut Francais du Petrole, v. 30, p. 743-777.

—— and D.H. Welte, 1978, Petroleum formation and occurrence: Springer-Verlag.

Ungerer, P., F. Behar, and D. Discamps, 1981, Tentative calculation of the overall volume of organic matter; implications for primary migration: Bergen, Proceedings of the 10th International Meeting of Organic Geochemistry.

Vail, P.R., R.M. Mitchum, Jr., and S. Thompson, III, 1977a, Seismic stratigraphy and global changes of sea level, in C. Payton, ed., Seismic stratigraphy—applications to hydrocarbon exploration: AAPG Memoir 26, p. 83-97.

——, ——, and ——, 1977b, Relative changes of sea level from coastal onlap, in C. Payton, ed., Seismic stratigraphy—applications to hydrocarbon exploration: AAPG Memoir 26, p. 36-71.

Waples, D.W., 1980, Time and temperature in petroleum formation; application of Lopatin's method to petroleum exploration: AAPG Bulletin, v. 64, p. 916-926.

Watts, A., and M.S. Steckler, 1981, Subsidence and tectonics of Atlantic type continental margins: Oceanologica Acta, N° SP., Colloque C3, Géologie des Marges Continentales, 26éme C.G.I., p. 143-154.

——, J.R. Cochran, and G. Selzer, 1975, Gravity anomalies and flexure of the lithosphere; a three-dimensional study of the great metear seamount, northeast Atlantic: Journal of Geophysical Research, v. 80, p. 1391-1398.

——, G.D. Karner, and M.S. Steckler, 1982, Lithospheric flexure and the evolution of sedimentary basins: Philosophical Transactions of the Royal Society of London, A305, p. 249-281.

—— and W.B.F. Ryan, 1976, Flexure of the lithosphere and continental margins basins: Technophysics, v. 36, p. 25-44.

—— and M.S. Steckler, 1979, Subsidence and eustasy at the continental margin of eastern North America: Washington, D.C., American Geophysical Union Maurice Ewing Symposium Series, v. 3, p. 218-234.

Welte, D.H., and M.A. Yükler, 1981, Petroleum origin and accumulation in basin evolution; a quantitative model: AAPG Bulletin, v. 65, p. 1387-1396.

Woodside, M., 1971, The thermal conductivity of porous media: Journal of Applied Physics, v. 32.

Zhuze, T.P., and Bourova, 1977, Influence des différents processus de la migration primaire des hydrocarbues sur la composition des pétroles dans les gisements, in J. Goni and E. Campos, eds., Advances in organic geochemistry: Madred, ENADIMSA, p. 493-499.

APPENDIX 1: BACKSTRIPPING METHOD[1]

Reconstruction of Burial

The volume of solid matrix Vs in a porous sediment of volume V with porosity $\emptyset$ may be written as: $Vs = (1-\emptyset)V$. When dealing with the evolution of the thickness Δh of slice of sediments during its burial, Perrier and Quiblier (1974) have shown that the thickness Δh is related to the porosity/depth relationship $\emptyset(Z)$ and the thickness of the solid matrix: Δh_s in the following form:

$$\Delta h_s = \int_{Z}^{Z+\Delta h} \{'1 - \emptyset(Z)\}\, dz \qquad (1)$$

Assuming Δh_s remains constant during the burial history, the Δh variations with depth may be computed after formula (1).

---

[1]The equations used in the backstripping method have been taken from Perrier and Quibler (1974) concerning the reconstitution of burial of a slice of sediments and were inspired from Watts and Steckler (1981) and Timoschenko et al (1959) about the tectonic subsidence computation.

Tectonic Subsidence

The tectonic subsidence Z_T is related to the total subsidence Z and the deflection A due to the load of sediments by:

$$Z_T = Z - A$$

Assuming local isostatical compensation, the deflection A will be:

$$A = S\left(\frac{\rho s - \rho w}{\rho m - \rho w}\right) + \Delta SL \frac{\rho w}{\rho m - \rho w}$$

(from Watts and Steckler, 1981).

Assuming regional isostasy, the deflection A(x) along the horizontal axis x is obtained by the equation of behavior of a thin elastic plate.

The deflection w(x) due to a point load P, located at x_o is given by:

$$D\frac{d^4 w(x)}{dx^4} + (\rho m - \rho w)g\; w(x) = \rho\delta(x - x_o)$$

(from Timoshenko et al, 1959)

with δ: Dirac function

$$D = \frac{ET e^3}{12(1 - \sigma^2)}$$

The total deflection A(x) is obtained by:

$$A(x) = \underset{\text{point load}}{\Sigma w(x)}$$

the point load P being given by:

$$P(x_o) = \left[S(x_o)\frac{\rho s(x_o) - \rho w}{\rho m - \rho w} + \Delta SL\frac{\rho w}{\rho m - \rho w}\right]\Delta x_o$$

with Δx_o: width of an elementary sedimentary column.

with S = sediment thickness corrected for compaction (m)
    ρs = mean density of sediments (kg cu m)
    ρm = mean density of mantle (kg cu m)
    ρw = mean density of water (kg cu m)
    ΔSL = sea level relative to the present day (m)
    g   = average gravity ($m.s^{-2}$)
    D   = flexural rigidity (N − m)
    E   = Young's Modulus (Pa)
    Te = elastic thickness (m)
    σ = Poisson's ration (no dimension).

APPENDIX 2: THERMAL TRANSFER IN SEDIMENTS[2]

The model including heat transfer by conduction and convection due to the expulsion of water during compaction has been studied in detail by Perrin (1983).

The variation with time of the amount of heat is equal to the heatflow due to conduction and the heat transported by the fluid. The corresponding differential equation is:

$$(\rho c^*)\frac{\delta T}{\delta t} = -(\rho c^* \vec{V}_s + \rho c_e \vec{u})\mathrm{gr\vec{a}d}\; T + \mathrm{div}(\lambda^* \mathrm{gr\vec{a}d}\; T)$$

$$\frac{\partial \emptyset}{\partial t} + \mathrm{div}\vec{V}_e = 0$$

$$\frac{\partial \emptyset}{\partial t} + \mathrm{div}(\emptyset - 1)\vec{V}_s = 0$$

$$\vec{u} = \emptyset(\vec{V}_e - \vec{V}_s).$$

The heat capacity of the bulk sedimen ρc^* being:

$$\rho c^* = \rho ce + \rho cs(1 - \emptyset).$$

The thermal conductivity of the bulk sediment * is computed after the semi-empirical formula (see Woodside and Messmer, 1961).

$$\lambda^* = \lambda s\left(\frac{\lambda e}{\lambda s}\right)^\emptyset$$

$\vec{u}$ = velocity of filtration of the fluid with respect to the solid matrix ($m.s^{-1}$)
V_e = velocity of expulsion of the fluid ($m.s^{-1}$)
V_s = velocity of burial of the solid ($m.s^{-1}$)
ρce = heat capacity of the fluid ($J\, kg^{-1}\, K^{-1}$).
ρcs = heat capacity of the solid matrix ($J\, kg^{-1}.K^{-1}$)
ρe = thermal conductivity of the fluid ($W\, m^{-1}\, °C^{-1}$)
ρs = thermal conductivity of the solid matrix ($W\, m^{-1}\, °C^{-1}$)
T = temperature (°K)
t = time (s)
$\emptyset$ = porosity (no dimension).

The heat flow $\emptyset$ coming out of the basement is computed from a thermal model of lithosphere.

The corresponding equation of conductive heat transfer through the lithosphere is:

$$\frac{\delta T}{\delta t} = K\frac{\partial^2 T}{\partial Z^2} + \frac{A}{\rho c}$$

[2]The equations of the thermal model of heat transfer by conduction and convection within the sediments are taken from Combarnous and Bories (1970). The conductive heat transfer theory is developed in Carslaw and Jaeger (1965).

with K = diffusivity of the lithosphere $(m^2.s^{-1})$
ρc = thermal capacity $(J.kg^{-1}.k^{-1})$

The thermal generation in the lithosphere is computed after the empirical formula of Lachenbruch (1965).

$$A = Ao\ \exp\left(-\frac{Z}{D}\right)$$

Ao = heat generation at the surface of the continental crust (Wm^{-3})
D = parameter related to the decrease of enrichment in the continental crust (around 10 km).

Both thermal models have been solved using a finite difference procedure (Perrin, 1983; Carnahan et al, 1969).

APPENDIX 3: OIL FORMATION AND MIGRATION MODEL[3]

Hydrocarbon Formation

The degradation of organic matter is described as in the Tissot and Espitalie model (1975) by a series of six parallel chemical reactions obeying a kinetic of order 1 and Arrhenius's law:

$$Q = \sum_{k=1}^{6} (\xi_{ko} - \xi_k)$$

$$\frac{d\xi k}{\xi k} = A_k e^{-E_k/RT_{dt}}$$

with Q = quantity of hydrocarbons formed $(mg/g\ C_{org}$ initial)
ξ_{ko} = initial oil potential relative to reaction $k (mg/g\ C_{org}$ initial)
ξ_k = residual oil potential relative to reaction k $(mg/g\ C_{org}$ initial)
A_k = reaction coefficient (s^{-1})
E_k = activation energy (cal/mole)
R = perfect gas constant $(R = 2$ cal/mole K)
t = time (s)
T = temperature (K)

The values of parameters ξ_{ko}, A_k, and E_k used are fixed for each type of organic matter whose degradation is to be described, on the basis of classification by Tissot et al (1974).

Representation of Fluid Flows (Water and Oil)

The formulation of fluid flows in porous environments is based on Darcy's law adapted to the diphasic case by the use of relative permeabilities:
For water:

$$\vec{V}_e = \frac{-KK_{re}\rho_e g}{\mu_e}\vec{grad}\,H_e$$

For the hydrocarbon phase, assumed to be unique:

$$\vec{V}_h = \frac{-KK_{rh}\cdot\rho_h g}{\mu_h}\vec{grad}\,H_h$$

with K = intrinsic permeability (m)
$\vec{V}_e, \vec{V}_h$ = filtration velocity of water and hydrocarbons (m/s)
μ_e, μ_h = dynamic viscosities of water and hydrocarbons (Pa.s)
K_{re}, K_{rh} = relative permeabilities of water and hydrocarbons (no dimension)
ρ_e, ρ_h = densities of water and hydrocarbons (kg/m)
g = gravity acceleration (9.81 m/s)

$$H_e = \frac{P}{\rho_e g} - Z \text{ head (water)}$$

$$H_h = \frac{P_h}{\rho_h g} - Z \text{ head (hydrocarbons)}$$

P, P_h = pressure of water and oil (Pa)
Z = depth (m).

Characteristic intrinsic permeabilities of the porous environment are calculated according to porosity by the Kozeny-Carman formula, introducing an anisotropy coefficient:

$$K_x = \frac{0\cdot2\ \emptyset^3}{(1-\emptyset)^2 S_0^{\,2}}$$

$$K_y = \lambda K_x$$

with K_x, K_y = horizontal and vertical intrinsic permeabilities (sq m)
$\emptyset$ = porosity (no dimension)
S_0 = specific area of rock matrix (sq m/cu m)
λ = anisotropy coefficient, less than 1 (no dimension).

Once the permeability coefficients for each element of the grid have been calculated, those applicable to each boundary are derived at by interpolation.

Relative permeabilities K_{re} and K_{rh} are assumed to depend on oil saturation by curves similar to those used for modelling fluid movement in reservoirs (Figure 15).

Capillarity Equation

$$P_h - P = P_c$$

P_h = pressure in the oil phase
P = pressure in the water phase
P_c = capillary pressure.

---

[3]The equations used for the model have been taken from Tissot and Espitalie (1975) concerning the oil generation, and from Marle (1972) and Scheidegger (1960) about the fluid flow calculations. We also were inspired from Smith (1971) for the description of compaction.

The value of P_c is computed from a mean radius of access to pores R (in meters) characteristic of each lithology, by the standard formula:

$$P_c = \frac{2\gamma}{R}$$

We selected for γ a value of 30.10^{-3} Newton/n and values of R such that capillary pressure in the clays is approximately 2 MPa.

Representation of Compaction

The equation of effective stress is used to express stresses between the solid matrix and fluids:

$$S_g = \sigma + P$$

with S_g = total stress (P_a)
σ = effective stress (P_a)
P = pressure (P_a)

As we could not do otherwise, we considered only the vertical component of stress, given by the following relation which expresses the weight of overlying sediments:

$$S_g = \int_0^z \rho g dz$$

with z = depth of point considered (m)
ρ = total density of rock (kg/cu m)
g = gravity acceleration.

Total stress S_g is then termed geostatic load. To describe sediment compaction we shall write, like Smith (1971), that effective stress σ is a function of porosity:

$$\sigma = s_k(\emptyset)$$

The index k signifies that the function s_k differs from one lithology to another (Figure 16).

The concept of compaction irreversibility is introduced by linking σ, s_k, and $\emptyset$ as soon as values of $\emptyset$ stop decreasing continuously in a grid unit and tend to increase above the $\emptyset_{min}$ value then reached, using an empirical relation which is easy to calculate. This relation expresses the fact that once values of $\emptyset$ increase, they no longer follow the initial $\sigma = s_k$ curves but, rather, much steeper ones which meet the $\sigma = s_k$ curves at a tangent at the point where the $\emptyset_{min}$ value is reached (Figure 16).

The density for the water (ρ_e) as well as for oil (ρ_h) have been assumed to be constant.

The density and viscosity values used correspond to a low viscosity oil containing dissolved gas:

ρ_h = 750 kg/cu m (density)
$\mu_h = ae^{b/T}$ (dynamic viscosity according to Amdrade formula in Latil, 1975).

T is the temperature, a and b are selected so as to obtain a viscosity of 16.10^{-3} Pa.s at 40°C and $1.15.10^{-3}$ Pa.s at 160°C.

Water viscosity according to temperature is calculated by the standard Bingham formula (Latil, 1975).

Algorithm and Boundary Conditions

The geological section under investigation from the beginning of subsidence up to the present time is represented by a finite element grid with conditions at the limits as follows: 1) at the base of the grid, which it is assumed represents an impermeable bedrock affected by subsidence, zero flow and displacement specified; and 2) at the upper boundary of the grid, hydrostatic pressure and overburden by sediments of uniform 2,000 kg/cu m density, vertical displacements free.

The number of grid units in the upward direction increases with evolution, an extra row is added when the depth of the top of the last grid level exceeds the initial thickness of the following level. The number of grid units is constant breadthwise. Using the equations set forth earlier, water and hydrocarbon flow velocities can be calculated at each moment. These velocities are used at each iteration to produce the equations of mass balance of water hydrocarbons the solid matrix for each element of the grid, which in turn enable us to compute the pressures, the porosities, the saturations and the nodes depths through a finite difference method.

APPENDIX 4: THERMODYNAMIC MODEL OF SOLUBILITY OF HC[4]

The thermodynamic equilibrium is expressed by the equation of the fugacities in the gas phase and in the liquid phase, for each component:

$$f_i' = f_i''$$

The fugacities are expressed as follows:

$$f_i' = f_i°(P, T)x'_i\gamma'_i(P, T, s_j)$$

with $f_i°(P, T)$ = fugacity of pure component i at temperature T and pressure P
x'_i = molar percentage of i in phase '
$\gamma'_i(P, T, x_j)$ = activity coefficient of i in phase '.

The activity coefficients γ'_i are computed, introducing the solubility parameters:

$$\log \gamma'_i = \frac{V_i(\delta_i - \bar{\delta'})^2}{RT}$$

with δ_i = solubility parameter of component at P and T
$\bar{\delta'}$ = average solubility parameter in phase '
V_i = co-volume of component i
R = molar gas constant (R = 2 cal/mole K).

Coupled to the mass balance equations, the equations above are sufficient for the determination of the concentrations of each component in both phases for a mixture at given pressure and temperature.

---

[4]The equations of the model were partially taken from Hildebrand (1929).

248

Reprinted by permission of the International Human
Resources Development Corporation. Published as
Chapter 9 in *Geochemistry in Petroleum Formation*, by D.
W. Waples, pp. 121–154, Boston, 1985.

Predicting Thermal Maturity

INTRODUCTION

Measured maturity values for possible source rocks are invaluable because they tell us much about the present status of hydrocarbon generation at the sample location. In most cases, however, measured maturity data are of limited value in exploration. Part of this problem is a consequence of the limitations we face in attempting to obtain reliable maturity measurements. In some areas there are no well samples available; indeed, in frontier basins there may not be a single well within tens or hundreds of miles.

Even in maturely explored basins the samples available for analysis often do not give a representative picture of maturity in the basin (fig. 1.2). Furthermore, maturity measurements can only tell us about present-day maturity levels. If our measurements indicate that a rock has already passed through the oil-generation window, we still have no clue as to when oil generation occurred, nor do we know at what depth or temperature it occurred. These considerations are important when we want to compare timing of generation, expulsion, and migration with timing of structure development or trap formation.

In order to circumvent these difficulties, methods have been developed for calculating maturity levels where measurements are not available. Among the most popular models are those of Tissot (1969), Lopatin (1971), and Hood et al. (1975). They have been discussed elsewhere (Waples, 1984a).

The common thread running through all these models is the assumption that oil generation depends upon both

249

the temperature to which the kerogen has been heated and the duration of the heating. This assumption is a logical and defensible one, for it is in keeping with the predictions of chemical–kinetic theory.* The Arrhenius equation (eq. [9.1]) gives the exact dependence of the reaction-rate constant k on the activation energy E_a and the temperature T.

$$k = A\exp(-E_a/RT) \qquad (9.1)$$

The preexponential factor A is a constant, the exact value of which depends upon the particular reaction under consideration, and R is the universal gas constant.

Several workers have calculated activation energies for the process of oil generation. The values thus obtained for E_a are in the range of 11,000 to 14,000 calories per mole (cal/mol) (Tissot, 1969; Connan, 1974). Because activation energies in this range are far lower than one would expect for the breaking of carbon–carbon or carbon–oxygen bonds (40,000 to 60,000 cal/mol), many workers have interpreted the low values of E_a as proof of the importance of mineral catalysis in oil generation.

In one respect the catalysis hypothesis appears to be reasonable, because catalysts lower activation energies by providing alternative, lower-energy pathways. One problem with the catalyst idea, however, is that no known catalysts are capable of lowering the activation energies to 14,000 cal/mol. Most catalytic effects are far less dramatic.

In 1975 Jüntgen and Klein presented a much more plausible explanation for the low calculated values for activation energies. They pointed out that hydrocarbon generation involves the simultaneous occurrence of many distinct reactions, and that the overall rate of hydrocarbon generation should depend upon the sum of the rates of all the parallel chemical reactions that produce hydrocarbon molecules. When these individual reactions are summed, and the overall reaction scheme is treated mathematically as though it were a single reaction, the calculated activation energy turns out to be much lower than the activation energy of any of the individual reactions (fig. 9.1). This

* Price (1982) has represented a minority opinion that time is of no importance in hydrocarbon generation. I do not believe that such a view is tenable in light of either empirical evidence or theoretical considerations.

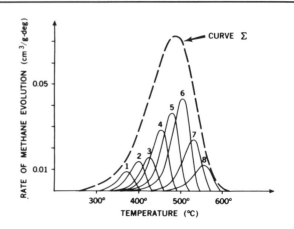

FIGURE 9.1 *Rate of methane evolution from coal for eight parallel reactions (numbered curves) as a function of pyrolysis temperature during programmed-temperature pyrolysis. A Gaussian distribution is assumed for the initial concentrations of the eight reactants. The following activation energies were assumed (kcal/mol):*

(1) = 48	*(2) = 50*	*(3) = 52*	*(4) = 54*
(5) = 56	*(6) = 58*	*(7) = 60*	*(8) = 62*

A-factors are taken as 10^{15}/min for each reaction. Curve Σ, representing the sum of the eight parallel reactions, has a pseudo-activation energy of 20 kcal/mol and an A-factor of 10^4/min. After Jüntgen and Klein, 1975, by permission of Erdöl und Kohle, Erdgas, Petrochemie.

calculated activation energy is a mathematical construct rather than a true activation energy. The calculated activation energy is thus best referred to as a "pseudo-activation energy," because the low values do not actually describe any single hydrocarbon-generating reaction.

If we accept that the Arrhenius equation (when equipped with an appropriate pseudo-activation energy) adequately describes the process of hydrocarbon generation, we must also accept that both time and temperature play roles. These two factors are interchangeable: a high temperature acting over a short time can have the same effect on maturation as a low temperature acting over a longer period. Nevertheless, early efforts to take both time and temperature into account in studying the process of hydrocarbon generation were only partially successful because of the mathematical difficulties inherent in allowing both time and temperature to vary independently. In 1971, however, N.V. Lopatin in the Soviet Union described a simple method by which the effects of both time and temperature could be taken into account in calculating the thermal maturity of organic material in sediments. He developed a "Time–Temperature Index" of maturity (TTI) to quantify his method.

Lopatin's original work was greeted with some enthusiasm and much criticism in his homeland. Several of the problems that subsequently surfaced could be attributed to the poor quality of the data with which Lopatin originally calibrated his model. Despite a few minor difficulties, however, Lopatin's basic idea has much merit, and it has been generally well received in the West.

Lopatin's method allows one to predict both where and when hydrocarbons have been generated and at what depth liquids will be cracked to gas. It has even been suggested that maturity models are more accurate than measured data for determining the extent of petroleum generation (Yükler and Kokesh, 1984). In addition, TTI values have been used to estimate the extent of diagenesis of inorganic minerals (Siever, 1983; Schmoker, 1984).

In this chapter you will learn how to carry out maturity calculations using Lopatin's method and how to use Lopatin's method in exploration. We shall also look at some of the other maturity models that are available.

CONSTRUCTION OF THE GEOLOGICAL MODEL

One of the advantages of Lopatin's method is that the required input data are very simple and easy to obtain. We need data that will enable us to construct a time stratigraphy for the location of interest and to specify its temperature history. Time–stratigraphic data are usually available as formation tops and ages obtained by routine biostratigraphic analysis of cuttings. If no well data are available, a time stratigraphy can sometimes be constructed using seismic data, especially if the seismic reflectors can be tied to well data. If no subsurface data are available, estimates can be made, perhaps from thicknesses of exposed sections nearby.

BURIAL-HISTORY CURVES

Implementation of Lopatin's method begins with the construction of a burial-history curve for the oldest rock layer of interest. An example is shown in figure 9.2, which was constructed from the time stratigraphy for the Tiger well

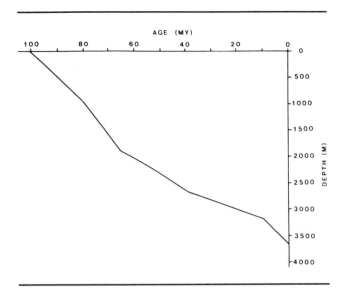

FIGURE 9.2 *Burial-history curve for the deepest datum (100-my-old rock) in the Tiger well constructed from the time–stratigraphic data given in table 9.1.*

TABLE 9.1 *Time-stratigraphic data for the Tiger Well*

Age (my)	Depth (m)
0	0
10	500
38	900
65	1800
80	2800
100	3700

(table 9.1). In the Tiger well, sediment has accumulated continuously but at varying rates since deposition of the oldest rock 100 million years ago (mya). Today the rock is at a depth of 3700 m. The burial-history curve was constructed in the following way: two points, representing the initial deposition of the sediment and its position today, are marked on the age-depth plot (fig. 9.3).

The next step is to locate the first control point from the time–stratigraphic data in table 9.1. Neglecting compaction effects, by 80 mya the sediment had been buried to a depth of 900 m (fig. 9.4). Using the other control points from table 9.1, we can construct figure 9.5. Connecting

251

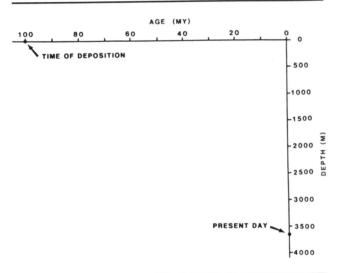

FIGURE 9.3 *First step in the construction of the burial-history curve in figure 9.2. Plot the two points corresponding to present-day depth of burial and the moment of deposition of the rock of interest.*

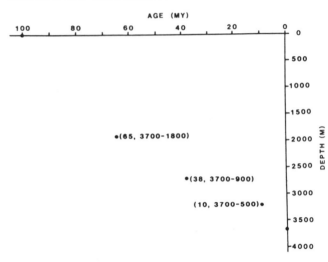

FIGURE 9.5 *Third step in the construction of the burial-history curve in figure 9.2. Plot all the remaining control points (age of the datum, thickness of rock presently separating it from the 100-my-old rock).*

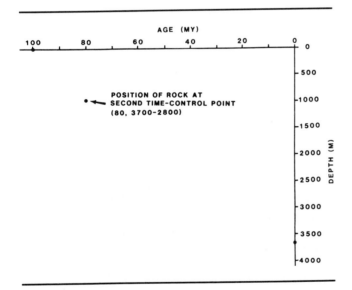

FIGURE 9.4 *Second step in the construction of the burial-history curve in figure 9.2. Plot the second time-control point, corresponding to the next oldest datum (in this case, 80 my). The depth of burial of the 100-my-old rock is given by the thickness of rock that separates the two datums at the present time (900 m in this case).*

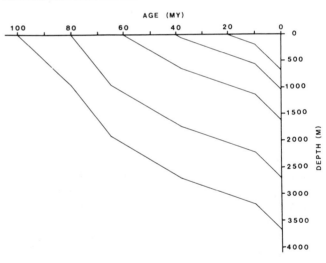

FIGURE 9.6 *Family of burial history curves for the Tiger well based on the time–stratigraphic data presented in table 9.1.*

252

the six dots in figure 9.5 completes the burial-history curve already shown in figure 9.2.

All of the shallower and younger horizons will have burial-history curves whose segments are parallel to those of the oldest horizon (fig. 9.6). This geometry is a direct consequence of ignoring compaction effects.

Burial-history curves are based on the best information available to the geologist. In cases where biostratigraphic data are available and deposition has been reasonably continuous, it is easy to construct burial-history curves with a high level of confidence. In cases where biostratigraphic data are lacking or where the sediments have had complex tectonic histories, a burial-history curve may represent only a rather uncertain guess. Nevertheless, if constructed as carefully as the data permit, burial-history curves represent our best understanding of the geological history of an area.

TEMPERATURE HISTORY

The next step is to provide a temperature history to accompany our burial-history curve. The subsurface temperature must be specified for every depth throughout the relevant geologic past. The simplest way to do this is to compute the present-day geothermal gradient and assume that both the gradient and surface temperature have remained constant throughout the rock's history. Suppose, for example, that the Tiger well was logged, and that a corrected bottom-hole temperature of 133° C was obtained at 3800 m. Suppose further that local weather records indicate a yearly average surface temperature of 19° C. Using these present-day data and extrapolating them into the past, we can construct the temperature grid shown in figure 9.7.

Where measured bottom-hole temperatures are not available, maps of regional geothermal gradients can be useful in estimating the gradient at a particular location. In many poorly explored areas, temperature profiles will be based largely on guesswork.

There are numerous other variations that can be employed in creating temperature grids. For example, we can change surface temperatures through time without altering the geothermal gradient (fig. 9.8). Causes for such events could include global warming and cooling or local climatic variations resulting from continental drift or elevation changes.

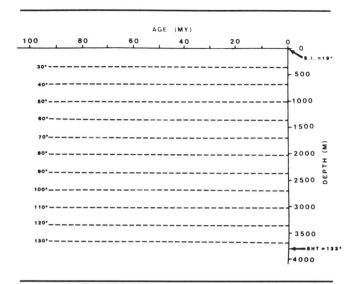

FIGURE 9.7 *Subsurface-temperature grid that assumes a constant surface temperature (19° C) and geothermal gradient (3° C/100 m) during the last 100 my. Isotherms are spaced every 10° C for convenience in calculating maturity.*

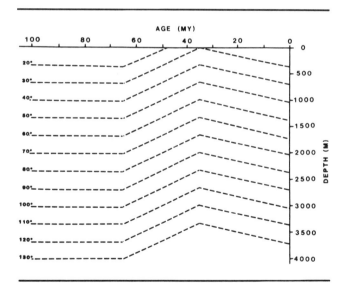

FIGURE 9.8 *Subsurface-temperature grid that assumes a constant geothermal gradient (3° C/100 m) but a variable surface temperature during the last 100 my.*

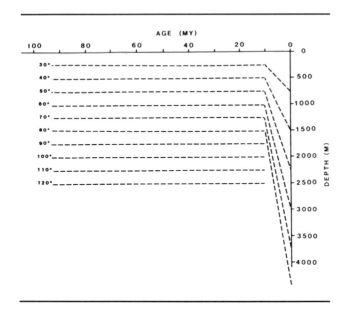

FIGURE 9.9 *Subsurface-temperature grid that assumes a constant surface temperature but a geothermal gradient that dropped dramatically during the last 10 my after being constant for 90 my.*

In other cases the surface temperature remains constant, but the geothermal gradient varies in response to heating or cooling events. The example in figure 9.9 represents a location at which rapid sediment accumulation in the last ten million years has lowered the geothermal gradient, resulting in subsurface temperatures that are anomalously low compared to the "normal" ones that dominated previously. More complicated temperature histories (for example, fig. 9.10) are also possible. "Dogleg" gradients can be used to reflect changes in thermal conductivities caused by variations in lithology (fig. 9.11).

There is no theoretical limit to the complexity that can be introduced into our temperature histories. Given adequate data or an appropriate model on which to base complex temperature reconstructions, we are limited only by our own creativity. In most cases, however, the data necessary for highly sophisticated temperature reconstructions are simply not available.

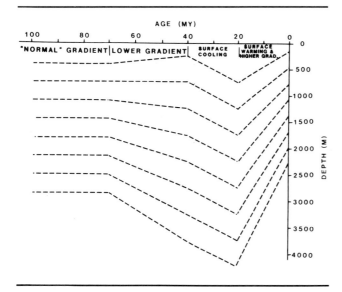

FIGURE 9.10 *Complex subsurface-temperature grid arising from a variety of changes noted at the top of the diagram.*

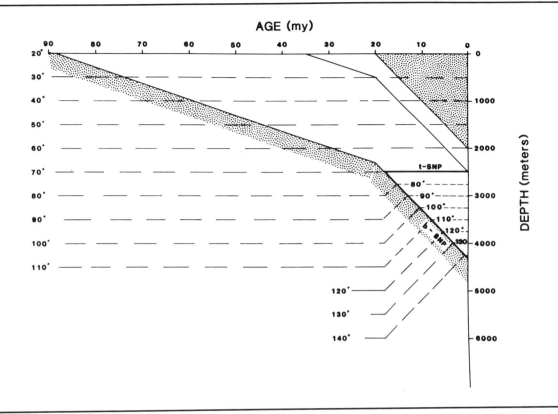

FIGURE 9.11 *"Dog-leg" geothermal gradient that arose as a result of development of overpressuring (SNP) in a massive shale section sandwiched between two sandy sections.*

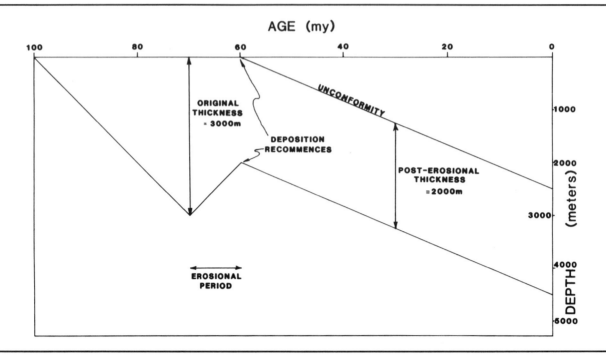

FIGURE 9.12 *Loss of 1000 m of section by erosion during an uplift event lasting from 70 mya to 60 mya. Individual burial-history curves remain parallel, but the distance between the two lines which bracket the erosion decreases by 1000 m.*

SPECIAL CONSIDERATIONS ABOUT BURIAL-HISTORY CURVES

The most common complicating factor in constructing burial-history curves is erosional removal. Erosion is indicated in a burial-history curve by an upward movement of the curve. If deposition resumes later, the burial-history curve again begins to trend downward (fig. 9.12).

Whenever erosional removal occurs, the resultant thinning of the section must be represented in the entire family of burial-history curves. The individual segments of each of the burial-history curves in a family will remain parallel, but the distances between them will be reduced (fig. 9.12).

Faulting can be dealt with by considering the hanging wall and footwall as separate units having distinct burial histories. If part of the section is missing as a result of faulting, burial-history curves for both hanging wall and footwall can be represented on a single diagram (fig. 9.13). If, however, some part of the section is repeated as a result of thrusting (fig. 9.14), two separate diagrams should be used for the sake of clarity (fig. 9.15).

The effects of thrusting on thermal maturity are not well understood. If thrusting is rapid compared to the rate of thermal equilibration between thrust sheets, the movement of hot rocks from the bottom of the overthrusted slab over cool rocks at the top of the underthrusted slab will affect organic maturation by causing important perturbations in subsurface temperatures (Furlong and Edman, 1984). Studies by Edman and Surdam (1984) and Angevine and Turcotte (1983) in the Overthrust Belt of Wyoming indicate that a slow-equilibration model is superior to a simple model invoking rapid thermal equilibration (fig. 9.16). However, more work is required before we will understand fully how thrusting influences hydrocarbon generation and destruction.

Van Hinte (1978) has utilized "geohistory diagrams," which are similar to burial-history curves (fig. 9.17). Geohistory diagrams take sea level as the datum, how-

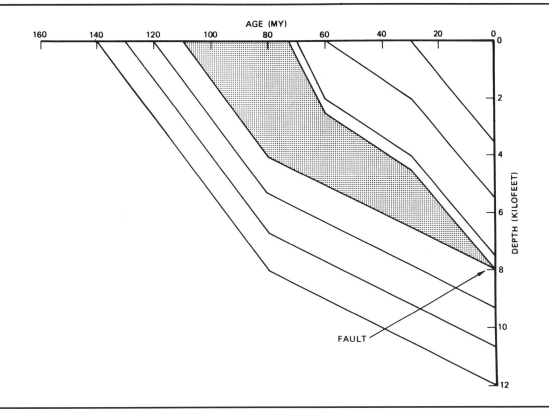

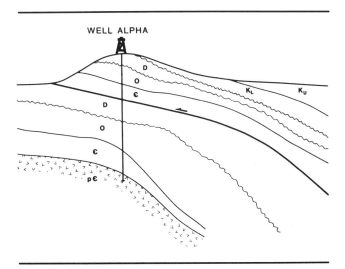

FIGURE 9.13 *Juxtaposition of burial-history curves for hanging wall and footwall. Shaded area represents missing section.*

FIGURE 9.14 *Repeated section in Well Alpha where thrusting has occurred.*

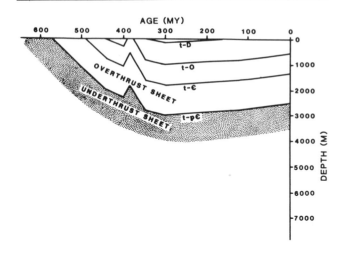

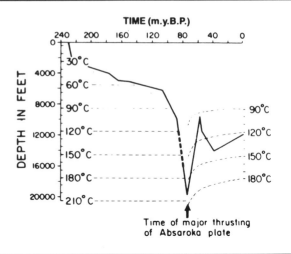

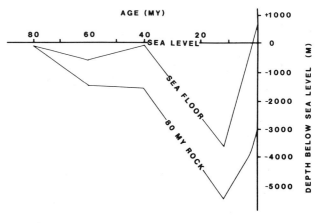

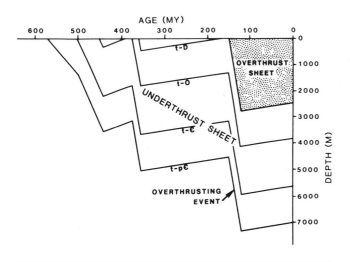

FIGURE 9.15 *Burial-history curves for overthrusted (top) and underthrusted sheets in Well Alpha. For the sake of clarity two diagrams must be used because section is repeated rather than missing as in figure 9.13.*

FIGURE 9.16 *Perturbation in subsurface-temperature grids in hanging wall and footwall of thrust faults. Reprinted by permission of the American Association of Petroleum Geologists from Edman and Surdam, 1984.*

FIGURE 9.17 *Geohistory diagram (burial-history curve using sea level as the datum).*

258

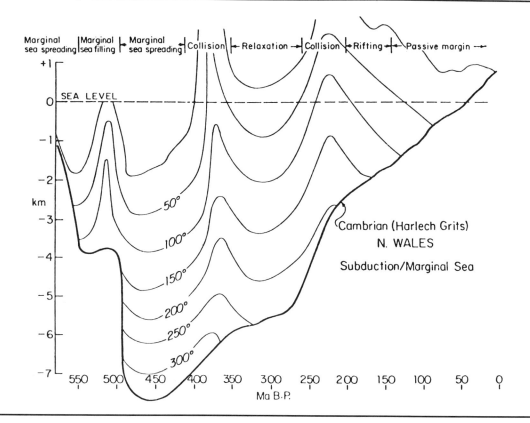

FIGURE 9.18 *Geohistory diagram for north Wales showing the complex pattern of isotherms when uplift occurs without rapid erosion. Reprinted by permission of the Canadian Society of Petroleum Geologists from Siever and Hager, 1981.*

ever, whereas burial-history curves use the sea floor. Geohistory diagrams permit one to see clearly the water depth as it evolves through time, and thus can be combined with models for depositional environments and organic facies. They also allow one to assess relationships between basin subsidence and sediment supply.

Geohistory diagrams are valuable and should be included in the geochemical analysis of basins. They are not as well suited as burial-history curves for assessing thermal maturity, however, because it is more complicated to develop a subsurface temperature grid where the datum is sea level (fig. 9.18). Furthermore, in some cases geohis-

tory diagrams can easily be misinterpreted. For example, in figure 9.17 the indicated rock subsided rapidly between 40 and 10 mya, but despite the steepness of its trajectory, very little overburden was added because of sediment starvation in the basin.

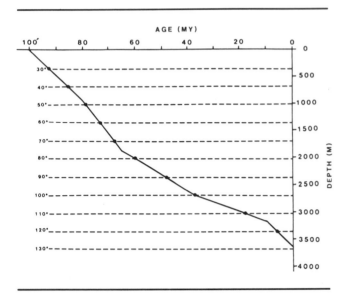

AGE (MY)

TABLE 9.2 *Relation between rock temperature, the index value* n, *and the* γ-*Factor*

Temperature Range (°C)	n	Temperature-Factor (γ)
30–40	−7	$\frac{1}{128}$
40–50	−6	$\frac{1}{64}$
50–60	−5	$\frac{1}{32}$
60–70	−4	$\frac{1}{16}$
70–80	−3	$\frac{1}{8}$
80–90	−2	$\frac{1}{4}$
90–100	−1	$\frac{1}{2}$
100–110	0	1
110–120	1	2
120–130	2	4
130–140	3	8
140–150	4	16
150–160	5	32
160–170	6	64

FIGURE 9.19 *Juxtaposition of the burial-history curve and simplest subsurface-temperature grid for the 100-my-old rock in the Tiger well. Dots mark the beginning and end points of each temperature interval.*

CALCULATION OF MATURITY

Once the burial-history curves and temperature grids have been constructed, we must put them together. Figure 9.19 shows the superposition of the simplest temperature grid from figure 9.7 over the burial-history curve for the oldest rock from the Tiger well (fig. 9.2). Intersections of the burial-history curve with each isotherm are marked with dots. These dots define the time and temperature intervals that we shall use in our calculations. *Temperature intervals* are defined by isotherms spaced 10° C apart. A *Time interval* is the length of time that the rock spent in a particular temperature interval. Total maturity is calculated by summing the incremental maturity added in each succeeding temperature interval.

Now we can carry out the maturity calculations. Chemical reaction-rate theory states that the rate of a reaction occurring at 90° C (a reasonable average for oil generation) and having a pseudo-activation energy of 16,400 cal/mol will approximately double with every 10° C increase in reaction temperature. Lopatin (1971) assumed that the rate of maturation followed this doubling rule. Testing of

his model by Waples (1980) and the successful application of Lopatin's method in numerous published examples have confirmed the general validity of Lopatin's assumption.

In order to carry out maturity calculations conveniently, we need to define both a time factor and a temperature factor for each of the temperature intervals shown in figure 9.19. Lopatin defined each time factor simply as the length of time, expressed in millions of years, spent by the rock in each temperature interval.

The temperature factor, in contrast, increases exponentially with increasing temperature. Lopatin chose the 100°–110° C interval as his base and assigned to it an index value $n = 0$. Index values increase or decrease regularly at higher or lower temperatures, respectively (table 9.2). Because the rate of maturation was assumed to increase by a factor of two for every 10° C rise in temperature, for any temperature interval the temperature factor, which Lopatin called γ, was given by equation (9.2).

$$\gamma = 2^n \qquad (9.2)$$

The γ-factor thus reflects the exponential dependence of maturity on temperature.

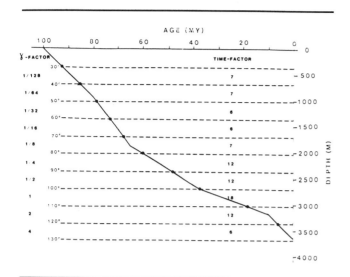

AGE (M.Y)

FIGURE 9.20 *Temperature factors (γ-factors) and time factors for the 100-my-old rock in the Tiger well.*

TABLE 9.3 *Summary of a TTI calculation for the 100-my-old rock in the Tiger Well*

Temp. Interval	Temp. Factor	Time Factor	Interval TTI	Total TTI	Time (m.y.BP)
30°–40°C	1/128	7	.05	.05	86
40°–50°	1/64	7	.11	.16	79
50°–60°	1/32	6	.19	.35	73
60°–70°	1/16	6	.38	.73	67
70°–80°	1/8	7	.88	1.6	60
80°–90°	1/4	12	3.0	4.6	48
90°–100°	1/2	12	6.0	10.6	36
100°–110°	1	18	18.0	28.6	18
110°–120°	2	12	24.0	52.6	6
120°–130°	4	6	24.0	76.6	0

Multiplying the time factor for any temperature interval by the appropriate γ-factor for that temperature interval gives a product called the Time–Temperature Index of maturity (TTI). This interval-TTI value represents the maturity acquired by the rock in that temperature interval. To obtain total maturity, we simply sum all the interval-TTI values for the rock. Maturity always increases; it can never go backward because interval-TTI values are never negative. Furthermore, even if a rock cools down, maturity continues to increase (albeit at a slower rate) because γ is always greater than zero.

A good analogy can be drawn between oil generation and baking. If we put a cake in a cold oven and turn the oven on, the cake will bake slowly at first but will bake faster and faster as the temperature rises. If we turn off the oven but leave the cake inside, baking will continue, although at increasingly slower rates, as the oven cools down. On the other hand, if we forget about the cake when the oven is hot and let it burn, we cannot "unburn" it, no matter how much or how rapidly we cool it down.

The first step in calculating TTI is illustrated in figure 9.20, where the time factors and γ-factors for each temperature interval are shown on the burial-history curve.

In table 9.3 interval-TTI values and total-TTI values up to the present day are calculated.

It is also possible to determine the total-TTI value at any time in the past simply by stopping the calculation at that time. For example, if we wanted to know the TTI for the 100-my-old rock in the Tiger well during the early Oligocene (36 mya), we would add up the interval-TTI values for the first 64 my of the rock's history. From the calculation in table 9.3 we can see that 36 mya the total TTI was 10.6.

FACTORS AFFECTING THERMAL MATURITY

Because maturity is affected by both baking time and baking temperature, the specific burial history of a rock can strongly affect its maturity. Figure 9.21 shows four of the many paths by which an 80-my-old rock could have reached a present burial depth of 3000 m. In figure 9.21A the rock was buried at a constant rate for its entire 80-my history. In figure 9.21B burial was very slow during the first 70 my of the rock's existence, but quite rapid in the last 10 my. Figure 9.21C shows rapid burial during the first 20 my, followed by a nonerosional depositional hiatus for the last 60 my. In Figure 9.21D 40 my of rapid burial to a depth of 4000 m was followed by a hiatus lasting 30 my and, finally, by 10 my of uplift and erosion.

TTI values differ appreciably among these four scenarios. Maturities for the rocks in figure 9.21 are di-

261

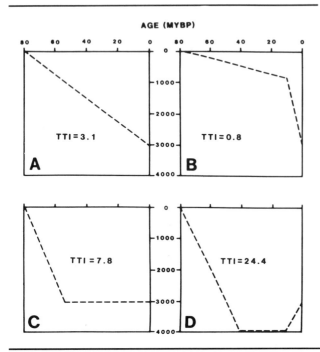

FIGURE 9.21 *Four possible burial-history curves for a rock of 80-my age that is presently buried to a depth of 3000 m. TTI values were calculated assuming a constant geothermal gradient of 2° C/100 m and an average surface temperature of 19° C.*

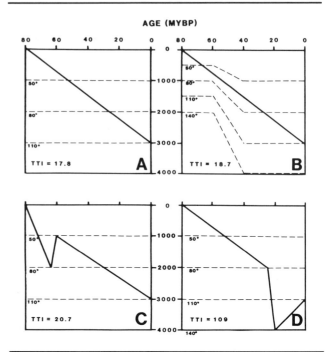

FIGURE 9.22 *Important thermal events that occur early in a rock's history do not affect its maturity very much, as shown by three possible combinations of varying burial and geothermal history (A, B, and C). In contrast, the maturity in model (D) is much higher because the thermal event (extra burial) occurred when the rock was near its maximum paleotemperature.*

rectly related to the amount of time spent near the maximum burial depth and temperature. The calculated maturity in figure 9.21D is by far the highest, because the temperature reached and maintained at 4000 m was much higher than that achieved in the examples where no erosion occurred.

However, when important events occur early in the history of a rock, their impact on maturity is much less. For example, figure 9.22 illustrates two events that will increase maturity: a higher geothermal gradient (fig. 9.22B) and extra burial preceding later uplift and erosion (fig. 9.22C). However, the higher geothermal gradient was only in force early in the rock's history when its temperature remained low because of shallow burial. Thus the TTI in figure 9.22B is only slightly higher than in the control example (fig. 9.22A). One implication of this exercise is that source-rock maturity in rifted basins,

where high geothermal gradients are mainly present in the early stages of basin evolution, will not be unusually high unless the source rock was deposited during the very early rifting stage. Cohen (1985) has cited such an example in the Recôncavo Basin of Brazil.

Similarly, the additional burial that occurred in figure 9.22C between 70 and 60 mya has not greatly affected present-day maturity because subsequent burial to much greater depths and higher temperatures has swamped out the slight increase in maturity resulting from that early event. A brief residence near 80° C obviously is not very important compared to a longer time spent near 110° C. Figure 9.22D shows, however, that burial to depths and temperatures greater than present-day conditions, followed by erosion, will affect the maturity very strongly (compare TTI of 109 in fig. 9.22D with 17.8 in fig. 9.22A). Changes in the geothermal gradient when a rock is near its maximum depth of burial will also influence maturity strongly.

The effect of igneous intrusions is difficult to treat accurately with maturity models because our present models have not been calibrated for such high temperatures. Furthermore, there is substantial uncertainty about the temperatures actually achieved in the sedimentary rocks and the rate at which thermal anomalies decay. Nevertheless, the effects of igneous events can be detected by finding discrepancies between measured and predicted maturity values. Kettel (1983), for example, deduced the existence and shape of an ancient heat anomaly in the southern North Sea by showing that vitrinite-reflectance values were significantly higher than normal subsurface heating could account for.

Bond (1984) has used Lopatin's method in a very sophisticated manner to model maturation in the San Juan Basin of New Mexico. The maturity history there has been complicated greatly by an intense thermal event during the Neogene and by many thousands of feet of uplift and removal. Nevertheless, Bond was able to develop a satisfactory model that fit the measured data.

INTERPRETATION OF TTI VALUES

Uncalibrated TTI values obviously are of little value; to be useful they must be compared in some way with measured maturity values. Lopatin's (1971) original calibration was shown to be in error by Waples (1980), who proposed a revised version of the TTI-R_o correlation (table 9.4). Subsequent work has not strongly questioned the TTI calibration within the oil-generation window, but Katz et al.(1982) showed that beyond the oil-generation window Waples's correlation is probably incorrect.

Other complicating factors have arisen. Recent investigations that have improved our understanding of the oil-generation window indicate that Waples's choice of $R_o = 0.65\%$ as the threshold for oil generation is almost certainly too high. Moreover, different kerogen types have different oil-generation thresholds (see the section *Potential Problems with Maturity Calculations*). These various problems must be dealt with before we can construct a correlation between TTI and oil generation. Table 9.5 shows a proposed correlation between TTI and hydrocar-

TABLE 9.4 *Correlation of TTI values with vitrinite reflectance*

% Vitrinite Reflectance (R_o)	TTI
0.40	<1
0.50	3
0.60	10
0.65	15
0.70	20
0.85	40
0.90	50
1.00	75
1.15	110
1.35	180
1.50	300
1.75	500
2.00	900
2.50	2,700
3.00	6,000
4.00	23,000
5.00	85,000

NOTE: Reprinted, by permission, from D. W. Waples, 1980, Time and temperature in petroleum formation: application of Lopatin's method to petroleum exploration: Bulletin of the American Association of Petroleum Geologists, v. 64, pp 916–926, table 4.

TABLE 9.5 *Correlation of selected TTI values with vitrinite-reflectance values and hydrocarbon generation*

TTI	R_o	Generation
1	0.40	Condensate from resinite
3	0.50	From S-rich kerogen
10	0.60	early
15	0.65	
20	0.70	oil
50	0.90	peak
75	1.00	
180	1.35	late
900	2.00	wet gas
		dry gas

263

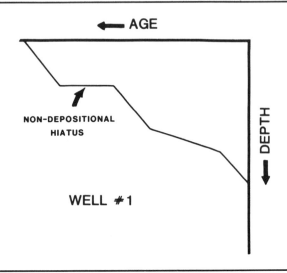

FIGURE 9.23 *Initial proposed burial-history model for Well #1. The model includes an extensive nonerosional depositional hiatus.*

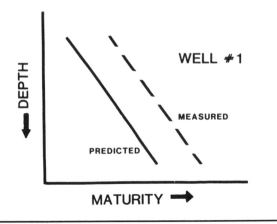

FIGURE 9.24 *Poor correlation between measured and predicted maturity data for Well #1.*

bon generation based on our best present understanding of catagenesis and hydrocarbon formation. The onset of oil generation is shown to vary from about TTI = 1 for resinite to TTI = 3 for high-sulfur kerogens to TTI = 10 for other Type II kerogens, to TTI = 15 for Type III kerogens.

Where both TTI values and measured maturity data are available, the two should always be compared. Such comparisons can often teach us important, hitherto undiscovered, facts about the geology of the area under consideration. For example, suppose that for Well #1 we have constructed the burial-history curve shown in figure 9.23. This proposed burial history includes a depositional hiatus but no erosional removal. When we compare our calculated maturities with measured ones obtained from well samples however, the correlation (fig. 9.24) is rather poor, with calculated maturity values being consistently lower than the measured ones. Assuming that the measured data are reliable, how can we explain the consistent underestimation of maturity in our calculations?

In general, there are two possible explanations if calculated maturities do not agree with reputable measured ones. One explanation is that the paleotemperatures used

in the model were incorrect. This situation can arise easily when our present-day temperature data are poor or when past events (such as igneous activity) created a thermal regime quite different from the present one. A second explanation, which is the one we shall adopt in the present example, is that we have made a poor estimate of the amount of erosional removal. In the case of Well #1, we postulated a nonerosional hiatus. Because our calculated maturity values were too low, it seems likely that substantial removal occurred during the time of the unconformity. The measured maturity level was therefore higher than our model predicted. On the basis of these data we can revise our geological model and include enough erosional removal (fig. 9.25) to bring measured and calculated maturities into agreement.

Because the effects of underestimating paleotemperature are about the same as the effects of underestimating erosion, one might think that it would be difficult to decide which was the better explanation for discrepant data. In practice, however, it is usually easy to distinguish between the two possibilities on the basis of local geology.

For example, Magoon and Claypool (1983) chose among three possible thermal histories by comparing predicted and measured maturities for the Inigok-1 well on the Alaskan North Slope. The thermal history finally chosen (fig. 9.26) was rather complex but justified on the

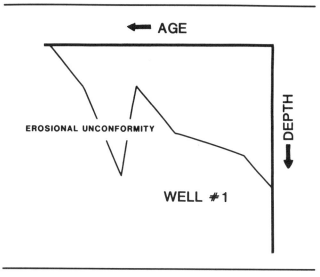

FIGURE 9.25 *Revised burial-history model for Well #1 based on the poor correlation with measured maturity data. The hiatus has been reinterpreted as an erosional unconformity in which a substantial amount of section was removed.*

basis of specific events on the North Slope that affected geothermal gradients.

One important question is "How well should we expect measured and calculated maturities to agree?" In the C.O.S.T. #1 well, drilled off the Texas Gulf Coast, samples were taken and analyzed carefully, and time-stratigraphic and temperature data were determined with a high level of confidence. There is nevertheless a discrepancy of about 1000 ft at both the top and bottom of the oil-generation window between measured and calculated maturities (fig. 9.27). Why is the fit not better?

The vitrinite-reflectance values obviously form a very nice trend, and thus appear at first to be high-quality data. However, closer examination of the data indicates that vitrinite was actually rather sparse in the well; no more than 31 particles were analyzed in any sample, and many samples contained fewer than 10 vitrinite particles. We should therefore be cautious in ascribing the misfit between measured and predicted maturities solely to deficiencies in our thermal model. In fact, in many other cases we easily obtain a near-perfect fit.

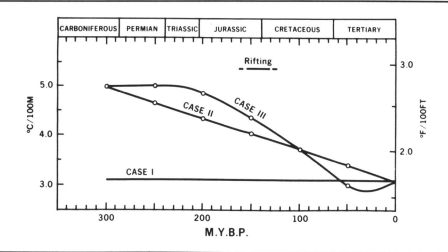

FIGURE 9.26 *Three possible thermal models for the Inigok-1 well, Alaskan North Slope. Case I represents a constant geothermal gradient; Case II shows a gradual decrease in gradient since the Carboniferous; and Case III, which gave the best fit between measured and predicted maturities, is approximately a combination of I and II. From Magoon and Claypool (1983).*

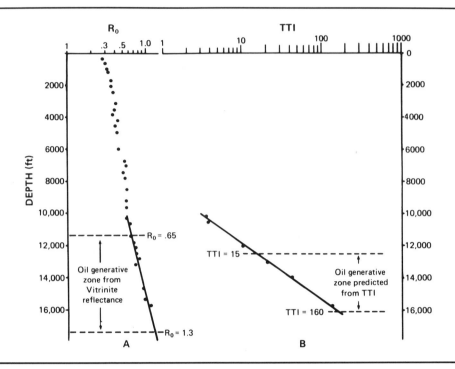

FIGURE 9.27 *Comparison of measured vitrinite-reflectance data with TTI values, and the oil-generation windows delineated by each data set for the C.O.S.T. #1 well, Texas offshore Gulf Coast.*

On the basis of a fair amount of experience, I consider about 500 ft as the intrinsic accuracy of Lopatin's method. If we accept such a limitation, then where reliable measured and calculated maturities do not agree within 500 ft, the original model should be modified to bring measured and calculated maturities within the acceptable limits.

Skeptics might argue that an uncertainty of 500 ft is unacceptably large, and maturity modeling is therefore of little value. The obvious answer to this objection is that we must use maturity modeling only where an error of 500 ft would be acceptable. For most applications such an uncertainty is not a problem. After all, we live happily with uncertainties about whether oil generation commences at vitrinite-reflectance values of 0.5%, 0.6%, or even 0.7%. We even blissfully ignore the uncertainty of whether vitrinite reflectance is a valid measure of oil generation at all. Our doubts about thermal modeling are really no more serious.

APPLICATIONS TO HYDROCARBON PRESERVATION

The preservation deadline for oil has been the subject of investigation for many years. In 1915 David White proposed his famous "Carbon-Ratio Theory," which showed that the oil deadline correlates with coal rank. It follows that if we can predict coal rank (itself a measure of thermal maturity), we can predict the oil deadline. The potential of thermal modeling for this purpose is obvious.

Waples (1980) suggested the applicability of Lopatin's method in defining hydrocarbon deadlines and proposed TTI values for several deadlines. Further empirical work suggests that a slight modification of those original estimates is appropriate (table 9.6), but even these new values are still somewhat uncertain. The deadlines proposed in table 9.6 are expressed at the 80% confidence

TABLE 9.6 *Hydrocarbon deadlines (80% confidence level) correlated with vitrinite-reflectance and TTI values*

TTI	R_o	Expected Hydrocarbons (80% Probability)
50	0.90	normal oil
75	1.00	normal-light oil
180	1.35	condensate-wet gas
500	1.75	wet gas
900	2.00	dry gas

level; that is, at higher TTI values heavier hydrocarbons could occur, but these occurrences would be rather rare.

Despite the apparent success of Lopatin's method in predicting oil deadlines, at least one important problem remains. The kinetic parameters used by Lopatin were defined in order to best describe the overall process of oil generation, which comprises many distinct chemical reactions. Many of the cleavage reactions in oil generation involve carbon-heteroatom bonds, which have low activation energies compared to breaking carbon–carbon bonds.

Cracking of crude oil, in contrast, includes a more homogeneous set of reactions, most of which involve cleavage of carbon–carbon bonds. One would therefore expect the pseudo-activation energy for oil destruction to be higher than that for oil generation. Furthermore, the temperatures at which extensive cracking occurs are probably about 20° to 30° C higher than typical temperatures of oil generation. The difference in kinetic parameters for generation and destruction of oil requires that we use different γ-factors for the two processes.

The best solution to uncertainties about the appropriateness of any of our thermal-maturity models for cracking would be to select a new γ-factor on the basis of careful empirical studies of known hydrocarbon occurrences. Unfortunately, no such study has been published. Until such research is carried out, the best we can do with Lopatin's method is to utilize the same kinetic parameters as for oil generation. We can take comfort in the fact that, despite this important oversimplification, TTI values appear to correlate well with observed hydrocarbon deadlines.

Predicting the gas deadline is far more difficult because destruction of methane is an oxidative process rather than cracking. The rate of methane oxidation is therefore dependent on the availability of suitable oxidizing agents as well as on temperature. Oxidizing agents are apparently rather rare in sands, but in carbonates there are often substantial amounts of sulfate and elemental sulfur present. Barker and Kemp (1982) have shown that methane in the presence of anhydrite is unstable at high temperatures. Furthermore, the economic dry-gas deadline is often controlled more by reservoir quality than by methane stability. For these reasons there is little chance that maturity calculations will be of value in locating the dry-gas deadline.

APPLICATIONS TO EXPLORATION

MATURITY THROUGH TIME

The earliest application of Lopatin's method to exploration was published by Zieglar and Spotts (1978) in an analysis of the hydrocarbon production and future potential of the Central Valley of California. Using the as-yet-unpublished approach of Waples (1980), they drew iso-TTI lines on burial-history curves to show the development of maturity through time and to compare the thermal histories of various depocenters within the basin (fig. 9.28).

The simplest way to put iso-TTI lines on burial-history curves is to begin by selecting the TTI values of interest. Useful TTI values might include those for the onset and end of oil generation or those for the liquid deadline. Then starting with the deepest horizon in the burial-history-curve family and beginning with deposition of that sediment and working toward the present, one calculates interval-TTI values and sums them until the desired TTI value is reached. The point at which the desired TTI was reached is marked on the burial-history curve. The process is repeated for each member of the burial-history-curve family. Then all the points having the same TTI value are connected with a line. If TTI values corresponding to the onset and end of oil generation are plotted, the resulting maturity band (fig. 9.29) defines the oil-generation window through time.

Figures like 9.29 have proven extremely useful, particularly in understanding the timing of hydrocarbon genera-

267

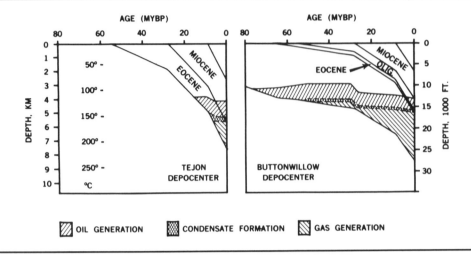

AGE (MYBP)

TEJON DEPOCENTER

BUTTONWILLOW DEPOCENTER

OIL GENERATION   CONDENSATE FORMATION   GAS GENERATION

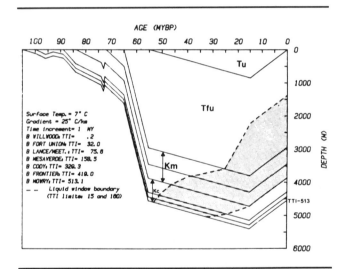

FIGURE 9.29 *Family of burial-history curves for a well in the Big Horn Basin, Wyoming, showing the evolution of the oil-generation window through time. Tu = undifferentiated Tertiary; Tfu = Fort Union Formation; Km = Lance-Meeteetse formations; Kc = Cody-Frontier formations. Reprinted by permission of the Rocky Mountain Association of Geologists from Hagen and Surdam, 1984.*

FIGURE 9.28 *Hydrocarbon-generation histories in the deepest parts of the Tejon and Buttonwillow depocenters, southern Great Valley, California. Reprinted by permission of the American Association of Petroleum Geologists from Zieglar and Spotts, 1978.*

tion. Their only weakness is that because each diagram represents a single place on the earth, a series of such figures must be used if an area having a variety of maturity histories is being considered.

CROSS SECTION AND PLAN VIEW

Cross sections with isomaturity lines on them have been used to give a more regional picture of maturation. Figure 9.30 shows isomaturity lines superimposed on cross sections across the Otway Basin, southeastern Australia, at four times from 60 mya to the present. The gradual movement of maturity upward through the stratigraphic sequence is clear. In such diagrams the possibilities for migration of hydrocarbons at various times can be examined and correlated with structural events that regulate migration pathways. TTI values can, of course, be contoured instead of reflectance values.

268

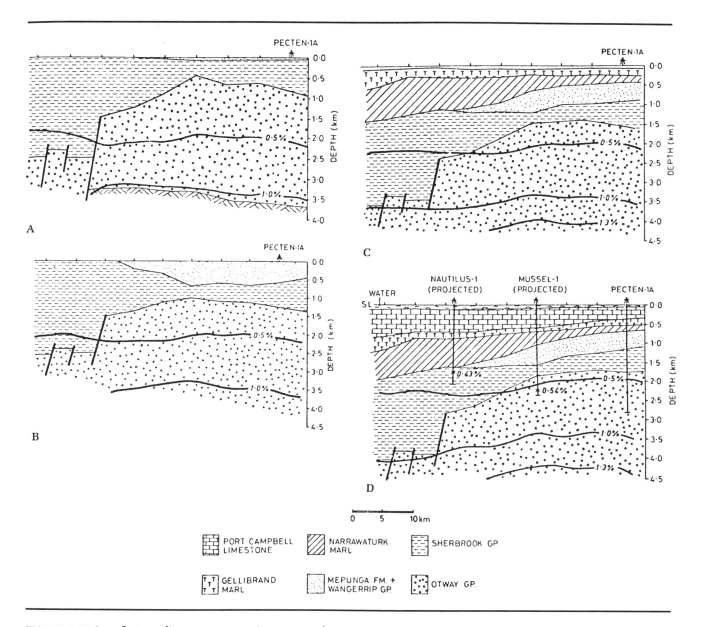

FIGURE 9.30 *Isoreflectance lines on cross sections across the Otway Basin of southeastern Australia calculated at four different times: 60 mya (A), 40 mya (B), 20 mya (C), and present-day (D). Reprinted by permission of the American Association of Petroleum Geologists from Middleton and Falvey, 1983.*

DEADLINES

One of the potentially most important applications of thermal modeling to exploration is in predicting the oil deadline. The economics of oil and gas exploration are often very different; indeed, gas is of no value at all in some remote regions. Knowledge of oil deadlines (limits below which economic accumulations are unlikely to occur) is therefore very important in exploration.

Preservation deadlines can be easily shown on burial-history curves, in plan view, or on cross sections, in the same manner as hydrocarbon-generation limits. Even a very preliminary estimate of the liquid-hydrocarbon deadline can be extremely valuable in formulating a drilling program.

COMPUTERIZATION

Virtually all the calculations and plots discussed thus far can be carried out by a computer. Computerization greatly speeds up the maturity-calculation process and makes it possible to modify the input data quickly and easily. Numerous companies have developed their own maturity-calculation programs, and some consulting companies offer them for sale. Some of the programs are more sophisticated than the basic program we have described, and can include such features as decompaction. Others are less flexible than hand calculations because they limit the number of uplift events or changes in subsurface temperatures.

The main advantages of a computerized system are its speed in calculating and plotting results and its ease in making revisions in the geological or geothermal model. These are important conveniences that make thermal modeling more palatable to many geologists.

There are, however, some disadvantages to carrying out maturity calculations entirely by computer. One potential problem is that the technology can easily become a "black box" to a geologist. If a geologist simply is required to put in age, depth, and temperature data, he or she may not understand at all how these data affect the calculations. This ignorance will lead to a lack of appreciation of the factors affecting maturation.

The second problem is more severe. One of the main benefits of carrying out maturity calculations using Lopatin's method lies in the construction of the burial-history curves themselves. These curves allow a geologist to bring the concept of time into geology in an effective and unusual way. Frequently, in fact, the burial-history curves themselves prove to be more interesting than the TTI values. When maturity calculations are carried out by a computer, the burial-history curves are often deemphasized compared to TTI values. Furthermore, when the computer constructs a burial-history curve, the geologist is not as involved and will not learn as much as if he or she had constructed it by hand.

Computerization is desirable but, in my view, should only be used after a geologist has mastered drawing the burial-history curves and calculating and plotting maturity by hand. Thus prepared, the geologist will be able to derive maximum benefit from data created by the computer.

COMPARISON OF SEVERAL MATURITY MODELS

Those wishing to carry out maturity calculations will have at their disposal several alternative models from which to choose. Lopatin's method is probably the most widely used at the present time, but also available are several other published models, as well as proprietary, unpublished versions developed within individual companies.

The most readily accessible alternative model for most geologists is the LOM (Level of Organic Metamorphism) method developed by Shell (Hood et al., 1975). LOM values are calculated by considering only the time that the rock spent within 15° C of its maximum paleotemperature. These time and temperature values are then used to compute LOM from the nomograph shown in figure 9.31. For example, a rock whose maximum paleotemperature was 120° C, and which spent 40 my between 105° and 120° C, would have reached LOM = 10 at the present day. The nomograph also shows the pseudo-activation energy required to reach each LOM value (in this case 21,000 cal/mol). LOM values have been calibrated to various measured maturity parameters (fig. 9.32). A LOM value of 10 corresponds to R_o = 0.8%, near peak oil generation.

LOM has an advantage over Lopatin's method in that calculations are simpler, since one need not reconstruct in detail the complete burial history of a rock. LOM therefore oversimplifies the influence of temperature on maturity. In many cases this assumption is quite acceptable, but in other cases it is not.

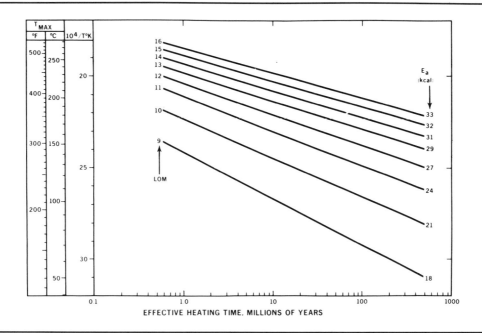

EFFECTIVE HEATING TIME, MILLIONS OF YEARS

FIGURE 9.31 *Nomograph used to calculate LOM by knowing the maximum paleotemperature of a rock and its "effective heating time," the length of time the rock spent within 15° C of that temperature. One finds the effective heating time on the x-axis and the maximum paleotemperature on the y-axis. The LOM value and pseudo-activation energy are found at those coordinates. Reprinted by permission of the American Association of Petroleum Geologists from Hood et al., 1975.*

Probably the greatest weakness, however, is that a geologist will miss out on many of the additional benefits of using Lopatin's method. The LOM method does not allow one to discuss timing of generation in any convenient way. Furthermore, LOM does not require construction of a burial-history curve, itself a valuable learning device. Despite these shortcomings, the LOM method has achieved moderate popularity and appears to be satisfactory.

The earliest published model for calculating the extent of hydrocarbon generation is that developed at the French Petroleum Institute (Tissot, 1969; Tissot and Espitalié, 1975). The Tissot-Espitalié model is highly mathematical,

fully computerized, and rather formidable. Application thus far has been mainly by their French colleagues.

The advantages of the Tissot-Espitalié model are that, because of its sophisticated mathematical foundation, it may be the most accurate in assessing maturity. Furthermore, because the model assumes that the kinetics of oil generation and destruction are different, in describing oil destruction it is almost certainly superior to Lopatin's. The Tissot-Espitalié model also allows timing of generation to be represented, although in a slightly different format than Lopatin's model (fig. 9.33).

One of the model's main weaknesses is that geological input is minimal. A geologist using the program is therefore unlikely to derive the benefits that come from constructing burial-history curves. A second problem is that the model appears to underestimate maturity. Much higher temperatures and longer geologic times are required for oil generation than in other models. Maturities calculated by the Tissot-Espitalié model in the Michigan Basin, for example, are much lower than those predicted by other models (Nunn et al., 1984).

Advocates of the Tissot-Espitalié model have been

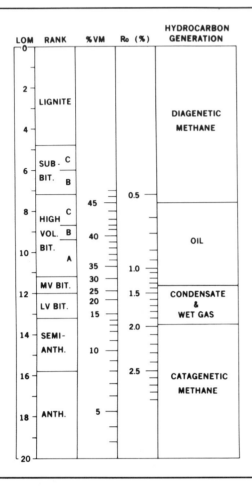

FIGURE 9.32 *Correlation of LOM with various coal-rank parameters. Reprinted by permission of the American Association of Petroleum Geologists from Hood et al., 1975.*

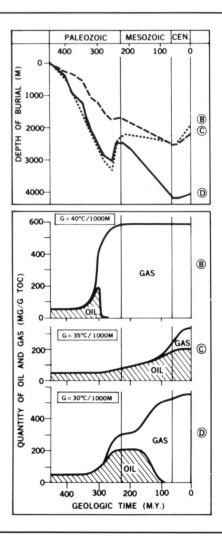

FIGURE 9.33 *Burial-history curves and hydrocarbon generation at three locations in the Illizi Basin, Algeria, predicted using the model of Tissot (1969). Reprinted by permission of the World Petroleum Congresses from the Proceedings of the Ninth World Petroleum Congress, Applied Science Publishers, from Tissot et al., 1975.*

vocal in their criticism of Lopatin's method. Their main complaint is that Lopatin's assumption of a doubling of reaction rate for each 10° C increase in temperature is unrealistic, since the actual activation energies are on the order of 50,000 cal/mol rather than 15,000. This argument would be valid and telling if hydrocarbon generation were a homogeneous reaction, but the analysis of Jüntgen and Klein discussed earlier (fig. 9.1) refutes their claim. In fact, the Tissot-Espitalié model is very similar to Lopatin's, because Lopatin's model uses a single kinetic

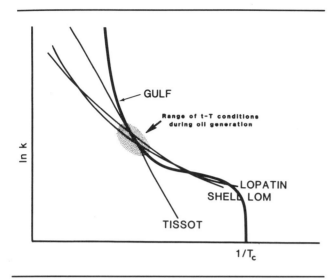

FIGURE 9.34 *Arrhenius plot showing why the various thermal models all seem to work adequately despite their fundamental differences. Because the range of times and temperatures we normally observe for oil generation in nature is narrow, the available data are tightly clustered. Consequently, it is possible to draw many very different lines (representing the various models) through the data. Only if we obtain new data that increase our range of time and temperature values will we be able to choose among these models.*

parameter (equivalent to curve Σ in fig. 9.1), whereas the Tissot-Espitalié model inputs each of the individual members of the family (curves 1–8 in fig. 9.1).

The most unusual model published recently was developed at Gulf (Toth et al., 1983). Their work indicated that changes in vitrinite reflectance in the North Sea were best explained by a pseudo-activation energy of 50 cal/mol, a value far too small for a bond-breaking chemical reaction. This result suggests that increases in vitrinite reflectance may be related to changes in the relative positions of portions of the kerogen molecule rather than to chemical reactions. Despite its oddness, the Gulf model has been used with apparent success by its developers.

Most privately developed models with which I am familiar are similar to Lopatin's method. Their main differences involve the particular ways in which they use time and temperature in the maturation process. At the

present time it is not possible to select one of the models (Lopatin, LOM, Tissot-Espitalié, Gulf, unpublished) as the best, either from a theoretical point of view or on the basis of empirical data. The process of hydrocarbon generation is so complex and chemically variable that any model will only be an approximate description of the actual system.

All of the models mentioned worked adequately for their developers within the data sets used to develop and test those models. We may therefore assume that they also will work well in the future for other data sets. The reason that the various models, which differ substantially among themselves in their descriptions of the kinetics of hydrocarbon generation, can all work satisfactorily is that the range of times and temperatures that occur in natural settings is quite small: perhaps 70° C (from 70° to 140° C) and generally from 5 to 50 million years. In determining kinetic parameters in the laboratory, a chemist normally tries to observe reaction rates over a much wider range of conditions in order to distinguish as carefully as possible between the contributions of the time and temperature variables.

The small ranges of time and temperature data available from oil-generation studies mean that measured data points cluster in a small area on an Arrhenius plot, which is a typical device for determining reaction kinetics (fig. 9.34). Given the natural variations and experimental uncertainties always associated with geological samples, the data do not lend themselves to a unique interpretation. There are in fact many ways that we can draw a straight line through our data. Figure 9.34 shows schematically how the various laboratories have selected a variety of interpretations. Until we obtain data giving us a broader range of times and temperatures, we shall be unable to select a "best" model among those proposed.

The implications of this dilemma are fortunate for explorationists. Basically, whichever model one is using can be defended on the basis of both theory and experimental data, and therefore is acceptable. There is at the present time no superior model; one should select a model on the basis of availability, convenience, and ancillary features (permitting discussion of timing, degree of integration with geology, visual output, etc.). Perhaps at some time in the future, one model will emerge as superior to the others on the basis of its ability to explain measured data, but for now the best model for each geologist is that which is most practical.

POTENTIAL PROBLEMS WITH MATURITY CALCULATIONS

The most obvious errors in maturity calculations will come from inaccuracies in time and temperature data. In actuality, time data are seldom a problem. First, the dependence of maturity on time is linear, so even a rather large error in baking time will not produce a catastrophic change in maturity. Secondly, we usually have excellent control on rock ages through micropaleontology. Age calls are often made within a million years, and can be even better in Cenozoic rocks. Only in cases where micropaleontological dating was not or could not be carried out might we anticipate possible problems with time.

Temperature, in contrast, is the single most important cause of uncertainty and error in maturity calculations. The sensitivity of maturity to temperature is clearly indicated by the exponential dependence of maturity on temperature predicted by the Arrhenius equation.

Furthermore, our uncertainties about the true values of subsurface temperatures are much greater than about time. Present-day subsurface temperatures are difficult to measure accurately. Most logged temperatures are too low and require correction. Various methods have been developed for this purpose (see Yükler and Kokesh, 1984 for a brief discussion), but there is no guarantee of their accuracy in any particular case.

Even if we could measure present-day subsurface temperatures with perfect accuracy, however, we still would have to extrapolate the present somehow into the past. In many cases, where present-day temperatures are maximum paleotemperatures, even an inaccurate extrapolation into the past may not cause significant problems. In other cases, however, particularly where Paleozoic rocks are involved, an accurate interpretation of the ancient geothermal history may be critical. In such cases we should be very careful about using predicted maturities unless we have some independent confirmation of the validity of our model from a comparison with measured maturity data.

A question of some concern comes from the previously mentioned fact that most of the maturity models treat all types of kerogen identically. Despite experimental evidence indicating that different kerogens decompose to yield hydrocarbons at different levels of maturity (see chapter 4), Lopatin's model, the LOM model, and the unpublished models with which I am familiar do not utilize different kinetic parameters for the various kerogen types. The only published model that does consider different reaction kinetics for the various kerogen types is that of Tissot and Espitalié. This modification was possible because of the algebraic nature of the Tissot-Espitalié model, but it has not yet been adapted to graphical approaches like Lopatin's.

The Tissot-Espitalié model, therefore, has a clear theoretical advantage over the others in considering oil generation from different types of kerogen. However, the practical implications of this theoretical advantage are relatively modest, for two reasons. First, most oil is probably generated from Type II macerals. Type I kerogens are very rare. Type III kerogen is common, but because the woody and cellulosic components themselves generate little oil, we need not worry much about them. (Most oil derived from Type III kerogens probably comes from Type II material within the Type III matrix.) The most important distinction, therefore, would be among the kinetics of oil generation from the various Type II macerals (resinite, exinite, cutinite, etc.). The Tissot-Espitalié model does not make this distinction, and therefore fails to make full use of its theoretical advantage.

Secondly, the other models are, in a crude but probably satisfactory fashion, able to take different kerogen types into consideration simply by adjusting the thresholds for generation from different kerogen types (table 9.5). For example, liptinite might begin to be strongly converted to bitumen at TTI = 10, whereas an equal intensity of generation from resinite might be reached at TTI = 1. This approach distorts somewhat the theoretical foundation of our chemical–kinetic approach, but it is probably acceptable in a practical sense, given the other uncertainties under which we labor. Thus, the problem of dealing with different kerogen types can be addressed by all the models, albeit in different ways.

CONCLUSIONS

Models for predicting thermal maturity have been developed to aid in understanding the hydrocarbon-generation and -preservation histories of sedimentary basins. Applications include defining the hydrocarbon-generation window, determining timing of generation, and defining deadlines for liquid hydrocarbons. Calibration and much testing of these models have shown them to be reliable and accurate enough for routine use in exploration programs, both in frontier and maturely explored areas.

Some of the models employed are also capable of aiding the geologist in reconstructing the history of an area. By comparing calculated maturity levels with measured values, one can discover erroneous assumptions about geologic or geothermal histories. Construction of burial-history curves is usually enlightening.

The numerous published and unpublished maturity models all appear to give satisfactory correlations with measured data, despite substantial differences in the ways in which they interchange time and temperature. We cannot yet say with certainty that one of the models is mathematically superior to the others. Choice of a suitable model at the present time is best made on the basis of availability, convenience, and integrability with geology. Lopatin's methodology, with a variety of relatively minor variations in the time–temperature interrelationship, is the most widely used technique today.

Although maturity calculations are often carried out by hand, the utility of the models can be increased by using computers to perform calculations and plot the results. One must guard against the computerized version becoming a "black box" technique, however. One can only take full advantage of maturity-modeling technology by maintaining a sound geological foundation for the model. Output data will only be as good as the geological model responsible for those data.

SUGGESTED READINGS

Bond, W. A., 1984, Application of Lopatin's method to determine burial history, evolution of the geothermal gradient, and timing of hydrocarbon generation in Cretaceous source rocks in the San Juan Basin, northwestern New Mexico and southwestern Colorado, in J. Woodward, F. F. Meissner, and J. L. Clayton, eds., *Hydrocarbon Source Rocks of the Greater Rocky Mountain Region:* Denver, Rocky Mountain Association of Geologists, pp. 433–447.

de Bremaecker, J.-Cl., 1983, Temperature, subsidence, and hydrocarbon maturation in extensional basins: a finite element

model: Bulletin of the American Association of Petroleum Geologists, v. 67, pp. 1410–1414.

Edman, J. D. and R. C. Surdam, 1984, Influence of overthrusting on maturation of hydrocarbons in Phosphoria Formation, Wyoming–Idaho–Utah Overthrust Belt: Bulletin of the American Association of Petroleum Geologists, v. 68, pp. 1803–1817.

Snowdon, L. R., 1979, Errors in extrapolation of experimental kinetic parameters to organic geochemical systems: Bulletin of the American Association of Petroleum Geologists, v. 63, pp. 1128–1138.

van Hinte, J. E., 1978, Geohistory analysis—application of micropalentology in exploration geology: Bulletin of the American Association of Petroleum Geologists, v. 62, pp. 201–222.

Waples, D. W., 1980, Time and temperature in petroleum formation: application of Lopatin's method to petroleum exploration: Bulletin of the American Association of Petroleum Geologists, v. 64, pp. 916–926.

———, 1984a, Thermal models for oil generation, in J. Brooks and D. Welte, eds., *Advances in Petroleum Geochemistry, Volume 1:* London, Academic Press, pp. 7–67.

Zieglar, D. L. and J. H. Spotts, 1978, Reservoir and source bed history of Great Valley, California: Bulletin of the American Association of Petroleum Geologists, v. 62, pp. 813–826.

PRACTICE PROBLEMS

1. The Black Well was drilled off the Louisiana Gulf Coast. It penetrated 1000 ft of Pleistocene sediments, 3500 ft of Pliocene, and 11,000 ft of upper Miocene before being abandoned at 16,150 ft in the middle Miocene. The corrected bottom-hole temperature was 270° F. A plausible average surface temperature is 68° F. Construct a family of burial-history curves for the well and calculate the present-day TTI at total depth.

2. Calculate present-day TTI at 3000 m in the Red Well, assuming a constant geothermal gradient through time. Find when the rock at 3000 m began to generate oil (TTI = 10). Determine when each of the strata began to generate oil.

Time-stratigraphic data:

Age (my)	Depth (m)
0	0
2	500
38	1200
65	2700
80	3000
100	4000

Temperature data:

Present-day average surface temperature:	15° C
Corrected BHT (4200 m):	141° C
Estimated surface temperature at end Cretaceous:	25° C

3. Calculate present-day TTI for a rock at 3000 m in the Beige Well using both maximum and minimum scenarios for Tertiary removal.

Time-stratigraphic data:

Age (my)	Depth (m)
30	0
38	300
65	1400
80	1700
100	3000

Erosional removal is estimated to have begun about 5 mya and probably comprises between 500 and 2000 m.

Temperature data:

Present-day average surface temperature:	10° C
Corrected BHT (3000 m):	70° C

4. The Ultraviolet Well is spudded in Paleocene sediments. At a depth of 1500 ft, micropaleontology indicates the rocks to be of Maestrichtian age. The following Upper Cretaceous boundaries are noted:

Maestrichtian–Campanian	1807 ft
Campanian–Santonian	2002 ft
Santonian–Coniacian	2360 ft
Coniacian–Turonian	2546 ft
Turonian–Cenomanian	3017 ft

The Cenomanian is 480 ft thick and overlies 1000 ft of Kimmeridgian-age shale. Total depth is reached at 6120 ft in Middle Jurassic rocks.

Evidence from related sections indicates that the Paleocene was originally about 3000 ft thick and that no other Cenozoic sediments were ever deposited. Total original thickness of the Kimmeridgian is thought to be 1500 ft. It is also believed that 500 ft of Lower Cretaceous sediments were deposited before uplift and erosion began.

Assuming a surface temperature of 10° C and a geothermal gradient of 2° F/100 ft, draw a burial-history curve for the section penetrated and calculate maturity for the Kimmeridgian shale.

5. Analyze the timing of oil generation in the Pink Well, drilled in the Midcontinent region of the United States. The geothermal gradient was found to be 1.0° F/100 ft, and the surface temperature today is about 59° F. Time–stratigraphic data are given in the following table. No unconformities are recognized within the Paleozoic. Erosional removal since the Permian probably totals about 2000 ft.

Tops	Age (my)	Period	Depth (ft)
Permian	230	Permian	0
Virgil	280	L. Carboniferous	7,000
Missouri	288	"	8,000
Des Moines	296	"	11,000
Atoka	304	"	13,000
Morrow	309	"	18,500
Mississippian	320	E. Carboniferous	21,000
Kinderhook	340	"	23,000
Sylvan	425	Ordovician	25,500
Arbuckle	470	"	27,500

6. You have been asked to evaluate an undrilled prospect in a remote area that is available in an expensive farm-in deal. Because of the high operations cost, upper management has decided that gas and condensate are not economical. Your responsibility is to make a recommendation regarding the nature of hydrocarbons that might be present in the prospect. The following geological summary is available to you.

"A regional study of the area suggests the probable presence of a thin, rich, oil-prone source rock at about 4300-m depth near the prospect. The source rock is thought to be about 300 my old. No other source rocks were noted. Highly fractured carbonates overlie the source rock; they are in turn overlain at 2750 m by a sandstone of excellent reservoir quality. The reservoir is sealed by a thick salt layer. No other reservoirs are anticipated.

"The basin filled at a generally uniform rate from about 300 mya to 100 mya. At that time nearby orogenic activity caused the first traps to be formed during a gradual 1200-m uplift lasting until 40 mya. From 40 mya to the present about 500 m of additional burial occurred.

"Nearby well control indicates that a geothermal gradient of 3.65° C/100 m and a surface intercept of 15° C are reasonable for the area. The traps at the prospect location formed slightly prior to the beginning of erosional removal in the basin and have retained integrity to the present."

Utilizing the principles of hydrocarbon generation and preservation, evaluate the prospect.

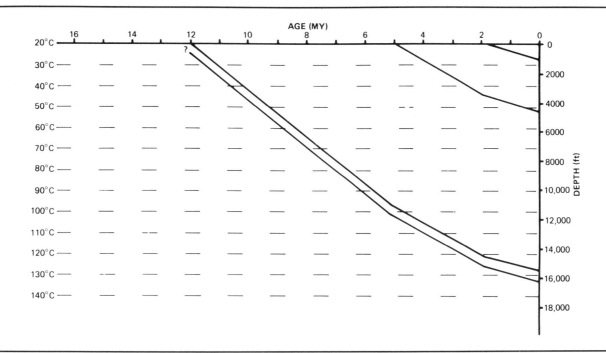

FIGURE 9.35 *Family of burial-history curves and subsurface-temperature grid for the Black well.*

SOLUTIONS TO PRACTICE PROBLEMS

1. Because we have only a single bottom-hole temperature and no information with which to reconstruct a detailed temperature history of the Black well, it would be satisfactory to assume that the present-day geothermal gradient is constant throughout the section and that it has remained constant since the middle Miocene. Calculation of the geothermal gradient gives 202° F/16,150 ft, or 1.24° F/100 ft. After converting to Celsius and inverting, we have a thickness of 1,442 ft for every 10°-C temperature interval. These data permit construction of the subsurface-temperature grid (fig. 9.35).

The time–stratigraphic data permit a complete definition of three horizons: the Pleistocene–Pliocene boundary, the Pliocene–upper Miocene boundary, and the boundary between upper Miocene and middle Miocene. Using the dates of 1.8, 5.0, and 12.0 my for these three events, the three members of the family of burial-history curves for the Black Well can be drawn (fig. 9.35). A fourth burial-history curve corresponding to the rock at total depth can also be partially drawn, but because the

exact age of that rock is not known, the curve cannot be taken back to the time of deposition. Fortunately, however, the temperature during this period of uncertainty was very low and did not affect maturity appreciably.

Calculation of maturity is now easy. Using the temperature factors given in table 9.2, we find the TTI at 16,150 ft to be about **17.7.**

2. The present-day temperature profile was constructed from the present-day surface temperature of 15° C and the gradient of 3° C/100 m (calculated from (141 − 15)/42). The spacing of the 10° isotherms is therefore 333 m. At the end of the Cretaceous the gradient was the same as the present, but the surface temperature was 10° higher. In the absence of any more information, let us assume that cooling proceeded at a constant rate from the end of the Cretaceous (65 mya) to the present. Furthermore, let us assume that during the Late Cretaceous the temperature did not change. With these assumptions we can construct the subsurface-temperature grid shown in figure 9.36.

The family of five burial-history curves for the Red Well is shown in figure 9.36 superimposed on the subsurface-temperature grid. TTI values can now be calculated by defining the length of time spent in each temperature interval. An ex-

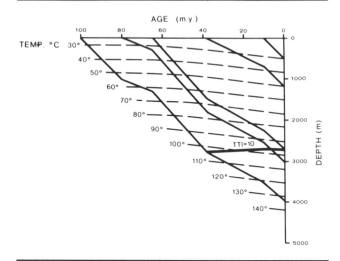

FIGURE 9.36 *Family of burial-history curves and subsurface-temperature grid for the Red well. Isomaturity line represents the beginning of oil generation at TTI 10.*

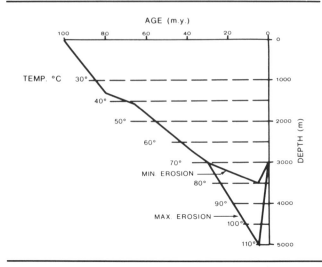

FIGURE 9.37 *Burial-history curves for a 100-my-old rock, assuming two different erosional models (500 m or 2000 m of removal) commencing 5 mya in the Beige well.*

TABLE 9.7 *Calculation of TTI at the present day for the rock now at 3000 m in the Red Well*

Temp. Interval (°C)	γ-Factor	Time Factor (my)	Interval TTI	Total TTI
30–40	$\frac{1}{128}$	10	.08	.08
40–50	$\frac{1}{64}$	6	.09	.17
50–60	$\frac{1}{32}$	7	.22	.39
60–70	$\frac{1}{16}$	7	.44	.83
70–80	$\frac{1}{8}$	11	1.38	2.21
80–90	$\frac{1}{4}$	15	3.75	5.96
90–100	$\frac{1}{2}$	11	5.50	11.46
100–110	1	4	4.00	15.46

ample of a calculation for the stratum now at 3000 m is given in table 9.7.

The present-day TTI calculated from table 9.7 for the horizon now at 3000 m is about 15.5. This horizon reached a TTI of 10, corresponding to the onset of oil generation, about 8 mya (table 9.7 and fig. 9.36). The 4000-m horizon reached TTI 10 about 39 mya, while the rock now at 2700 m has not quite reached maturity (fig. 9.36).

3. Construction of the subsurface-temperature grid for the Beige Well is simple because there are no changes in gradient or surface temperature through time (fig. 9.37). Two burial-history curves are constructed in figure 9.37, corresponding to minimum and maximum possible removals. The curves are identical from 100 mya to 30 mya. Between 30 mya and 5 mya, when uplift and erosion began, they are different. In the case of minimum erosion only 500 m of sediment was deposited during those 25 my, whereas in the case of maximum erosion 2000 m had to be deposited. Removal began 5 mya in both cases; the difference was the amount of removal.

Present-day TTI values reflect the differences in burial depth and maximum paleotemperature for the two models. The maximum-erosion case has a TTI of 14.8 today, whereas the minimum-erosion case is only 5.5.

4. Draw the temperature grid as in problem #1 after converting Fahrenheit to Celsius; then consider the problem of the unconformities. The last unconformity is at the surface, where Paleocene sediments are exposed. Loss of 1500 ft of Paleocene sediment has occurred in the last 55 my. In the absence of any evidence to the contrary, it can be assumed that erosion has occurred at a constant rate throughout that time interval. The temperature and depth of burial therefore have also decreased at a uniform rate.

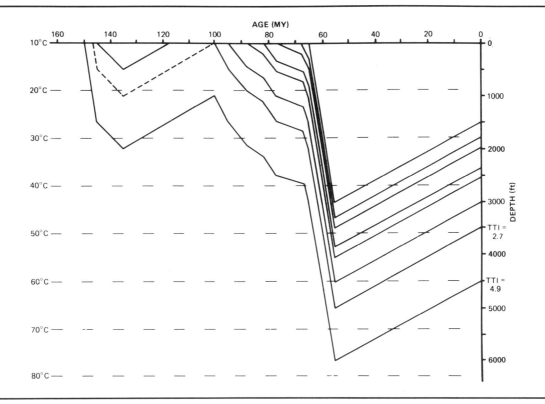

FIGURE 9.38 *Family of burial-history curves and subsurface-temperature grid for the Ultraviolet well.*

The first unconformity is also erosional. The section has lost 500 ft of Kimmeridgian and 500 ft of Lower Cretaceous. The exact time interval represented by the lost Lower Cretaceous rocks is not known, but this lack of knowledge is not serious because all events occurred when the bed of interest (the Kimmeridgian) was at a very low temperature.

Uplift in a basin is usually preceded by a slowing of the subsidence and sediment-accumulation rates. Assume that the Lower Cretaceous sediments were deposited at a slower rate than the Kimmeridgian sediments. In this case a rate of 500 ft in 10 my was selected. If we also assume that erosion occurred at a constant rate between 135 mya and 100 mya, the horizon lines can be completed as shown in figure 9.38. The dashed line represents the top of the uneroded Kimmeridgian.

TTI values can be calculated for the bottom and top of the uneroded Kimmeridgian. None of the Kimmeridgian turns out to be thermally mature; TTI values lie between 2.7 and 4.9 and therefore correspond to vitrinite-reflectance values between 0.50% and 0.55% (table 9.4).

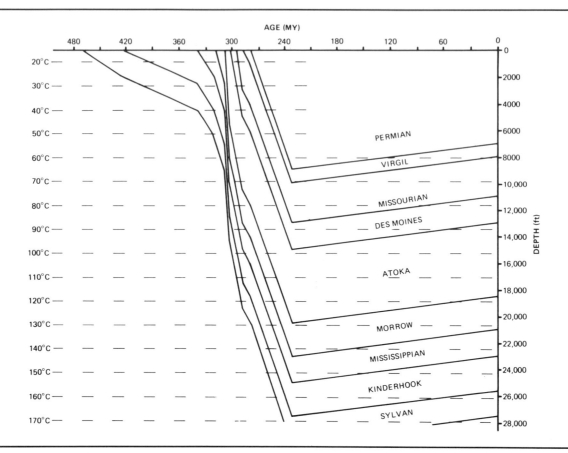

FIGURE 9.39 *Family of burial-history curves and subsurface-temperature grid for the Pink well.*

5. The family of burial-history curves constructed from the data is shown in figure 9.39. In order to discuss the timing of oil generation, we must put isomaturity lines on the burial-history curves. Let us assume that the kerogen present is Type III, which probably begins to generate oil a bit later than Type II—say at $R_o = 0.65\%$ (TTI = 15). If the end of oil generation is taken as TTI = 160, then these values define the oil-generation window. In order to draw the isomaturity lines, we merely find the points on each burial-history curve where TTI is 15 and 160. Then we connect all the TTI = 15 points with one line and all the TTI = 160 points with another (fig. 9.40).

It is evident from figure 9.40 that, at the present time, the interval from 7600 ft to 14,000 ft is within the oil-generation window. Temperatures in this interval range from 58° to 93° C. The relatively low temperatures of the oil-generation window are a direct consequence of the long time that these sediments have spent in the subsurface.

In the distant past the oil-generation window was deeper and hotter because the sediments had not been baked so long. The unusually high temperatures (115°–140° C) and great depths of burial (18,000–22,500 ft) required for oil generation during the Late Paleozoic from the Sylvan Formation were a result of the rapidity with which the overlying sediments were deposited.

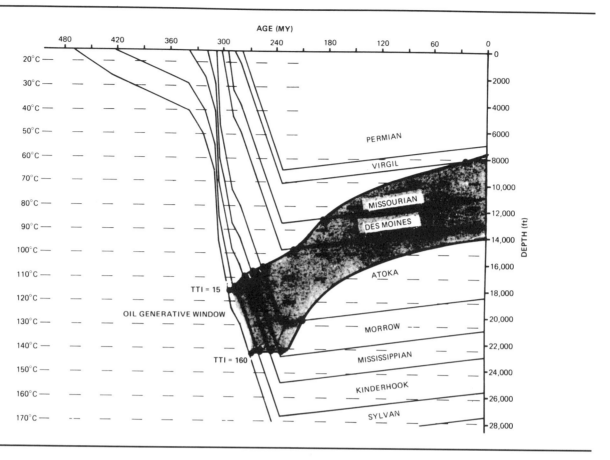

FIGURE 9.40 *Isomaturity lines superimposed on burial-history curves for the Pink well.*

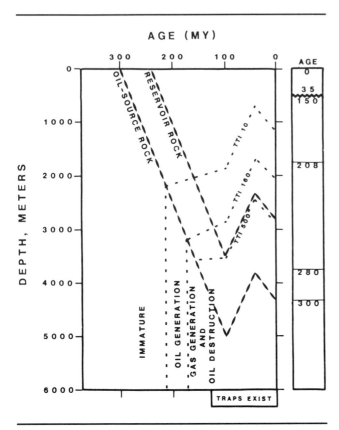

FIGURE 9.41 *Burial-history curves and isomaturity lines for farm-in prospect, showing that oil generation in the source rock predated trap formation by many tens of millions of years and that the maturity level of the reservoir rock today is too high for oil to be preserved. This is therefore a gas prospect.*

6. Burial-history curves for the source rock and reservoir rock are shown in figure 9.41. The subsurface-temperature grid is not shown. Maturities of both source and reservoir were calculated and isomaturity lines defining the oil-generation window (TTI = 10 to TTI = 160) and limit for condensate preservation (TTI = 500) drawn on the burial-history curves. The source rock generated its oil around 200 mya, long before any trapping mechanism existed in the prospect. After trap formation (liberally interpreted as starting as early as 120 mya) the only product being generated by the source rock was gas.

Furthermore, the maturity of the reservoir today is very high (TTI nearly 500), indicating that even if oil had somehow found its way into the traps, it would not have been preserved. Wet gas is the most likely hydrocarbon product from the prospect. In view of the economics of the play and the mandate from management, your recommendation would be to reject the farm-in offer.

The American Association of Petroleum Geologists
Bulletin, v. 65, No. 8 (August 1981), p. 1387–1396.

Petroleum Origin and Accumulation in Basin Evolution—A Quantitative Model[1]

D. H. WELTE and M. A. YUKLER[2]

ABSTRACT

Basin data—geologic, geophysical, geochemical, hydrodynamic, and thermodynamic—can be combined for quantified hydrocarbon prediction. A three-dimensional, deterministic dynamic basin model can be constructed to calculate all the measurable values with the help of mass- and energy-transport equations and equations describing the physical and/or physicochemical changes in organic matter as a function of temperature. Input data consist of heat flux, initial physical and thermal properties of sediments, paleobathymetric estimates, sedimentation rate, and amount and type of organic matter. Subsequently, the model computes pressure, temperature, physical and thermal properties of sediments, maturity of organic matter, and the hydrocarbon potential of any source rock as a function of space and time.

Thus the complex dynamic processes of petroleum formation and occurrence in a given sedimentary basin can be quantified. For example, hydrocarbon potential maps for any given source rock and any geologic time slice of the basin evolution can be provided as computer printouts. The computer model can be applied to any stage of an exploration campaign and updated as more information becomes available.

INTRODUCTION

Increasing demand and decreasing supply of hydrocarbons require increased activities in petroleum exploration and an improvement in the exploration success ratio. New oil and gas fields must be found and explored areas should be reassessed for additional oil and/or gas pools.

The systematic search for petroleum accumulations started toward the end of the 19th century with the acceptance of the "anticlinal theory." The vertical movement of petroleum in a static medium was considered. Studies on multiphase flow systems of gas, oil, and water resulted in the "hydrodynamic theory." The search focused on the detection of suitable subsurface structures which could host petroleum accumulations. Geophysical methods have been developed and improved to help locate these structures. However, the timing and the amount of petroleum generation were seldom considered. Later, particularly over the last 10 years,

organic geochemical studies supplied the most needed chemical data on the generation, migration, and accumulation of petroleum. From these data, new concepts were developed on the temperature-time dependence of petroleum generation and on the other complex processes of migration and accumulation.

Because of the enormous amount of data required to describe this complex basin system, quantification of the processes was not possible. Therefore, only qualitative or semiquantitative studies were made and presented as case studies. With the invention of large and fast computers, quantitative studies can be initiated.

Until now statistics have played a major role in exploration for hydrocarbons. The success ratio was directly related to the complexity of the system studied and how well it fitted a predetermined frequency distribution. Our studies show that most processes in a sedimentary basin are not only time-dependent but are also strongly interrelated to a degree never considered in the past. A small error in the determination of one process can result in a totally erroneous answer. On the basis of experience gained in basin studies, we, therefore, chose a three-dimensional, dynamic deterministic model to quantify the previously mentioned processes. This approach has been successfully tested with existing sedimentary basins. For reasons of confidentiality no details on the areas studied can be given.

SYSTEM CONCEPT AND SIMULATION

To understand the complex natural phenomena of petroleum occurrences in sedimentary basins, surface and subsurface samples are systematically collected and analyzed. The various processes are described and the interrelations are determined. Thereafter one tries to comprehend the "system" in which these processes occur. Once an understanding of the system is reached, practical goals are sought which go beyond a purely scientific or theoretical description of the qualitative nature of the problem. To reach these practical goals a quantification is needed to answer questions as to when and how much petroleum was formed and where it has accumulated.

About 1890 the "anticlinal theory" was widely accepted as the controlling principle of petroleum accumulation. The main idea was that petroleum is driven in a water-saturated environment by buoyancy forces,

[1]Manuscript received, December 24, 1980; accepted, February 11, 1981.
[2]Institute for Petroleum and Organic Geochemistry (ICH-5), KFA-Julich,

P. O. Box 1913, D-5170 Julich, Federal Republic of Germany.
Financial support for the study came from German Federal Ministry for Research and Technology (BMFT) Grant No. ET 3070 B. M. Radke gave valuable advice with respect to hydrocarbon generation curve.

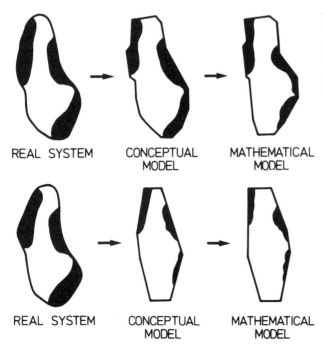

REAL SYSTEM   CONCEPTUAL MODEL   MATHEMATICAL MODEL

REAL SYSTEM   CONCEPTUAL MODEL   MATHEMATICAL MODEL

FIG. 1—Steps in development of model of real-world system: a, conceptual and mathematical models successfully represent real system; b, owing to poor understanding of real system, models do not represent it.

which are vertical, and accumulates in crestal positions of anticlines or, in general, in the highest local positions to which it can migrate in any structure. At the beginning of the 20th century the "hydraulic theory" of oil and gas accumulations was developed (Munn, 1909a, b; Shaw, 1917; Mills, 1920; Rich, 1921, 1923, 1931, 1934; Illing, 1938a, b, 1939). Hubbert (1940, 1953) further developed the hydraulic theory with a mathematical basis and this theory then received wide acceptance. These theories, however, addressed only the question of where accumulations were located.

With the acceptance of the anticlinal theory geologists started to search for anticlines or for "geologic highs" in the subsurface. These structures commonly are not visible from the surface and, therefore, geophysics became an important tool in petroleum exploration.

Gravimetric, magnetic, and geoelectric methods were developed, applied, and improved to satisfy the needs of oil exploration. The invention of refraction and reflection seismics in the 1920s was a big step forward in the search for petroleum. Every improvement in geophysical methods resulted in better understanding of the subsurface formations and structures and enabled detection of series of new finds. In the past 20 years remote sensing techniques were developed and their use has become popular. However, all these methods are concerned only with the possible location of a petroleum accumulation.

Therefore, it is clear that to answer not only the questions "where," but also, "why, when, and how much," we have to establish the generation, migration, and accumulation processes quantitatively. In the past few decades, extensive studies on sediments and petroleum have shown that petroleum originates from finely disseminated organic matter buried within sediments that have been subjected to elevated temperatures (about 50°C and higher). A sediment may be considered a good source rock if it meets certain criteria. These include the amount of organic matter, both soluble (bitumen) and insoluble (kerogen), the type of kerogen, and the maturity of the organic matter (Tissot and Welte, 1978). Identification of specific source rocks is dependent also on the correlation of the composition of extractable hydrocarbons and nonhydrocarbons.

The release of petroleum compounds from kerogen and their movement within and through the pores of a source rock are defined as primary migration. The movement of petroleum expelled from a source rock through the wider pores of more permeable and porous carrier and reservoir rocks before final emplacement, being mainly controlled by buoyancy, is called secondary migration. Petroleum may then be collected in reservoir rocks in various types of structural or stratigraphic traps and form accumulations.

The science of organic geochemistry or petroleum geochemistry is now adequately developed to be applied quantitatively to problems of the generation, migration, and accumulation of petroleum. During the past decade organic geochemistry has become a useful tool in petroleum exploration. New and rapid methods have been developed to fulfill the main requirements of petroleum exploration—assessment of source rock potential, source rock maturity, and source rock/oil correlation (Welte et al, 1981).

The system in which the generation, migration, and accumulation of petroleum occurs is the three-dimensional dynamic geologic framework. The geometry and location of the system are determined from paleogeography and seismic information about the basin. The sediment inputs (type, source, rate, etc), depositional environments, paleobathymetric estimates, mineralogic changes, and tectonic movements are determined from geology. The direction and rate of fluid movement and hydraulic properties of fluids and sediments are determined from hydrodynamics. The direction and rate of heat flow and thermal properties of fluids and sediments are determined from thermodynamics. Finally, type, amount, maturity, and generation potential of organic matter are determined from organic geochemistry. A "qualitative" determination of these processes and the relations among them have been obtained through numerous temporal and spatial observations in the field and in the laboratories. Now, we unite all these data and their interrelations to comprehend the real system which is called the conceptual model (Fig. 1a). When our determination of the processes and the relations among them is incorrect or incomplete the conceptual model does not represent the real system (Fig. 1b).

Once an understanding of the system is reached, we try to make quantitative petroleum exploration predictions, that is, answer the questions, why, when, how much, and where. This requires the simulation of the various processes in the system by either physical or mathematical models. Because the complex nature of

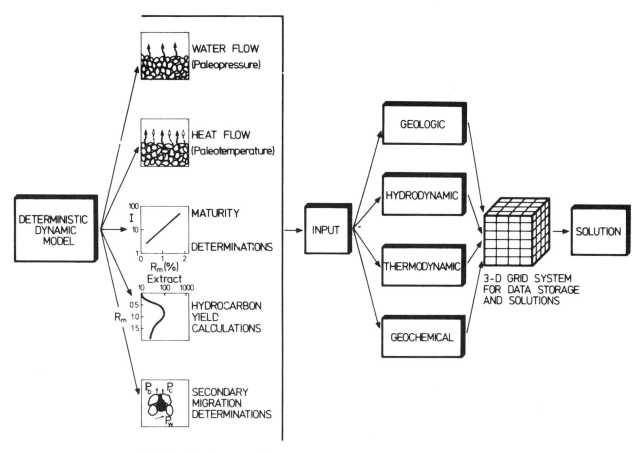

FIG. 2—Development of three-dimensional deterministic dynamic model.

the problem is unsuitable for a physical model, a mathematical model should be used.

Simulation is a class of techniques that involves setting up a model of a real system and then performing experiments on the model. In this sense it is a simplification of the real world and, thus, leads to errors—conceptual and mathematical. A mathematical model is based on the conceptual model and therefore inherits the errors from the conceptual model. These errors are due to poor knowledge of the real system and its behavior (Fig. 1a, b). Incorrect original input will result in errors in the conceptual model.

Certain assumptions are made in the mathematical formulation of a real system. The assumptions and simplications made in the mathematical formulation and/or in the solution techniques lead to mathematical errors. These errors should be quantitatively determined so that one can obtain reliable answers from simulation studies. Sensitivity analysis aids in the computation of such errors (Yukler, 1976, 1979).

Petroleum generation, migration, and accumulation occur in a three-dimensional framework as a function of time, and are controlled by very complex and inter-related mechanisms (Tissot and Welte, 1978; Yukler et al, 1978; Welte and Yukler, 1980). Furthermore, the distribution and the characteristics of organic matter do not show any fixed pattern. Under these conditions we have chosen a three-dimensional deterministic dynamic

model. This model can be used at any stage during an exploration program. It is applicable to new areas, as well as to old prospects with abundant data where it provides a more quantitative and detailed fresh appraisal.

Figures 2, 3, and 4 illustrate the flow chart of the three-dimensional dynamic deterministic model. The details are explained in the following.

QUANTITATIVE BASIN ANALYSIS—THREE DIMENSIONAL MODEL TO SIMULATE GEOLOGIC, HYDRODYNAMIC, AND THERMODYNAMIC DEVELOPMENT

A "qualitative" understanding of petroleum generation, migration, and accumulation has been obtained from numerous temporal and spatial reconstructions of basin histories and laboratory studies. As a result, various theories and hypotheses have been developed to explain these complex phenomena. In a normal qualitative approach the validity of these theories and hypotheses are demonstrated by case histories. Unfortunately, case histories which do not prove the validity of the theories and hypotheses are either not considered or lead to speculations. To overcome these difficulties or to improve these misleading approaches, a "quantitative" evaluation of the total system and mechanisms prevailing therein is required. Only then can the validity

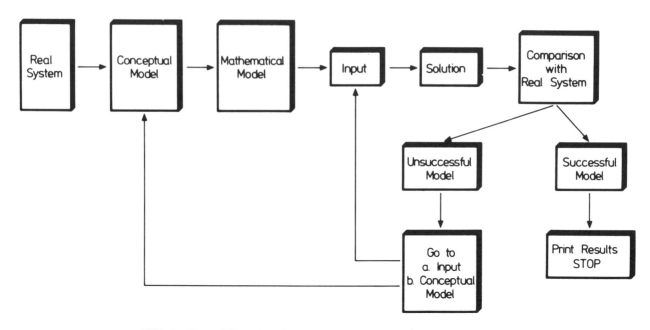

FIG. 3—General flow chart for three-dimensional quantitative basin model.

of the theories and the hypotheses applied be effectively checked.

Knowledge of paleopressures and paleotemperatures during the evolution of a sedimentary basin is important for the solution of many problems such as (1) stratigraphic and structural development; (2) changes in physical properties of fluids (density and viscosity) and sediments (compaction, permeability, porosity, etc); (3) changes in thermal properties of fluids and sediments (heat capacity, thermal conductivity, etc); (4) mineralogic changes affected by temperature and pressure; (5) generation, migration, and accumulation of petroleum; (6) fluid flow mechanism, such as fluid flow directions and rates and determination of abnormal fluid pressure zones.

Gibson (1958) studied the excess pressures assumed to be generated by a moving boundary condition, such as continuous sedimentation, and developed an equation to compute the consolidation of a clay layer. Using Gibson's equation, Bredehoeft and Hanshaw (1968) examined pressure-producing mechanisms in a basin. Sharp and Domenico (1976) used the same equation to compute energy transport in compacting sedimentary sequences. Bishop (1979) determined compaction of thick, abnormally pressured shales using the same equation, but added the effect of external loading (Taylor, 1948). All these quantitative studies are carried out in one dimension which requires the highest degree of symmetry in a three-dimensional framework. However, Gibson's equation cannot accurately determine compaction of sediments, for it does not handle the compressibility of sediments rigorously. A new equation for fluid flow in sediments with moving boundary conditions (sedimentation, compaction, and erosion) was derived by Yukler et al (1978). The hydraulic head (or pore pressure) in sediments can be computed in three-dimensions (x, y, and z) and as a func-

tion of time with the following equation (Welte et al, 1980), where the inflow-outflow is equal to the net accumulation due to grain and fluid compressibility plus the net accumulation due to the change in sediment density, change in rate of sedimentation, and change in water depth:

$$\frac{1}{\rho}\left[\frac{\partial}{\partial x}\rho K\frac{\partial h}{\partial x} + \frac{\partial}{\partial y}\rho K\frac{\partial h}{\partial y} + \frac{\partial}{\partial z}\rho K\frac{\partial h}{\partial z}\right]$$
$$= S_s\frac{\partial h}{\partial t} + \alpha\left[-(L-z)\frac{\partial\partial_s}{\partial t} - (\partial_s - \partial_w)\frac{\partial H}{\partial t} - \gamma_w\frac{\partial H}{\partial t}\right] \quad (1)$$

Here,

L	= length
M	= mass
T	= time
h	= hydraulic head, L
H	= water depth, L
L	= sedimenth thickness, L
K	= hydraulic conductivity, L/T
S_s	= storativity, 1/L
t	= time, T
x, y, z	= three orthogonal vectors
α	= compressibility of solid skeleton, LT^2/M
γ_s	= specific weight of bulk sediment, M/L^2T^2
γ_w	= specific weight of fluid, M/L^2T^2
ρ	= density of water, M/L^3

The term with the compressibility of fluid is neglected, since the error is found to be negligible.

The heat flow equation for the simultaneous transfer of heat both by conduction and convection (due to water flow) was introduced by Stallman (1963):

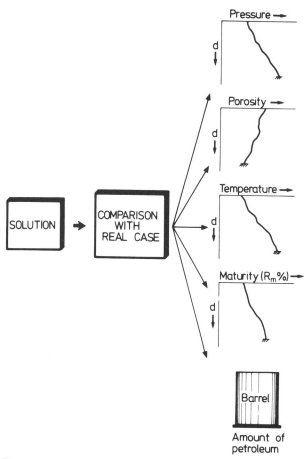

FIG. 4—**Comparison of model results with selected real values.**

$$\frac{\partial}{\partial x} \, K \, \frac{\partial T_m}{\partial x} + \frac{\partial}{\partial y} \, K \, \frac{\partial T_m}{\partial y} + \frac{\partial}{\partial z} \, K \, \frac{\partial T_m}{\partial y} -$$

$$\underset{\text{conduction}}{}$$

$$\rho_w C_{pw} \left[\frac{\partial}{\partial x} \, V_x T_m + \frac{\partial}{\partial y} \, V_y T_m + \frac{\partial}{\partial z} \, V_z T_m \right] + Q$$

$$\underset{\text{convection}}{} \qquad \underset{\text{source/sink}}{}$$

$$= \rho_{ws} C_{ws} \, \frac{\partial T}{\partial t} \qquad\qquad (2)$$

$$\underset{\text{net accumulation}}{}$$

where

E	= energy
°C	= temperature in degrees Celsius
C_{pw}	= specific heat of fluid, E/M°C
C_{ws}	= specific heat of bulk sediment, E/M°C
K	= thermal conductivity, E/LT°C
Q	= sink (−) or source (+) term, E/L³T
T_m	= temperature, °C
V_x, V_y, V_z	= fluid flow in x, y, and z directions, respectively, L/T
ρ_w	= density of fluid, M/L³
ρ_{ws}	= density of bulk sediment, M/L³

All physical and thermal parameters of the system are pressure- and temperature-dependent and are, therefore, always recalculated with each computation of pressure and temperature by a special iterative technique (Yukler et al, 1978). The preceding equations are also integrated from the bottom of a sedimentary unit to the top and the resulting equations are solved by a suitable numerical analysis technique (Yukler, 1976).

QUANTITATIVE APPRAISAL OF HYDROCARBON POTENTIAL

The generation and emplacement of petroleum is a time-dependent, dynamic process linked to the evolution of a sedimentary basin. Hence, the quantitative appraisal of the hydrocarbon potential is integrated into the three-dimensional deterministic dynamic basin model (Figs. 2, 3) which is constructed according to the regional geologic framework.

Basic data for the assessment of the hydrocarbon potential of a basin consist of the regional geology, hydrodynamics, and geothermics and knowledge on generation, migration, and accumulation of petroleum. The identification of source rocks throughout the basin and the amount, type, and maturity of their organic matter are important parameters used for this part of the model (Welte and Yukler, 1980). A decisive role is played by the regional geothermal gradients and their variation with time.

Effect of Temperature on Maturation of Organic Matter

Thermal evolution of source rocks changes many physical and chemical properties of the organic matter. The changes in these properties are used as indicators for maturation. The parameters most commonly used in petroleum exploration are optical examination of kerogen, physicochemical analysis of kerogen, and chemical analysis of extractable bitumen (Tissot and Welte, 1978). All these measurements and studies are done on the present end products of thermal evolution. Our main objective is to determine changes in maturity as a function of time to the present and then to compare the computed end results with the observed values.

Thermal evolution of source rocks is directly related to the geologic and hydrodynamic processes. Studies by Yukler et al (1978) showed that temperature and pressure distributions are interrelated and depend on the geologic development of a basin. Hence, temperature not only affects the maturity of organic matter, but also affects the physical and thermal properties of fluids and sediments. This means that such parameters as density, viscosity, porosity, permeability, and thermal conductivity, as well as compaction of sediments and the specific yield of hydrocarbons are influenced.

Heat flow in a sedimentary system is composed of two components. There is heat transport by conduction and by convection (eq. 2). Heat distribution by convection is directly related to fluid movement. The direction and rate of fluid flow are determined from hydraulic

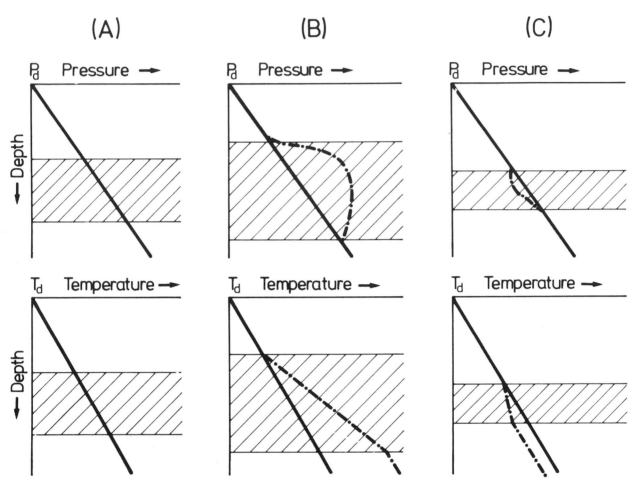

FIG. 5—Pressure and temperature distribution in normal, abnormally high, and abnormally low pressured systems (A, B, C, respectively). P_d and T_d are pressure and temperature at arbitrary depth d. Solid lines show changes in abnormally high and low pressure systems.

head and hydraulic conductivity in the system. Water, however, has a low thermal conductivity, but a high specific heat capacity with respect to sediments. Therefore, water is a poor heat conductor, but has a large heat-holding capacity. These properties of water are pressure- and temperature-dependent.

Pressure distribution in the deposited sediments is influenced by the hydrodynamic properties of sediments, especially by porosity and permeability, and by the changes in the overburden pressure due to additional sedimentation, erosion, and changes in water levels. Increase in the overburden pressure on a sedimentary unit, that is, the increase in load, is supported by water pressure and by grain-to-grain stress distribution. With increase in overburden pressure the sedimentary unit compacts because of the bleed-off of compaction water and as a result of solid-matrix and water compressibility. As the unit compacts, the water leaves the system and the overburden pressure is supported by the grain-to-grain stress distribution and by the pressure in the remaining water in the unit. If the permeability in the sedimentary unit is not sufficient to release water with respect to increase in overburden pressure, for example,

in fine-grained sediments such as shales, most of the water stays in the unit and the overburden pressure is then mainly supported by the water pressure. This is termed an abnormally high pressure medium. Other factors, such as aquathermal pressuring and generation of hydrocarbons, may also contribute to pressure increases. When the amount of water escaping from the system is abnormally high owing to high permeability such as in coarse sands, the overburden is supported mainly by the solid skeleton, and the pore space in the system decreases sharply. This is called an abnormally low pressure medium. As a result there is a sharp increase in the temperature gradient on top of an abnormally high pressure medium and a sharp decrease in the temperature gradient on top of an abnormally low pressure medium (Fig. 5). The abnormally high-pressured sedimentary unit represents an insulator because of low thermal conductivity of the water-filled pore spaces. As the water cannot easily move out of the system, the heat coming into the unit is largely stored in the water of the unit and results in a sharp increase in temperature gradient. In an abnormally low pressure medium, pore space is reduced so thermal conductivity

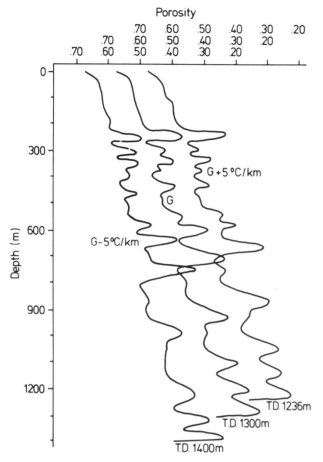

FIG. 6—Effect of heat flow on porosity versus depth relation (from Yukler et al, 1978).

is kept higher than in a normal pressure medium (hydrostatic) and specific heat capacity is kept low. Hence, the abnormally low pressure medium represents a conductor.

From 0 to 100°C, changes in density of water can be neglected, whereas decrease in viscosity with increasing temperature is important (Paaswell, 1967). Hydraulic conductivity of a medium is the ability of a medium to transmit water and is defined as the product of permeability of the medium and the specific weight of the fluid divided by the viscosity of the fluid. Therefore, decrease in viscosity with increase in temperature will yield higher hydraulic conductivities which will result in higher water flow rates. Higher water flow rates will yield higher compaction and changes in temperature, pressure, and physical and thermal properties versus depth relations. Figure 6 illustrates the changes in porosity versus depth relations with changes in heat flux (Yukler et al, 1978).

The system concept which is formulated by the conceptual model and the resulting mathematical model in this study allows us to determine the temperature distribution in a sedimentary basin with greater accuracy than any other indirect method available. With the knowledge of the spatial and temporal distribution of temperature, the thermal evolution of the organic

matter can be determined.

Our mathematical model allows us to develop, at least, a time-dependent three-dimensional relative-temperature-distribution pattern even in unexplored basins without any well data. This relative pattern can easily be gauged as soon as samples are available to measure physical properties of the sediments, temperature, and vitrinite reflectance.

Determination of Maturity of Organic Matter from Lopatin's Method

Lopatin (1971) analyzed the Ruhr coals and especially the coal seams in the Munsterland-1 borehole. In coalification reactions the reaction rate, in general, doubles with every increase of 10°C in temperature. Lopatin mathematically analyzed the geologic and petrologic findings of the Munsterland borehole and introduced the temperature-time index (I) which is the sum of the products of the effective geologic heating time (G), and the temperature correction factor (T):

$$I = T_1G_1 + T_2G_2 + \ldots + T_nG_n \ldots \quad (3)$$

Subsequently, a correlation equation between vitrinite reflectance, $R_m\%$, and the temperature-time index was found,

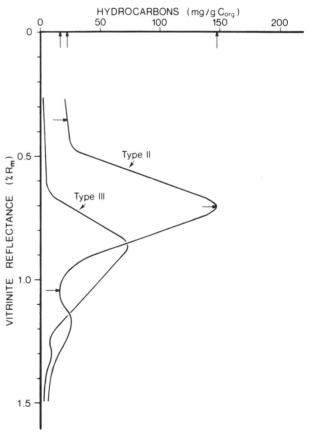

FIG. 7—Hydrocarbon generation curve for type II and type III kerogen.

289

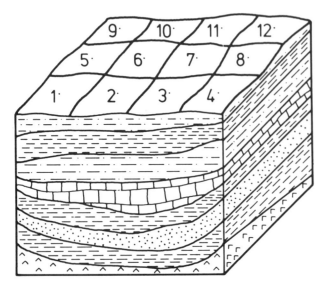

FIG. 8—General three-dimensional illustration of study area.

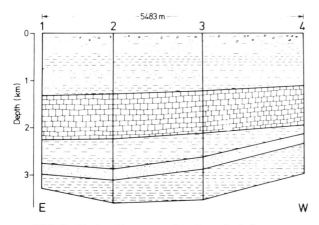

FIG. 9—East-west cross section through study area.

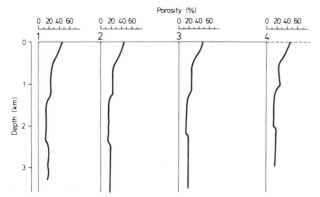

FIG. 10—Porosity versus depth relations at four locations along cross section in Figure 9.

$$R_m\% = 1.301 \lg I - 0.5282 \ldots \quad (4)$$

From simultaneous solutions of equations 1 and 2, the temperature is computed as a function of time and space. Then T and G values are determined. From equation 3 the temperature-time index is computed and replaced in equation 4. The vitrinite reflectance is then calculated as a function of space and time.

Lopatin's method is used as a first approach. It needs further improvement to be applicable to the different types of organic matter and at different levels of maturation. Despite its "oversimplified" approach, however, Lopatin's method in our experience has worked well.

Quantification of Hydrocarbon Generation Potential

At the next step the calculated vitrinite reflectance values will be used to determine the amount of hydrocarbons to be expected from possible source rocks within the three-dimensional sedimentary basin as a function of time.

For this purpose a plot of a hydrocarbon generation curve for type II and type III kerogen is given in Figure 7. In the example treated in this paper the source rock in question contains organic matter of type II kerogen. The peak hydrocarbon generation for this curve is shown at a vitrinite reflectance value of 0.70%. In this connection it must be realized that the hydrocarbon generation curves represent observed maximum values of hydrocarbons which are not always reached. Because hydrocarbon generation curves also have other deficiencies due to migrational phenomena, parameters such as "kerogen to hydrocarbon conversion" should be employed to appraise a source rock. Nevertheless, for the time being, hydrocarbon generation curves as given in Figure 7 are used for calculations of the hydrocarbon potential.

Let's assume that three different vitrinite reflectance values ($R_m\%$) of 0.35, 0.70, and 1.04% are computed for three different source rocks. These vitrinite reflectance values will result in ratios of hydrocarbons/organic carbon of about 22, 147, and 16 mg/g, respectively (Fig. 7). In this way, prior to any exploratory drilling and before any actual source rock analyses are available a relative distribution of the regional hydrocarbon potential for a given source rock can be established. At a later stage, when rock samples are available the hydrocarbon potential can be quantified more accurately. Along this line, with the assumption or estimation of expulsion efficiency, amount of petroleum probably expelled from source rocks can be given.

PRIMARY AND SECONDARY MIGRATION OF HYDROCARBONS

As a first approach we assumed that primary migration is mainly a pressure-driven hydrocarbon-phase movement. Primary migration, therefore, is a normal process that occurs in any mature source rock accompanying the generation of hydrocarbons. It takes place through available pores or by microfracturing of dense source rocks. Hydrocarbon-phase migration is in agreement with the empirical geologic and geochemical data (Tissot and Welte, 1978).

There are three major parameters that control secondary migration and the subsequent formation of oil and gas pools. These are buoyancy, capillary pressure, and hydrodynamics. The buoyancy is computed by subtracting the petroleum density from the formation water density and multiplying by the height of the petroleum column. At present, in the absence of reliable data, the estimation of the height of a petroleum column is very problematic. The pore sizes in the capillary pressure equation for sandy layers are computed from Berg's (1975) equation. The interfacial tension is corrected for temperature as given by Schowalter (1979). The pore pressures or hydrodynamic conditions are computed from equation 1. With the combination of all these parameters, as a first approach, possible secondary migration directions of petroleum and traps most likely to contain petroleum can be indicated.

THREE-DIMENSIONAL MODEL APPLICATION

The model has been applied successfully to real basins as shown by the example in Figure 8. The objectives of this study are to determine the geologic history, paleopressure, paleotemperature and generation, migration, and accumulation of the hydrocarbons in a given sedimentary basin. On the basis of all the available data we have an understanding of the basin, that is, the conceptual model. The mathematical model is based on the conceptual model.

Figure 8 shows continuous sedimentation of a sequence of different sediments (shale, sandy shale and sand, shale and sandy shale, limestone, sandy shale, shale and sandy shale layers) as an input. The sedimentation occurred on top of a practically impermeable basement during a total time span of 195 m.y. The lowest shale unit is the source rock and the overlying sand layer is the reservoir rock. The sedimentation took place in shallow water (20 to 100 m) except for carbonates where water depth was approximately 500 m. The structural pattern is also very simple with some normal faults that have very small throws. The faults are discontinuous and are mainly at the edges of the basin owing to local stress anomalies.

Initial porosities (pore volume/bulk volume) of 0.62, 0.40, 0.28, and 0.54 are assumed for shale, sand, limestone, and sandy shale, respectively. Permeability versus porosity relations are determined from available literature data and from core analysis. The heat flux is 1.1×10^{-6} cal/cm^{-2}/sec^{-1} and the average initial temperature at the sediment-water interface is 15°C. Initial physical and thermal parameters are chosen depending on the lithologic descriptions (Yukler et al, 1978). The errors in the initial estimates of sediment input and heat flow are minimized as discussed by Yukler et al (1978).

A time step of 100,000 years is chosen with a vertical

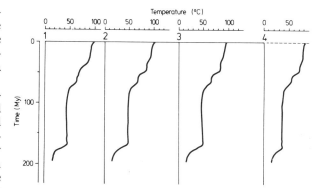

FIG. 11—Temperature-time relations at four locations along cross section in Figure 9.

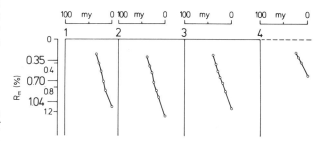

FIG. 12—Evolution of maturity of organic matter as function of time at four locations along cross section in Figure 9.

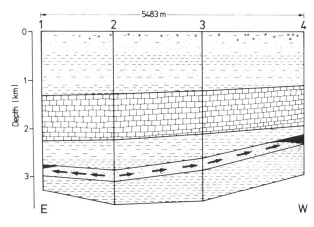

FIG. 13—Possible secondary migration directions and accumulation of petroleum. Arrows show directions as determined from model study (same cross section as in Fig. 9).

grid interval of 10 m and lateral grid interval of 1,000 m. The initial time step can be very small and can be increased by a certain factor at successive time steps based on the numerical analysis technique used (Halepaska and Hartman, 1972; Yukler, 1976).

The basin studied is 3,384 km^2 with a maximum sedi-

291

ment thickness of 4,500 m, generating 1.5228×10^6 node points. Equations 1, 2, 3, and 4 are solved for this grid network and the pore pressure, temperature, physical, and thermal properties, maturation of organic matter, and amount of hydrocarbons generated are determined at each grid point as a function of time. These results are then used to determine possible secondary migration directions. This latter information is related to the existence of traps. The solution of this problem can be obtained by a fast computer with large memory. We used an IBM 370/168 for this problem.

Figure 9 illustrates an east-west cross section computed by the model through the center of the sedimentary basin. For simplicity and discretion, we will present the model results only along this cross section. The porosity versus depth relation is shown in Figure 10 at locations 1, 2, 3, and 4. The low porosity and permeability of the limestone layers decrease the subsurface fluid movement into or out of the formations below it, thus forming a closed system. Therefore, petroleum generated in this system cannot migrate out of it. This is important for mass balance studies on the generation of petroleum. The temperature-time relation for the source rock at locations 1, 2, 3, and 4 is shown in Figure 11. From Lopatin's method as described previously, the maturation of the organic matter, in terms of vitrinite reflectance (R_m), is calculated as a function of time for the source rock (Fig. 12). The time of peak oil generation, where R_m is about 0.7, is 20, 25, and 23 m.y. at locations 1, 2, and 3, respectively. Vitrinite reflectance of 0.7 is not reached at location 4. Figure 13 is a simplified example showing the computed secondary migration directions and possible locations of oil accumulations.

CONCLUSIONS

A computer model to quantify generation, migration, and accumulation of oil has been developed and applied to existing sedimentary basins. The model results were computed with all available data on sedimentary thicknesses, pressure, temperature, porosity, density, thermal conductivity, specific heat, maturity of organic matter, and amount of petroleum in place. Necessary corrections in sedimentation rates, initial physical and thermal parameters, heat flux, and coefficients of the maturity equation were made using sensitivity equations. The allowable errors were ± 8% in physical and thermal parameters, ± 2°C in temperature, ± 10% in maturity, and ± 15% of petroleum in place.

REFERENCES CITED

Berg, R. R., 1975, Capillary pressure in stratigraphic traps: AAPG Bull., v. 59, p. 939-956.
Bishop, R. S., 1979, Calculated compaction states of thick abnormally pressured shales: AAPG Bull., v. 63, p. 918-933.
Bredehoeft, J. D., and B. B. Hanshaw, 1968, On the maintenance of anomalous fluid pressure. I. Thick sedimentary sequences: Geol. Soc. America Bull., v. 79, p. 1097-1106.

Gibson, R. E., 1958, The progress of consolidation in a clay layer increasing in thickness with time: Geotechnique, p. 172-182.
Halepaska, J. C., and F. W. Hartman, 1972, Numerical solution of the 3-dimensional heat flow, in Short papers on research in 1971: Kansas Geol. Survey Bull. 204, pt. 1, p. 11-13.
Hubbert, M. K., 1940, The theory of ground-water motion: Jour. Geology, v. 48, p. 785-944.
_____ 1953, Entrapment of petroleum under hydrodynamic conditions: AAPG Bull., v. 37, p. 1954-2026.
Illing, V. C., 1938a, The migration of oil, in A. E. Dunstan et al, eds., The science of petroleum, v. 1: London, Oxford Univ. Press, p. 209-215.
_____ 1938b, An introduction to the principles of the accumulation of petroleum, in A. E. Dunstan et al, eds., The science of petroleum, v. 1: London, Oxford Univ. Press, p. 218-220.
_____ 1939, Some factors in oil accumulation: Inst. Petroleum Jour., v. 25, p. 201-225.
Lopatin, N. V., 1971, Temperature and geologic time as factors in coalification: Akad. Nauk. Uzb. SSR Izv. Ser. Geol., v. 3, p. 95-106.
Mills, R. van E., 1920, Experimental studies of subsurface relationship in oil and gas fields: Econ. Geology, v. 15, p. 398-421.
Munn, M. J., 1909a, Studies in the application of the anticlinal theory of oil and gas accumulation—Sewickley quadrangle, Pa.: Econ. Geology, v. 4, p. 141-157.
_____ 1909b, The anticlinal and hydraulic theories of oil and gas accumulation: Econ. Geology, v. 4, p. 509-529.
Paaswell, R. E., 1967, Thermal influences on flow from a compressible porous medium: Water Resources Research, v. 3, p. 271-278.
Rich, J. L., 1921, Moving underground water as a primary cause of the migration and accumulation of oil and gas: Econ. Geology, v. 16, p. 247-371.
_____ 1923, Further notes on the hydraulic theory of oil migration and accumulation: AAPG Bull., v. 7, p. 213-225.
_____ 1931, Function of carrier beds in long-distance migration of oil: AAPG Bull., v. 15, p. 911-924.
_____ 1934, Problems or the origin, migration, and accumulation of oil, in Problems of petroleum geology: AAPG, p. 337-345.
Schowalter, T. T., 1979, Mechanisn of secondary hydrocarbon migration and entrapment: AAPG Bull., v. 63, p. 723-760.
Sharp, J. M., Jr., and P. A. Domenico, 1976, Energy transport in thick sequences of compacting sediments: Geol. Soc. America Bull., v. 87, p. 390-400.
Shaw, E. W., 1917, The absence of water in certain sandstones of the Appalachian oil fields: Econ. Geology, v. 12, p. 610-628.
Stallman, R. W., 1963, Computation of ground-water velocity from temperature data, in Methods of collecting and interpreting ground-water data: U.S. Geol. Survey Water Supply Paper 1544-H, p. 36-46.
Taylor, D. W., 1948, Fundamentals of soil mechanism: New York, John Wiley and Sons, 510 p.
Tissot, B., and D. H. Welte, 1978, Petroleum formation and occurrence: Berlin, Springer-Verlag, 538 p.
Welte, D. H., and M. A. Yukler, 1980, Evolution of sedimentary basins from the standpoint of petroleum origin and accumulation—an approach for a quantitative basin study: Organic Geochemistry, v. 2, p. 1-8.
_____ et al, 1981, Application of organic geochemistry and quantitative analysis to petroleum origin and accumulation—An approach for a quantitative basin study, in G. Atkinson and J. J. Zuckerman, eds., Origin and chemistry of petroleum: Elmsford, N.Y., Pergamon Press, p. 67-88.
Yukler, M. A., 1976, Analysis of error in groundwater modelling: PhD thesis, Univ. Kansas, 182 p.
_____ 1979, Sensitivity analysis of groundwater flow systems and an application to a real case, in D. Gill and D. F. Merriam, eds., Geomathematical and petrophysical studies in sedimentology, computery and geology, v. 3: Elmsford, N.Y., Pergamon Press, p. 33-49.
_____ C. Cornford, and D. H. Welte, 1978, One-dimensional model to simulate geologic, hydrodynamic and thermodynamic development of a sedimentary basin: Geol. Rundschau, v. 67, p. 960-979.

PREDICTIVE STRATIGRAPHY

American Association of Petroleum Geologists Memoir
26, *Seismic Stratigraphy—Applications to Hydrocarbon
Exploration*, edited by Charles E. Payton, copyright 1977,
pp. 63–81.

Seismic Stratigraphy and Global Changes of Sea Level, Part 3: Relative Changes of Sea Level from Coastal Onlap [1]

P. R. VAIL, R. M. MITCHUM, JR.,[2] and S. THOMPSON, III[3]

Abstract Relative changes of sea level can be determined from the onlap of coastal deposits in maritime sequences. The durations and magnitudes of these changes can be used to construct charts showing cycles of the relative rises and falls of sea level. Such charts summarize the history of the fluctuations of base level that control the distribution of the sequences and the strata within them.

A relative rise of sea level is indicated by coastal onlap, which is the landward onlap of littoral and/or nonmarine coastal deposits. The vertical component, coastal aggradation, can be used to measure a relative rise, but it should be adjusted for any thickening due to differential basinward subsidence. During a relative rise of sea level, a transgression or regression of the shoreline, and a deepening or shallowing of the sea bottom may take place. A common misconception is that transgression and deepening are synonymous with a relative rise, and that regression and shallowing are synonymous with a relative fall. A relative stillstand is indicated by coastal toplap; intermittent stillstands between rapid rises are characteristic of a cumulative rise. A relative fall of sea level is indicated by a downward shift in coastal onlap from the highest position in a sequence to the lowest position in the overlying sequence. After a major relative fall of sea level, the shelf tends to be bypassed, and the coastal onlap may be restricted to the apex of a fan at the basin margin.

Seismic sections provide the best means of determining the onlap and toplap patterns within the depositional sequences, and well control can provide the determinations of coastal and marine facies. Each cycle is plotted on a chart in chronologic order, dating and measuring the relative rise by increments of coastal aggradation, dating any relative stillstands by the duration of coastal toplap, and dating and measuring the relative fall by the downward shift of coastal onlap. Seismic examples illustrate the procedures and some of the problems encountered.

INTRODUCTION

Relative changes of sea level can be determined from the onlap of coastal deposits in depositional sequences. Relative stillstands can be determined from coastal toplap. As shown in Part 2 (Vail et al, this volume), seismic sections provide the best means of recognizing onlap and toplap patterns within the depositional sequences. Well control can provide data for the distinction between coastal and marine facies within the sequences.

Determinations of the durations and magnitudes of relative changes of sea level and the durations of the relative stillstands are needed to construct charts showing cycles of relative rises, stillstands, and falls of sea level. Such charts summarize the history of the fluctuations of sea level,

which is the effective base level during the deposition of most maritime sequences and during subsequent erosion. These fluctuations control the distribution of the sequences and the strata within them, and the extent of the unconformities and correlative conformities along the sequence boundaries.

This study examines concepts of relative changes of sea level, some of the depositional patterns related to these changes, and how regional charts showing cycles of relative rise, stillstand, and fall of sea level may be constructed. The following paper (Part 4, Vail et al, this volume) will show how charts of regional cycles are used to determine global cycles.

CYCLES OF RELATIVE CHANGE OF SEA LEVEL

A *relative change of sea level* is defined as an apparent rise or fall of sea level with respect to the land surface. Either sea level itself, or the land surface, or both in combination may rise or fall during a relative change. A relative change may be operative on a local, regional, or global scale. This paper will work with conceptual models showing relative changes and stillstands on a regional scale. Simultaneous relative changes in three or more widely spaced regions around the globe are interpreted as global changes of sea level (Part 4, Vail et al, this volume).

A *cycle of relative change of sea level* is defined as an interval of time during which a relative rise and fall of sea level takes place. A cycle may be recognized on a local, regional, or global scale, but this paper will deal with the regional ones. Most regional cycles are eventually determined to be global, but even those that are not are useful in regional stratigraphic studies.

A cycle of relative rise and fall of sea level typically consists of a gradual relative rise, a period of stillstand, and a rapid relative fall of sea level (Fig. 1). In detail, the gradual cumulative rise consists of a number of smaller scale rapid rises and stillstands. Such a small-scale event is a *paracycle*, defined as a relative rise and stillstand of

[1]Manuscript received, January 6, 1977; accepted, June 13, 1977.

[2]Exxon Production Research Co., Houston, Texas 77001.

[3]New Mexico Bureau of Mines and Mineral Resources, Socorro, New Mexico 87801.

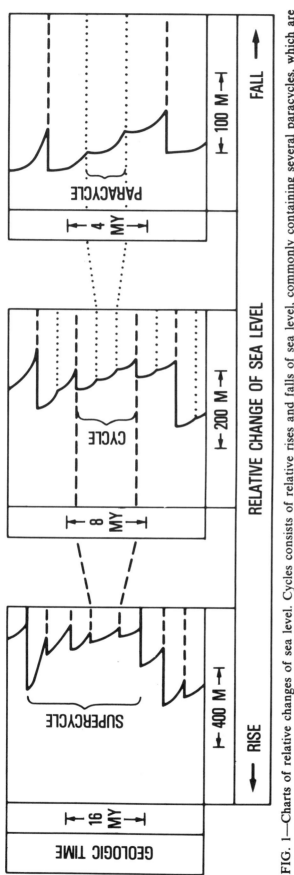

FIG. 1—Charts of relative changes of sea level. Cycles consists of relative rises and falls of sea level, commonly containing several paracycles, which are smaller scale pulses of relative rises to stillstands. Several cycles usually form a higher order cycle (supercycle) with pattern of successive rises between major falls. Note asymmetry of gradual rises and abrupt falls at each scale.

sea level, followed by another relative rise with no significant fall intervening. These small-scale events commonly are not detected with seismic data, but are more readily recognized with data observed in outcrops, cores, and well logs.

Commonly, a set of several regional or global cycles will form a distinctive pattern consisting of successive rises to higher relative positions of sea level, followed by one or more major relative falls to a lower position. This set forms a cycle of higher order (Part 4, Vail et al, this volume). Such a set of cycles is termed a supercycle.

A hierarchy of supercycle, cycle, and paracycle reflects relative changes of sea level of different orders of magnitude. As seen in Figure 1, a supercycle contains several cycles, and a cycle may contain several paracycles. These relations are shown on a regional or global cycle curve.

RELATIONS OF DEPOSITIONAL SEQUENCES TO CYCLES AND PARACYCLES OF RELATIVE CHANGES OF SEA LEVEL

Our basic operational stratigraphic unit is the depositional sequence, defined as a stratigraphic unit composed of a relatively conformable succession of genetically related strata and bounded at its top and base by unconformities or their correlative conformities (Part 2, Vail et al, this volume). If a sufficient sediment supply is available, one or more depositional sequences are deposited during one cycle of relative rise and fall of sea level.

If a cycle contains a continuous relative rise to stillstand of sea level, only one sequence is likely to be deposited during that time. The abrupt fall at the end of the cycle tends to produce an unconformity that will separate the sequence from the overlying one of the next cycle.

If a cycle contains two or more paracycles, at least two sequences are likely to be deposited during the cycle. The boundary between the sequences would be marked most commonly by downlap of the overlying sequence, although toplap of the underlying sequence may be present.

Two or more sequences may be deposited during a cycle or a paracycle. After a rapid rise of sea level, a surface of non-deposition may be developed before the progradational deposits of the stillstand are laid down. The surface should be marked by downlap of the overlying progradational deposits. Frazier (1974) recognized such surfaces in defining depositional episodes during the Pleistocene of the Gulf of Mexico. Each sequence of transgressive sandstones is overlain by a sequence of upward coarsening, progradational strata.

295

The smaller scale sequences deposited during paracycles or in shorter pulses during development of delta lobes may be too thin to be recognized with seismic data alone; outcrops and/or close well control with adequate cores may be needed. Stratal terminations are commonly subtle and depositional in nature.

In the construction of charts of relative changes of sea level, a simple one-to-one relation of depositional sequences to sea-level cycles should not necessarily be expected. Each sequence is a building block in the regional stratigraphic framework; and comprehensive analyses of the strata, their facies, and their ages are needed to determine if one sequence or a set of them represents one cycle of relative change of sea level.

In general, the greater the sea-level fall the easier it is to recognize sequence boundaries by onlap, downlap, and truncation. If structuring occurs, erosion of the tilted strata during a succeeding lowstand commonly causes spectacular angular unconformities.

Maritime and Hinterland Sequences

Distribution of the strata and facies is controlled directly by relative changes in sea level in some depositional sequences and indirectly or not at all in others. The primary consideration is whether a sequence was deposited in a maritime or hinterland environment.

A *maritime sequence* is a depositional sequence that consists of genetically related coastal and/or marine deposits. The coastal facies of a maritime sequence especially is controlled by the position of sea level as a base level. Shallow marine facies are partially controlled by sea level, but the deep marine facies are not directly controlled by it. Therefore, cyclical changes in the relative position of sea level exert a major control on the landward extent of maritime depositional sequences.

A *hinterland sequence* is one that consists entirely of nonmarine deposits laid down at a site interior to the coastal area, where depositional mechanisms are controlled indirectly or not at all by the position of sea level. Although our experience with hinterland sequences is limited, they seem to be deposited independently of the maritime sequences, and therefore are omitted from the discussion of relative changes of sea level.

INDICATORS OF RELATIVE CHANGES OF SEA LEVEL

The most reliable stratigraphic indicators of relative changes of sea level are the depositional limits of onlap and toplap (Part 2, Vail et al, this volume) within the coastal facies of maritime sequences. With adequate paleobathymetric control, marine facies may be used; however, the deep-marine control generally is not adequate. Other methods may be employed to help measure relative changes of sea level, but there are pitfalls in using alone such phenomena as transgression/regression of the shoreline and deepening/shallowing of the sea bottom.

The next sections discuss some basic concepts dealing with the stratigraphic indicators used to determine relative changes of sea level in maritime sequences. For purposes of illustration, diagrams show deposits of terrigenous clastics, but the models apply also to carbonates and other rock types. Parts of these concepts are modified from the works of Weller (1960, p. 498-501), van Andel and Curray (1960), Curray (1964), and others. Pitman (1977) discussed interrelations of eustatic changes of sea level, tectonic movements, and rates of sediment supply.

Relative Rise of Sea Level

A *relative rise of sea level* is an apparent sea level rise with respect to the underlying initial depositional surface (Fig. 2) and is indicated by coastal onlap. It may result from (1) sea level actually rising while the underlying initial surface of deposition subsides, remains stationary, or rises at a slower rate; (2) sea level remaining stationary while the initial surface of deposition subsides; or (3) sea level falling while the initial surface of deposition subsides at a faster rate. Additionally, *coastal onlap* is the progressive landward onlap of littoral and/or nonmarine coastal deposits in a given maritime sequence. Coastal deposits may be determined by paleoecology or sedimentology.

During a relative rise of sea level, where the sedimentary supply is sufficient, coastal deposits progressively onlap the underlying initial surface of deposition. The process is unable to build much above sea level, which approximates effective depositional base level. Without the rise of effective base level, the depositional site would be unable to accommodate the sediment, and each increment of coastal deposition would be terminated laterally before it could onlap the depositional surface.

A relative rise of sea level can be measured most accurately where littoral deposits (those laid down between low and high tide) onlap the underlying depositional surface. However, nonmarine coastal deposits (those laid down on the coastal plain above high tide) most commonly onlap the surface; they may build a few meters above sea level, introducing a small error in the

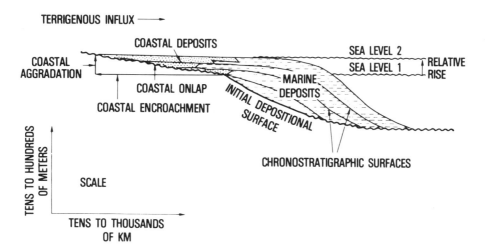

FIG. 2—Coastal onlap indicates a relative rise of sea level. Relative rise of base level allows coastal deposits of a maritime sequence to aggrade and onlap initial depositional surface.

LOW TERRIGENOUS INFLUX — TRANSGRESSION

COASTAL AGGRADATION

COASTAL ONLAP

COASTAL ENCROACHMENT

INITIAL DEPOSITIONAL SURFACE

FINAL

INITIAL

RELATIVE RISE

HIGH TERRIGENOUS INFLUX — REGRESSION

COASTAL ONLAP

INITIAL DEPOSITIONAL SURFACE

FINAL

INITIAL

RELATIVE RISE

BALANCED TERRIGENOUS INFLUX — STATIONARY SHORELINE

COASTAL ONLAP

INITIAL DEPOSITIONAL SURFACE

FINAL

INITIAL

RELATIVE RISE

NONMARINE COASTAL DEPOSITS

LITTORAL DEPOSITS

MARINE DEPOSITS

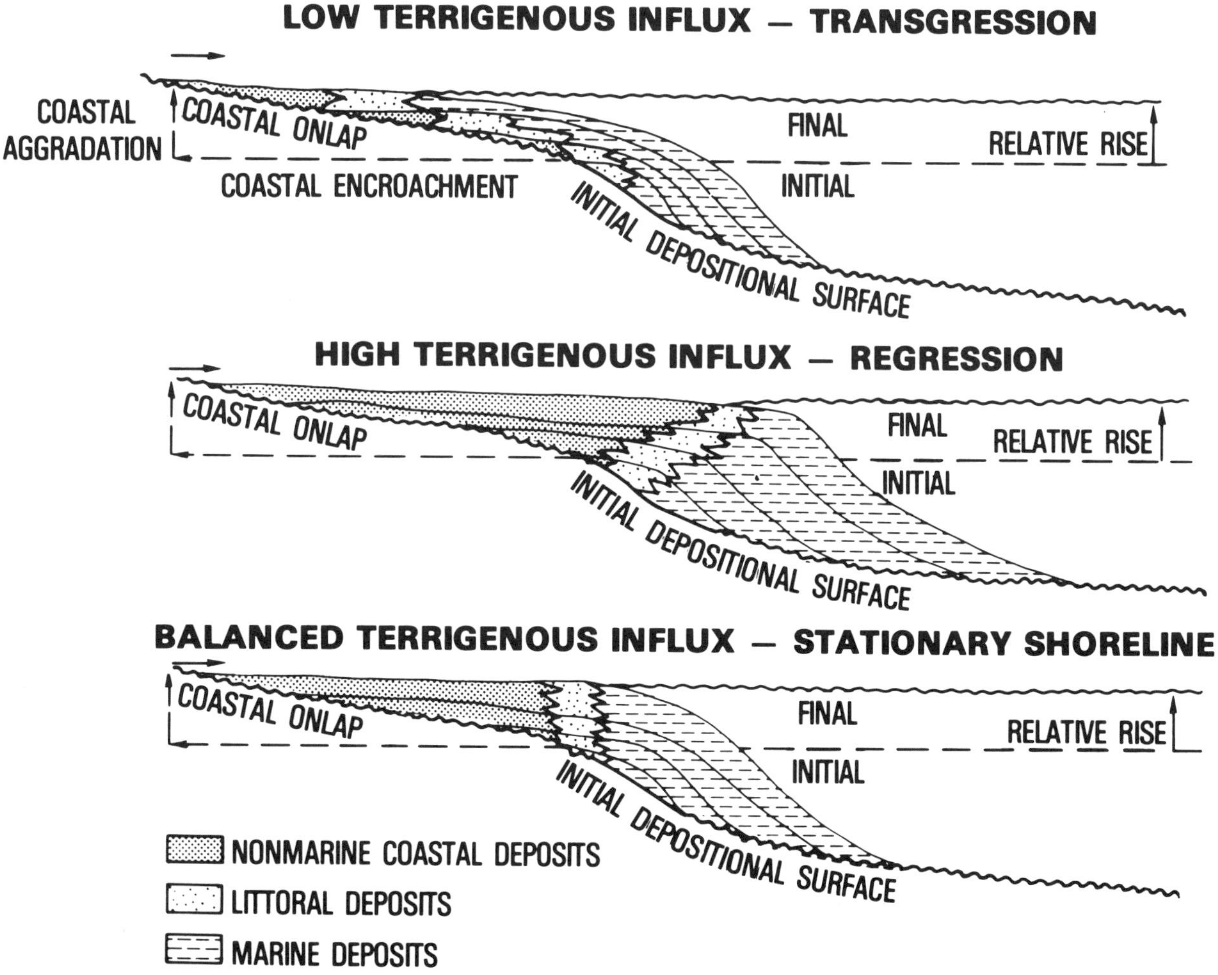

FIG. 3—Transgression, regression, and coastal onlap during relative rise of sea level. Rate of terrigenous influx determines whether transgression, regression, or stationary shoreline is produced during relative rise of sea level.

estimate of the relative rise. On coastal plains that are hundreds of kilometers wide, the error may be several tens of meters.

A relative rise of sea level may be measured on a stratigraphic cross section provided that structural deformation is not too intense. A seismic section is excellent for this purpose. Either the vertical or the horizontal components of coastal onlap can be used, and are termed coastal aggradation and coastal encroachment, respectively (Fig. 2). However, measurements of coastal aggradation should be adjusted for any distortion due to basinward thickening of strata that may have resulted from differential basinward subsidence. Likewise, measurements of coastal encroachment need to be adjusted for variations in the slope of the underlying initial depositional surface that would distort measurements of sea level rise. Other variables, such as compaction, may need to be considered in individual cases.

The onlapping coastal deposits of a particular sequence may have been removed by erosion at a given locality. In some cases the missing strata may be restored to a projection of the underlying initial deposition surface, sometimes with the help of isolated erosional remnants, but generally it is better to search the region for a section in which the coastal onlap is preserved.

Where a relative rise of sea level is more rapid than the rate of deposition, the result may be on-

lap of marine strata (marine onlap) instead of coastal onlap, and paleobathymetric control will be needed to help measure the relative rise. Assuming that structural movements are not complicated, the amount of the relative rise may be approximated by measuring the vertical component of marine onlap (marine aggradation) if the paleobathymetry remains constant. If not, the amount of rise may be estimated by determining the marine aggradation plus any amount of deepening, or minus a lesser amount of shallowing. Because paleobathymetric measurements are given in intervals of several hundreds of feet, the measurement of relative rise with marine onlap is only an approximation and should be checked against measurements of coastal onlap of the same unit in other areas.

During a relative rise of sea level, a transgression or regression of the shoreline, and deepening or shallowing of the sea bottom, may take place. Marine transgression and regression during a relative rise of sea level are illustrated in Figures 3 and 4. A transgression of the shoreline is indicated by landward migration of the littoral facies in a given stratigraphic unit, and a regression is indicated by seaward migration of the littoral facies. Instead of transgression or regression, the shoreline may be stationary. Similarly, a deepening of the sea bottom (Fig. 4) is indicated by evidence of increasing water depth, and a shallowing

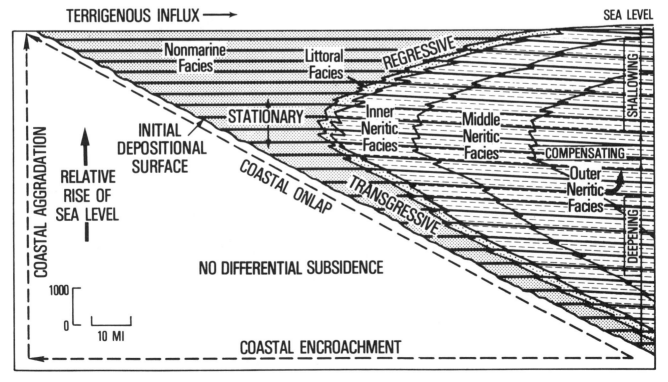

FIG. 4—Coastal onlap with marine transgression and regression. During relative rise of sea level, littoral facies may be transgressive, stationary, or regressive, and neritic facies may be deepening, compensating, or shallowing.

298

of the sea bottom is indicated by evidence of decreasing water depth. Instead of deepening or shallowing, the sea bottom may be compensating. Any of the above may take place during a relative rise of sea level. A common misconception is that transgression and deepening are synonymous with a relative rise, and that regression and shallowing are synonymous with a relative fall of sea level.

Although it is true that either transgression or deepening may indicate at least a part of a relative rise of sea level, neither can be assumed to indicate the entire rise. A transgression may be terminated by an increase in terrigenous clastic supply to produce a stationary shoreline or a regression while the relative rise of sea level continues (Grabau, 1924, p. 724; Weller, 1960, p. 500; van Andel and Curray, 1960; Curray, 1964). A bathymetric deepening may also be terminated by an increase in sediment supply to produce a shallowing sea bottom while the relative rise continues. Moreover, the distribution of sediment may be such that the littoral facies is regressive while a starved marine section is deepening during a relative rise. Because regression and shallowing can occur during a relative rise, stillstand, or fall of sea level, they cannot be indicative of any one of them. However, regression is most common during a relative rise or stillstand.

On a seismic section through the coastal facies of a maritime sequence, a relative rise of sea level can be recognized by onlapping reflections. Transgressions and regressions of the shoreline are recognized with more difficulty by lateral changes in reflection characteristics, such as amplitude, frequency, and wave form, indicating changes from coastal to marine facies. This point is illustrated in the following example.

Northwestern Africa Example

On a seismic section from the continental margin of northwestern Africa (Fig. 5), a major relative rise of sea level is indicated by coastal onlap. Also, marine transgression and regression are indicated by landward and seaward migration of the littoral facies during this rise. A large sedimentary wedge, Triassic through Tertiary in age, thickens seaward into the Atlantic. Boundaries of depositional sequences (shown by the heavy black lines) have been interpreted from the onlapping and downlapping reflection terminations (shown by arrows). Paleontologic control in wells indicates the two major sequences to be Lower and Upper Cretaceous. Coastal and marine deposits are also identified from well control. Individual seismic reflectors can be traced from interpreted marine deposits on the left into coastal

deposits on the right and finally to the point of onlap against the underlying unconformity. Shown on the section are an Early Cretaceous transgression, stationary shoreline, and regression, followed by a Late Cretaceous transgression and regression. Both these cycles of transgression occurred during a relative rise of sea level documented by continuous coastal onlap during Early Cretaceous and much of Late Cretaceous time.

Relative Stillstand of Sea Level

A *relative stillstand of sea level* is an apparently constant position of sea level with respect to the underlying initial surface of deposition, and is indicated by coastal toplap. It may result if both sea level and the underlying initial suurface of deposition actually remain stationary, or if both rise or fall at the same rate.

During a relative stillstand of sea level, where the sedimentary supply is sufficient, deposition in the coastal environment is hindered in any attempt to build above the effective base level, and the strata are prevented from onlapping the initial depositional surface (Fig. 6). The result is coastal toplap—the toplap (Part 2, Vail et al, this volume) of coastal deposits in a depositional sequence. Each unit of strata laps out in a landward direction at the top of the unit, but the successive terminations lie progressively seaward. Toplap can be recognized on seismic sections, and the evidence of coastal deposits can be determined from paleoecology or sedimentology using well data.

If a relative stillstand occurs after a rise that is more rapid than the rate of deposition, the result may again be marine onlap. In such a case, the stillstand may be reflected in the paleobathymetry as a shallowing at the same rate as the marine aggradation, again assuming no differential structural movements or other complications.

A cumulative relative rise of sea level that occurs over several million years commonly is characterized by shorter pulses of sea-level rise alternating with intervals of stillstand. In general, the rapid rises are more frequent and are of greater magnitude in the early part of the cumulative rise, and the stillstands are more frequent and last longer in the later part of the cumulative rise. This gradual diminishing of the rapid rises slows down the rate of the cumulative rise with time.

The cyclic pulses consisting of alternations of rapid rises and stillstands are called paracycles (Fig. 1). They are recognized frequently as smaller scale depositional sequences on detailed well-log or outcrop sections, but commonly are too small to be recognized on seismic sections.

Minor surfaces displaying toplap may be produced by periods of rapid excess deposition of

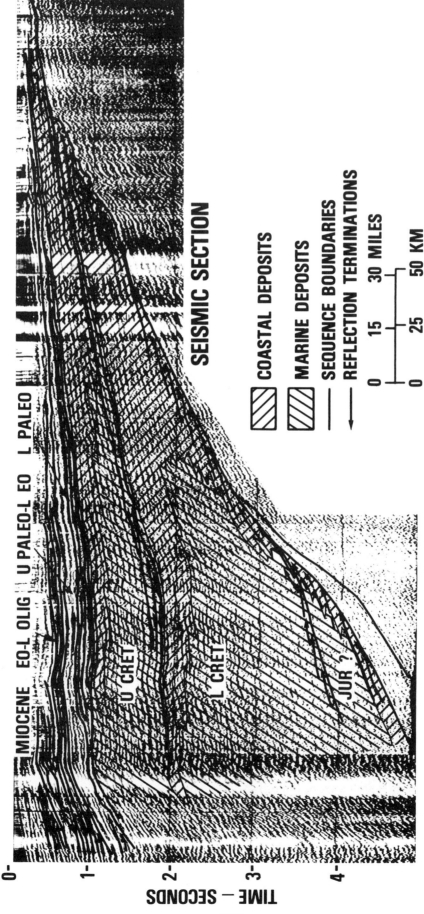

FIG. 5—Offshore West Africa seismic example. Coastal onlap with transgression and regression.

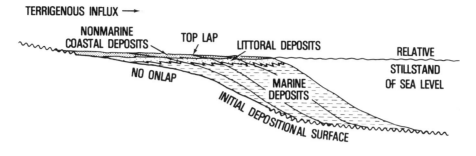

FIG. 6—Coastal toplap indicates relative stillstand of sea level. With no relative rise of base level, nonmarine coastal and/or littoral deposits cannot aggrade, so no onlap is produced; instead, by-passing produces toplap.

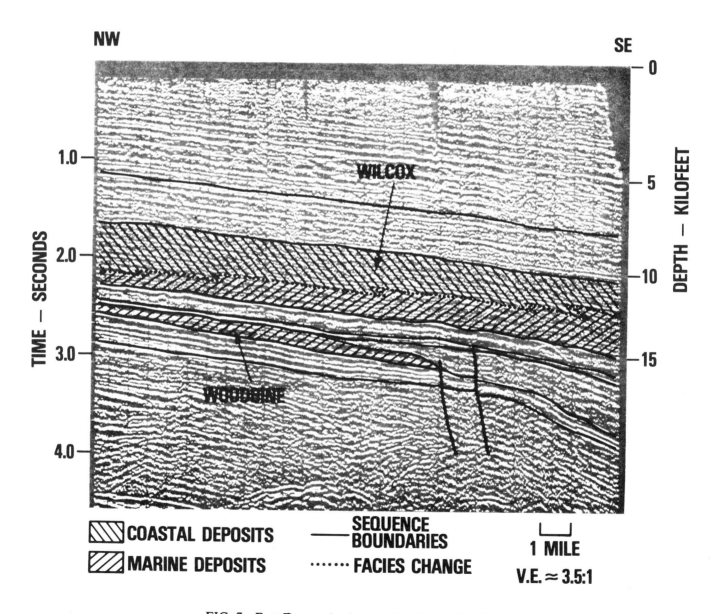

FIG. 7—East Texas seismic example of coastal toplap.

clastic sediments during a relative rise. These indicators of relative stillstand may be localized to areas of abnormally high depositional rates, such as in deltaic lobes. These local paracycles may or may not be related to more regional or global stillstands.

Texas Gulf Coast Example

Two examples of relative stillstand marked by coastal toplap are shown on the seismic section of Figure 7. These are from the Woodbine and Wilcox stratigraphic units of eastern Texas. Coastal and marine deposits were identified from wells.

The Woodbine (middle Cenomanian) forms a depositional sequence that consists mostly of shale in this area and lies basinward from the famous East Texas field. High-amplitude, slightly divergent, continuous reflectors mark the top and the base of the Woodbine sequence. A series of more steeply dipping reflections between these two strong reflections forms an oblique progradational configuration indicating depositional clinoforms (Part 6, Vail et al, this volume). These steeply dipping reflectors terminate by toplap immediately beneath the reflection marking the base of the overlying sequence. Thin deltaic sandstones at this toplap position produce gas in the area (Part 5, Vail et al, this volume). This Woodbine example closely resembles the diagrammatic example in Figure 6.

The Wilcox (Paleocene-Eocene) shows a variation of this toplap pattern. Reflections above and below the Wilcox are parallel. Within the Wilcox, a number of reflections dip at a greater angle than the bounding reflectors, indicating prograding clinoforms. To the northwest (left), reflections from the oblique prograding clinoforms show toplap below a stratigraphic horizon in the lower third of the Wilcox. To the southeast (basinward), the toplap pattern occurs at progressively higher stratigraphic levels. This pattern is interpreted as an alternation of relative stillstands and relative rises of sea level.

Relative Fall of Sea Level

A *relative fall of sea level* is an apparent fall of sea level with respect to the underlying initial surface of deposition, indicated by a downward shift of coastal onlap. It may result: (1) if sea level actually falls while the initial surface of deposition rises, remains stationary, or subsides at a slower rate; (2) if sea level remains stationary while the surface is rising; or (3) if sea level rises while the surface is rising at a faster rate.

A *downward shift of coastal onlap* is a shift downslope and seaward from the highest position of coastal onlap in a given maritime sequence to the lowest position of coastal onlap in the overlying sequence. In Figure 8a, the downward shift occurs between the highest coastal onlap of unit 5 in sequence A and the lowest coastal onlap of unit 6 in sequence B. The patterns of onlap indicate a relative rise of sea level during deposition of sequence A, then an abrupt relative fall to the position of unit 6 in sequence B, followed by another rise during deposition of sequence B.

In the seismic examples we have studied, fall of sea level is indicated by such an abrupt shift, and appears to occur as one event, rather than a succession of events. Where accurately dated, each fall is "rapid," occurring within a million years or less.

A stratification pattern, presented earlier by Weller (1960, Fig. 189B) and reproduced in Figure 8b, shows a series of units prograding at successively lower levels during a relative gradual fall of sea level. An example is the lower Gallup Formation in New Mexico, where several small falls in a short period of time produce a pattern similar to that of Figure 8b. A cross section showing these features in the lower Gallup is shown in Campbell (1977). These features are commonly beyond seismic resolution. In addition, the areas where we have seismic and well control are commonly areas of thick sedimentary deposits. In such areas, regional subsidence may proceed at a faster rate than any gradual fall of sea level and a relative rise is produced. Therefore, an actual fall in sea level, especially a gradual one, might not be detected by using coastal onlap as a criterion.

To measure a relative fall of sea level, the initial difference in elevation is determined between the highest coastal onlap in the underlying sequence (unit 5 in sequence A, Fig. 8a) and the lowest coastal onlap in the overlying sequence (unit 6 in sequence B). Difficulties encountered in the actual measurement are: (1) the underlying unit commonly is eroded during a relative fall, and if so, it must be approximated by restoration; (2) differential basinward subsidence that took place during the fall and the ensuing rise must be corrected; and (3) marine (instead of coastal) onlap is commonly encountered in the oldest beds of the overlying unit and accurate paleobathymetric determinations are needed to determine a relative fall of sea level.

After a relative fall, marine onlap again may result if the relative rise is more rapid than the rate of deposition. Paleobathymetric shallowing in deep marine strata may be helpful in measuring actual falls of sea level, where there are no structural complications.

a) DOWNWARD SHIFT IN COASTAL ONLAP INDICATES RAPID FALL

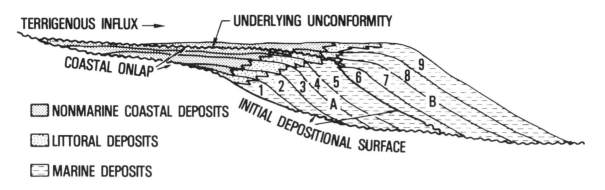

TERRIGENOUS INFLUX →

UNDERLYING UNCONFORMITY

COASTAL ONLAP

1 2 3 4 5 6 7 8 9

A

B

INITIAL DEPOSITIONAL SURFACE

⬛ NONMARINE COASTAL DEPOSITS

▨ LITTORAL DEPOSITS

▱ MARINE DEPOSITS

b) DOWNWARD SHIFT IN CLINOFORM PATTERN INDICATES GRADUAL FALL

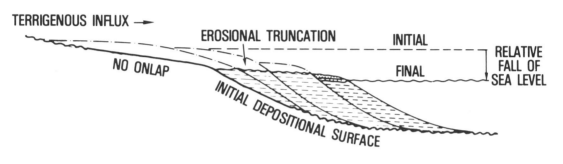

TERRIGENOUS INFLUX →

EROSIONAL TRUNCATION

INITIAL

FINAL

RELATIVE FALL OF SEA LEVEL

NO ONLAP

INITIAL DEPOSITIONAL SURFACE

FIG. 8—Downward shift of coastal onlap indicates relative fall of sea level. With relative fall of base level, erosion is likely: deposition is resumed with coastal onlap during subsequent rise. (a) Downward shift in coastal onlap indicates rapid fall observed in all cases studied so far. (b) Downward shift in clinoform pattern (after Weller, 1960), indicates gradual fall; but has not been observed on seismic data.

Because a major fall is so difficult to measure accurately, it rarely can be plotted quantitatively with any degree of confidence.

After a major relative fall of sea level, the shelf tends to be bypassed, and the area of coastal onlap may be restricted to the apex of a fan at the basin margin. The block diagrams in Figure 9 show differences in stratal patterns during periods of highstand and lowstand of sea level. A marked change in the depositional pattern of deep-marine strata commonly occurs as a result of a major relative fall of sea level.

Figure 9a illustrates an idealized pattern developed during a highstand of sea level when shallow seas cover much of the shelf. Deposition occurs in clinoform lobes that prograde across the shallow shelf; with sufficient sediment supply, the progradation continues into deep water as shown in this illustration. Coarse clastics tend to be trapped on the shelf, and finer material is transported to the toes of the clinoforms.

Figure 9b illustrates the depositional pattern after a major fall of sea level below the shelf edge. The shelf is exposed to subaerial erosion, and rivers tend to bypass the shelf and deposit directly onto the slope. During the ensuing rise of sea lev-

el, any coastal onlap occurs near the sediment source; marine onlap may be produced if the sediments are funnelled through a submarine canyon. Most of the onlap commonly occurs in marine deposits at the proximal edge of a submarine fan, as shown in Figure 9b. This type of marine onlap may extend to very deep water, where detailed bathymetric control is needed to calculate the magnitude of relative changes of sea level. However, the overall pattern is diagnostic of a major relative fall of sea level and is common along continental margins and in deep-marine basins.

San Joaquin Basin (California) Example

A seismic section from San Joaquin basin, California (Fig. 10), shows the upper Tertiary section divided into nine depositional sequences; the depositional environments are identified generally. In the middle Miocene sequence, consisting of the lower part of the Fruitvale formation, a prominent shelf-edge is evident, and the sequence is thickest on the shelfward side to the east. The sequence thins basinward along well-developed clinoforms. This sequence is very similar to the basinward part of sequence A in Fig. 8a.

303

The upper Miocene sequence consists of the upper Fruitvale–McLure shale and Santa Margarita Sandstone. It closely resembles sequence B (in Fig. 8a) as it onlaps against the underlying middle Miocene sequence; and it laps out completely at the middle Miocene shelf edge. The sequence thins as it progrades into the basin. Well control indicates marine onlap in the lower part of the sequence. The upper part of the sequence is composed of the Santa Margarita Sandstone which was deposited in a shallow-marine to deltaic environment. Therefore, coastal onlap is demonstrated in the upper part of the sequence.

Such coastal onlap of a sequence restricted to the basin indicates a major relative fall of sea level. Because this sequence overlies a widespread shelf-type sequence, and because there is no significant difference in age between them, the fall must have been relatively rapid.

North Sea Example

Several examples of a shift in depositional patterns with relative falls of sea level are shown in Figures 11 and 12. These are overlapping seismic sections showing the Tertiary of the North Sea.

The two sections together are more than 160 km long, and vertical exaggeration is about 20 to 1. Depositional sequences are dated and the environments of the strata are determined from well control. Shelf sediments pass into basinal deposits from right to left.

The sequences that are widespread on the shelf are interpreted as highstand deposits, and the sequences that lap out against the slope or lower part of the shelf are interpreted as lowstand deposits. Highstand sequences are thickest near the shelf edge and extend long distances across the shelf to the right. The overall depositional pattern is that of progradation and downlap in a basinward direction. Prominent highstand sequences were deposited in late Paleocene–early Eocene, early Oligocene, middle Miocene, and early Pliocene times, as shown on the chart of relative change of sea level in Figures 11 and 12.

In contrast, lowstand sequences are thickest in the basin to the left and thin rapidly shelfward by marine onlap. Lowstand sequences were deposited in middle Paleocene, early middle Eocene, middle and late Oligocene and early Miocene, late Miocene, and late Pliocene times. The low-

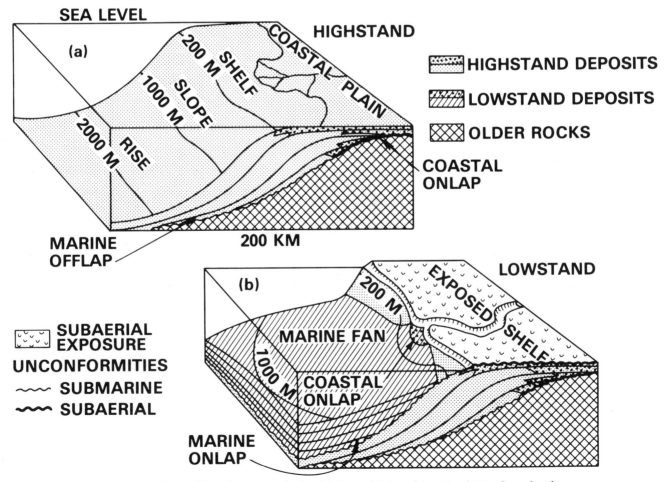

FIG. 9—Depositional patterns during highstand (a) and lowstand (b) of sea level.

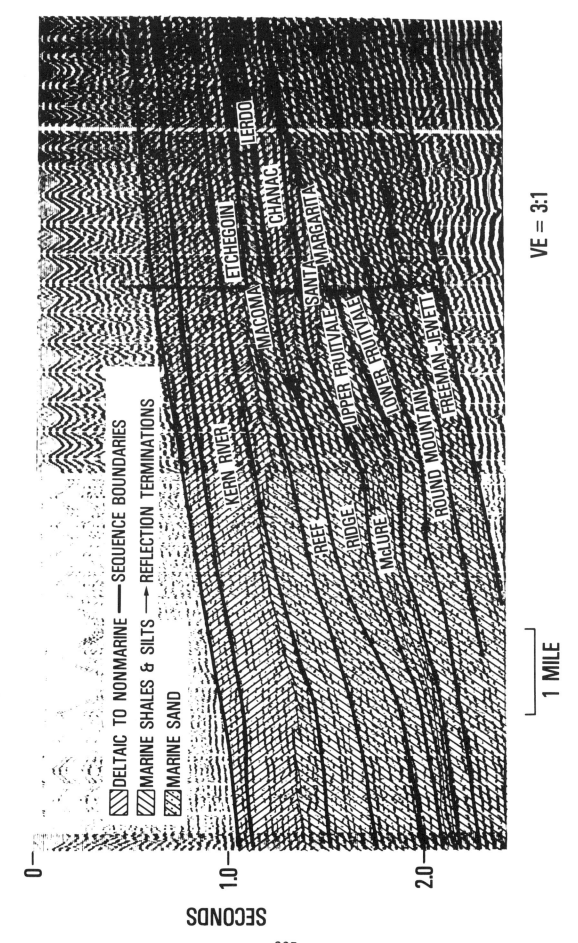

FIG. 10 — San Joaquin basin (California) seismic example of downward shift in coastal onlap.

VE = 3:1

1 MILE

SECONDS

ETCHEGOIN
LERDO
CHANAC
MACOMA
SANTA MARGARITA
KERN RIVER
UPPER FRUITVALE
LOWER FRUITVALE
ROUND MOUNTAIN
FREEMAN–JEWETT
REEF
RIDGE
McLURE

DELTAIC TO NONMARINE — SEQUENCE BOUNDARIES
MARINE SHALES & SILTS → REFLECTION TERMINATIONS
MARINE SAND

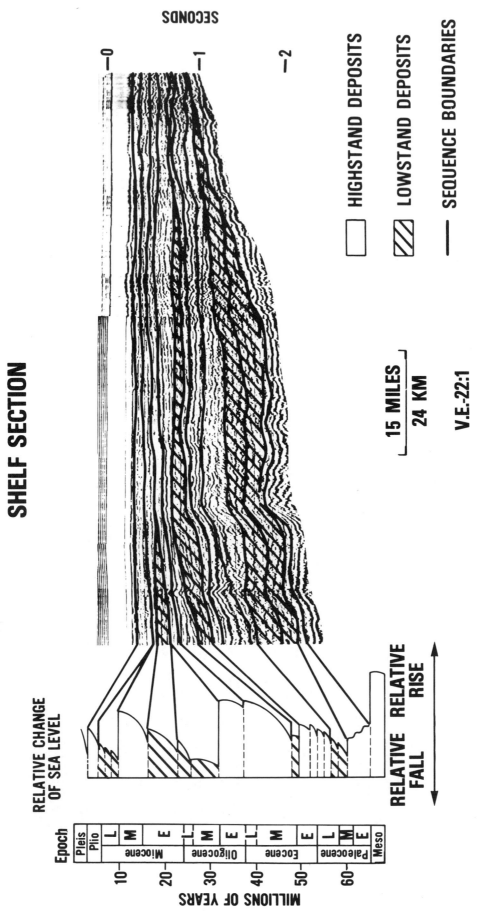

FIG. 11—North Sea Tertiary shelf seismic example of depositional patterns during highstands and lowstands of sea level.

306

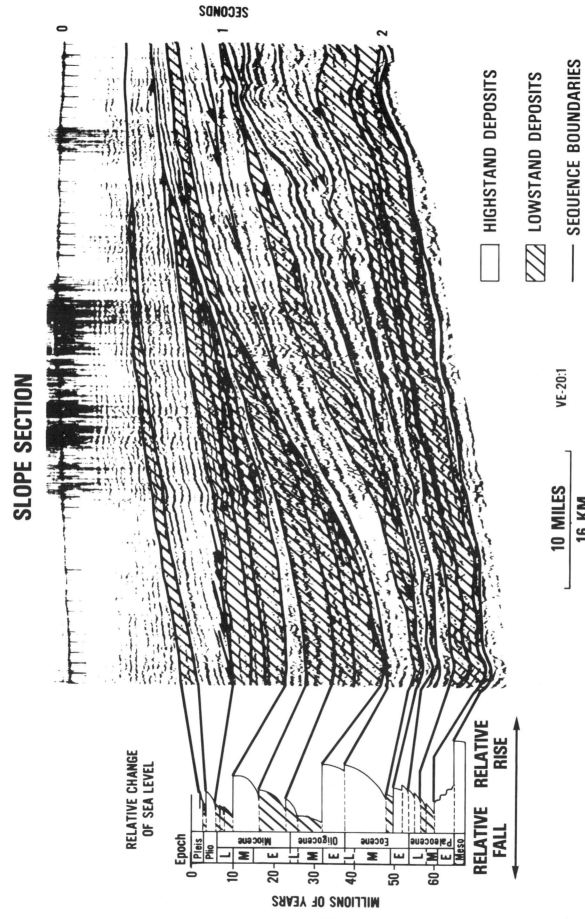

SLOPE SECTION

FIG. 12—North Sea Tertiary slope seismic example of depositional patterns during highstands and lowstands of sea level.

HIGHSTAND DEPOSITS

LOWSTAND DEPOSITS

SEQUENCE BOUNDARIES

VE-20:1

10 MILES
16 KM

RELATIVE CHANGE OF SEA LEVEL

RELATIVE RISE

RELATIVE FALL

Epoch

Pleis
Plio
Miocene
Oligocene
Eocene
Paleocene
Meso

L M E L M E L M E L M E

SECONDS

MILLIONS OF YEARS

stand sequences of late-middle and late Oligocene age are missing over much of north-central Europe. Pleistocene units are not discussed here.

CONSTRUCTION OF REGIONAL CURVES OF RELATIVE CHANGES OF SEA LEVEL

The preceding sections showed how to recognize cycles of relative rise, stillstand, and fall of sea level, using coastal onlap, toplap, and other stratigraphic criteria. This section describes how to plot curves of these cycles for a given region. A *region*, in the sense used here, can be an entire oceanic domain, a continental margin, an inland sea, or a strip of coast. The main requirement is that the sequences of strata had some original physical continuity, so that the regional charts are based on strata that have a definite relation in both time and space.

A sea-level curve may be compared with other regional data by plotting it on a chronostratigraphic correlation chart for the region (Part 7, Vail et al, this volume). Such a combination shows the relations of sea-level changes to geologic age, distribution of depositional sequences, unconformities, facies and environment, and other information.

The next paper of this suite (Part 4, Vail et al) will show how to compare several sea-level curves from different regions and produce a global chart of relative changes of sea level. The global chart may be used in an undrilled region to predict age and general depositional characteristics of sequences. As data are acquired, the regional chart can be improved, and comparisons with the global chart may show regional anomalies.

Procedure

The procedure for constructing a sea-level curve will be illustrated diagrammatically and also with a seismic example from northwestern Africa. The curve is plotted with respect to geologic time and shows the relative rises, stillstands, and falls of sea level determined in the region. A regional grid of stratigraphic cross sections is needed to draw the curve; these sections should include the shelf edge and show the most complete record of coastal onlap in the region. Seismic sections are excellent for this purpose if they have good reflection quality and sufficient well control, and if structural deformation has not been too complicated.

Figure 13 diagrammatically illustrates the steps in constructing a sea-level curve. The first step is to analyze the maritime sequences (A to E in Fig. 13a). This step includes determinations of sequence boundaries, ages, areal distributions, and the presence of coastal onlap and toplap. On seismic sections the reflections representing strata within each sequence are traced to the proximal depositional limits of onlap or toplap. Environmental control is added to distinguish coastal and marine facies, and available age control is used to determine the geologic-time range of each sequence. The same procedures may be followed using subsurface well logs or surface sections, as long as the stratification surfaces are traced accurately.

The second step is to construct a chronostratigraphic correlation chart (Fig. 13b) of the sequences (Part 2, Vail et al, this volume). This chart plots the information shown on the stratigraphic cross section (Fig. 13a) against geologic time. The geologic-time ranges and relative areal distributions of the sequences are shown on this type of display.

After determining and plotting the ages of the depositional sequences, the third step is to identify the cycles of relative rise and fall of sea level, to measure the magnitudes of the rises and falls, and to plot them and the stillstands with respect to geologic time (Fig. 13c). Coastal aggradation is the best measure of a relative rise, and a downward shift of coastal onlap is the best measure of a relative fall of sea level. Coastal toplap indicates a relative stillstand. Where the onlap is in marine deposits, the paleobathymetry may sometimes be used to determine a relative change of sea level. If the record of the onlap is obscure, or has been removed by erosion, it is preferable to search for more complete sections elsewhere in the region.

In this example, relative rises and falls are determined with measurements of coastal aggradation, the vertical component of coastal onlap (Fig. 13a). Each increment of coastal aggradation is plotted on a geochronologic chart (Fig. 13c). Beginning with the oldest unit, sequence A, the first increment of coastal aggradation is 100 m, and the time interval is from 26 to 24 m.y. before present. Similarly, the successive increments of 150, 100, and 50 m are determined to give a total of 600 m of coastal aggradation during the deposition of sequence A. The measurements of coastal aggradation are made as closely as possible to the underlying unconformity to minimize the effect of differential basinward subsidence.

The unconformity at the top of sequence A shows erosional truncation of some units. However, the youngest coastal onlap in sequence A is dated 17-18 m.y., approximately the same as the oldest coastal onlap in sequence B. Evidence for a significant stillstand is missing. The relative fall of sea level at the end of sequence A occurred in less than one million years. The magnitude of the fall, from the highest coastal onlap in sequence A to

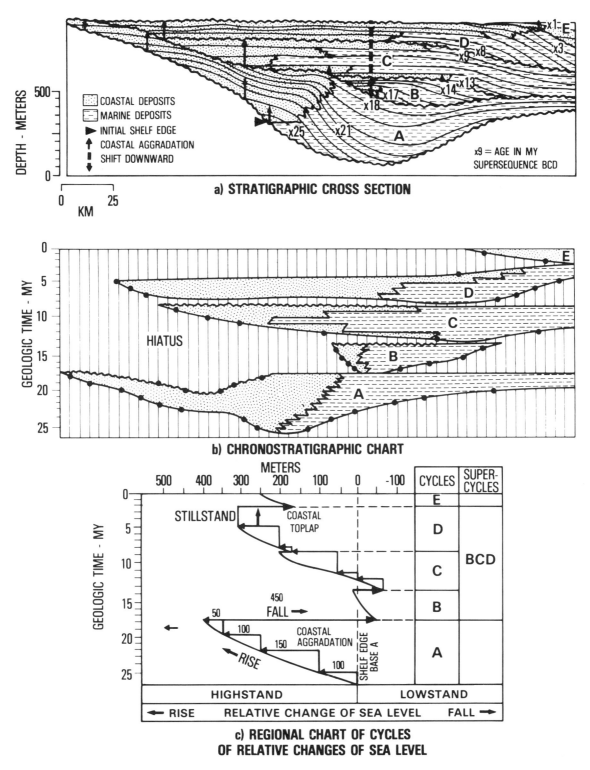

FIG. 13—Procedure for constructing regional chart of cycles of relative changes of sea level.

the lowest coastal onlap in sequence B, is measured as 450 m (heavy dashed line, Fig. 13a). Because the stratal surfaces are parallel over the area of measurement in sequences B, C, and D, no correction for differential basin subsidence is necessary.

Repeating the steps, the cycles of relative changes of sea level are plotted on Figure 13c

from onlap patterns of sequences B, C, D, and E on Figure 13a. In the upper part of sequence D, coastal toplap indicates a relative stillstand of sea level.

In this example, each cycle of relative change of sea level is asymmetrical, with a slow rise to stillstand and a rapid fall (Fig. 13c). This asymmetry of cycles has been observed in the investigations to date. The cycles as a group also show a pattern of asymmetry. Cycle A represents a highstand, cycles B, C, and D begin with a lowstand and gradually rise to a highstand, and cycle E rapidly falls to a lowstand. A higher order cycle (supercycle) BCD is recognized by the progressive rise and fall within the asymmetrical pattern.

A *highstand* is defined as the interval of time when sea level is above the shelf edge, and a *lowstand* as the interval of time when sea level is below the shelf edge. A *comparative* lowstand may be recognized as when sea level is at its lowest position on the shelf during the deposition of a series of sequences, or at its lowest position but when a shelf edge is not evident. Knowledge of these times can be helpful in predicting sedimentary events. For example, the highstands are the most likely times for trapping terrigenous clastics in deltas on the shelf, and the lowstands are the most likely times that the clastics will be funneled through submarine canyons or other notches in the shelf edge and deposited in submarine fans in the basin.

As shown in Fig. 13c, the times when sea level is at the shelf edge may be plotted as arrows on the curves. Then the highstands are to the left of the arrows and the lowstands are to the right. In preparing a group of regional cycle charts, we orient them with the relative rises plotted toward the left (as in Fig. 13c) to correspond to the general convention of showing landward on the left. However, in a given region, the relative rises of sea level are plotted to correspond to the landward direction on the stratigraphic section. Thus a section with landward on the right may be used to construct a chart with relative rises plotted toward the right, as in Figures 11 and 12. When charts for several regions are compared, they all should be oriented the same way.

Together, the stratigraphic section (Fig. 13a), the chronostratigraphic correlation chart (Fig. 13b), and the sea-level cycle curve (Fig. 13c) summarize the relations between sequences, correlation, and cycles of relative changes in sea level. Such a combination provides a summary of the geology within a regional stratigraphic framework. The regional cycles of relative change of sea level shown on the geochronologic chart can be compared with curves of global relative changes of sea level (Part 3, Vail et al, this volume).

In many regional seismic stratigraphic studies, practical problems such as the distribution and quality of data, erosion, or structural displacement of stratigraphic units, make a complete analysis of coastal onlap impractical or impossible. In such areas, a partial curve may be plotted for the cycles represented by data, and the remainder of the regional curve may be inferred from the global cycle chart.

Northwestern Africa Example

Figures 14 and 15 illustrate the procedure for constructing a curve of relative changes of sea level from seismic data for an offshore region in northwestern Africa. Figure 14 shows the same seismic section used in Figure 2, but the environmental interpretation was deleted and the measurement of coastal aggradation was added. From well control, we know the geologic age, and that the onlapping strata were deposited in a coastal environment. Figure 15 shows the chronostratigraphic correlation chart and the chart of relative changes of sea level.

The first step in constructing the curve is to determine the magnitudes of the relative rises from coastal aggradation. On the seismic section (Fig. 14), note where the onlapping reflections in a given sequence are parallel, and where they begin to diverge as a result of differential basinward subsidence. Measure increments of coastal aggradation where the reflections are parallel to avoid adjustments where they diverge. Beginning at the base of the sequence, measure the first increment as the vertical component of coastal onlap from the base of the sequence up to the highest parallel reflection. Then trace this reflection laterally to its point of onlap and measure the next increment at that point. Repeat the process until the top of the sequence is reached. Sum the lengths of the vertical components; seismic times must be converted to depth—the summation gives the total coastal aggradation and thus the total relative rise in sea level.

On this section, we measured 1,100 m of coastal aggradation by the incremental method. Note that the total thickness of the Cretaceous (measured where the reflections are divergent on the left side of the section) is approximately 5,000 m and thus is much greater than the aggradation because of differential basinward subsidence. Incremental measurements of aggradation are made near the point of onlap to avoid such problems with stratal thickening.

In the next step, relative falls are determined from the downward shift in coastal onlap. A prominent downward shift occurs at the Tertiary-Cretaceous boundary between the highstand position of the youngest Upper Cretaceous coastal strata and the oldest coastal onlap of the overlying Paleocene strata. Unfortunately the highest coastal onlap of the Cretaceous strata is not seen on the seismic section. However, much of the relative fall can be determined indirectly by measuring the subsequent rise in early Tertiary time. Differential subsidence occurring during the fall; or

310

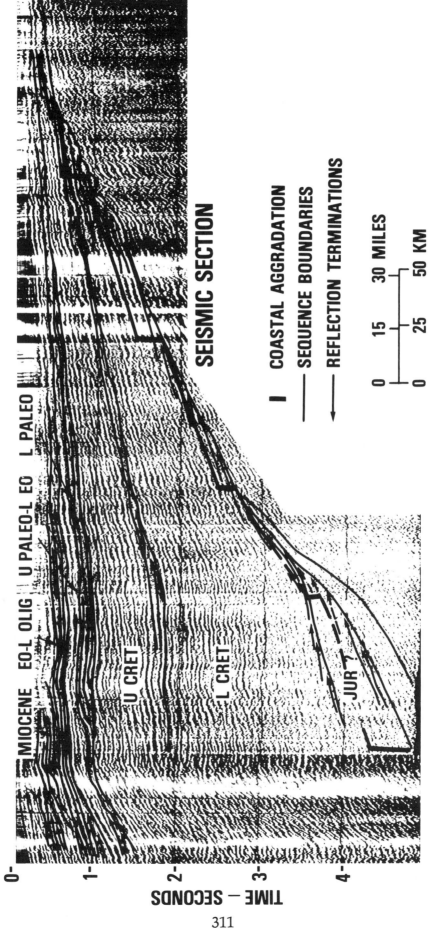

MIOCENE EO-L OLIG U PALEO-L EO L PALEO

U CRET

L CRET

JUR ?

SEISMIC SECTION

| COASTAL AGGRADATION
—— SEQUENCE BOUNDARIES
⟶ REFLECTION TERMINATIONS

```
0        15        30 MILES
|---------|---------|
0        25        50 KM
```

TIME — SECONDS

FIG. 14—Offshore West Africa example of seismic stratigraphic interpretation.

311

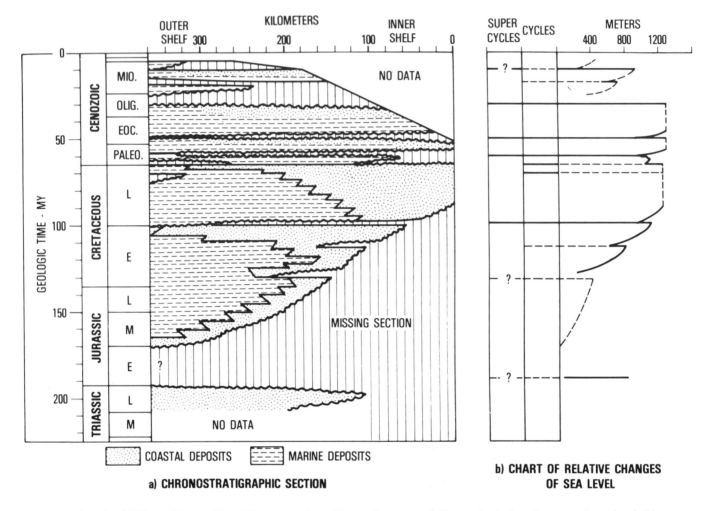

FIG. 15—Offshore West Africa. Chronostratigraphic section (a) and chart of relative changes of sea level (b).

before the rise, would make the subsequent rise greater than than the fall. The estimates of the relative falls are shown on Fig. 15b.

Such complications are encountered so often that accurate measurements of relative falls are rarely obtained. Nevertheless, the best approximations are given on the sea level curves so the magnitudes of relative change of sea level can be expressed to some degree in quantitative terms. Additional data and further research on this problem may yield more accurate results.

The combination of the chronostratigraphic section (Fig. 15a) and the chart of relative changes of sea level (Fig. 15b) provides a summary of the geologic history of the area based on the seismic section (Fig. 14).

REFERENCES CITED

Campbell, C. V., 1977, Depositional model of beach shoreline—Gallup Sandstone (Upper Cretaceous),

northwestern New Mexico: New Mexico Bur. Mines and Mineral Resources Circular (in press).

Curray, J. R., 1964, Transgressions and regressions, in R. L., Miller, ed., Papers in marine geology (Shepard commemorative vol.): New York, Macmillan Co., p. 175-203.

Frazier, D. E., 1974, Depositional episodes—their relationship to the Quaternary stratigraphic framework in the northwestern portion of the Gulf basin: Texas Univ. Bur. Econ. Geology Geol. Circ. 74-1, 28 p.

Grabau, A. W., 1924, Principles of stratigraphy: New York, D. G. Seiler, 1,185 p.

Pitman, W. C., III, 1977, Relationship between sea level changes and stratigraphic sequences: Geol. Soc. America Bull., (in press).

van Andel, T. H., and J. R. Curray, 1960, Regional aspects of modern sedimentation in northern Gulf of Mexico and similar basins, and paleogeographic significance, in F. P. Shepard, F. B. Phleger, and T. H. van Andel, eds., Recent sediments, northwest Gulf of Mexico: AAPG, p. 345-364.

Weller, J. M., 1960, Stratigraphic principles and practices: New York, Harper and Brothers, 725 p.

American Association of Petroleum Geologists Memoir
26, *Seismic Stratigraphy—Applications to Hydrocarbon
Exploration*, edited by Charles E. Payton, copyright 1977,
pp. 83–97.

Seismic Stratigraphy and Global Changes of Sea Level, Part 4: Global Cycles of Relative Changes of Sea Level. [1]

P. R. VAIL, R. M. MITCHUM, JR.,[2] and S. THOMPSON, III[3]

Abstract Cycles of relative change of sea level on a global scale are evident throughout Phanerozoic time. The evidence is based on the facts that many regional cycles determined on different continental margins are simultaneous, and that the relative magnitudes of the changes generally are similar. Because global cycles are records of geotectonic, glacial, and other large-scale processes, they reflect major events of Phanerozoic history.

A global cycle of relative change of sea level is an interval of geologic time during which a relative rise and fall of mean sea level takes place on a global scale. A global cycle may be determined from a modal average of correlative regional cycles derived from seismic stratigraphic studies.

On a global cycle curve for Phanerozoic time, three major orders of cycles are superimposed on the sea-level curve. Cycles of first, second, and third order have durations of 200 to 300 million, 10 to 80 million, and 1 to 10 million years, respectively. Two cycles of the first order, over 14 of the second order, and approximately 80 of the third order are present in the Phanerozoic, not counting late Paleozoic cyclothems. Third-order cycles for the pre-Jurassic and Cretaceous are not shown. Sea-level changes from Cambrian through Early Triassic are not as well documented globally as are those from Late Triassic through Holocene.

Relative changes of sea level from Late Triassic to the present are reasonably well documented with respect to the ages, durations, and relative amplitudes of the second- and third-order cycles, but the amplitudes of the eustatic changes of sea level are only approximations. Our best estimate is that sea level reached a high point near the end of the Campanian (Late Cretaceous) about 350 m above present sea level, and had low points during the Early Jurassic, middle Oligocene, and late Miocene about 150, 250, and 200 m, respectively, below present sea level.

Interregional unconformities are related to cycles of global highstands and lowstands of sea level, as are the facies and general patterns of distribution of many depositional sequences. Geotectonic and glacial phenomena are the most likely causes of the sea-level cycles.

Major applications of the global cycle chart include (1) improved stratigraphic and structural analyses within a basin, (2) estimation of the geologic age of strata prior to drilling, and (3) development of a global system of geochronology.

INTRODUCTION

Cycles of relative change of sea level on a global scale are evident throughout Phanerozoic time. The evidence is based on the fact that many regional cycles determined on different continental margins are simultaneous and that the relative magnitudes of the changes generally are similar. Concepts and methods of determination of rela-

tive changes of sea level and regional cycles were given previously (Part 3, Vail et al, this volume). In this paper are presented charts of global cycles, the methods for constructing the charts from a modal average of correlative regional cycles based on seismic stratigraphy, and our estimates of the actual magnitudes of the sea-level changes.

Because the global cycles are records of geotectonic, glacial, and other large-scale processes, they reflect major events of Phanerozoic history. The timing and relative importance of these events are indicated by charts of the cycles. Such a composite record offers a means of subdividing Phanerozoic time into significant geochronologic units based on a single criterion.

Fairbridge (1961) summarized the historical development of concepts of sea-level change on a global scale, including the classic works of Haug (1900), Suess (1906), Stille (1924), Grabau (1940), Umbgrove (1942), Kuenen (1940, 1954, 1955), Arkell (1956), and others. These pioneer investigations laid the foundation for later work including ours. However, some developments have confused "transgressions and regressions" of the shoreline with "rises and falls" of sea level. Grabau (1924) recognized this problem. The charts we present in this paper show relative and eustatic rises and falls of sea level on a global scale, and differ from charts that show transgressions and regressions of the shoreline.

GLOBAL CYCLES

Figures 1 through 3 are charts of relative changes of sea level on a global scale. The vertical axis of each chart is scaled in millions of years (Ma, after Van Hinte, 1976 a, b), with standard periods and epochs plotted alongside. The horizontal axis shows relative positions of sea level and is scaled from 1.0 to 0.0, with 1.0 being the maximum relative highstand (65 Ma) and 0.0 being the minimum relative lowstand (30 Ma). Relative rises of sea level are plotted toward the left, and relative falls toward the right. The present position of sea level is extended through

[1]Manuscript received, January 6, 1977; accepted, June 13, 1977.

[2]Exxon Production Research Co., Houston, Texas 77001.

[3]New Mexico Bureau of Mines and Mineral Resources, Socorro, New Mexico 87801.

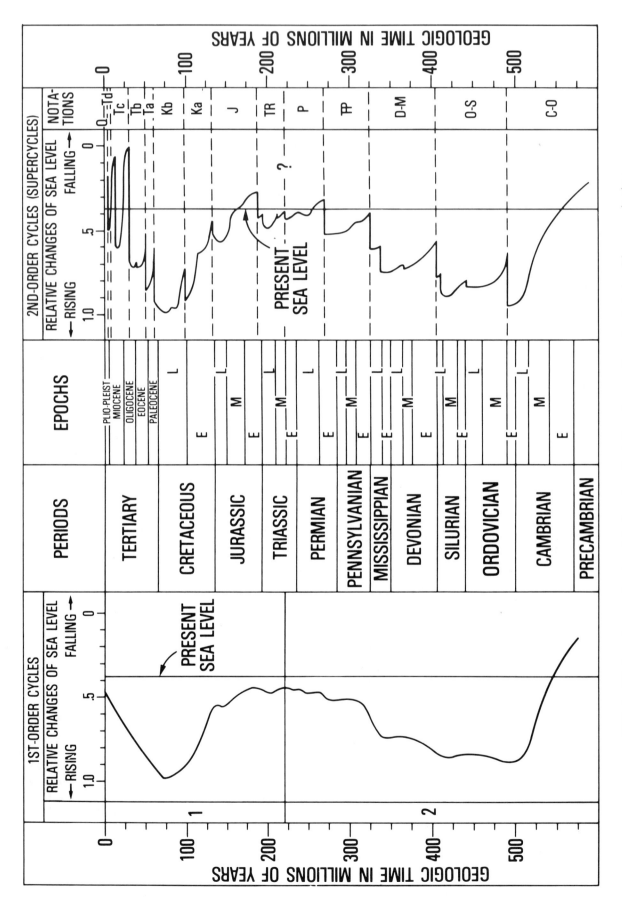

FIG. 1—First- and second-order global cycles of relative change of sea level during Phanerozoic time.

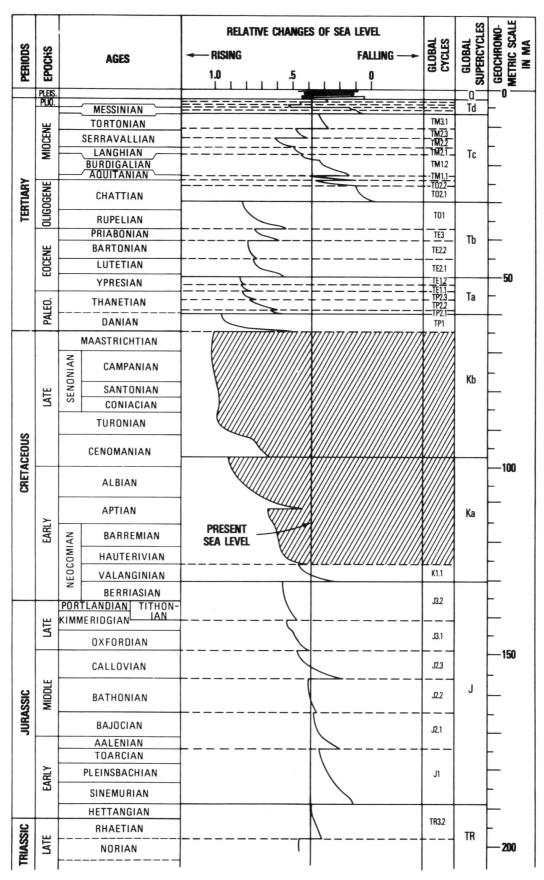

Jurassic - Cretaceous time scale after Van Hinte 1976 a, b

FIG. 2—Global cycles of relative change of sea level during Jurassic-Tertiary time. Cretaceous cycles (hatchured area) have not been released for publication.

315

Phanerozoic time as a vertical reference line, although it rarely can be related to measurement of ancient changes of sea level.

A *global cycle* of relative change of sea level is an interval of geologic time during which a relative rise and fall of sea level takes place on a global scale. Intermittent stillstands (and therefore paracycles) may occur in any part of the cycle, but tend to predominate after the major part of the rise has taken place and before the fall begins.

The global cycle charts (Figs. 1-3) show cycles of three orders of magnitude. The older of the two first-order cycles (Fig. 1) occured from Precambrian to Early Triassic, with a duration of over 300 million years; the younger first-order cycle occurred from middle Triassic to the present within a duration of about 225 million years. The durations of the 14 second-order cycles (Fig. 1) range from 10 to 80 million years. Over 80 third-order cycles (Figs. 2, 3), not including late Paleozoic cyclothems, have durations that range from approximately 1 to 10 million years. In this paper we do not show the third-order cycles for the pre-Jurassic and Cretaceous periods. Pre-Jurassic cycles are not included because documentation comes mainly from North America with only limited data from other continents. Cretaceous cycles have not been released for publication.

Trends of rise and fall of sea level reveal a marked asymmetry in the second- and third-order cycles. The relative rise generally is gradual and the relative fall generally is abrupt. In the first-order cycles, the cumulative falls tend to be more gradual and the curves are relatively symmetrical. Although the ages and durations of the first-, second-, and third-order cycles are fairly accurate, the amplitudes of the relative changes are only approximations.

Figure 1 shows first- and second-order cycles of the entire Phanerozoic. No distinct boundary occurs between the two first-order cycles, but the best dividing point appears to be between Early and Middle Triassic.

Figure 2 is a chart of relative changes of sea level on a global scale during Jurassic-Triassic time. The second- and third-order cycles are more clearly shown at this expanded time scale. The horizontal scale is the same as that of Figure 1 (see Part 7, Vail et al, this volume).

Figure 3 is a chart of the second- and third-order cycles during Tertiary and Quaternary time. A scale expanded from that of Figure 2 is needed to show the third-order cycles during these times; however, even at this scale all the cycles of glacial events in late Quaternary time are not clearly shown. The ages and durations of the Cenozoic third-order cycles have the best documentation, based mainly on zones of planktonic

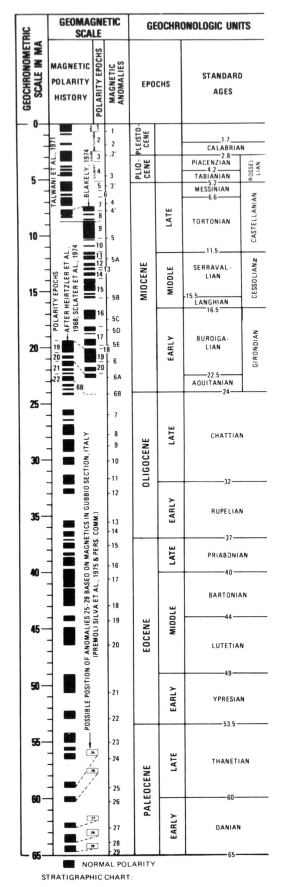

316

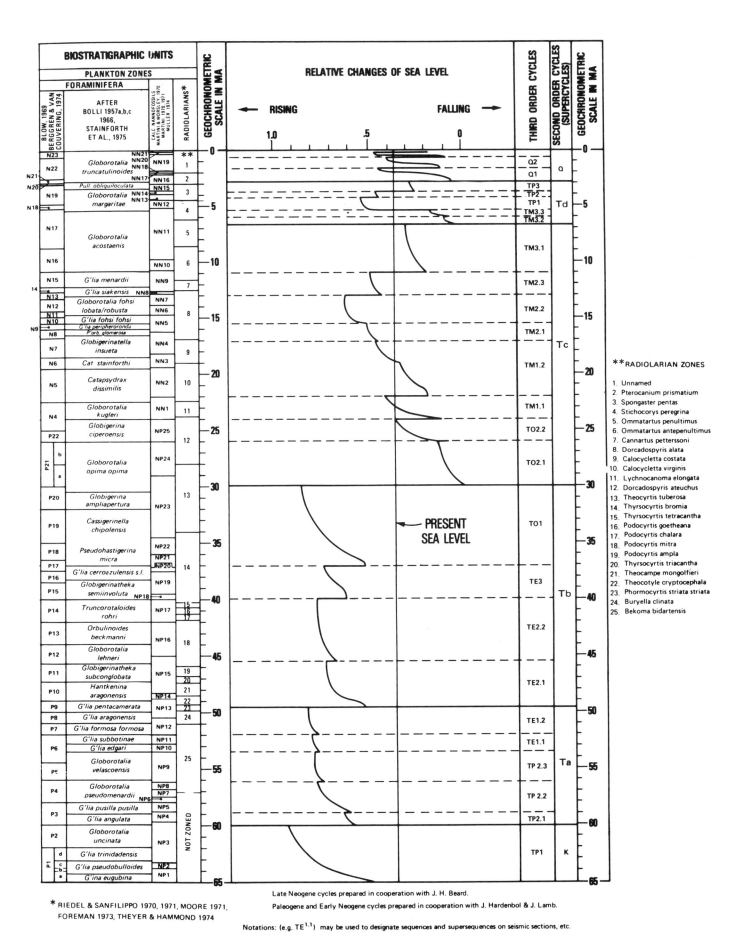

FIG. 3—Global cycles of relative change of sea level during Cenozoic time. Basic references for the stratigraphic part of the chart are Hardenbol (unpublished after Ryan et al, 1974) and Hardenbol and Berggren (1977).

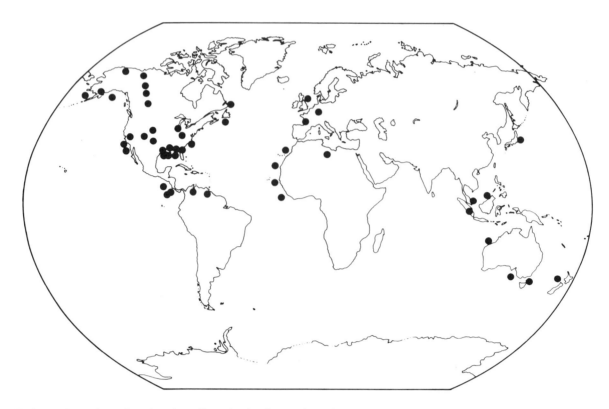

FIG. 4—Location of regional studies of seismic stratigraphy used in construction of global cycle chart for Phanerozoic time.

Foraminifera that have been tied to the geochronometric scale. Amplitudes of the relative changes are determined mainly from seismic stratigraphic analysis of grids of seismic sections tied to well data.

On the right side of the global cycle charts (Figs. 1-3), are columns containing notations that are useful for identifying stratigraphic units on seismic and stratigraphic sections. We identified depositional sequences according to their ages and their relations with cycles of relative changes of sea level. For example, the supersequence corresponding to the Jurassic supercycle is identified as J. It in turn is subdivided into J1, J2, or J3 corresponding to the Early, Middle, or Late Jurassic epoch in which the sequence occurs. Where more than one cycle of sea-level change occurs within a given epoch, such as the Middle Jurassic (J2), we used the notation J2.1 and J2.2, depending on the number of cycles. If we wanted to identify more than one sequence within a single cycle, we used the notation J2.1A, J2.1B, etc.

It also is important to identify sequence boundaries, especially those that are unconformities. This is because a given unconformity may truncate strata with a wide range of ages and also may be onlapped by strata of different ages. We identified an unconformity as pre- the oldest overlying sequence at the point where the surface approaches conformity. For example, an unconformity that becomes conformable at the base of the Callovian would be identified as pre-Callovian (pre-J2.3).

CONSTRUCTION OF GLOBAL CYCLE CHART

Figure 4 is a world map that shows the general locations of the regional grids of seismic data used in the construction of the global cycle charts (Figs. 1-3). The seismic sections, supplemented by well control and other geologic data, provided the regional stratigraphic framework needed to measure and date the relative changes of sea level. These studies of seismic stratigraphy are drawn from all continents except Antarctica, and provide a representative worldwide sample for Jurassic and younger cycles. However, data are concentrated in areas of petroleum exploration where sedimentary sections are relatively thick and subsidence rates are high, with resultant higher rates of relative rise than would be recognized in thinner sections. Pre-Jurassic cycles are determined primarily on the evidence from North America, with supporting data from other continents.

The global cycle chart is simply a modal average of correlative regional cycles from many areas around the world. The construction of regional cycle charts is explained in Part 3 (Vail et al, this volume). A global cycle can be approximated

318

with a modal average of three or more correlative regional cycles from different continents. The more continents represented, the greater the accuracy of the resulting chart. Some obviously provincial effects such as orogenic deformation, excessive tectonic subsidence, or excessive sedimentary loading may distort the average amplitudes of sea level changes and should not be used without adjustment. Ages of the cycles in these regions generally correspond to the global pattern.

Figure 5 is an example of how four regional cycles are correlated and averaged to construct global cycles (column 5). In this example Cenozoic cycles of specific regions on four continents are used. Although many regional differences in the four curves are obvious, ages and durations of the cycles generally are correlative and amplitudes of the relative changes generally are similar. The global cycle curve is not based on these four regional curves alone, but from those of other areas shown in Figure 4.

The main bases of comparison are the ages of the major relative falls of sea level. For example, the pre-late Miocene fall (10.8 Ma) occurs on all three charts where data are present and is indicated by a major unconformity in all these regions. A pre-middle late Oligocene fall (30 Ma) can be recognized on all charts. The latest early Eocene fall (49 Ma) and the mid-Paleocene fall (60 Ma) are recognized in all three regions where Eocene and Paleocene deposits are found. After major falls are correlated and charted, ages of individual cycles are charted. These are quite similar in all four areas where data are available. Some differences in age relations may be explained by local differences in paleontologic techniques of age dating. Some cycles are not recognized with seismic data because the stratigraphic section is too thin, such as within the Tertiary of northwestern Africa.

Determination of average amplitudes of cycles is the least quantitative step in this procedure. With the exception of the Gippsland basin, shapes of the curves relating to the first- and second-order cycles generally show an overall fall with large fluctuations at supercycle boundaries. The Gippsland basin curve shows an anomalous overall rise probably related to the geotectonic history of Australia. Amplitudes of third-order cycles are charted, giving greatest weight to regions where most complete data are available. Although the first-order cycle of the Gippsland basin is anomalous, the third-order cycles fit the global pattern well.

Accuracy of the global cycle chart depends on the quality and quantity of the regional charts that are used to construct it. On the charts in this report (Figs. 1-3), the ages, durations, and relative amplitude of the global cycles represent a relatively high level of accuracy, but measurement of the actual amplitudes is still a major problem. Direct measurement is made difficult by: (1) a sometimes wide difference range in thickness of coastal onlap for the same global cycle from various regions; (2) practical difficulties in making complete regional onlap analyses such as lack of control, erosion of critical portions of coastal onlap, or structural complications; (3) necessity of inferring coastal onlap from other facies relations where onlap for portions of the curve can not be measured directly; and (4) difficulties in measuring relative falls of sea level from seaward shifts in coastal onlap. For these reasons, the horizontal scale on the global charts showing amplitude of relative rises and falls is not calibrated in meters, but with a relative scale normalized on the maximum range of sea level positions of the curve. The highest position of sea level, occurring at the end of the Cretaceous (65 Ma), is set at 1.0 and the lowest position, at the mid-Oligocene (30 Ma), is set at 0.0. In the example cited (Fig. 5), each regional curve has been normalized according to this pattern. Where the regional curves do not include the Late Cretaceous high or the Oligocene low, they are normalized by making the best fit of the overlapping portions of the curves. If a given Upper Cretaceous or mid-Oligocene regional cycle has an anomalous magnitude, the regional curve is normalized by making a best fit to the other curves with a less distorted portion of the curve.

ESTIMATION OF EUSTATIC CHANGES OF SEA LEVEL

Global cycle curves (Figs. 1-3) summarize relative changes of sea level as described above. However, these curves include large-scale subsidence that must be discounted to determine the eustatic curve. An estimate of the true eustatic curve (Fig. 6c) has been made for Jurassic to Holocene time by calibrating the global cycle curve (Fig. 2) with the work of Pitman (in press), Sleep (1976), and J. H. Beard (personal commun.). These authors calculated quantitative values for the position of sea level during parts of Cretaceous, Tertiary, and late Neogene time. Their results conform to our preliminary estimates of eustatic change.

Pitman (in press) presented a curve of sea level change from Late Cretaceous to late Miocene time (Fig. 6a). His calculations, based only on rates of seafloor spreading and resultant volumes of midocean ridges, show a cumulative fall from

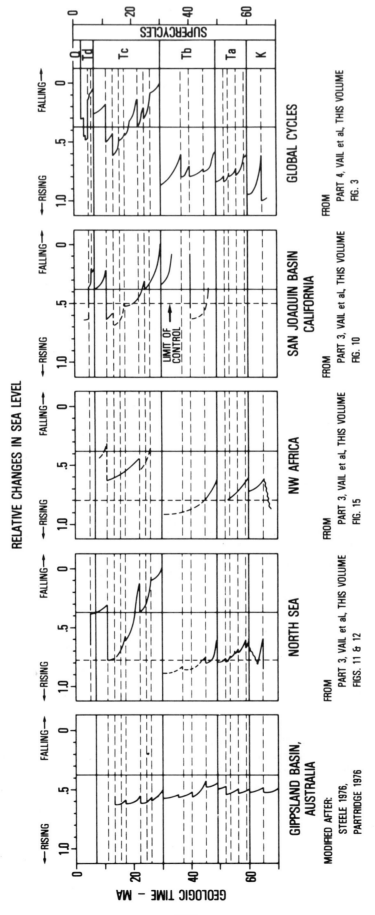

FIG. 5—Correlation of regional cycles of relative change of sea level from four continents and averaging to construct global cycles.

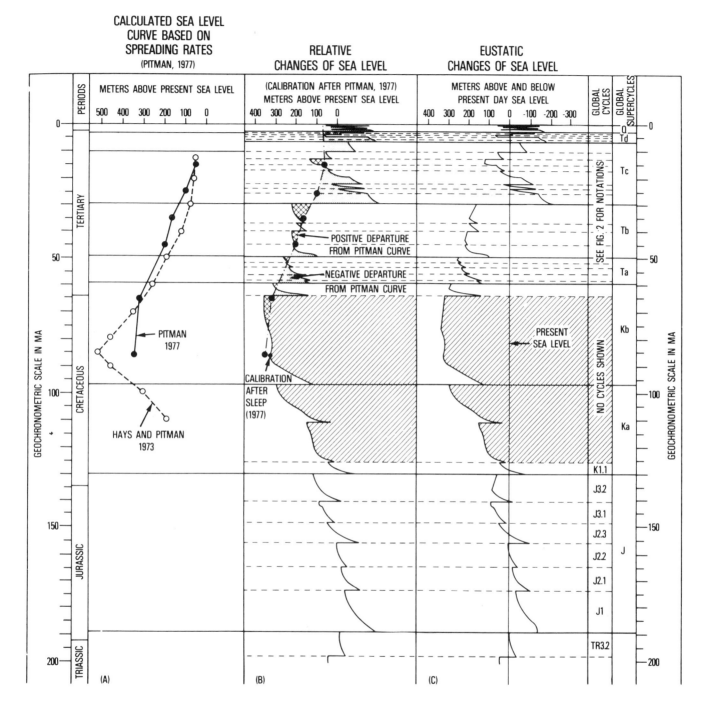

FIG. 6—Estimation of eustatic changes in sea level from Jurassic to Holocene: **a.** Pitman's (in press) and Hays and Pitman's (1973) calculated sea-level curves based on rates of seafloor spreading and volumes of midocean ridges; **b.** Pitman's (in press) curve from (**a.**) overlain on global curve of relative changes of sea level; and **c.** best estimate of eustatic changes of sea level, calibrated from Pitman's (in press) curve.

321

350 to 60 m above the present position of sea level; he explained the remaining 60 m as water retained in present ice caps. His curve matches closely the Tertiary part of our Triassic-Holocene first-order cycle curve (Fig. 1). Sleep (1976) suggested sea level in the late Turonian to be 300 m above the present, based on work in the Precambrian shield area of Minnesota. J. H. Beard (personal commun.) estimated late Neogene eustatic changes by relating them to known Pleistocene eustatic changes using paleontology and seismic sequence analysis. This work provided a calibration for the youngest part of our chart.

Pitman's curve and our first-order cycle curve are very similar. However, our second- and third-order cycles show significant departures, both positive (higher) and negative (lower), from Pitman's curve (Fig. 6b). Where our curve is higher than Pitman's, his curve indicates that an overall gradual eustatic fall of sea level may be in progress, but our analysis of onlap indicates a relative rise. This discrepancy may be due to the fact that in our study areas with relatively thick sedimentary columns, subsidence occurs at a greater relative rate than sea level falls, producing a relative rise of sea level (see Part 3, Vail et al, this volume). During such a relative highstand, Pitman's curve (showing a gradual fall) is considered more representative of eustatic change than is ours.

The negative departures of our second- and third-order cycle curves from Pitman's curve are rapid falls of sea level probably caused by glacial withdrawals of water not evaluated by Pitman (see later section on causes of global cycles). Subsequent deglaciation produces a rise in sea level that diminishes in rate until the rises reach the position of the eustatic fall shown on Pitman's curve. As discussed later, glaciation is not documented in the geologic record prior to Oligocene within our last first-order cycle, so additional evidence of glaciation or of some other cause is needed to explain some of the rapid changes, especially the falls, of the second- and third-order cycles in the Early Tertiary and Mesozoic.

Although not covered by Pitman's curve, a general first-order rise from Early Jurassic to Late Cretaceous is indicated from our curve. If glaciation or some other factor produced the rapid second- and third-order falls, the subsequent rises should not cross the first-order curve. Therefore, no gradual eustatic falls are indicated except possibly during the Tithonian and Berriasian.

Our best estimate of true eustatic changes from Jurassic to Holocene is shown on Fig. 6c. Amplitudes of the changes are calibrated in meters with respect to present sea level, based on our curves,

those of Pitman (in press), and data from Sleep (1976) and J. G. Beard (personal commun.). A more accurate curve for the Triassic through Early Cretaceous segment can be made when the first-order rise of sea level for that part of the curve has been calculated from rates of seafloor spreading or from some other long-term cause.

GLOBAL HIGHSTANDS AND LOWSTANDS AND MAJOR INTERREGIONAL UNCONFORMITIES

Table 1 lists the major global highstands and lowstands (or comparative lowstands) of sea level during Phanerozoic time. They are separated by major global falls of sea level. These falls are associated with major interregional unconformities.

A *global highstand* is an interval of geologic time during which the position of sea level is above the shelf edge in most regions of the world. Conversely, a *global lowstand* is an interval during which sea level is below the shelf edge in most regions. A *comparative lowstand* occurs when sea level is at its lowest position on the shelf between periods of highstand.

Global highstands of sea level (Part 3, Vail et al, Fig. 9, this volume) are characterized by widespread shallow marine to nonmarine deposits on the shelves and "starved" basins. If the supply of terrigenous sediment is abundant, delta lobes may prograde over the shelf edge into deep water. Global lowstands are characterized by erosion and nondeposition on the shelves, and deposition of deep marine fans in the basins. After a major fall of sea level to a global lowstand, a major interregional unconformity commonly is developed by subaerial erosion and nondeposition on shelves and basin margins, and by periods of nondeposition or shifts in depositional patterns in the deep-water parts of the basin.

The two first-order global cycles of sea-level change (Fig. 1) may be generally described in terms of global lowstands and highstands. During the older first-order cycle, sea level rose from a lowstand in late Precambrian time to a highstand during a long interval from Late Cambrian to Mississippian, and gradually fell to a lowstand which was at a broad minimum through Permian and Early Triassic time. In the cycle from Triassic to the present, cumulative rises built to a highstand peak in Late Cretaceous followed by cumulative falls to lowstand with many fluctuations.

CAUSES OF GLOBAL CYCLES

According to Fairbridge (1961), a eustatic change of sea level on a global scale may be produced by a change in the volume of sea water, by a change in the shape of the ocean basins, or by a combination of both. A change in the volume of seawater may be produced by glaciation and deglaciation, or by additions of juvenile water from magmatic sources (Rubey, 1951, p. 1137-1138), volcanos, or hot springs (Egyed, 1960). A change in the shape of the ocean basins may be produced by geotectonic mechanisms or sedimentary filling of the basins.

Among these factors, only geotectonic mechanisms appear to be of sufficient duration and magnitude to account for the first-order and most of the second-order cycles. Glaciation and deglaciation probably account for many third-order cycles and some of the second-order cycles, especially those in the late Neogene. Other unidentified causes may produce the rapid changes evident in second- and third-order cycles, or may work in combination with geotectonics and/or glaciation to accentuate or diminish the changes. Pitman (in press) discussed the effect of abrupt changes in rates of seafloor spreading on onlap patterns along continental shelves.

Changes in volume or elevation of midocean ridges, which are related to changes in the rate of

seafloor spreading, appear to produce significant changes in the shape of ocean basins (Hallam, 1963, 1969, 1971; Menard, 1964; Russell, 1968; Wise, 1972, 1974; Flemming and Roberts, 1973; Rona, 1973a, b; Hays and Pitman, 1973; and Rona and Wise, 1974; Pitman, in press). Volumetric changes along subduction zones are more difficult to quantify and evaluate and have not been treated quantitatively to our knowledge. Carey (1976) suggested that earth expansion may be the major cause for the rapid changes.

Pitman (in press) stated that, except for glacial effects, volumetric change in the midocean ridges related to change in rate of seafloor spreading is potentially the fastest and volumetrically the most significant way to change sea level. According to his calculations, sea level has fallen steadily but at varying rates from the Late Cretaceous, due to a contraction in size of oceanic ridges related to decreasing rates of seafloor spreading. At the same time, passive continental margins of the Atlantic and other ocean basins have subsided tectonically at decreasing rates, following a pre-

Table 1. Global Highstands and Lowstands of Sea Level and Associated Major Interregional Unconformities During Phanerozoic Time.

SEA LEVEL HIGHSTANDS	MAJOR GLOBAL SEA LEVEL FALLS	SEA LEVEL LOWSTANDS
	PRE-LATE PLIOCENE & PRE-PLEISTOCENE (3.8 & 2.8 MA)	LATE PLIOCENE - EARLY PLEISTOCENE
EARLY & MIDDLE PLIOCENE		
	PRE-LATE MIOCENE & PRE-MESSINIAN (10.8 & 6.6 MA)	LATE MIOCENE
MIDDLE MIOCENE		
	PRE-MIDDLE LATE OLIGOCENE (30 MA)	MIDDLE LATE OLIGOCENE
LATE MIDDLE EOCENE & EARLY OLIGOCENE		
	PRE-MIDDLE EOCENE (49 MA)	EARLY MIDDLE EOCENE
LATE PALEOCENE - EARLY EOCENE		
	PRE-LATE PALEOCENE (60 MA)	MID-PALEOCENE
CAMPANIAN & TURONIAN		
	PRE-MIDDLE CENOMANIAN (98 MA)	MID-CENOMANIAN
ALBIAN - EARLIEST CENOMANIAN		
	PRE-VALANGINIAN (132 MA)	VALANGINIAN
EARLY KIMMERIDGIAN		
	PRE-SINEMURIAN (190 MA)	SINEMURIAN
NORIAN & MIDDLE GUADALUPIAN		
	PRE-MIDDLE LEONARDIAN (270 MA)	MID-LEONARDIAN
WOLFCAMPIAN & EARLIEST LEONARDIAN		
	PRE-PENNSYLVANIAN (324 MA)	EARLY PENNSYLVANIAN
OSAGIAN & EARLIEST MERAMECIAN		
	PRE-DEVONIAN (406 MA)	EARLY DEVONIAN
MIDDLE SILURIAN		
	PRE-MIDDLE ORDOVICIAN (490 MA)	EARLY MIDDLE ORDOVICIAN
LATE CAMBRIAN & EARLY ORDOVICIAN		
		EARLY CAMBRIAN & LATEST PRECAMBRIAN

dictable thermal cooling curve (Sclater et al, 1971). The general model can be used to explain the cumulative first-order fall since the Late Cretaceous.

There may be a general correspondence of times of orogenic movement and volcanism with times of second-order sea-level highstands (Fig. 1). In general, high rates of seafloor spreading should be associated with relatively shallow ocean floors, highstand flooding of continental and basin margins, and greater subduction. Increased subduction should tend to produce increased volcanism and orogeny from continent-continent collisions. Such orogenic episodes should have fairly long terms of occurrence associated with durations of second-order highstands. However, pronounced angular unconformities associated with rapid falls of sea level can give impressions of short periods of orogeny. If short-term orogenies can be documented, they may be related to third-order cycles.

First-order cycles show a possible overall relationship to patterns of seafloor spreading rates and orogeny. For example, rifting and continental pull-apart dominate times of major sea-level rise in Cambrian and Jurassic–Early Cretaceous, and orogenies dominate intervals of falling sea level within each first-order cycle.

Many workers have noted unconformities which occur simultaneously in many regions of the globe (Stille, 1924; Arkell, 1956; Sloss, 1963, 1972; Gussow, 1963; Vail and Wilbur, 1966; Moore et al, 1974; and Dennison and Head, 1975). Most of these unconformities are coincident with major relative falls of sea level on our charts (Figs. 1-3). The unconformities that correspond to major sea-level falls at the end of the second-order cycles are shown on Table 1.

We concur with Sleep (1976) that unconformities caused by major eustatic changes of sea level modify the history of subsidence within cratonic basins and continental margins. Where thermal contraction of the lithosphere controls basin subsidence, basins should continue to subside even during times of eustatic lowstands. Sleep calculated that significant unconformities in the geologic record could be produced by eustatic falls of sea level, with amplitudes of those presented on Figure 6c, even in rapidly subsiding basins. Sleep's work supports our contention that interregional unconformities are not primarily due to uplift of the continental interior or continental margins, but are primarily caused by erosion or non-deposition during eustatic changes in sea level.

Glaciation and deglaciation are the only well understood causal mechanisms that occur at the relatively rapid rates of third-order cycles (Pit-

man, in press). Rates of geotectonic mechanisms related to seafloor spreading are too slow. Glaciation has been documented in the Pleistocene, late and early Miocene, and to some degree in the late Oligocene, but there is no evidence of glaciation at the times of many other lowstands. Other evidence for climatic changes, such as from oxygen isotopes (Savin, in press; Savin and Douglas, in press; Fischer and Arthur, in press) and other faunal studies (Haq and Lohmann, 1976) show that lowstands generally represent climatically cool conditions, and highstands represent climatically warm conditions.

Other evidences of cyclicity which correlate in general with global cycles include frequency of unconformities in deep sea cores and cycles of faunal diversity (Fischer and Arthur, in press), and changes of calcite compensation depth (van Andel, 1975, in press).

In summary, the cause for the first-order and some second-order cycles may be related to geotectonic mechanisms. Some of the second- and third-order cycles can be explained by glaciation. The empirically observed rapid falls of sea level at the ends of the third-order cycles remain unexplained where evidence for glaciation is not known.

APPLICATIONS

Major applications of global cycles fall into three categories: (1) improved stratigraphic and structural analyses incorporating the effects of sea-level changes, (2) estimation of geologic age ahead of the drill, and (3) development of a global system of geochronology.

In regional stratigraphic studies, after analysis of seismic sequences and regional sea-level changes is begun (Parts 2, 3, 7, Vail et al, this volume) comparison of regional and global sea level curves can aid prediction of age of sequences for which control is lacking and fill gaps in regional sea-level curves. Correlation of regional curves with times of unconformities, lowstands, and highstands on global curves aids prediction of depositional facies and distribution of sequences (Parts 3, 9, 10, Vail et al, this volume). Moreover, departures of the regional curve from the global curve indicate anomalous regional effects such as tectonic subsidence or uplift.

Estimation of geologic age of strata prior to drilling is a seismic-stratigraphic technique commonly applied in areas of sparse or no well control. Where wells are present and biostratigraphic zones are determined, they can be tied to seismic sequences for accurate age dating throughout the area of the seismic grid. If there is no well control within the grid, geologic age can be inferred by

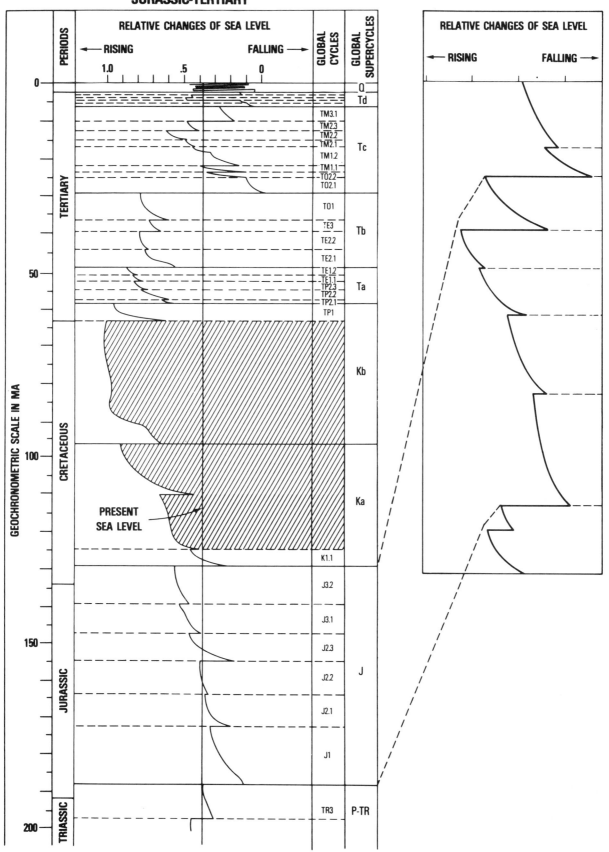

FIG. 7—Estimating geologic age prior to drilling.

building a regional chart of relative changes of sea level from seismic data and matching it with the global chart (Fig. 7). Accuracy can be improved with information from outcrops or distant wells that help to establish the general age of strata that are known to be present in the basin.

One of the greatest potential applications of the global cycle chart is its use as an instrument of geochronology. Global cycles are geochronologic units defined by a single criterion—the global change in the relative position of sea level through time. Determination of these cycles is dependent on a synthesis of data from many branches of geology. As seen on the Phanerozoic chart (Fig. 1), the boundaries of the global cycles in several cases do not match the standard epoch and period boundaries, but several of the standard boundaries have been placed arbitrarily and remain controversial. Using global cycles with their natural and significant boundaries, an international system of geochronology can be developed on a rational basis. If geologists combine their efforts to prepare more accurate charts of regional cycles, and use them to improve the global chart, it can become a more accurate and meaningful standard for Phanerozoic time.

REFERENCES CITED

Arkell, W. J., 1956, Jurassic geology of the world: London, Oliver and Boyd Ltd., 806 p.

Carey, S. W., 1976, The expanding earth—developments in geotectonics, part 10: Amsterdam, Elsevier, 470 p.

Dennison, J. M., and T. W. Head, 1975, Sea level variations interpreted from the Appalachian basin Silurian and Devonian: Am. Jour. Sci., v. 275, p. 1089-1120.

Egyed, L., 1960, On the origin and constitution of the upper part of the earth's mantle: Geol. Rundschau, v. 50, p. 251-258.

Fairbridge, R. W., 1961, Eustatic changes in sea level, in L. H. Ahrens, et al, eds., Physics and chemistry of the earth: London, Pergamon Press, v. 4, p. 99-185.

Fischer, A. G., and M. A. Arthur, (in press), Secular variations in the pelagic realm, in H. E. Cook, and P. Enos, eds., Basinal carbonate sediments: SEPM Spec. Pub. no. 25.

Flemming, N. C., and D. G. Roberts, 1973, Tectono-eustatic changes in sea level and sea floor spreading: Nature, v. 243, p. 19-22.

Foreman, H. P., 1973, Radiolaria of Leg 10 with systematics and ranges for the families *Amphipyndacidae, Artostrobiidae* and *Theoperidae*: deep sea drilling project, leg 10, in Initial reports of the DSDP, v. 10: Washington (U.S. Govt. Printing Office) p. 407-474.

Grabau, A. W., 1924, Principles of stratigraphy: New York, A. G. Seiler, 1,185 p.

———— 1940, The rhythm of the ages: Peking, Henri Vetch Pub., 56 p.

Gussow, W. C., 1963, Metastacy, in D. C. Mungan, ed., Polar wandering and continental drift: SEPM Spec. Pub. 10, p. 146-169.

Hallam, A., 1963, Major epeirogenic and eustatic changes since the Cretaceous and their possible relationship to crustal structure: Am. Jour. Sci., v. 261, p. 397-423.

———— 1969, Tectonism and eustacy in the Jurassic: Earth-Sci. Rev., v. 5, p. 45-68.

———— 1971, Mesozoic geology and the opening of the North Atlantic: Jour. Geology, v. 79, p. 129-157.

Haq, B. U., and G. P. Lohmann, 1976, Early Cenozoic calcareous nannoplankton biogeography of the Atlantic Ocean: Marine Micropaleontology, v. 1, no. 2, p. 119-194.

Hardenbol, J., and W. A. Berggren, (1977), A new Paleogene numerical time scale: AAPG Studies Geology, no. 6, in press.

Haug, E., 1900, Les geosynclinaux et les aires continentales: Soc. Geol. France Bull., Ser. 3, v. 28, p. 617-711.

Hays, J. D., and W. C. Pitman, 1973, Lithospheric plate motion, sea level changes, and climatic and ecological consequences: Nature, v. 246, p. 18-22.

Kuenen, Ph. H., 1940, Causes of eustatic movements: 6th Pacific Sci. Cong., Proc., v. 2, p. 833-837, Berkeley, Univ. Calif. Press.

———— 1954, Eustatic changes of sea-level: Geologie en Mijnbouw, v. 16, p. 148-155.

————1955, Sea level and crustal warping; crust of the earth—a symposium: Geol. Soc. America, v. 62, p. 193-204.

Menard, H. W., 1964, Marine geology of the Pacific: New York, McGraw-Hill, 271 p.

Moore, T., 1971, Radiolaria, deep sea drilling project, leg 8, in Initial reports of the DSDP, v. 8: Washington (U. S. Govt. Printing Office) p. 727-748.

Moore, T. C., Jr., et al, 1974, Cenozoic hiatuses in pelagic sediments, in E. Siebold and W. R. Riedel, eds, Marine plankton and sediments; 3rd Plankton Conference (Kiel) Proc.

Partridge, A. D., 1976, The geologic expression of eustacy in the early Tertiary of the Gippsland basin: APEA Jour., v. 16, p. 73-79.

Pitman, W. C., (in press), Relationship between sea level change and stratigraphic sequences: Geol. Soc. America Bull.

Riedel, W. R., and A. Sanfilippo, 1970, Radiolaria, deep sea drilling project leg, Leg 4, in Initial reports of the DSDP, v. 4: Washington (U.S. Govt. Printing Office) p. 503-575.

————1971, Cenozoic radiolaria from the western tropical Pacific, deep sea drilling project, leg 7, in Initial reports of the DSDP, v. 7: Washington (U.S. Govt. Printing Office) p. 1,529-1,672.

Rona, Peter A., 1973a, Relations between rates of sediment accumulation on continental shelves, sea-floor spreading, and eustacy inferred from the central North Atlantic: Geol. Soc. America Bull., v. 84, p. 2851-2872.

————1973b, Worldwide unconformities in marine sediments related to eustatic changes of sea level: Nature Phys.-Sci., v. 244, p. 25-26.

———— and D. U Wise, 1974, Symposium: global sea level and plate tectonics through time: Geology, v. 2, p. 133-134.

Rubey, W. W., 1951, Geologic history of sea water: Geol. Soc. America Bull., v. 62, p. 1111-1147.

Russell, K. L., 1968, Oceanic ridges and eustatic changes in sea level: Nature, v. 218, p. 861-862.

Savin, S. M., (in press), The history of the earth's surface temperature during the past hundred million years: Annu. Rev. Earth and Planetary Sci., v. 5.

———— and R. Douglas, (in press), Changes in bottom-water temperatures in the Tertiary and its implications: Geology.

Sclater, J. G., R. N. Anderson, and M. L. Bell, 1971, Elevation of ridges and evolution of the central eastern Pacific: Jour. Geophys. Research, v. 76, p. 7888-7916.

Sleep, N. H., 1976, Platform subsidence mechanisms and "eustatic" sea-level changes: Tectonophysics, v. 36, p. 45-56.

Sloss, L. L., 1963, Sequences in the cratonic interior of North America: Geol. Soc. America Bull, v. 74, p. 93-113.

————1972, Synchrony of Phanerozoic sedimentary-tectonic events of the North American craton and the Russian platform. Sect. 6, in Stratigraphie et sedimentologie: 24th Int. Geol. Cong. (Montreal).

Stille, H., 1924 Grundfragen der vergleichenden Tektonik: Berlin, Borntraeger.

Suess, E., 1906, The face of the earth: Oxford, Clarendon Press, v. 2, 556 p.

Theyer, F., and S. R. Hammond, 1974a, Paleomagnetic polarity sequence and radiolarion zones, Brunhes to polarity Epoch 20: Earth and Planetary Sci. Letters, v. 22, p. 307-319.

———— 1974b, Cenozoic magnetic time scale in deep-sea cores—completion of the Neogene: Geology, v. 2, no. 10, p. 487-492.

Umbgrove, J. H. F., 1942, The pulse of the earth: The Hague, Nijhoff 179 p.

Vail, P. R., and R. O. Wilbur, 1966, Onlap, key to worldwide unconformities and depositional cycles (abs.): AAPG Bull., v. 50, p. 638.

van Andel, Tj. H., 1975, Mesozoic-Cenozoic calcite compensation depth and the global distribution of carbonate sediments: Earth and Planetary Sci. Letters, v. 26, p. 187-194.

———— (in press), An eclectic overview of plate tectonics, paleogeography, and paleoceanography, in Historical biogeography, plate tectonics and the changing environment: 37th Biology Colloquium, Oregon State Univ. Press.

Van Hinte, J. E., 1976a, A Jurassic time scale: AAPG Bull., v. 60, p. 489-497.

———— 1976b, A Cretaceous time scale: AAPG Bull., v. 60, p. 498-516.

Wise, D. U., 1972, Freeboard of continents through time: Geol. Soc. America Mem. 132, p. 87-100.

———— 1974, Continental margins; freeboard and volumes of continents and oceans through time, in C. A. Burke and C. L. Drake, eds., The geology of continental margins: New York, Springer-Verlag, p. 45-58.

American Association of Petroleum Geologists Memoir
36, *Interregional Unconformities and Hydrocarbon
Accumulation*, edited by John S. Schlee, copyright 1984,
pp. 129–144.

Jurassic Unconformities, Chronostratigraphy, and Sea-Level Changes from Seismic Stratigraphy and Biostratigraphy

P.R. Vail
J. Hardenbol
Exxon Production Research Company
Houston, Texas

R.G. Todd
Esso Exploration and Production Norway
Stavanger, Norway

Seventeen global unconformities and their correlative conformities (sequence boundaries) subdivide the strata of the Jurassic and earliest Cretaceous into genetic depositional sequences produced by 16 eustatic cycles. These 16 cycles make up the Jurassic supercycle. Eight of the global unconformities are both subaerial and submarine (Type 1), and are believed to have been caused by rapid eustatic falls of sea level. Nine of the unconformities are subaerial only (Type 2), and are believed to be related to slow eustatic falls of sea level. In addition, 16 marine condensed sections (starved intervals) have been identified. These condensed sections are interpreted to be related to rapid eustatic rises of sea level. Unconformity recognition is locally or regionally enhanced by periodic truncation of folded and faulted strata during sea-level lowstands and onlap onto topographic highs during sea-level highstands, but we find no evidence that the tectonics caused the global unconformities. The 16 eustatic cycles that make up the Jurassic supercycle correspond to 16 global chronostratigraphic intervals that subdivide Jurassic strata into a series of genetic depositional sequences, which are recognizable from seismic, well, and outcrop data.

The Jurassic unconformities and the stratal and facies patterns between them are caused by the interaction of basement subsidence, eustatic changes of sea-level, and varying sediment supply. Detailed analyses of the sediments with seismic stratigraphy and well data permit quantification of the subsidence history and reconstruction of paleoenvironment and sea-level changes through time. The integrated use of seismic stratigraphy and biostratigraphy provides a better geologic age history than could be obtained by either method alone. Paleobathymetry, sediment facies, and relative changes of sea level can be interpreted from seismic data and confirmed, or improved on, by well control. Geohistory analysis based on geologic time-depth diagrams provides a quantitative analysis of total basin subsidence. When this subsidence is corrected for compaction and sediment loading, the tectonic subsidence and long-term eustatic changes may be determined. Short-term, rapid changes of sea level can be demonstrated from seismic, well, and outcrop data. The stratigraphic resolution of these changes rarely allows exact quantification of their magnitude, but a minimum rate of sea-level change often can be determined.

INTRODUCTION

A representative seismic line from the Inner Moray Firth in the North Sea (Figure 1) illustrates a seismic sequence interpretation of the Jurassic. Black lines indicate seismic sequence boundaries (unconformities or their equivalent conformities), arrows mark representative reflection terminations, numbers (132) indicate the age of the sequence boundaries determined where they are conformable, and symbols (J3.2) identify the sequence. Seismic stratigraphic studies tied to well control in over 60 areas located all over the globe indicate that any regional grid of reflection seismic profiles with reasonable data quality can be interpreted in a similar manner. Not only are sequence boundaries always present, but their ages dated in areas where they become conformable are the same in different basins around the world.

This paper discusses the Jurassic supercycle and shows how eustatic changes of sea level cause the global similarities of unconformity age, coastal onlap, and marine-condensed sections. It also shows how shoreline transgressions and regressions relate to these criteria and why transgressions and regressions are not necessarily globally synchronous.

The material for this paper is from a series of previously published papers: Vail et al (1977), Vail and Mitchum (1979), Vail and Hardenbol (1979), Vail et al (1980), and Vail and Todd (1981), and Hardenbol, Vail, and Ferrer (1981).

AGE, TYPES, AND CAUSES OF JURASSIC GLOBAL UNCONFORMITIES AND CONDENSED SECTIONS

Seventeen global unconformities and their correlative conformities (sequence boundaries) subdivide the strata of the Jurassic and earliest Cretaceous into genetic sequences

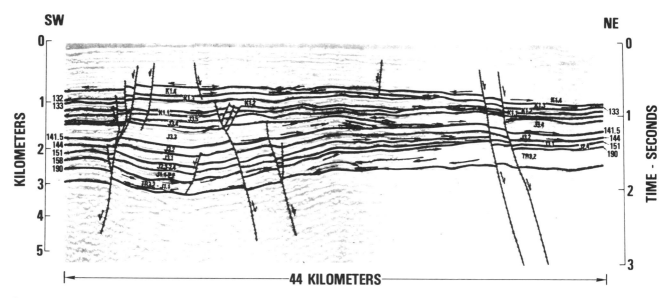

Figure 1: Seismic section northeast of Beatrice Field, Inner Moray Firth, North Sea, United Kingdom, showing seismic sequences of the Jurassic supersequence. Numbers (132) refer to the ages of the sequence boundaries where they are conformable. Symbols (J3.2) identify the individual sequences. Seismic section courtesy of Geophysical Company of Norway (United Kingdom) Ltd. (GECO).

produced by 16 eustatic cycles. These 16 cycles make up the Jurassic supercycle (Vail et al, 1977, part 8) (Table 1). Eight of the global unconformities are both subaerial and submarine (Type 1) and are believed to have been caused by rapid eustatic falls of sea level. Nine of the unconformities are subaerial only (Type 2) and are believed to be related to slow eustatic falls of sea level. In addition, 16 marine-condensed sections (starved intervals) have been identified. These condensed sections are interpreted to be related to rapid eustatic rises of sea level (Vail and Todd, 1981). Unconformity recognition is enhanced locally or regionally by periodic truncation of folded and faulted strata during sea-level lowstands and by onlap onto topographic highs during sea-level highstands, but we find no evidence that the tectonics caused the global unconformities. Table 1 is our latest summary of the ages of Type 1 and Type 2 unconformities and condensed sections. Of special significance is the fact that, by more careful age dating, we have discovered that the unconformities we previously placed at the base of the Sinemurian, Callovian, Oxfordian, Kimmeridgian, and Valanginian (Vail and Todd, 1981) are all older and occur within or at the base of the late portion of the preceding age. Figure 2 shows the relation of unconformity age and type and condensed interval age to the relative changes of coastal onlap and eustatic changes of sea level in the Jurassic supercycle.

Unconformities, marine condensed sections, coastal onlap, and marine transgressions-regressions are controlled by the interaction of these variables: subsidence, eustatics, and sediment supply. Empirical observations from seismic stratigraphic studies indicate that unconformities, condensed sections, coastal onlap, and marine transgression-regression are related as shown diagrammatically on Figure 3 and discussed below.

Unconformity is defined (Bates and Jackson, 1980, p. 675) as the structural relationship between rock strata in contact characterized by a lack of continuity in deposition, and cor-

responding to a period of nondeposition, weathering, or especially erosion (either subaerial or subaqueous) prior to the deposition of the younger beds, and often (but not always) marked by absence of parallelism between strata. We use the term unconformity for a surface representing a significant time gap with erosional truncation (subaerial or subaqueous) and/or subaerial exposure. Marine surfaces with significant hiatuses but without evidence of erosion are not unconformities, according to this usage. This definition also restricts the usage of unconformity as used in Vail et al, 1977. Angularity of the underlying strata is commonly due to erosion and truncation of tilted strata, although sedimentary bypass may produce a similar effect referred to as *toplap* (Vail et al, 1977, part 2). Discordance of the overlying rocks with an unconformity is commonly caused by terminations of overlying strata by sedimentary lapouts. These lapouts are termed onlap or downlap, depending upon whether they lap out in an updip or downdip direction, respectively, at the time of deposition (Figure 3).

Global unconformities are unconformities that are present at one place or another in all sedimentary basins with a sea-level base level at the time of deposition, and become conformable at the same geologic age. They are not necessarily unconformable surfaces everywhere throughout each basin. In other words, although a global unconformity may be present within a basin, it may be characterized by extensive conformable areas in that basin as well.

Type 1 (subaerial-submarine) unconformities are typified by downward shifts of coastal onlap (defined below), commonly below the shelf edge, typically producing subaerial exposure of the shelf, valley entrenchment, and the initiation of canyon cutting along the shelf edge (Figures 4A and 5). Fluvial sediments and lowstand deltas commonly fill the entrenched valleys, and submarine fans and slope-front fill deposits accumulate in deep water basins. Lowstand evaporite deposits and regional truncation of folded or faulted

Table 1

Type		Age	Sequence Identification	Van Hinte (1976) Time Scale in MA	Relative Magnitude		
					Maj.	Med.	Min.
Subaerial/Submarine	1	latest Berriasian	-	132	X		
Condensed Section	CS	mid Late Berriasian	K1.3	132.5			X
Subaerial	2	early Late Berriasian	-	133			X
Condensed Section	CS	basal Late Berriasian	K1.2	133.5			X
Subaerial	2	mid Early Berriasian	-	134			X
Condensed Section	CS	basal Berriasian	K1.1	135		X	
Subaerial/Submarine	1	within Tithonian	-	138		X	
Condensed Section	CS	within Tithonian	J3.5	138.5		X	
Subaerial/Submarine	1	within Tithonian	-	139		X	
Condensed Section	CS	within Tithonian	J3.4	139.5		X	
Subaerial/Submarine	1	within Tithonian	-	140	X		
Condensed Section	CS	basal Tithonian	J3.3	141		X	
Subaerial/Submarine	1	mid Late Kimmeridgian	-	141.5		X	
Condensed Section	CS	basal Kimmeridgian	J3.2	143	X		
Subaerial	2	mid Late Oxfordian	-	144		X	
Condensed Section	CS	basal Middle Oxfordian	J3.1	147	X		
Subaerial	2	basal Late Callovian	-	151		X	
Condensed Section	CS	mid Callovian	J2.4	153	X		
Subaerial/Submarine	2	mid Early Callovian	-	155			X
Condensed Section	CS	basal Callovian	J2.3	156			X
Subaerial	2	basal Late Bathonian	-	158	X		
Condensed Section	CS	basal Bathonian	J2.2	165		X	
Subaerial	2	basal Late Bajocian	-	168		X	
Condensed Section	CS	basal Bajocian	J2.1	171	X		
Subaerial/Submarine	1	mid Early Aalenian	-	173		X	
Condensed Section	CS	mid Toarcian	J1.5	175		X	
Subaerial	2	mid Early Toarcian	-	177			X
Condensed Section	CS	basal Toarcian	J1.4	178			X
Subaerial/Submarine	1	mid Late Pliensbachian	-	179		X	
Condensed Section	CS	basal Late Pliensbachian	J1.3	180		X	
Subaerial	2	basal Pliensbachian	-	184			X
Condensed Section	CS	mid Late Sinemurian	J1.2	185		X	
Subaerial/Submarine	1	latest Hettangian	-	190		X	

Table 1: Jurassic supercycle global unconformities (sequence boundaries) and condensed sections.

strata are characteristic in the proper geological setting. Type 1 unconformities are typified by both subaerial and submarine erosion and are formed when the rate of eustatic sea-level fall exceeds the rate of subsidence at the shelf edge.

Type 2 (subaerial) unconformities are characterized by downward shifts of coastal onlap to a position at or landward of the shelf edge, subaerial exposure of the landward portion of the shelf, and no evidence of canyon cutting (Figures 4B and 5). Lowstand evaporite deposits are characteristic in the proper geologic setting. Type 2 unconformities have only subaerial erosion, but their equivalent submarine conformities can commonly be recognized by submarine onlap resulting from shifts in the sites of deposition. Type 2 unconformities are formed when the rate of fall of eustatic sea-level is less than the rate of subsidence at the shelf edge, but exceeds the rate of subsidence on the inner portion of the shelf.

The chronostratigraphic significance of an unconformity is that all the rocks below the unconformity are older than the rocks above it. The ages of the strata immediately above and below the unconformity differ geographically according to the areal extent of erosion or nondeposition. The dura-tion of the hiatus associated with an unconformity differs correspondingly, but the unconformity itself is a chronostratigraphic boundary because it separates rocks of different ages, and no chronostratigraphic surfaces cross it. Although many chronostratigraphic surfaces may merge along an unconformity, none actually cross the unconformity. For these reasons, unconformities are not diachronous but are time boundaries that may be assigned a specific geologic age dated in those areas where the hiatus is least and/or where the rocks above and below become conformable. By careful correlation of unconformities and their correlative conformities, a sedimentary interval can be divided into genetic depositional sequences bounded by these unconformities. A depositional sequence is a chronostratigraphic interval because it contains all the rocks deposited during a given interval of geologic time limited by the ages of the sequence boundaries where they are conformities. The global Jurassic unconformities and their correlative conformities subdivide the Jurassic supersequence (all sediments deposited within the Jurassic supercycle) (Figure 2) into a series of genetic depositional sequences with global chronostratigraphic significance. Our studies indicate that the most

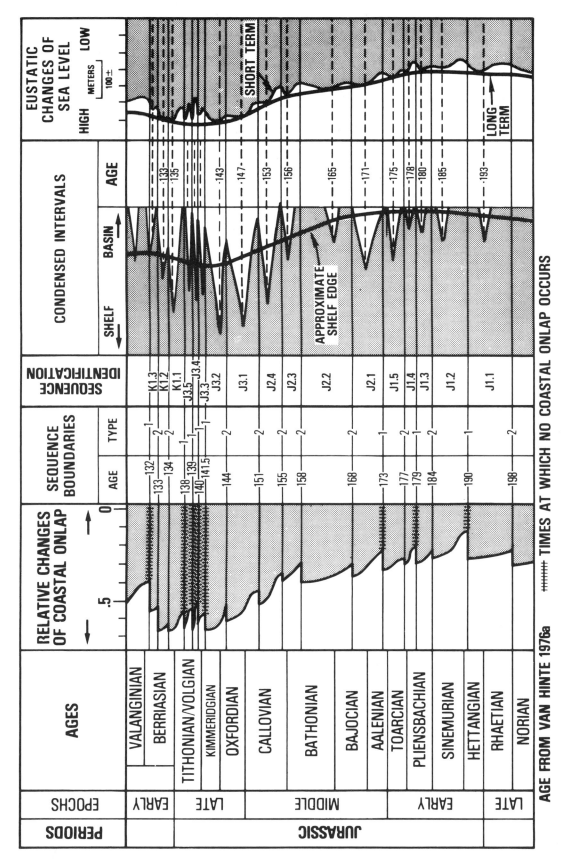

Figure 2: Global cycle chart of the Jurassic supercycle showing age and type of sequence boundaries (unconformities and their correlative conformities), sequence identification, age of condensed intervals and charts showing relative changes of coastal onlap and eustatic changes of sea level.

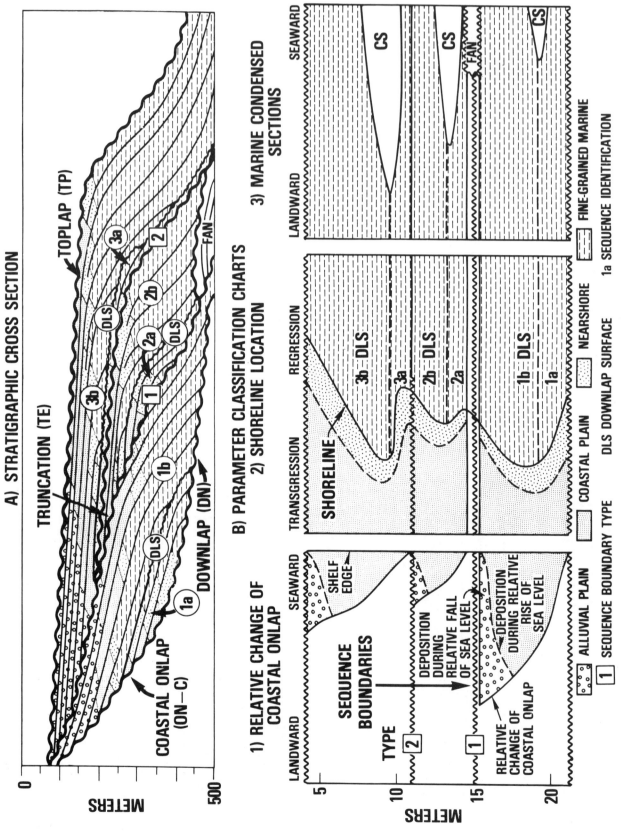

Figure 3: Diagrammatic stratigraphic cross section and chart showing parameters used to make the global cycle chart of the Jurassic supercycle. The stratigraphic cross section (A) shows the distribution of boundary types, downlap surfaces (condensed sections), and facies of three idealized sequences. The parameter classification chart (B) shows the relation between the parameters of the three sequences in time. CS means condensed section.

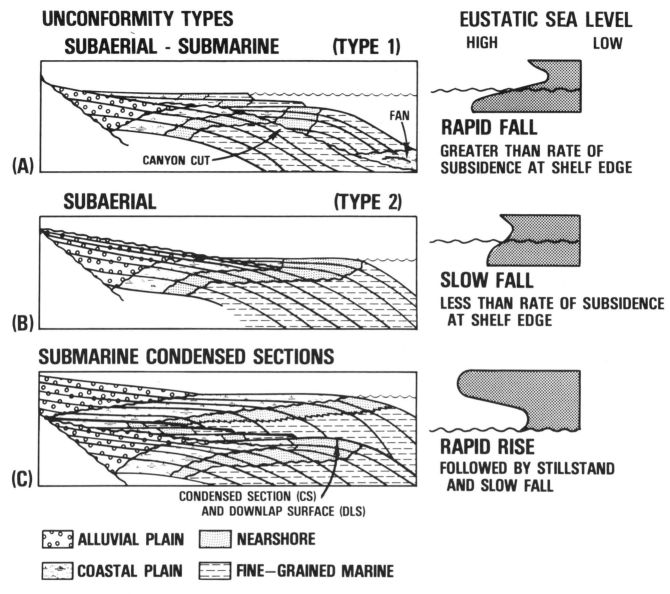

UNCONFORMITY TYPES

SUBAERIAL - SUBMARINE (TYPE 1)

(A)

CANYON CUT

FAN

SUBAERIAL (TYPE 2)

(B)

SUBMARINE CONDENSED SECTIONS

(C)

CONDENSED SECTION (CS)
AND DOWNLAP SURFACE (DLS)

EUSTATIC SEA LEVEL

HIGH   LOW

RAPID FALL
GREATER THAN RATE OF
SUBSIDENCE AT SHELF EDGE

SLOW FALL
LESS THAN RATE OF SUBSIDENCE
AT SHELF EDGE

RAPID RISE
FOLLOWED BY STILLSTAND
AND SLOW FALL

ALLUVIAL PLAIN   NEARSHORE

COASTAL PLAIN   FINE—GRAINED MARINE

Figure 4: Diagrammatic charts showing the relation between Type 1 and Type 2 unconformities and submarine condensed sections to eustatic sea-level changes.

widespread chronostratigraphic unit at the time of deposition is near the top of each sequence, as shown on Figure 5. The upper portions of clastic sequences characteristically show the greatest landward onlap of the alluvial plain, and at the same time they exhibit major seaward progradation.

In most basins of the world, Type 1 unconformities are characterized by both subaerial and submarine erosion. However, there are occasional exceptions where the shelf margin is locally subsiding so rapidly that its rate exceeds that of the fall of sea level, and the unconformity has the characteristics of a Type 2. An example is a shelf margin controlled by a rapidly subsiding growth fault, such as in the Neogene of the Gulf of Mexico. Similarly, Type 2 unconformities may locally show Type 1 characteristics in areas where structural activity causes stability or uplift so that sea level may locally fall below the shelf edge.

Condensed sections are thin marine stratigraphic intervals characterized by very slow depositional rates (less than 1 cm/1000 yrs). They are commonly associated with thin but continuous zones of burrowed or lightly lithified beds (hardgrounds, omission surfaces). They are conspicuous by the presence of scattered pebbles or high concentrations of volcanic ash, glauconite, phosphate, or radioactive minerals, and are commonly linked to marine hiatuses or faunal mixing horizons (Figures 4C and 5). In vertical deepening-upward sections, a condensed section commonly marks the greatest paleowater depth. However, in areas of rapid subsidence the greatest paleowater depths may be above the condensed section. Condensed sections should not be confused with a winnowed interval associated with high-energy removal of fine sediment.

A condensed section develops when the rate of relative

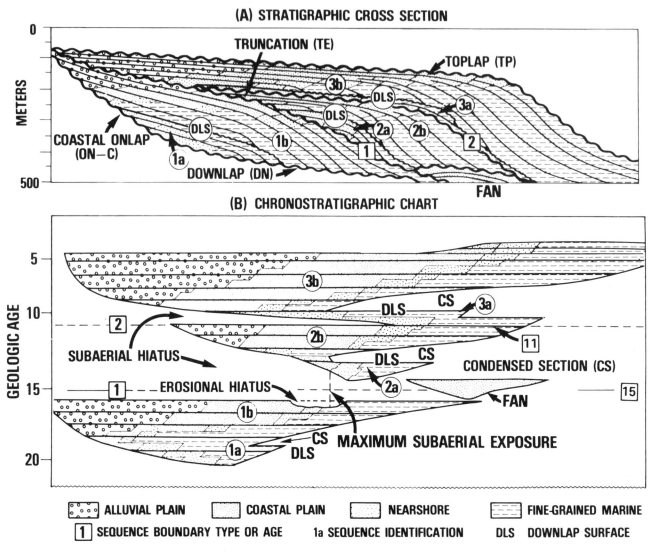

(A) STRATIGRAPHIC CROSS SECTION

TRUNCATION (TE)

TOPLAP (TP)

3b

DLS

3a

DLS

COASTAL ONLAP
(ON–C)

DLS

2a

2b

1b

1

2

1a

DOWNLAP (DN)

FAN

(B) CHRONOSTRATIGRAPHIC CHART

3b

DLS

CS

3a

2

2b

11

SUBAERIAL HIATUS

DLS

CS

1

EROSIONAL HIATUS

2a

CONDENSED SECTION (CS)

FAN

15

1b

CS

MAXIMUM SUBAERIAL EXPOSURE

1a

DLS

▨ ALLUVIAL PLAIN   ▨ COASTAL PLAIN   ▨ NEARSHORE   ▨ FINE-GRAINED MARINE

1 SEQUENCE BOUNDARY TYPE OR AGE   1a SEQUENCE IDENTIFICATION   DLS DOWNLAP SURFACE

Figure 5: Diagrammatic stratigraphic cross section and chronostratigraphic chart showing the distribution of sequence boundary types, downlap surfaces (condensed sections), and facies of three idealized sequences in depth and time.

rise (eustatic rise plus subsidence) of sea level is significantly greater than the rate of accumulation, causing the depositional site to shift landward and resulting in low sedimentation rates (starved conditions) seaward of the depositional site. The marine hiatuses associated with condensed sections tend to disappear landward into essentially conformable successions developed at the site of deposition (Figures 3 and 5). The age of a condensed section within a given depositional sequence tends to be synchronous globally but may differ slightly from basin to basin with changes in rates of deposition and subsidence.

Downlap surfaces are submarine surfaces characterized seismically by a downlap over a concordant pattern and are commonly associated with a marine hiatus (Figures 4C and 5). Downlap surfaces associated with condensed sections mark the change from the end of transgression to the start of regression as the rate of eustatic rise decreases, and sediments begin to prograde out over the old starved surface.

The downlap represents deep-marine clinoform toes developed as prograding progresses seaward. Downlap surfaces may also be present above submarine fans.

Coastal onlap refers to the progressive landward encroachment of the coastal (littoral, coastal plain, or alluvial plain) deposits of a given sequence (Vail et al, 1977, part 3, Figure 3 and 5). All sequences exhibit almost continuous coastal onlap throughout the time of their deposition. Periodically, there is a major downward (basinward) shift in the coastal onlap pattern at the times of Type 1 or Type 2 unconformities. This abrupt downward shift in coastal onlap gives a chart of relative changes of coastal onlap the sawtooth character observed in Figure 2. The almost continuous presence of coastal onlap indicates that most sediments are deposited during a relative rise of sea level. Either eustatic sea level is rising, or the depositional surface is subsiding, or the two are combining to create space for the accommodation of sediments. Type 1 and Type 2 uncon-

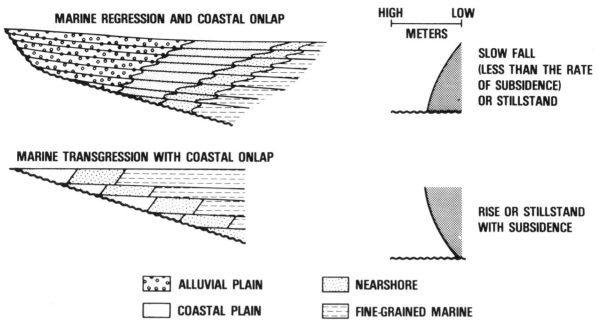

STRATIGRAPHIC AND FACIES PATTERNS

MARINE REGRESSION AND COASTAL ONLAP

MARINE TRANSGRESSION WITH COASTAL ONLAP

INDICATED EUSTATIC CHANGES OF SEA LEVEL

HIGH   LOW
METERS

SLOW FALL
(LESS THAN THE RATE
OF SUBSIDENCE)
OR STILLSTAND

RISE OR STILLSTAND
WITH SUBSIDENCE

ALLUVIAL PLAIN   NEARSHORE

COASTAL PLAIN   FINE-GRAINED MARINE

Figure 6: Diagrammatic charts showing the relation between transgression-regression and eustatic sea-level changes.

formities mark those exceptional periods of time when the rate of eustatic sea-level fall is greater than that of subsidence, producing relative falls of sea level and globally synchronous downward shifts of coastal onlap. During the periods of time involved in Type 1 relative falls, and prior to the beginning of the following relative rises, the only sediments deposited in significant volumes are submarine fans, which have no corresponding coastal onlap. Thus, the downward shift in coastal onlap appears only as a gap or discontinuity in the coastal onlap record and is instantaneous in terms of geologic time (Figures 3 and 5).

In previous papers (Vail et al, 1977; 1980), we directly equated relative changes of coastal onlap with relative changes of sea level. Pitman (1978) and M.T. Jervey (personal communication) demonstrated that changes observed in coastal onlap cannot be related directly to relative changes of sea level in the upper part of the cycle, as we have done in the past. We now refer to our sawtoothed global charts as relative changes of coastal onlap. Ideally, a chart of coastal onlap should be constructed by plotting the successive positions of the upper limit of the coastal plain (paludal and deltaic deposits). When an alluvial plain is associated with the coastal plain, the relative sea-level boundary approximates the facies change between the coastal plain and the alluvial plain (M.T. Jervey, personal communication, Figure 3). Unfortunately, seismic stratigraphic techniques do not always permit identification of this facies change between coastal and alluvial plains. Consequently, the charts developed from studies of seismic sections include data from both the coastal plain and the alluvial plain with its slightly greater alluvial dips. This greater dip is the source of the differences between the coastal onlap chart and the relative change-of-sea-level chart. We recognized this factor in Vail et al (1977, part 3), but did not take it into account when interpreting relative changes of sea level. The two charts, however, are similar. Charts of relative changes of coastal onlap versus distance typically show abrupt sawtooth shifts from widespread to restricted, whereas charts of relative changes of sea level versus distance commonly show more gradual shifts.

Transgression is defined as the landward displacement of the shoreline indicated by landward migration of the littoral facies in a given stratigraphic unit. Regression is the seaward displacement of the shoreline (Vail et al, 1977, part 3, Figures 3 and 4). Our studies indicate that transgressions and regressions are *not* necessarily globally synchronous. Other writers have reached similar conclusions (Yanshin, 1973). There is, however, a general tendency for certain periods of time to be dominated by transgressive deposits, while in others the prevalent trend is regressive. Times of maximum transgression tend to be more globally synchronous than times of regression, because of the cumulative effects of sealevel rise and subsidence. In general, transgression results when eustatic sea level rises and regression results when eustatic sea level falls at a slower rate than subsidence (Figure 6). We consider an interval transgressive if the overall tendency is to deepen upward, even though there may be a series of stacked, retrogradational regressive units.

We believe that cycles of relative changes of coastal onlap are more useful than transgressive-regressive cycles for defining stratigraphic intervals. The lows in the sawtooth pattern mark times of global unconformities that subdivide the section into depositional sequences genetically related to cycles of coastal onlap. The cycles define chronostratigraphic intervals, which include all strata deposited during the time period between the points where the unconformi-

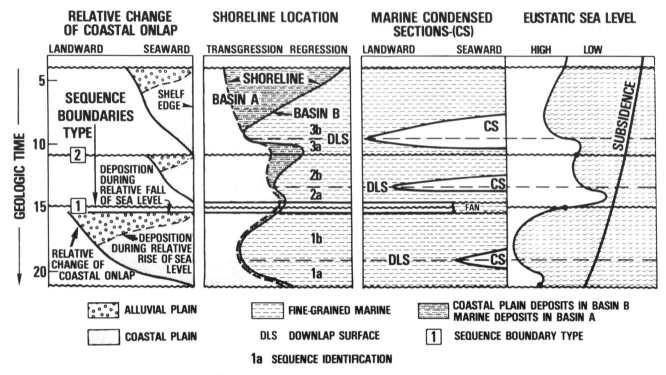

RELATIVE CHANGE OF COASTAL ONLAP	SHORELINE LOCATION	MARINE CONDENSED SECTIONS-(CS)	EUSTATIC SEA LEVEL

Figure 7: Summary chart showing relation of sequence-boundary types, relative changes of coastal onlap, relative changes of sea level, transgression-regression and marine condensed sections to eustatic changes of sea level and subsidence.

ties become conformable. Stratigraphic units bounded by unconformities or their equivalent conformities are defined as sequences (Vail et al, 1977, part 2). They have the advantage of being genetic units and are ideal for facies analysis and paleogeographic reconstructions. Marine condensed sections occur within sequences, and can be used to subdivide sequences.

Cycles of transgression and regression do not necessarily coincide with cycles of relative change of coastal onlap. In general, the maximum point of transgression occurs in the middle of a cycle of coastal onlap, and the maximum point of regression occurs after the downward shift of coastal onlap and thus above the associated global unconformity (Figures 3 and 5). As stated previously, this relationship does not always exist because of the variations of sediment supply. In general, transgressions are more rapid than regressions.

The stratigraphic relations among global unconformities, condensed sections, coastal onlap, relative changes of sea level, marine transgressions and regressions, subsidence, and eustatic changes of sea level are diagrammed in Figure 7. In general, as we discuss later in this report, the tectonic subsidence along most passive margins is long-term and gradually decreases in rate, because it is related to a thermal decay curve. It does not change rapidly enough to cause regional unconformities. Tectonic subsidence patterns differ from region to region, and are not globally synchronous. We therefore conclude that the many synchronous unconformities and abrupt changes in stratigraphy observed in basins globally are caused by eustatic sea-level changes, superimposed on regional tectonic and sedimentary regimes

that change at much slower rates.

Type 1 and Type 2 unconformities (Figure 7) are globally synchronous because when they are dated precisely they coincide in time with the inflection points of curves showing eustatic falls of sea level (Vail and Todd, 1981; M.T. Jervey, personal communication). An inflection point also marks the time of downward shift of coastal onlap. Global cycles of coastal onlap are chronostratigraphic intervals that can be dated precisely because they include all the rocks deposited during the interval of time between the inflection points of eustatic falls of sea level.

Truncation below unconformities caused by erosion during a relative fall of sea level has a very different pattern depending on whether or not tectonic activity (folding, faulting, uplift) is taking place during the time of the fall. If no tectonic activity other than subsidence is taking place, the truncation should be greatest at the shelf edge and should progress shelfward by headward erosion and downcutting of river valleys (Figure 8a). In general, valley and canyon truncation dies out landward away from the shelf edge, the overlying and underlying strata are generally parallel, and very detailed stratigraphic correlation is necessary to identify the truncation and onlap patterns associated with the canyon and valley fills. Farther shelfward, as truncation effects die out, the unconformities must be identified by coastal onlap and related facies patterns (Figure 8a). If folding, faulting, uplift, or large differential subsidence is taking place during the relative fall, erosion of the tilted beds produces an easily identifiable truncation unconformity (Figure 8b). In general, most global unconformities are not associated with tectonic activity.

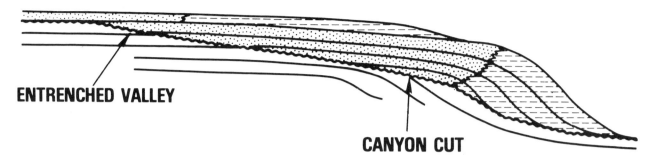

(A) RAPID SEA-LEVEL FALL WITH BASIN SUBSIDENCE

ENTRENCHED VALLEY

CANYON CUT

(B) RAPID SEA-LEVEL FALL DURING TECTONISM

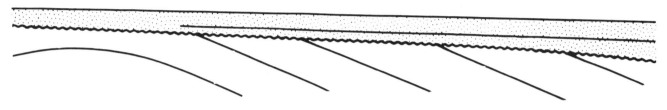

TRUNCATION OF TILTED STRATA

NEAR SHORE   FINE GRAINED MARINE

Figure 8: Diagrammatic sections showing stratal patterns associated with unconformities. Section A shows a typical headward erosion pattern associated with a Type 1 unconformity formed during basin subsidence. Section B shows the expression of a Type 1 unconformity formed during a period of tectonic uplift and folding.

The causes of global eustatic changes of sea level in the Jurassic (as shown on Figure 2) are unknown. Most workers agree, however, that the long-term changes in sea level, such as the overall rise in the Jurassic, result from changes in volume of the ocean basins due to such processes as changes in rates of sea-floor spreading (Hallam, 1963; Pitman, 1978). The causes of rapid changes of sea level (greater than 10 cm/1000 years), however, are problematical. The only known mechanism that produces rates of this magnitude is glaciation (Donovan et al, 1979), but conclusive evidence is lacking for glaciers of sufficient size to have caused the large and rapid changes of sea level we observe in the Jurassic and other Phanerozoic strata.

In summary, eustatic changes of sea level are the primary cause of major abrupt vertical changes in stratigraphy at the time of deposition and they are major factors controlling lateral changes. Eustatic changes of sea level produce interregional or "global" unconformities not only for continental margins, but for interior basins and deep-oceanic sites. As such, they provide a means to make exact correlations from the deep sea basins across the shelf into interior basins. Figure 2 shows our best estimate of eustatic sea-level changes during the Jurassic supercycle and how they relate to the ages of Type 1 and Type 2 regional unconformities, relative changes of coastal onlap, and age and relative magnitude of condensed sections. Transgressions and regressions are not included because of their global variability. Our studies

show that Type 1 and Type 2 regional unconformities are globally synchronous when dated precisely at the age they become conformable. The age of conformity corresponds to the time of the most rapid rate of eustatic fall of sea level (inflection point on the sea-level curve), as shown in Figures 4 and 7 (M. T. Jervey, personal communication).

Ages of condensed sections may differ slightly with changes in rate of deposition and subsidence. Downward shifts of coastal onlap, however, are synchronous globally, because they occur at inflection points on the sea-level curve at the times of highest rates of eustatic fall. The inflection points also mark the most accurate dates to establish the synchrony of the global unconformities.

When the coastal onlap patterns of restricted and widespread sequences of the same age are compared in several areas globally, the patterns tend to be similar in areas where the subsidence pattern is continuous and uninterrupted over the interval studied. If stability or uplift occurs, the onlap patterns are distorted. Such distorted patterns are useful for indicating periods of tectonism. However, the ages of the sequence boundaries remain constant in the distorted sections. Shoreline transgressions and regressions are the most variable of the parameters discussed; however, sequences tend to be mostly transgressive in the lower part of the coastal onlap cycle below the downlap surface, and mostly regressive in the upper part of the cycle above the downlap surface.

STRATIGRAPHIC NOMENCLATURE

A major operational problem is how to name and refer to unconformities (sequence boundaries) and sequences once they have been identified. Our experience is that the sequence boundaries can be marked with colored pencils to facilitate identification as a grid of seismic or well and outcrop sections is correlated. After the areal continuity of the sequence boundaries and condensed sections has been verified by seismic loop ties or well-log correlation, the geological ages of the strata are determined by means of paleontology. At this point, we believe that information is conveyed precisely by naming the unconformities after the oldest strata known to occur above the unconformity (e.g., the latest Berriasian unconformity) and by informally naming the sequences between the unconformities with symbols representative of age (e.g., Kl. 3). If precise age dating cannot be achieved because a significant hiatus is present throughout the study area or because the paleontologic data are poor or lacking, the age of the closest global unconformity is used and the error range is indicated on a summary chronostratigraphic chart. For reference purposes we designate unconformities on seismic or well-log sections by the age of their correlative conformities in million of years (m.y.). It is important to note which time scale is used, since the age in millions of years of a particular faunal zone may vary from one scale to another and may change in the future. The Jurassic global chart (Figure 2 and Table 1) shows the symbols we use for the global cycles and the ages of the global unconformities we have identified in the Jurassic supercycle. The numerical age is also useful for input into quantified stratigraphic studies, such as geohistory analysis referred to later in this paper.

Formal formation names for stratigraphic units serve a very useful descriptive purpose, but remember that formations are lithostratigraphic units and may transgress time. For proper facies and paleogeographic interpretation, the lithostratigraphy must be placed in the proper chronostratigraphic framework or serious interpretation errors will result. We also believe that naming unconformities after orogenic events is misleading. In the North Sea, for example, two major unconformities are called the Middle and Late Cimmerian. We prefer to designate the respective unconformities as latest Bathonian (158) and latest Berriasian (132), believing that more accurate stratigraphic information is thereby provided. Our studies indicate that these particular unconformities were caused by falls of eustatic sea level superimposed on longer term tectonic movements, and were not caused directly by orogenic events.

PROCEDURE FOR INTERPRETATION OF UNCONFORMITIES, CHRONOSTRATIGRAPHY, COASTAL ONLAP, TECTONIC SUBSIDENCE, AND EUSTATIC CHANGES OF SEA LEVEL

The procedure used for charting the Jurassic unconformities, chronostratigraphy, coastal onlap, tectonic subsidence, and eustatic changes of sea level consists of the following five steps:

1) Developing chronostratigraphic framework by determining sequence boundaries for a study region.

2) Dating sequence boundaries and making a representative chronostratigraphic chart for a study region.

3) Making regional chart of relative changes of coastal onlap for study regions and combining them to make a global onlap chart.

4) Determining tectonic subsidence and long-term eustatic changes of sea level with geohistory analysis corrected for loading and compaction.

5) Estimating short term eustatic changes of sea level.

Developing Chronostratigraphic Framework By Determining Sequence Boundaries for a Study Region

Sequence boundaries are interpreted most accurately by using a combination of reflection seismic data, well control, and outcrop information. Figure 5A is a diagrammatic example of a geologic cross section showing chronostratigraphic surfaces, unconformities, and facies. To interpret unconformities and sea-level changes, it is critical that we make stratal correlations in sufficient detail to recognize truncation, lapout, prograding, and other patterns associated with different depositional environments. Seismic data tied to well control or closely spaced well logs with detailed marker-bed correlations are ideal for this purpose.

Figure 4 illustrates the three types of surfaces discussed previously: Type 1 and Type 2 unconformities, and condensed sections. In general, we find that if we can trace these types of surfaces over a regional grid of seismic and well data, they will correspond to one of the global unconformities or condensed sections listed in Table I. The 15 m.y. unconformity is a Type 1; the 11 m.y. unconformity is a Type 2. Condensed sections are present at 9, 13, and 18 m.y. (Figure 5B).

The condensed sections (CS) or downlap surfaces (DLS) shown schematically within sequences 1, 2, and 3 (Figure 5) overlie transgressive deposits and thus occur within a sequence between the bounding global unconformities. In deep-water depositional sites such as ocean basins, such condensed intervals or downlap surfaces commonly directly overlie both types of unconformities in areas where lowstand deposits are absent.

We have found that unconformities and condensed sections can be recognized best on reflection seismic sections by identifying onlap, downlap, and truncation patterns and dating them with well control (Vail et al, 1977, parts 2 and 6; 1980). Because seismic reflections are produced largely by stratal surfaces and unconformities with significant velocity-density (impedance) contrasts (Vail et al, 1977, part 5), reflection patterns portray stratal configurations within the limits of seismic resolution.

Unconformities characterized by large hiatuses, such as those along basin margins, commonly extend laterally into areas where deposition has been more nearly continuous and where hiatuses gradually decrease as the unconformities approach conformity (Figure 5). A conformable surface is a chronostratigraphic horizon and must be traced with its correlative unconformity to define completely the sequence it bounds. In this way, the three-dimensional sequence framework bounded by unconformities and their correlative conformities is completely defined for subsequent seismic facies and structural analysis (Vail et al, 1977, parts 2 and 7). As shown on Figure 4, Type 1 and Type 2 unconformities

338

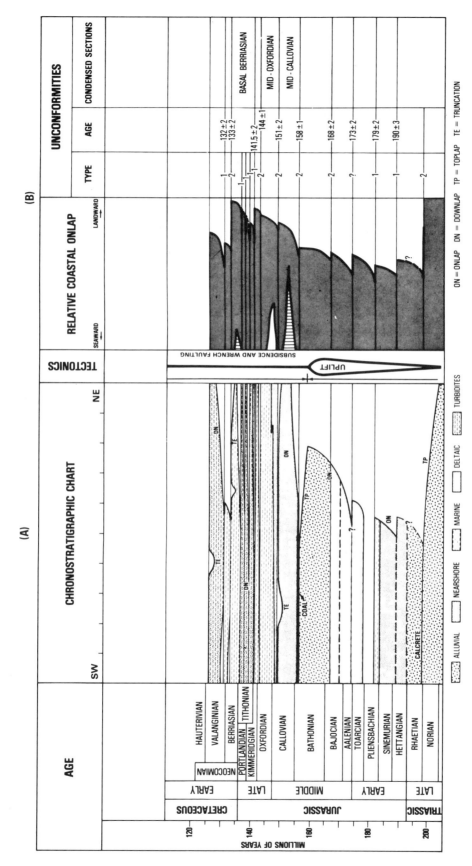

Figure 9: Charts showing chronostratigraphy, relative changes of coastal onlap, age and type of unconformities and condensed sections (from the Inner Moray Firth in the North Sea). This chart is discussed in detail in Vail and Todd, 1981.

339

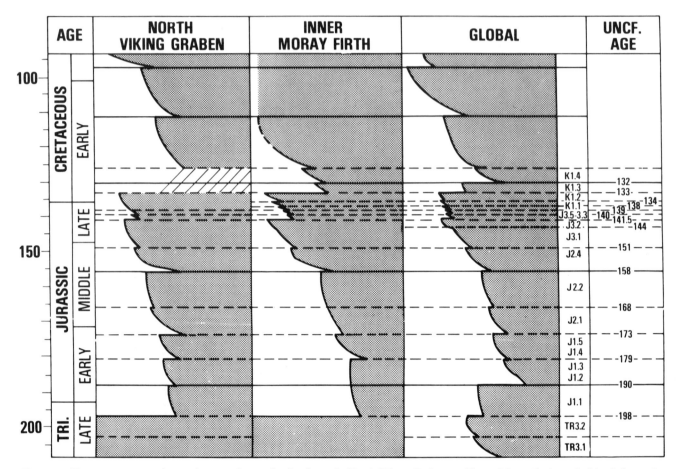

Figure 10: Charts comparing relative changes of coastal onlap from the North Viking Graben and Inner Moray Firth, in the North Sea with the global chart.

become most conformable at the point of downward shift of coastal onlap. Locally, Type 1 unconformities have the least hiatus on the upper slope landward of the submarine fan. We give the submarine fan the same age as the unconformity, since we believe it formed during the most rapid eustatic fall of sea level, but we carry the base of the fan as our principal mapping surface.

Dating Depositional Sequence Boundaries and Making a Representative Chronostratigraphic Chart for a Study Region

An unconformity is dated most accurately by establishing the ages of overlying and underlying strata at points along the unconformity where the hiatus is least. The ideal point is where the surface becomes a conformity. In many cases, however, conformity is not reached and the point of minimum hiatus is used. Dating is accomplished using biostratigraphy from wells and outcrop, and comparing with local and global onlap charts.

Figure 5B is an example of how a chronostratigraphic chart is made from a geologic cross section. The chart portrays the sequences in geological time, and thus emphasizes the genetic nature of the facies within the sequences and the abrupt discontinuities between sequences. Chronostrati-

graphic charts also emphasize the periods of time when no sediments are present because of nondeposition or erosion. Figure 9 is a chronostratigraphic chart for the Inner Moray Firth seismic section shown on Figure 1. This chart is discussed in detail by Vail and Todd (1981).

Making Regional Charts of Relative Changes of Coastal Onlap for Study Regions and Combining Them to Make a Global Onlap Chart

Charts of relative changes of coastal onlap are derived from geological cross sections or interpreted seismic sections (such as Figure 1) and from the chronostratigraphic charts (such as Figure 9). The procedure for constructing charts of relative changes of coastal onlap is discussed in Vail et al (1977, parts 3 and 4). Figure 10 shows how we combine charts of relative changes of coastal onlap to produce the global chart. In this example, unconformity age and relative patterns of coastal onlap are compared among the North Viking Graben, the Inner Moray Firth, and the global chart (see Vail and Todd, 1981). Unconformities of the same age and type are added to the global chart when they have been identified in sedimentary basins on at least three different continents. Patterns of relative changes of coastal onlap are included on the global chart only from areas undergoing

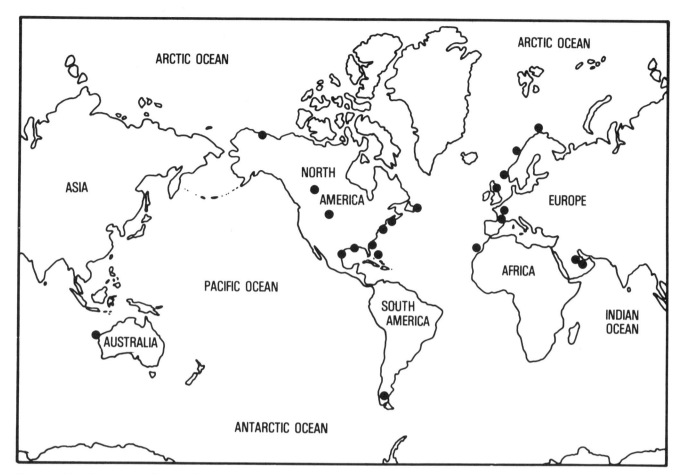

Figure 11: Map showing distribution of seismic stratigraphic and/or well studies of the Jurassic supercycle.

uniform subsidence, because the coastal onlap pattern in a given area is influenced by a variable tectonic and subsidence history. Figure 11 shows the areas in which we have made seismic stratigraphic and well-log studies used in the construction of the global Jurassic coastal onlap chart (Figure 2).

Determining Tectonic Subsidence and Long-Term Eustatic Changes of Sea Level with Geohistory Analysis (Van Hinte, 1978) Corrected for Loading and Compaction (Hardenbol, Vail, Ferrer, 1981)

Sea-floor subsidence, eustatic changes of sea level, and sediment supply are the principal interacting variables in a marine basin. This interaction is recorded in the sedimentary section in the form of sequences filling the basin. Geohistory analysis provides a technique for depicting the interaction of eustatic changes of sea level and sea-floor (tectonic) subsidence and relating these factors to the sedimentary record. This technique separates the total subsidence of an area into individual components due to the effects of sediment and water-column loading, compaction, and tectonic activity (e.g., thermal cooling, basement faulting, and tectonic loading).

Also separated are the apparent vertical movement effects of eustatic changes of sea level. By quantification of the individual effects of sediment compaction, sediment loading,

and subsidence, the magnitudes of the long-term eustatic changes can be established.

An example of this geohistory analysis from offshore northwestern Africa is described in Hardenbol, Vail, and Ferrer (1981). It shows that in the early Cenomanian, sea level could have been nearly 300 m (984 ft) higher than present sea level.

The eustatic sea-level values obtained from geohistory analysis show close agreement with previous results obtained by different methods by Vail et al (1977, Figure 8, part 4), Hayes and Pitman (1973), Pitman (1978), and Kominz (this volume). A significant difference exists, however, with the much smaller values obtained by Watts and Steckler (1979), who used a similar method based on the subsidence history of the Atlantic margin of the North American east coast. Possibly their assumption of a single thermal contraction event for the thermo-tectonic subsidence of the Atlantic margin is not valid. Thermo-tectonic subsidence histories for Georges Bank and Baltimore Canyon C.O.S.T. wells suggest increased tectonic activity in the Early Cretaceous and middle Tertiary.

Estimating Short-Term Changes of Eustatic Sea Level

The pattern of short-term eustatic changes of sea level is derived from a combination of several data sets (Figure 12).

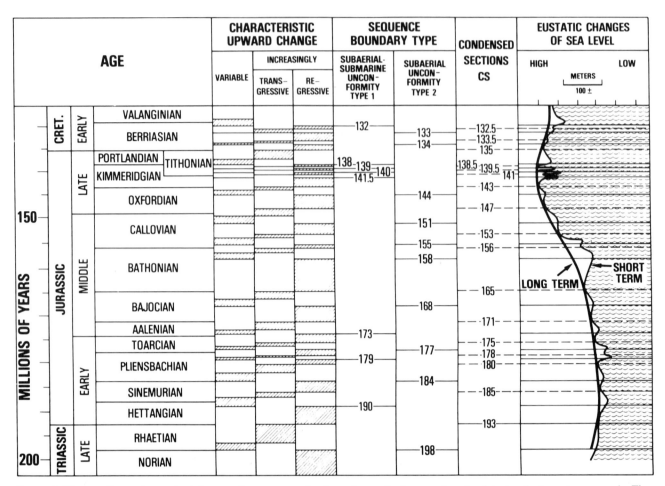

Figure 12: Diagram illustrating methodology used to construct the chart of eustatic changes of sea level during the Jurassic supercycle. The ages of Type 1 and Type 2 unconformities and condensed sections are plotted along with the intervals of time when the Jurassic sediments tend to be increasingly transgressive or regressive upward.

The ages of Type 1 and Type 2 unconformities and condensed sections are plotted, along with the intervals of time when Jurassic sediments tend to be increasingly transgressive or regressive upward. From these are derived the direction and slope of the short-term eustatic curve based on the patterns shown in Figures 4 and 6. The final eustatic sea-level curve is then derived from the configuration of the short-term changes superimposed upon the independently determined envelope of the long-term eustatic changes.

The absolute magnitudes of the rapid falls of sea level are difficult to calculate, but studies indicate that the 132-million-year and 140-million-year drops are the greatest and may approximate the magnitudes of the Pleistocene changes. The other changes appear to be less, as shown on Figures 2 and 12. These magnitudes are estimated from the downward shift of coastal onlap below the shelf edge and from the depth of subaerial erosion obtained from geohistory reconstructions.

CONCLUSIONS

Our experience indicates that analysis of stratigraphy based on depositional sequences is the most accurate approach for interpreting geologic age, depositional environment, lithofacies, and paleogeography. Not only does the approach subdivide strata into genetic depositional units, but it places them in a global context controlled by short-term variations in eustatic sea level. Sequences can be readily recognized from seismic interpretation, well-log and outcrop correlations, and facies studies. Interpretations based on lithofacies and biostratigraphy could be misleading unless they are place within a context of detailed stratal chronostratigraphic correlations. This paper illustrates how sequence concepts may be applied to the Jurassic.

ACKNOWLEDGMENT

This paper represents the joint effort of many Exxon scientists who contributed to the development, documentation, and testing of the concepts. We are especially indebted to R. M. Mitchum and M. A. Uliana, whose critical review of the manuscript provided many important improvements. In addition, we thank Exxon Production Research for permission to release this information.

REFERENCES CITED

Bates, R.E., and J.A. Jackson, 1980, eds., Glossary of geology, second edition: Falls Church, Virginia, American Geological Institute, 794 p.

Donovan, D.T., et al, 1979, Rates of marine transgressions and regressions: Journal of the Geological Society of London, v. 136, p. 187-193.

Hallam, A., 1963, Eustatic control of major cyclic changes in Jurassic sedimentation: Geological Magazine, v. 100, p. 444-450.

Hardenbol, J., P. R. Vail, and J. Ferrer, 1981, Interpretating paleoenvironments, subsidence history and sea-level changes of passive margins from seismic and biostratigraphy: Oceanologica Acta No. SP, p. 3-44.

Hayes, J.D., W.C. Pitman, III, 1973, Lithospheric plate motion, sea-level changes and climatic and ecological consequences: Nature, v. 246, p. 18-22.

Kominz, M.A., in press, Oceanic ridge volumes and sea level change - an error analysis, in J.S. Schlee ed., Interregional unconformities: AAPG Memoir 36, this volume.

Pitman, W.C., 1978, Relationship between sea-level change and stratigraphic sequences: Geological Society of American Bulletin, v. 89, p. 1389-1403.

Vail, P.R., and J. Hardenbol, 1979, Sea-level changes during the Tertiary: Oceanus, v. 22, p. 71-79.

———, and R.M. Mitchum, 1979, Global cycles and sea-level change and their role in exploration: Bucharest, Romania, Proceedings of the Tenth World Petroleum Congress, v. 2, Exploration Supply and Demand, p. 95-104.

———, and R.G. Todd, 1981, North Sea Jurassic unconformities, chronostratigraphy and sea-level changes from seismic stratigraphy: Proceedings of the Petroleum Geology Continental Shelf, Northwest Europe, p. 216-235.

———, et al, 1980, Unconformities of the North Atlantic: Philosophical Transactions of the Royal Society of London, A 294, p. 137-155.

———, et al, 1977, Seismic stratigraphy and global changes of sea level, in Seismic stratigraphy—applications to hydrocarbon exploration; AAPG Memoir 26, p. 49-212.

Van Hinte, J.E., 1976a, A Jurassic time scale: AAPG Bulletin, v. 60, p. 489-497.

———, 1976b, A Cretaceous time scale: AAPG Bulletin, v. 60, p. 498-516.

———, 1978, Geohistory analysis - application of micropaleontology in exploration geology: AAPG Bulletin, v. 62, p. 201-222.

Watts, A.B., and M.S. Steckler, 1979, Subsidence and eustacy at the continental margin of eastern North America, in M. Talwani, W.F. Hay, and W.B.F. Ryan, eds., Deep drilling results in the Atlantic ocean; continental margins and paleoenvironment: American Geophysical Union, Maurice Ewing Series, No. 3, p. 218-234.

Yanshin, A., 1974, The so-called global transgressions and regressions: International Geology Review, v. 16, p. 617-646 (English translation of Russian article).

The American Association of Petroleum Geologists Bulletin
V. 64, No. 8 (August 1980) P. 1158-1178, 11 Figs., 1 Table

Assessing Oil and Gas Plays in Facies-Cycle Wedges[1]

DAVID A. WHITE[2]

Abstract Oil and gas potentials of formations in frontier areas can be assessed by comparison with formations in corresponding parts of facies-cycle wedges documented in producing areas. The transgressive-regressive facies-cycle wedge is a body of rock bounded above and below by regional unconformities or the tops of major nonmarine tongues. The ideal wedge includes, from base to top, facies successions from nonmarine, to coarse (sandstone or grain carbonate), to fine (marine shale or micrite), to coarse, and back to nonmarine. Five types of potential coarse reservoir plays, representing the wedge base, middle, top, edge, and subconformity positions, are identified by their distinctive vertical facies successions within this cycle.

Different play types have different risks that can affect assessments. Least risky of sandstone plays are the transgressive wedge base, which is capped by marine shale that commonly is both a good source and seal, and the wedge middle, which is both underlain and overlain by thick shale. Most risky are the regressive wedge-top sandstones with no thick shale seal above and a typically poor-source shale beneath, and the wedge edge, which has no thick marine shales whatever. Subunconformity plays, which include any of the other wedge parts truncated beneath another wedge, have intermediate risks.

Carbonate plays, with the exception of the wedge top, in general are riskier than their corresponding sandstone plays, probably because of more poorly developed porosity and permeability in the carbonate rocks. The exceptional wedge-top carbonate rocks have excellent leached porosity and anhydrite caps. The wedge classification contrasts different types of reefs for exploration purposes. Transgressive wedge-base reefs are encased in the overlying shale or micrite and are apt to be localized over unconformity topography. Wedge-middle reefs are both capped and underlain by shale, and many are localized by depositional topography. Typical regressive wedge-top reefs, which also may form at the edge of depositional slopes, overlie shale or micrite and commonly are capped by anhydrite and red beds.

Comparison based on wedge position considers the basic differences in source-reservoir-seal relations. Wedge position alone, however, cannot reflect all the other critical controls of oil and gas occurrence, such as source richness and maturation, reservoir quality, or trap capacity. Large variations in productivity thus can occur within any single play type. Accordingly, assessment procedures for new plays have three key steps: (1) selecting look-alike productive plays of the same wedge position; (2) scaling the potential hydrocarbon yields to compensate for obvious differences in thickness or areal extent; and (3) risking the results for the other factors that might render the new plays nonproductive.

This study is based on stratigraphic cross sections from 80 producing basins of the free world. Selected examples are from the Eastern Venezuela, Alberta, Gulf Coast, Permian, and Paradox basins.

INTRODUCTION

The transgressive-regressive facies-cycle wedge provides a good framework for geologic compari-
sons for assessments of undiscovered oil and gas potentials. Assessment by geologic analogy, or comparison of new areas with old producing ones, is a common approach to resource appraisal. There are three critical steps in such assessments: (1) comparisons should reflect similarities in as many of the basic geologic controls of oil and gas occurrence as possible; (2) assessors should scale these comparisons, using factors such as barrels per unit area or volume, to normalize size differences between compared areas; (3) assessors should estimate the risk that such comparisons with productive areas might be wrong. Dry holes are, after all, the most common results of exploration.

This study aims at providing a systematic approach for making these assessment comparisons and is based on more than 100 stratigraphic cross sections that show the settings of some 2,000 major oil and gas fields in 200 wedges within 80 producing basins of the free world. A major field is defined as one having ultimately recoverable reserves in excess of 50 million bbl of oil, or 300 Bcf

[1]A preliminary version of this paper was read before the Association at Houston, Texas, April 2, 1979. Manuscript received, September 17, 1979; accepted, February 1, 1980.

[2]Exxon Production Research Co., Houston, Texas 77001.

Many people participated in this project—particularly, C. V. Campbell, H. M. Gehman, T. A. Konigsmark, and R. O. Wilbur. Other contributors are S. Boggs, Jr., H. A. Garaas, G. C. Grender, W. M. Jordan, G. T. MacCallum, W. V. Naylor, R. A. Page, R. W. Prather, W. P. Ryman, J. A. Sawyer, J. R. Shouldice, P. Stauft, L. M. Toohey, D. G. Van Delinder, P. R. Vogt, and W. J. Wood. Cross sections presented are based on the much more detailed work of J. H. Wright (eastern Venezuela); J. Y. Smith and P. Purcell (Alberta Cretaceous); F. E. Grinstead and G. E. Welder (southeast Texas); F. A. Johnson, Jr., and others (west Texas); and R. M. Mitchum (Utah-Colorado). I thank C. V. Campbell, H. M. Gehman, and D. O. Smith for reviewing the manuscript, and Exxon Production Research Co. for permission to publish. Grender et al (1974, p. 491) published a brief description of the use of the wedge in quantitative stratigraphic studies.

Article Identification Number
0149-1423/80/B008-0001$03.00/0

of gas, or any equivalent combination of oil and gas. Only a few cross sections can be included here. The selected five sections illustrate wedge characteristics, the methodology, and a sampling of the exploration implications. The other studied basins are listed in Appendix 1.

Although the discussion centers on the stratigraphic controls of oil and gas, most of the field examples are in structural traps within the stratigraphic settings described.

WEDGE ANALYSIS

A facies-cycle wedge (Fig. 1) is a body of sedimentary rock bounded above and below by regional unconformities or the tops of major nonmarine tongues. The ideal wedge represents a transgressive-regressive cycle of deposition including, from base to top, the vertical succession of facies from nonmarine to coarse-textured to fine-textured to coarse-textured and back to nonmarine.

Wedge analysis is based on the vertical succession of three generalized facies groupings. *Fine facies* include marine shale, marl, micritic limestone, chalk, and salt; organic-rich types of these shales and limestones are the likely source rocks for oil and gas. *Coarse facies* include sandstones, siltstones, grain limestones, and dolomites interbedded or in contact with the fine marine facies; such coarse-textured rocks are the typical reservoirs for oil and gas. *Nonmarine and associated facies* are not interbedded with the fine marine rocks; they typically include interbedded mixtures of nonmarine shale, red shale, sandstone, conglomerate, and anhydrite. Anhydrite is in this grouping because its typical lagoonal or restricted-shelf environment of deposition is generally associated with nonmarine red-bed environments of the regressive parts of cycles. Grain limestones and dolomites are also classed as "nonmarine and associated" facies if they are not interbedded or in contact with the fine marine facies. The key to the groupings is the presence or absence of the fine-facies interbeds, rather than the nature of the coarse facies. The groupings thus emphasize the positions of potential marine source rocks for oil and gas, although they do not rule out nonmarine sources.

The analysis rests on Walther's (1894, p. 987-993) observation that the vertical succession of facies commonly is the same as the lateral order of their depositional environments. (See Middleton, 1973, and Woodford, 1973, for reviews of Walther's ideas.) Facies are distinctive, coeval, adjacent rock units. Different marine facies form concurrently on the seafloor in laterally contiguous environments of deposition. If the shoreline transgresses landward, the coexisting environ-

ments also shift, and the offshore deposits progressively cover the previous products of the nearer shore environments. The vertical succession of facies is reversed if the shoreline regresses seaward, and nearshore deposits come to overlie the offshore ones.

The presence of a fine facies or a nonmarine facies either above or below a prospective coarse reservoir facies defines the four basic wedge parts and related oil and gas plays (Fig. 2). The distinctive vertical facies succession of the transgressive wedge base is fine over coarse over nonmarine; of the wedge middle, fine over coarse over fine; of the regressive wedge top, nonmarine over coarse over fine; and of the wedge edge, nonmarine over coarse over nonmarine. Subunconformity plays, a fifth significant type, occur in any truncated wedge part in which the oil or gas is in direct contact with an overlying unconformity. A play is a group of geologically similar prospects having basically the same source-reservoir-trap controls of oil and gas; in this method of analysis the primary groupings by similar reservoir-facies types may be further subdivided by trap types or other geologic similarities.

Oil and gas follow the facies. Transgressive and regressive deposits have distinctive patterns of facies changes and corresponding progressions of oil and gas fields. In a marine depositional basin, *centerward* is the direction in which nonmarine facies change laterally within strata to marine facies. *Marginward* is the opposite direction. In the transgressive or *up-to-margin pattern* of the wedge-base boundary (Fig. 1), the facies changes occur progressively farther marginward in successively younger strata. Oil and gas fields, in either structural or stratigraphic traps, are apt to occur in an up-to-margin progression, that is, in progressively younger reservoirs marginward. Levorsen (1934) illustrated such a progression for the Cherokee sandstones of eastern Oklahoma. In the regressive or *up-to-center pattern* of the wedge top, the facies changes occur progressively farther centerward in successively younger strata; the oil and gas fields may occur in an up-to-center progression. The up-to-center pattern is analogous to the seaward "climb" of Lowman (1949, p. 129), a term he used to describe the Tertiary rocks of the United States Gulf Coast. In the zigzag or stairstep facies-boundary patterns, the "treads" are master stratal (chronostratigraphic) surfaces, and the "risers" represent the facies changes within strata. The terms up-to-margin and up-to-center objectively describe facies boundaries and progressions of oil and gas fields, even where the specific relations to transgressive or regressive shorelines are obscure.

To avoid confusion with cyclothems and the

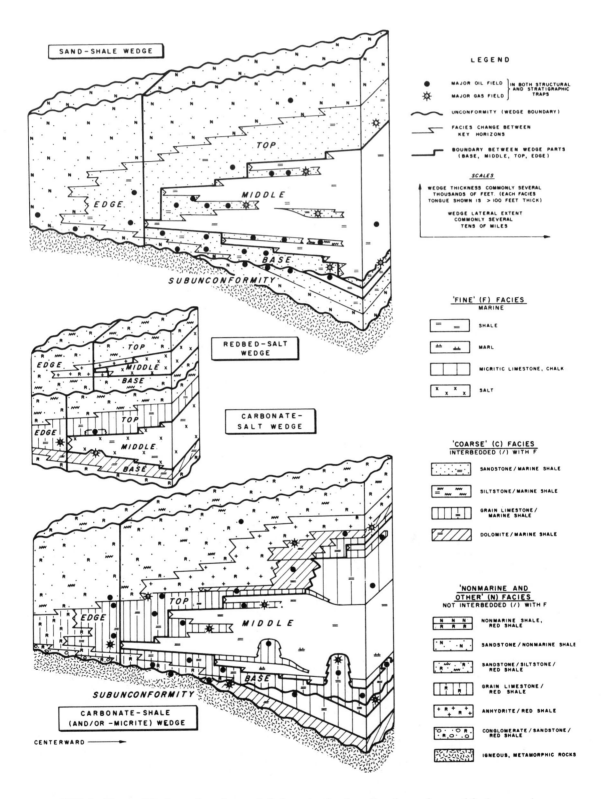

FIG. 1—Types of facies-cycle wedges and their parts, showing oil and gas plays and facies groupings.

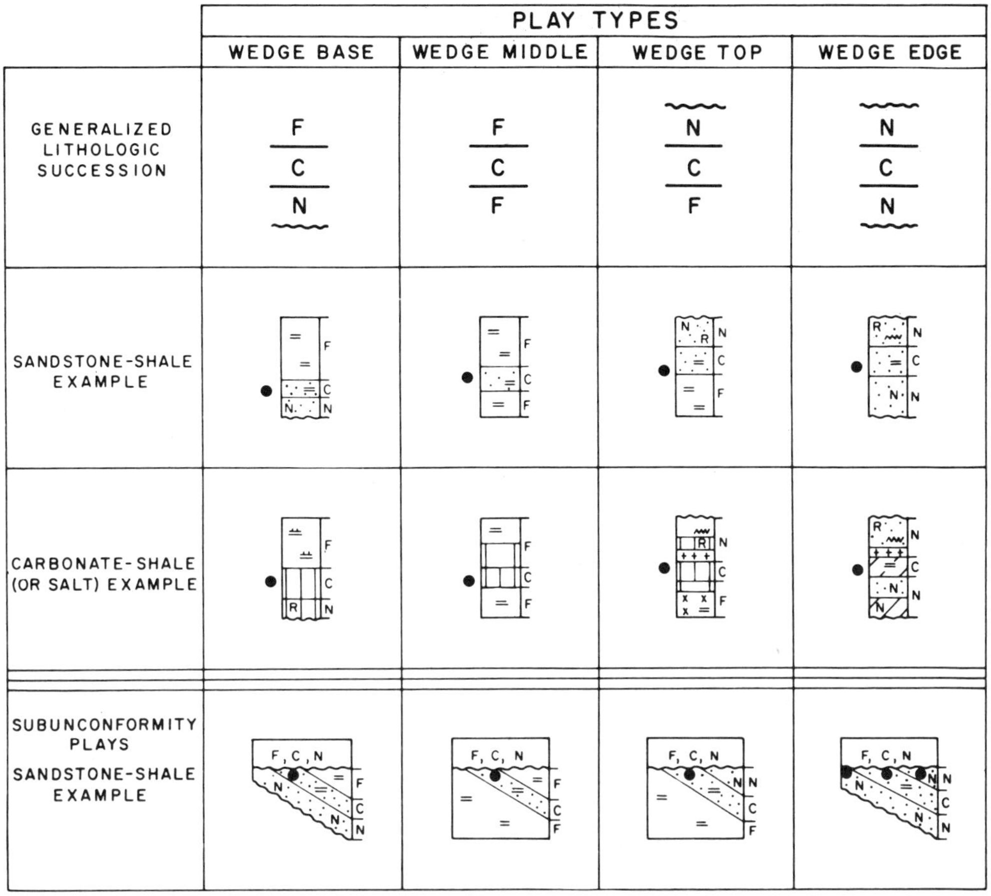

FIG. 2—Vertical facies successions defining play types in wedges. See Figure 1 for lithologic symbols. Facies units shown are at least 100 ft (30 m) thick. Thinner units are classed as interbeds.

deposits of other small-scale pulses, wedge analysis recognizes only the facies units that are at least 100 ft (30 m) thick. Thinner units are treated as interbeds. Vertical facies successions in single wells or outcrop columnar sections can define wedges objectively. A knowledge of lateral relations, as reconstructed from cross section, may be required, however, to determine the original wedge parts if key facies units are eroded away. For example, where the overlying fine facies is missing, a truncated wedge-base coarse facies looks like an edge. At some places, as shown in the upper right of Figure 3, a coarse facies directly underlying an unconformity appears to be a wedge-top play but could be a truncated wedge-middle one. Interpretations of original wedge positions are used in classifying subunconformity plays (Fig. 2).

In scale the typical wedge falls between the unconformity-bounded interregional sequence of Sloss (1963, p. 93) and the depositional sequence of Mitchum et al (1977, p. 53) or the "interthem" of Chang (1975, p. 1549). Although at places bounded by regional unconformities, wedges fundamentally are facies-defined bodies (Fig. 3), whereas depositional sequences are recognized by the geometry of their bedding surfaces. Wedges change as the facies change, as do the prospects for oil and gas. As shown on Figure 3, two wedges may merge into one where a bounding unconformity or regressive nonmarine tongue terminates toward the center of the depositional basin. Toward basin margins, the wedge base, middle, and top combine to form the wedge edge beyond the termination of the last tongue of fine facies. Wedges are apt to be 1,000 ft (300 m) or more thick and to extend 20 mi (32 km) or more laterally.

OIL AND GAS PLAY SUMMARY

About 95% of the free world's oil and gas reservoirs considered in this study occurs in the coarse

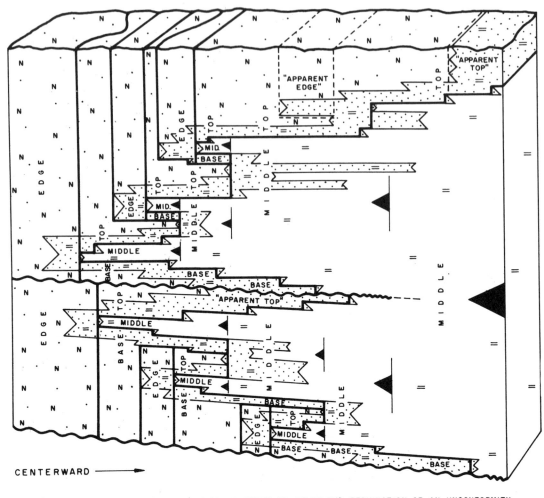

WEDGE SYMBOLS PLOTTED VERTICALLY ABOVE OR BELOW THE TERMINATION OF AN UNCONFORMITY OR NONMARINE TONGUE SEPARATING TWO WEDGES THAT CENTERWARD COMBINE TO FORM ONE GREATER WEDGE

FIG. 3—Relations of multiple wedges. See Figure 1 for lithologic symbols. To interpret this figure, imagine that all wedge parts must be defined by facies successions observed in single well drilled at any given place.

facies. These are the reservoirs most closely associated with fine marine source rocks. Most of the other 5% occurs in the nonmarine facies. Excluding the Middle East, the percentages in coarse and nonmarine facies are 90 and 10, respectively. Even though they are less common, nonmarine occurrences are important exploration targets, and their importance will increase in the future. Also, a few significant fields are in fractured fine-facies reservoirs.

The wedge-play types describe the five basic hydrocarbon source-reservoir-seal relations for the coarse facies. The main source possibility of the coarse wedge-base reservoir is in the overlying fine facies, which also commonly provides an ex-

cellent seal. The wedge-middle reservoir has fine source facies both above and below, the overlying one doubling as a seal. The chief potential source of the wedge top is beneath; thin, tight interbeds within the upper coarse or nonmarine facies provide uncertain seals, except where anhydrite is present. The wedge edge can have only a laterally equivalent body of fine source facies, and its seal is as uncertain as that of the wedge top. In addition to overlying and underlying sources, each of the preceding play types may have both interbedded and laterally equivalent sources. Subunconformity plays are more complex. The oil and gas in any truncated wedge part may be capped by any one of the three facies types. The main source

Table 1. Hydrocarbon Distribution by Play Types in Wedge*

	Base	Middle	Top	Edge	Subunconformity
Sandstone	23	25	8	8	13
Carbonate	4	5	5	4	5

*As percent of total studied oil-equivalent, ultimately recoverable reserves for free world excluding Middle East.

facies may lie either above or below the unconformity.

For the areas studied, excluding the Middle East, about three-fourths of the total oil and gas is in sandstones and one-fourth is in carbonate rocks (Table 1). The two most productive plays are the wedge-base and wedge-middle sandstones, which are two to three times as prolific as the other sandstone plays. Hydrocarbons are more evenly distributed throughout the carbonate plays. If the Middle East were included in Table 1, however, wedge-middle sandstones and wedge-top carbonate rocks would dominate all other plays. Excluding the Middle East, gas accounts for about 45% of the oil-equivalent reserves in both sandstone and carbonate reservoirs. Wedge-edge and subunconformity sandstones are the only plays that have unusually low gas contents, probably because of leakage.

These distributions of total hydrocarbons are strongly affected by a few giant occurrences; for this reason the numbers in Table 1 are not as directly usable for ordinary assessment purposes as the frequency-of-production data discussed later in the section on risking. The following examples of wedges show that wedge analysis is generally applicable in the regional setting, but they also illustrate that more detailed considerations are essential in exploring for oil and gas. The examples serve to document at least in part the conclusions in the succeeding summary of play characteristics.

EXAMPLES OF SANDSTONE WEDGES

Three examples of sandstone-shale wedges include one that thickens centerward from the area of sediment supply, one that thins centerward, and multiple wedges that initially thicken but ultimately thin centerward, as all wedges bordering open oceans must do. The first two examples are the settings of the greatest heavy-oil and bitumen accumulations on earth. Figure 4 locates the cross sections of Figures 5 through 7.

Tertiary, Eastern Venezuela

The Oligocene-Miocene-Pliocene cycle consists of the productive wedge-base Oficina sandstone, the wedge-middle Freites shale, and the wedge-top La Pica and Las Piedras sandstones and nonmarine facies (Fig. 5). The cross section runs near the Oficina group of fields on the south flank of the Eastern Venezuela basin. The basin lies between the Guiana shield on the south and the mobile belt of the Serranía del Interior on the north (Fig. 4A).

The Oficina sandstones onlapped during southward transgression toward the Guiana shield, which was the main source of clastics (Renz et al, 1958). Centerward was north, and the wedge-base boundary has an overall up-to-margin pattern (Fig. 5). During the late Freites–La Pica regression, however, sand also came eastward from the Anaco uplift. The wedge top has an up-to-center pattern eastward, so this north-south section (Fig. 5) is parallel with facies strike. Most stratigraphic intervals, particularly those of the wedge middle, thicken centerward in this area.

Most oil and gas pools are in the wedge base, but a few are in wedge-middle tongues extending from the base. The main pay zones of 22 major fields are stratigraphically projected onto Figure 5. Most pools are trapped on the high (north) sides of south-dipping normal faults, in combination with sandstone pinch-outs and cross faults (Hedberg et al, 1947). The number of pay zones per field ranges from 20 to 150. Shale interbeds within the wedge base form top seals for most pools. The Freites shale changes to sandstone just south of the cross section, and at the end of this main shale tongue is a wedge edge in which most of the Orinoco bitumen and heavy-oil belt lies. The Orinoco belt contains about 1 trillion bbl in place, with the main concentrations on four structual highs (Demaison, 1977).

For several possible reasons, the La Pica and Las Piedras sandstones of the wedge top are not productive here. (1) Many faults die out upward in the Freites, so there are fewer fault traps above, although lenticular sandstones are abundant. (2) There is no massive seal over the wedge top, although many tight interbeds may provide local seals. (3) Hydrocarbon maturation within the wedge top is unlikely, except at the trough of the structural basin a few miles to the north. (4) Last, and probably most important, is the lack of organic-rich source beds. The Miocene shales of the Freites are green and apparently lean, whereas the Oligocene shales associated with the wedge base are dark gray and apparently organic-rich. Sandstone wedge bases generally seem to be asso-

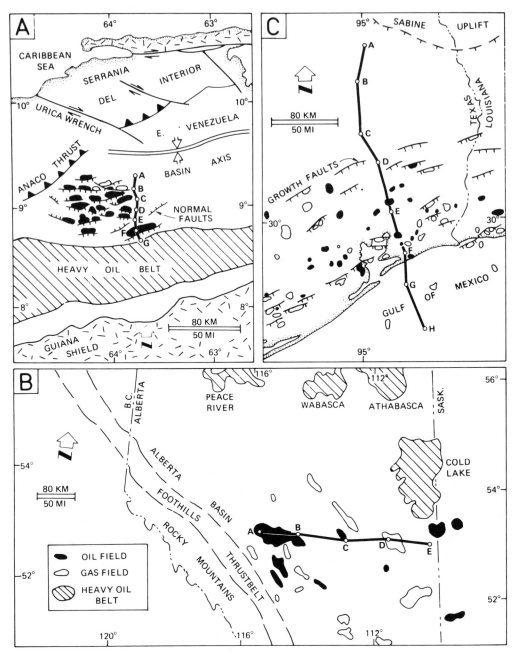

FIG. 4—Locations of cross-section examples of sandstone wedges: **A**, Tertiary, eastern Venezuela, Figure 5; **B**, Mesozoic, Alberta, Figure 6; **C**, Tertiary, Texas Gulf Coast, Figure 7.

ciated with better source rocks than are the wedge tops.

Mesozoic, Alberta

The Cretaceous cycle consists of the wedge-base Mannville sandstones, the wedge-middle Colorado shale and intertongued Viking and Cardium sandstones, and the wedge-top Belly River sandstones and Edmonton-Paskapoo nonmarine facies (Fig. 6). The Alberta basin lies between the Rocky Mountains on the west and the Canadian shield on the east (Fig. 4B).

The Early Cretaceous sea transgressed from the north; Mannville clastics came both from the Ancestral Rockies on the west and from the Canadian shield on the east (Jardine, 1974). Basal Mann-

350

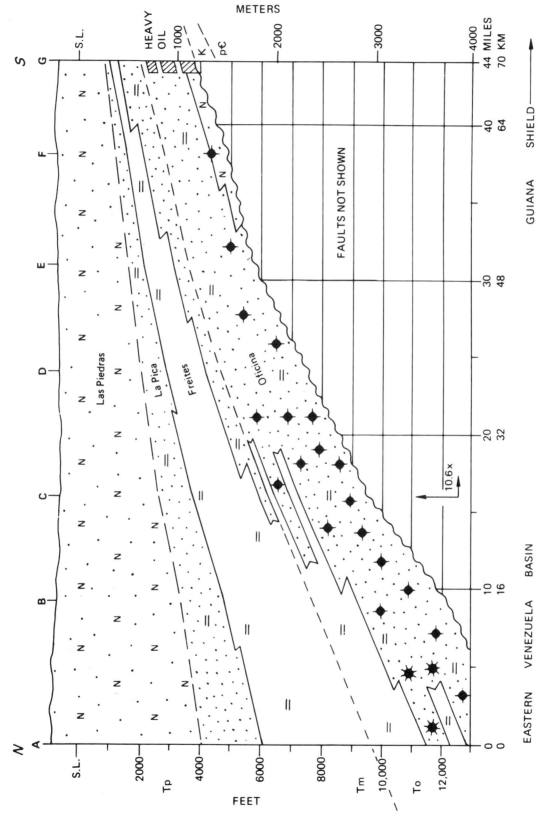

FIG. 5.—Cross section of Tertiary sandstone-shale wedge, eastern Venezuela. See Figure 1 for lithologic symbols and Figure 4A for location. Symbols for major-field producing zones on section show various combinations of oil and gas.

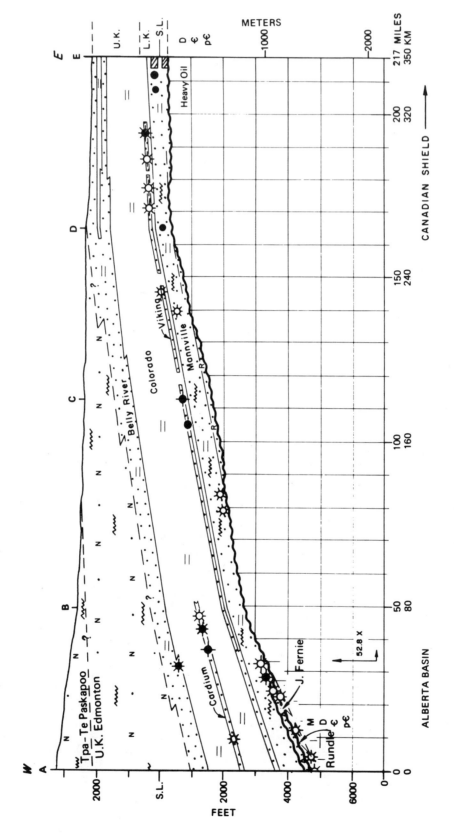

FIG. 6—Cross section of Mesozoic sandstone-shale wedge, Alberta. See Figure 1 for lithologic symbols and Figure 4B for location.

ville beds onlap the topography of the pre-Cretaceous unconformity. A sea transgressing from the south joined that from the north at the end of Mannville deposition. The western source of clastics dominated this area thereafter (Fig. 6), and the Colorado shales thin centerward (eastward). Greater subsidence toward the western source area reversed the eastward depositional dips of the Viking, Cardium, and Belly River sandstones, forming ideal updip stratigraphic traps at the eastern pinch-outs. After the Belly River regression with its up-to-center facies pattern, the ongoing flood of clastics from the west kept marine waters out of the area despite continued subsidence.

Major oil and gas production from the wedge-base Mannville unit is widespread. Either local interbeds or the basal Colorado shales provide seals and possible oil sources. Correlative sandstones on the north and east, including the McMurray sands on the Athabaska River, contain nearly 1 trillion bbl of bitumen and heavy oil in place (Energy, Mines and Resources Canada, 1977). A common theory is that biodegradation and water-washing have altered normal oil migrating from Lower Cretaceous shales updip toward the shield (Deroo et al, 1974). Stratigraphically trapped hydrocarbons are present in the wedge-middle Viking and Cardium sandstones, the latter containing the giant Pembina field (well A, Fig. 6). As a probable result of relatively poor adjacent oil source rocks, only one small major pool occurs in the wedge-top Belly River sandstones; the uppermost Colorado shales are lighter gray and apparently less organic-rich than those beneath.

Several subunconformity fields in Jurassic and Mississippian reservoirs are also shown on Figure 6. This gas is trapped at updip lobes of pre-Cretaceous scarps sealed by locally tight basal Mannville sediments. Possible organic-rich source beds are present both above and below the pre-Cretaceous unconformity, and some subunconformity hydrocarbons may have migrated into Mannville deposits.

This section again illustrates the hydrocarbon richness of the sandstone wedge base and middle as compared to the top.

Tertiary, Texas Gulf Coast

The Gulf Coast Upper Cretaceous and Tertiary rocks contain several depositional cycles (Fig. 7). Separate transgressive-regressive episodes center around the main wedge-middle shale tongues of the Midway, Crockett, Jackson, Anahuac, and middle and upper Fleming. Wedge parts for such compound cycles are defined in Figure 3. In the Anahuac wedge, for example, the richly productive wedge base consists of the *Marginulina* and uppermost Frio sands overlying the Frio nonmarine beds. Above the wedge-middle Anahuac *(Heterostegina)* shales are the wedge-top *Discorbis* sands. South or centerward of the end of the nonmarine facies, all productive Frio sands are classed as part of a great wedge middle.

The several wedge bases (Woodbine, uppermost Wilcox, uppermost Yegua, and uppermost Frio-*Marginulina)* generally have up-to-margin facies patterns, whereas the wedge tops (lower Wilcox, lower Yegua, lower Frio, and lower Fleming-*Discorbis)* have marked up-to-center zigzags. Most units here thicken centerward, at many places abruptly across growth faults that are not shown on Figure 7. Farther south, all units must thin again down depositional slopes toward the Gulf of Mexico. Figure 4C shows the section location.

The Sabine uplift was active at the end of Early Cretaceous time, when much of the north rim of the Gulf basin was elevated and eroded. The transgressive Upper Cretaceous Woodbine sandstone was locally eroded off the crest of the uplift, but the overall transgression culminated in the broad inland sweep of Midway seas. The subsequent history was dominantly regressive with transgressive punctuations; the depocenter of each succeeding unit shifted progressively farther gulfward.

Producing belts follow generally an up-to-center progression (younger reservoirs gulfward) in accordance with the dominantly regressive deposition of the favorable coarse facies. At the transgressive breaks at the top Yegua and top Frio, however, producing belts follow gentle up-to-margin progressions. Traps are salt domes (both piercement and deep-seated) and anticlinal roll-overs at growth faults, together with local sand pinch-outs. Most major fields producing from the Fleming nonmarine facies are on piercement domes; the oil apparently was displaced upward from its regular habitat. Of all oil and gas in the Gulf basin, about half comes from wedge middles, the other half being about equally divided between bases and tops. Wedge edges, present mainly in Jurassic–Lower Cretaceous rocks, are essentially nonproductive. Sources of Tertiary hydrocarbons are matters of controversy, but most geochemists believe some contributions are from shales at considerable depths (Curtis, 1979).

EXAMPLES OF CARBONATE WEDGES

Two examples of carbonate wedges show: (1) carbonate wedge-base reefs sealed by shale, and wedge-top reefs capped by marginal anhydrite;

353

and (2) a wedge in which the middle or central sediment is chiefly salt rather than shale. Figure 8 locates the cross sections of Figures 9 and 10.

Paleozoic, West Texas

The Pennsylvanian-Permian cycle consists of wedge-base Bend-Strawn-Canyon grain carbonate rocks and reefs, wedge-middle Strawn through Guadalupe shales and micritic limestones, and wedge-top Cisco through Ochoa grain limestones and dolomites, anhydrites, and red beds (Fig. 9). The cross section runs eastward from the Eastern shelf of the Midland basin across the Bend arch to the edge of the Fort Worth basin (Fig. 8A). Limestone textures are not differentiated on Figure 9, although coarser grain sizes probably predominate except for the Leonard-Guadalupe deep-water limestones of the Midland basin proper.

During the Strawn transgression, basal beds of the Caddo Limestone onlapped the post-Mississippian unconformity from both sides toward the crest of an old ridge near well G, Figure 9. The main clastic sources on the east and south were the eroding mountains of the Ouachita structural belt. Remote from the clastic source, wedge-base reefs (wells D, G, Fig. 9) started to grow above the old hills and scarps, or above gentle arches (Vest, 1970). Reef buildups kept pace with submergence as the basin tilted toward the west. Clastics from the mountains filled and overran the Fort Worth basin and thinned centerward into the "starved" Midland basin (Adams et al, 1951). Deposition shifted regressively westward in the form of successively younger depositional slopes (Van Siclen, 1958; Adams, 1962) in an up-to-center pattern. Typically, red beds were deposited on the flat, shallow undaforms (Rich, 1951) behind carbonate banks whose debris trailed down the depositional slopes or clinoforms. Dark shale and turbidites rested on the deep-water fondoforms. These sediments eventually swept over and killed the reef organisms that had been building mounds from the Caddo Limestone.

Massive Leonard dolomite reefs apparently were localized during the regression at the undaform edge of the uppermost Wolfcamp Coleman Junction limestone. Red beds and anhydrite formed back of these reefs, whereas the Dean and Spraberry turbidites filled the deep depositional basin in front. The Midland basin was nearly filled when it was covered by upper Guadalupe dolomite–anhydrite–red-bed facies backing the Capitan reef of the Central Basin platform. Ochoan and Triassic red beds completed the basin fill.

Wedge-base Canyon reefs capped by shale are major producers of oil, and a few fields are present in the wedge-base sheet limestones. Several wedge-middle sandstones and siltstones, notably the Spraberry turbidites (well A, Fig. 9), also contain major fields. Wedge-top Leonard dolomite reefs are also productive, as are the overlying San Andres dolomites capped by anhydrite and red beds. The Leonard regressive reefs typically have good porosity only in the narrow core; the immediate backreef facies updip consist of evaporites and micrograined dolomites that form the seals (Hills, 1972). Porosity in the overlying San Andreas dolomite is greater and much more widespread, as is typical of such leached units capped by anhydrite. In this general area, reservoirs in every wedge position could have been fed by hydrocarbon-source shales and micritic limestones of the wedge middle.

Oil and gas are more uniformly distributed throughout carbonate wedges than they are in sandstone ones, and at many places the carbonate wedge top is the best play. Wedge-base and wedge-top reefs have distinctly different seals and topographic controls of localization.

Paleozoic, Utah-Colorado

The Pennsylvanian-Permian rocks of the Paradox basin provide an example of a carbonate-salt wedge. Lower Hermosa and Molas carbonate rocks form the wedge base; Paradox salt and black shale beds form the wedge middle; and upper Hermosa carbonate rocks and Permian red beds and arkoses form the wedge top. Wedge edges are present at both ends of the cross section (wells A, E, Fig. 10), marginward of the facies changes from wedge-middle salts to carbonates. The cross section extends from the Tyende saddle northeast across the Paradox basin to the Uncompahgre uplift (Fig. 8B). The section does not show the structural complexities either of the salt anticlines or of the uplift.

Early Pennsylvanian seas transgressed the eroded Mississippian surface. In the line of cross section (Fig. 10), well D is nearest the center of the depositional basin, and in both directions from this well the wedge-base carbonates have up-to-margin facies boundaries with the overlying salts. With continued subsidence the area of salt deposition spreads to the maximum outline shown on the index map (Fig. 8B). During each successive step of this spreading, contemporaneous normal marine carbonates, backed by red beds, were deposited. At its maximum extent the Paradox salt basin evidently was rimmed by tectonically controlled shelf areas. Access to open marine waters may have been across relatively low sills on the northwest, west, and southeast (Wengerd, 1958, 1962). Soon afterward the Uncompahgre uplift became more active and began

354

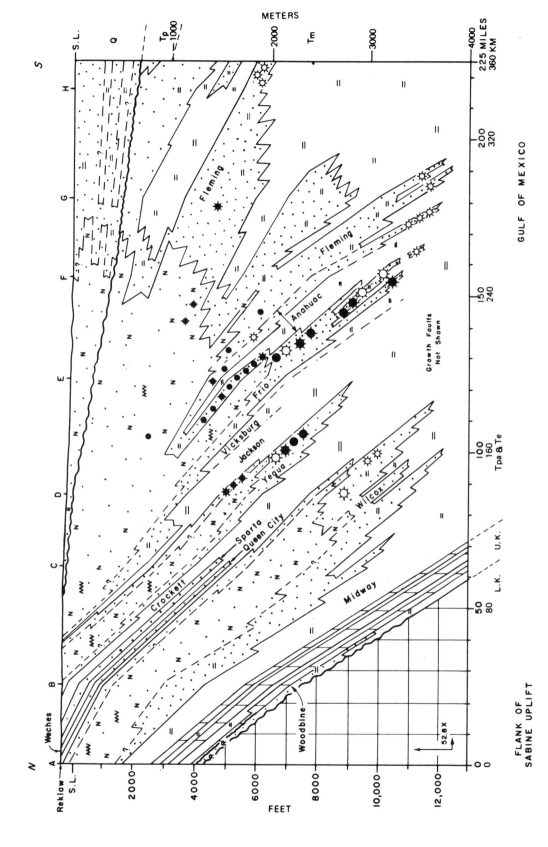

FIG. 7—Cross section of Tertiary sandstone-shale wedges, Texas Gulf Coast. See Figure 1 for lithologic symbols and Figure 4C for location. Larger oil and gas symbols each represent two or more major fields.

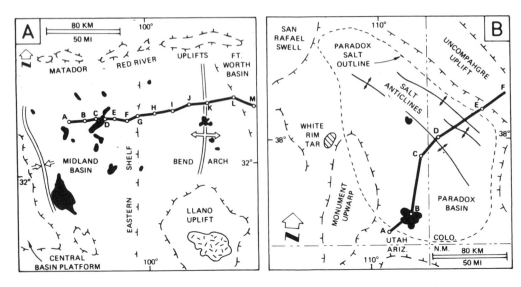

FIG. 8—Locations of cross-section examples of carbonate wedges: **A**, Paleozoic, west Texas, Figure 9; **B**, Paleozoic, Utah-Colorado, Figure 10.

to shed arkose into the salt basin. Marginal carbonates, including the productive Aneth bank, encroached the shrinking salt basin from the other side. Regressive or up-to-center facies patterns developed from both sides of the basin toward well D (Fig. 10).

Within the salt section itself, minor cycles of increasing and then decreasing salinity are reflected by the upward succession from black shale to dolomite to anhydrite to thick halite to (rarely) potassium salts, followed by the reverse succession (Herman and Barkell, 1957). Salt flow and growth of piercement anticlines were activated in Late Pennsylvanian time by differential loading as the Cutler arkose spread across the basin (Shoemaker et al, 1958).

The Aneth field produces from an algal bank in the wedge top. The wedge edge has a few minor fields (e.g., Barker Dome). The wedge base has only minor production and generally is tight. Most Pennsylvanian carbonate production of whatever wedge position is limited to the area of interbedded black shales, which are the presumed source rocks and seals. At the Northwest Lisbon field, the subunconformity Mississippian reservoir is in fault contact with the salt and black shale beds. The unconformity-bounded Cambrian and Upper Devonian–Mississippian sequences are both present as wedge edges (Fig. 10). Tar in the Permian White Rim sandstone may have originated in the laterally equivalent Phosphoria Formation on the northwest (Demaison, 1977).

PLAY CHARACTERISTICS

Wedge-Base Plays

The fine facies overlying the wedge base has an up-to-margin pattern and commonly provides an excellent hydrocarbon source and seal. Transgressive shales and micritic limestones, deposited in deepening water under reducing conditions and covered by more offshore deposits, are apt to be organic-rich. Basal reservoir beds typically onlap the underlying unconformity, the pinch-outs forming potential stratigraphic traps. Traps may also form where the coarse reservoir facies changes marginward to tight nonmarine facies. If depositional dips are later reversed, however, possible stratigraphic traps form along the coarse-to-fine facies boundary. The bases of many wedges are relatively thin and at places are missing entirely. Special exploration techniques include mapping of paleotopography, paleostructure, and units onlapping the underlying unconformity (MacKenzie, 1972).

Sandstone wedge bases are apt to be channel or shoreline deposits interbedded with marine shales. Most major fields in the nonmarine parts of wedge bases are close to the coarse facies. Sand distribution commonly is influenced by the topography of the underlying unconformity (Busch, 1959). Examples of productive wedge-base sandstones are the Pennsylvanian Burbank and Bartlesville of Oklahoma (Levorsen, 1934), Cretaceous Mannville of Alberta (Fig. 6), Cretaceous Sarir of Libya (Sanford, 1970), Cretaceous Wood-

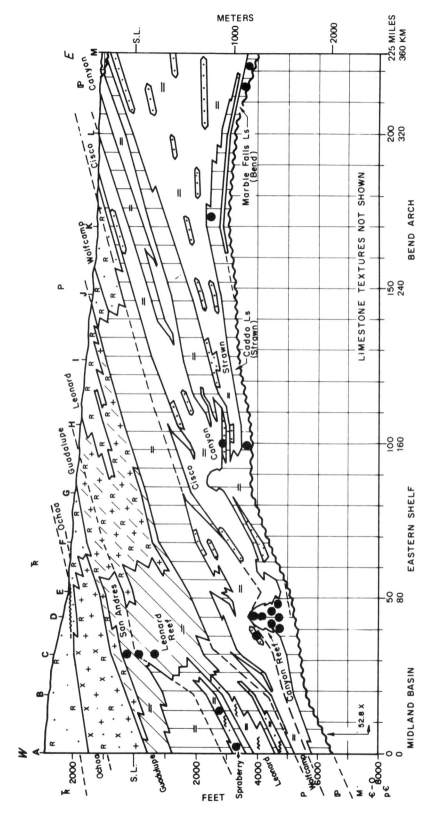

FIG. 9—Cross section of Paleozoic carbonate-shale wedge, west Texas. See Figure 1 for lithologic symbols and Figure 8A for location.

357

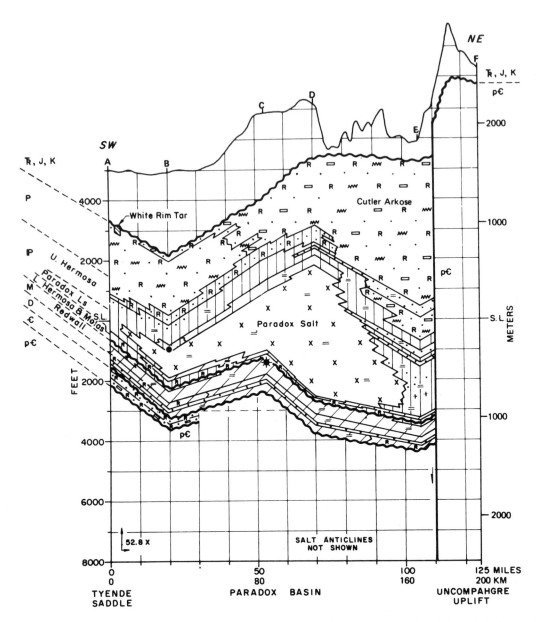

FIG. 10—Cross section of Paleozoic carbonate-salt wedge, Utah-Colorado. See Figure 1 for lithologic symbols and Figure 8B for location.

bine-Tuscaloosa of the United States Gulf Coast (Fig. 7), Oligocene Oficina of eastern Venezuela (Fig. 5), Oligocene-Miocene Lakat of central Sumatra (Soeparjadi et al, 1975), and Miocene Pyramid Hill of the Bakersfield arch, California (California Div. Oil and Gas, 1973).

Wedge-base carbonate reefs are encased by the overlying shale or micrite. Many such reefs appear to be localized over hills or scarps or arches on the underlying unconformity. Carbonate sheets are also productive. Productive wedge-base

carbonate rocks include the Ordovician Trenton of the Cincinnati arch (Lockett, 1968), Devonian deep-basin Beaverhill Lake and Leduc reefs of Alberta (Hemphill et al, 1970), Pennsylvanian Canyon reefs of west Texas (Fig. 9), Cretaceous Sligo of northern Louisiana (Rainwater, 1971), and Oligocene Porquero reefs of northern Colombia (Olsson, 1956).

Wedge-Middle Plays

With fine facies both above and below, wedge-middle plays have optimum source-reservoir-seal relations. Oil and gas occur in lenticular reservoirs enclosed in fine facies, or in tongues extending into the wedge middle from the base, edge, or top. The lower and upper segments of such tongues commonly have up-to-center and up-to-margin facies patterns, respectively. Regional reversal of depositional dips makes ideal stratigraphic traps (Fig. 6). The overlying transgressive fine facies is apt to be more organic-rich than the underlying regressive one. Wedge-middle plays are more likely to be associated with marked depositional slopes and interval thickenings than are other play types. Special exploration techniques include mapping the depositional topography that may control the distribution of reservoir rocks.

Sandstone reservoirs represent many depositional environments; nearshore and shelf sands may occur on the high side (undaform of Rich, 1951) and turbidites on the low side (fondoform) of the depositional slope (clinoform). Turbidites are most prevalent in this play type. At a few places, fractured shales, commonly cherty or calcareous, produce major amounts of oil or gas. Examples of the many productive wedge-middle sandstones are the Devonian Bradford of Pennsylvania (Newby et al, 1929), Permian Spraberry of west Texas (Fig. 9), Cretaceous Burgan and Zubair of the Middle East (Owen and Nasr, 1958), Cretaceous Viking and Cardium of Alberta (Fig. 6), Miocene Forest-Cruse of Trinidad (Bitterli, 1958; Michelson, 1976), Miocene-Pliocene Puente-Repetto of Los Angeles basin (Mayuga, 1970), and the centerward parts of many Tertiary tongues of the United States Gulf Coast (Fig. 7).

Wedge-middle carbonate reefs are both capped and underlain by shale or micrite, and many are localized by depositional topography at undaform edges or by differential compaction or by contemporaneous structural growth. Sheet carbonates, consisting mostly of limestone rather than dolomite, also produce, although porosity development is uncertain and difficult to predict. Wedge-middle carbonate rocks include the Devonian Leduc-Rimbey reef chain of Alberta (Belyea, 1964), the Jurassic Uwainat and Cretaceous Thamama of the Middle East (Hajash, 1967), and the Paleocene Heira and Ruaga of Libya (Terry and Williams, 1969; Bebout and Pendexter, 1975).

Wedge-Top Plays

The boundary between the wedge-top coarse facies and the underlying fine facies has an up-to-center pattern. Regressive shales and micritic limestones, deposited in shallowing water under oxidizing conditions and covered by more near-shore deposits, are apt to be organic-poor. Unless anhydrite is present, the only seals may be thin interbeds of tight rocks within the coarse or nonmarine facies. Stratigraphic traps occur where reservoirs change marginward to tight nonmarine rocks, or, if depositional dips are reversed, centerward at the change to fine facies (Fig. 6). Wedge tops commonly are very thick, although they may be truncated by the overlying unconformity. Special exploration techniques include mapping of underlying depositional topography and the distribution of thin interbedded seals.

Wedge-top sandstones contain only about one-third as much oil and gas as either wedge-base or wedge-middle sandstones (Table 1, which excludes Middle East plays). The contrast is especially striking for the tops normally include a far greater volume of sediment and are more thoroughly drilled than the bases. Wedge-top sand bodies of many types occur chiefly in thick deltaic complexes that become more marine downward. In the Gulf Coast Tertiary, although the bulk of production comes from wedge-middle tongues, the wedge tops contain some fields (Fig. 7). Other productive sandstone wedge tops are the Mississippian Chester of Illinois (Bond et al, 1971), Cretaceous Belly River of Alberta (Fig. 6), Cretaceous Pictured Cliffs of the San Juan basin (Weimer, 1960; Brown, 1973), Miocene lower Palembang of south Sumatra (Koesoemadinata, 1969), and Miocene Chanac of the San Joaquin Valley (California Div. Oil and Gas, 1973).

The wedge top is the most prolific carbonate play, if Middle East plays are included. The best situation is where porous limestones and dolomites are capped by anhydrite. Extensive leaching commonly contributes greatly to the porosity. Anhydrite forms behind barriers ranging from low oolite banks to massive regressive-reef complexes. Many such reefs, including the Permian Leonard and Guadalupe of west Texas (Fig. 9), are localized on the upper edges of clinoforms dropping off into the wedge middle. Examples of productive wedge-top carbonate rocks capped by anhydrite are the Mississippian Mission Canyon of the Williston basin (Craig, 1972), Permian Chase of western Oklahoma and Kansas (Pippin, 1970), Jurassic Smackover of the Gulf Coast (Ottmann et al, 1976), Jurassic Arab of Saudi Arabia (Steineke et al, 1958), Cretaceous Rodessa of the Gulf Coast (Rainwater, 1971), and Miocene Asmari of Iran (Hull and Warman, 1970). These reservoirs apparently have tapped, either along clinoforms or along fractures, good source rocks deeper within the wedge middle. The efficiency of accumulation must be due to the excellent anhydrite seals.

Wedge-Edge Plays

The most productive wedge-edge plays are on uplifts or platforms that affected sedimentation, where the reservoir facies lie close to laterally equivalent bodies of fine source facies. Unless anhydrite is present, seals are uncertain. Chances for finding major oil or gas diminish with increasing distance marginward from the end of the wedge-middle fine facies. Special exploration techniques include mapping such terminations.

Productive wedge-edge sand bodies typically are channel or shoreline deposits interbedded with shallow-water marine shale. Most known wedge-edge sandstone reserves are in Venezuela, in both coarse and nonmarine reservoirs (Miocene Lagunillas, Oligocene Oficina at Anaco, Miocene-Pliocene Quiriquire) on various uplifts of the Maracaibo and Eastern Venezuela basins (Miller et al, 1958; Renz et al, 1958). Tar seals are important in these occurrences, and much of the gas apparently has leaked out.

A special wedge-edge category includes nonmarine sandstones intertongued with lacustrine beds such as the organic-rich oil shales. These nonmarine units internally have their own reservoirs, sources, and the equivalents of all wedge parts; they require a more detailed facies analysis than that presented here. Lacustrine examples are the Jurassic Precipice of the Surat basin, Australia (Power and Devine, 1970); Cretaceous Chubutiano, San Jorge basin, Argentina (Criado Roque et al, 1959); Eocene Green River, Uinta basin, Utah (Osmond, 1965); Eocene to Miocene Kenai, Cook Inlet, Alaska (Calderwood and Fackler, 1972); and Miocene Tipam, Assam basin, India (Bhandari et al, 1973).

Carbonate wedge-edge examples are dominated by the Permian Chase of the Amarillo uplift and by the Permian Leonard-Guadalupe of the Central Basin platform. Tectonism provides a direct control of sedimentation (Ball, 1972). Most of the reservoirs have anhydrite caps.

Subunconformity Plays

Uplifts are particularly favorable for subunconformity plays, although many regional truncations are richly productive. Only half of all productive subunconformity reservoirs studied are capped by thick shale. The other half are sealed by tight interbeds within sandstone-shale, grain carbonate, or red-bed facies of the overlying wedge base or edge. About 40% of subunconformity fields are in truncated wedge edges. The source of oil and gas for some of these fields, as well as for the few in basement rocks, must lie above the unconformity. For other fields, however, including many from other wedge positions, the most likely organic-rich source lies below the unconformity. Special techniques for exploring subunconformity plays are mapping paleogeology, paleotopography, paleostructure, and the extents of both top and bottom seals for the reservoirs (Chenoweth, 1972; Rittenhouse, 1972).

Subunconformity sandstone reservoirs of truncated wedge bases, middles, and tops are the ones most commonly capped by thick shale. Most wedge-edge and basement reservoirs are unconformably overlain by sandstone-shale facies. Examples of subunconformity sandstone plays are the wedge-base Cretaceous Woodbine of the East Texas field (Minor and Hanna, 1941); wedge-middle Eocene, Lake Maracaibo (Miller et al, 1958); wedge-top Mississippian Chester of Illinois (Bond et al, 1971); and the wedge-edge Cambrian-Ordovician Amal, Libya (Roberts, 1970). Probably because of leakage, gas is not common in subunconformity sandstone reservoirs.

Subunconformity carbonate reservoirs are mainly limestones in truncated wedge bases, middles, and tops, and dolomites in truncated wedge edges. About 30% of the carbonate reservoirs are capped by other carbonate rocks. Examples of subunconformity carbonate plays are the wedge-base Triassic Baldonnel, British Columbia (Barss et al, 1964); wedge-middle Mississippian Meramec-Osage, Kansas (Euwer, 1968); wedge-top Mississippian Rundle, Alberta (Fig. 6); and wedge-edge Ordovician Ellenburger, Central Basin platform, New Mexico and west Texas (Hill, 1971).

ASSESSMENT BY GEOLOGIC ANALOGY

Assessment by analogy assumes that the oil and gas contents of untested areas will be similar to those of tested productive areas if the geologic settings are similar. The three essential steps are (1) to make comparisons based on geologic similarities, (2) to scale these comparisons to compensate for differences in thickness or area, and (3) to evaluate the risk that other geologic differences may make the untested area nonproductive. Unfortunately, the differences commonly are more significant than the perceived similarities, so the risking step is vital.

Making Comparisons

There are many ways to make geologic comparisons; the important thing is to ensure that comparisons of similarities are based on many key controls of hydrocarbon source, reservoir, and trap. Comparison based on wedge position is a good place to start, because it considers the basically different source-reservoir-seal relations. The comparing process should not stop there, however, for a lot of geologic variation can occur within any single play type.

Assessors can use columnar sections for initial comparisons of new and old areas. Within wedge parts, the vertical succession of lithologies, the reservoir objectives, and the geologic ages should be as nearly alike as possible. It is ideal if columnar well sections are available from many producing areas. In assessing a new wedge-base reef play, for example, such sections could provide examples of the most similar wedge-base reefs already productive. Comparisons can be further strengthened by reference to complete cross sections and by inclusion of other factors such as structural-trap types or measured source-rock characteristics. Vertical profiles can also be used for more detailed facies analysis (Visher, 1965).

Scaling

After a good look-alike is selected, the assessor should compensate for differences in the sizes of the compared areas. Various scaling factors have been used (White and Gehman, 1979) to normalize differences in facies thickness or areal extent. If thicknesses are about the same, areal yields in barrels per square mile or kilometer taken from the producing area can be applied to the area of the new play. If thicknesses are different, it may be better to use a volumetric yield such as barrels per cubic mile of total play sediment. If this volume is overweighted with an excess of either coarse or fine facies, however, the results may be very misleading (Hedberg, 1975). This problem can be partly solved by using the volume of only that facies which is in critically short supply (White et al, 1975)—that is, using only the sand volume where shale is overabundant, or only the shale volume where sand is oversupplied.

Where data are available, assessors can scale by using the potential pay thicknesses and productive areas of prospects, multiplied by look-alike yields in barrels per acre foot. Another scaling approach is to use numbers and sizes of fields. Still another is the extrapolation of discovery rates, such as barrels per drilled foot; in this method the untested area generally is a direct extension of the tested one, and large amounts of historical data are required.

Risking

The third assessment step should allow for any geologic risk that a new play will be dry. The comparing and scaling steps tacitly assume that the play will be productive. The risking step recognizes that this assumption may be wrong.

Risking is essential because it is generally impossible to make the perfect comparison in which all the key controls of oil and gas occurrence are identical. These controls are many, and each is required for production. The list of essentials includes adequate trap closure, seal, and timing with respect to migration; sufficient reservoir thickness and porosity; suitable organic source-rock quality, maturation, and migration plumbing; preservation of hydrocarbons from flushing, cooking, or biodegradation; and permeability, and fluid viscosity adequate for recovery. If any factor is missing or inadequate, any chance of the new play being productive is wiped out. Discounts for geologic risk can be applied to scaled assessments as described by White and Gehman (1979).

Different wedge positions have different overall risks. Figure 11 summarizes the chances of major-field production for a worldwide sample of 1,150 plays in 80 basins including the Middle East. Each basin contains at least one major field in one productive play, although most plays are nonproductive of major fields. Of all wedge-base sandstone plays studied, for example, 60% are productive, and the other 40% represent the risk that wedge-base sandstones will not produce. Including completely nonproductive basins in the sample would decrease the heights of the bars on Figure 11 but presumably would not change the relative rankings.

The most productive (least risky) sandstone plays are those of the wedge base and middle. These reservoirs have the best positions relative to sources and seals. Sandstones of the wedge top and edge are much more risky; out of every 100 plays, only 15 are productive, probably because of poor sources and seals. Subunconformity plays, which include truncated representatives of all other play types, come close to striking the average productive chance of all the others.

With the exception of the wedge top, carbonate plays generally are riskier than their corresponding sandstone plays. Particularly in the wedge

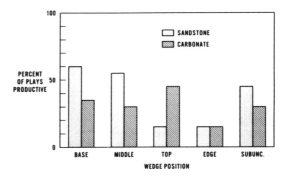

FIG. 11—Relative play chances for 1,150 studied plays in 80 productive basins. Productive play by definition contains at least one major field.

base and middle, more poorly developed porosity and permeability in the carbonate rocks seem to be the main reasons. Many carbonate wedge tops, however, have excellent leached porosity and very effective anhydrite seals. Like the sandstones, the carbonate wedge edge is very risky, and the subunconformity situation is about the average of the others.

Of all studied plays, 35% are productive. Taken separately, 38% of sandstone and 30% of carbonate plays are productive.

These world averages provide general guides to risking levels. Productive chances actually assigned to any specific new play, however, should depend as much as possible on its own geologic merits. One of the chief challenges in petroleum geology is finding exceptions to the rules.

In summary, the reliability of assessments of future oil and gas potentials depends on three things: (1) good comparisons that incorporate a lot of geology; (2) scaling that makes the comparisons more accurate; and (3) risk evaluation that faces reality without sacrificing the optimism essential to exploration.

APPENDIX I. LIST OF STUDIED BASINS

Cross sections, generally oriented perpendicular to depositional strike, were drawn from company and published data at standardized scales, chiefly using well control, through the main producing areas of the following basins.

AFRICA
Algeria: Touggourt, Ghadames, Ilizzi
Angola: Cuanza
Gabon: Gabon Coastal
Libya: Sirte
Niger: Niger Delta
Egypt: Suez

ASIA-AUSTRALIA
Australia: Amadeus, Gippsland, Surat
Burma: Burma
East Pakistan: Bengal
India: Assam, Cambay
Indonesia: Barito, North Java, Central Sumatra, North Sumatra, South Sumatra
Malaysia: Brunei-Sabah
Middle East: Persian Gulf
Pakistan: Indus

EUROPE
Austria: Vienna
France: Aquitaine
Italy: Bradanic, Caltanissetta, Po
Netherlands, etc: North European

NORTH AMERICA
Canada: Alberta, Mackenzie, Williston
Mexico: Mucuspana, Saline, Tampico
United States (western): Big Horn, Cook Inlet, Cuya-
ma, Denver, Green River, Los Angeles, North Slope, Paradox, Piceance, Powder River, Sacramento, Salinas, San Joaquin, San Juan, Santa Maria, Uinta, Ventura, Wind River
United States (central and eastern): Anadarko, Appalachian, Ardmore, Arkoma, Delaware, Fort Worth, Gulf Coast, Illinois, Michigan, Midland, Palo Duro

SOUTH AMERICA
Argentina: Cuyo, Magallanes, Neuquén, San Jorge
Bolivia: Tarija
Brazil: Recôncavo, Sergipe-Alagoas
Colombia: Lower Magdalena, Middle Magdalena
Ecuador: Southwest Ecuadorian
Peru: Northwest Peruvian, Oriente
Venezuela: Barinas, Eastern Venezuela (Guárico, Maturín, and Trinidad subbasins), Falcón, Maracaibo

REFERENCES CITED

Adams, J. E., 1962, Foreland Pennsylvanian rocks of Texas and eastern New Mexico, in C. C. Branson, ed., Pennsylvanian System in the United States: AAPG, p. 372-384.

———— et al, 1951, Starved Pennsylvanian Midland basin: AAPG Bull., v. 35, p. 2600-2607.

Ball, M. M., 1972, Exploration methods for stratigraphic traps in carbonate rocks, in Stratigraphic oil and gas fields: AAPG Mem. 16, p. 64-81.

Barss, D. L., E. W. Best, and N. Meyers, 1964, Triassic, in R. G. McCrossan and R. P. Glaister, eds., Geological history of western Canada: Alberta Soc. Petroleum Geologists, p. 113-136.

Bebout, D. G., and C. Pendexter, 1975, Secondary carbonate porosity as related to early Tertiary depositional facies, Zelten field, Libya: AAPG Bull., v. 59, p. 665-693.

Belyea, H. R., 1964, Upper Devonian, in R. G. McCrossan and R. P. Glaister, eds., Geological history of western Canada: Alberta Soc. Petroleum Geologists, p. 66-88.

Bhandari, L. L., R. C. Fuloria, and V. V. Sastri, 1973, Stratigraphy of Assam Valley, India: AAPG Bull., v. 57, p. 642-654.

Bitterli, P., 1958, Herrera subsurface structure of Penal field, Trinidad, B.W.I.: AAPG Bull., v. 42, p. 145-158.

Bond, D. C., et al, 1971, Possible future petroleum potential of region 9—Illinois basin, Cincinnati arch, and northern Mississippi embayment, in Future petroleum provinces of the United States: AAPG Mem. 15, p. 1165-1218.

Brown, C. F., 1973, A history of the development of the Pictured Cliffs Sandstone in the San Juan basin of northwestern New Mexico, in J. E. Fassett, ed., Cretaceous and Tertiary rocks of the southern Colorado Plateau: Four Corners Geological Soc. Mem., p. 178-184.

Busch, D. A., 1959, Prospecting for stratigraphic traps: AAPG Bull., v. 43, p. 2829-2843.

Calderwood, K. W., and W. C. Fackler, 1972, Proposed stratigraphic nomenclature for Kenai Group, Cook Inlet basin, Alaska: AAPG Bull., v. 56, p. 739-754.

California Division of Oil and Gas, 1973, Generalized cross section, southern San Joaquin Valley, in California oil and gas fields, v. 1.

Chang, K. H., 1975, Unconformity-bounded stratigraphic units: Geol. Soc. America Bull., v. 86, p. 1544-1552.

Chenoweth, P. A., 1972, Unconformity traps, in Stratigraphic oil and gas fields: AAPG Mem. 16, p. 42-46.

Craig, L. C., 1972, Mississippian System, in Geologic atlas of the Rocky Mountain region: Rocky Mtn. Assoc. Geologists, p. 100-110.

Criado Roque, P., et al, 1959, The sedimentary basins of Argentina: 5th World Petroleum Cong., New York, Proc., sec. 1, p. 883-899.

Curtis, D. M., 1979, Source of oils in Gulf Coast Tertiary—why look a gift horse in the mouth? (abs.): AAPG Bull., v. 63, p. 437-438.

Demaison, G. J., 1977, Tar sands and supergiant oil fields: AAPG Bull., v. 61, p. 1950-1961.

Deroo, G., et al, 1974, Geochemistry of the heavy oils of Alberta, in Oil sands—fuel of the future: Canadian Soc. Petroleum Geologists Mem. 3, p. 148-167.

Energy Mines and Resources Canada, 1977, Oil sands and heavy oils—the prospects: Rept. EP77-2, 36 p.

Euwer, R. M., 1968, Glick field, Kiowa and Comanche Counties, Kansas, in Natural gases of North America: AAPG Mem. 9, p. 1576-1587.

Grender, G. C., L. A. Rapoport, and R. G. Segers, 1974, Experiment in quantitative geologic modeling: AAPG Bull., v. 58, p. 488-498.

Hajash, G. M., 1967, The Abu Sheikhdom—the onshore oilfields history of exploration and development: 7th World Petroleum Cong., Mexico City, Proc., v. 2, p. 129-139.

Hedberg, H. D., 1975, The volume-of-sediment fallacy in estimating petroleum resources, in Methods of estimating the volume of undiscovered oil and gas resources: AAPG Studies in Geology, no. 1, p. 161.

———— L. C. Sass, and H. J. Funkhouser, 1947, Oil fields of Greater Oficina area, central Anzoategui, Venezuela: AAPG Bull., v. 31, p. 2089-2169.

Hemphill, C. R., R. I. Smith, and F. Szabo, 1970, Geology of Beaverhill Lake reefs, Swan Hills area, Alberta, in Geology of giant petroleum fields: AAPG Mem. 14, p. 50-90.

Herman, G., and C. A. Barkell, 1957, Pennsylvanian stratigraphy and productive zones, Paradox salt basin: AAPG Bull., v. 41, p. 861-881.

Hill, C. S., 1971, Future petroleum resources in pre-Pennsylvanian rocks of north, central, and west Texas and eastern New Mexico, in Future petroleum provinces of the United States: AAPG Mem. 15, p. 738-751.

Hills, J. M., 1972, Late Paleozoic sedimentation in west Texas Permian basin: AAPG Bull., v. 56, p. 2303-2322.

Hull, C. E., and H. R. Warman, 1970, Asmari oil fields of Iran, in Geology of giant petroleum fields: AAPG Mem. 14, p. 428-437.

Jardine, D., 1974, Cretaceous oil sands of western Canada, in Oil sands—fuel of the future: Canadian Soc. Petroleum Geologists Mem. 3, p. 50-67.

Koesoemadinata, R. P., 1969, Outline of geologic occurrence of oil in Tertiary basins of west Indonesia: AAPG Bull., v. 53, p. 2368-2376.

Levorsen, A. I., 1934, Relation of oil and gas pools to unconformities in the Midcontinent region, in W. E. Wrather and F. H. Lahee, eds., Problems in petroleum geology (Sidney Powers memorial volume): AAPG, p. 761-784.

Lockett, R., 1968, Production of gas in northern Cincinnati arch province, in Natural gases of North America: AAPG Mem. 9, p. 1716-1745.

Lowman, S. W., 1949, Discussion, in Sedimentary facies in geologic history: Geol. Soc. America Mem. 39, p. 125-130.

MacKenzie, D. B., 1972, Primary stratigraphic traps in sandstones, in Stratigraphic oil and gas fields: AAPG Mem. 16, p. 47-63.

Mayuga, M. N., 1970, Geology and development of California's giant—Wilmington oil field, in Geology of giant petroleum fields: AAPG Mem. 14, p. 158-184.

Michelson, J. E., 1976, Miocene deltaic oil habitat, Trinidad: AAPG Bull., v. 60, p. 1502-1519.

Middleton, G. V., 1973, Johannes Walther's law of the correlation of facies: Geol. Soc. America Bull., v. 84, p. 979-988.

Miller, J. B., et al, 1958, Habitat of oil in the Maracaibo basin, Venezuela, in L. G. Weeks, ed., Habitat of oil: AAPG, p. 601-640.

Minor, H. E., and M. A. Hanna, 1941, East Texas field, Rusk, Cherokee, Smith, Gregg, and Upshur Counties, Texas, in A. I. Levorsen, ed., Stratigraphic type oil fields: AAPG, p. 600-640.

Mitchum, R. M., Jr., P. R. Vail, and S. Thompson, III, 1977, The depositional sequence as a basic unit for stratigraphic analysis, in Seismic stratigraphy—applications to hydrocarbon exploration: AAPG Mem. 26, p. 53-62.

Newby, J. B., et al, 1929, Bradford oil field, McKean County, Pa., and Cattaraugus County, N.Y., in Structure of typical American oil fields: AAPG, v. 2, p. 407-442.

Olsson, A. A., 1956, Colombia, in Handbook of South American geology: Geol. Soc. America Mem. 65, p. 293-326.

Osmond, J. C., 1965, Geologic history of site of Uinta basin, Utah: AAPG Bull., v. 49, p. 1957-1973.

Ottmann, R. D., P. L. Keyes, and M. A. Ziegler, 1976, Jay field, Florida—a Jurassic stratigraphic trap, in North American oil and gas fields: AAPG Mem. 24, p. 276-286.

Owen, R. M. S., and S. N. Nasr, 1958, Stratigraphy of the Kuwait-Basra area, in L. G. Weeks, ed., Habitat of oil: AAPG, p. 1252-1278.

Pippin, L., 1970, Panhandle-Hugoton field, Texas-Oklahoma-Kansas, in Geology of giant petroleum fields: AAPG Mem. 14, p. 204-222.

Power, P. E., and S. B. Devine, 1970, Surat basin, Australia—subsurface stratigraphy, history, and petroleum: AAPG Bull., v. 54, p. 2410-2437.

Rainwater, E. H., 1971, Possible future petroleum potential of Lower Cretaceous, western Gulf basin, in Future petroleum provinces of the United States: AAPG Mem. 15, p. 901-926.

Renz, H. H., et al, 1958, The Eastern Venezuelan basin, in L. G. Weeks, ed., Habitat of oil: AAPG, p. 551-600.

Rich, J. L., 1951, Three critical environments of deposition, and criteria for recognition of rocks deposited in each of them: Geol. Soc. America Bull., v. 62, p. 1-20.

Rittenhouse, G., 1972, Stratigraphic-trap classification, *in* Stratigraphic oil and gas fields: AAPG Mem. 16, p. 14-28.

Roberts, J. M., 1970, Amal field, Libya, *in* Geology of giant petroleum fields: AAPG Mem. 14, p. 428-437.

Sanford, R. M., 1970, Sarir oil field, Libya—desert surprise, *in* Geology of giant petroleum fields: AAPG Mem. 14, p. 449-476.

Shoemaker, E. M., J. E. Case, and D. P. Elston, 1958, Salt anticlines of the Paradox basin, *in* A. F. Sanborn, ed., Guidebook to the geology of the Paradox basin: Intermtn. Assoc. Petroleum Geologists, p. 39-59.

Sloss, L. L., 1963, Sequences in the cratonic interior of North America: Geol. Soc. America Bull., v. 74, p. 93-114.

Soeparjadi, R. A., et al, 1975, Exploration play concepts in Indonesia: 9th World Petroleum Cong., Tokyo, Proc., v. 3, p. 51-64.

Steineke, M., R. A. Bramkamp, and N. J. Sander, 1958, Stratigraphic relations of Arabian Jurassic oil, *in* L. G. Weeks, ed., Habitat of oil: AAPG, p. 1294-1329.

Terry, C. E., and J. J. Williams, 1969, The Idris "A" bioherm and oilfield, Sirte basin, Libya, *in* P. Hepple, ed., The exploration for petroleum in Europe and North Africa: London, Inst. Petroleum, p. 31-48.

Van Siclen, D. C., 1958, Depositional topography—ex-

amples and theory: AAPG Bull., v. 42, p. 1897-1913.

Vest, E. L., 1970, Oil fields of Pennsylvanian-Permian Horseshoe Atoll, west Texas, *in* Geology of giant petroleum fields: AAPG Mem. 14, p. 185-203.

Visher, G. S., 1965, Use of vertical profile in environmental reconstruction: AAPG Bull., v. 49, p. 41-61.

Walther, J., 1894, Einleitung in die Geologie als historische Wissenschaft: Jena, Verlag von Gustav Fischer, 1055 p.

Weimer, R. J., 1960, Upper Cretaceous stratigraphy of the Rocky Mountain area: AAPG Bull., v. 44, p. 1-20.

Wengerd, S. A., 1958, Pennsylvanian stratigraphy, southwest shelf, Paradox basin, *in* A. F. Sanborn, ed., Guidebook to the geology of the Paradox basin: Intermtn. Assoc. Petroleum Geologists, p. 109-133.

——— 1962, Pennsylvanian sedimentation in Paradox basin, Four Corners region, *in* Pennsylvanian System in the United States: AAPG, p. 264-330.

White, D. A., and H. M. Gehman, 1979, Methods of estimating oil and gas resources: AAPG Bull., v. 63, p. 2183-2192.

——— et al, 1975, Assessing regional oil and gas potential, *in* Methods of estimating the volume of undiscovered oil and gas resources: AAPG Studies in Geology, no. 1, p. 143-159.

Woodford, A. O., 1973, Johannes Walther's law of the correlation of facies—discussion: Geol. Soc. America Bull., v. 84, p. 3737-3740.

The American Association of Petroleum Geologists Bulletin
V. 62, No. 2 (February 1978), P. 201-222, 20 Figs.

Geohistory Analysis—Application of Micropaleontology in Exploration Geology[1]

J. E. VAN HINTE[2]

Abstract "Geohistory analysis" is the use of quantitative stratigraphic techniques to unravel and portray geologic history. Quantification of routine stratigraphic well information is possible as a result of recent advances in microbiostratigraphy that allow exploration paleontologists to determine geologic ages in terms of millions of years, and to express depositional environments in terms of water depth. Geohistory diagrams and numerical stratigraphic techniques, such as the calculation of rates of sediment accumulation and rates of subsidence or uplift (with and without corrections for consolidation), are useful in widely different aspects of petroleum exploration. Knowledge of rates and timing of vertical movements is of local importance in distinguishing between different kinds of movements; these data should constitute a parameter in any structural classification. In a more general sense, such knowledge is a key to understanding basin evolution and plate tectonics.

INTRODUCTION

Notably increased capability to zone Mesozoic and Cenozoic sedimentary sections and to estimate the water depth at which beds were deposited has been developed recently for a multiplicity of microfossil groups. The use of numbers for paleontologic conclusions (e.g., age in years and water depth in meters) allows for a new, numerical stratigraphy.

"Standard" biostratigraphy zonations now are tied with greater precision to the relative geologic time scale and to a linear numerical time scale. Consequently, paleontologists can assign geologic ages accurately to most fossiliferous sections and can express these ages in megayears (1 megayear = 1 Ma = 1 m.y.). Seismic stratigraphers can do likewise, provided that their framework has been calibrated by paleontologic data. Placing the geologic record of an area in a linear chronostratigraphic setting brings the sequence of events into a perspective that facilitates the interpretation of geologic processes.

Recent summaries of time scales applied in micropaleontologic dating are given by Berggren (1972) for the Cenozoic and by van Hinte (1976a, b) for the Jurassic and Cretaceous. The typical micropaleontologic biochronologic unit comprises about 1 Ma for the Cenozoic and about 2 Ma for the Mesozoic. This resolution is adequate for most practical purposes, but, when needed, combined zonations and additional biohorizons often can increase resolution. Within the next decade, the biochronology for several groups of age-significant microfossils is expected to be better known and to be interrelated more precisely, so the scales might approach a resolution of a few hundred-thousand years.

Paleo-water-depth determination is a highly complex art. The quantity and quality of knowledge on benthic microfossils from a greater variety of environments are increasing steadily; more importantly, we now can obtain additional ecologic insight from diversity trends, distribution patterns, etc. of benthic microfossils. Recent oceanographic research has provided more reliable data on the depth distribution of plankton in the water column and has uncovered useful details on depth-related phenomena such as calcium carbonate dissolution. Further, when dealing with frontier material, paleontologists have to attempt to incorporate available knowledge on paleogeog-

[1]Manuscript completed, 1975; received, June 4, 1976; revised and accepted, May 18, 1977.

[2]Esso Production Research—European, 213, Cours Victor-Hugo, 33321 Bègles, France.

I gratefully remember the stimulating work with W. A. Berggren, A. S. Laughton, K. Perch-Nielsen, and A. Ruffman on board the *Glomar Challenger* during Leg 12 of the Deep Sea Drilling Project which led to my first geohistory diagrams and this paper.

With one exception, the examples used in this paper are from recently drilled exploration wells for which no precise location can be given; they are discussed to demonstrate the approach rather than the local geology. Paleontologic work was by J. L. Lamb at Exxon Production Research Co. and by D. O. LeRoy, S. Grossman, R. L. Mullins, and others at Exxon Company U.S.A. I thank these colleagues for making this information available and their companies for permitting its use in schematic form.

D. H. Horowitz (Exxon Production Research Co.) kindly introduced me to his computer program for sediment thickness restoration, which is designed on the principles discussed in Horowitz (1976).

I thank the management of Exxon Production Research Co. for suggesting that I write this paper, and for allowing me time to complete it and to select examples freely from my company reports.

Article Identification Number
0149-1423/78/B002-0003$3.00/0

raphy and paleooceanography of the area. Although a certain amount of circular reasoning or even guessing is hard to avoid, a reasonable estimate of paleo-water depth usually can be made. Fortunately, in areas of intensive exploration where the explorationist needs more details, consideration of local geologic factors and paleontologic experience can make for sharper estimates.

Numerical geologic time, coupled with direct geologic measurements (e.g., interval thickness) and interpreted data (e.g., paleo-water depth), permits the expression of significant stratigraphic conclusions in quantitative terms. For example, we can now determine rates of fill, subsidence, or uplift, and can estimate the time span of unconformities, the amount of removed section, or the rate of fluid expulsion. Also, well data can be presented graphically to portray the timing and magnitude of geologic events and processes, and to permit (by extrapolation) estimates on unknown parts of sections.

Geohistory analysis, that is, the use of numerical stratigraphic techniques to unravel and portray geologic history, has important practical po-tential in the interpretation of geologic events and in the determination of geologic controls of oil and gas that are rate or age dependent. The use of numbers rather than a multitude of names affords more disciplined geologic analysis and reasoning, facilitates computer applications, and, perhaps foremost, increases communication between members of an exploration team with different backgrounds such as micropaleontologists and engineers. This paper demonstrates the construction of a geohistory diagram and gives some schematized exploration examples. Concepts and applications of numerical stratigraphy are discussed in the Appendix.

GEOHISTORY DIAGRAM

In his book "Géologie du Bassin de Paris," Lemoine (1911, Figs. 136, 137) summarized the geologic history of the Paris basin in a "schéma des mouvements d'ascension et de descente de la surface" (diagram of rising and descending movements of the surface) combined with a "schéma de la descente générale du fond" (diagram of the general descent of the basement; Fig. 1). Burollet

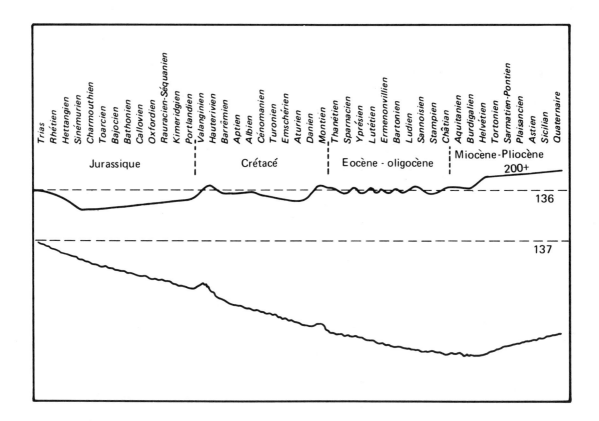

FIG. 1—Qualitative geohistory diagram for Paris basin as constructed by Lemoine (1911, Figs. 136, 137). Numbers 136 and 137 are added to Lemoine's figure, indicating two baselines he used: 136 shows water depth or elevation; 137 shows subsidence and uplift.

366

(1956, Figs. 85-87) applied Lemoine's approach to summarize stratigraphic conclusions for different sections in central Tunisia, entitling his diagrams "variations de la profondeur des mers en Tunisie centrale en relation avec les mouvements de tréfonds" (variations of the depth of the seas in central Tunisia in relation with the movements of the basement; Fig. 2). A similar diagram was published by Fülöp (1971).

Diagrams such as these are an excellent method of portraying stratigraphic data and conclusions and provide a direct means of reading the timing and magnitude of geologic events. Quantification of paleontologic conclusions on age and environment now permits considerable refinement and the routine use of linear scales in such portrayals. Geohistory diagrams thus were developed and have been used in our work since 1970.

Construction

The mathematical derivation of the sediment accumulation and subsidence rates is given in the Appendix. The data shown in Figures 10 and 19 can be used to construct a time/depth plot with a linear time scale as the horizontal axis and a depth scale as the vertical (Fig. 3). First, the paleobathymetry at the time of the datum levels is determined from the paleodepth curve (solid triangles, Fig. 19); these water depths then are plotted against the linear scales (solid triangles, Fig. 3). Next, the ages at total depth (TD) and at the unconformity are determined by extrapolation of sedimentary-accumulation rates, and their paleobathymetry is plotted (open triangles). Finally, the cumulative thickness for each level is plotted on the graph as solid and open circles. The circles connected by broken lines represent present thicknesses, and the solid lines are restored thicknesses. The slope of the lines shows the uncorrected subsidence rate (uRs) and the true subsidence rate (Rs), respectively. The Rs line is the subsidence path for level TD, but it does not represent tectonic subsidence (R_T), unless TD is at the top of basement or the sediment below TD was completely compacted at TD time. The "true" subsidence history of any chosen level or unit of a well section can be drawn in the diagram, as illustrated by the shaded path of unit E. The slope of the oldest segment of the Rs line = Rs_o for unit A, and the slope of the last part of the uRs line = Rs_o for unit H.

Philipp (1961) and later German petroleum geologists, following a personal suggestion by A. Roll, constructed burial history curves (*Zeit-teufenkurve, Absenkingskurve*) in connection with studies on maturation, migration, and porosity prediction. They commonly used a linear time scale, and these plots were very similar to geohistory diagrams except that paleo-water depth was not taken into account and interest was focused on geochemical and physical rather than geologic processes.

Applications

The inclusion of a geohistory diagram in a paleontlogic well report has the following obvious advantages.

1. The construction itself forces the biostratigrapher to be aware of the geologic implications of his conclusions. He sees, for example, where more data are needed, what the consequences are of reporting an unconformity, and which of his interpretations do, and which do not, make geologic sense. For example, when a displaced shallow-water fauna is not recognized as such, it will plot too high on the diagram, and consequently the subsidence curve will show an uplift "kick."

2. Comparisons between wells and areas are possible at a glance.

3. The diagram forms a convenient linear time-depth frame for the geologist to plot other parameters, such as heat-flow data, porosity changes, and mineralizations.

4. Generalized diagrams are most informative models of basin evolution, tectonic style, continental-margin history, etc.

5. Extrapolation of the curves fills gaps in cases of fragmentary information.

Most applications listed in the appendix under sediment accumulation and subsidence rates can be portrayed in geohistory diagrams and the same conclusions can be achieved by simple graphic extrapolation. Thus, anomalies in the tectonic subsidence rate (R_T; see Appendix) because of eustatic sea-level changes, dating errors, or displaced faunas, appear as conspicuous deflections in the subsidence curve.

A computer program can be written to machine generate all rate parameters and to plot a geohistory diagram on a routine basis. Needed input data are numerical geologic age, present drill depth and interpreted water depth of data points ("tops"), shale percentage of units, and pressure gradient.

EXAMPLES OF GEOHISTORY ANALYSIS

Interpretation of Presumed Unconformity

Paleontologic analysis of two Gulf Coast wells shows a marked break in paleobathymetry (Fig. 4), which raises the question of whether this represents an unconformity and, if so, how much section is missing. No markers were found at the

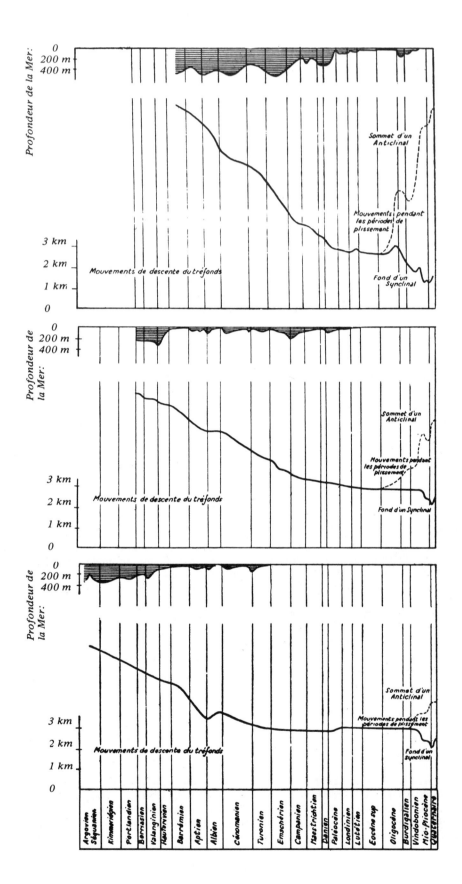

FIG. 2.—Qualitative geohistory diagrams for central Tunisia (from Burollet, 1956, Figs. 85-87). *Profondeur de la Mer* is depth of the sea; *mouvements de descente du tréfonds* is descending movements of basement; *sommet d'un anticlinal* is top of an anticline; *mouvements pendant les périodes de plissement* is movement during periods of folding.

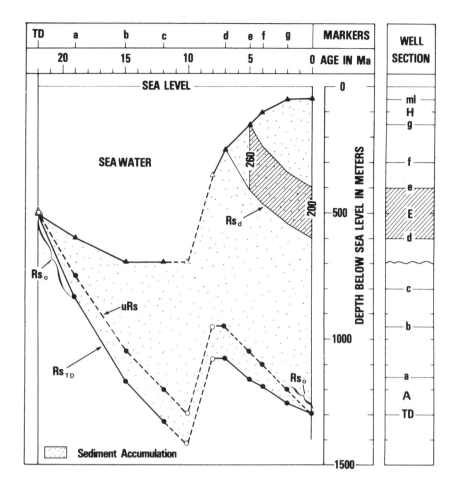

FIG. 3—Geohistory diagram for hypothetical well of Figures 10 and 19 (Appendix). Upper curve shows water-depth history for location; lower curves show subsidence and uplift at that location. Shaded path shows subsidence and reduction in thickness of unit E. For further explanation, see text.

presumed unconformity (Fig. 4a). Extrapolation of the uncorrected rate of sediment accumulation (uR) above the break (calculated from 1.7 and 2.0 Ma) and below the break (calculated from 3.1 and 2.8 Ma) suggests 2.72 Ma and 2.75 Ma as the ages at the break. The difference of 30,000 years is negligible, and we conclude that the paleobathymetry break in well 1 represents a eustatic change in sea level of 700 ft (213 m). In well 2, however, the water-depth break is only 250 ft (76 m; Fig. 4b). The ages at the break are 2.2 and 2.8 Ma. The difference of 600,000 years is a considerable hiatus in this young section.

A schematic reconstruction for the time of sea-level drop is shown in Figure 4c. Sediments were being deposited in 1,000 ft (350 m) of water at well 1, so the 700-ft (200 m) drop in sea level did not necessarily disrupt sedimentation. At well 2,

however, lowering of sea level by only 350 ft (107 m) would have exposed the shelf and caused an unconformity. We may ask how much section was eroded at well 2. The sea-level drop occurred at −2.75 Ma or somewhat later if the calculation of 30,000 missing years for well 1 is correct, so at well 2 at least 50,000 pre-event years are not represented. With sediment accumulating (uR) below the break at a rate of 25 cm/1,000 years, substitution in equation 2 of the Appendix reveals that the equivalent of a present thickness of 12.5 m (about 40 ft) was deposited and subsequently eroded.

Depiction of Structural Style and Timing

The directness with which structural style and timing can be portrayed is illustrated in Figure 5. An Exxon Company U.S.A. paleontologic study

369

of three wells drilled offshore from Louisiana in an area underlain by mobile shale (Fig. 5a) provides age and environmental interpretations from which geohistory diagrams were constructed (Fig. 5b-d); the major sea-level drop at 2.8 Ma was taken into account.

It is readily apparent that sites 1 and 2 continuously subsided at a rapid rate during the past 4 Ma. Site 3, however, remained high while the surrounding area subsided; later this site even showed some uplift as a response to rapid loading. During the past 0.5 Ma, site 3 has subsided at a rate more than 1.5 times greater than the regional rate at the other two sites. The cause of this seemingly excessive rate may be interpreted as either fluid release from the shales or shale withdrawal. Regardless of which interpretation is accepted by the regional geologist, the paleontologist has provided him with a clear picture from which to start.

Smaller differences in the slope of subsidence curves (Fig. 5) are hard to evaluate visually; therefore, when different wells are compared, bar diagrams are useful (Fig. 6). Local sharp changes in rates shown by these graphs probably are due to movements and changes in density anomalies within the sediment column. For example, the initial high rates at site 1 clearly are due to growth-fault activity (Fig. 5a) and the high rate between −1 Ma and −2 Ma at site 2 could be caused by withdrawal and be coincident with the uplift of site 3.

Apart from changes in rates that occur at the different locations at different times, all three sites simultaneously show an acceleration in the average sediment accumulation and subsidence rates for the latest 0.5 Ma. This acceleration also was observed in other wells in the Gulf of Mexico and probably is a regional phenomenon related to massive deposition of glacial detritus from the continent. Although the rates are high at all three sites, they are low at sites 1 and 2 compared with site 3 (Fig. 6). One could imagine that (1) rapid regional loading and subsidence possibly caused development of growth faults that functioned as fluid escape routes, enabling the shale dome to

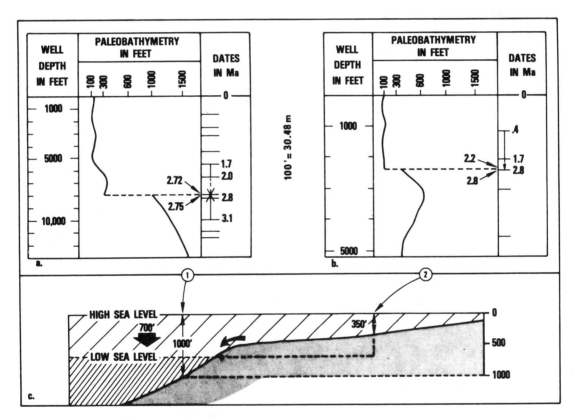

FIG. 4—Paleobathymetry curves and numerical age of markers for parts of two offshore Gulf wells **a**, well 1; **b**, well 2. Ages at break in water depth (2.72 and 2.75 Ma in well 1; 2.2 and 2.8 Ma in well 2) are derived from extrapolation of sediment accumulation rates. **c**, Schematic reconstruction of bathymetric profile at time of sea-level lowering to illustrate different effects of drop.

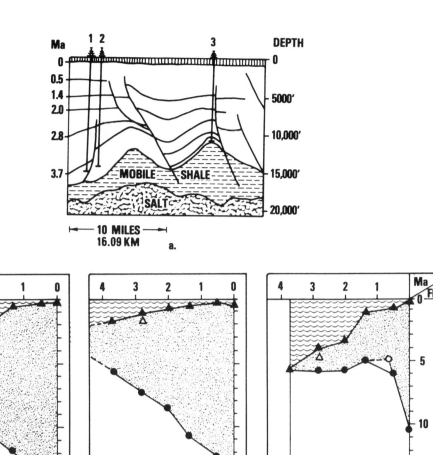

SEAWATER   SEDIMENT ACCUMULATION

FIG. 5—a, Cross section in Mississippi delta area of offshore Louisiana based on seismic data and three wells. Ma numbers are megayears before present (1,000 ft = 304.8 m).

b-d, Geohistory diagrams for wells show that dome is suspended in sediment column where its top remains at about same level while surrounding area subsides.

collapse, or (2) redistribution of load (prograding depocenter) initiated the growth of younger domes that are withdrawing shale from the older ones. In either case, accelerated regional subsidence under the gigantic sediment load seems especially favorable here for petroleum accumulation; it brought sediments into warmer levels and triggered the collapse of overpressure, thus accelerating hydrocarbon maturation and initiating large-scale fluid migration.

Some Implications of Correcting for Compaction

The first example dealt with calculations made on beds at comparable burial depths and, there-fore, uncorrected values of sediment accumulation rate (uR) could be used without significant errors being introduced. In the second example, the use of uR and uRs (uncorrected subsidence rate) served the illustrative purpose and allowed for general conclusions but would have been unacceptable for quantitative estimates and predictions. The following example illustrates some aspects of geohistory analysis of wells for which lithology and pressure data are available so that corrections for compaction can be made.

Geohistory diagrams of seven wells drilled in one area offshore Louisiana (Fig. 7) are based on marker points that are both fossil tops and elec-

371

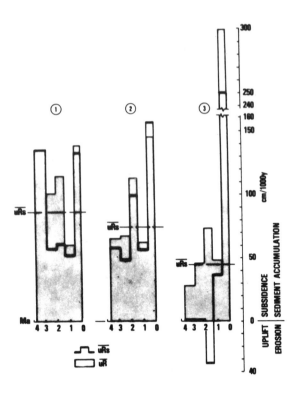

FIG. 6—Bar diagrams showing rates of vertical movement and sediment accumulation for wells 1 to 3 of Figure 5. Horizontal axis is linear time scale, and vertical axis shows rates of sediment accumulation and subsidence in cm/1,000 years. Width of each bar is determined by spacing of time markers. Mean rates of subsidence (uRs) for latest 4 Ma suggest that site 3 will continue to subside at higher than regional rate to catch up with "normal" area. That is, shales still are overpressured unless difference has been made up by sediment added from withdrawal toward site 3.

tric-log picks. The uRs line (uncorrected subsidence using present thickness) diverges from the Rs_{TD} line ("true" subsidence rate for TD using restored thickness) in the older part of the section and converges in the younger part; hence uRs and Rs_{TD}, respectively, in turn are the better estimates for the tectonic subsidence rate R_T. A still better estimate for the tectonic subsidence can be obtained in the following way. From regional geologic and geophysical observations, a thick sedimentary section is known to be present below TD. Under normal conditions, compaction of an interval now 10,000 ft (3,048 m) below TD would have been practically complete at TD time; thus subsidence of that interval after TD time is a measure of tectonic subsidence. To obtain the lower R_T curves (Fig. 7), we assumed a normally pressured 10,000-ft unit below TD with 80%

shale. When the tectonic subsidence curves were "smoothed," three anomalies became apparent in all wells (vertically shaded areas in Fig. 7). Differences in shape and size of these areas between the wells probably are due to inaccuracies in the paleontologic conclusions, but their presence is real and in our opinion reflects eustatic changes in sea level. The first change is in the order of 700 ft (213 m); later ones are considerably smaller. (The first change appears too large in Figure 7 because the "interpreted tectonic subsidence" line has been drawn on large increments, and because the −2.2 Ma control point is not well defined and possibly older than 2.2 Ma.) The resolution of our example does not permit determination of the beginning and end of low or high stands of sea level. Research focused toward that goal, however, may reveal such detail. Another general conclusion that could be drawn from this example is that the rate of net-sediment accumulation (averaged for the seven wells) during the past 3.7 Ma was 58 cm/1,000 years, whereas the tectonic subsidence rate for the same period was 51 cm/1,000 years.

Geohistory of Continental Margins

The schematic cross sections of Figure 8a and b show tectonostratigraphic models of continental margins: a passive (or block-faulted, or divergent, or Atlantic-type) margin in Figure 8a (Vogt, 1970; Bott, 1971), and an active (or compressional, or convergent, or Pacific) margin with a trench, an underthrust ridge, and a fore-arc basin in Figure 8b (Seely et al, 1974). The well symbols give the approximate drill locations of actual exploration wells. Schematic geohistory diagrams for these wells (Fig. 8c, d) reveal at a glance the marked differences in their geologic histories.

The passive margin subsided at a rate of about 20 cm/1,000 years during the early Early Cretaceous when shallow-water sedimentation more or less kept up with subsidence. During the late Early Cretaceous and the Late Cretaceous, subsidence slowed, but, nevertheless, the sea deepened because sedimentation slowed even more. A sudden (within the time equivalent of one planktonic foraminiferal biozone) rapid subsidence occurred at the end of the Cretaceous–beginning of the Tertiary, followed thereafter by only slight subsidence; the area simply filled with sediment. Different paleontologists may interpret the paleobathymetry somewhat differently, which may change slightly the rates and magnitudes of the movements; however, the general trend is apparent: early rapid subsidence that decreases, then accelerates briefly before stopping. A similar pattern has been observed in several wells drilled near the edge of the continent, whereas more

372

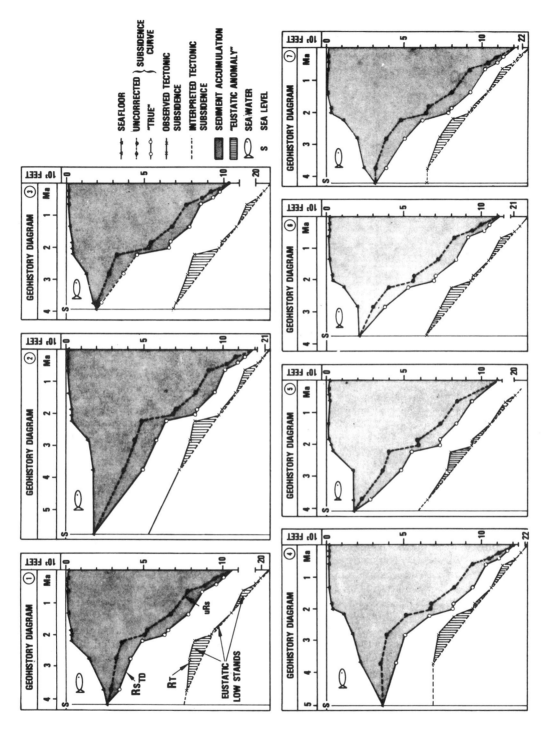

FIG. 7—Geohistory diagrams showing past 6 to 4 Ma of subsidence and shoaling at seven well sites offshore Louisiana. Lower part of vertical scales is interrupted. 10^3 ft = 304.8 m.

373

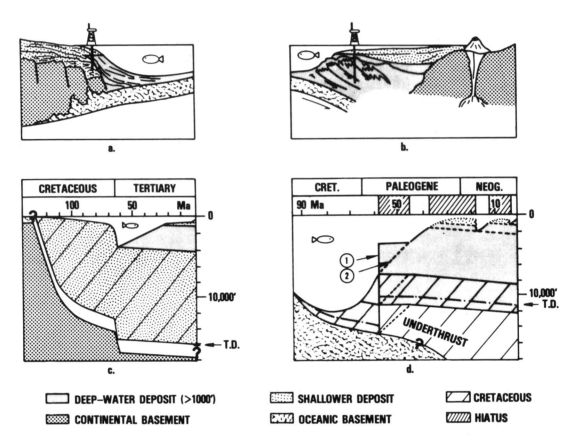

DEEP-WATER DEPOSIT (>1000')   SHALLOWER DEPOSIT   CRETACEOUS

CONTINENTAL BASEMENT   OCEANIC BASEMENT   HIATUS

FIG. 8—Tectonostratigraphic models: **a**, for passive continental margin; **b**, for active continental margin. Schematic geohistory diagrams for typical well sections: **c**, on passive continental margin (as well at **a** did not reach basement, vertical and horizontal scales are interrupted to show its supposed subsidence path); **d**, on active continental margin illustrates two of three alternative interpretations for uplift and shoaling path during first major hiatus: (1) "instantaneous" uplift; (2) gradual uplift. Stippled bathymetry lines at right suggest one possibility of what may have happened during times not represented because of later hiatuses.

landward wells show the same trend but without the sudden brief acceleration in subsidence. The subsidence curve of these wells closely resembles that of oceanic crust (Fig. 9) except that initial rates are higher (the more so with more sediment accumulating) and, if the sedimentation rate continues to be high, substantial subsidence ceases later than would be expected from the oceanic-subsidence curve. Confirming the suggestion by Vogt and Ostenso (1967), it seems that passive-margin fault blocks move downward only (not up) and that the rate of their subsidence depends merely on distance from the spreading center (cooling) and isostatic response to loading (Sleep, 1971; Keen and Keen, 1973). Where geohistory diagrams reveal diversions from this general pattern, the activity of other forces (e.g., wrenching) may be concluded, or, generally on a smaller scale, such diversions may be apparent and the result of eustatic sea-level changes.

The geohistory diagram is entirely different for a well drilled on an active margin (Fig. 8d). Here, deep-water sediments accumulated on a normally subsiding oceanic seafloor until about 58 m.y. ago when they were uplifted abruptly as a result of underthrusting that accompanied subduction (Seely et al, 1974). The rate of uplift cannot be interpreted from the well data because of a hiatus of −58 to −43 Ma. Three possibilities exist (Fig. 8d): (1) uplift occurred "instantaneously" 58 Ma ago and was followed by a period of nondeposition because of increased current velocities over the new seafloor high; (2) uplift was spasmodic or gradual during the entire period of the hiatus (with a minimum average rate of about 10 cm/1,000 years), and increased current velocities over the rising seafloor prevented sediment accumulation, or if sediments were deposited, they were eroded subsequently; or (3) uplift occurred rapidly but later than −58 Ma, and pre-event

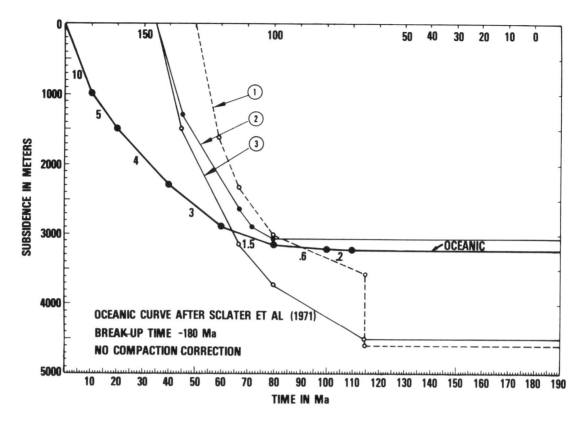

FIG. 9—Passive-margin descent compared with oceanic lithogenetic subsidence curve. Oceanic subsidence curve of Sclater et al (1971) has been drawn in increments of 10 and 20 Ma; average subsidence rate is given in cm/1,000 years for each increment. Curves 1 to 3 show subsidence as interpreted from wells at shelf edge of older Atlantic. Well 1 is same as that used in Figure 8c; curves 2 and 3 are preliminary conclusions on two West African offshore wells. Breakup time for this part of Atlantic, and consequently beginning of oceanic subsidence, is placed arbitrarily at −180 Ma.

sediments have been eroded. Whatever actually occurred, the paleontologic conclusions point out that uplift (underthrusting) apparently took place within a specific time span upon which the location had become part of the continent and thereafter subsided slowly under an increasing sediment load. The two younger hiatuses in this particular well (stippled parts in Fig. 8d) may be the result of erosion caused by eustatic sea-level drops or by changes in current regime. A second active-margin exploration well for which a paleontologic report was available showed the same basic pattern.

The two examples (Fig. 8) illustrate that geohistory diagrams are useful for depicting sedimentary history and structural timing. Because the diagrams characterize tectonic settings, they will be particularly useful in basins having more complex histories than the two straightforward margins discussed here. Recent developments in global tectonics have emphasized large-scale horizontal movements and related the geologic history of large areas to their plate positions. It now seems opportune to resolve the more subtle vertical movements that are of immediate importance to the explorationist. A classification of vertical movements according to their rates will have predictive value in undrilled areas of known tectonic setting; conversely, geohistory diagrams (see Fig. 8) can serve as models from which one can recognize the former plate positions of, for example, known Paleozoic sections.

REFERENCES CITED

Berggren, W. A., 1972, A Cenozoic time-scale—some implications for regional geology and paleogeography: Lethaia, v. 5, p. 195-215.

Bott, M. H. P., 1971, Evolution of young continental margins and formation of shelf basins: Tectonophysics, v. 11, p. 319-327.

Burollet, P. F., 1956, Contribution à l'étude stratigraphique de la Tunisie Centrale: Tunisia Serv. Mines, Ann. Mines et Géologie, no. 18, 345 p.

Fülöp, J., 1971, Les formations jurassique de la Hongrie, in Colloque du jurassique Méditerranéen: Hungary, Magy. Allami Földt. Intéz., Évk., v. 54, no. 2, p. 31-61.

Horowitz, D. H., 1976, Mathematical modeling of sediment accumulation in prograding deltaic systems: 9th Internat. Sedimentol. Cong. Proc., Pergamon Press (in press).

Keen, M. J., and C. E. Keen, 1973, Subsidence and fracturing on the continental margin of eastern Canada: Canada Geol. Survey Paper 71-23, p. 23-42.

Lemoine, P., 1911, Géologie du bassin de Paris: Paris, Librairie Sci. Hermann et Fils.

Philipp, W., 1961, Struktur- und Lagerstättengeschichte des Erdölfeldes Eldingen: Deutsch. Geol. Gesell. Zeitschr., v. 112, p. 414-482.

Roll, A., 1974, Langfristige Reduktion der Mächtigkeit von Sedimentgesteinen und ihre Auswirkung—eine Übersicht: Geol. Jahrb., series A, no. 14, 76 p.

Ruffman, A., and J. E. van Hinte, 1973, Orphan Knoll—a "chip" off the North American plate: Canada Geol. Survey Paper 71-23, p. 407-449.

Sclater, J. G., R. N. Anderson, and M. L. Bell, 1971, Elevation of ridges and evolution of the central eastern Pacific: Jour. Geophys. Research, v. 76, p. 7888-7915.

Seely, D. R., P. R. Vail, and G. G. Walton, 1974, Trench slope model, in C. A. Burk et al, eds., The geology of continental margins: New York, Springer-Verlag, p. 249-260.

Sleep, N. H., 1971, Thermal effects of the formation of Atlantic continental margins by continental break up: Royal Astron. Soc. Geophys. Jour., v. 24, p. 325-350.

Van Hinte, J. E., 1976a, A Jurassic time scale: AAPG Bull., v. 60, p. 489-497.

——— 1976b, A Cretaceous time scale: AAPG Bull., v. 60, p. 498-516.

Vogt, P. R., 1970, Magnetized basement outcrops on the south-east Greenland continental shelf: Nature, v. 226, p. 743-744.

——— and N. A. Ostenso, 1967, Steady state crustal spreading: Nature, v. 215, p. 810-817.

APPENDIX

SEDIMENT ACCUMULATION RATES

Determine uR, Rf, and "R"

Assignment of numerical ages to paleotops or other marker horizons enables the calculation of the average rate ("R") at which the sedimentary rock accumulated during the time interval bounded by the markers. If the time span (A) of the interval is expressed in megayears and the thickness (T) in meters, then

$$\text{"R"} = \frac{T}{10A} \text{ cm/1,000 years.} \qquad (1)$$

In the example of Figure 10, 100 m of sediment accumulated during the most recent 2 megayears (unit H, Fig. 11) at an average accumulation rate of $100/(10 \times 2)$ = 5 cm/1,000 years, or "R" = 5. In the same example, 200 m of sediment between markers a and b was deposited during 4 Ma (unit B), also at an average of 5 cm/1,000 years.

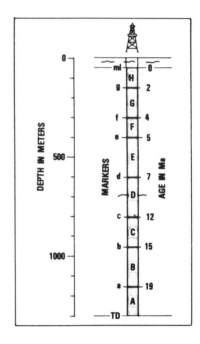

FIG. 10—Hypothetical offshore well section. Markers can be tops of index fossils, log picks, or other horizons with demonstrated value as time markers in area. Units between markers are indicated by capital letters.

MARKER	UNIT	THICKNESS IN M			ACCUMULATION RATES CM/1000y			φ
		T_p	T_o	T_N	R_u	R_f	R	
ml								
g	H	100	100	54	5.00	5.00	2.70	.459
f	G	150	155	89	7.50	7.75	4.45	.409
e	F	100	115	63	10.00	11.50	6.30	.373
d	E	200	220	132	10.00	11.00	6.69	.342
c	D	200	230	139	4.00	4.60	2.78	.306
b	C	150	180	108	5.00	6.00	3.60	.277
a	B	200	255	150	5.00	6.38	3.75	.252
TD	A	150	185	116				.230

FIG. 11—Thickness and sediment-accumulation rates for well units shown in Figure 10. T_p is present thickness; T_o is initial thickness; T_N is net thickness. Initial and net thicknesses were derived using Figure 12 and 13. Zero depth is at mudline (ml), not at sea level.

Rates of deposition for units B and H are not the same, however, because unit B is deeply buried and consequently is more consolidated than unit H. Thus, equation 1 should be modified to express the uncorrected rate (uR) as

$$uR = \frac{T_p}{10A} \text{ cm}/1{,}000 \text{ years,} \qquad (2)$$

where T_p is present thickness. In other words, units should be restored to their original thickness (T_o) to obtain comparable rates of fill (Rf):

$$Rf = \frac{T_o}{10A} \text{ cm}/1{,}000 \text{ years.} \qquad (3)$$

Rates so calculated, however, still do not fully represent the sedimentation process because thick units (as a whole) have lower initial porosity than thin units, inasmuch as their own load results in increased compaction downward. Consequently, neither uR nor Rf can be used to compare rates of sediment accumulation for units having different initial thicknesses and depths of burial. A "pure" rate of sediment-mass accumulation can be calculated from the net thickness (T_N, porosity = 0) of a unit:

$$R_{\phi = o} = \frac{T_N}{10A} \text{ cm}/1{,}000 \text{ years.} \qquad (4a)$$

If present porosity is p, this expression can be written as:

$$R_{\phi = o} = \frac{T_p (1-p)}{10A} \text{ cm}/1{,}000 \text{ years.} \qquad (4b)$$

Of course, accumulation rates also are comparable if they all are converted to the same shale porosity; for instance, if porosity = a,

$$R_{\phi = a} = \frac{T_N}{10A (1-a)} \text{ cm}/1{,}000 \text{ years,} \qquad (4c)$$

or, if present porosity = p,

$$R_{\phi = a} = \frac{T_p (1-p)}{10A (1-a)} \text{ cm}/1{,}000 \text{ years.} \qquad (4d)$$

Equations 4a and 4b are the special case when a = 0.

Assume that the example of Figure 10 is a normally pressured clay-shale section. Then, an estimate for the porosity of the units (from Fig. 12) gives the net thickness (T_N), which in turn (using Fig. 13) gives an estimate of the initial thickness (T_o; Fig. 11). For example, the porosity of the 100-m-thick unit F is estimated as 0.373 (Fig. 12), and $T_N = (1 - 0.373) \times 100 \text{ m} = 63 \text{ m}$ (or 207 ft). The dashed line on Figure 13 shows the initial thickness to be 375 ft (115 m). Values derived

from the nomogram of Figure 14 are considerably higher (see Fig. 15), which suggests that, if available, local empirical data should be used as a basis for correction.

Figure 11 lists the three kinds of average accumulation rates calculated separately using equations 2, 3, and 4a, respectively. The differences in rates obtained by the three methods of calculation emphasize the need to apply rate equations consistently with due regard for assumptions inherent in each.

The volume reduction (consolidation, compaction) is difficult to generalize for sandstone and limestone because sorting, solution, crushing, and cementation play varying roles, and their timing and intensity are unknown.

These reductions are not considered in the present general discussion. However, they should not be ignored in detailed studies (Chilingarian and Wolf, 1975; Shinn et al, 1977).

Applications

Dating a section while it is being drilled, and therefore knowing the sedimentation rates, has distinct advantages for well-site geologists. Application of this knowledge depends, of course, on the particular geologic or technologic questions involved. A few of the more common well-site applications are discussed here.

1. It is possible to make a prediction of age or thickness ahead of the bit. By using equation 2, we can calculate estimates of the sedimentation rate and answer such questions as, "At what depth will we reach the top of the Miocene?" or "What is the age of the reflector at -500 m?" For example, assume that drilling is at -450 m as shown in Figure 10, and the question is: "Will the reflector at -500 m be the expected top of unit D?" We are 50 m below marker e, and our latest uR is 10 cm/1,000 years; the top of D, by definition, is 2 Ma older than e. Hence, substituting in equation 2 gives 10 $= T_p/(10 \times 2)$, so $T_p = 200$ m, which indicates that we must drill $200 - 50 = 150$ m before the top of D is reached. This depth, of course, is 100 m deeper than the reflector. Then, what age is the reflector? Its depth below marker e is 100 m $= T_p$; substituting in equation 2 gives 10 = 100/10A, so A = 1 Ma. Because e is 5 Ma old, the reflector marks beds that are 5 + 1, or 6 Ma old.

2. The same kind of extrapolation permits an estimate of the age at total depth in the absence of markers. In our example of Figure 10, uR above the lowest marker is 5 cm/1,000 years. Substituting in equation 2 gives 5 = 150/10A from which A = 3 Ma, and the age at total depth can be estimated to be 19 + 3, or 22 Ma. In the same way, the age can be estimated for any particular level below total depth (such as a seismic reflector).

The basic assumption in these extrapolations is that no change in sedimentation rate occurs below (or above, in case of upward extrapolation) the last marker. This is a weak assumption, but comparisons with nearby wells may determine its reliability. Of course, extrapolation across different lithologic or seismic facies and across unconformities should be avoided.

3. A third application is to estimate the time span of missing section ("hiatus" of Mitchum et al, 1974; Vail et al, 1974; "lacuna" of Wheeler, 1958). The age of beds directly above and below an unconformity rarely is

φ	DEPTH METER	AVERAGE UNIT φ PER 50 M	PER 100 M	PER 200 M
.490	0	.4730	.4598	.4384
		.4465		
.436		.4265	.4170	
		.4070		
.400	200	.3920	.3855	.3735
		.3790		
.373		.3670	.3615	
		.3560		
.351	400	.3463	.3415	.3323
		.3368		
.332		.3275	.3230	
		.3185		
.314	600	.3098	.3055	.2973
		.3013		
.297		.2930	.2890	
		.2850		
.281	800	.2773	.2735	.2663
		.2698		
.266		.2625	.2590	
		.2555		
.252	1000	.2488	.2455	.2393
		.2423		
.239		.2360	.2330	
		.2300		
.227	1200	.2243	.2215	.2163
		.2188		
.216		.2135	.2110	
		.2085		
.206	1400	.2035	.2010	.1963
		.1985		
.196		.1938	.1915	
		.1893		
.187	1600	.1848	.1825	.1783
		.1803		
.178		.1760	.1740	
		.1720		
.170	1800	.1680	.1660	.1623
		.1640		
.162		.1603	.1585	
		.1568		
.155	2000			

φ	DEPTH METER	AVERAGE UNIT φ PER 50 M	PER 100 M	PER 200 M
.155	2000	.1533	.1515	.1483
		.1498		
.148		.1465	.1450	
		.1435		
.142	2200	.1405	.1390	.1360
		.1375		
.136		.1345	.1330	
		.1315		
.130	2400	.1287	.1275	.1250
		.1263		
.125		.1238	.1225	
		.1213		
.120	2600	.1188	.1175	.1150
		.1163		
.115		.1138	.1125	
		.1113		
.110	2800	.1088	.1075	.1050
		.1063		
.105		.1038	.1025	
		.1013		
.100	3000	.0990	.0980	.0960
		.0970		
.096		.0950	.0940	
		.0930		
.092	3200	.0910	.0900	.0880
		.0890		
.088		.0870	.0860	
		.0850		
.084	3400	.0835	.0830	.0820
		.0825		
.082		.0815	.0810	
		.0805		
.080	3600	.0795	.0790	.0780
		.0785		
.078		.0775	.0770	
		.0765		
.076	3800	.0755	.0750	.0740
		.0745		
.074		.0735	.0730	
		.0725		
.072	4000			

FIG. 12—Shale porosity at increasing burial depth (according to Roll, 1974, Table 1). Left column indicates porosity at 100-m intervals. Other porosity columns record average porosity for units 50, 100, and 200 m thick.

known precisely from marker forms. Rather than simply guessing at what the magnitude of the unconformity might be within certain reasonable limits, we now can calculate an estimated age at the unconformity by extrapolating uncorrected rates of sediment accumulation (uR) downward and upward, respectively, from the nearest markers. Because of an unconformmity at −700 m in the section of Figure 10, the average rate of sediment accumulation for unit D (as given on Figure 11) is quite meaningless. What to do about it? An unconformity can be considered as the record of a geologic event. Thus, the section directly below an unconformity most likely was part of the sedimentary regime that prevailed prior to the event, and the section directly above an uconformity was part of the sedimentary regime that established itself after the event took place. From upward extrapolation of the uncorrected accumulation rate (uR = 5 cm/1,000 years for unit C), the age at the unconformity is 12 − 2 = 10 Ma derived from substituting in equation 2, 5 = 100/10A, A = 2 Ma. From downward extrapolation of uR (= 10 cm/1,000 years for unit E), the age at the unconformity is 7 + 1 = 8 Ma [10 = 100/10A, A = 1 Ma]. Hence, 10 − 8 = 2 Ma is missing.

If more wells have been drilled in the area, we may, for example, know that elsewhere the age span −7 to

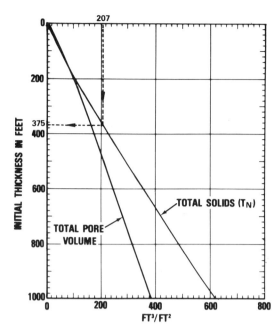

FIG. 13—Relation between total pore volume, net thickness (T_N), and initial thickness (T_o) of shales (from Gretener and Labute, 1969, Fig. 2). 100 ft = 30.48 m.

−12 Ma is missing, and somewhere else −7.5 to −9, and at a third location −8 to −10.5, etc. By combining this information, we can obtain an idea of when the event that caused the unconformity took place. In our example (Fig. 16), it occurred about 8.5 Ma ago. Irregularities in the time span of the gap, like those in Figure 16, may suggest that we are dealing with a tectonic event. To the contrary, one would expect a lacuna to decrease regularly in an offshore direction, suggesting that the event is a downward movement of base level (Barrell, 1917; Eicher, 1968) owing to a eustatic drop in sea level or other cause. Graphic extrapolation of such a situation may give a sharper estimate of the timing of the event (Fig. 17).

4. Once the age of an event is known, we can estimate the amount of removed section ("erosional hiatus" of Mitchum et al, 1974; "erosional vacuity" of Wheeler, 1958) by extrapolation of the pre-unconformity accumulation rate. We saw previously that in the case shown in Figure 10, sedimentation continued until 8.5 Ma ago, that is, 3.5 Ma after −12 Ma (marker c). Where uR = 5, 3.5 Ma of deposition will have resulted in 175 m of sediment accumulation, calculated $5 = T_p/(10 \times 2.5)$, where $T_p = 175$ m. Of this 175 m, 100 m is now preserved; consequently 75 m has been eroded.

5. If an abrupt change in water depth occurs in a section, the procedure of point 3 can be followed downward and upward toward the break. Such a paleo-water-depth break most likely records an eustatic change in sea level if no time gap is found; it could or could not represent such a change if an unconformity is present.

Examples 1 to 5 are applications at the well site where the initial thickness (T_o) or net thickness (T_N) cannot be

determined because sandstone/shale ratios and shale porosities are unknown and uncorrected rate values (uR) must be used. This procedure is not as inaccurate as it may seem, because in each case the rates are extrapolated from one unit to its adjacent unit only; hence, compaction differences are mostly minor and errors tend to be small and irrelevant with respect to the accuracy of the method. The same holds for regional comparisons of uR's of similar formations at more or less equivalent burial depths, although in these cases the use of corrected values is more appropriate. Corrections for compaction are imperative, however, in other regional and in general applications.

6. Contour maps of rate-of-fill (Rf) values outline depocenters and other trends in the depositional environment for selected time intervals. Rate-of-fill maps have an advantage over isopach maps because a correction is made for distortion caused by regional variation in burial depth of mapped units.

Rf values, however, are unit-thickness dependent, so that $R_{\phi-a}$ has to be used to compare information on units of widely different initial thicknesses. Lithologic variation further complicates the picture. Thus, an area where initially a relatively thin but sandy section was deposited may appear as a "thick" on an R-contour map if the unit is less sandy elsewhere.

7. An R-contour map, therefore, may be similar to a sand-percentage map. The R map has the advantage that even though no sand is detected, trends show up that may lead to discovery of sand outside the mapped area. Further, an R map of, for example, a reservoir time-stratigraphic unit from offshore Louisiana can be compared directly with one from Nigeria, regardless of differences between the two regions in age, thickness, or sand percentage.

8. The corrected average rate of sediment accumulation (R) is a meaningful sedimentary parameter for general studies. For example, in order to improve predictions on the presence of source rocks and their possible yield, one could investigate relations between hydrocarbon source potential (amount and type of organic matter) and accumulation rate of shales (Rogers and van Hinte, in press). Or, for example, regional R maps for large time slices can be compared with hydrocarbon occurrences. Depending on results of such studies, R may be recognized as a useful parameter in quantitative assessment techniques.

In any case, R is a highly informative number that should be considered in the description and classification of sediments. It ranges from >0.1 cm/1,000 years for red, deep-sea clay (Berger, 1974) to several meters/1,000 years for some deltaic deposits (cf. Fig. 6). Anomalously low average numbers for a certain kind of sediment suggest the presence of unconformities; anomalously high numbers also can have important geologic implications that one might not recognize otherwise.

SUBSIDENCE RATES

Paleobathymetry

Irreversibility of the evolution of the organic world is the basis for paleontologic dating of sediments. Organisms, however, not only occurred during a particular time but also lived in certain environments having their

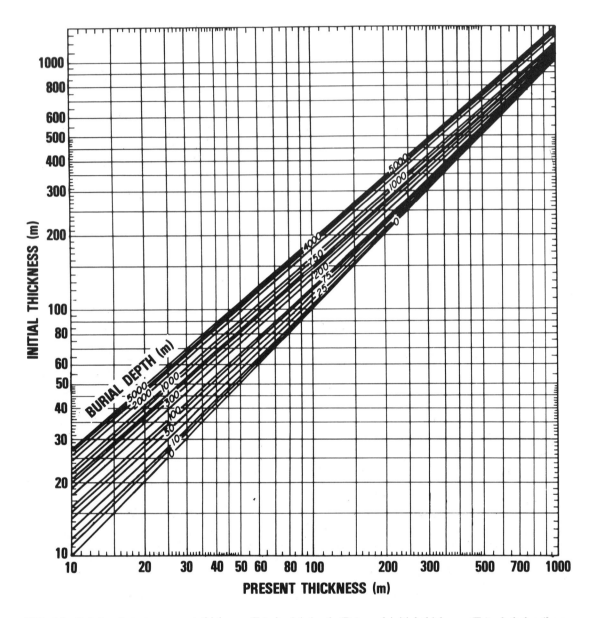

FIG. 14—Relation between present thickness (T_p), burial depth (D_p), and initial thickness (T_o) of shales (from Perrier and Quiblier, 1974, Fig. 1).

own habitat. Paleontologic well reports, therefore, should include conclusions on both age and environment of deposition of penetrated sections. In the present discussion, one aspect of the environment—paleobathymetry—is most important.

Although the distribution of some species of assemblages seems to be related to depth per se (hydrostatic pressure), availability of direct depth indicators is limited so that paleobathymetric conclusions commonly are obtained indirectly.

Depending on the material at hand, a paleontologist may try one or more of the several available ways to determine the marine depositional environment—quali-

tatively, by comparison with the modern occurrence of certain species or assemblages and by recognition of ecophenotypic trends (benthos and plankton), or, quantitatively, by determining relative frequencies (e.g., plankton/benthos, arenaceous/calcareous forams, percentages of radiolarians or ostracods), species number, species dominance, species diversity, coiling ratio, sex ratio, ontogenetic age, and, for deep-water deposits, solution facies. Thus, he obtains an idea about one or a combination of environmental factors which may be chemical (salinity, pH, availability of oxygen, nutrients, carbon dioxide), physical (temperature, substratum, energy level, light penetration, turbidity, circulation, barriers), or biological (intracommunity, intercommunity, plants). Many of these factors can be depth indicators, and the better we know the paleo-oceanographic conditions, the better we can determine to what degree "cool," "dark," or "euxinic" means "deep."

Paleo-water-depth estimates can be quite accurate in areas where the recent faunal distribution, fossil record, and regional geologic history are well known (e.g., the Gulf Coast). Elsewhere, water-depth estimates may have to be somewhat less precise; the older the sediments the more speculative the paleodepth determinations become. The accuracy with which paleo-water-depth determinations can be made in frontier areas is indicated in Figure 18. The resolution may seem too general to be of practical use, but it should be realized that, although individual determinations may be inconclusive, a succession of determinations (e.g., based on side-wall cores in a well) shows a trend through which a line can be drawn establishing with some confidence relative changes in water depth between parts of the section.

In Figure 19, paleo-water-depth interpretations have been added to the data of Figure 10, with the length of the horizontal lines indicating the uncertainty with which a conclusion is drawn for each individual core. Despite the uncertainties, a trend emerges, and a paleontologist's best-guess "paleobathymetry curve" can be drawn. Then, an estimate of the bathymetry at each marker time can be read from the curve.

Qualitative and quantitative methods similar to those mentioned can be applied to associations of palynomorphs in nonmarine deposits. The results may indicate the paleo-elevation of the deposit or of its hinterland. In the following discussion, however, we treat only negative elevation, that is, depth below sea level, or water depth.

Determine Subsidence Rates

Assume that during the past 2 megayears the water depth did not change at the location of Figure 10. This would mean that in order to accommodate the sediments of unit H, the site must have subsided 100 m with respect to sea level. If, however, shoaling or deepening had occurred, the amount of subsidence would have been, respectively, less than or greater than the sediment thickness (see Bandy and Arnal, 1960). What holds for the amount of subsidence is also valid for the rate, which can be written as

$$``Rs" = ``R" - \frac{(W+E)}{10A} \text{ cm}/1,000 \text{ years}, \quad (5)$$

in which W is the change in water depth, and E is the eustatic sea-level change during deposition of the unit. Deepening from 0 to −50 m would mean that W = −50; in the same sense a eustatic rise in sea level of 20 m would make E = −20.

In the preceding example, "Rs" is 5 cm/1,000 years. That is, during the time of deposition of the unit, its base went down (subsided) at an average velocity of 5 cm/1,000 years. To obtain a comparable subsidence rate for older units, the thickness prior to compaction, or initial thickness T_o, is used.

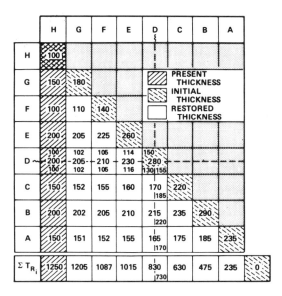

FIG. 15—Present thickness, initial thickness, and restored thickness for units in hypothetical offshore well (Fig. 10). Letters across top of diagram indicate time at end of unit deposition; letters at left indicate unit. For example, at marker-g time (top of unit G), unit D was 205 m thick and the total section was 1,205 m thick. Numbers were obtained from nomogram of Figure 14. Compare with Figure 11 and note differences.

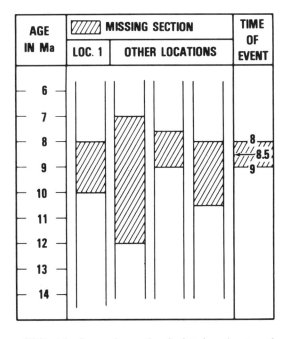

FIG. 16—Comparison of calculated estimates of missing section at various locations in one area. Such comparison permits reliable estimate of timing of event causing unconformity.

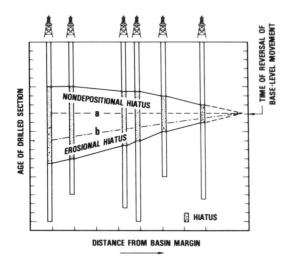

$$Rs_O = \frac{T_o - (W+E)}{10A} = Rf - \frac{(W+E)}{10A} \text{ cm}/1{,}000 \text{ years.} \quad (6)$$

Use of present thickness (T_p) for units other than the highest one would give uncorrected rates that are lower than Rs_o:

$$uRs = \frac{T_p - (W+E)}{10A} = uR - \frac{(W+E)}{10A} \text{ cm}/1{,}000 \text{ years.} \quad (7)$$

Subsidence at or near the surface is not caused exclusively by subcrustal forces but includes the effects of loading: (1) compaction of the underlying section ("pseudosubsidence" of Roll, 1974), and (2) isostatic adjustment. The calculated Rs_o value, therefore, applies only to the base of a unit. Deeper levels at that time subsided at a slower rate, because lower section was being less compacted, and higher units, of course, did not exist. The calculated uRs value is the rate at which all levels subsided during that time and is "correct" only if the entire section consists of noncompactible sediment. Fortunately, more and more empirical compaction data are becoming available for estimating the subsidence rate of any chosen level in a compacting section (Rs_n). Knowing the compaction also allows for distinguishing the "tectonic" subsidence rate (R_T), which may be entirely a result of isostasy or of isostasy together with other basement-involved and/or basement-uninvolved structuring.

For example, Figure 15 gives, step by step, the loading effect of the deposition of each new unit on the section below it. The estimates are made using the no-

FIG. 17—Derivation of time of reversal in downward base-level movement that caused unconformity of basinward-decreasing magnitude. Drawing line a as boundary between erosional and nondepositional hiatus assumes that base-level drop occurred within limits of time-stratigraphic resolution and appears as "instantaneous;" line b assumes that base level moved downward and upward at about same rate; first assumption generally is preferred (Frazier, 1974; Mitchum et al, 1974). Wheeler (1958) used terms "lacuna," "hiatus," and "erosional vacuity," respectively, for hiatus, nondepositional hiatus, and erosional hiatus. Broken lines at right of wells are extrapolation.

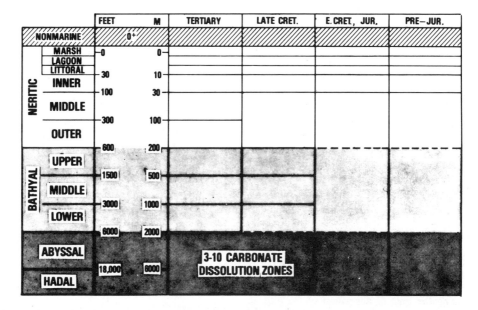

FIG. 18—Approximate micropaleontologic paleobathymetric resolution using forams, ostracods, and microfacies. Pre-Tertiary sections deposited under particular conditions (e.g., carbonate shelf) may be interpreted with greater precision than shown.

mogram of Figure 14 and assuming 100% shale. We now can read the restored thickness (TR) for any part of the section at any of the marker times. For example, 5 Ma ago, at time e, unit D was 230 m thick (TR_{De} = 230 m), and the entire section at that time was 155 + 210 + 160 + 230 + 260 = 1,015 m thick ($\Sigma\ TR_{ie}$ = 1,015 m, i = a through e). That is, this is the depth of burial of TD at time e, or D_{TD}. The actual subsidence of a particular level (n) can be expressed as its increase in burial depth minus the water-depth change, so that

$$Rs_n = \frac{\Delta D_n - (W+E)_n}{10A}\ cm/1,000\ years. \qquad (8)$$

"Rate of subsidence" stands for "average rate of subsidence with respect to sea level" if E is excluded from the form, and for "average rate of subsidence with respect to the geoid" if E is included in the form.

Differences between the subsidence rates for a particular unit are significant and have to be taken into account, as shown in Figure 20, which lists the uRs (uncorrected rate of subsidence), Rs_o (initial rate of subsidence), Rs_n (rate of subsidence of level n), Rc (rate of compaction during an interval), and the data needed to calculate these values.

Rs_{TD} is the tectonic rate of subsidence (R_T) if the bottom of the well is in basement, but usually Rs_{TD}

results largely from compaction of sediments underlying the penetrated well section. In that case, one has to make an assumption as to the lithology (shale percentage) and thickness of the underlying sediments in order to calculate an estimate for the tectonic subsidence rate.

$$R_T = Rs_B, \qquad (9)$$

for B = top of basement or top of compacted section, or

$$R_T = Rs_n - pRs_n, \qquad (10)$$

in which pRs is the rate of "pseudosubsidence" or compaction of the section below n.

The rate of compaction (Rc) for any unit at any chosen time can be calculated from data like those in Figure 15. For example, during the 2 Ma of deposition of unit G, the thickness of unit B decreased from 205 to 202 m, that is, 3/(10×2) = 0.15 cm/1,000 years (Rc = 0.15). During the same time, unit E shrank from 225 to 205 m, that is, 20/(10×2) = 1 cm/1,000 years (Rc = 1).

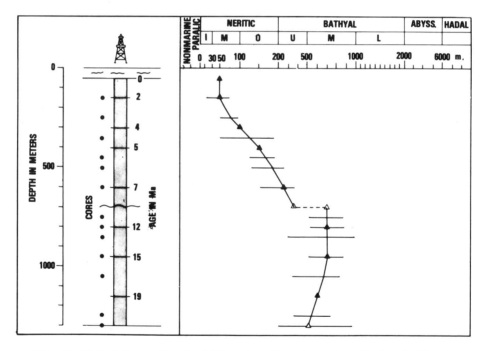

FIG. 19—Well of Figure 10 with addition of position of core data (dots) and with paleobathymetric interpretations for core samples given as horizontal lines in right-hand part of figure. Length of horizontal bars indicates confidence of each environmental interpretation. A paleontologist drawing a paleobathymetric curve through horizontal bars commonly is guided by data on faunal distribution obtained from examination of cuttings and does not necessarily use midpoints of bars in constructing his model. Triangles in figure indicate which single-figure paleobathymetric estimates were read from model for dated levels of well.

$$Rc = \frac{TR_n - TR_{n-1}}{10A} \text{ cm}/1{,}000 \text{ years}, \qquad (11)$$

hence,

$$pRs = \frac{\Sigma(TR_n - TR_{n-1})}{10A}. \qquad (12)$$

In the absence of data on the section below TD, uRs and Rs_{TD} values can be used to estimate R_T. For the older part of the section, uRs is smaller than Rs_{TD}, the basal units being thinner than they were at earlier times. For the younger part of the section, uRs is larger than Rs_{TD} because the compaction of the section below TD becomes less important whereas the thickness of the younger units on which uRs is based increases. Below the point where uRs > Rs_{TD} reverses the uRs < R_{TD},

uRs is the best estimate for R, whereas for the section younger than the reversal point, Rs_{TD} is the best estimate. If drilling continues, that is, deeper section is being drilled, the reversal point will move down (get older) until it reaches a level where compaction is complete (or basement is reached). From there down, Rs_{TD} = R_T.

The average rate at which each unit of Figure 20 shrank during the last 2 Ma (i.e., during deposition of unit H) is shown in the figure. Unit H also compacted during its deposition, and the numbers in parentheses are based on the assumptions that present average porosity of the unit is 50% and that porosity of the upper inches was 80%. For this 100% shale example, it is obvious that Rc decreases with depth. In nature, variation in shale percentage can be the reason that the deeper units of a section compact at a higher rate than the shallower units.

In summary, five "average rates of subsidence" are distinguished: Rs_o during an interval is the rate at

	UNIT	AGE	T_p	T_o	WATER DEPTH W	uRs $\frac{T_p-(W+E)}{10A}$	Rs_o $\frac{T_o^*-(W+E)}{10A}$	FOR TD D_R ΔD	FOR TD $\frac{Rs}{\Delta D-(W+E)}{10A}$	FOR e D_R ΔD	FOR e $\frac{Rs}{\Delta D-(W+E)}{10A}$	DURING ML ΔTR $(TR_{ml}-TR_g)$	DURING ML Rc_{ML}
1250		0			50			1250		350			
	H	2	100	100	0	5.0	5.0	45	2.25	60	3.0	(150)	(7.5)
1150		2			50			1205		290			
	G	2	150	180	50	5.0	9.0	118	3.4	150	5.0	30	1.5
1000		4			100			1087		140			
	F	1	100	140	50	5.0	14.0	72	2.2	140	9.0	10	1.0
900		5			150			1015		0			
	E	2	200	260	100	5.0	13.0	185	4.25			5	.25
700		7			250			830					
		(1)	(100)	(150)	100	0	5.0	100	0				
600	D	(8)(10) (2)	200	280	350/700	-5.0 <-17.5	-1.4 <-17.5	730	-5.0 <-17.5			5	.10
		(2)	(100)	(155)	0	5.0	7.75	100	5.0				
500		12			700			630					
	C	3	150	220	0	5.0	7.33	155	5.16			2	.06
350		15			700			475					
	B	4	200	290	-100	7.5	9.75	240	8.5			2	.05
150		19			600			235					
	A	(3)	150	235	-100	8.33	11.16	235	11.16			1	.03
		(22)			500			0					

FIG. 20—Different "average rates of subsidence" and average rate of compaction. Numbers needed to calculate rates for hypothetical well shown in Figure 10 are given. T_p = present thickness; T_o = initial thickness; W = change in water depth; E = eustatic sea-level change; A = duration of unit; uRs = uncorrected subsidence rate; Rs_o = initial rate of subsidence; Rs_{TD} = rate of subsidence at base of section; D = burial depth (= Σ TR, in which TR is restored thickness); Rc = rate of compaction. Ages at unconformity in unit D are discussed in paragraph on applications of sediment accumulation rates.

which the base of the rock record subsides; Rs_n during an interval is the rate at which an older level n subsides; uRs during an interval is the rate at which all levels of a section would subside if no compaction took place; R_T during an interval is the rate at which the top of the fully compacted section or of basement subsides; and pRs during an interval is the rate at which a level subsides because of compaction of the older section. A sixth rate, Rc, is the rate of compaction of a unit as a result of loading by a later unit.

Applications

The use of subsidence and compaction rates as parameters has countless applications in exploration and general geology. The following are a few of the possible uses.

1. A map of Rs_o can be used to predict facies distribution. Areas with the highest surface subsidence are potentially lowest topographically and will be sand prone if a sand source is available. Rs_o contours may further outline areas affected by growth faults or diapirs. Rs_o contours do not necessarily conform to those of sediment accumulation rates (Rf).

2. Estimation of R_T values from well data allows conclusions on magnitude and timing of structuring to be drawn with unprecedented precision. A suite of maps portrays the dynamics of subsurface geometry.

3. Maps of R_T values distinguished between regional, basement-involved subsidence and anomalies caused by movements within the sediment column (growth faults, shale or salt domes). A suite of R_T anomaly maps shows the dynamics of such structures.

4. Subsidence rates are a useful parameter in the characterization and early recognition of structural styles and, on a larger scale, in basin classification and characterization of plate-tectonic setting.

5. Rs_n values give the proper history of vertical movement of a level. Thus, an Rs_n contour map of a reservoir sandstone can be made for a particular time interval. Superimposing this map on a paleosurface map ($D_{n-1} + W_{n-1}$) shows the direction of buoyant flow during that time.

6. A contour map of compaction rates (Rc) for a particular unit during a selected (younger) time interval is like a piezometric contour map. The pattern of fluid movement will have been away from the high compaction rates and toward the low rates. If hydrocarbons were present in the formation, then the fluid movement pattern will have been the migration pattern. Also, if the compacting formation itself were immature, hydrocarbons that came from elsewhere will have had to travel those routes. A superimposed paleostructure map shows which structures could have been functioning as traps and which more likely were being drained at that time.

7. R_T values are fundamental data in geophysical (geodynamic) studies on crustal thickness, isostasy, and, in succession, on properties and processes below the Moho.

8. Noting, while drilling, that subsidence rates are significantly lower than regional rates is a warning that drilling is over a dome of either salt or shale and may penetrate overpressured formations.

9. At present, absolute values for eustatic sea-level variations are unknown, and in calculating rates of vertical movement, we assume that E = O. Kuenen (1939) and more recently Wise (1972, p. 92) have shown that "for 80% of post-Precambrian time oscillations have remained within ± 60 m of normal freeboard." (Wise's "normal freeboard" is 20 m above present sea level.) If this is true, then even maximum oscillations will not affect our vertical movement rate numbers so long as we are dealing with high average rates over long periods of time. But, the lower the rate and/or the shorter the time interval considered, the more significant the eustatic effect will be. For example, a very large eustatic change of 100 m will give a significant apparent vertical movement of 10 cm/1,000 years when a time interval of 1 Ma is considered, but an insignificant 1 cm/1,000 years when a time interval of 10 Ma is used. Conversely, the rates can give us an idea about the eustatic changes. Assume that at a particular location averaged over 20 Ma no vertical movement took place and that we have a good age determination for every 2 Ma. Then, variations found in the rates of tectonic subsidence (R_T) calculated over the smaller time increments are the result of variations in sea level (or of wrong age or paleobathymetry determinations). Or, imagine that basement regularly subsided at 10 cm/1,000 years except for some brief periods of uplift and accelerated subsidence. Again, the anomalies probably are due to eustatic sea-level changes or to errors. When anomalous R_T values are found consistently for the same time interval in different areas, they most likely are not the result of errors in individual age or paleobathymetry determinations and, therefore, indicate eustatic sea-level changes or general errors in the linearity of the time scale used. Consequently, with increasing accuracy of the time scale, the recognition of R_T anomalies will improve our knowledge of the time and magnitude of eustatic sea-level changes. Accumulating this empirical knowledge in a quantitative sea-level plot similar to the qualitative curve of Umbgrove (1939, p. 128) will give a curve from which E can be read.

APPENDIX REFERENCES CITED

Bandy, O. L., and R. E. Arnal, 1960, Concepts of foraminiferal paleoecology: AAPG Bull., v. 44, p. 1921-1932.

Barrell, J., 1917, Rhythms and the measurement of geologic time: Geol. Soc. America Bull., v. 28, p. 745-904.

Berger, W. H., 1974, Deep-sea sedimentation, in C. A. Burk and C. L. Drake, eds., The geology of continental margins: New York, Springer-Verlag, p. 213-241.

Chilingarian, G. V., and K. H. Wolf, eds., 1975, Compaction of coarse-grained sediments, 1: Developments in Sedimentology, no. 18A, 552 p.

Eicher, D. L., 1968, Geologic time: Englewood Cliffs, N.J., Prentice-Hall Inc., 149 p.

Frazier, D. E., 1974, Depositional episodes: their relationship to the Quaternary stratigraphic framework in the northwestern portion of the Gulf Basin: Texas Univ. Bur. Econ. Geology Geol. Circ. 74-1, 28 p.

Gretener, P. E., and G. J. Labute, 1969, Compaction—a discussion: Bull. Canadian Petroleum Geology, v. 17, p. 296-303.

Kuenen, Ph., 1939, Quantitative estimations relating to

eustatic movements: Geologie en Mijnbouw, v. 18, p. 194-201.

Mitchum, R. M., P. R. Vail, and J. B. Sangree, 1974, Regional stratigraphic framework from seismic sequences (abs.): Geol. Soc. America Abs. with Programs, v. 6, p. 873.

Perrier, R., and J. Quiblier, 1974, Thickness changes in sedimentary layers during compaction history; methods for quantitative evaluation: AAPG Bull., v. 58, p. 507-520.

Rogers, M. A., and J. E. van Hinte, in press, Organic geochemical evaluation of some Tertiary deep water sediments from the Atlantic North of 45° North: Deep Sea Drilling Proj. Supp. Rept., North Atlantic.

Roll, A., 1974, Langfristige Reduktion der Mächtigkeit von Sedimentgesteinen und ihre Auswirkung—eine Übersicht: Geol. Jahrb., series A, no. 14, 76 p.

Shinn, E. A., et al, 1977, Limestone compaction: an enigma: Geology, v. 5, p. 21-24.

Umbgrove, J. H. F., 1939, On rhythms in the history of the earth: Geol. Mag., v. 76, p. 116-129.

Vail, P. R., R. M. Mitchum, and S. Thompson, 1974, Eustatic cycles based on sequences with coastal onlap (abs.): Geol. Soc. America Abs. with Programs, v. 6, p. 993.

Wheeler, H. A., 1958, Time-stratigraphy: AAPG Bull., v. 42, p. 1047-1063.

Wise, D. U., 1972, Freeboard of continents through time, in Studies in earth and space sciences: Geol. Soc. America Mem. 132, p. 87-100.

Reprinted by permission of the Society of Economic
Paleontologists and Mineralogists. Published in *Tectonics
and Sedimentation*, Society of Economic Paleontologists
and Mineralogists Special Publication 22, p. 1–27.

PLATE TECTONICS AND SEDIMENTATION

WILLIAM R. DICKINSON
Stanford University, Stanford, California

ABSTRACT

The theory of plate tectonics offers a fresh opportunity to interpret the evolution of sedimentary basins in terms of changing plate interactions and shifting plate junctures. Although plate-tectonic theory lays primary emphasis on horizontal movements of the lithosphere, large vertical movements are also implied in response to changes in the thickness of crust, in the thermal condition of lithosphere, and in the isostatic balance of lithosphere over asthenosphere. As thick sedimentation requires either an initial depression or progressive subsidence to proceed, the auxiliary vertical movements largely control the evolution of sedimentary basins. Ancillary geographic changes related to the governing horizontal movements also affect patterns of sedimentation strongly.

The geosynclinal terminology used prior to the advent of plate tectonics is inadequate to describe fully the plate-tectonic settings of sedimentary basins. Basins can be described instead in terms of the type of substratum beneath the basin, the proximity of the basin to a plate margin, and the type of plate juncture nearest to the basin. Intraplate settings of oceanic or continental character contrast with zones of plate interaction, which include those of divergent, convergent, and transform motions and within each of which the underlying crustal structure is or may be complex. The evolution of a sedimentary basin thus can be viewed as the result of a succession of discrete plate-tectonic settings and plate interactions whose effects blend into a continuum of development.

Oceanic basins contain an assemblage of diachronous facies whose relations are controlled by thermal subsidence of the lithosphere as it moves away from midoceanic rises. Rifted continental margins undergo successive stages of structural evolution as the following features are formed: prerift arch, rift valley, proto-oceanic gulf, narrow ocean, and open ocean. Sedimentary phases related to each stage are components of the rifted-margin prism of strata that masks the continent-ocean interface beneath a continental terrace-slope-rise association or a progradational continental embankment. Marginal fracture ridges along marginal offsets and aulacogens along failed arms of triple junctions locally break the continuity of rifted-margin prisms. Sedimentary basins associated with arc-trench systems where oceanic lithosphere is consumed include trenches beyond the subduction complex beneath the trench slope break, forearc basins in the arc-trench gap, intra-arc basins within the magmatic arc, and interarc basins or retroarc basins in the backarc area. Interarc basins are oceanic basins between a migratory intraoceanic arc and a remnant arc, whereas retroarc basins rest on continental basement adjacent to a foreland fold-thrust belt behind a continental margin arc. Peripheral basins adjacent to suture belts formed by crustal collision occur in an analogous foreland setting between orogen and craton, but in front of a colliding magmatic arc. Retroarc basins and peripheral basins both imply partial subduction of continental lithosphere. Intracontinental basins include infracontinental types, beneath which incipient continental separation gave rise to crust of transitional thickness, as well as supracontinental types.

INTRODUCTION

The theory once termed the *new global tectonics* (Isacks and others, 1968), which postulates a segmented and mobile lithosphere, is no longer new. Most geologists apparently accept its main tenets as valid, together with the corollary concepts of continental drift and seafloor spreading, transform faults (Wilson, 1965) and subduction zones (White and others, 1970), although some geologists, notably Beloussov (1970) and Meyerhoff (Meyerhoff, 1972), have challenged these fundamental concepts. *Plate tectonics* (McKenzie, 1972a; Dewey, 1972) has become an alternate designation for the new global tectonics because the discrete spherical caps of essentially rigid lithosphere inferred to be in relative motion with respect to one another, and to the softer and weaker asthenosphere beneath, are commonly called *plates* (McKenzie and Parker, 1967). The characteristic patterns of lateral motion of these surficial slabs, curved to conform to the spherical outline of the earth, were described by Morgan (1968) as motions of crustal blocks. He indicated, however, that these fundamental tectonic entities are actually blocks of *tectosphere,* a layer thicker than crust in the ordinary sense of the layer above M and essentially synonymous with the lithosphere of others. Because of the large lateral dimensions of the main intact pieces of lithosphere, which are of the order of only 100 km thick, the passage of time has favored usage of the word plate, rather than block, as the basic descriptor.

As a comprehensive theory that purports to explain the global distribution of all belts of tectonic deformation within the crust as the loci of the boundaries or junctures between plates of lithosphere, plate tectonics has the flavor of a fresh paradigm that must be accepted or rejected almost in its entirety with only modest allowances made for deviant behavior. The evident tectonic complexity of the earth admittedly forces the recognition that unusually broad zones of deformation occur along some plate junctures (Atwater, 1970), that the motions of some small plates are controlled partly by the interaction of adjacent large plates (McKenzie, 1970, 1972b; Roman, 1973a, 1973b), and that intraplate deformation is possible on a limited scale or to a limited degree (Sykes, 1970; Doyle, 1971). These adjustments within the framework of plate-tectonic theory dilute its elegance somewhat but do not challenge its fundamental premises. Moreover, the history of mountain belts is better illuminated by plate tectonics than by any preceding theory (Coney, 1970; Dewey and Bird, 1970a).

Concepts derived from plate tectonics are used here as the basis for a discussion of general relations between tectonics and sedimentation. Plate tectonics offers fresh ways to explain the evolution of sedimentary basins, and many concepts of the past can be discarded or must be modified to conform to the new point of view. The development of plate-tectonic interpretations and models of sedimentary basins thus entails the mental exercise of changing outworn interpretations and unjustified conclusions without denying established facts. Application of plate-tectonic analysis to the evolution of a specific sedimentary basin also requires the uniformitarian assumption that present styles of plate-tectonic behavior are useful keys to plate-tectonic behavior during the time span represented by the evolution of the basin.

Unfortunately, there seem to be no clearcut means yet to judge when plate-tectonic behavior of the modern sort began, or whether somewhat different forms of plate interactions prevailed at different times in the past. Events that could conceivably mark times of tectonic transition when plate tectonics could have been initiated or could have undergone some change in kind include (a) the breakup of Pangaea starting roughly a quarter of a billion years ago (Dietz and Holden, 1970), (b) the formation of the oldest recognized blueschist belts (Ernst, 1972) and ophiolite sequences (Burke and Dewey, 1972) about a half billion years ago, (c) a null in the reported frequency of radiometric dates for orogenic granitic and metamorphic rocks at about three-quarters of a billion years ago, (d) the development of the oldest lasting cratons during the Precambrian, or (e) the formation of the first cratonic nuclei deep in the Precambrian (see also Burke and Dewey, 1973).

Perhaps the most revolutionary facet of plate-tectonic theory as applied to sedimentary basins is the startling light it sheds on the tempo of major geologic events. At ordinary rates of spreading along midoceanic rises and of plate consumption at trenches (Le Pichon, 1968; Chase, 1972), oceanic basins 5000 km wide can form or, once formed, can disappear within 50 to 100 my, a span of time representing only one or two periods of the standard geologic column. It follows that no sedimentary basin with a long history of deposition is likely to have remained in the same plate-tectonic setting throughout its evolution. Realization of this principle is a vital guard against oversimplified versions of local geologic history in terms of plate tectonics. From a plate-tectonic standpoint, the Phanerozoic alone is an immense span of time, nominally long enough to open and close an ocean as broad as the Atlantic five or ten times!

Among the many things that might be written about plate tectonics and sedimentation, this paper discusses the following topics: (a) the vertical movements of lithosphere that are inherent in plate tectonics and required to set the *conditions for sedimentation*, (b) the *ancillary effects* of horizontal movements of lithosphere described as continental drift and seafloor spreading, (c) the problem of translating the *basin terminology* employed by geosynclinal theory into terms compatible with plate-tectonic theory, (d) the main *plate-tectonic settings* important for sedimentation, and (e) the gross outlines of *basin evolution* implied by the concept of plate tectonics.

CONDITIONS FOR SEDIMENTATION

Thick sedimentation in a given place implies the prior existence of a deep hole into which sediment can be dumped or progressive subsidence of the substratum to accommodate successive increments of strata. The formation of either kind of sediment trap on a large scale requires pronounced vertical movements of the earth's crust. Plate-tectonic theory as a geometric paradigm to explain tectonic patterns lays special emphasis instead on grand horizontal translations of lithosphere with its capping of crust. However, major vertical motions of crust and lithosphere are required to accompany the horizontal motions by any feasible geologic interpretations of the mechanisms of plate motions and interactions. The vertical motions are related to changes in crustal thickness, in thermal regime, and in the conditions for isostatic balance. These three facets of plate-tectonic theory postulate inherent vertical motions of an order and on a scale that no previous tectonic theory can match in overall scope. Despite its quite proper formal emphasis on horizontal

388

translations of lithosphere, plate-tectonic theory thus also affords the best theoretical basis yet devised to account for grand vertical movements of crust and lithosphere. From the standpoint of sedimentation, therefore, it is a mistake to view the emphasis on horizontal motions as a potential weakness of plate-tectonic theory. The ancillary vertical motions induced by plate interactions are fully equal to the demands of data on sedimentary basins and their provenances.

Crustal thickness.—New oceanic crust and lithosphere is formed continually at spreading centers along midoceanic rise crests and within marginal or interarc basins behind migrating arc structures. When continental blocks are rifted apart by extensions of spreading centers, thin oceanic crust of igneous origin is thus created adjacent to thick continental crust by submarine volcanism and associated intrusions. Frequency curves of crustal elevation show that the floor of such a new oceanic basin should stand typically about 4 km below the mean surface level of the two continental fragments formed by the rift. The process of continental separation can thus form a new crustal depression capable of serving as a receptacle for sediment, and in principle such a sediment trap can form adjacent to any part of a continental block as potential provenance. In detail, continental separation involves the development of a belt of transitional crust and lithosphere between each continental fragment and the adjacent oceanic basin. The width of the transitional region is not well known, but is probably 100 to 250 km wide in typical cases. Prior to sedimentary loading, the transitional crust will presumably stand at elevations intermediate between those of the continental block on one side and the oceanic basin on the other. Off Norway, Talwani and Eldholm (1972) described a broad transitional region in water depths of 1000 to 2500 m.

Studies of incipient continental separations suggest that two types of processes contribute to the development of belts of *transitional crust* of intermediate thickness (fig. 1):

(1) Attenuation of continental crust by stretching is accomplished by extensional faulting at upper crustal levels accompanied probably by pseudoplastic flowage at deep crustal levels (Lowell and Genik, 1972); sediment deposited on this type of transitional crust will rest upon basement rocks of continental character, but not upon a continental block of ordinary crustal thickness.

(2) Crust with oceanic affinities but unusual thickness forms where sedimentation contemporaneous with volcarism within an incipient rift depression helps construct a crustal profile of mingled sedimentary and igneous components in a complex of lavas, dikes, sills, and sediments (D. G. Moore, 1973); sediment deposited later on this type of transitional crust will rest upon a substratum of oceanic character having a crustal profile that may be of nearly continental thickness.

Plate interactions related to the consumption of lithosphere at arc-trench systems, rather than its creation at rise crests, can also produce crust of anomalous thickness that differs from both the normal oceanic value of 5 to 10 km and the normal continental value of 30 to 40 km. Such anomalous crust can be either of fundamentally oceanic or of fundamentally continental character in terms of its rock components. *Anomalous crust,* in the sense the term is used here, can form by either of two basically different mechanisms that are tectonically linked only as two contrasting facets of the geologic processes that operate within arc-trench systems (fig. 2):

(1) Igneous materials are added to the crustal structures of magmatic arcs, either as volcanic components of the surficial edifice or as intrusive components within the crustal roots of the arcs. Presumably by this process, the crustal thickness beneath intraoceanic arcs like Tonga-Kermadec (Shor and others, 1971, fig. 3) and the Marianas (Murauchi and

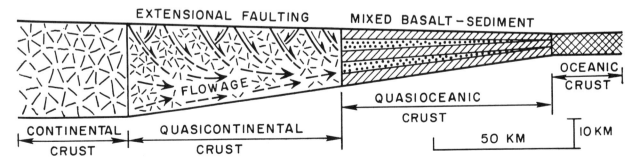

FIG. 1.—Schematic diagram at true scale to illustrate concepts of quasicontinental and quasioceanic types of transitional crust along a rifted continental margin (see text for discussion). Either type of transitional crust may be dominant to the near exclusion of the other type in specific cases (*e.g.,* see Lowell and Genik, 1972, figs. 3, 5 for quasicontinental and Moore, 1973a, fig. 10 for quasioceanic).

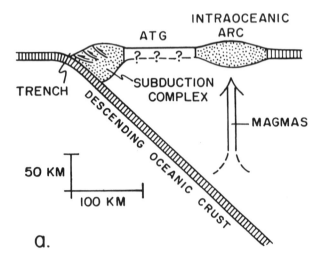

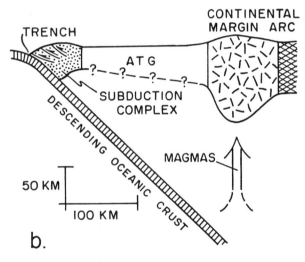

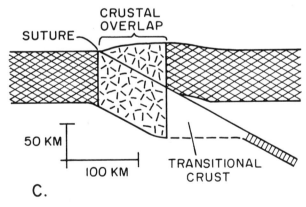

Fig. 2.—Schematic diagram at true scale to illustrate concepts of paraoceanic and paracontinental types of anomalous crust (see text for discussion): *a,* intraoceanic arc-trench system; *b,* continental margin arc-trench system; *c,* suture belt formed by crustal collision when continental block on descending plate encountered subduction zone. Oceanic crust is ruled and continental crust is cross-hatched; paraoceanic crust formed of overthickened oceanic elements is stippled and paracontinental crust formed of overthickened continental elements is jackstrawed. ATG denotes crust of variable and uncertain thickness within arc-trench gap.

others, 1968, fig. 3), whose deep underpinnings are probably oceanic crust, can be increased from the normal oceanic range to a thickness of 12 to 15 km, or perhaps even to 15 to 25 km, as argued by Markhinin (1968) for the Kuriles. Beneath the continental margin arc of the central Andes (James, 1971), the unusually thick crust of nearly 75 km likely also includes major contributions of magmatic rock injected from below into pre-existing continental crust during arc activity.

(2) Oceanic crustal slabs are stacked tectonically in subduction zones to produce thickened crust, or subduction can cause actual overlapping of continental blocks. In California, the subduction complex of the Franciscan assemblage (Ernst, 1970) is a structurally scrambled terrane of oceanic materials (Hamilton, 1969). Included are ophiolitic scraps, deformed seamounts, oceanic pelagites, and turbidite graywackes in mélanges (Hsu, 1968) and thrust slices with a total apparent tectonic thickness of 20 to 30 km (Hamilton and Myers, 1966). No basement rocks of continental character have been detected either within or beneath the complex. In Tibet, the unusual crustal thickness of as much as 75 km has been attributed by Dewey and Burke (1973) to crustal thickening that was effected by essentially doubling up the continental crust. A northern extension of the Indian subcontinent apparently was carried beneath the Tibetan plateau from a subduction zone marked now by the suture belt of ophiolitic mélanges and other deformed oceanic materials along the Indus line between the Himalayas and the Trans-Himalayan ranges.

Changes in crustal thickness within arc-trench systems where lithosphere is consumed thus tend uniformly, although in diverse fashion, to produce thicker crust. This trend fosters isostatic uplift and thus, potentially, the creation of elongate highlands as major sources of sediment. Intraplate crust can also thicken with time—certainly in oceanic regions and possibly in continental ones. In the oceans, the construction of volcanic chains like that of Hawaii on top of previously formed lithosphere can roughly double the thickness of the crust locally. By contrast, continental separations and arc migrations promote crustal attenuation and produce thinner crust associated with newly formed lithosphere. From considerations of the isostatic balance of crust, taken in isolation from other factors, spreading centers are thus generators of sites of potentially thick sedimentation.

Thermal regime.—Plate-tectonic behavior involves convective motions of asthenosphere and lithosphere (Elsasser, 1971), regardless of whether some sort of triggering perturbation of

390

the system is induced primarily by tidal forces governed by astronomic relations (Bostrom, 1971; Knopoff and Leeds, 1972; G. W. Moore, 1973). Convectional overturn of mantle material causes relative uplift and subsidence of the surface of the lithosphere in places whose locations are partly independent of local crustal thickness.

The magnitude of the thermal effect is best understood for the elevation of oceanic crust, which stands at shallow depths beneath active rise crests and at progressively greater depths down the flanks of the rises (Sclater and others, 1971). As the age of the oceanic crust can be inferred from the correlation of magnetic anomalies, rates of subsidence can be estimated empirically. The crests of midoceanic rises have depths of 2.5 to 3 km, but all ocean floors that are roughly 75 my old and lack much sediment cover have a depth of about 5.5 km; oceanic crust of intermediate age stands at intermediate depths related in a regular fashion to age. Rates of subsidence are initially almost 100 m/my but decline with time towards a figure of 10 m/my. The observed subsidence can be explained well simply by the thermal contraction of a cooling lithosphere that is about 100 km thick. The calculations assume that isostatic compensation takes place at the base of the slab of lithosphere where it is in contact with the asthenosphere. Various assumptions for the conductivity and basal temperature of a slab of lithosphere 75 to 100 km thick allow subordinate contributions to crustal elevation from phase changes in the slab and from convective bulging of the asthenosphere beneath the slab.

Thermal tumescence along intracontinental rifts prior to continental breakup, and the succeeding thermal decay following continental separation, also cause major uplift and subsidence of continental basement rocks (Sleep, 1971). An initial thermal uplift of the order of 1 to 2.5 km can be inferred, and is matched well by the observed domal uplift of 1.5 km during the late Tertiary in central Kenya along the East African rift system (Baker and others, 1972). Crustal thinning by erosion of the arched region along an incipient rift may contribute significantly to the net crustal attenuation associated with continental separations (Hsu, 1972). The duration of purely thermal subsidence along a fresh continental margin newly formed by rifting is unlikely to persist for more than 100 my, and for typical continental ruptures the major effects probably occur within 50 my while the continental margin is within 1000 km of the rise crest (fig. 3).

Potential activators of crustal uplift and subsidence traceable to changing thermal regimes

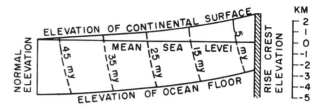

Fig. 3.—Schematic diagram to illustrate subsidence of a rifted continental margin as it moves away from a rise crest. Dashed lines show successive idealized positions of the continental slope at the intervals of time indicated. Width of diagram is about 1000 km for a half-spreading rate of 2 cm/yr; effects of sedimentation are ignored.

include the poorly understood hotspots, whose positions Morgan (1972) ascribed to fixed advective plumes or columns of hot material rising from the deep mantle. If hotspots form bumps on the upper boundary of the asthenosphere, the motions of mobile lithosphere may cause parts of the plates to bob up and down as they cross over the sites of the underlying hotspots (Menard, 1973a). Epeirogenic warping amounting to hundreds of meters vertically, and with wavelengths of the order of 1000 km or more, may be attributable to such a phenomenon. The eventual impact of recent analyses (Burke and others, 1973; Molnar and Atwater, 1973) showing that all supposed hotspots cannot be fixed in position relative to one another is not yet clear. In discussing midplate tectonics, Turcotte and Oxburgh (1973) have offered alternative explanations for the origin of the linear island and seamount chains whose relations the concept of hotspots purports to explain. With fixed hotspots, the unidirectional extension of these volcanic chains is interpreted as a result of the passage of plates of lithosphere over fixed hotspots below. However, the same general effects can be achieved in theory by postulating the development of propagating cracks in the lithosphere induced by thermal stresses from the cooling of slabs of lithosphere and by membrane stresses from changes in the radii of curvature of spherical caps of lithosphere as they change latitude on the globe. Regardless of how the hotspot controversy is resolved, the possibility of epeirogenic warping of lithosphere in irregular patterns as plates pass over a bumpy asthenosphere remains open (Menard, 1973b), unless the boundary between lithosphere and asthenosphere is assumed arbitrarily to be uniformly smooth.

Isostatic balance.—In the past, isostatic reasoning commonly has been applied by assuming the base of the crust at M to be the level of compensation. To the extent that slabs of litho-

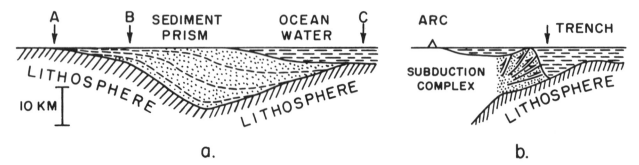

Fig. 4.—Schematic diagrams to illustrate subsidence of substratum by flexural bending of lithosphere under surficial loading: *a*, downbowing of continental margin owing to load of sediment prism deposited offshore (after Walcott, 1972, fig. 2) where A is inland line of flexure and B is initial edge of continental block before marginal subsidence (note that substratum at B is depressed to a depth at roughly the same level as that of normal ocean floor at C); *b*, hypothetical downbowing of ocean floor offshore from load of subduction complex stacked tectonically in subduction zone associated with trench beneath which oceanic substratum is depressed prior to final descent into the mantle (after Hamilton, 1973, fig. 1 and note added in proof); vertical exaggeration 10 ×.

sphere are literally rigid, the base of the lithosphere is a more appropriate level of compensation to choose. In reality, perhaps, given the limited strength of rock masses, assumption of partial compensation at both those levels, and likely at others as well, may prove the most useful stance to adopt. In any case, our past tendency to think of isostasy in terms of crustal balance alone is invalid.

If we may speak, nevertheless, of crustal isostasy in isolation from the broader context, then depressions associated with plate consumption and subduction zones are anisostatic, in the sense that the elevation of the top of the crust is not there a function of crustal thickness and density alone. Instead, the overall motion of a descending slab of lithosphere compels the crustal elements in the top tier of the slab to follow downward. Where oceanic lithosphere is consumed, the trenches are 2.5 to 5 km deeper than the floors of the adjacent oceanic basins; despite this marked difference in elevation, the crustal profile beneath the oceanward slope of the trench is demonstrably the same as in the open ocean. Beneath the axis of the trench, the ponding of turbidites may even increase the thickness of the crust there, although in cases where this effect is dominant a bathymetric trench may not be present as an expression of subduction.

McKenzie (1969) has argued convincingly that the presence of continental blocks prevents plate consumption by simple gravimetric resistance to plate descent. Even so, the attempted subduction of a continental margin, or of continental lithosphere generally, though arrested at a stage short of actual plate consumption, may be able to accomplish appreciable anisostatic subsidence, as that term was used above with reference to oceanic trenches. The local subsidence of a continental surface standing initially near sea level to nearly oceanic depths along a linear belt seems conceivable if the excess depth of 2.5 to 5 km noted for trenches can be extrapolated to this roughly analogous setting. Even if conditions of deep water were never attained, a linear belt of thick sediments deposited on subsiding continental basement might develop as a result of partial subduction. Such a region would presumably appear in the geologic record as a mobile pericratonic fringe bordering an otherwise stable craton.

Walcott (1972) has shown also that flexural bending of lithosphere under sedimentary loading of oceanic and transitional crust just offshore from a rifted continental margin can cause marked subsidence of continental basement along the adjacent edge of the continental block (fig. 4a). As sediments accumulate off the continental margin, the isostatically compensated lithosphere sags downward, and the upper surface of the continental block tilts seaward. The depressed belt along the continental margin may be 200 km wide inward from the initial continental edge to the line of no vertical displacement within the continental block, and the substratum at the initial edge of the continent can become buried beneath as much as 4 km of sediment deposited as an elongate wedge thickening seaward within the depressed belt. Landward of the depressed belt is a gentle linear upwarp (not shown on fig. 4a) parallel to the axis of the linear sag in the lithosphere offshore.

Elevation changes related to crudely analogous flexures of the lithosphere may occur around arc-trench systems in response to tectonic loading represented by the buildup of tectonically stacked subduction complexes. Hamilton (1973) argues that the weight of a seaward-thinning wedge of mélanges forming the subduction complex bows down the descending plate of lithosphere oceanward of the subduction zone (fig. 4b). This action would tend to reduce the angle of plate descent near the surface, as accretionary lateral growth of the welt of mélanges covered the initial site of plate consumption, and perhaps to increase the depth of the trench as a result of the sag developed in the lithosphere.

A complication of uncertain significance colors all detailed considerations of the behavior of lithosphere. Several kinds of data, and especially those on terrestrial heat flow, suggest that continental lithosphere is thicker than oceanic lithosphere and perhaps as much as twice as thick (Scalater and Francheteau, 1970). If so, important questions about the origin of continental lithosphere and about the motions of plates of lithosphere over the asthenosphere are raised. In any case, rigorous treatment of the isostatic balance of lithosphere over asthenosphere clearly cannot be attempted until possible variations in the thickness of the lithosphere are better known.

ANCILLARY EFFECTS

The sedimentary record is only partly a result of paleotectonic conditions suitable for sedimentation. Paleogeographic relations govern to a large extent the nature of the sediment accumulated in a given place at a given time. The factors that govern geographic variations with time are largely ancillary side effects of plate tectonics. The main influences of this kind are related to changes in latitude, changing patterns of geography, and eustatic changes.

Research on paleomagnetism (McElhinny, 1973) indicates that drifting continents have changed latitude drastically during the course of geologic history. Unless one supposes a wholly uniform climate from equator to pole at times in the past, this conclusion implies that each continental block has moved through fundamentally different climatic belts during its history. In general, reconstructed paleolatitudes also cross each continental block in different directions for different times. It follows that an adequate analysis of any sedimentary basin must include the recovery of the paleolatitudes of the basin for the times of interest. Where long periods of time are represented by the sedimentary sequence, a graph showing the trend of changing paleolatitudes for the center of the basin, or for the ends of an elongate basin, may well prove to be essential for a full interpretation of sedimentation.

Changing geographic patterns arising from continental drift may exert important influences on the distribution of potential sediment sources. Patterns of oceanic circulation should also be affected, as well as patterns of atmospheric circulation related to rain shadows and other important effects. Unfortunately, a full assessment of these types of influences on sedimentation in a particular basin at specific times in the past must await the development of a sequential atlas of paleogeography on a globally integrated basis. For much of the Phanerozoic our knowledge is still inadequate to shape this goal even with respect to the positions of all the parts of the present continental blocks. We may never be able to reconstruct well the configurations of the floors of vanished ocean basins, which may have harbored rises for which no clear evidence remains.

Eustatic changes in sea level stemming from ice storage in polar regions are probably modulated, in the long view, by the movement of continental blocks into and out of positions where they can support large glaciers; distributions of other continents so as to block latitudinal circulation in the oceans probably also favor extensive glaciation on the continents located at high latitudes (Crowell and Frakes, 1970). The remarkable display of cyclic sedimentation in the late Paleozoic sequences deposited at low paleolatitudes on coastal plains and in epeiric seas of North America and Europe can be ascribed tentatively to fluctuations of the glaciation at high paleolatitudes in Gondwana (e.g., Wanless and Shepard, 1936).

The glacial explanation of eustatic effects relies upon changing volumes of ocean water coupled with constant volume of the ocean basins. In recent years, various authors (*e.g.*, Valentine and Moores, 1972) have speculated that changes in the globally integrated spreading rate along midoceanic rises can cause eustatic effects by changing the volume of the ocean basins while the volume of ocean water remains constant. The root of the idea rests upon the principle that oceanic crust subsides with age; therefore, if the mean age of the oceanic crust changes with time, the mean depth will also vary. Evaluation of the effect is difficult (e.g., Johnson, 1971), both because we lack sufficient data to estimate globally integrated past spreading rates closely on an areal basis and because the available worldwide data on areal flooding of continental blocks at specific times in the past remains partly equivocal (but see the paper by Sloss and Speed in this volume for a fresh synthesis and a unique interpretation). The whole question of continental freeboard through time is discussed provocatively by Wise (1972).

BASIN TERMINOLOGY

Any field of human inquiry requires a sort of code of simple words or phrases to denote complex concepts. Without this aid to brevity all communication becomes too tedious to pursue. When the underlying framework of concepts changes, the established code begins to lose meaning. Something of this sort has occurred in the past few years as the geosynclinal theory for sedimentary basins and orogeny has given way to plate-tectonic theory. Although the rocks, which are the substance of the geologic record that we discuss, remain the same, the way that we view their evidence has changed. During the present period of transition in concepts, there are two fundamentally opposed ways to proceed with description. One course is the adoption of a wholly new terminology for sedimentary basins. The other course is the adaption of the old terminology to reflect the new concepts. In practice, the most likely path of thinking is a middle course that blends the two approaches by borrowing from the old where convenient and inventing the new where necessary. In practice, also, a quick consensus of views is unlikely, for many thoughtful workers will offer conflicting terminological schemes, each with its own flavor, strengths, and weaknesses (e.g., Mitchell and Reading, 1969; Dewey and Bird, 1970; Dickinson, 1971a).

The most challenging obstacle to a clean translation from geosynclinal to plate-tectonic terminology stems from the two meanings of the word *basin* in geological science. In one sense a basin is merely a bathymetric or topographic depression, but in a more significant sense a basin is the prism of rock forming a thick sedimentary succession. Various types of bathymetric basins can be related readily to the current global pattern of plate tectonics, but sedimentary basins can be related to plate tectonics only by deductive reasoning that postulates the historical dimensions of plate interactions. On the other hand, the existence of sedimentary basins is the starting point for geosynclinal theory, and their relationship to bathymetric basins must be inferred from inductive reasoning.

The dominant theme of geosynclinal theory is the recognition of parallel and adjacent miogeosynclinal and eugeosynclinal belts (Kay, 1951). Although several other kinds of geosynclines have been named, their designations arise as extensions of terminology to encompass sequences whose history does not conform to the ruling concepts for the recognition of these two basic stratigraphic elements of orogenic belts. In general, miogeosynclinal terranes are characterized by clearcut depositional contacts with continental basement and by a lack or paucity of turbidites and interstratified volcanic rocks. By contrast, eugeosynclinal terranes are characterized by equivocal contact relationships with continental basement and by an abundance of turbidites and volcanic rocks within the sedimentary sequence. As a first approximation, the former can thus be interpreted as thick accumulations of strata on the margins of continental blocks and the latter, as strata formed somewhere within an adjacent oceanic basin including its island arcs. This loose approach to the translation of geosynclinal terminology into terms compatible with plate tectonics is not entirely satisfactory. It does not allow for the considerable sophistication of geosynclinal theory in full flower, and results in the unnecessary lumping of things that the full geosynclinal terminology accords different status. Nor does it meet the need to relate various types of sedimentary basins to different kinds of plate interactions, rather than just to the two main kinds of substratum.

PLATE-TECTONIC SETTINGS

In terms of plate tectonics, the settings of basins can be described with reference to three fundamental factors: (1) the type of crust and lithosphere that serves as substratum for the basin; (2) the proximity of the basin to a plate margin, and (3) the type of plate juncture or junctures nearest to the basin.

Types of substratum.—In terms of immediate substratum, normal *continental* crust and standard *oceanic* crust are clearly two end members. For the *transitional* crust discussed in an earlier section, the term *quasicontinental* is applied here to the type characterized by attenuated continental basement, and the term *quasioceanic* is applied to the type characterized by an overthickened profile of plainly oceanic elements including both igneous and sedimentary materials (see fig. 1). The terms *paracontinental* and *paraoceanic* are applied, respectively, to *anomalous* crust of previously continental or quasicontinental, and previously oceanic or quasioceanic, character where crustal thickening has occurred by the addition of igneous materials in magmatic arcs and hotspot chains or through processes of tectonic stacking related to subduction zones (see fig. 2). There is inherent ambiguity between arc-related and subduction-related subtypes of paracontinental and paraoceanic anomalous crust because both magmatic arcs and subduction zones can migrate with respect to the intervening sliver of lithosphere (Dickinson, 1973). The different categories of crust are largely conceptual at present, for the operational means to distinguish between them with geophysical observations remain elusive for the most part. Moreover, the terms are unlikely to suffice as a full catalogue of significantly different types of crust and associated lithosphere. They would prove to be especially inadequate, and perhaps misleading to some extent, if it develops that significant exchange of substance can occur between crust and mantle, or between lithosphere and asthenosphere, in settings that lie within intact plates.

Proximity to junctures.—The degree of proximity of a basin to a plate margin must be understood in relative rather than absolute terms. The point is the extent to which tectonic effects related directly to plate interactions influence the setting of the basin. The thermal decay of the lithosphere as it moves away from a spreading center is one such effect, which will be confined within a certain distance from the rise crest depending upon the spreading rate. Similarly, the locus of arc magmatism along a certain trend parallel to the associated trench will occur at a distance that depends upon such parameters as the rate of plate consumption and the dip of the inclined seismic zone. In broad terms, basin settings can thus be divided into *intraplate* settings as opposed to *zones of plate interaction.*

Plate junctures.—There are three basic varieties of plate junctures: (1) *divergent,* where old lithosphere separates at spreading centers, and new lithosphere is built along midoceanic rise crests to fill the opening gap by accretion of material to the retreating edges of the separating plates; (2) *convergent,* where plate consumption carries old lithosphere downward into the asthenosphere along inclined seismic zones, and the processes that operate in the subduction zones and magmatic arcs modify the lithosphere of the overriding plate by adding both magmatic and tectonic increments to its profile; and (3) *transform,* where two plates slide past one another along a lateral fault zone without either forming new lithosphere or destroying old lithosphere.

The three kinds of junctures are geometric end members and are analogous, in terms of the geometry of strain indicated, to the three familiar classes of faults: normal (extensional), reverse-thrust (contractional), and strike-slip (lateral). There are variants of the three types of plate junctures where motions oblique to the trends of the junctures occur (Dickinson, 1972). Obliquity of convergence, as indicated independently by slip vectors determined for earthquakes and by calculations of relative plate motions from correlations of magnetic anomalies at sea, is common along modern subduction zones. Obliquity of divergence, however, is commonly resolved into a rectilinear lattice of rise crests and connecting transforms for apparently mechanical reasons (Lachenbruch and Thompson, 1972). Along transforms where the relative motion of the plates has a component of divergence, the same mechanical tendency evidently fosters a similar rectilinear lattice of transform segments connected by short rise segments. Along transforms where the relative motion of the plates has a component of convergence, the combined effect has been called transpression (Harland, 1971).

Some incipient plate junctures may become inactive before developing the full characteristics of their class. For example, an aborted divergent juncture within a continent might never undergo sufficient plate separation to develop a fully oceanic crust and lithosphere. A sedimentary basin with an unexposed floor of transitional crust might well form above the site of such a feature. Although clearly intracontinental in its setting, such a basin can be described as *infracontinental* as opposed to *supracontinental* basins deposited on a full complement of continental basement. Similarly, plate consumption might be arrested at some stage of partial subduction before a magmatic arc was developed. As the confident application of plate-tectonic logic depends upon the recognition of the full display of geologic features characteristic of each type of plate juncture, such partially developed junctures are apt to foster ambiguous interpretations.

Combined parameters.—Considering jointly the parameters of crustal substratum, proximity to plate interaction, and type of plate juncture, the gross settings of sedimentary basins can be grouped into a hierarchy consistent with plate tectonics. Initially, intraplate settings are contrasted with settings within zones of plate interaction, which include zones of divergent, convergent, and transform motion. For each of the four broad classes of plate settings so derived, the nature of the crustal substratum may vary, and subclasses of the settings can be recognized on the basis of the variations:

(a) For intraplate settings, the substratum need not be normal continental or standard oceanic in nature, for transitional or anomalous crust inherited from extinct plate junctures can be present.

(b) Zones of divergence include both intra-continental and intraoceanic types, although the two commonly are merely sequential stages in the evolution of a single plate juncture responsible also for the formation of transitional crust during intermediate stages of its evolution.

(c) Zones of convergence include types, or phases of development, in which either oceanic or continental (or transitional) crust is drawn into the subduction zone, and in parallel also include types in which the anomalous crust of the arc-trench system develops from crust of either oceanic or continental (or transitional) character initially; consequently, arc-trench systems embrace multiple settings with diverse overall relations.

(d) Transform zones include three basic types in which the two plates sliding past one another are both oceanic, both continental, or one continental and one oceanic, but transitional or anomalous crustal blocks may also be involved; moreover, each of the three basic types includes two variants where some auxiliary motion is either divergent or convergent.

The most important classes of geosynclines represent the net development of sedimentary successions over spans of time long enough for the plate-tectonic settings of the sites of deposition to change. Typical steps in the evolution of different classes of geosynclines thus apparently represent characteristic patterns in the evolution of plate interactions and the consequent changes in plate-tectonic setting. Examples of apparently anomalous geosynclinal evolution then represent the results of less common sequences of plate interactions with the different changes in plate-tectonic setting implied thereby.

BASIN EVOLUTION

Geosynclinal theory is forced by its inductive basis to address the analysis of basin evolution in terms of type examples. Where deviations from supposed norms of evolutionary trends occur, the theory is unable to offer clear insights. Plate tectonics, by contrast, approaches the problem of basin evolution in terms of alternate sequential patterns of plate interactions. Given the overall framework of varied plate motions, deductive reasoning from the theory has the potential to shed some insight on quite unfamiliar evolutionary trends, as well as to explain in coherent fashion a range of events that might issue in different circumstances from any particular stage of the evolutionary development of a given type of basin. By treating basin evolution as a function of plate motions and interactions, plate tectonics thus broadens the scope of theoretically controlled analysis and reduces the need for wholly intuitive suggestions.

Plate tectonics as a framework of thinking thus precludes the possibility of a neat catalogue of basin types. Each basin, seen in a developmental perspective of space and time, to some extent partakes of a unique flavor in principle. The constants in the equation of basin evolution are the types of plate interactions and settings, but the order in which they may be arranged in space and time is variable within wide limits. Geosynclinal theory, by contrast, assumes the overall trend of development as the standard, and views the incremental events that occur during evolution as variable within wide limits. In an analogical but very real sense, geosynclines as conceived by classic theory thus have an ontogeny, or life history, driven by tendencies akin in their effects to a vital force. Plate tectonics casts a more prosaic light on sedimentary basins, but allows naturally for more shadings of behavior without the need to modify any of its fundamental tenets.

From the standpoint of plate tectonics, the evolution of sedimentary basins is incidental to the formation and consumption of lithosphere. The major perturbations of a stable and level earth's surface are related to the opening of oceanic basins accompanied by the rifting and fragmentation of continental blocks, and to the closing of oceanic basins accompanied by the collision and assembly of continental blocks. The principal trends of basin evolution can thus be described as they pertain to the following realms of interplay between tectonics and sedimentation: (a) *oceanic basins* underlain by oceanic lithosphere, (b) *rifted continental margins* along the transitional interface between oceanic and continental lithosphere, (c) *arc-trench systems* where oceanic lithosphere is consumed beneath island arcs or continental margins, (d) *suture belts* where continental blocks are juxtaposed by crustal collision, and (e) *intracontinental* basins in the interior of continental blocks. For none of these realms of interplay should one assume invariant modes of evolution, but a discussion of each in order does afford the means to indicate the salient relationships of plate tectonics to basin evolution.

Oceanic Basins

Depending upon their size, the distribution of divergent plate junctures within them, and the distribution of convergent plate junctures within or around them, oceanic basins may lie at any given time either wholly within zones of plate interaction or in wholly intraplate settings; most typically, they lie partly within zones of plate interaction but otherwise in an intraplate setting. Each segment of oceanic crust and lithosphere typically experiences the following succession of plate-tectonic settings in order: (1) the zone of plate interaction along a divergent plate juncture where the oceanic substratum is formed; (2) the intraplate setting of a deep oceanic basin; and (3) the zone of plate interaction along a convergent plate juncture where the bulk of the oceanic lithosphere is consumed while variable and uncertain proportions of the oceanic crustal elements are caught up in the subduction zone. During either the initial or final phases of evolution in zones of plate interaction, selected segments of the oceanic crust may be subjected also to deformation along transform plate junctures associated with the divergent or convergent plate junctures.

Ignoring features related to rifted continental margins and to arc-trench systems, the principal settings of oceanic facies controlled by tectonic relations are the following (fig. 5): (a) rise crests where the layered igneous succession of ophiolite sequences (Vine and Moores, 1972) are formed along the trends of the spreading centers, (b) rise flanks where the oceanic substratum gradually subsides as it cools in moving away from spreading centers, and (c) deep basins beneath which the thermal contraction of lithosphere is essentially complete. This gross picture must be modified to allow for special conditions of shearing along active transforms near rise crests and for sharp topographic contrasts across and along fracture zones that mark the inactive extensions of transforms down the rise flanks. The outline of settings may also be inappropriate in detail for the oceanic basins generated by spreading centers within marginal seas or interarc basins behind migrating island arcs.

Nevertheless, the ideal triad of rise crest, rise flank, and deep basin serves to emphasize the main characteristic trends of evolution for an oceanic basin (see fig. 5). The igneous components of the ophiolite sequence formed at the spreading center are the first of a series of diachronous facies typical for oceanic sequences. The pelagic sediment that covers the igneous portion of the ophiolite sequence has a stratigraphy with facies relationships that reflect changing water depths (Berger, 1973). While the rise crest and flanks are above the carbonate compensation depth, calcareous sediment is deposited, but is succeeded by siliceous sediment deposited lower on the rise flanks and in the

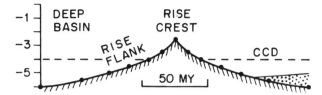

FIG. 5.—Sketch to illustrate principal settings of oceanic facies on transverse profile of typical mid-oceanic rise (depths after Sclater and others, 1971). Vertical scale is in km of water depth, but horizontal scale is in my because lateral dimensions of a mid-oceanic rise are dependent upon the spreading rate. CCD (dashed line) is typical level of carbonate compensation depth (Berger, 1973). Abyssal plain of turbidites indicated schematically, without showing isostatic compensation of substratum, by stippled area on right.

deep basins. In basins that tap turbidity currents from landmasses, the pelagic layers are covered eventually by turbidites of abyssal plains.

Where the oceanic basin changes latitude during its history or otherwise encounters different oceanic provinces, complexities are introduced into the diachronous sedimentary succession (Frakes and Kemp, 1972; Heezen and others, 1973). For example, owing to the high equatorial productivity of calcareous plankton, the carbonate compensation depth is lowered significantly below the lysocline in a narrow belt along the equator. This phenomenon has potential consequences for an oceanic sedimentary sequence formed on one side of the equator as a doublet of rise-crest calcareous pelagites overlain by siliceous pelagites reflecting later deposition in deeper water. If the segment of the oceanic basin bearing this doublet then crosses the equator, its transit may be marked by the deposition of a layer of equatorial calcareous pelagites. After the segment of the basin has moved away from the equator into the other hemisphere, it will then carry two calcareous-siliceous doublets. The two successive calcareous horizons, each overlain gradationally by siliceous sediment, record the times of positioning at the rise crest and equatorial transit, respectively. Whether details of this kind can ever be read clearly from the deformed oceanic facies of orogenic belts is a moot question at present, but avenues for inquiry are surely open.

A special set of oceanic facies is associated with islands and seamounts built as isolated mounds or in linear chains across oceanic regions. The thick volcanic piles themselves may be capped by reefoid sediment and flanked by archipelagic aprons of volcaniclastic turbidites derived locally. In certain instances, widespread and thick carbonate platforms like those in the

Bahamas may also be built within oceanic regions, probably on the quasioceanic crust of marginal fracture ridges (see below) or the submarine ridges of hotspot-generated island-seamount chains.

One of the most remarkable corollaries of plate-tectonic theory is the inference that all the old oceanic crust—igneous rocks and sediments alike older than the present ocean floors—has been placed into one of three non-oceanic repositories: (1) the mantle, into which crustal materials capable of pressure-induced inversion to suitably dense phases could be swept together with the bulk of the plates of lithosphere consumed through time by descent into or through the asthenosphere; (2) subduction complexes, into which crustal materials could be scraped from the tops of descending plates and thus welded by accretion to the flanks of continental and island-arc crust; or (3) magmatic arc structures, into the roots of which crustal materials melted from the upper levels of descending plates could be fed from below. Given the ages currently estimated for the present ocean floors, this inference means that the presumably immense bulk of all the turbidites in all the subsea fans and abyssal plains of all pre-Jurassic oceanic basins have met one of those fates, of which the second seems the most likely at present.

Rifted Continental Margins

Rifted continental margins form in pairs when continental separations occur along divergent plate junctures, and form singly when magmatic arc structures are rifted away from the margins of continental blocks by spreading behind the arcs. In the former case of simple continental separation, each rifted continental margin presents the juxtaposition of a high-standing continental block with sediment sources against a newly formed oceanic basin to serve as a sediment sink. The resulting sedimentation forms a characteristic sedimentary prism spanning the interface between continental and oceanic crust. Different components of the prism, here called *rifted-margin prism,* rest on continental, transitional, and oceanic crust. The prism thus contains strata of both miogeosynclinal and eugeosynclinal affinities. The nearshore assemblage of mainly paralic and shelf facies resting on continental basement has been aptly termed the miogeocline (Dietz and Holden, 1967) in recognition of the fact that these strata form, in transverse section, a wedge thickening seaward toward the shelf edge in existence at the time of deposition. Similarly, the offshore assemblage of turbidites and other deposits in deep water near the foot of the continental slope can be termed the eugeocline in analogous recognition of the asymmetric form of the thick accumulation, whose site is controlled by the position of the continental margin. However, as these latter deposits may grade imperceptibly into those of the broad oceanic basin nearby, the designation eugeocline is commonly less useful in practice than the term miogeocline.

The rifted-margin prism, when completed, includes a number of distinctive sedimentary phases within a complex assemblage of deposits. Each of the phases reflects either deposition in a particular plate-tectonic setting during the time when the rifted continental margin still lay within the zone of plate interaction along a divergent plate juncture, or else deposition during a particular stage in the growth of the prism during the time when the rifted margin was later in an intraplate setting. Variations arise within the total sedimentary assemblage as rates of spreading during different continental separations vary in relation to rates of sediment delivery to the rifted margins. In principle, the process of accumulation of a rifted-margin prism can also be terminated during any given phase of sedimentation by orogeny. Such orogeny may be related either to the activation of an arc-trench system along the continental margin, beneath which the offshore oceanic lithosphere thus begins to be consumed, or to crustal collision with an arc-trench system that approaches the continental margin by consuming the intervening oceanic lithosphere offshore (Dickinson, 1971b). By assuming that the sedimentation of a rifted-margin prism can be arrested at any stage in the growth of its successive depositional phases by several kinds of orogeny, a broad spectrum of individual geosynclinal developments can be accommodated within the same conceptual framework of plate tectonics. Important complications in the succession of depositional phases within rifted-margin prisms are introduced also by the presence locally of marginal offsets of the continental blocks involved and by the failed or aborted arms of triple junctions distributed along the trend of a rift belt. The marginal offsets may give rise to marginal fracture ridges and the triple junctions, to aulacogens.

The series of plate-tectonic settings that mark the successive stages of the evolution of rifted continental margins can be denoted loosely by the following five terms: pre-rift arch, rift valley, proto-oceanic gulf, narrow ocean, and open ocean (see also Schneider, 1972). The five gradational stages of structural evolution are

398

associated with depositional phases whose strata are intercalated locally as contemporaneous facies. The successive phases of deposition may form markedly diachronous facies along any rifted-margin prism, for the geometry of plate tectonics requires most continental separations to proceed as wedge-like openings, rather than as instantaneous separations along the whole length of given rift belts (Dickinson, 1972).

Pre-rift arch.—During the thermal arching that precedes and accompanies incipient rifting, peralkaline volcanism is characteristic (fig. 6a). This activity is apparently not uniform along the rift belt but is concentrated near the crests of broad domal uplifts, from 250 to 1250 km across, that are spaced like beads with centers at intervals of roughly 1000 to 2000 km along the trend of the rift belt (LeBas, 1971). The balance between the rate of accumulation of such volcanics and the rate of erosion of the thermal arches that they crown is uncertain, but relations in Africa and South America adjacent to the South Atlantic suggest that erosion of the thermal arch is the dominant effect areally. In that region, uplifted terranes of Precambrian basement are prominent along both coasts between the ocean and extensive inland basins in which Paleozoic and Mesozoic strata are preserved on both continents (Burke and others, 1971).

Rift valley.—When sufficient crustal extension affects the arched region, rift valleys begin to form as grabens and half-grabens (fig. 6b). Probably these develop first within the domal uplifts, but later they extend as an essentially continuous branching network along the full trend of the rift belt. In the rift valleys, continental redbeds are intercalated with volcanics that continue to erupt through the growing system of crustal fractures (Scrutton, 1973). Broad regions to either side of the eventual zone of rupture between the separating continents apparently can be scarred by extensional faulting during this time. Large-scale extensional faulting has offset continental basement rocks across a belt 100 to 250 km wide west of the axial rift

zone in the modern Red Sea (Hutchinson and Engels, 1972), and the Triassic basins of the Appalachian region lie as much as 250 to 500 km inland from the present continental slope, which can be taken as marking roughly the line of Jurassic continental separation.

Proto-oceanic gulf.—As continued crustal distension induces subsidence along the zone of incipient continental rupture despite continued thermal effects, the floors of the main rift valleys become partially or intermittently flooded to form proto-oceanic gulfs. Restricted conditions in these basins, which are probably still rimmed by uplifts that block delivery of clastic sediment, promote the deposition of evaporites in suitable climates (fig. 6c). Immense thicknesses, as much as 5 to 7.5 km, of evaporites are present in the subsurface beneath parts of the Red Sea (Lowell and Genik, 1972; Hutchinson and Engels, 1972). Buried salt layers that feed extensive diapir fields are present off many North Atlantic coasts (Pautot and others, 1970). Extensive evaporites are known also from coastal basins on both sides of the South Atlantic, where they are apparently correlative and represent dismembered portions of the same elongate and initially continuous evaporite basin (Reyment, 1972). The proto-oceanic evaporites are presumably deposited mainly on transitional crust, probably in most instances of the quasicontinental variety representing attenuated continental basement. Deposition on oceanic crust, or as part of the sedimentary component of quasioceanic crust, could conceivably occur if thermal uplift along the rift belt were sufficiently pronounced during and just after full continental rupture.

Narrow ocean.—Once new oceanic crust and lithosphere begin to form along the belt between two separated continental blocks, fully oceanic conditions are attained in the structural sense (fig. 7). In the geographic sense, however, a distinction can be drawn between narrow oceans and open oceans. In narrow oceans, sediment delivery to a given oceanic site from both bounding continental blocks could con-

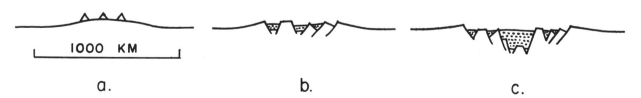

a.   b.   c.

FIG. 6.—Sketches to illustrate successive pre-oceanic stages in evolution of rifted continental margins in cross-section (vertical exaggeration 25× except on dips of faults): *a*, pre-rift thermal arch shown about 750 km across with idealized volcanoes capping it; *b*, rift valley system with terrestrial sediments ponded locally across a belt about 500 km wide; *c*, proto-oceanic gulf with thick saline deposits shown within a belt about 250 km wide. Stipples indicate sediment.

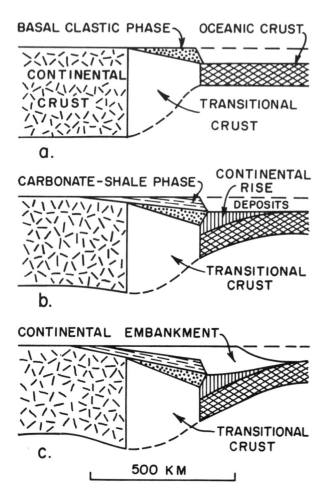

BASAL CLASTIC PHASE — OCEANIC CRUST

CONTINENTAL CRUST

TRANSITIONAL CRUST

a.

CARBONATE-SHALE PHASE — CONTINENTAL RISE

DEPOSITS

TRANSITIONAL CRUST

b.

CONTINENTAL EMBANKMENT

TRANSITIONAL CRUST

c.

500 KM

Fig. 7.—Idealized diagrams to illustrate successive depositional phases in evolution of rifted-margin prism along continent-ocean interface with sea level shown as dashed line (vertical exaggeration 10×): *a*, basal clastic phase of miogeocline deposited during thermal subsidence of transitional crust, within which earlier deposits of rift-valley redbeds, proto-oceanic evaporites, rift lavas, etc. are not differentiated; *b*, carbonate-shale phase of miogeocline deposited during shelf-slope-rise configuration of rifted continental margin; *c*, progradational continental embankment. See text for discussion.

ceivably occur, although the tendency of the spreading center to form an elevated midoceanic rise would tend to divide the oceanic basin into two halves with different sediment sources for beds deposited from bottom-hugging turbidity currents. If turbidity currents could reach or cross the actual spreading center, the net effect would be to continue forming transitional crust of the quasioceanic type.

More important in a narrow ocean is the fact that the transitional crust along and adjacent to the attenuated continental margins would continue to subside thermally as a closing stage of the plate interaction that formed the rifted continental margins. This period of subsidence probably persists along a rifted continental

margin for perhaps 50 to 75 my, during which it proceeds independently of sedimentary loading. Its influence would tend to eliminate the uplifted belts that previously acted to bar sediment delivery to the rifted continental margin. This early thermal subsidence of the substratum beneath the belt of quasicontinental transitional crust is probably the factor that induces rather rapid accumulation of thick clastics as a basal phase of typical miogeoclines (fig. 7a). Such strata form a basal wedge, thickening seaward, as the oldest areally continuous deposits in the outer or oceanward parts of both the Appalachian and Cordilleran miogeoclines (King, 1969, p. 11-12; Stewart, 1972), in both of which the basal clastic phase is latest Precambrian and earlier Cambrian in age.

Open ocean.—When the strictly thermal subsidence of a rifted continental margin is complete, the margin is left in an intraplate setting and facing an open ocean. At this point, the drowned belt of transitional crust along the margin is already complex geologically. The attenuated continental basement rocks are faulted and covered locally by continental clastics, volcanics, and evaporites concentrated in varying degree within downfaulted blocks. Across this compound substratum, the basal clastic phase deposited during and closely following the main thermal subsidence is draped as a wedge of marine and paralic strata built upward to form an isostatically balanced continental terrace in the initial configuration of that feature. From this point onward, any further subsidence apparently is the result of sedimentary loading of crust offshore from the shelf break at the edge of the continental terrace.

The continued evolution of the rifted margin prism can be described using the terminology of Dietz (1963) for continental terrace, slope, rise, and embankment. The continental terrace, upon which shelf and paralic sedimentation dominates, extends to the slope break at the shelf edge, from which the continental slope leads down to deep water where the continental rise of turbidites accumulates along the edge of oceanic crust (fig. 7b). Bending of the lithosphere caused by the loading of the continental rise (Walcott, 1972) causes the continental terrace to tilt progressively seaward. This process enables the continental terrace to receive successive wedges of strata that thicken rather uniformly from a nearly common landward hinge zone toward the shelf edge (*e.g.*, Rona, 1973). The flexure at the hinge zone lies perhaps 100 to 250 km from the shelf edge. As subsidence of this kind is not linked directly to sedimentary loading of the continental terrace itself, erosional episodes on the shelf may produce disconformities within the shallow marine and shoreline deposits of the accumulating terrace wedge. These miogeoclinal strata, deposited more slowly than the underlying basal clastic phase, are probably represented by the succeeding carbonate-shale phase (*e.g.,* King, 1969) of the lower Paleozoic section in the Appalachian and Cordilleran miogeoclines. For the Appalachians, a carbonate platform marginal to the continent is recognized clearly for this interval (Rodgers, 1968), and similar strata appear in the Cordilleran case (Armstrong, 1968b).

The continental slope beyond the shelf break is largely a region of sediment bypass that serves for the transit of turbidity currents headed from shallower water toward the continental rise. Sediment thicknesses beneath the terrace-slope-rise association thus give an hourglass effect in section, with the pinched region of thin strata lying along the continental slope. Available data suggest that sediment thicknesses beneath the shelf break at the outer edge of the continental terrace, and also those beneath the continental rise, can reach at least 5 km.

If clastic sedimentation along a rifted continental margin is voluminous enough, upward construction of the continental rise and outward construction of the continental terrace lead to the development of a progradational continental embankment (fig. 7c). This type of feature is discussed here as a sequel to the stage of development represented by the terrace-slope-rise association, but appropriate relations among the details of structural development of transitional crust, the rate of thermal subsidence, and the timing and rate of sediment delivery could blur the distinction between the two stages of development in some instances. The continental slope on the front of a continental embankment becomes a constructional feature owing to wholesale progradation of the continental edge. Shelf break and slope thus advance seaward from the region of transitional crust until both reach a position above fully oceanic crust, as Dietz (1963) suggests for the Texas coast. The top of the embankment receives mainly fluvial and paralic sediments while the frontal slope and toe receive mainly turbidites. Immense thicknesses of sediment are possible for continental embankments; at least 12.5 km of sediment are present beneath the Texas coast and Walcott (1972) suggests that thicknesses of 17.5 km could be attained by simple isostatic subsidence of oceanic crust and lithosphere. By analogy with the deep-water Niger delta, the structure of the continental embankment in the region beyond the edge of the pre-existing continental terrace can be inferred to include three main depositional phases (Burke, 1972): a basal phase of sandy turbidites deposited near the toe of the embankment, a middle phase of mainly shaly rocks deposited on the advancing frontal slope of the embankment, and an upper phase of largely sandy paralic strata deposited along the prograding outer edge of the top of the embankment.

Marginal offsets.—On the floor of the modern Atlantic Ocean, the major transforms that offset the crest of the midoceanic rise are extensions of fracture zones whose extremities at the flanks of the ocean appear to coincide with abrupt offsets of the adjacent continental margins (Le Pichon and Hayes, 1971; Le Pichon and Fox, 1971). To some extent, therefore, the gross shape of the rectilinear trellis of rise segments and connecting transforms in the ocean is inherited from the shape of the initial rupture formed by continental separation. The marginal offsets were transform fault zones, rather than extensional rift zones, during continental separation. The edges of the continental blocks along the trends of the marginal offsets thus underwent a different early evolution than the edges that face toward the center of the ocean and are masked now by rifted-margin prisms of the type just discussed (fig. 8). Strike-slip along the marginal offsets during continental separation would disrupt and displace segments of the earlier phases of any sedimentary accumulations that might form along those parts of the continental edges. More important, however, is the fact that the structural character of the transitional crust along the marginal offsets is likely to be different in kind (Francheteau and Le Pichon, 1972).

During continental separation, continental margins along the marginal-offset transforms are swept by the butt ends of incipient midoceanic rise segments (see fig. 8). Although the full thermal and petrologic effects of this process are unclear, the lateral transit of the end of a rise segment along a marginal offset apparently leads to the formation of a distinctive feature termed a marginal fracture ridge (Le Pichon and Hayes, 1971; Le Pichon and Fox, 1971). Where fully developed, marginal fracture ridges extend along the marginal offset itself and at least that far again out to sea along the same trend. They are formed probably in part by crumpling and shearing along the slip zone, and in part by overthickened quasi-oceanic crust formed by exceptional leakage of igneous materials from the regions near the butt

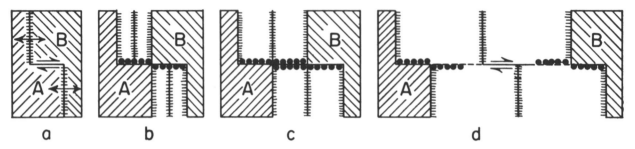

FIG. 8.—Diagram after LePichon and Hayes (1971) showing in plan view the development of marginal fracture ridges during continental separation of continental fragments A and B shown joined prior to separation in *a* and progressively farther apart in *b, c, d*. Barred lines are spreading centers linked by a transform and hachured lines show positions of normal rifted-margin prisms. Heavy dots are marginal fracture ridges along marginal offsets and offshore. Dashed lines in *d* are fracture zones along same trend.

ends of the migrating rise segments. Thick sedimentation along marginal offsets is probably delayed by prolonged thermal uplift, and clastic sedimentation may be inhibited locally by high-standing marginal fracture ridges. However, the marginal offsets are subject to prolonged thermal uplift and marginal fracture ridges may later actually promote abnormally thick sedimentation of biogenic sediments in the oceanic realm along offshore trends in line with marginal offsets. The Bahama platform, an elongate accumulation of 5 km of mainly shallow-water carbonates above perhaps 15 km of quasioceanic crust (Dietz and others, 1970) may reflect such a phenomenon, although other means of generating the quasioceanic crust beneath the carbonate platform have also been suggested.

Aulacogens.—The term aulacogen is applied here in the usage of Hoffman (and others, 1974) as adapted from the Russian literature, in which the term was devised for a class of features initiated mainly in the later Precambrian (Salop and Scheinmann, 1969). Aulacogens are elongate sedimentary basins that extend, as gradually narrowing wedges or pie slices in plan view, from the margins of cratons toward the interiors of cratons. The sedimentary sequences of aulacogens are mainly similar in general nature to facies equivalents in platformal sequences of the cratons adjacent on both sides, but are much thicker. Aulacogens are thought to evolve from semioceanic gashes formed at re-entrants in rifted continental margins during continental separations (fig. 9).

The overall geometry of RRR and RRF triple junctions (McKenzie and Morgan, 1969) is attractive as an explanation for the tectonic setting of aulacogens. Such a triple junction has two or three spreading centers as arms. The Afar region linking the Red Sea, Gulf of Aden, and East African rift systems is a modern example. Burke and others (1971) and Grant (1971) argue that the Benue Trough, which

extends into Africa from the head of the Gulf of Guinea, was temporarily one spreading arm of a triple junction in the Cretaceous. When continuation of motion along the other two arms opened the South Atlantic, the Benue arm was aborted in an incipient stage of development. Cretaceous and younger sediments beneath the trough are more than 5 km thick for at least 500 km along its axis. Their accumulation was probably accommodated by the subsidence of transitional crust beneath the trough. Thermal subsidence following the failure of the Benue spreading arm to continue into a fully oceanic configuration probably served as a trigger to initiate sedimentation, which then forced further isostatic subsidence under sedimentary loading. The location of a long-lived Benue depression also apparently controlled the course of the Niger River and the position of its delta. The delta itself is evidently a local continental embankment containing sediment some 10 km thick built into deep water beyond the initial continental margin off the seaward end of the aulacogen.

In North America, the best example of an aulacogen containing Phanerozoic strata is the Anadarko-Ardmore basin, which extends inland nearly 500 km from the southern margin of the Paleozoic continent. Ham and Wilson (1967) describe the section in the elongate basin as 10 to 12 km of Paleozoic strata overlying at least 2 km of Cambrian volcanics. Late Paleozoic deformation and coarse clastic sedimentation within the basin was greatest toward its open end, and was probably related to the Ouachita orogeny along the nearby continental margin.

Arc-Trench Systems

Arc-trench systems are the characteristic geologic expression of convergent plate junctures (Dickinson, 1970). As recognized plainly by Kay (1951), the volcaniclastic rocks of vol-

402

canic island chains built along magmatic arcs are prominent within many eugeosynclinal terranes. Eugeosynclinal terranes also include the subduction complexes associated with trenches, where oceanic strata are mingled tectonically as they are detached from the tops of slabs of lithosphere descending beneath the flanks of arc-trench systems. Sedimentary sequences that accumulate on the flanks of magmatic arcs, which stand as positive topographic features during arc activity, receive a variety of geosynclinal designations locally, depending on details of their relationships to various types of substratum and also upon the nature of the strata themselves. A full discussion of the evolution of arc-trench systems requires an essential focus on magmatism and metamorphism, but the emphasis here is solely upon the facets of behavior that affect the associated sedimentary basins (Dickinson, 1974).

Arc-trench systems include the following five major morphotectonic elements (*e.g.*, Dickinson, 1973): (1) the *trench,* a bathymetric deep floored by oceanic crust; (2) the *subduction zone* beneath the inner wall of the trench and the trench slope break marking the top of the inner wall; (3) the arc-trench gap, a belt within which a *forearc basin* may occur between the trench slope break and the magmatic arc; (4) the magmatic arc, within which *intra-arc basins* may occur; and (5) the backarc area, within which may lie either an *interarc basin* floored by oceanic crust and separated from the rear of the arc by a normal fault system, or a *retroarc basin* floored by continental basement and separated from the rear of the arc by a thrust fault system.

Sedimentation in the various types of basins noted for the different elements of arc-trench systems is contemporaneous with both volcanism and plutonism along the magmatic arc, and with metamorphism both in the cool subduction zone and in the hot roots of the magmatic arc. Faulting and other deformation in the subduction zone, within the magmatic arc, and in the backarc area is also contemporaneous with sedimentation. Although sequential phases of sedimentation can doubtless be recognized for each of the kinds of sedimentary basins noted in the various morphotectonic settings, the areal contrast in facies among the various kinds of sedimentary sequences forms the most important and regular genetic pattern. This pattern of distinctive and parallel sedimentary terranes, coupled with their igneous and metamorphic associates, can be used as a means to identify the petrotectonic assemblages that form the geologic record of past arc-trench systems.

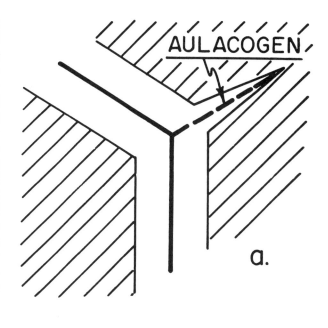

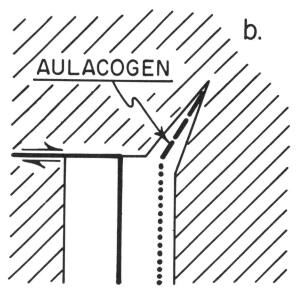

Fig. 9.—Diagrams showing in plan view alternate mechanisms for the development of aulacogens at re-entrants in rifted continental margins (continental blocks shaded, plate junctures that continue active shown as solid heavy lines, failed spreading centers along axis of aulacogen shown as dashed heavy line): *a,* aulacogen as failed arm of formerly stable RRR triple junction (spreading directions changed along the two arms that continued spreading when motion stopped along the failed arm); *b,* aulacogen as failed arm of inherently unstable RRF triple junction (after Grant, 1971).

Thick sedimentation associated with arc-trench systems is best discussed, therefore, in relation to subduction zones, forearc basins, intra-arc basins, interarc basins, and retroarc basins (fig. 10). The progressive development of these features presumably will continue until the plate consumption that fosters an arc-trench

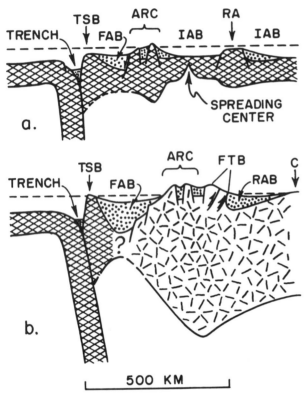

FIG. 10.—Idealized diagrams to illustrate tectonic settings of sedimentary basins (stippled) associated with arc-trench systems. Dashed line is sea level. Vertical exaggeration is 10×; note apparent steep angles of descent of plates beneath trench although true angles depicted are 60 degrees (*a*) and 30 degrees (*b*). Oceanic and paraoceanic crust is crosshatched; continental and paracontinental crust is jackstrawed. Trench slope breaks (TSB) lie above subduction zone complexes at thresholds of arc-trench gaps within which forearc basins (FAB) are shown. For intraoceanic arc (*a*), active or frontal island arc (ARC) is shown with a marine intra-arc basin, and remnant arc (RA) stands between two interarc basins (IAB), one active (left) and one inactive (right) but both with volcaniclastic wedges along one flank. For continental margin arc (*b*), volcanic highlands (ARC) are shown with a terrestrial intra-arc basin, a flanking intermontane lowland, and a foreland fold-thrust belt (FTB) above zone of partial crustal subduction lies between magmatic arc and retroarc basin (RAB) on depressed crust of pericratonic foreland adjacent to margin of interior craton (C).

system is terminated, commonly by crustal collision to form a suture belt.

Subduction zones.—Seaward from the trench in typical arc-trench systems is a broad upwarp of the ocean floor marking the flexure of the lithosphere as it bends to descend beneath the arc-trench system. The inner slope of this outer arch is the gentle outer slope of the trench, and is scarred in places by normal faults reflecting local extensional deformation of the ophiolite sequence represented by the igneous oceanic crust and its sediment cover. The trench is a bathymetric deep immediately adjacent to the

tectonic front of the subduction zone, which begins at the base of the steep inner wall of the trench. On the trench floor, variable thicknesses of turbidites are ponded above the sediment layers rafted tectonically into the trench from the open ocean floor. Transport by turbidity currents within a trench is mainly longitudinal along the trench axis (von Huene, 1972; Marlow and others, 1973), although the initial entry of sediment into the trench may occur at intervals along the inner wall as well as from the ends of the trench (Ross, 1971; J. C. Moore, 1973). Where the rate of sediment delivery to a trench is high enough in relation to the rate of plate consumption, the trench may be filled with sediment and the trench site covered by subsea fans that mask the position of the tectonic front of the subduction zone (Silver, 1969). It is fair to infer that the volume of locally deposited turbidite sediment incorporated within the nearby subduction complex is thus in some measure inversely proportional to the bathymetric depth of the associated trench. An empty trench leaves little evidence in the geologic record.

Recent data leave little doubt that the steep inner wall of the trench is underlain almost directly by deformed and uplifted oceanic strata with only a local cover of undeformed sediment (Karig, in press). This material is interpreted here as a subduction complex of mélanges and crumpled beds sliced by thrusts and including ophiolitic scraps. The tectonic top of the subduction complex is assumed to lie at or just beneath the sea bottom at the trench slope break, a bathymetric transition point located at the top of the inner wall of the trench. The mass of the subduction complex is inferred to grow by the accretion of successive increments of oceanic crustal materials that either are jammed against its seaward flank at the trench axis or scraped into its basal levels from the top of the slab of lithosphere that descends beneath the subduction zone. It is important to note that these materials thus added to the subduction complex include not only indigenous trench turbidites deposited nearby, but potentially also include samplings of all the turbidites deposited over extensive areas of the ocean floor from sources wholly exotic to the arc-trench system into whose flank they are incorporated.

As oceanic materials are stacked tectonically within subduction zones, net uplift of the subduction complex must occur even while subduction continues, and should be dramatic when subduction ceases for any reason. The condition of a subduction complex where exposed to view on land is thus never the initial condition. Always there is the overprint of deformation during uplift, which must amount to a minimum of 5 to 10 km if the depths of modern trenches are representative. Mass movement of material off the steep inner wall of the trench may in time recycle some materials back through the process of subduction. We are still at a loss to understand fully the complex structures of mélanges and thrust-bounded slabs in subduction complexes (Suppe, 1972). It seems clear, however, that their great apparent thicknesses are tectonic, rather than stratigraphic.

Forearc basins.—The topographic and bathymetric configuration within the arc-trench gap between the trench slope break and the volcanic front is highly varied. The elevation of the threshold at the trench slope break is evidently controlled by the elevation of the top of the subduction complex, which may be emergent as islands or may lie at depths as great as 2 to 3 km. Different arc-trench gaps contain, singly or in combination, such diverse geographic elements as mountainous uplifts, longitudinal troughs, transverse submarine slopes, shallow shelves, deep-marine terraces or plains, and terrestrial plains or valleys. In a number of modern arc-trench gaps, thick sequences of largely undeformed sediments attest to progressive subsidence to develop forearc basins as the term is used here, and sequences interpreted as forearc basins in the geologic record attain thicknesses of 5 to 12 km. Subsidence may be related to the descent of a dense slab of lithosphere beneath the arc-trench gap.

The sedimentary sequences of forearc basins rest on a substratum of variable and partly uncertain character. On the arc flank of the arc-trench gap the substratum may include eroded igneous rocks, both plutonic and volcanic, of the magmatic arc. On the trench flank of the arc-trench gap the substratum may include parts of the subduction complex. Beneath the center of a forearc basin the substratum may be para-oceanic crust made of previously accreted elements of a subduction complex that broadens by growing seaward with time, or may be oceanic or transitional crust that existed before the arc-trench system was activated (*e.g.*, Grow, 1973). Thus, in general, forearc basins are commonly successor basins (King, 1969) in the sense that they overlie older, deformed elements or orogenic belts.

Forearc basins receive sediment mainly from the extensive nearby arc structures, where not only volcanic rocks but also plutonic and metamorphic rocks exposed by uplift and erosion may serve as sources. Sources may also include local uplands along the trench slope break or within the arc-trench gap itself. Facies grada-

tions may presumably occur between strata of forearc basins and volcaniclastic beds of the volcanic arcs, but prominent normal fault zones commonly bound the basins on the arc side (Karig, in press). On the trench side, tectonic gradation into the disrupted strata of the subduction zone presumably occurs locally. In several instances, however, nearly intact ophiolite sequences that underlie continuous sedimentary sequences of inferred forearc basins are in sharp fault contact with the adjacent subduction complexes. This circumstance implies little or no transfer of material into the subduction zone from the part of the forearc basin now preserved. Instead, the forearc basins appear to have wholly overridden the subduction zones. This relation holds for the late Mesozoic Great Valley sequence of California where faulted against the coeval Franciscan complex (Bailey and others, 1971), for the late Paleozoic and early Mesozoic western marginal facies of New Zealand where faulted against the coeval eastern axial facies or Torlesse Group (Landis and Bishop, 1972; Blake and Landis, 1973), and for the early Tertiary succession of the central Burmese lowland where faulted against the Indoburman flysch terrane (Brunnschweiler, 1966).

By inference from the bathymetry of modern forearc basins, and from the sedimentology of older sequences inferred to have been deposited in similar settings, forearc basins may contain a variety of facies. Shelf and deltaic or terrestrial sediments, as well as turbidites with either transverse or longitudinal paleocurrents, may occur in different examples. The local bathymetry is presumably controlled by the elevation of the trench slope break, the rate of sediment delivery to the forearc basin, and the rate of basin subsidence acting in combination. Various facies patterns and successive depositional phases probably can occur in different cases.

Intra-arc basins.—Magmatic arcs include both intraoceanic and continental margin types (Dickinson, 1974). Intraoceanic arcs include those with only paraoceanic crust built by magmatic additions to oceanic crust and those underlain at depth by a sliver of continental crust detached as part of a migratory arc structure. Continental margin arcs include island arcs backed by shallow epicontinental seas as well as those standing along the edges of landmasses. Fault-bounded extensional basins that occur within many magmatic arcs may be related to local volcano-tectonic subsidence, or to arching that accompanies uplift of paracontinental crust, or to the development of an incipient interarc basin. Volcaniclastic strata are

405

characteristic, but may include a range of types from terrestrial redbeds to turbidites, as well as various intermediary facies. Local sources of sediment within the arc structure are typical, but low-standing island arcs located near continental blocks may accumulate clastic strata from external sources as well; these have been termed basinal arcs (Berg and others, 1972).

Backarc areas.—The distinction made here between interarc and retroarc basins in the backarc area reflects the existence of two distinct variants of arc-trench systems (see fig. 10). Both types of basins are related indirectly to the convergent plate junctures to which the trenches and their associated subduction zones are related directly. The contrast between the tectonic settings of interarc and retroarc basins seemingly stems, therefore, from influences other than simple plate interactions at convergent plate junctures. The key control is apparently the relative motion of the plate of lithosphere in the backarc area with respect to the underlying asthenosphere (Coney, 1971; Dickinson, 1972; Hyndman, 1972; Wilson and Burke, 1972; Wilson, 1973).

The lithosphere is apparently not wholly intact across the region beneath the magmatic arc owing to thermal softening from the high heat flux. The narrow belt of lithosphere beneath the arc-trench gap can thus be viewed as a separate narrow plate. Where the lithosphere behind the arc has a component of motion, relative to underlying asthenosphere, away from the magmatic arc, then the arc structure may split. An interarc basin underlain by newly formed oceanic crust built by a backarc spreading center then opens between the active or frontal arc and a remnant arc (Karig, 1972), which may be of either intraoceanic or continental margin type. This mode of behavior is characteristic of eastward-facing island arcs in the western Pacific region (Karig, 1970, 1971a). Where the lithosphere behind the arc has a component of motion, relative to underlying asthenosphere, toward the magmatic arc, then partial subduction of continental lithosphere beneath the rear of the arc structure is assumed here to occur (e.g., Coney, 1972). A fold-thrust belt thus develops in the backarc area as cover rocks are stripped off descending basement. The resulting highlands shed debris into a downbowed retroarc basin along a belt that can be termed pericratonic between the continental margin arc and the craton. This mode of behavior is characteristic of the westward-facing Andean arc, which is flanked on the east by the Subandean fold-thrust belt, beyond which are the Subandean sedimentary basins that lie between the Andes

and the craton (Ham and Herrerra, 1963; Sonnenberg, 1963).

Conceivably, the contrasting behavior of eastward-facing and westward-facing arc-trench systems may reflect the different tectonic regimes, respectively extensional and contractional, induced in barkarc areas by the postulated net westward drift of lithosphere with respect to asthenosphere as a result of tidal influences (G. W. Moore, 1973). By implication, arc-trench systems with a roughly east-west orientation might experience no marked deformation in backarc areas, and hence might display neither interarc nor retroarc basins.

Interarc basins.—The sedimentary record of interarc basins is not well documented, but their global abundance at present suggests that eugeosynclinal terranes of the past probably contain numerous examples. It must be inferred that some ophiolitic sequences of orogenic belts represent oceanic crust formed as the floors of interarc basins, rather than in open oceans. If there are significant differences between the igneous rocks of the two kinds of ophiolitic sequences, the distinction is not yet established.

The sedimentary strata in modern interarc basins include distinctive turbidite aprons of volcaniclastic beds shed backward from the rifted rear sides of migratory frontal arcs (Karig, 1970, 1971a, 1972). These turbidite wedges appear to rest almost directly on the igneous oceanic crust with little or no intervening pelagites present. Beyond the interarc spreading centers sedimentation varies markedly. Where a given interarc basin is bounded on the side away from the arc-trench system by a submerged remnant arc, no effective source of clastic sediment is present and oceanic pelagites accumulate. Where successive remnant arcs with paraoceanic crust are calved in succession from migratory frontal arcs, a broad oceanic region is formed in which the only thick sedimentary accumulations are turbidite wedges stranded behind each submerged remnant arc.

On the other hand, where an interarc basin forms by disruption of a continental margin arc, one side of the interarc basin is a form of rifted continental margin along which some variant of a rifted-margin prism can be formed (Mitchell and Reading, 1969) beside a marginal sea (Karig, 1971b; Packham and Falvey, 1971; Moberly, 1972). It may be argued that the pattern of parallel facies belts associated with such a continental margin fringed by migratory intraoceanic arcs lying offshore faithfully reproduces the classic miogeosyncline-eugeosyncline couple. If so, extreme horizontal motions

406

of lithosphere may be unnecessary assumptions to explain the juxtaposition of diverse terranes within orogenic belts. The rifted continental margin on the inner side of the interarc basin is interpreted then as the miogeosynclinal belt, whereas the adjacent interarc basin, the offshore island arc, and the open ocean beyond together represent the complex tectonic elements of the eugeosynclinal belt. Although attractive, the analogy harbors a potential fallacy. Only if the substratum beneath the supposed miogeoclinal wedge includes igneous rocks representing part of the geologic record of the earlier stages of arc evolution prior to arc migration can the analogy be defended in detail. As most miogeoclinal wedges appear to rest on truncated continental basement considerably older than the base of the miogeoclinal wedge, this logical requirement of the analogy does not appear to be met in typical orogenic belts.

Retroarc basins.—The sedimentary record of retroarc basins includes fluvial, deltaic, and marine strata as much as 5 km thick deposited in terrestrial lowlands and epicontinental seas along elongate pericratonic belts between continental margin arcs and cratons. Where the magmatic arcs stand along continental margins that have grown seaward by tectonic accretion, some retroarc basins may be successor basins in the sense of resting upon previously deformed terranes. Sediment dispersal into and across retroarc basins is mainly transverse, in a gross sense, from highlands on the side toward the magmatic arc, although contributions from the craton are also present. The deposits of retroarc basins are thus exogeosynclinal in the sense that debris is shed toward the craton from sources within marginal orogenic belts.

The sources of sediment may include the magmatic arc itself, but commonly the principal sources are uplifted strata in the fold-thrust belt formed by partial subduction behind the arc. Such was the case for the Cretaceous retroarc basin of the interior and Rocky Mountain region of North America (Weimer, 1970). The main highland sources were uplands of folded and faulted pre-Mesozoic strata lying just west of the retroarc basin, but still east of the batholith belt that marks the position of the associated magmatic arc (Hamilton, 1969). Part of the subsidence in retroarc basins is probably in response to flexure of the lithosphere or other isostatic adjustments induced by the tectonic load of thrust sheets in the adjacent foreland fold-thrust belt (Price and Mountjoy, 1971). As the retroarc basin evolved, contractional deformation disrupted piedmont facies along the highland flank of the basin, and ultimately crumpled the flank of the basin fill within the fold-thrust belt (Armstrong, 1968a). Where the main sources of sediment are thus in the fold-thrust belt behind the arc, rather than within the magmatic arc, the nature of the sources depends upon the previous history of the continental margin. Where the magmatic arc arises following the initiation of plate consumption along a previously inactive continental margin draped with a rifted-margin prism, the sources are apt to be uplifted miogeoclinal strata.

The fold-thrust belts that parallel the orogenic margin of retroarc basins thus may be described commonly as foreland thrust belts (Coney, 1973). In this sense, the foreland is simply the cratonal or platformal interior of the continent, and the foreland basin is a retroarc basin. However, foreland basins in this same setting with respect to the continental interior may form as a result of crustal collisions in which a rifted continental margin with its rifted-margin prism encounters the main subduction zone associated with the trench of an arc-trench system. The designation of the foreland can thus be ambiguous with respect to the polarity of the arc-trench system responsible for the orogenic belt. So long as parts of a rifted-margin prism are thrust back toward the continental interior, and a pericratonic fringe of continental basement is drawn down by partial subduction to form an elongate basin parallel to the fold-thrust belt, the concept of a foreland to the orogenic belt is appropriate. Foreland basins formed by partial subduction of continental margins during crustal collisions are here termed peripheral basins as discussed in the next section. Designation of a given foreland basin as either a retroarc basin or a peripheral basin thus depends upon a knowledge of the sequence and timing of tectonic events in the adjacent orogen.

Suture Belts

The term suture belt is used here for the complexly deformed joins along which crustal blocks are welded together by the crustal collisions that occur when lithosphere bearing thick crustal blocks reaches a subduction zone along a convergent plate juncture where oceanic lithosphere was previously being consumed (fig. 11). Crustal collisions include a variety of types involving both intraoceanic and continental margin arc-trench systems (Dickinson, 1971c). In all cases, crustal collision involves juxtaposition of the tectonic elements of an arc-trench system, together with its variety of sedimentary basins, against other crustal blocks across the

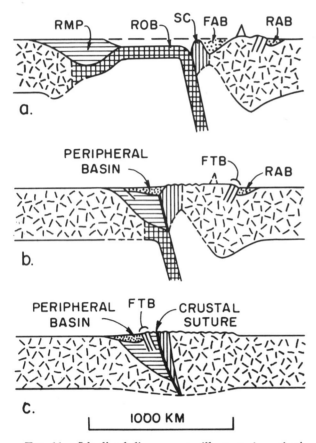

Fig. 11.—Idealized diagrams to illustrate hypothetical sequence of events, and associated sedimentary basins (stippled), during crustal collision between rifted continental margin (see fig. 7) on left and continental margin arc (see fig. 10b) on right: *a,* prior to collision; *b,* initial collision; *c,* final suturing (vertical exaggeration is 10✕). Oceanic and quasioceanic crust is crosshatched; continental, quasicontinental, and paracontinental substratum is jackstrawed. Symbols: RMP, rifted-margin prism (horizontal rules); ROB, remnant ocean basin (sea level as dashed line); SC, subduction complex (vertical rules); FAB, forearc basin; RAB, retroarc basin; FTB, foreland fold-thrust belts.

suture belt. Along a sutured join, deformed sedimentary sequences that were deposited on the ophiolitic basement of an open oceanic basin or an interarc basin can be caught between the sutured crustal blocks. Such sequences commonly appear to view after suturing as tectonically scrambled mélanges of ophiolitic scraps and oceanic facies. Such suture-belt mélanges may not be visible within a suture belt if the extent of subduction during crustal collision was sufficient to hide them beneath rocks of the overriding plate. Clear evidence of the sutured join may also be absent if contractional deformation during collision is sufficient to squeeze the suture-belt mélanges upward to tectonic levels that are later removed by erosion. These

two cases of obscured suture belts can be described as hidden sutures where telescoping thrust sheets cover the suture and as cryptic sutures where the materials caught in the suture are pressed out tectonically and lost by later erosion (e.g., Dewey and Burke, 1973).

Suture belts contain deformed examples of all the various types of sedimentary sequences discussed in connection with oceanic basins, rifted-margin prisms, and arc-trench systems. In addition, sedimentary basins of a unique type here termed *peripheral basins* with strata as thick as 5 km are also formed by processes related to collision (see fig. 11). As a continental crustal block is drawn toward a subduction zone just prior to crustal collision, bending of the lithosphere probably first causes extensional faulting analogous to that seen on the oceanic outer arch seaward of arc-trench systems. These faults might offset the strata of a rifted-margin prism in a sense similar to that of the earlier faults that were associated with continental rifting or with growth faulting during deposition of the prism. Later in the progress of crustal collision, the edge of the continental block is depressed by partial subduction to form a pericratonic or foreland basin peripheral to the suture belt on the plate being partly consumed. As the process of subduction is braked by crustal collision, the subsidence in this peripheral basin may be succeeded by marked uplift. As the peripheral basin is drained, a phase of evaporite deposition could conceivably ensue. The well-known sabhka deposition in the Persian Gulf may be an example of such deposition in a restricted seaway remaining along a belt parallel to the suture belt of the Zagros Crush Zone in Iran (Wells, 1969).

Perhaps the most characteristic deposits of peripheral basins are exogeosynclinal clastic wedges spread toward the craton as fluvial and deltaic strata shed from a suture belt involving the continental margin (Graham and others, in press). If the peripheral basin is deep enough, however, these deposits may be preceded by turbidites deposited on depressed continental or transitional crust, rather than oceanic crust (W. M. Neill, personal commun., 1973). Paleocurrent trends in the clastic wedges may be dominantly transverse to the orogenic trend, whereas those in the turbidites may be dominantly longitudinal to the orogenic trend. Clastic wedges of peripheral basins, as well as any clastic wedges shed toward the other side of highlands along the suture belt, may thus be termed molasse in many cases. The turbidites of peripheral basins, as well as the turbidites of oceanic basins or forearc basins caught within

408

the suture belt, may thus be termed flysch in many cases.

The evolution of suture belts forms an attractive explanation, though not the only one, for the tectonic relations of flysch and molasse (Graham and others, in press). In general, any completed suture belt will represent the end result of a sequential closure of a remnant ocean basin (Dickinson, 1972). Only if the shapes of colliding continental margins are mirror images of one another, and the vector of the relative plate motion causing crustal collision is exactly as required, can crustal collisions be synchronous along their whole length. In the general case, extensive suture belts must be diachronous in development as successive adjustments in plate motions and boundaries allow progressive welding of crustal blocks to proceed. A tectonic transition point between the segment already sutured (fig. 11c) and the segment yet to be sutured (fig. 11a) will migrate along the developing suture belt with time. Behind the transition point, orogenic highlands, clastic wedges and filled peripheral basins are characteristic. Ahead of the transition point, remnant ocean floor and incipient peripheral basins are present. As the drainage of orogenic highlands is commonly longitudinal, much of the sediment derived from the collision orogen will not be shed transversely as clastic wedges, but will be shed longitudinally into the remnant oceanic basin and deepening peripheral basins along tectonic strike. Many of the deposits that reflect erosion of the collision orogen thus will be incorporated later into the same orogenic belt as the tectonic transition point migrates along the growing suture belt. In this fashion, synorogenic flysch of turbidites with mainly longitudinal paleocurrents and postorogenic molasse of clastic wedges with largely transverse paleocurrents may be seen as the natural result of crustal collision to form suture belts.

Intracontinental Basins

Intracontinental basins are the most difficult type to treat constructively in terms of plate tectonics if basins related to apparently intracontinental orogenic belts like the Urals are excluded. Provided such orogenic terranes are interpreted as suture belts (Hamilton, 1970), the associated basins can be interpreted variously in terms of former oceanic basins, rifted continental margins, arc-trench systems, and collision orogens. Basins related to these kinds of features include foreland basins of both retroarc and peripheral types where the basin fill is supracontinental in the sense of resting on continental crust or on older rifted-margin prisms.

Basins bounded on all sides by anorogenic terranes forming a basement that is uniformly older than the basin fill are the intracontinental ones difficult to explain using principles of plate tectonics. For infracontinental basins (see above) the basement does not extend unmodified beneath the floor. Partial attenuation of continental basement along an aborted rift that never advanced beyond an incipient stage could lead to conditions permitting marked crustal subsidence locally, especially under sedimentary loading. Unfortunately, detection of transitional crust hidden beneath an infracontinental basin depends upon geophysical observations, for the basin fill permanently masks the substratum.

Presumably, infracontinental basins would tend to be elongate in many cases, but not necessarily in all. If antecedent or contemporaneous domal uplifts were distributed at intervals along the belt of partial crustal attenuation, as appears to be the pattern for early stages of continental rifting, then crustal thinning by stretching and erosion might be concentrated within relatively equant areas. As a result, the infracontinental basins that developed after thermal tumescence gave way to thermal decay might appear as apparently isolated and more or less round features distributed apparently at random across a continental block. The only clue to their essentially common origin might be a rough contemporaneity of development. The initial stages of a major continental separation may well involve extensive gashing of the continental block, while still joined, over a broad region that would lapse eventually into quiescence except where the rifting was fully established along a single trend. Dispersed infracontinental basins might then remain as a record of the widespread extent of incipient rifting.

Alternatively, a fundamentally elongate belt of incipient continental separation might be marked by a chain of isolated infracontinental basins linked only by intracontinental transforms. If the transform segments of the integrated tectonic system were masked by cover rocks or later deformation, the fundamental pattern might be difficult to detect by any means. There seems an especially strong possibility that successor basins might form along recently completed suture belts in this fashion as residual plate motions were resolved into translation along transforms roughly parallel to the suture belt.

None of these speculations touch upon the possibility of long-lived supracontinental basins with underpinnings of normal continental crust. The motion of a plate of lithosphere over a bumpy asthenosphere accounts well only for re-

409

versible epeirogenic warping and temporary subsidence. Note that this effect might affect local areas distributed in unpredictable fashion over a continental block, but that any local and temporary subsidence would occur as part of a wave of shbsidence. The passage of the lithosphere over a bump or depression on top of the asthenosphere thus might leave a sort of subtle track in the stratigraphic record of any epicontinental seas covering a continental block. Mechanisms for permanent subsidence of supracontinental basins on a large scale in truly intraplate settings are not apparent from plate-tectonic theory.

Summary

The preceding tentative classification of sedimentary basins in a plate-tectonic framework indicates that satisfactory alternatives to the geosynclinal terminology can be devised, and that points of correspondence between the two schemes of nomenclature can be appreciated. The discussion also indicates that direct equivalency between individual terms in the two sets cannot be expected. For example, eugeosynclines apparently contain the strata of oceanic basins, intra-arc basins, and interarc basins as modified by deformation in subduction complexes, magmatic arcs, and suture belts. On the other hand, rifted-margin prisms may include the superimposed strata of taphrogeosynclines, miogeosynclines, and paraliageosynclines. Exogeosynclinal foreland basins may be either retroarc basins or peripheral basins in the terminology suggested. Forearc basins and aulacogens have been described by some as epieugeosynclines and zeugogeosynclines, respectively, but others have applied different terms to analogous features and the same terms to different kinds of features. Such discordances in terminology are to be expected, given the dramatic change in frame of reference. Whatever terminology is used, progress in applying plate-tectonic theory to problems of sedimentation can come easily only if sedimentary basins are classified and discussed in a manner that is congruent with concepts of plate tectonics. In this paper, I have tried simply to find phraseology that would convey meaning now, without prejudice to either past or future usage.

REFERENCES CITED

ARMSTRONG, R. L., 1968a, Sevier orogenic belt in Nevada and Utah: Geol. Soc. America Bull., v. 79, p. 429–458.

——, 1968b, The Cordilleran miogeosyncline in Nevada and Utah: Utah Geol. and Mineralog. Survey Bull. 78, 58 p.

ATWATER, TANYA, 1970, Implications of plate tectonics for the Cenozoic tectonic evolution of western North America: Geol. Soc. America Bull., v. 81, p. 3513–3536.

BAILEY, E. H., BLAKE, M. C., JR., AND JONES, D. L., On-land Mesozoic crust in California Coast Ranges: U.S. Geol. Survey Prof. Paper 700-C, p. 70–81.

BAKER, B. H., MOHR, P. A., AND WILLIAMS, L. A. J., 1972, Geology of the eastern rift system of Africa: Geol. Soc. America Special Paper 136, 67 p.

BELOUSSOV, V. V., 1970, Against the hypothesis of sea-floor spreading: Tectonophysics, v. 9, p. 489–511.

BERG, H. C., JONES, D. L., AND RICHTER, D. H., 1972, Gravina-Nutzotin belt—tectonic significance of an upper Mesozoic sedimentary and volcanic sequence in southern and southeastern Alaska: U.S. Geol. Survey Prof. Paper 800-D, p. 1–24.

BERGER, W. H., 1973, Cenozoic sedimentation in the eastern tropical Pacific: Geol. Soc. America Bull., v. 84, p. 1941–1954.

BLAKE, M. C., JR., AND LANDIS, C. A., 1973, The Dun Mountain ultramafic belt—Permian oceanic crust and upper mantle in New Zealand: Jour. Research U.S. Geol. Survey, v. 1, p. 529–534.

BOSTROM, R. C., 1971, Westward displacement of lithosphere: Nature, v. 234, p. 536–538.

BRUNNSCHWEILER, R. O., 1966, On the geology of the Indoburman Ranges (Arakan Coast and Yoma, Chin Hills, Naga Hills): Jour. Geol. Soc. Australia, v. 13, p. 137–194.

BURKE, K. C. A., 1972, Longshore drift, submarine canyons, and submarine fans in development of Niger delta: Am. Assoc. Petroleum Geologists Bull., v. 56, p. 1975–1983.

——, AND DEWEY, J. F., 1972, Permobile, metastable, and plate-tectonic regimes in the Precambrian: Geol. Soc. America Abstracts with Programs, v. 4, p. 462.

——, DESSAUVAGIE, T. F. J., AND WHITEMAN, A. J., 1971, Opening of the Gulf of Guinea and geological history of the Benue Depression and Niger Delta: Nature Phys. Sci., v. 233, p. 51–55.

——, KIDD, W. S. F., AND WILSON, J. T., 1973, Relative and latitudinal motion of Atlantic hot spots: Nature, v. 245, p. 133–137.

——, AND DEWEY, J. F., 1973, An outline of Precambrian plate development, in Tarling, D. H., and Runcorn, S. K., Implications of continental drift to the earth sciences: Academic Press, N.Y., p. 1035–1045.

CHASE, C. G., 1972, The N plate problem of plate tectonics: Geophys. Jour. Roy. Astron. Soc., v. 29, p. 117–122.

CONEY, P. J., 1970, Geotectonic cycle and the new global tectonics: Geol. Soc. America Bull., v. 81, p. 739–748.

——, 1971, Cordilleran tectonic transitions and motion of the North America plate: Nature, v. 233, p. 462–465.

————, 1972, Cordilleran tectonics and North American plate motion: Am. Jour. Sci., v. 272, p. 603–655.
————, 1973, Plate tectonics of marginal foreland thrust-fold belts: Geology, v. 1, p. 131–134.
CROWELL, J. C., AND FRAKES, L. A., 1970, Phanerozoic glaciation and the causes of ice ages: Am. Jour. Sci., v. 268, p. 193–224.
DEWEY, J. F., 1972, Plate tectonics: Sci. American, May, p. 56–68.
————, AND BIRD, J. M., 1970a, Mountain belts and the new global tectonics: Jour. Geophys. Research, v. 75, p. 2625–2647.
————, AND BIRD, J. M., 1970b, Plate tectonics and geosynclines: Tectonophysics, v. 10, p. 625–638.
————, AND BURKE, K. C. A., 1973, Tibetan, Variscan, and Precambrian basement reactivation: products of continental collision: Jour. Geology, v. 81, p. 683–692.
DICKINSON, W. R., 1970, Relations of andesites, granites, and derivative sandstones to arc-trench tectonics: Rev. Geophysics and Space Physics, v. 8, p. 813–862.
————, 1971a, Plate tectonic models of geosynclines: Earth and Planetary Sci. Lettrs., v. 10, p. 165–174.
————, 1971b, Plate tectonic models for orogeny at continental margins: Nature, v. 232, p. 41–42.
————, 1971c, Plate tectonics in geologic history: Science, v. 174, p. 107–113.
————, 1972, Evidence for plate tectonic regimes in the rock record: Am. Jour. Sci., v. 272, p. 551–576.
————, 1973, Widths of modern arc-trench gaps proportional to past duration of igneous activity in associated magmatic arcs: Jour. Geophys. Research, v. 78, p. 3376–3389.
————, 1974, Sedimentation within and beside ancient and modern magmatic arcs, in Dott, R. H., Jr., and Shaver, R. H. (eds.): Soc. Econ. Paleontologists and Mineralogists Special Pub. 19, p. 230–239.
DIETZ, R. S., 1963, Wave-base, marine profile of equilibrium, and wave-built terraces: a critical appraisal: Geol. Soc. America Bull., v. 74, p. 971–990.
————, AND HOLDEN, J. C., 1966, Miogeoclines in space and time: Jour. Geology, v. 74, p. 566–583.
————, AND HOLDEN, J. C., 1970, Reconstruction of Pangaea, breakup and dispersion of continents, Permian to present: Jour. Geophys. Research, v. 75, p. 4939–4956.
————, HOLDEN, J. C., AND SPROLL, W. P., 1970, Geotectonic evolution and subsidence of Bahama platform: Geol. Soc. America Bull., v. 81, p. 1915–1928.
DOYLE, H. A., 1971, Australian seismicity: Nature Phys. Sci., v. 234, p. 174–175.
ELSASSER, W. M., 1971, Sea-floor spreading as thermal convection: Jour. Geophys. Research, v. 76, p. 1101–1112.
ERNST, W. G., 1970, Tectonic contact between the Franciscan melange and the Great Valley sequence, crustal expression of a late Mesozoic Benioff zone: ibid., v. 75, p. 886–902.
————, 1972, Occurrence and mineralogic evolution of blueschist belts with time: Am. Jour. Sci., v. 272, p. 657–668.
FRAKES, L. A., AND KEMP, E. M., 1972, Generation of sedimentary facies on a spreading ocean ridge: Nature, v. 236, p. 114–117.
FRANCHETEAU, JEAN, AND LEPICHON, XAVIER, 1972, Marginal fracture zones as structural framework of continental margins in South Atlantic Ocean: Am. Assoc. Petroleum Geologists Bull., v. 56, p. 991–1007.
GRAHAM, S. A., DICKINSON, W. R., AND INGERSOLL, R. V., in press, Himalayan-Bengal model for flysch dispersal in Appalachian-Ouachita system: Geol. Soc. America Bull.
GRANT, N. K., 1971, South Atlantic, Benue trough, and Gulf of Guinea Cretaceous triple junction: Geol. Soc. America Bull., v. 82, p. 2295–2298.
GROW, J. A., 1973, Crustal and upper mantle structure of the central Aleutian arc: ibid., v. 84, p. 2169–2192.
HAM, C. K., AND HERRERRA, L. J., JR., 1963, Role of Subandean fault system in tectonics of eastern Peru and Ecuador, in Childs, O. E. and Beebe, B. W. (eds.), Backbone of the Americas: Am. Assoc. Petroleum Geologists Mem. 2, p. 47–61.
HAM, W. E., AND WILSON, J. L., 1967, Paleozoic epeirogeny and orogeny in the central United States: Am. Jour. Sci., v. 265, p. 332–407.
HAMILTON, WARREN, 1969, Mesozoic California and the underflow of Pacific mantle: Geol. Soc. America Bull., v. 80, p. 2409–2430.
————, 1970, The Uralides and the motion of the Russian and Siberian platforms: ibid., v. 81, p. 2553–2576.
————, 1973, Tectonics of the Indonesian region: Geol. Soc. Malaysia Bull. 6, p. 3–10.
————, AND MYERS, W. B., 1966, Cenozoic tectonics of the western United States: Rev. Geophysics, v. 4, p. 509–549.
HARLAND, W. B., 1971, Tectonic transpression in Caledonian Spitzbergen: Geol. Mag., v. 108, p. 27–42.
HEEZEN, B. C., AND 11 CO-AUTHORS, 1973, Diachronous deposits, a kinematic interpretation of the post-Jurassic sedimentary sequence on the Pacific plate: Nature, v. 241, p. 25–32.
HOFFMAN, PAUL, BURKE, K. C. A., AND DEWEY, J. F., 1974, Aulacogens and their genetic relation to geosynclines, with a Proterozoic example from Great Slave Lake, Canada, in Dott, R. H., Jr., and Shaver, R. H. (eds.): Soc. Econ. Paleontologists and Mineralogists Special Pub. 19, p. 38–55.
HSU, K. J., 1968, Principles of melanges and their bearing on the Franciscan-Knoxville problem: Geol. Soc. America Bull., v. 79, p. 1063–1074.
————, 1972, The concept of the geosyncline, yesterday and today: Leicester Lit. Philos. Soc. Trans., v. 66, p. 26–48.
HUTCHINSON, R. W., AND ENGELS, G. G., 1972, Tectonic evolution in the southern Red Sea and its possible significance to older rifted continental margins: Geol. Soc. America Bull., v. 83, p. 2989–3002.
HYNDMAN, R. D., 1972, Plate motions relative to the deep mantle and the development of subduction zones: Nature, v. 238, p. 263–265.
ISACKS, BRYAN, OLIVER, JACK, AND SYKES, L. R., 1968, Seismology and the new global tectonics: Jour. Geophys. Research, v. 73, p. 5855–5899.
JAMES, D. E., 1971, Plate tectonic model for the evolution of the central Andes: Geol. Soc. America Bull., v. 82, p. 3325–3346.

Johnson, J. G., 1971, Timing and coordination of orogenic, epeirogenic, and eustatic events: *ibid.*, v. 82, p. 3263–3298.

Karig, D. E., 1970, Ridges and basins of the Tonga-Kermadec island arc system: Jour. Geophys. Research, v. 75, p. 239–254.

———, 1971a, Structural history of the Mariana island arc system: Geol. Soc. America Bull., v. 82, p. 323–344.

———, 1971b, Origin and development of marginal basins in the western Pacific: Jour. Geophys. Research, v. 76, p. 2542–2561.

———, 1972, Remnant arcs: Geol. Soc. America Bull., v. 83, p. 1057–1068.

———, in press, Crustal accretion in subduction zones.

Kay, Marshall, 1951, North American geosynclines: Geol. Soc. America Mem. 48, 143 p.

King, P. B., 1969, The tectonics of North America—a discussion to accompany the tectonic map of North America: U.S. Geol. Survey Prof. Paper 628, 94 p.

Knopoff, Leon, and Leeds, A., 1972, Lithospheric momenta and the deceleration of the earth: Nature, v. 237, p. 93–95.

Lachenbruch, A. H., and Thompson, G. A., 1972, Oceanic ridges and transform faults: their intersection angles and resistance to plate motion: Earth and Planetary Sci. Lettrs., v. 15, p. 116–122.

Landis, C. A., and Bishop, D. G., 1972, Plate tectonics and regional stratigraphic-metamorphic relations in the southern part of the New Zealand geosyncline: Geol. Soc. America Bull., v. 83, p. 2267–2284.

Le Bas, M. J., 1971, Per-alkaline volcanism, crustal swelling and rifting: Nature Phys. Sci., v. 230, p. 85–86.

Le Pichon, Xavier, 1968, Sea-floor spreading and continental drift: Jour. Geophys. Research, v. 73, p. 3661–3698.

———, and Fox, P. J., 1971, Marginal offsets, fracture zones, and the early opening of the North Atlantic: *ibid.*, v. 76, p. 6294–6308.

———, and Hayes, D. E., 1971, Marginal offsets, fracture zones, and the early openings of the South Atlantic: *ibid.*, v. 76, p. 6283–6293.

Lowell, J. D., and Genik, G. J., 1972, Sea-Floor spreading and structural evolution of southern Red Sea: Am. Assoc. Petroleum Geologists Bull., v. 56, p. 247–259.

Markhinin, E. K., 1968, Volcanism as an agent of formation of the earth's crust, *in* Knopoff, Leon, Drake, C. L., and Hart, P. J. (eds.), The crust and upper mantle of the Pacific area: Am. Geophys. Union Mon. 12, p. 413–423.

Marlow, M. S., Scholl, D. W., Buffington, E. D., and Alpha, T. R., 1973, Tectonic history of the central Aleutian arc: Geol. Soc. America Bull., v. 84, p. 1555–1574.

McElhinny, M. W., 1973, Palaeomagnetism and plate tectonics: Cambridge Univ. Press, 358 p.

McKenzie, D. P., 1969, Speculations on the consequences and causes of plate motions: Geophys. Jour. Roy. Astron. Soc., v. 18, p. 1–32.

———, 1970, Plate tectonics of the Mediterranean region: Nature, v. 226, p. 239–243.

———, 1972a, Plate tectonics, *in* Robertson, E. C. (ed.), The nature of the solid earth: McGraw-Hill, N.Y., p. 323–360.

———, 1972b, Active tectonics of the Mediterranean region: Geophys. Jour. Roy. Astron. Soc., v. 30, p. 109–185.

———, and Morgan, W. J., 1969, Evolution of triple junctions: Nature, v. 224, p. 125–133.

———, and Parker, R. L., 1967, The North Pacific, an example of tectonics on a sphere: Science, v. 216, p. 1276–1280.

Menard, H. W., 1973a, Depth anomalies and the bobbing motion of drifting islands: Jour. Geophys. Research, v. 78, p. 5128–5138.

———, 1973b, Epeirogeny and plate tectonics: Am. Geophys. Union Trans. (EOS), v. 54, p. 1244–1255.

Meyerhoff, A. A., and Meyerhoff, H. A., 1972, "The new global tectonics": major inconsistencies: Am. Assoc. Petroleum Geologists Bull., v. 56, p. 269–336.

Mitchell, A. H., and Reading, H. G., 1969, Continental margins, geosynclines, and ocean-floor spreading: Jour. Geology, v. 77, p. 629–646.

Moberly, Ralph, 1972, Origin of lithosphere behind island arcs, with reference to the western Pacific: Geol. Soc. America Mem. 132, p. 35–55.

Molnar, Peter, and Atwater, Tanya, 1973, Relative motion of hot spots in the mantle: Nature, v. 246, p. 288–291.

Moore, D. G., 1973, Plate-edge deformation and crustal growth, Gulf of California structural province: Geol. Soc. America Bull., v. 84, p. 1883–1906.

Moore, G. W., 1973, Westward tidal lag as the driving force of plate tectonics: Geology, v. 1, p. 99–100.

Moore, J. C., 1973, Cretaceous continental margin sedimentation, southwestern Alaska: Geol. Soc. America Bull., v. 84, p. 595–614.

Morgan, W. J., 1968, Rises, trenches, great faults, and crustal blocks: Jour. Geophys. Research, v. 73, p. 1959–1982.

———, 1972, Plate motions and deep mantle plumes: Geol. Soc. America Mem. 132, p. 7–22.

Murauchi, S., and 12 Co-authors, 1968, Crustal structure of the Philippine Sea: Jour. Geophys. Res., v. 73, p. 3143–3172.

Packham, G. H., and Falvey, D. A., 1971, An hypothesis for the formation of marginal seas in the western Pacific: Tectonophysics, v. 11, p. 79–109.

Pautot, Guy, Auzende, J-M., and Lepichon, Xavier, 1970, Continuous deep sea salt layer along North Atlantic margins related to early phase of rifting: Nature, v. 227, p. 351–354.

Price, R. A., and Mounjoy, E. W., 1971, The Cordilleran foreland thrust and fold belt in the southern Canadian Rockies: Geol. Soc. America Abstracts with Programs, v. 3, p. 404–405.

Reyment, R. A., 1972, The age of the Niger delta (West Africa): 24th Internat. Geol. Cong. Rept., Sec. 6, p. 11–13.

RODGERS, JOHN, 1968, The eastern edge of the North American continent during the Cambrian and Early Ordovician, *in* Zen, E-An, and White, W. S. (eds.), Studies of Appalachian geology, northern and maritime: Wiley Interscience, N.Y., p. 141–149.

ROMAN, CONSTANTIN, 1973a, Buffer plates where continents collide: New Scientist, 25 Jan, p. 180–181.

——, 1973b, Rigid plates, buffer plates, and subplates: Geophys. Jour. Roy. Astron. Soc., v. 33, p. 369–373.

RONA, P. A., 1973, Relations between rates of sediment accumulation on continental shelves, sea-floor spreading, and eustacy inferred from the central North Atlantic: Geol. Soc. America Bull., v. 84, p. 2851–2871.

ROSS, D. A., 1971, Sediments of the northern Middle America trench: *ibid.*, v. 82, p. 303–322.

SALOP, L. I., AND SCHEINMANN, Y. M., 1969, Tectonic history and structures of platforms and shields: Tectonophysics, v. 7, p. 565–597.

SCHNEIDER, E. D., 1972, Sedimentary evolution of rifted continental margins: Geol. Soc. America Mem. 132, p. 109–118.

SCLATER, J. G., AND FRANCHETEAU, JEAN, 1970, The implications of terrestrial heat flow observations on current tectonic and geochemical models of the crust and upper mantle of the earth: Geophys. Jour. Roy. Astron. Soc., v. 20, p. 509–542.

SCLATER, J. G., ANDERSON, R. N., AND BELL, M. L., 1971, Elevation of ridges and evolution of the central Pacific: Jour. Geophys. Research, v. 76, p. 7888–7915.

SCRUTTON, R. A., 1973, The age relationship of igneous activity and continental break-up: Geol. Mag., v. 110, p. 227–234.

SHOR, G. G., JR., KIRK, H. K., AND MENARD, H. W., 1971, Crustal structure of the Melanesian area: Jour. Geophys. Research, v. 76, p. 2562–2586.

SILVER, E. A., 1969, Late Cenozoic underthrusting of the continental margin off northernmost California: Science, v. 166, p. 1265–1266.

SLEEP, N. H., 1971, Thermal effects on the formation of Atlantic continental margins by continental breakup: Geophys. Jour. Roy. Astron. Soc., v. 24, p. 325–350.

SONNENBERG, F. P., 1963, Bolivia and the Andes, *in* Childs, O. E., and Beebe, B. W. (eds.), Backbone of the Americas: Am. Assoc. Petroleum Geologists Mem. 2, p. 36–46.

STEWART, J. H., 1972, Initial deposits in the Cordilleran geosyncline, evidence of a late Precambrian continental separation: Geol. Soc. America Bull., v. 83, p. 1345–1360.

SUPPE, JOHN, 1972, Interrelationships of high-pressure metamorphism, deformation, and sedimentation in Franciscan tectonics, U.S.A.: 24th Internat. Geol. Cong. Rept., Sec. 3, p. 552–559.

SYKES, L. R., 1970, Seismicity of the Indian Ocean and a possible nascent island arc between Ceylon and Australia: Jour Geophys. Research, v. 75, p. 5041–5055.

TALWANI, MANIK, AND ELDHOLM, OLAV, 1972, Continental margin off Norway; a geophysical study: Geol. Soc. America Bull., v. 83, p. 2575–3606.

TURCOTTE, D. L., AND OXBURGH, E. R., 1973, Mid-plate tectonics: Nature, v. 244, p. 337–339.

VALENTINE, J. W., AND MOORES, E. M., 1972, Global tectonics and the fossil record: Jour. Geology, v. 80, p. 167–184.

VINE, F. J., AND MOORES, E. M., 1972, A model for the gross structure, petrology, and magnetic properties of oceanic crust: Geol. Soc. America Mem. 132, p. 195–205.

VON HUENE, ROLAND, 1972, Structure of the continental margin and tectonism at the eastern Aleutian trench: Geol. Soc. America Bull., v. 83, p. 3613–3626.

WALCOTT, R. I., 1972, Gravity, flexure, and the growth of sedimentary basins at a continental edge: *ibid.*, v. 83, p. 1845–1848.

WANLESS, H. R., AND SHEPARD, F. P., 1936, Sea level and climatic changes related to late Paleozoic cycles: *ibid.*, v. 47, p. 1177–1206.

WELLS, A. J., 1969, The crush zone of the Iranian Zagros Mountains, and its implications: Geol. Mag., v. 106, p. 385–394.

WEIMER, R. J., 1970, Rates of deltaic sedimentation and intrabasin deformation, Upper Cretaceous of Rocky Mountain region, *in* Morgan, J. P. (ed.), Deltaic sedimentation, modern and ancient: Soc. Econ. Paleontologists and Mineralogists Special Pub. 15, p. 270–292.

WHITE, D. A., ROEDER, D. H., NELSON, T. H., AND CROWELL, J. C., 1970, Subduction: Geol. Soc. America Bull., v. 81, p 3431–3432.

WILSON, J. T., 1965, A new class of faults and their bearing on continental drift: Nature, v. 207, p. 343–347.

——, 1973, Mantle plumes and plate motions: Tectonophysics, v. 19, p. 149–164.

——, AND BURKE, K. C. A., 1972, Two types of mountain building: Nature, v. 239, p. 448–449.

WISE, D. U., 1972, Freeboard of continents through time: Geol. Soc. America Mem. 132, p. 87–100.

Reprinted by permission of National Academy Press.
Published in *Studies in Geophysics—Climate in Earth
History*, pp. 97–104, Washington, D. C., National
Academy Press, 1982.

Long-Term Climatic Oscillations
Recorded in Stratigraphy

ALFRED G. FISCHER
Princeton University

INTRODUCTION

The marine stratigraphic record reveals cyclic changes of various sorts, including periodic interruptions of deposition, change in the sedimentary constituents supplied, change in faunas and floras, and change in the nature of the depositional environment. Many of these changes are too general in character and in distribution to be attributed to local causes: they seem to reflect global changes in climate and their effects on the marine system.

These phenomena are discussed here in sequence of increasing period. Bedding phenomena visible at the outcrop level appear to correspond to climatic changes induced by the Earth's orbital perturbations—in the 20,000-500,000-yr range—the same forces that drove the glacial advances and retreats of the Pleistocene. Broader phenomena that must generally be synthesized from regional or global data suggest a possible climatic cycle in the 30-36 million years (m.y.) realm. This in turn appears to ride on an extremely long cycle (not necessarily of fixed period) that brought on alternation of "icehouse" and "greenhouse" climates: in the last 700 m.y., the Earth seems to have completed two and started on a third of these cycles.

CHANGES IN THE 20,000-500,000-YEAR (MILANKOVITCH) RANGE

21,000- and 43,000-Year Cycles

Sediments when viewed at the level of a roadcut or hillside are characterized by stratification. This is generally attributed to random fluctuations in the supply of sediment to, and removal of sediment from, a given depositional site. Such processes might be expected to produce a fairly random aggregation of thicker and thinner strata, yet many sedimentary "formations" show rather striking uniformity of bedding thickness. This is particularly true of many limestones, in which thicker beds of biogenically formed carbonate alternate with thin interbeds of shale, recording a simple oscillation cycle, as recognized by Gilbert (1895, 1900), Schwarzacher (1975), Fischer (1980, 1981), and others.

Gilbert (1895) attributed rhythmic bedding in the Cretaceous of Colorado to climatic influence of the axial precession having a period of about 21,000 yr. These variations in insolation were first worked out quantitatively by Milankovitch (1941) and have been revised by Berger (1980).

414

The timing of such rhythms may be approached in two different ways, from "below" or from "above." In varved sequences, in which beds are composed of presumed annual laminations, the duration of a bed in years should equal the number of varves within it. The alternative is to take a radiometrically well-dated interval, such as a stage or an epoch, and to divide its length by the number of beds found within it.

The first of these methods was applied to the Green River Oil Shale (Eocene) by Bradley (1929). Bradley did not actually count the varves in a bed—he determined the mean thickness of varves from thin sections and the mean thickness of beds from measurements on outcrop and found the ratio to be 21,000 : 1. Whereas this work needs independent verification, it seems to have confirmed Gilbert's hypothesis in principle.

The second method has been applied to Cretaceous pelagic and hemipelagic limestones by Arthur (1979a) and by Fischer (1980, 1981). Various problems arise with this approach. One is that few rhythmic bedding sequences and continuous exposures span the length of a stage, so that is becomes necessary in most sequences to extrapolate. Another is that in rival time scales—Obradovich and Cobban (1975) versus Van Hinte (1976)—some stages differ by a factor of 3, so that it becomes necessary to average the results from several stages. A series of 11 Cretaceous sequences from Colorado, France, and Italy (Fischer 1980, 1981) yielded raw averages ranging from 10,000 to 87,000 yr per bed. The eight shorter ones yielded a mean of 17,125 on the Obradovich-Cobban scale and 26,375 on the Van Hinte scale, for a combined mean of 21,750. Of the remaining three, one is poorly dated and the other two seem to lie in the vicinity of 50,000 yr and might be related to the 43,000-yr cycle in obliquity (Milankovitch, 1941; Berger, 1980).

Thus the existence of the 21,000-yr rhythm—and therewith of a precessional influence on Cretaceous sedimentary regimes—appears to be moderately well established. The case for a sedimentary record of the cycle in tilt, on the other hand, is not strong except in the deep-sea record (Arthur, 1979b).

100,000-Year Cycle

Simple bedding rhythms of the type discussed above tend to occur in sets, and while there is considerable variation in the number per set, statistical averages out of any one sequence usually yield a mean number of about five (Schwarzacher, 1975). This ratio holds for the Precambrian-Cambrian boundary beds in Morocco (Monninger, 1979); for Carboniferous limestones in Ireland, Triassic limestones in the Alps, Triassic lake deposits in New Jersey, and Jurassic limestones in southern Germany (Schwarzacher, 1975); and for five of the eight Cretaceous sequences studied (Fischer, 1980, 1981). If the Cretaceous sequences cited above are of precessional origin, then the "Schwarzacher bundles," which they compose, would seem to have a timing of about 100,000 yr. Furthermore, by analogy, it appears reasonable to interpret this bedding pattern, characteristic of various parts of the Phanerozoic, as a record of precessional cycles grouped into 100,000-yr sets.

This 100,000-yr rhythm is the strongest of the glacial rhythm signals in the Pleistocene marine record (Hays et al., 1976), and is attributed there to the orbital quasi-rhythm in eccentricity (Milankovitch, 1941; Berger, 1980). It is extremely tempting to consider the Schwarzacher bundles as the product of the precession coupled with eccentricity. Indeed, the precession can influence climate only by way of orbital eccentricity, so that a bundling of precessional beds into larger sets is virtually demanded by theory.

Cycles at the 500,000-Year Level

An example of long rhythms in stratal sequences is provided by the Permo-Carboniferous megacycles of Kansas (Moore et al., 1951; Heckel, 1977), in which terrigenous deposits at base and top separate a marine sequence characterized by a peculiarly patterned alternation of shales and limestones. Some 25 of these megacycles characterize the 10 m.y. of Missourian-Virgilian time, a mean duration of 400,000 yr.

In various other sequences, such as the Precambrian-Cambrian boundary beds in Morocco (Monninger, 1979), the Triassic lake deposits of New Jersey (Van Houten, 1964), and the Cretaceous and Eocene in central Italy, Schwarzacher bundles are in turn grouped into sets of four to six, representing about 0.5 m.y. each.

This cycle too appears to have a match in orbital perturbations, namely in a longer cycle of eccentricity, which emerges from Berger's (1980) calculations. This cycle has recently been recognized in the Pleistocene record by Briskin and Berggren (1975).

Multiple Pathways of Expression

Whereas the sediments in which these bedding patterns have been found are mainly limestone sequences, they include different depositional regimes, in which rhythmicity is induced by different factors. In the alpine Triassic, for example, the cause is a variation in sea level, leading to repeated emergence and submergence of carbonate banks (Fischer, 1964). In the late Cretaceous of central Italy, the setting is one of deep water throughout, and the rhythms reflect a change from carbonate deposition to clay deposition—either because the carbonate supply was reduced or because of carbonate dissolution on the seafloor (Arthur and Fischer, 1977). In the Cretaceous of Kansas, and in the Mid-Cretaceous (Aptian-Albian) of Italy and of the present ocean floor, some of the rhythmicity was produced by changes in oxygen content of bottom waters: the depositional sites oscillated between aerobic and anaerobic conditions (Arthur and Fischer, 1977; Arthur, 1979b; Fischer, 1980). In the Triassic lake deposits of the Newark rift (Van Houten, 1964), the rhythmicity resulted from changes in lake level and in the chemistry of the lake. The only common denominator for all of these changes is climate—climatic fluctuation so severe as to change sea level (presumably by growth of glaciers), the chemistry and behavior of the oceans, and the salinity of lakes.

Conclusions and Problems Regarding Cyclicity at the Milankovitch Level

The data summarized above have led me to conclude that climatic oscillations driven by the Earth's orbital perturbations have not been limited to the Pleistocene but have affected the Earth's climates through Phanerozoic time—the last 600 m.y. We have barely begun to recognize their record in the sediments and are far from having adequate descriptions, let alone understanding. There are suggestions that the nature of this record has changed with time. In the Late Pleistocene record, for example, the eccentricity signal is strongest, the obliquity signal next, and the precessional signal weakest. In the Late Carboniferous cyclothems of Kansas—another glacial time—eccentricity cycles—in particular, the 400,000-yr cycle—seem again to dominate the picture. At nonglacial times the precessional signal seems the strongest. Are there definite time changes in the kinds of rhythms—signals of changes in orbital character or of changes in the Earth's response to constant signals? Is there a solar factor in addition? We do not know the answers. What the record tells us is that different parts of the Earth recorded the climatic changes in different ways, and this in turn should serve to develop some understanding of the functioning of the Earth. Puzzling, for example, are the sharp sea-level changes suggested by the record for times generally thought to have been free of polar ice. Was there perhaps mountain glaciation on scales far beyond that of today? Such questions call for further studies.

CHANGES AT THE 30-MILLION-YEAR AND 300-MILLION-YEAR LEVELS

30-Million-Year Cycle

While historical geologists since Lyell have given lip service to the principle of uniformitarianism, in which the Earth is viewed as having developed in a gradual and steady manner, a majority of stratigraphers and tectonicists, going back to Cuvier and d'Orbigny, including Chamberlin, Grabau, and Umbgrove, have been impressed with the segmentation of geologic history into episodes. Some of these changes appear to be rhythmic, and one of the rhythms represented lies at the 30-36-m.y. level.

Dorman (1968) suggested a 30-m.y. cycle in global temperatures, based on oxygen isotope analyses of Cenozoic mollusks. Damon (1971) analyzed the record of marine transgressions and regressions from the continents and concluded that Phanerozoic sea level rose and fell with a periodicity of 36 ± 11 m.y. and that this bore some correspondence to periodicities in global mountain building and in regional plutonism.

Fischer and Arthur (1977) suggested that the Mesozoic-Cenozoic part of Earth history is logically subdivided not into four periods as currently practiced but into seven, with a mean duration of 32 m.y., corresponding essentially to Grabau's (1940) seven pulses: the Triassic, Liassic, "Jurassic," Comanchean, Gulfian or "Cretaceous," Paleogene, and Neogene.

Each of these corresponds to an expansion of organic diversity in the pelagic marine realm (development of polytaxy) followed by a decline to an "oligotaxic" state. This pattern appears in global counts of coexisting genera and species as well as in the structure of marine communities, in which the polytaxic state is characterized by the development of superpredators, while the crash leading to oligotaxy is accompanied by the spread of opportunistic generalists (Figure 9.1). Fischer and Arthur tentatively recognized some reflections of this cycle in marine temperature regimes (Figure 9.2), in the oxygenation of the oceans, in carbon isotope ratios, in the ups and downs of the calcite compensation depth, in the development of submarine unconformities, and in other factors.

Their overall conclusion was that oceanic structure and behavior have changed on a time scale of about 32 m.y., responding to some change in general climate: in polytaxic episodes the high latitudes were warmer and the temperature of the ocean mass as a whole was higher than during oligotaxy. However, these fluctuations ride on a much longer oscillation, which will be discussed below. During the last 100 m.y. the Earth has passed through three polytaxic episodes—that of the Late Cretaceous, that of the Eocene, and that of the Miocene—delimited by three oligotaxic ones—the Maastrichtian-Danian boundary crisis, the Oligocene crash, and the current decline. During this time we have experienced the "climatic deterioration" long recognized by the terrestrial paleobotanists (Dorf, 1970). Each episode of polytaxy has been merely a step back "up" in what has been a general "downward" trend toward colder high latitudes and colder ocean masses—a trend that finally culminated in the glacial episode in which we find ourselves now.

The history of pelagic diversity in the Paleozoic offers some support for the existence of the 30-m.y. cycle through the Paleozoic, but the precision of Paleozoic data remains marginal. For that matter, the existence of this cycle in the Mesozoic-Cenozoic is still a matter of debate; Hallam (Chapter 17), for example, finds no convincing evidence for the postulated Mid-Jurassic break, and some of the polytoxic episodes recognized by Fischer and Arthur are split by minor reductions in faunal diversity.

The general causes for the postulated 30-m.y. cycle remain uncertain. The pattern suggests that it was a minor modulation of the long (300-m.y.) greenhouse-icehouse cycle discussed below. I am therefore inclined to think that it, too, was engendered by changes (lesser ones) in atmospheric carbon dioxide pressure and that it, too, expresses imbalances between the rates at which carbon dioxide is added to the outer Earth by volcanism and withdrawn from it by weathering and sedimentation (see discussion below). Indeed, just as the long-range cycle is here attributed to first-order changes in volcanism and in sea level, so the 30-m.y. cycle seems to match shorter fluctuations in these factors (Damon, 1971).

300-Million-Year Level

In the last 700 m.y. the Earth has undergone three major episodes of glaciation, during which ice caps not only covered one or both of the polar regions but extended at times more than

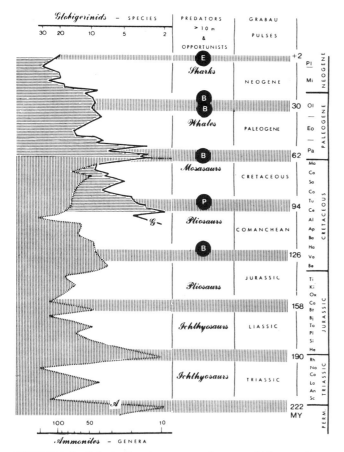

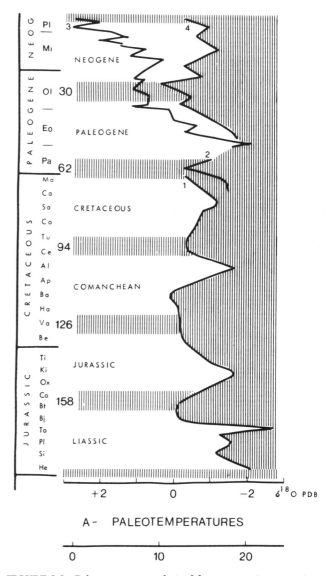

and short duration, was associated with the end of the Ordovician, about 435 Ma.

In between, the world seems to have lacked polar ice caps. The Paleogene of the Arctic region, for example, contains a warm temperate to subtropical forest assemblage including large trees, remains of amphibians, a wide range of reptiles, and, among the mammals, horses and monkeys (Koch, 1963; West and Dawson, 1978). Also, the marine molluscan fauna of western Greenland is distinctly subtropical (Kollmann, 1979).

This paleobotanical evidence for a once very different world is corroborated by the marine record: The present ocean's warm waters are confined to a thin surficial layer in the lower

FIGURE 9.1 Pelagic diversity, superpredators, and blooms of opportunists over last 220 m.y. On left, changes in global diversity. Genera of ammonites (A) and species of planktic (globigerinacean) foraminifera (G), plotted logarithmically. Episodes of increasing diversity are separated by biotic crises of varying magnitude. Crises of moderate and high intensity recur at intervals of approximately 32 m.y. (shaded bands, defining seven cyclic episodes or pulses of diversification: polytaxy). These essentially coincide with transgressive pulses of Grabau (right). Each polytaxic pulse brought superpredators exceeding 10 m in length, a role that has been successively filled by ichthyosaurs, pliosaurs, mosasaurs, whales, and sharks, as shown in middle. Superpredators are known only from stages opposite the names. Mid-Triassic ichthyosaurs: Cymbospondylus and Shastasaurus; Toarcian ichthyosaur: Stenopterygius; Oxfordian pliosaur: Stretosaurus; Albian pliosaur: Kronosaurus; Campanian-Maastrichtian mosasaurs: Hainosaurus and others; Eocene whale: Basilosaurus; Mio-Pliocene shark: Carcharodon megalodon. Biotic crises are accompanied by local mass-occurrences of single pelagic species, rare in normal biotas. These are interpreted as blooms of opportunists and have been plotted in black circles. B, Braarudosphaera, a coccolithophorid; P, Pithonella, a problematicum; E, Ethmodiscus rex, a giant diatom. From Fischer and Arthur (1977).

halfway to the equator (see Chapter 6 for summary and literature). The times of these first-order glaciations (Figure 9.3) are Late Precambrian [about 750-650 m.y. ago (Ma)], Late Carboniferous-early Permian (340-255 Ma), and Pleistocene-Recent, having commenced about 2 Ma and stretching on into the unknown future. Another glacial episode, of lesser vigor

FIGURE 9.2 Paleotemperatures derived from oxygen isotope ratios in calcitic fossil skeletons, assuming constant oxygen isotope ratios in seawater. 1, belemnites, northwestern Europe, uncorrected lat. 45-55°; 2, planktonic foraminifera, South Atlantic, uncorrected lat. 30-32°; 3, planktonic foraminifera, South Pacific, uncorrected lat. 47-52°; 4, planktonic foraminifera, tropical Pacific, uncorrected lat. 7-19°. From Fischer and Arthur (1977).

417

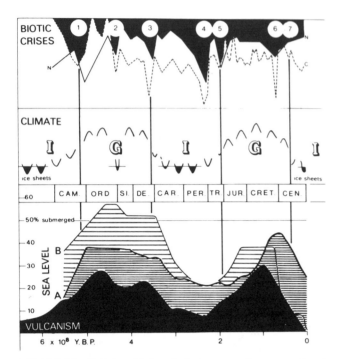

FIGURE 9.3 Relation of inferred climates to secular patterns in volcanism, sea level, and organic diversity. Volcanism: emplacement of plutons in North America, after Engel and Engel (1964). Sea level: A, first-order eustatic curve of Vail *et al.* (1977); B, compromise between North American and Russian records, constructed from Hallam (1977); the scale at left refers to this curve. Biotic record: N, Stehli *et al.*'s (1969) curve of disappearance of animal families; C, net gain-and-loss curve of Cutbill and Funnel (1967), overlap shaded. Inferred climatic states from Fischer (1981); minor oscillations (which may bring about growth of ice sheets, shaded) after Fischer and Arthur (1977). Diagram modified from Fischer (1981).

and middle latitudes. Bottom waters are close to the freezing point throughout the oceans, and the mean temperature of the oceanic water masses lies at about 3°C. While we do not have reliable measurements of paleotemperatures—oceanic or otherwise—for the ice ages of the Paleozoic and Precambrian, it seems likely that their oceans had a temperature and structure rather similar to that of today.

In contrast, studies of oxygen isotopes from Cretaceous oceanic deposits (e.g., Douglas and Savin, 1975) show that the temperatures of the bulk of the Cretaceous ocean masses lay in the vicinity of 15°C, i.e., that the mean temperature of the oceans averaged some 10° above that of the present ones. This suggests that bottom waters were not formed as today, by chilled polar surface waters mixed with a strong dash of meltwater from the polar ice caps. Several alternatives appear possible. Either lightly cooled surface waters descended in the high latitudes, to form a bottom water of simple origin, or, alternatively, bottom waters generated in the paraequatorial dry belts (horse latitudes) to form a deep warm layer, as yet not sampled. A likely compromise is that bottom waters arose from a mixture of both of these processes.

In short, it appears that the Cretaceous and Paleogene periods had a markedly different climate and ocean: tropical

temperatures were much as they are today, while the temperature gradients toward the poles (and, in the ocean, toward the bottom) were very much lower. This implies a more uniform distribution of energy received from the Sun. The most appealing mechanism for this is that of a climatic "greenhouse" (Budyko, 1977; Manabe and Wetherald, 1975), in which an enrichment of atmospheric carbon dioxide inhibits radiation losses of energy into space. Temperatures in the tropical seas would not rise appreciably, because of increased evaporation, but the water content of the atmosphere would increase, and the transport of heat to the higher latitudes would occur largely as latent heat of evaporation, released in the high latitudes by heavy rainfall.

Fischer (1981) has contrasted these climatic states of the Earth as the "icehouse state" and the "greenhouse state" (Figure 9.3). I view the history of the last 700 m.y. as a passage through two great icehouse-greenhouse cycles and into the beginning of a third. Associated phenomena include first-order changes in sea level, mean ocean temperatures, and oceanic aeration, possibly linked to changes in volcanicity and plate motions, as explained below. The transitions from one state to the other appear to be punctuated by the four major biotic crises. A special explanation, in this scheme, has to be invented for the glaciation and biotic crisis at the end of Ordovician time.

Associated Phenomena

Volcanism Long-term secular variations in global volcanism are not easily apprehended. Of the various volcanic processes, those associated with the generation of the oceanic lithosphere rank largest, but their pre-Jurassic record is lost by the recycling of the oceanic crust back into the mantle. On the continents, the great andesitic volcanic piles formed at convergent plate boundaries are largely lost to erosion. The best record of long-term volcanism is probably that of the granitic plutons that form the substructure of these belts (Fischer, 1981).

If we may take the emplacement of such plutons as indicative of volcanic activity in general, and if North America is representative, then a plot of the rate of "granite" emplacement in North America, through geologic time, should serve as an index to worldwide volcanicity. Such a plot, by Engel and Engel (1964) is shown at the bottom of Figure 9.3. It is essentially bimodal, showing one broad (and bifid) peak that matches the inferred greenhouse state of the Ordovician-Devonian and another, sharper peak that matches the Mesozoic greenhouse interval.

Sea-Level Change The eustatic curves by Vail *et al.* (1977) show the fluctuations of sea level relative to the continents. While this demonstrates a lively history of sea-level oscillations, its most generalized version—Vail's first-order curve or Hallam's (1977) curve (here modified by Fischer)—coincide with the inferred succession of icehouse and greenhouse states: the three major glaciations occur within the three lows of the curve, while the greenhouse states correspond to the highs. While glaciation itself drives sea level through oscillations because of withdrawal of water from the hydrosphere into ice caps, the longer persistence of these first-order lows shows them to have been a condition for the growth of ice sheets. The latter merely contributed feedback. For the reasons why high sea levels should coincide with greenhouse states, see below.

Oceanic Aeration   The ocean of our day is moderately well oxygenated throughout, so that animal life, dependent on aerobic respiration, is possible on almost all bottoms and throughout almost all of the water mass. Exceptions are encountered only in localized areas such as the Black Sea, the Cariaco Trench, and the Santa Barbara Basin and, seasonally, in certain tropical belts of upwelling. Because of this, bottom sediments are almost everywhere stripped of organic matter by scavengers and bacteria and are plowed and mixed in the process (Fischer and Arthur, 1977). Accumulation of petroleum source beds is at a minimum.

In contrast, during much of the past, black, organic-rich, finely laminated sediment was widely deposited on marine bottoms deprived of free oxygen. In my experience, widespread anaerobism of this sort peaked in the Ordovician to Devonian interval and again in the Jurassic and Cretaceous (Fischer and Arthur, 1977; Jenkyns, 1980). These times correspond to the greenhouse states. Berry and Wilde's (1979) alternative explanation, that the Paleozoic black shales are holdovers from the poorly oxygenated atmosphere of the Precambrian, fails to explain the earlier (Lower Cambrian) spread of highly oxygenated (red) fossiliferous marine sediments, as well as the recurrence of anoxia in the Mesozoic. Within these episodes, there was a waxing and waning of anoxia, on what Fischer and Arthur (1977) interpreted as a 30-m.y. cycle, as well as on the yet smaller scales of the Milankovitch cycles.

The causes for these variations in oceanic aeration are not resolved. We may think of the ocean in analogy to an organism that digests food: In anaerobic periods, the sea has indigestion. This may be brought on in one of two ways: either by a surfeit of organic matter supplied to it, which overwhelms its digestive capacity, or by a breakdown in the digestive system itself.

Fischer and Arthur (1977) ascribed mainly to the latter cause and linked aeration to the cooling of oceans, which (a) permits a given volume of water to absorb more oxygen while in contact with the atmosphere and (b) favors a vigorous marine circulation, shortening the residence time of water in the depths, between times of recharge at the surface. This explanation thus offers indirect evidence for a warming of seas between times of massive glaciation. On the other hand, the greenhouse state is likely to have increased organic production on the lands, owing to more plant growth in high latitudes, to more widespread rainfall, and to the greater availability of carbon dioxide. It seems likely that during greenhouse states the supply of organic matter from the lands to the oceans may have been greater than it is today. I have therefore come to believe that both factors worked together to promote oceanic anaerobism.

Punctuation by Biotic Crises

Two curves at the top of Figure 9.3 illustrate changes in faunal diversity revealed by the fossil record. N is Stehli *et al.*'s (1969) compilation of the disappearance of animal families; C is Cutbill and Funnel's (1967) analysis, depicting net gain and loss in invertebrate diversity. Six times of large-scale disappearance of taxa—first-order biotic crises, numbered 1-6—stand out in both curves. Of these, numbers 1, 3, and 5 coincide rather well with the boundaries between the climatic states suggested in the middle of the diagram. Number 2 coincides with the brief plunge into glaciation that occurred in the middle of the Early Paleozoic greenhouse. In a previous paper (Fischer, 1981) I sought to relate crises 4 and 6 to the climatic transitions as well, but this does not appear feasible. Both occur too soon, 4 before the breakup of Pangea, 6 before the greenhouse came to an end, as evidenced by the warm nature of the Arctic in Paleogene times (see above). Also, there is now strong evidence for an extraterrestrial origin of crisis 6 (Alvarez *et al.*, 1980). The transition to the Late Cenozoic icehouse state is marked, instead, by the Late Eocene-Oligocene biotic crisis, which does not show on Stehli *et al.*'s curve, is a second-order crisis in Cutbill's and Funnell's compilation, and was strongly developed in the marine realm (Fischer and Arthur, 1977).

Possible Causes

Elsewhere (Fischer, 1981), I have suggested that the fluctuations in atmospheric carbon dioxide content result from fluctuations in the rate of supply (from volcanism) and in the rate of withdrawal (by the linked processes of weathering and sedimentation) (Budyko, 1977; Holland, 1978). Both are linked to a major cyclic process—mantle convection. The hypothesis can only be outlined here.

While there is much uncertainty about the manner in which the lithospheric plates are driven, thermal convection of the mantle is generally taken to be the ultimate cause (e.g., Morgan, 1972). There is no reason to suppose that this process runs at a constant rate, and Fischer (1981) has proposed that the historical pattern in which episodes of continental dispersion and episodes of continental aggregation succeed each other results from a cyclicity in mantle convection in the following manner.

In one phase, the mantle is in a quiescent state, having few and slowly turning cells, which tend to sweep the continental masses together into one pangea. As a result of few cells, the total length of midoceanic ridge is relatively short; because of slow convection and slow spreading rates, the ridge is relatively narrow. It follows (Russell, 1968; Hays and Pitman, 1973) that the displacement of waters from the ocean basins by the ridge is small and that the continents stand high and dry. Contributions of carbon dioxide from the interior to the ocean-atmosphere system are at a low, because both basaltic volcanism in rift zones (which brings carbon dioxide from the mantle) and andesitic volcanism from subduction zones (which recycle lithospheric carbon back to the atmosphere) are minimal. At the same time, the large area of the lands implies that uptake of carbon dioxide by weathering is at a maximum. As a result, a high atmospheric content of carbon dioxide, inherited from a former state, cannot be maintained. Carbon dioxide pressure will drop until the rate of weathering slows and the rate of carbon dioxide withdrawal approaches the rate of volcanic addition. This low balance results in the development of the icehouse state. In this, the growth of a glacial armor over the lands further reduces weathering, providing a negative feedback that may help to explain why the Earth has never turned into a complete iceball.

Mantle convection increases, by development of more cells and of more vigorous convection. New rifts develop, and old ones spread more rapidly. As marine ridges grow in length as well as in width, and as continents spread out by rifting, seawater is displaced, flooding the continents. The increased volcanism in rift belts and in belts of plate convergence raises the output of carbon dioxide to the atmosphere-hydrosphere system. At the same time, flooding of the continents cuts carbon dioxide losses to the lithosphere by weathering. The net result is that atmospheric CO_2 pressure must rise, until the weathering rates, increased thereby, once again withdraw carbon dioxide at rates matching the volcanic addition. This high balance results in the greenhouse state.

The late Ordovician ice age, coming in the middle of a greenhouse episode, at a time of sea-level highs, and associated with a major biotic crisis, seems altogether exceptional and asks for special explanation. One possibility that comes to mind is that of a greenhouse that overshot, producing a cloud cover so dense as to reflect enough of the solar radiation to cool the Earth. The alternative is to find a geologically transient sink for carbon dioxide.

Conclusions and Problems

Global climates have alternated between states susceptible to widespread glaciation (icehouse states) and greenhouse conditions. Two such cycles have been completed in the 600 m.y. of Phanerozoic time. The reasons for them are not firmly established. While several authors (cf., Pearson, 1978) have suggested a tie to changes in insolation, related to the cycle, of galactic rotation, an apparent correlation with sea-level changes and volcanicity suggests an internal cause. This is here sought in hypothetical cycles of mantle convection, which drive sea levels and atmospheric carbon dioxide content by independent pathways linked by a feedback mechanism (weathering). Whereas the Phanerozoic record suggests a length of about 300 m.y. per cycle, a rigorous periodicity throughout Earth history is not implied, inasmuch as mantle behavior must change in a cooling Earth. Nevertheless, it seems likely that we are in the early part of an icehouse state.

Riding on this long cycle are a family of smaller climatic fluctuations, of which one seems to have a periodicity of perhaps slightly more than 30 m.y. In this one, too, the climatic effect may depend on changes in carbon dioxide, but the mechanisms remain obscure.

CONCLUSIONS

In summary, I suggest that the stratigraphic record holds evidence of a wide range of global changes in climatic state. Largest among these are the 150-m.y.(?) alternations between the major greenhouse and icehouse states. We know only the latter, and the traditional attempts to reconstruct the Mesozoic or the mid-Paleozoic world along strictly actualistic lines are grossly inadequate.

Upon this great cycle rides a smaller one, having a period somewhere around 30 m.y. to 36 m.y., which mimics the large one on a smaller scale and which is recorded in its effects on life. The oligotaxic times that happen to coincide with the turnover in the large cycle, from its greenhouse phase to its icehouse phase and vice versa, are particularly pronounced as some of the world's great biotic crises.

Smaller pre-Pleistocene climatic oscillations are seen at the 500,000-yr level, at the 100,000-yr level, at the 50,000-yr level, and at the 20,000-yr level, in round numbers. These appear to match similar periods in ice flux within the Pleistocene, which have been attributed to climatic effects of the Earth's orbital perturbations. These rhythmic events are recorded in a wide variety of sedimentary settings and record a multitude of pathways by means of which climatic change became expressed in sediments. We have only begun to recognize them and are far from any understanding of them.

In the normal course of development, we could expect to slide more deeply into the icehouse state for some millions of years to come, with continued gradual loss of species. The burning of fossil fuels may instead provide a brief brush with the greenhouse state within a generation. That would be a brief passage only, limited by the amounts of fossil fuels available. The effects on climate, I must leave to others more qualified. However, I believe that a full greenhouse of the kinds that existed in the mid-Paleozoic and Mesozoic would require the complete melting of the ice caps and the warming of the oceans as a whole. That would produce a major biotic crisis of the sort that brought about the partial or complete collapse of some of the world's organic communities in Devonian, Triassic, and Oligocene time. That event seems unlikely to occur for another 70 m.y., but even a brief brush with greenhouse conditions may upset the accustomed structure and behavior of atmosphere and oceans. It might thus have marked effects on the biosphere and on human life and history.

REFERENCES

Alvarez, L. W., W. Alvarez, F. Asaro, and W. V. Michel (1980). Extraterrestrial cause of the Cretaceous-Tertiary extinctions, *Science 208*, 1095-1108.

Arthur, M. A. (1979a). Sedimentologic and geochemical studies of Cretaceous and Paleogene pelagic sedimentary rocks: The Gubbio sequence, Dissertation, Princeton U., Part I, 174 pp.

Arthur, M. A., (1979b). North Atlantic Cretaceous black shales: The record at Site 398 and a brief comparison with other occurrences, in *Initial Reports of the Deep Sea Drilling Project Vol. 47, pt. 2*, U.S. Government Printing Office, Washington, D.C.

Arthur, M. A., and A. G. Fischer (1977). Upper Cretaceous-Paleocene magnetic stratigraphy at Gubbio, Italy. I. Lithostratigraphy and sedimentology, *Geol. Soc. Am. Bull. 88*, 367-371.

Berger, A. L. (1980). The Milankovitch astronomical theory of paleoclimates: A modern review, *Vistas in Astronomy 24*, 103-122.

Berry, W. B. N., and P. Wilde (1979). Progressive ventilation of the oceans—an explanation for the distribution of the Lower Paleozoic black shales, *Am. J. Sci. 278*, 257-275.

Bradley, W. H. (1929). The varves and climate of the Green River epoch, *U.S. Geol. Surv. Prof. Pap. 645*, 108 pp.

Briskin, M., and W. A. Berggren (1975). Pleistocene stratigraphy and quantitative oceanography of tropical core V-16 205, *Micropaleontol. Spec. Publ. 1*, 167.

Budyko, M. I. (1977). *Climatic Changes*, American Geophysical Union, Washington, D.C., 261 pp.

Cutbill, J. L., and B. M. Funnel (1967). Computer analysis of the fossil record, in *The Fossil Record*, W. B. Harlan *et al.*, eds., Geol. Soc. London, pp. 791-820.

Damon, P. E. (1971). The relationship between late Cenozoic volcanism and tectonism and orogenic-epeirogenic periodicity, in *Late Cenozoic Glacial Ages*, K. K. Turekian, ed., Yale U. Press, pp. 15-36.

Dorf, E. (1970). Paleobotanical evidence of Mesozoic and Cenozoic climatic changes, in *Proceeding of the North American Paleontological Convention, Vol. 2*, E. I. Yochelson, ed., Allen and Unwin, London, pp. 323-346.

Dorman, F. H. (1968). Some Australian oxygen isotope temperatures and a theory for a 30-million-year world temperature cycle, *J. Geol. 76*, 297-313.

Douglas, R. G., and S. M. Savin (1975). Oxygen and carbon isotope analyses of Cretaceous and Tertiary microfossils from Shatsky Rise and other sites in the North Pacific Ocean, in *Initial Reports of Deep Sea Drilling Project, Vol. 32*, U.S. Government Printing Office, Washington, D.C., pp. 509-521.

Engel, A. E. J., and C. G. Engel (1964). Continental accretion and the evolution of North America, in *Advancing Frontiers in Geology and Geophysics*, A. P. Subramaniam and S. Balakrishna, eds., Indian Geophysical Union, pp. 17-37e.

Fischer, A. G. (1964). The Lofer cyclothems of the alpine Triassic, *Kansas State Geol. Surv. Bull. 169*, Vol. l, pp. 107-150.

Fischer, A. G. (1980). Gilbert—bedding rhythms and geochronology, in *The Scientific Ideas of G. K. Gilbert*, E. I. Yochelson, ed., Geol. Soc. Am. Spec. Pap. 183, pp. 93-104.

Fischer, A. G. (1981). Climatic oscillations in the biosphere, in *Biotic Crises in Ecological and Evolutionary Time*, M. Nitecki, ed., Academic, New York, pp. 103-131.

Fischer, A. G., and M. A. Arthur (1977). Secular variations in the pelagic realm, in *Deep Water Carbonate Environments*, H. E. Cook and P. Enos, eds., Soc. Econ. Paleontol. Mineral. Spec. Publ. 25, pp. 18-50.

Gilbert, G. K. (1895). Sedimentary measurement of geologic time, *J. Geol. 3*, 121-127.

Gilbert, G. K. (1900). Rhythms and geologic time, *Am. Assoc. Adv. Sci. Proc. 49*, 1-19.

Grabau, A. W. (1940). *The Rhythm of the Ages*, Henry Vetch Publ. Co., Peking, 561 pp.

Hallam, A. (1977). Secular changes in marine inundation of USSR and North America through the Phanerozoic, *Nature 269*, 769-772.

Hays, J. D., and W. C. Pitman III (1973). Lithospheric motion, sea level changes and climatic and ecological consequences, *Nature 246*, 18-22.

Hays, J. D., J. Imbrie, and N. J. Shackleton (1976). Variations in the Earth's orbit: Pacemaker of the ice ages, *Science 194*, 1121-1132.

Heckel P. H. (1977). Origin of phosphatic black shale facies in Pennsylvanian cyclothems of midcontinent North America, *Am. Assoc. Petrol. Geol. Bull. 61*, 1045-1068.

Holland, H. D. (1978). *The Chemistry of Oceans and Atmospheres*, Wiley, New York, 351 pp.

Jenkyns, H. C. (1980). Cretaceous anoxic events: From continents to oceans, *J. Geol. Soc. 137*, 171-188.

Koch, D. E. (1963). Fossil plants from the lower Paleocene of Agatdalen (Angmartusuk) area, central Nuqssuaq Peninsula, northwest Greenland, *Medd. Groenl. 172*, 1-120.

Kollmann, H. (1979). Distribution patterns and evolution of gastropods around the Cretaceous-Tertiary boundary, in *Cretaceous-Tertiary Boundary Events*, Vol. II Proceedings, W. K. Christensen and T. Birkelund, eds., U. of Copenhagen, Copenhagen, Denmark.

Manabe, S., and R. T. Wetherald (1975). The effect of doubling the CO_2 concentration on the climate of a general circulation model, *J. Atmos. Sci. 32*, 3-15.

Milankovitch, M. (1941). *Kanon der Erdbestrahlung und seine Anwendung auf das Eiszeitenproblem*, Ed. Spec. Acad. Roy. Serbe, Belgrade, 633 pp., English translation, U.S. Department of Commerce.

Monninger, W. (1979). The section of Tiout (Precambrian/Cambrian boundary beds, Anti-Atlas, Morocco): An environmental model, *Arb. Palaeontol. Inst. Wuerzburg (Germany)*, 289 pp.

Moore, R. C., J. C. Frye, J. M. Jewett, W. Lee, and H. O'Connor (1951). The Kansas rock column, *Kansas Geol. Surv. Bull. 89*, 132 pp.

Morgan, W. J. (1972). Plate motions and deep mantle convection, in *Studies in Earth and Space Sciences*, R. Shagam, ed., Geol. Soc. Am. Mem. 132, 7-22.

Obradovich, J. D., and W. A. Cobban (1975). A time scale for the Late Cretaceous of the Western Interior of North America, in *The Cretaceous System in the Western Interior of North America*, W. G. E. Caldwell, ed., Geol. Assoc. Canada Spec. Pap. 13, pp. 31-54.

Pearson, R. (1978). *Climate and Evolution*, Academic, New York, 274 pp.

Russell, K. L. (1968). Oceanic ridges and eustatic changes in sea level, *Nature 218*, 861-862.

Schwarzacher, W. (1975). *Sedimentation Models and Quantitative Stratigraphy*, Developments in Stratigraphy, Vol. 19, Elsevier, New York, 377 pp.

Stehli, F. G., R. G. Douglas, and N. D. Newell (1969). Generation and maintenance of gradients in toxonomic diversity, *Science 164*, 947-949.

Vail, P. R., R. M. Mitchum Jr., and S. Thompson (1977). Seismic stratigraphy and global changes in sea level, part 4, in *Seismic Stratigraphy*, C. E. Peyton, ed., Am. Assoc. Petrol. Geol. Mem. 26, pp. 83-97.

Van Hinte, J. E. (1976). A Cretaceous time scale, *Am. Assoc. Petrol. Geol. Bull. 60*, 498-516.

Van Houten, F. B. (1964). Cyclic lacustrine sedimentation, Upper Triassic Lockatong Formation, central New Jersey and adjacent Pennsylvania, *Kansas Geol. Surv. Bull. 169*, 497-531.

West, R. M., and M. R. Dawson (1978). Vertebrate paleontology and Cenozoic history of the North Atlantic region, *Polarforschung 48*, 103-119.

The American Association of Petroleum Geologists Bulletin
V. 66, No. 6 (June 1982), P. 750-774, 12 Figs., 2 Tables

Upwelling and Petroleum Source Beds, With Reference to Paleozoic[1]

JUDITH TOTMAN PARRISH[2]

ABSTRACT

Upwelling zones are areas of persistent high organic productivity in the oceans and represent one type of setting for the deposition of petroleum source beds. Upwelling currents are driven by winds associated with the major features of atmospheric circulation, and the locations of ancient upwelling zones can be predicted from global atmospheric circulation models. Qualitative circulation models for the past are now made possible by the availability of good reconstructions of past continental positions and paleogeography. Atmospheric circulation was modeled on paleogeographic reconstructions for the following Paleozoic stages: Franconian (Late Cambrian), Llandeilo-Caradoc (Late Ordovician), Wenlock (Late Silurian), Emsian (Early Devonian), Visean (Early Carboniferous), Westphalian (Late Carboniferous), and Kazanian (Late Permian).

Four types of persistent upwelling currents are recognized. Coastal upwelling currents (e.g., the Peru Current) are the most familiar and are the types usually cited as petroleum source-bed settings. However, three additional types of upwelling have also been important in the past, especially during times of high sea-level stands. They are associated with divergence under stable atmospheric low-pressure systems and comprise: (1) symmetrical divergence at the equator, (2) symmetrical divergence in wide oceans at about 60° lat. (e.g., around Antarctica), and (3) radial divergence in more restricted oceans at about 60° lat. (e.g., around the southern tip of Greenland).

To the extent that the data allow reliable conclusions, the distribution of Paleozoic petroleum source beds appears to correspond closely to the distribution of predicted upwelling zones. Coupled with sea level models, the upwelling model can be particularly powerful, because while sea-level models predict likely times of source-bed deposition, the upwelling model contributes information on the sites of source-bed deposition.

INTRODUCTION

The preservation, maturation, migration, and trapping of oil are critical stages in determining the location of economically valuable deposits, but the distribution of oil is also limited by the original distribution of the organic matter from which it was derived. Organic matter is not randomly dispersed over the earth's surface. It has long been recognized that the production of marine algal organic matter in particular is concentrated in the small areas of the world's oceans occupied by persistent upwelling currents, where essential nutrients are recycled into the photic zone. As it is generally accepted that oil is derived primarily from algal organic matter (Dott and Reynolds, 1969; Philippi, 1974; Tissot and Welte, 1978, and references therein, especially Chapter 4), researchers have suggested repeatedly that upwelling zones are excellent settings for petroleum source-bed deposition (Trask, 1932; Brongersma-Sanders, 1948; Dott and Reynolds, 1969; Dow, 1978; Demaison and Moore, 1980). However, this hypothesis has not been used extensively in predicting source-bed distribution. Insofar as we recognize that upwelling commonly occurs off the west coasts of continents, and that the present west coasts of continents have not had very different orientations in the past, we have been able to attribute certain source sequences to upwelling. Good examples are the middle Miocene Monterey Formation of California and the Upper Permian Phosphoria Formation of the Rocky Mountains, both of which were deposited on west-facing coasts. The identification of these as upwelling deposits, however, has been a matter of hindsight rather than foresight. Continental drift reconstructions have opened the way to prediction of petroleum source regions based solely on the west-facing coast model. Sheldon (1964) and Freas and Eckstrom (1968) have used this model to explain the distribution of phosphorite deposits, also related to upwelling. The key to using the upwelling model to predict areas of source-bed deposition is to be able to construct global models of upwelling for the past.

Persistent, large-scale upwelling currents are driven by winds connected with major atmospheric circulation systems, such as the subtropical high-pressure systems off California and Peru. Therefore, one must predict upwelling from models of atmospheric circulation (Parrish et al, 1979). Using recent paleogeographic reconstructions by Scotese et al (1979) and following the lead of Hansen in Ziegler et al (1977), nonnumerical models of atmospheric circulation

[1] Manuscript received, Feb. 17, 1981; revised and accepted. Feb. 12, 1982.

[2] Dept. of Geophysical Sciences, University of Chicago, Chicago, IL 60637. Present address: U.S. Geological Survey, Denver, CO 80225.

The petroleum source-bed data used in this paper were largely taken from source-bed geochemistry files supplied by Amoco Production Co., Tulsa. The writer is grateful to Amoco Production Co. for making the files available.

Kirk S. Hansen contributed significantly to the construction of the circulation models. I have, however, modified the original maps that we worked on. I have benefited greatly from discussions with several people, especially Gerard Demaison and James A. Momper. These men, along with P. N. Froelich, E. J. Barron, and R. P. Sheldon, read earlier versions of the manuscript and made many useful comments. I am particularly grateful to the reviewers, H. D. Hedberg and W. W. Hay, for their suggestions. Errors in this paper are entirely my own.

This work was supported by a grant to A. M. Ziegler from Amoco International Oil Co. Rebecca Curtis drafted the figures and Gwen Pilcher and Barbara Shuey typed the manuscript.

were constructed for several geologic ages in the Paleozoic. The locations of major upwelling currents were predicted from the circulation models. To test the models, which were constructed on theoretical grounds, the distribution of the predicted upwelling zones and the location of petroleum source beds were compared.

The purpose of this paper is to present the atmospheric circulation and upwelling models and to analyze their usefulness and limitations for predicting past areas of source-bed deposition. In addition, I will review some of the basic principles of atmospheric circulation and upwelling, to aid in the understanding of how the models were constructed.

PREVIOUS WORK

Distribution of Source Beds

The first effort to put the distribution of petroleum into some kind of global framework was the monumental study by Trask (1932), in which he analyzed several thousand samples of aquatic sediments to determine the present global distribution of organic matter. He concluded that upwelling zones would be favorable settings for the deposition of source beds, based on the quantity of organic matter present.

In 1948, Brongersma-Sanders published a paper discussing in great detail the upwelling zone off South-West Africa, specifically the phenomenon of mass mortality of fishes caused by plankton blooms, its connection to upwelling, and its effects on the sediment. In that paper, she also suggested a connection between upwelling and petroleum source beds, which were developed in subsequent papers (e.g., Brongersma-Sanders, 1966).

In a recent paper, Demaison and Moore (1980) also considered the present distribution of organic matter in an attempt to construct a general model for the distribution of petroleum source beds. Their model deals with the requirement of anoxia for the qualitative preservation of oil-prone organic matter and is the first to put the distribution of anoxic environments into a global and predictive framework. Demaison and Moore (1980) included upwelling zones as one of the four anoxic environments they discussed, along with large anoxic lakes, anoxic silled basins, and anoxic open-ocean basins. They showed that the existence of each of these four environments is at least partially controlled by climate, and therefore potentially predictable by climatic-oceanographic models.

In a brief but comprehensive overview, Dow (1978) summarized the salient aspects of source-bed deposition on continental slopes and rises. He cited high productivity partly due to upwelling and a higher proportion of algal organic matter among the important features that favor a source potential for slope and rise sediments.

While the distribution of source beds through time and space is the major concern here, several other studies attempting to put the distribution of petroleum into a global context should be mentioned. These studies were concerned primarily with the disposition of oil fields, which may not correspond in location and age to the sources of the oil. Nevertheless, they were the first to take paleogeography and paleoclimatology into account in explaining the distribution of petroleum. Irving and Gaskell (1962), Deutsch (1965), and Irving et al (1974) considered the distribution of oil in the context of drifting continents. Irving and Gaskell's study (1962) was a remarkable early effort that used paleomagnetic techniques to orient continents to their paleolatitudinal positions. They observed that the fields tended to occur at relatively low paleolatitudes. Later paleomagnetic studies, including that of Deutsch (1965), have modified but not contravened this observation.

In the early 1970s, paleomagnetic techniques had improved and the amount of data available had expanded. These advances made feasible the more detailed study by Irving et al (1974), in which they plotted numerous source beds and reservoirs against paleolatitude. Their data suggest that source-bed distribution is as dependent on the distribution of land area as on paleolatitude. Irving et al (1974) proposed a correlation between times of rapid sea-floor spreading, development of petroleum basins, and volcanism. They recognized high plankton productivity as key to the distribution of source beds, although they erroneously attributed high productivity to increased volcanism (see Pisciotto and Garrison, 1981, p. 103; Calvert, 1966a). Irving et al (1974) were also early proponents of the hypothesis that source-bed deposition is linked to transgressions, a hypothesis later promoted by Arthur and Schlanger (1979) and Summerhayes (1981), among others. Irving et al (1974) did not explain how all these processes—sea-floor spreading, volcanism, etc—are related to the observed paleolatitudinal distribution of oil.

Although none of these studies resulted in a general predictive model for the distribution of source beds, they highlighted some of the key factors. Most consistently mentioned was the importance of high organic productivity in at least some settings for source-bed deposition. High productivity is the most predictable factor, temporally and geographically, controlling source-bed deposition, hence the attention to it in this study.

Paleoclimatic Models

There are three approaches to the modeling of atmospheric circulation and associated paleoclimatic patterns. An outline of these approaches illustrates some of the problems encountered in paleoclimatic modeling and defines the context for this study.

Models based on direct measurements.—For the Late Cretaceous and later, isotopic data on sea-surface paleotemperatures are available (e.g., Savin et al, 1975), and paleogeographic and biogeographic data are better than are those available for earlier periods. Thus, for the latter part of the Phanerozoic, fairly reliable quantitative data on important climatic parameters can be entered into existing numerical climate models derived from our understanding of present atmosphere and ocean dynamics (CLIMAP Project Members, 1976; Moore et al, 1980; Barron et al, 1981). Not only do we know something about sea-surface temperature for the late Phanerozoic, but also ice sheet extent and albedo. Albedo is related to the distribution of vegetation as well as ice, and Barron et al (1980) stressed its importance as a variable in climatic modeling. Unfortunately, data are so scattered for the Paleozoic that the direct approach is presently unfeasible.

Inference models.—A second, less quantitative, approach is similar in principle to the first. Diversity gradients (Fischer, 1960) and climatically controlled deposits such as evaporites and reefs (Rosen, 1975; Heckel and Witzke, 1979) can be used to infer paleoclimatic patterns, provided good paleogeographic reconstructions are available. Although this approach is enlightening to a degree, it involves circular reasoning. The hypothesis is that one can use the distribution of sediments and organisms to infer paleoclimatic patterns. There is no rigorous independent test of the hypothesis, only the general one that ancient sediment and biogeographic patterns were similar to modern ones, although an internally consistent reconstruction involving many types of data, such as that by Heckel and Witzke (1979), can be convincing. However, one can only loosely apply modern climatic patterns to the past. Ancient, particularly Paleozoic, continental positions were very different than they are now, and this surely had a profound effect on the climatic patterns. Furthermore, we often have an overly simplistic view of even modern climatic patterns. A commonly overlooked feature of present climate, for example, is the asymmetry of the climatic zones on opposite sides of the oceans (Parrish and Ziegler, 1980). At the same latitude, the ocean can be as much as 18° F (10°C) warmer on one side than on the other side. For example, in July the Atlantic is 73°F (23°C) at 20°S off eastern South America and 61°F (16°C) off southwest South Africa at the same latitude. A more extreme asymmetry must have prevailed in the Late Permian. During that time, there was an extraordinarily long fetch for the equatorial current, which, when it reached the western side of Tethys, would have been forced into two Gulf Stream-type currents flowing north and south. These currents would have carried very warm water into high latitudes, thereby creating a wide tropical zone on that side of Pangea. The Late Permian world was characterized by the greatest asymmetry ever, with tropical faunas extending to 40 to 50° lat. north and south in eastern Pangea (Ziegler et al, 1981). As Humphreville and Bambach (1979) pointed out, this asymmetry misled Stehli (1970) when he interpreted the distribution of the tropical Tethyan brachiopods of the Late Permian.

Another pitfall of the approach using paleoclimatic indicators to infer paleoclimate is the temptation to group data over too long a time interval. For example, evaporites tend to form in the dry subtropics around 30° lat., and as a continent moves through that latitude it might have evaporites deposited over much of its area. The evaporites would be diachronous, however, and the use of evaporites from an entire period might lead to the modeling of a dry zone that is too wide, although a wide dry zone might result in other circumstances (e.g., see Robinson, 1971). This type of problem prompted Meyerhoff (1970) to conclude that the distribution of paleoclimatic indicators does not support continental drift.

Theoretical approach.—A third approach is to separate the modeling from the data, an approach that is appropriate where limited paleoclimatic data are available. This approach is more rigorous in a number of ways than is the previous one. The modeling must be based on the paleogeography and the more general principles of atmospheric circulation. One then predicts from the climate models the distribution of paleoclimatically controlled sediments or organisms. The available date on paleoclimatic indicators thus independently test the climate models. Ziegler et al (1981) used this approach, and it is the one employed in this study. The paleoclimatic indicators used to test the models here are source beds, and companion papers to this one explore the distribution of phosphorites (Ziegler et al, 1979; Parrish and

Humphreville, 1981; Parrish and Ziegler, 1981), which are also related to upwelling (Sheldon, 1964; McKelvey, 1967; Burnett, 1977), and evaporites and coals (Ziegler and Parrish, 1981; Parrish et al, in press).

ATMOSPHERIC CIRCULATION

Two physical processes govern atmospheric circulation, (1) the distribution and exchange of heat over the earth's surface and (2) the rotation of the earth. The exchange of heat between the equator and the poles provides the primary driving force behind atmospheric circulation. The rotation of the earth and the differential heating between land and ocean control different aspects of the general circulation pattern. In the following sections, I discuss the form and cause of several features of atmospheric circulation that constitute present-day circulation and that have been incorporated in the models of atmospheric circulation for the Paleozoic. These references have been particularly helpful to the formulation of the following discussion: Perry and Walker (1977), Lorenz (1967), Petterssen (1969), Barry and Chorley (1970), and Riehl (1978).

Zonal Circulation

The primary heat exchange path over the earth's surface is between the equator and the poles, and in discussing that path, one can demonstrate the importance of the earth's rotation. The surface of the earth receives more sunlight per unit area at the equator than it does at the poles, creating a temperature gradient that provides energy for atmospheric circulation. Heated air rising at the equator flows poleward while cool air sinking at the poles flows equatorward. However, the rotation of the earth does not permit direct heat exchange between the poles and the equator. The effect of rotation, called the Coriolis effect, deflects equatorward-moving air to the west and poleward-moving air to the east. The present rotational speed is such that air rising at the equator is deflected eastward in a curving path until it is flowing nearly parallel with latitude at about 30°. Air sinking at the poles is deflected increasingly westward until it is moving parallel with latitude at about 60°. As the temperature gradient still requires heat exchange between equator and poles, there are intermediate cells, called Ferrell cells, between about 30 to 60° lat. in each hemisphere. The dynamics of Ferrell cells and the heat exchange they effect require that the equatorial air sink at 30°, completing the so-called Hadley cells, and the polar air rise at 60° (Fig. 1). Rising air causes low barometric pressure and sinking air causes high barometric pressure, so the pressure regime of the circulation is as follows: high pressure at the poles and at 30° lat. in each hemisphere, low pressure at the equator and at 60° lat. (Fig. 1). The wind and pressure regime is zonal, that is, parallel with latitude. The zonal aspect of atmospheric circulation has been recognized since the 1800s (Lorenz, 1967) from wind patterns on the oceans. As our understanding of atmospheric dynamics has increased, the zonal model has needed little improvement.

With regard to the zonal component of atmospheric circulation, I made the following assumption in constructing the models, that the present zonal pattern, described above, was the same in the past as it is now. Two points bear on this problem.

Rotation.—The earth's rotation was certainly faster in the past than it is now (J. W. Wells, 1963; Scrutton, 1965, 1978; Mohr, 1975), having been slowed through time by tidal friction (Sundermann and Brosche, 1978). The effect of a more rapid rotation would be to compress the climatic zones equatorward until at some rotational speed the heat exchange between the equator and each pole would require more than three zonal cells to be effected. Recent modeling experiments (Hunt, 1979) suggest that the rotational speed would have to be much higher than that postulated for the Paleozoic in order for the present zonal pattern to break down. The rotational speed for the Paleozoic is thought to have been 15 to 20% faster than it is today (Scrutton, 1978), amounting to a compression of about 4 to 6° lat. in each hemisphere. In the maps presented here, I have not attempted to model this compression of the climatic zones. As the error in positioning the continents is 5 to 10° (C. R. Scotese, personal commun.) and the modeled positions of the atmospheric highs and lows are also subject to error, I felt that little would be gained by trying to take the zonal compression into account. In any case, the predicted upwelling zones would still be in their indicated positions or very close to them.

Obliquity.—I have also assumed that the tilt of the earth's rotational axis relative to the plane of the ecliptic was the same as it is today. Milankovitch (1941), Hays et al (1976), A. L. Berger (1976, 1978, 1980), and others have recognized that variations in the earth's orbital parameters, obliquity among them, could have significant impact on paleoclimate. However, the cycles resulting from these variations are quite short on a geologic time scale. The longest cycle recognized so far is about 413,000 years and related to orbital eccentricity. This cycle might be evident in the Pleistocene deep-sea record (Briskin and Harrell, 1980), but our chances of seeing its effects in the distant past are almost nil, largely due to imprecision in correlation.

A. L. Berger (1976) has calculated variations in the earth's orbital parameters for the last 5 m.y. During this time, obliquity varied with a slight cyclical "wobble," never by more than about 3° from its present value. While there is no physical or astronomical basis for assuming that obliquity has been very different from this pattern, large-scale variations in obliquity have been proposed in the geologic literature (Wolfe, 1978, 1980; G. E. Williams, 1975, 1976) to explain some apparently anomalous thermal events. In order to explain the high-latitude distribution of some apparently warm-climate plants in the Eocene, Wolfe (1978, 1980) suggested that the rotational axis was more nearly perpendicular to the plane of the ecliptic than it is now. His reasoning was that if obliquity were near zero, the equator-to-poles gradient would be much reduced. However, this may not be the case. With an obliquity of zero, the poles would be slightly warmer than their present winter temperatures, but slightly cooler than their present summer temperatures. Indeed, the radiative heat loss from the poles might result in their being very cold continually through the year (A. L. Berger, personal commun.). This is a simplistic view that ignores the effects of geography, but it is possible that the average temperature gradient cannot be reduced by changes in obliquity. It is therefore likely that the reduced gradient in the Eocene and at other times in the past resulted from a combination of other processes, such as differing continental positions and the wider extent of epicontinental seas. On the other hand, if it can be demonstrated that there was no seasonality during these times, then obliquity reduction will have to be considered.

G. E. Williams (1975, 1976) proposed an obliquity of 90° for the late Precambrian diamictites, which he and others have equated with tillites, in areas that apparently lay in low latitudes at the time of deposition of the sediments. The thick diamictites alternate stratigraphically with thick dolomites, which G. E. Williams (1976) took as evidence that the climate alternated between very cold and very warm. All other things being equal, if the rotational axis lies in the plane of the ecliptic, the yearly insolation cycle at the "solstices" and "equinoxes" would consist of the following: (1) continuous illumination of one pole and the equator on the terminator, receiving very little sunlight, (2) day-night cycle similar to present one, with the insolation and geographic equators coinciding as during the modern spring and autumn equinoxes and little sunlight at the poles, (3) continuous illumination of the other pole, the equator again receiving little sunlight, and (4) normal day-night cycles as in (2). The glaciers would have to form during the solstices (1 and 3) and survive the equinoxes (2 and 4) each year. The alternation between dolomite and tillite is also difficult to explain in this context. Therefore, an obliquity of 90° is in itself probably insufficient to explain the apparent equatorial glaciations during the Precambrian. Proposed solutions of this problem and the problem of warm poles have rarely included analyses of the effects of different paleogeographies (an exception is the work of Barron et al, 1981), and the causes and effects of ice-free polar regions are not well understood. The solutions to these problems await further research.

It is difficult to overemphasize the importance of the zonal aspect of atmospheric circulation, particularly for modeling

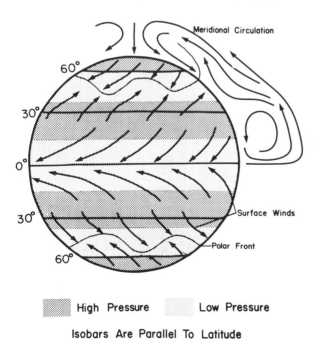

High Pressure   **Low Pressure**

Isobars Are Parallel To Latitude

FIG. 1—Distribution of surface winds and atmospheric pressure on a homogeneous earth.

425

past circulation. Today's world is, in a sense, a "worst case" for zonal circulation, since the major continents trend predominately north-south across the climatic zones. Yet the zonal aspect of circulation is evident throughout the world's oceans. The zonal pattern predominates in the wide Pacific Ocean and is fully developed in the circum-Antarctic Southern Ocean. Furthermore, the positions of the major high and low pressure cells, which form in response to the presence of continents, generally conform to the belts of high and low pressure predicted by the zonal model. How zonal circulation is modified into cellular circulation is explained in the following section.

Cellular Circulation and Seasonality

Continents disrupt the zonal circulation because the thermal properties of land and water differ. Land acquires and loses heat more rapidly than does water and, as a result, land temperatures fluctuate more than do ocean temperatures. Therefore, a temperature contrast usually exists between a continental landmass and its adjacent oceans. Seasonality is important because some cellular circulation features are created and destroyed on a seasonal basis. The latitudinal positions of the continents are also important, because at mid-latitudes the contrast between land and sea temperatures is greater, thus enhancing the cellular nature of the circulation.

Permanent high-pressure cells occur over the poles because those areas receive the least sunlight per unit area and are therefore continually cold. The northern and southern polar cells are quite different, however, because the distribution of land and sea at each pole is different. At the south pole, the high-pressure cell is well defined owing to the centering of the cold land over the pole and the contrast of the temperature regime of the land with the relatively warmer adjacent ocean. The south polar high pressure and the surrounding oceanic low-pressure belt are most intense during the southern winter, when the polar land is the coldest. In contrast, the north polar high-pressure cell is more diffuse, particularly during the northern winter, when it is displaced toward the relatively colder Asian and North American continents.

The zonal model predicts a belt of low pressure at about 60° lat. north and south. This predicted low-pressure belt is very well developed in the Southern Hemisphere, where an ocean girdles the earth at just the right latitude. This belt persists year round without disruption, although its latitudinal position shifts a few degrees with the seasons. The low pressure is strongest during the southern winter, when the temperature contrast between Antarctica and the Southern Ocean is greatest, and local intense cells develop during that season in the shallow embayments in the coastline.

In the northern hemisphere, the 60° low-pressure belt is continuous only in the northern summer, and then it is very weak. During the northern winter, this low-pressure belt is expressed as two strong low-pressure cells, one centered at the southern tip of Greenland, the other centered south of the Aleutians. These cells develop over oceanic areas in response to the cooling of the adjacent continents.

A belt of high pressure is predicted for about 30° lat. north and south by the zonal model. As in the previous case, the actual pattern is closer to the prediction in the Southern Hemisphere, owing to the lesser continentality there than in the Northern Hemisphere. The high-pressure zones develop best during the winters in both hemispheres. During the summers, low-pressure cells developing over the heated continents break the high-pressure zones into cells. This effect is particularly marked in the Northern Hemisphere, where a very intense low-pressure cell develops over southwestern Asia. The intensity of this cell creates a large temperature contrast in the Northern Hemisphere, resulting in a concomitant increase in the intensity of the adjacent high-pressure cells over the oceans.

Low pressure is expected and observed at the equator. Local intensifications of the low pressure over the relatively warmer continents track the thermal equator, being centered slightly south of the equator in January and north of the equator in July. The positions of all the zonal bands shift as much as 10° as the earth moves between the summer and winter solstices. The shift is particularly noticeable in the more stable zonal elements (i.e., the circum-Antarctic low, the equatorial low, and the oceanic subtropical high-pressure cells). The shift is related to the shift of the thermal equator, whose position is also partly determined by continentality. Note that the thermal equator lags behind the insolation equator, which has a total shift of 46° from tropic to tropic.

A final feature of the present circulation, the Asian monsoon, illustrates particularly well the importance of seasonality. The Asian continent combines three features that are important to the establishment of the seasonally alternating circulation pattern known as the monsoon. Asia is a large continent, which means that the central region becomes hotter in the summer and cooler in the winter because it is isolated from the ameliorating influences of the oceans. Second, Asia has a very high mountain range, the Himalayas, on the south, and highlands, comprising several mountain ranges, out to the coast on the east. During the winter, these mountains shield the interior of the continent from the ameliorating influences of the Pacific and Indian Oceans, permitting the development of a very intense high-pressure cell. In the summer, the Himalayas channel and intensify the ascending air of the low-pressure cell that develops over Afghanistan and Pakistan, and they also protect this hot region from cooler air masses to the north. The result is the development of strongly contrasting seasonal circulation. The importance of the monsoon lies in two features. One is that the monsoon accounts for the greatest disruption of the zonal pattern, a disruption that is total in the vicinity of the continent during the northern summer. The second feature is that winds flowing into the summer low-pressure cell and out of the winter high-pressure cell and the oceanic currents driven by those winds cross the equator. Indeed, the summer low has the effect of intensifying the high pressure cell in the southern Indian Ocean. Everywhere else on earth, as predicted by the zonal model, the circulation in the two hemispheres is more or less independent.

ATMOSPHERIC CIRCULATION AND UPWELLING

Although upwelling currents occur on all scales and are caused by a variety of processes, the ones of particular interest to petroleum geologists are large-scale, persistent, and wind-driven. The most familiar example of this type of upwelling current is the Peru Current. The Peru Current is part of a coastal upwelling system that is highly productive, containing one of the world's richest fisheries. The shelf sediments under the Peru Current are rich in organic matter (Manheim et al, 1975) and the sediments are oxygen-poor to anoxic (Goering and Pamatmat, 1971), thus appearing to be excellent petroleum source bed precursors. The Peru Current is typical of modern upwelling systems, occurring on the west coast of a continent and centered at about 30° lat., shifting position north and south with the seasons. For this discussion, however, the key feature of the Peru upwelling system is that it is driven by winds associated with the permanent subtropical high-pressure cell over the southeastern Pacific. These winds blow northward, parallel with the coast, and not offshore. The fact that the wind parallels the coastline is common to all the major coastal upwelling systems, and provides a springboard for the discussion of upwelling mechanisms. For additional discussion of upwelling, see the excellent review by Smith (1973).

Upwelling Dynamics

According to an account by Perry and Walker (1977), in 1902, Fridtjof Nansen published a paper in which he described the motion of ice floes blown over the surface of the sea by winds. Nansen noticed that the ice did not move directly with the wind, but at an angle 20 to 40° to the right of the wind direction. He concluded that this was due to the Coriolis effect. Subsequently, his conclusion was confirmed mathematically by Ekman (1905). In summary, a freely moving particle set into motion on the earth's surface does not maintain a straight path, but is deflected, owing to the rotation of the earth. This deflection is called the Coriolis effect, described in a previous section. To understand the consequences of the Coriolis effect on wind-driven water, imagine a column of water comprising several hypothetical horizontal layers. As the wind blows over the water, it sets into motion the topmost layer of water, which flows at an angle to the right of the wind. This layer of water is in frictional contact with the layer below it, which moves when the surface layer moves. However, the second layer is also deflected by the Coriolis effect, and consequently moves in a path that is at an angle to the direction of motion of the surface layer and therefore at a greater angle to the wind direction than that of the surface layer. The second layer sets into motion the next deeper layer and so on until, at some point below the surface, water is moving in the opposite direction from the wind, and still deeper, water is moving in the same direction. Since much of the frictional energy is lost as heat, each successive layer of water moves slower than the one above it until at some depth no motion occurs that can be attributed to the wind at the surface. The motion and speed of each layer of water can be represented by a vector and together the vectors describe a spiral, called the Ekman spiral. Adding all these vectors, one finds that the net transport of water is perpendicular to the wind direction, to the right in the Northern Hemisphere and to the left in the Southern Hemisphere. This is called Ekman transport, and the water from the point of zero motion to the surface is called the Ekman layer. A comparison of good global atmospheric and oceanic circulation maps will reveal the global pattern of Ekman transport and its importance. Knowing this, we can describe two types of upwelling zones, coastal upwelling and open-ocean divergence.

Coastal Upwelling

Petroleum geologists are most familiar with coastal upwelling because oil source bed precursors occur under such upwelling systems. Coastal upwelling that results in long-term high productivity requires steady wind that prevails from the proper direction, determined by the Coriolis effect. In the Northern Hemisphere, the net transport of wind-driven water is 90° to the right of the wind direction, meaning that the wind must blow parallel to the coast with the coast at the left. The net transport of water is then offshore, and the water moved away from the shore is replaced from below. In the Southern Hemisphere, the wind must blow with the coast on the right, because the net transport of water is to the left of the wind.

The major coastal upwelling zones today occur off western North America, western South America, and northwestern and southwestern Africa (Wooster and Reid, 1963). Three important coastal upwelling zones do not occur in the usual setting of a continental west coast: one off northern South America and Yucatan (Curl, 1960; Richards, 1960; Fukuoka et al, 1968; Corredor, 1979) and seasonal upwelling zones off Somalia and possibly western India. Together, these seven coastal upwelling zones constitute three major types: meridional, zonal, and monsoonal (Fig. 2). Meridional upwelling occurs on west-facing, longitudinally oriented coastlines at about 30° lat. Zonal coastal upwelling occurs on north- or south-facing coastlines that are situated at the proper latitude relative to the major zonal wind systems, described in previ-

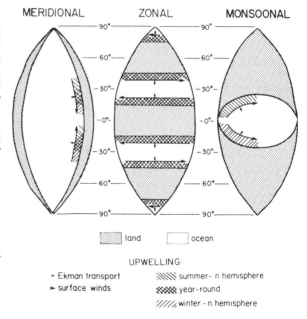

FIG. 2—Schematic diagram of coastal upwelling settings. Meridional coastal upwelling is associated with subtropical high pressure cells (e.g., off California). Zonal upwelling is driven by zonal winds (see Fig. 1) discussed in text—this type occurs off northern Venezuela, for example. Monsoonal upwelling is seasonal and driven by cross-equatorial winds associated with monsoons. This type is represented by Somalian upwelling. Models from K. S. Hansen and A. M. Ziegler, unpub.

427

ous sections. Monsoonal upwelling occurs when monsoonal winds blow parallel with a coastline in the proper direction, as do the summer monsoonal winds off Somalia (Swallow, 1980 et seq.) and western India (Banse, 1968; Sharma, 1978; cf. Taft, 1971). Upwelling may be partly or largely responsible for the anoxia in the Indian Ocean which Demaison and Moore (1980) cited as an example of an anoxic open-ocean environment, rather than as an upwelling environment.

Zonal coastal upwelling has the potential to be the most extensive (Fig. 2), not being limited in length by the Coriolis effect and the confines of the zonal climate pattern as are meridional upwelling currents (Kindle and O'Brien, 1974). This type of upwelling zone has not been considered before in the petroleum literature because attention has been directed to meridional upwellings, which are the predominant type in the modern world, owing to the preponderance of meridionally oriented coastlines. Zonal coastlines were more common during certain periods in the earth's history. The zonal upwelling zones in Figure 2 are schematic and represent the situation that would exist if the zonal wind pattern were undisturbed by coastline irregularities or unfavorable continental configurations. At no time was this perfect situation realized, so in order to identify past zonal upwelling areas, it is necessary first to model the atmospheric circulation.

As Demaison and Moore (1980) pointed out, not all upwelling zones are presently anoxic environments. For example, the northwest African upwelling is associated with a pronounced oxygen minimum but not anoxia. However, anoxic sediments of Cretaceous (Tissot et al, 1980; Lancelot et al, 1977), Eocene (Lancelot et al, 1977), and possibly Miocene (Lancelot et al, 1977; Diester-Haass and Schrader, 1979) age were deposited there. Indeed, variability is characteristic of upwelling zones, not only in the amount of organic carbon preserved, but also in the distribution of siliceous organisms, phosphorite, and glauconite, which are also considered characteristic of upwelling sediments. The causes of variability in upwelling sediments are not readily apparent and are probably complex. For example, they may be partly due to climatic changes and changes in sea level.

Open-Ocean Divergences

The second type of upwelling is open-ocean divergence, of which there are three types: cellular (radial), zonal, and equatorial (both symmetrical). These three types are similar in that they all occur under atmospheric low-pressure systems; they differ only in the configuration of the winds that create them. Low-pressure systems occur where a body of air is rising relative to the surrounding air, meaning that air flows into the low-pressure system from the high-pressure areas around it. The winds do not flow directly toward the center of the low-pressure system, however, but rather spiral into it owing to the Coriolis effect and friction with the earth's surface, so that the wind vectors are nearly parallel with the isobars (Fig. 3). In the Northern Hemisphere, the winds spiral counterclockwise into low-pressure cells. Because the net transport of wind-driven water is 90° to the right of the wind direction in the Northern Hemisphere, water is driven directly away from the center of the low-pressure system (Fig. 3). This water is replaced from below, resulting in upwelling. Where the low-pressure systems are persistent on at least a seasonal time scale, they are associated with high

productivity. In a low-pressure belt that lies within one hemisphere, such as the one around Antarctica, the winds comprising the westerlies and the polar easterlies converge and rise from opposite directions (Fig. 3; see also Fig. 4). Again, water is transported 90° to the wind direction, away from the center of the low pressure, and upwelling occurs. The third type of divergence occurs at the equator (Fig. 3; see also Fig. 4). There, winds converge from the same direction, but because the Coriolis effect changes sign across the equator, water is driven away from the equator to the north in the Northern Hemisphere and to the south in the Southern Hemisphere. The system is complicated by the presence of the equatorial counter-currents, but the divergence and higher productivity are characteristic (e.g., Sverdrup et al, 1942, Figs. 172, 198). The Coriolis effect goes to zero at the rotational equator, so exceptionally steady and strong

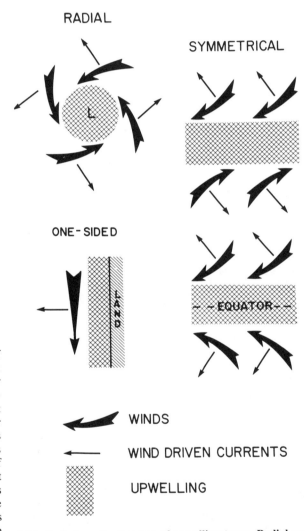

FIG. 3—Schematic diagram of upwelling types. Radial and symmetrical upwelling occurs under stable, low atmospheric pressure systems. Except for equatorial area, diagram illustrates wind and water currents for northern hemisphere. Southern hemisphere wind patterns would be mirror images of ones pictured here, with water transported to left of wind (facing downwind).

tradewinds are required to maintain upwelling of the vigor required to affect productivity. The tradewinds are the strongest on the eastern sides of oceans, and equatorial upwelling often results in higher productivity there (e.g., in the Pacific, Fig. 4). Higher productivity is usually not noted in the equatorial regions on the western sides of oceans (Fig. 4), although there is some evidence that during glacial time at 18,000 years B.P. and at times during the late Mesozoic and Cenozoic, strong equatorial divergence extended all the way across the Pacific (Moore et al, 1980; Leinen, 1979; K.J. Miskell, personal commun.).

The productivity associated with divergence upwelling has been overlooked as a possible source for petroleum source beds, probably because today such divergences occur only over very deep ocean (Fig. 4), and insufficient organic matter for oil generation is preserved in deep-ocean sediments no matter how productive the overlying surface water is. Three factors prevent the preservation of organic carbon in deep-ocean sediments. First, organic matter must settle through thousands of meters of water, and thus is more likely to be utilized by organisms living in the water column. Second, the sedimentation rate in the deep ocean is low and the sediments are not buried deeply enough to generate oil. Third, deep waters of the open oceans are now well oxygenated and may have been through much of Phanerozoic time (see discussion), thus making the preservation of organic matter in the sediments even less likely. The high productivity in modern ocean divergences is reflected in a higher content of biogenic silica in the underlying sediments (e.g., Lisitzin, 1972, Fig. 146). Anoxia in the productive zone, caused by the abundance of organic matter, does not extend to the bottom sedi-

ments in the deep ocean as it does, for example, over the shallower shelf off Peru.

The fact that today divergences are unfavorable settings for source-bed deposition does not mean that they always were. During periods in the earth's history when the continents were flooded, and under the right circumstances, divergence could have occurred in the epeiric seas. In that case, the organic matter resulting from the high productivity would have had a better chance of being preserved, owing to the relative shallowness of the water, the possibility of a higher sedimentation rate, and the likelihood that the oxygen minimum layer would impinge on the bottom sediments.

Channeled-Flow Upwelling

Although upwelling is generally not the result of winds blowing water offshore, as is commonly believed, there is a situation in which the water is effectively blown ahead of the wind. This occurs where winds are channeled down an elongated narrow seaway, resulting in high productivity at the upwind end. The modern example is the upwelling in the Gulf of California. Productivity is highest when the winds prevail from the north and parallel the long axis of the gulf. However, while there is an exchange of gulf water, flowing out on the surface, and Pacific water, flowing in at depth, in detail the water motion is spiral with upwelling on the mainland side and downwelling on the peninsular side, particularly at the head of the gulf. When winds prevail from the opposite direction, the pattern is reversed. A similar situation exists in the Gulf of Thailand, but is less striking because the Gulf of Thailand is broader (Emery and Niino, 1963). Both

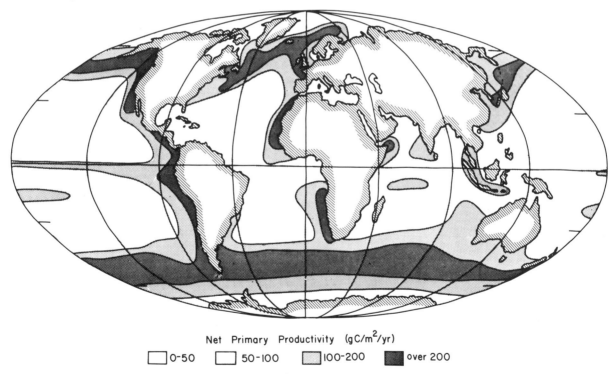

Net Primary Productivity (gC/m²/yr)

☐ 0-50   ☐ 50-100   ▨ 100-200   ■ over 200

FIG. 4—Present productivity of marine waters (from Leith, 1964). Areas of organic carbon productivity over 200 gC/m²/yr are all areas of upwelling. Some areas of high productivity result from two types of upwelling. For example, Peruvian coastal upwelling and East Pacific equatorial upwelling blend to form continuous high productivity belt.

the Gulf of Thailand and the Gulf of California have organic-rich sediments, the Gulf of California sediments being siliceous and anoxic as well (Calvert, 1966a, b). Organic matter in the Gulf of Thailand may be primarily of terrestrial origin, brought in by the Nam Chao Phraya River.

Upwelling on East-Facing Coasts

Dynamic upwelling (McKelvey, 1967) constitutes a type of upwelling that is not wind-driven divergence, but related to the dynamics of currents. Although it results in persistent high productivity, its importance to source-rock deposition is uncertain. Dynamic upwelling is most apparent in the strong western boundary currents such as the Gulf Stream and the Kuroshiro Current, and in certain small oceanic gyres, of which the Costa Rica Dome is the best-known example. In the Gulf Stream, dynamic upwelling is manifested as a very narrow (<0.6 mi; <1 km) band of relatively fresh, highly productive water bounding the left (facing down-current) side of the Gulf Stream (Yentsch, 1974). Because the productive area is so small, it seems unlikely that it could result in a significantly higher amount of organic carbon in the underlying sediments. Moreover, the productive area occurs over relatively deep water, although this may not always have been the case (Pinet and Popenoe, 1980). Tissot et al (1979, 1980) found that marine organic matter was relatively scarce in Cretaceous rocks off eastern United States, and this may be evidence that dynamic upwelling is relatively unimportant, although this analogy cannot be taken too far because Cretaceous paleogeography was different. It should be noted here that the high productivity on the Grand Banks is not due to Gulf Stream dynamic upwelling but to high latitude divergence. The Grand Banks sea bottom is near the surface and the water column is turbulent, hence little organic matter is preserved in the sediments despite the high productivity.

Average winds for east coasts flow parallel with the coast in the proper direction for upwelling. However, the winds are too variable to permit the initiation of persistent coastal upwelling currents (Saunders, 1977), although the winds do augment topographically induced upwelling (Blanton et al, 1981). It is conceivable, however, that under the more vigorous circulation regimes of glacial epochs, the winds might have been steadier, so that coastal upwelling could occur. This might explain, for example, the extreme southern penetration of cold-water isotherms off eastern Asia 18,000 years B.P. (Moore et al, 1980) and the Pacific-wide high productivity events in the middle Miocene. A somewhat similar suggestion was made by K. Bryan to Moore et al (1980, p. 240).

Bathymetry and Upwelling

Although atmospheric circulation is the primary driving force for upwelling, other factors also determine the location and configuration of upwelling zones. Coastline position has been dealt with in a previous section; a second paleogeographic constraint is bathymetry. Upwelling does not draw water from the deep ocean, but only from depths of a few tens to a few hundreds of meters, depending on the type of upwelling and the depth and configuration of the underlying sea floor (Wooster and Reid, 1963; Sverdrup et al, 1942). The actual depth required for upwelling over a shallow shelf

depends on the thicknesses of the Ekman layers, which include the layer at the surface containing water set into motion by wind friction and a similar layer of retarded motion caused by friction with the sea bottom (see discussion of upwelling dynamics). Water at the surface constitutes the outward flow in an upwelling situation, and water at the bottom constitutes the return flow. As long as the two Ekman layers are separated, upwelling is possible. When they meet, turbulence from surface to bottom ensues.

The thicknesses of the Ekman layers vary. Cited depths for the upper Ekman layer are 33 to 492 ft (10 to 150 m; Smith et al, 1971; O'Brien, 1971; Wooster, 1969; Mooers et al, 1976), with 164 to 328 ft (50 to 100 m) representing the typical depth. The depth generally employed by numerical modelers is 164 ft (50 m). The lower Ekman layer is usually considered to be quite thin, although O'Brien (1971) gave a figure of 328 ft (100 m). Transitory upwelling has been observed in water as shallow as 33 ft (10 m). Svansson (1975) stated categorically that upwelling by Ekman transport will not occur if the total water depth is less than twice the Ekman layer thickness. A good figure for the minimum depth required for upwelling is 164 ft or 50 m (R. Garvine, personal commun.). The shoreward limit of the upwelling system is not the shoreline itself but the bathymetric contour marking the depth where turbulence takes over from Ekman transport. The "upwelling shoreline" and the geographic shoreline are quite close together in most modern upwelling regions, but they may not have been in the past, where very shallow depth gradients resulted in widely spaced depth contours.

As nutrients are consumed, productivity decreases away from the site of upwelling, although upwelled water may remain on the surface from some distance (e.g., Ryther et al, 1971; Barber and Smith, 1981). Therefore, the location of the shoreward limit to the upwelling is important in determining the area in which source beds are most likely to have been deposited. No attempt was made in this study to map paleobathymetric contours on the paleogeographic base maps used here, save to identify the shelf break (Scotese et al, 1979), and adding global contours was beyond the present scope of the project. Therefore, I have assumed, for the purposes of this study, that the "upwelling shoreline" and the geographic shoreline are the same, recognizing that this assumption will surely lead to some error.

METHODS

Circulation Map Construction

Up to a point, the algorithm for the construction of the circulation maps is relatively simple and predicated on the principles described in the previous sections. Beyond that point, the procedure is specific to the paleogeography. Because the paleogeography is different for each time period and often not directly analogous to the modern, a degree of fine-tuning was required. A brief summary of the most general procedural steps follows, with examples and some exceptions to illustrate the fine-tuning. The reader should bear in mind, while reading the following lists, that each procedural step is subject to modification dependent on the paleogeography. It must also be remembered that the circulation maps are topographic maps in a sense, and the isobars must be read as contours. To some extent, the isobars illus-

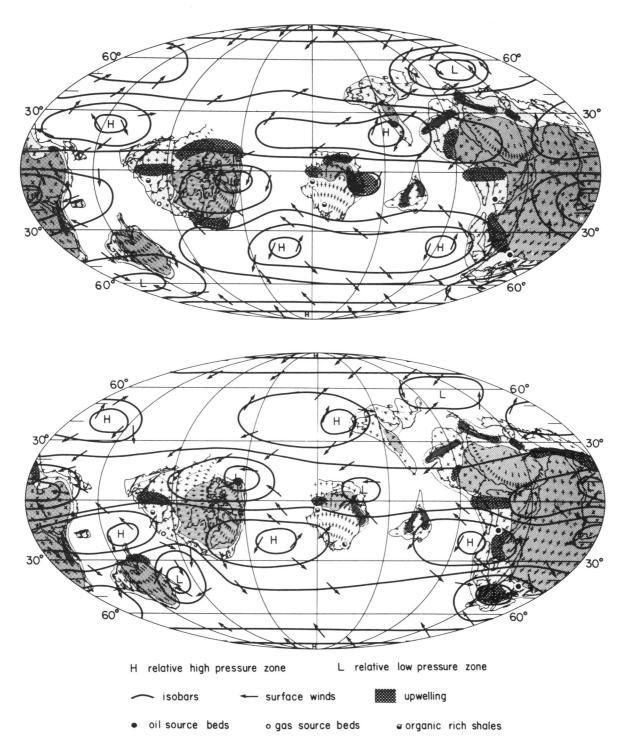

FIG. 5—Atmospheric circulation and upwelling during Franconian (Late Cambrian) time. Top = northern winter; bottom = northern summer. Base maps are those of Scotese et al (1979), and are fully described there. Large continent is Gondwana, consisting of Africa, South America, Antarctica, Australia, India, and pieces of Eurasia. North America is in left center, rotated clockwise 90° relative to its present position. Western Europe lies to south. Remaining three small continents are Siberia (upside down relative to its present position), Kazakhstan (part of USSR), and China. Dark shading = highlands; medium shading =lowlands; light shading = continental shelf.

431

trated were chosen arbitrarily out of a continuum of contours. No numerical values are implied, and isobar density indicates relative intensities only.

Placement of oceanic subtropical high-pressure cells.— The centers of the cells were placed between 40° lat. in the summer and 20° lat. in the winter. Generally, the centers were predicted to have occurred in the eastern parts of the ocean basins in question. An exception is, in a narrow basin, the high may be pulled toward the center of the basin if the continent on the east is relatively small or if the continents on either side were large enough to have cooled off significantly in the winter. The latter is illustrated by the winter circulation

in the present North Atlantic; the former is how the central southern subtropical high in the Franconian (Late Cambrian) northern summer (Fig. 5) was modeled. In very wide ocean basins, subtropical highs were predicted to have formed in the western parts of the basins near the continents, as modeled, for example, off the east coast of Siberia in the Visean (Early Carboniferous; Fig. 10).

Placement of subpolar oceanic low-pressure cells.—The centers of the low-pressure cells were placed near coastlines, where land-sea contrasts would have been greatest. The cells were made more intense in the winter and when they would have been permitted by the continental configuration to occur

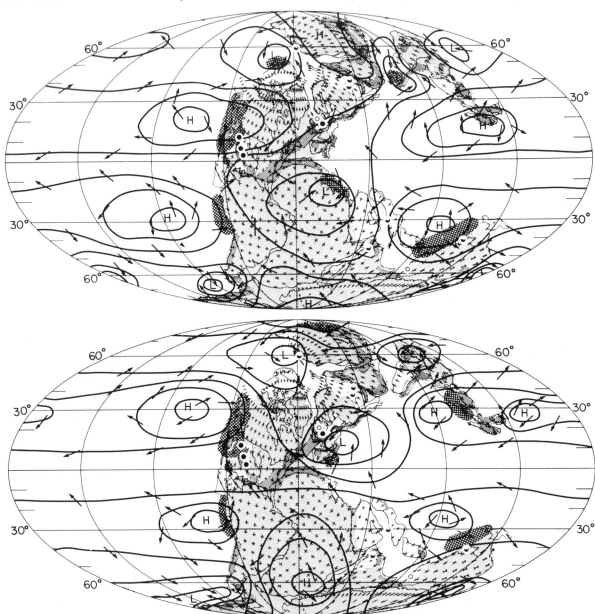

FIG. 6—Atmospheric circulation and upwelling during Kazanian (Late Permian) time. Top = northern winter; bottom = northern summer. Symbols and shading as in Figure 5. Base maps from Scotese et al (1979). Main continent is Pangea, with China to northeast. In south, Pangea consists of South America, Africa, India, Antarctica, and Australia; a piece of Eurasia lies off Africa. In north, Pangea consists of North America, western Europe, Kazakhstan, and Siberia.

in the normal zones of low pressure near 60° (e.g., off southwestern Pangea in the Late Permian Kazanian northern summer; Fig. 6).

Placement of continental lows.—The intensity of a continental low is dependent on the size of the continent and the positions of mountain ranges. Like an oceanic low, a continental low will be more intense if the continental configuration constrains it to an area that is already low in the general zonal circulation. The centers of lows were placed equatorward of 40° in the summer. The distance from the center of the low to the equator depends on the amount of continent above the equator (e.g., contrast the predicted summer lows over Gondwana in the northern summer and the northern winter, in the Franconian, Fig. 5). Generally, the lows were placed slightly to the east, for example, in the Kazanian maps (Fig. 6), where the adjacent marine air would have been warmer and thus would not have ameliorated the heating of the continents.

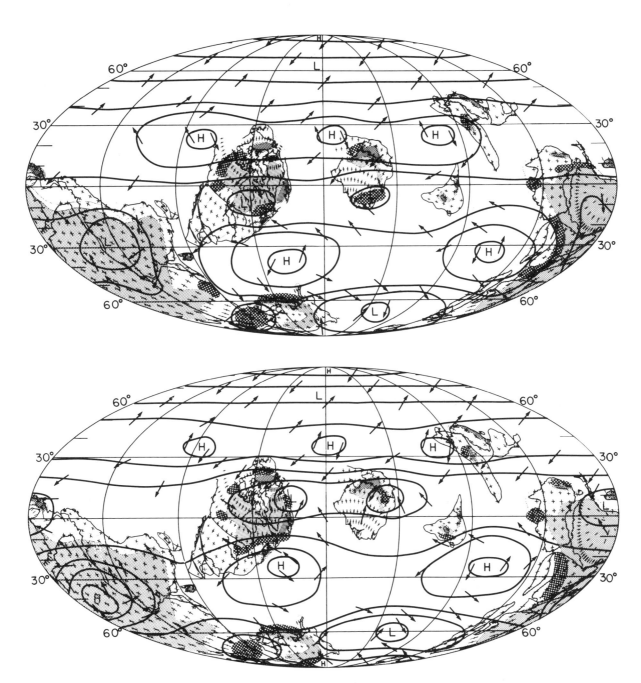

FIG. 7—Atmospheric circulation and upwelling during Llandeilo-Caradoc (Late Ordovician) time. Top = northern winter; bottom = northern summer. Symbols and shading as in Figure 5. Base maps from Scotese et al (1979).

433

Placement of continental highs.—Like a continental low, the intensity of a continental high depends on the size of the continent and whether it is in one of the zones of high pressure predicted by the zonal model. Continental highs were placed higher than 30° lat. A normal polar high was predicted if the pole was far from a shoreline, either in an ocean or over a continent. The high was predicted to be displaced away from the pole by a distance related to the proximity of the shoreline nearest the pole and by the size of the continent. For example, in the present world, the strongest "polar" high in the northern winter is the intense high pressure over Asia. This point is best illustrated by the predicted development of the winter highs over Gondwana (later southern Pangea) from the Ordovician onward in Figures 7 to 11.

Wind directions.—Surface winds run almost parallel with the isobars except near the equator. The angles between the arrows and the isobars on the maps were exaggerated for clarity.

Upwelling zones.—Upwelling zones were illustrated when

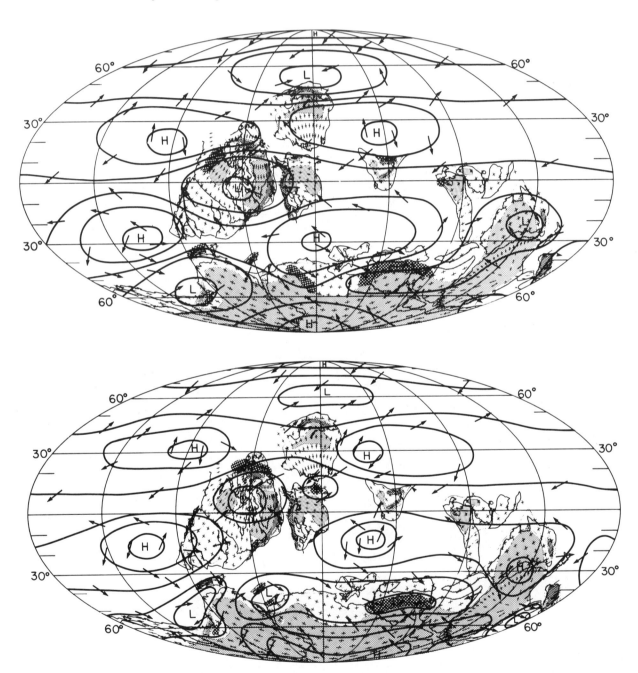

FIG. 8—Atmospheric circulation and upwelling for Wenlock (Late Silurian) time. Top = northern winter; bottom = northern summer. Symbols and shading as in Figure 5. Base maps from Scotese et al (1979).

434

they would have occurred over or near the continental shelves (i.e., excluding the oceanic divergences over deep ocean). Where a low pressure that would have created a divergence was predicted over a shelf, for example, off southwestern Gondwana in the Franconian, upwelling was illustrated (Fig. 5). Equatorial divergences were indicated only for the western sides of the continents, in keeping with the observation in the present world that equatorial divergence is more vigorous in that position. Coastal upwelling was predicted wherever the winds were predicted to have blown parallel with the coast in the proper direction. The exception is in a Gulf Stream situation, where in the present world, the winds are apparently too variable to give consistent upwelling (see discussion of upwelling on east coasts). East coast upwelling was predicted, however, from the monsoonal circulation in the Westphalian (Late Carboniferous; Fig. 11) and Kazanian (Fig. 6).

Source Beds

Source-bed data for this study were provided by Amoco Production Co. and supplemented with data from the litera-

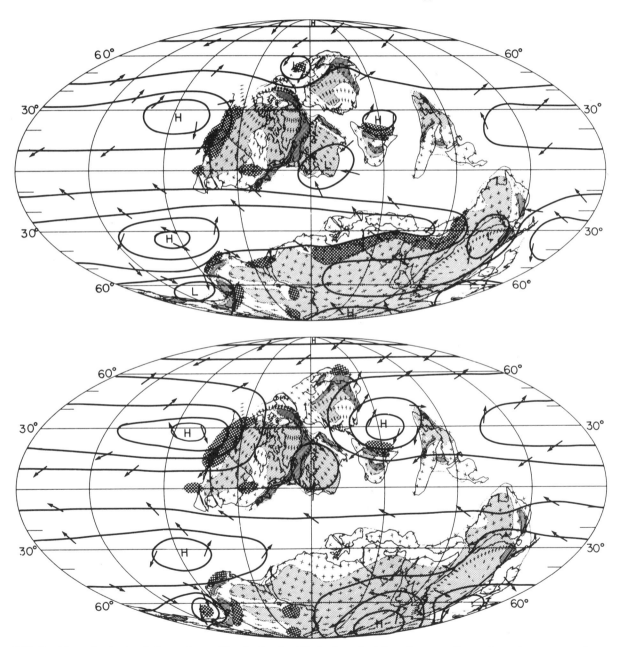

FIG. 9—Atmospheric circulation and upwelling for Emsian (Early Devonian) time. Top = northern winter; bottom = northern summer. Symbols and shading as in Figure 5. Base maps from Scotese et al (1979). Emsian paleogeography is presently under revision by Ziegler et al.

435

ture. The data were divided into three categories. The first category includes oil source beds and the second, gas source beds. Each of these two categories includes potential as well as effective sources. The third category is organic-rich shales (ORS), which includes beds identified as sources on qualitative grounds and beds containing greater than 1% total organic carbon (TOC), where additional geochemical information was lacking. In the literature, there was either little direct information on the source type or conflicting interpretations of the information available, necessitating the assignment of most literature data to the ORS category. Criteria for the assignment of source beds to the first and second categories included the following: (1) TOC greater than or equal to 0.4% (Momper, 1978); (2) position of the bed on an H/C-O/C plot (Tissot et al, 1974; greater than 1.0% TOC generally indicates oil source, although it can be lower, depending on thermal maturity); (3) percent elemental carbon (less than 75% indicates immature beds, greater than 90% indicates advanced diagenesis; Harwood, 1977); (4) ratio of bitumen to TOC (greater than 0.05 indicates

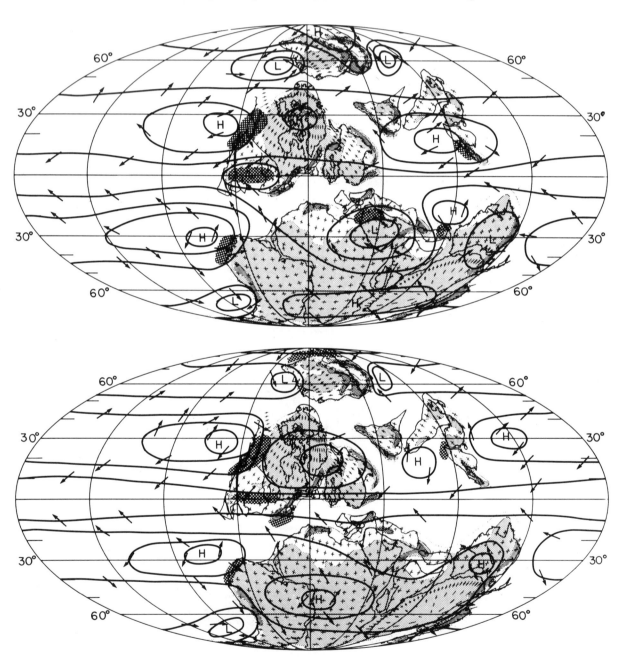

FIG. 10—Atmospheric circulation and upwelling for Visean (Early Carboniferous) time. Top = northern winter; bottom = northern summer. Symbols and shading as in Figure 5. Base maps from Scotese et al (1979). Visean paleogeography is presently under revision by Ziegler et al.

probable oil sources, depending on thermal maturity; R. J. Harwood, personal commun.); (5) kerogen type (structured or gas-prone versus amorphous or oil-prone; Harwood, 1977). Kerogen type was not always available and some subjective judgment was required in a few cases. Beds with as little as 0.4% TOC were included only if there was additional geochemical information regarding their source potential. While the use of a double standard (1.0% TOC for ORS, 0.4% TOC for others) is not ideal, exclusion of the latter data would have depleted the data set of classifiable

(i.e., gas versus oil) source beds. Bitumen/TOC and percent elemental carbon were used only as guides in interpreting H/C-O/C.

Source-bed data were plotted on maps produced at the University of Chicago under the direction of A. M. Ziegler. Some of these maps (Scotese et al, 1979) were the global paleogeographic maps used as the base maps for the atmospheric circulation and are represented in Figures 5 to 11. The rest of the maps (Scotese, 1979) had continental positions only, but for the entire Phanerozoic in 20 m.y. intervals.

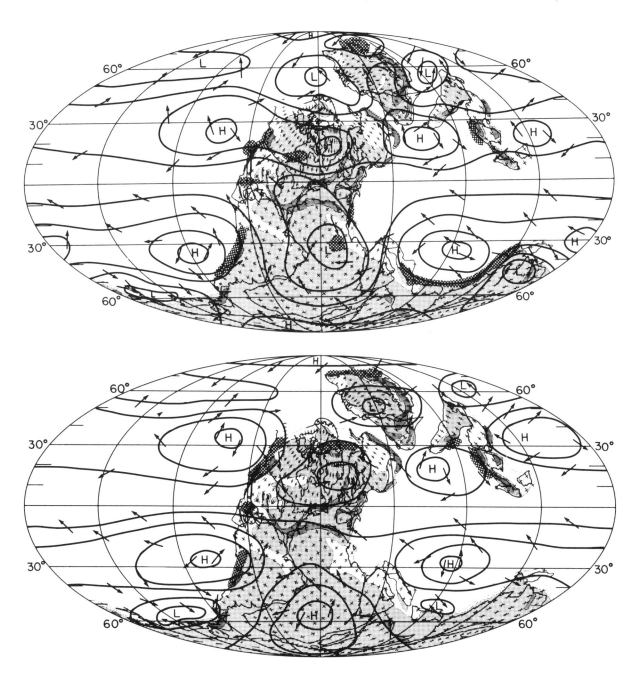

FIG. 11—Atmospheric circulation and upwelling for Westphalian (Late Carboniferous) time. Top = northern winter; bottom = northern summer. Symbols and shading as in Figure 5. Base maps from Scotese et al (1979).

437

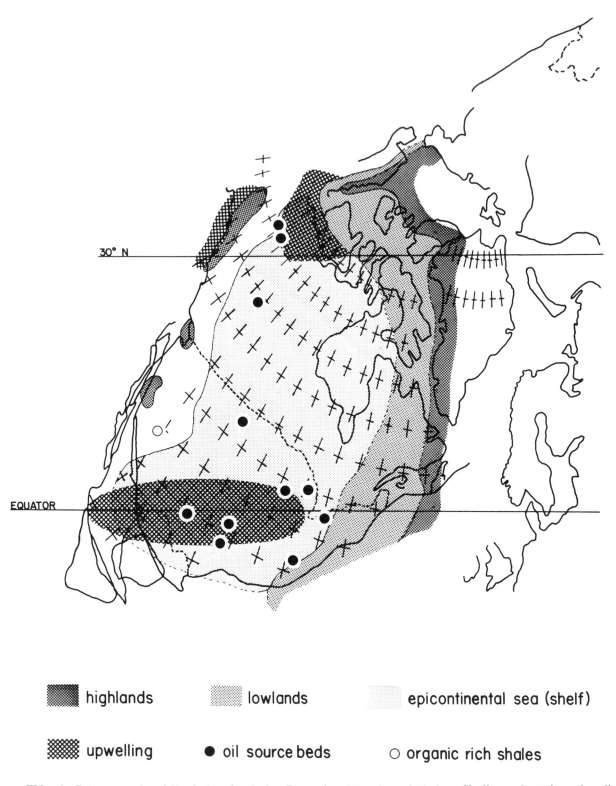

30° N

EQUATOR

| highlands | lowlands | epicontinental sea (shelf) |

upwelling   ● oil source beds   ○ organic rich shales

FIG. 12—Paleogeography of North America during Famennian (latest Devonian) time. Shading and continental outline reconstruction after Scotese et al (1979; see Fig. 5 caption). Upwelling is shown for uniform depth over whole shelf (see text). Upwelling probably also occurred between 30 and 15° on the northwestern margin, as the shelf just east of that margin was probably very shallow.

Space does not permit the inclusion of the latter group of maps. Two sets of circulation maps (Figs. 5, 6) have source-bed data plotted to illustrate typical data distributions.

The paleogeographic maps were being constructed when the present project was started, and the time intervals were chosen with other problems, such as biogeography, in mind. Therefore, the intervals for which we have paleogeographic maps are not always the most ideal for source beds. For example, the Emsian stage was that chosen for the Devonian, but far more Devonian source beds are of Famennian age. Therefore, source-bed data were plotted on continental reconstruction maps for time intervals not represented by the paleogeographic maps. Because of the importance of the Famennian-early Tournaisian to petroleum geology, a Scotese et al (1979)-style paleogeographic map of North America for that time was constructed (Fig. 12).

One of the difficulties encountered in this study was that precise ages were rarely given for source beds. Usually the period was reported, and sometimes the epoch, but rarely the age during which the source bed was deposited. While it was occasionally possible to infer a more precise age for the bed from paleontologic data or from the stratigraphic nomenclature, often such information was not available, particularly for sequences outside North America. Except for the Cambrian, I did not consider beds that could not be dated at least to the epoch level. However, since the level of resolution for the maps was at the age level, there are probably some data points that are not plotted on the correct maps. Although using such data potentially introduces error, rejecting them would deplete the data base, so a compromise was necessary. The small number of Cambrian data points necessitated plotting them on one map.

Four hundred twenty-three individual determinations on source beds were collected, of which the vast majority were from North America. Since many data were from closely spaced locations, there was the danger of weighting the analysis of the distribution of the source beds too heavily in favor of the regions where the data were clustered. In analyzing the data, I used as a single data point all the analyses within a 5° square longitude and latitude. This approach is preferable for several reasons. First, the individual source-bed analyses were often redundant in that they indicated the source quality of several sites in the same stratigraphic unit. While it is important to know if a single, laterally extensive unit is of source quality throughout its geographic range, there is no advantage in a global study to knowing that the unit is of source quality everywhere within, say, a few kilometers, since an upwelling zone will span several tens to hundreds of kilometers. Rather, the information of interest is that the upwelling did result in source beds in that area. It is useful, on the other hand, to indicate a single extensive unit by several data points if they define a broad region of source-quality beds, because the geographic extent of the upwelling zone and source beds would coincide under ideal conditions. Second, most data came from North America, and the test of the upwelling model is biased toward the circulation over that continent through time. Therefore, any reduction of North American data that does not alter the broad pattern sought here will result in a more honest assessment of the model's validity. Since upwelling sediments should generate oil, oil sources are more interesting for the purposes of this study and gas sources more interesting than

ORS. Therefore, where geochemical analyses resulted in more than one type within a 5° square, the whole square was counted as oil if any oil sources were present, gas if any gas sources but no oil sources were present, and ORS if that was the only type of source in that square.

RESULTS

A total of 126 5° squares contained source-bed data. Forty-eight (38%) of the squares contained oil sources, 19 (15%) contained gas (but no oil) sources, and 59 (47%) were only ORS. For comparison, of the 423 individual data on source beds that were clustered into the 5° squares, 86 (20%) were oil, 47 (11%) were gas, and 290 (69%) were ORS. From this, one can get a sense of the effects of clustering the data.

The data were divided into four categories: (1) those explainable by upwelling, (2) those that might be explainable by upwelling, (3) those that do not fall into a modeled upwelling zone, and (4) those formed in nonmarine areas. Nonmarine source beds were not included in the statistics because they are not expected to be related to upwelling. These data are summarized in Table 1.

Over a third (36.5%) of the marine data fell into the modeled upwelling zones. A few of the individual determinations from areas with modeled upwelling zones were gas-source data points, indicating the presence of some terrestrial organic matter. There are several possible explanations for this mixture. First, the paleogeography may not be accurate enough, so that deposits that appeared to be marine on paleogeographic grounds may in fact not have been. Second, the interpretation of the geochemical analyses could have been inaccurate, or the organic matter may have been degraded by chemical or biochemical processes, becoming no longer representative of original generating capability (J. A. Momper, personal commun.). Third, regions that contained both gas and oil sources could have been assembled from data on beds that were not precisely coeval, so that the regions were in fact terrestrial at one time and marine during a period shortly before or after; alternatively, marine and terrestrial deposits might have been too closely associated vertically and/or laterally for resolution. Since it was impossible, within the context of this study, to interpret each analysis, these various possibilities for the distribution of some source types were not determined.

Of the data used, 17.5% occurred in areas that might have had upwelling under certain conditions, including the following. First, zonal coastal upwelling zones are partly dependent on the latitudinal position of the shoreline (see Fig. 2) and divergences, particularly the equatorial divergence, are also related to latitude. The positions of the continents on the maps were determined largely from paleomagnetic data, and there can be errors in the placement of the continents. Source regions that lay within 5° latitude of an upwelling zone were therefore included in this category. Bathymetry also affects upwelling, and a few cases occurred where source regions would have fallen into upwelling zones if the 164 ft (50 m) isobath (the effective shoreline for upwelling) were remote from the geographic shoreline.

The potential importance of the positions of the continents and the bathymetry is illustrated by the two coeval points in Siberia in the Cambrian. Neither of these points occurs in a modeled upwelling zone in its present position. The northern

Table 1. Distribution of Source-Bed Types in Relation to Modeled Upwelling Zones as Percent of Total Individual Analyses and Total 5° Squares*

	Oil Sources		Gas Sources		Organic-Rich Shales		All	
	i.a.	5°	i.a.	5°	i.a.	5°	i.a.	5°
Sources explainable by upwelling	80.2	70.8	36.2	30.3	41.0	20.3	48.5	36.5
Sources possibly explainable by upwelling	16.3	22.9	4.3	27.3	8.4	18.6	9.4	17.5
Sources not explainable by upwelling	2.3	4.2	2.1	3.0	7.2	11.9	5.7	7.1
Nonmarine sources	1.2	2.1	57.4	100.0	43.4	49.2	36.4	38.9

i.a. = individual analyses (n=423)
*See text for explanation

5°=5° squares longitude-latitude (n=126)

zone occurs near the paleocoastline, which is too close to the equator for zonal upwelling to occur (the Coriolis effect is too weak). However, if Siberia were 5° farther south than plotted on Figure 6, zonal upwelling would have occurred along that coastline, and the northern source would have been deposited in the resulting high-productivity area. With Siberia in the indicated position, it would be possible to account for the southern source region by upwelling if the 164 ft (50 m) isobath were 10 to 15° south of the geographic shoreline. Then, however, the northern source could not be accounted for by upwelling. Finally, if Siberia lay 5° north of the indicated position, the continent would straddle the equator and both sources could be accounted for by equatorial divergence over the craton.

A few sources (7.1%) that apparently were marine could not be accounted for by upwelling. Most of these lay in settings where upwelling could not have occurred, for example, marginal marine (deposits of mixed marine and terrestrial origin of the USSR) or in restricted basins (e.g., the Copper Shale in the North Sea region). Bathymetry and basin configuration aside, however, these sources did not occur in areas where upwelling could be expected.

About a third of the data (38.9%) were nonmarine, as determined from the paleogeography and/or from the presence of gas sources coupled with the absence of oil sources. Forty-nine 5° squares were classed as nonmarine. Of these, two-thirds were ORS (30) and most of the rest were gas (19). Only one point was oil, and was apparently lacustrine. This distribution fits nicely with the hypothesis (Tissot et al, 1974) that terrestrial organic matter gives rise to gas rather than oil.

Statistical Analysis

Oil source beds were likely to have been deposited on ancient continental shelves whether or not the organic matter was derived from upwelling. Therefore, it is reasonable to presume that some sources would have been deposited in areas corresponding to the modeled upwelling zones even if there were no genetic relationship. For example, if 20% of

the continental shelf area is included in the modeled upwelling zones, then about 20% of the sources would be expected to fall into the modeled upwelling zones, even if the sources were randomly distributed with respect to upwelling. Therefore, using chi-square and related statistics, I tested the null hypothesis that the source beds were randomly distributed. In order to test the hypothesis, one must have some way of estimating the expected correspondence between upwelling zones and source beds. If the beds are randomly distributed with respect to upwelling, the proportion of beds occurring in upwelling zones should approximate the proportion of shelf area covered by upwelling. The area covered by upwelling was determined for each of the time intervals for which paleogeographic maps were available. An average value for upwelling areas was used for some of the statistical tests. As discussed earlier, the bathymetric profile of the continental shelf is critical to the position and, in some cases, the existence of upwelling zones. In measuring the area covered by the modeled upwelling zones, I took the conservative approach of including all the shelf area that could have contained the locus of upwelling, even though the upwelling itself should have been confined to a smaller region at any instant in time. However, I did not include areas not originally modeled, that is, areas that might have had upwelling depending on latitude and bathymetry but which otherwise were not in a modeled upwelling zone. The average value for upwelling area/total shelf area for the Paleozoic was 0.2363. The source-bed data were grouped several different ways, and the statistical tests performed on each group. The groups included marine data only and all Paleozoic data lumped together or data from individual intervals. The results are summarized in Table 2.

The correspondence between the distribution of upwelling zones and source beds was highly significant, with $p \ll 0.005$, for all Paleozoic data taken together. This was true whether only marine data were considered or whether nonmarine data were included with those not accounted for by upwelling. This later test was performed because of the uncertainties in the classification of organic matter as marine

Table 2. Results of Statistical Tests of Correlation Between Upwelling and Source-Bed Distribution (5° Squares)

	Sources Explainable by Upwelling[a] (Total Marine Sources)	Shelf Area Occupied by Modeled Upwelling Zones (%)	Statistic[b]
Paleozoic	46 (77)	23.63[c]	55.61**
Cambrian	2 (6)	28.64	0.32
Late Ordovician	3 (6)	38.59	0.43
Silurian[d]	1 (4)	25.82	0.70
Early Devonian[d]	3 (8)	15.35	0.11
Middle Devonian	5 (7)	15.35	0.001**
Late Devonian[e]	13 (17)	15.35	3.3×10^{-8}**
Early Mississippian[d]	3 (5)	12.25	0.02*
Late Mississippian	4 (4)	12.25	2.3×10^{-4}**
Early Pennsylvanian	3 (3)	21.13	0.009*
Late Pennsylvanian[d]	2 (2)	21.13	0.04*
Early Permian	3 (4)	23.66	0.04*
Late Permian[d]	4 (11)	23.66	0.005*

* Indicates results significant at 5% level.

** Indicates results significant at 0.5% level.

[a] Does not include sources possibly explainable by upwelling (see text, Table 1).

[b] For all Paleozoic data together, the statistic is chi-square. For individual periods, numbers are the probabilities calculated directly.

[c] Mean of shelf areas for the individual periods.

[d] Times for which paleogeography was determined by Scotese et al (1979).

[e] Paleogeography determined for North America (this paper); shelf area not reconsidered for this time.

or nonmarine, a classification based on paleogeography, which could only be extrapolated from the paleogeographic maps to the intervening time intervals, and on geochemical criteria.

Taking the Paleozoic periods individually, the results were also usually significant. Some of these tests are not so reliable, however, as the ones on the lumped data because the sample sizes for some of the individual periods are very small, making the statistical results less meaningful. Furthermore, estimates of shelf area covered by upwelling cannot be made for the time intervals for which the paleogeography has not been worked out. In these cases, the areas measured from the known intervals are used for the intervening ones (e.g., the Emsian estimate was used for the Middle and Late Devonian). The results of the analyses of the individual time periods are also in Table 2. When only source beds falling into modeled upwelling zones were considered, only a few periods showed a distribution of source beds significantly different from random with respect to upwelling, within the constraints of small sample sizes (Table 2). When, however, the beds that might be attributable to upwelling were included with those that were attributable to upwelling, then almost all periods showed a significant correspondence between the modeled upwelling zones and the distribution of the source beds (not shown in Table 2).

DISCUSSION

If a source bed lies in a modeled upwelling zone, that does not prove that the organic matter in that bed resulted from upwelling. A full understanding of the relation between up-

welling and source-bed deposition will require more detailed studies on the local paleogeography, associated sediments, and origin of organic matter. The work of Sheldon et al (1967) on the Phosphoria Formation indicates the importance of knowing local paleogeography. They demonstrated the shelf-edge setting of the Phosphoria Formation, thereby illustrating the resemblance between the settings for the Phosphoria and the modern Peruvian upwelling system. Wardlaw (1980) has contributed to the confirmation of the Phosphoria as an upwelling deposit by showing that the times of maximum phosphorite deposition were also times when the seawater was cooler. Both phenomena characterize vigorous upwelling. The sediments in an upwelling deposit are also characteristic. Phosphorite, siliceous oozes, and organic matter are commonly associated and result from upwelling processes (Diester-Haass and Schrader, 1979; Burnett, 1977; Calvert, 1979). A. M. Ziegler (unpub.) calls this association "bioproductite." This association is found, for example, off Peru and southwest Africa. Siliceous oozes are characteristic of the Gulf of California upwelling. Fully developed bioproductites occur in the Miocene Monterey Formation in California and in the Phosphoria Formation.

It is important to know the type of organic matter occurring in the source beds. An abundance of organic matter does not necessarily indicate upwelling. For example, Tissot et al (1979, 1980) pointed out that although Late Cretaceous sediments on both sides of the North Atlantic contain about the same amount of organic matter, the organic matter is predominately terrestrial in origin on the west side and of marine origin on the east, off Africa. This has also been demonstrated for the present (Emery and Milliman, 1978). Al-

441

though the sediments are marine on both sides of the Atlantic, they clearly were influenced by different climatic and oceanographic regimes.

The Famennian (latest Devonian-earliest Mississippian) black shales of North America illustrate the interrelationship of paleogeography, sediment types, and organic matter. Since the North American paleogeography for Famennian time was reconstructed specially for this project (Fig. 12), it is appropriate to use this particular example to demonstrate this interrelationship.

North America was situated on the equator in the Famennian in such a way that the locus of equatorial divergence would have been over Mexico, New Mexico, northern Texas, Oklahoma, Illinois, and Michigan. The most vigorous upwelling would have been southwest of the eastern highlands (present orientation), that is, over Texas, Kansas, and Oklahoma. The sediments near the highlands would have experienced less vigorous upwelling and some input of terrestrial organic matter. The degree of upwelling influence would also have decreased to the northwest, toward the Williston basin, as the shelf became shallower in that direction.

The sediment distribution predicted from this upwelling pattern would be as follows. The richest sources should be in the southwest and midwest, and that would also be the most likely site for chert and phosphorite deposition. Shales to the northeast would be somewhat less rich, with less chert and phosphorite, and would contain some terrestrial organic matter washed in from the highlands. Shales in the Williston region would be similar to the northeastern ones, but with less terrestrial input. This distribution is precisely the one seen among the Woodford, Chattanooga, Bakken, New Albany, and related shales, which are at least partly correlative. The Woodford contains abundant chert (Amsden et al, 1967; Galley, 1971), and it and the New Albany are phosphatic at their tops (Hendricks et al, 1947, cited in the *Lexique Stratigraphique;* Collinson et al, 1967). Furthermore, thick chert accumulations, at least partly Late Devonian in age, occur in that region and in Mexico—the Caballos and Arkansas Novaculites and the La Yerba Formation (McBride and Thomson, 1970; Goldstein and Hendricks, 1953; Amsden et al, 1967; Lopez-Ramos, 1969). The distribution of these thick chert accumulations exactly traces the line of the Famennian equator, with the westward restoration of Mexico along the Monterey-Torreon shear zone (de Cserna, 1970), a rotation accepted by Scotese et al (1979). The Chattanooga Shale also contains chert (J. S. Wells, 1971) and abundant terrestrial plant remains (Conant and Swanson, 1961) and is the source for much gas in the Appalachian basin (Ray, 1971). Other Late Devonian-Early Mississippian black shales from the Appalachian basin show an east to west trend of carbon isotopic composition from nonmarine in eastern New York to fully marine in western Pennsylvania and West Virginia (Maynard, 1981). The organic matter in the New Albany Shale in Illinois is mostly marine amorphous (Barrows et al, 1979). The Bakken Formation is a rather thin unit, but has proved to be a source for oil in the Williston basin (J. A. Williams, 1974).

The interpretation of the North American black shales as normal marine and further as upwelling deposits precludes the shallow-water interpretation for the novaculites proposed by Folk (1973) and for the black shales, especially the Chattanooga (Conant and Swanson, 1961). Folk's (1973) interpre-

tation rested heavily on his textural evidence, which is far from conclusive, and on the presence of tree trunks. The trunks are explained by the climatic regime, which was wet in the eastern highlands and included equatorial winds that would have carried any floating debris away from the highlands and west over Texas. Several other lines of evidence, discussed in detail by McBride and Thomson (1970), point toward a deeper water origin for cherts and shales. Of particular importance is the predominance of radiolarians, which are open ocean organisms that tend to concentrate in tropical high-productivity zones (Lisitzin, 1972, p. 151). The Chattanooga Shale also contains conodonts and radiolarians. Conant and Swanson's (1961) other arguments for the shallow water were contradictory (e.g., in one place they said the Chattanooga contains no shallow-water sedimentary structures and in another they claimed it does), and they lacked the information, gained since the publication of their paper, regarding the placement of nearby terrains (i.e., Florida). Finally, the black shales of northwestern Canada (e.g., the Canol Formation) lay in a coastal upwelling zone analogous to the California Current, and were quite independent of the equatorial upwelling that created the black shales farther south. In short, the evidence for the deep-water origin of the North American Famennian black shales still predominates. Furthermore, the distribution of organic matter, siliceous rocks, and phosphatic layers within those shales suggest that the high concentration of organic matter resulted from upwelling.

Not all source beds were deposited in upwelling zones. The best-known alternate type of setting for source bed deposition is the Black Sea, which represents an important class of basins characterized by anoxic bottom waters. Demaison and Moore (1980) cogently discussed the role of anoxic basins of the Black Sea type. In their model, they pointed out the importance of having the silled basin in a relatively humid climate, as is the Black Sea, as opposed to a dry climate, such as that influencing the Mediterranean. The humid climate permits the establishment of the permanent salinity stratification of the water column that characterizes the Black Sea and leads to the anoxia of its bottom waters. A possible ancient example is interior North America in the Late Pennsylvanian. Heckel (1977) attributed the black shales there to upwelling; however, this is unlikely because the many islands scattered through the area (his Fig. 6) would have prevented the development of the required divergent flow. The shales could fit nicely into the Black Sea model of Demaison and Moore (1980), however, in that a continuous influx of fresh water from the wet equatorial lands in the east could have created a persistent salinity stratification. By linking the development of anoxia to paleoclimatic regimes, Demaison and Moore (1980) have opened the way for global prediction of the likelihood of anoxic basin formation.

Black shales are geographically extensive in certain intervals of the Paleozoic, although no compilation has been made to document their precise extent. In order to explain these widespread deposits, Berry and Wilde (1978) proposed a model of successive oxygenation (ventilation) events, beginning in the Precambrian. They argued that the early ocean had to have been completely anoxic, and they set up a model to explain the timing and mechanism of oxygenation. They tentatively proposed a ventilation cycle beginning with the Precambrian glaciation, continuing through the Late Ordovician glaciation, and concluding in the Middle Devonian or possibly later. Oxygen was introduced, according to their model, by mixing at the surface and in cold, saline bottom waters generated by ice formation. During interglacial times, the entire water column below the mixed zone became dysoxic through mixing of the oxygenated bottom waters and the overlying anoxic waters; dysoxia was less with each successive interglacial. Interestingly, Berry and Wilde (1978) worked upwelling of anoxic waters into their model to explain mass kills of graptolites. It would be instructive to compare the distribution of graptolite-rich rocks to the distribution of modeled upwelling zones presented here.

The problem in evaluating Berry and Wilde's (1978) model lies in the question of rates. The present turnover rate for the ocean, including bottom waters, is only 1,600 years (Broecker, 1974). There were at least two major glacial episodes during the Precambrian (Frakes, 1979), each presumably lasting at least as long as the Pleistocene, although the imprecision of dating Precambrian rocks precludes precise determination of the episodes' timing. If one considers only half of the Pleistocene as glacial (van Eysinga, 1975; bear in mind that the present is interglacial and bottom waters are being generated), then the ocean has turned over about 500 or 600 times in the past 2 m.y. With those turnover rates, it is difficult to imagine that the oceans could have remained anoxic into the Paleozoic, even allowing for the necessity of many cycles to complete oxygenation and for lower initial atmospheric oxygen values (but see Schopf, 1978). Nevertheless, it should be noted that in the conservative treatment of the data presented here (Table 2), upwelling and oil source-bed distribution do not correlate until the Middle Devonian, about the time by which Berry and Wilde (1978) suggested complete oxygenation had taken place.

Leggett et al (1981) proposed an alternative model for the deposition of Paleozoic black shales. Concentrating on British sequences, they investigated the temporal distribution of Cambrian-Silurian black shales from the viewpoint of Fischer and Arthur's (1977) cycles of oxic and anoxic oceans. They concluded that the lower Paleozoic black shales in Britain do not follow the 32 ma cycle postulated by Fischer and Arthur (1977) for the Mesozoic and Cenozoic, although an irregular, longer wavelength cycle is evident. Leggett et al (1981) offered a variety of explanations for the pattern seen, and recognized that the limited scope of their investigations might be partly responsible for the pattern they observed. The cyclicity seen by Leggett et al (1981) is not reflected in the data presented here.

A related problem is that of oceanic anoxic events. The occurrence of anoxic events has been attributed to relative stagnation of the oceans in times of low equator-to-poles temperature gradient (Schlanger and Jenkyns, 1976). Slowing of ocean currents might permit a widening and intensification of the ocean-wide oxygen minimum, which now occurs with varying intensities between 328 ft (100 m) and about 4,922 ft (1,500 m), depending on location. This is an intuitively attractive idea, but the evidence from cores is conflicting. For example, Prell et al (1980) concluded that in the Indian Ocean only the West Australian Current became stronger during the last glacial maximum and that the Somalian upwelling was weaker than it is today. Sancetta (1978) saw a strengthening of circulation with the initiation of glaciation, but no apparent further change as glaciation progressed. Diester-Haass and Schrader (1979) and W. H. Berger et al (1978) postulated an intensification of circulation off northwest Africa during the Neogene and Holocene, coinciding with cooling events. However, Diester-Haass and Schrader (1979) did not see an intensification of the southwest African upwelling in the Neogene; moreover, they stressed the importance of other factors (e.g., sea level) in complicating the interpretation of the strength of upwelling at any given locality.

Oceanic anoxic events would result in increased preservation of organic matter on all the continental margins, meaning that the algal organic matter produced would be preserved. This is significant because productivity is higher over continental shelves in general than in the low-productivity areas of the mid-latitude central ocean. However, as in a normal ocean, upwelling zones will result in the richest sources, even if upwelling occurred during times of anoxia. In analyzing the effects of upwelling during anoxic events, it is necessary to classify source beds by kerogen type (as did Tissot et al, 1979, 1980; Summerhayes, 1981; and as was attempted here) and by richness (not considered in detail here) in order to see the effects of upwelling.

There is real potential in combining the predictive strengths of the upwelling models with the information provided by sea-level and related curves (e.g., Vail et al, 1977). The upwelling model provides information about the general location of high-productivity areas and intensified oxygen-minimum zones. Sea-level curves, together with detailed paleobathymetric profiles of the continental margin, would provide the means for identifying the precise location of the high productivity and the distribution of anoxic bottom sediments. For example, a shallow shelf gradient will result not only in a locus of upwelling that is removed from the geographic shoreline, but will also give rise to an extensive anoxic zone (see also Jenkyns, 1980). This explanation also applies to the correlation of source-bed deposition and transgressions. Sea-level rise would permit the migration of upwelling and oxygen-minimum zones onto the cratons. Finding the locus of upwelling in an epeiric sea permits an explanation of the distribution of source beds in epicontinental seas. The distribution of source beds is usually disjunct in such situations, even in vertically and laterally continuous sequences, and the oxygen-minimum expansion model does not explain this disjunct distribution, where the upwelling model might.

The importance of the results reported here is twofold. First, the predicted upwelling zones can be used to identify regions that ought to be examined for source beds. The models provide information about the geographic extent of the likely source region and also about the age one might expect the source bed to be. Clearly, one cannot explore for oil on the basis of this one piece of evidence alone. Rather, the present models and similar ones for the Mesozoic and Cenozoic that are forthcoming (Parrish, 1981; Parrish and Curtis, in press) can contribute strength to one link in the chain of events, all of which must be learned for a particular area, that gives rise to source beds. Second, these models, and models similar to the ones discussed by Demaison and Moore (1980), can together provide an understanding of the more subtle global controls on the distribution of oil. The upwelling model cannot supplant all the previous models, such as the oceanic anoxic event model. Rather, in combination with previous models, it can contribute to our overall conception of the distribution of oil source beds.

The results presented here are preliminary. One way to improve the models is to reconstruct the atmospheric circulation using a modified version of one of the existing numerical global circulation models. Using such models requires such a large commitment of time and money, however, that it seemed better to construct qualitative models first, however crude they might be, to assess the models' potential. With the results of this study being positive, it is worthwhile pursuing the use of numerical models, and efforts are under way to do that (e.g., Barron and Washington, 1981). The models presented here can also be improved by work done on some of the problems the models have helped identify, namely studies of local paleogeography, associated sediments (e.g., phosphorites, Parrish and Ziegler, 1981; siliceous sediments, Parrish et al, in prep.), and others.

A formal prediction of the proportion of sources that should be explainable by the upwelling models was not possible at the outset of this study. One could not expect all source beds to be deposited in upwelling zones, as discussed above, and uncertainties about continental positions and orientation, paleogeography, and correlation frustrate the most careful attempts to construct global models or explain global patterns (Gray and Boucot, 1979). That over a third of the source beds lie in modeled upwelling zones is, therefore, an encouraging result. The conclusion that the distribution of some oil source beds is related to the distribution of ancient upwelling zones is further strengthened, although not proved, by the statistical significance of the correlation between source bed and upwelling distribution. It is hoped that this study will spark related studies that eventually will lead to a comprehensive view of the distribution of petroleum source beds.

SELECTED REFERENCES

Amsden, T. W., et al, 1967, Devonian of the southern Mid-Continent area, United States, in D. H. Oswald, ed., International symposium on the Devonian System: Alberta Soc. Petroleum Geologists, p. 913-932.

Arthur, M. A., and S. O. Schlanger, 1979, Cretaceous "oceanic anoxic events" as causal factors in development of reef-reservoired giant oil fields: AAPG Bull., v. 63, p. 870-885.

Banse, K., 1968, Hydrography of the Arabian Sea shelf of India and Pakistan and effects on demersal fishes: Deep-Sea Research, v. 15, p. 45-79.

Barber, R. T., and R. L. Smith, 1981, Coastal upwelling ecosystems, in A. R. Longhurst, ed., Analysis of marine ecosystems: New York, Academic Press, p. 31-68.

Barron, E. J., and W. M. Washington, 1981, Modeling the Cretaceous climate using realistic geography; simulations with the NCAR Global Circulation Model (abs.): Geol. Soc. America Abs. with Programs, v. 13, p. 404.

———— J. L. Sloan, and C. G. A. Harrison, 1980, Potential significance of land-sea distribution and surface albedo variations as a climatic forcing factor; 180 m.y. to the present: Palaeogeography, Palaeoclimatology, Palaeoecology, v. 30, p. 17-40.

———— S. L. Thompson, and S. H. Schneider, 1981, An ice-free Cretaceous? results from climate model simulations: Science, v. 212, p. 501-508.

Barrows, M. H., R. M. Cluff, and R. D. Harvey, 1979, Petrology and maturation of dispersed organic matter in the New Albany Shale Group of the Illinois basin: 3rd Eastern Gas Shales Symposium Proc. (METC/SP-79/6), p. 85-114.

Barry, R. G., and R. J. Chorley, 1970, Atmosphere, weather, and climate: New York, Holt, Rinehart, and Winston, 320 p.

Berger, A. L., 1976, Obliquity and precession for the last 5,000,000 years: Astronomy and Astrophysics, v. 51, p. 127-135.

———— 1978, Long-term variations of caloric insolation resulting from the earth's orbital elements: Quaternary Research, v. 9, p. 139-167.

———— 1980, The Milankovitch astronomical theory of paleoclimates; a modern review, in A. Beer, K. Pounds, and P. Beer, eds., Vistas in astronomy: London, Pergamon Press, v. 24, p. 103-122.

Berger, W. H., L. Diester-Haass, and J. S. Killingley, 1978, Upwelling off north-west Africa; the Holocene decrease as seen in carbon isotopes and sedimentological indicators: Oceanologica Acta, v. 1, p. 3-7.

Berry, W. B. N., and P. Wilde, 1978, Progressive ventilation of the oceans—an explanation for the distribution of the lower Paleozoic black shales: Am. Jour. Sci., v. 278, p. 257-275.

Blanton, J. O., et al, 1981, The intrusion of Gulf Stream water across the continental shelf due to topographically-induced upwelling: Deep-Sea Research, v. 28, p. 393-405.

Briskin, M., and J. Harrell, 1980, Time-series analysis of the Pleistocene deep-sea paleoclimatic record: Marine Geology, v. 36, p. 1-22.

Broecker, W. S., 1974, Chemical oceanography: New York, Harcourt, Brace, Jovanovich, 214 p.

Brongersma-Sanders, M., 1948, The importance of upwelling water to vertebrate paleontology and oil geology: Koninkl. Nederlandse Akad. Wetensch Verh. Afd. Natuurk., v. 45, p. 1-112.

———— 1966, The fertility of the sea and its bearing on the origin of oil: Adv. Sci., v. 23, p. 41-46.

Burnett, W. C., 1977, Geochemistry and origin of phosphorite deposits from off Peru and Chile: Geol. Soc. America Bull., v. 88, p. 813-823.

Calvert, S. E., 1966a, Origin of diatom-rich, varved sediments from the Gulf of California: Jour. Geology, v. 74, p. 546-565.

———— 1966b, Accumulation of diatomaceous silica in the sediments of the Gulf of California: Geol. Soc. America Bull., v. 77, p. 569-596.

———— 1979, Environment of phosphorite deposition off Namibia, in W. C. Burnett and R. P. Sheldon, eds., Report on the marine phosphatic sediments workshop: Honolulu, East-West Center, p. 8-9.

CLIMAP Project Members, 1976, The surface of the ice-age earth: Science, v. 191, p. 1131-1144.

Collinson, C., et al, 1967, Devonian of the north-central region, United States, in D. H. Oswald, ed., International symposium on the Devonian System: Alberta Soc. Petroleum Geologists, p. 933-972.

Conant, L. C., and V. E. Swanson, 1961, Chattanooga Shale and related rocks of central Tennessee and nearby areas: U.S. Geol. Survey Prof. Paper 357, 91 p.

Corredor, J. E., 1979, Phytoplankton response to low level nutrient enrichment through upwelling in the Columbian Caribbean basin: Deep-Sea Research, v. 26A, p. 731-741.

Curl, H., Jr., 1960, Primary production measurements in the north coastal waters of South America: Deep-Sea Research, v. 7, p. 183-189.

de Cserna, Z., 1970, Mesozoic sedimentation, magmatic activity, and deformation in northern Mexico, in K. Seewald and D. Sundeen, eds., The geologic framework of the Chihuahua tectonic belt; a symposium in honor of Ronald K. Deford: West Texas Geol. Soc., p. 99-117.

Demaison, G. J., and G. T. Moore, 1980, Anoxic environments and oil source bed genesis: AAPG Bull., v. 64, p. 1179-1209.

Deutsch, E. R., 1965, The paleolatitude of Tertiary oil fields: Jour. Geophys. Research, v. 70, p. 5193-5203.

Diester-Haass, L., and H.-J. Schrader, 1979, Neogene coastal upwelling history off northwest and southwest Africa: Marine Geology, v. 29, p. 39-53.

Dott, R. H., Sr., and M. J. Reynolds, 1969, Sourcebook for petroleum geology: AAPG Mem. 5, 471 p.

Dow, W. G., 1978, Petroleum source beds on continental slopes and rises: AAPG Bull., v. 62, p. 1584-1606.

Ekman, F. W., 1905, On the influence of the earth's rotation on ocean currents: Arkiv f. Matem., Astr. o. Fysik, Stockholm, v. 2, no. 11, 53 p.

Emery, K. O., and J. D. Milliman, 1978, Suspended matter in surface waters: influence of river discharge and of upwelling: Sedimentology, v. 25, p. 125-140.

———— and H. Niino, 1963, Sediments of the Gulf of Thailand and adjacent continental shelf: Geol. Soc. America Bull., v. 74, p. 541-554.

Fischer, A. G., 1960, Latitudinal variations in organic diversity: Evolution, v. 14, p. 64-81.

———— and M. A. Arthur, 1977, Secular variations in the pelagic realm: SEPM Spec. Pub. 25, p. 19-50.

Folk, R. L., 1973, Evidence for the peritidal deposition of the Devonian

Caballos Novaculite: AAPG Bull., v. 57, p. 702-725.

Frakes, L. A., 1979, Climates throughout geologic time: New York, Elsevier, 310 p.

Freas, D. H., and C. L. Eckstrom, 1968, Areas of potential upwelling and phosphorite deposition during Tertiary, Mesozoic, and late Paleozoic time (C-XXIII), in Proceedings of the seminar on sources of mineral raw materials for the fertilizer industry in Asia and the Far East: United Nations Mineral Resources Development Ser., v. 32, p. 228-238.

Fukuoka, J., A. Ballester, and F. Cervigon, 1968, An analysis of hydrographical condition in the Caribbean Sea (III)—especially about upwelling and sinking: Studies in Oceanography, p. 145-149.

Galley, J. E., 1971, Summary of petroleum resources in Paleozoic rocks of region 5—north, central, and west Texas and eastern New Mexico, in Future petroleum provinces of the United States—their geology and potential: AAPG Mem. 15, p. 726-737.

Goering, J. J., and M. M. Pamatmat, 1971, Denitrification in sediments of the sea off Peru: Investigaciones Pesqueras, v. 35, p. 233-242.

Goldstein, A., Jr., and T. A. Hendricks, 1953, Siliceous sediments of Ouachita facies in Oklahoma: Geol. Soc. America Bull., v. 64, p. 421-442.

Gray, J., and A. J. Boucot, 1979, Historical biogeography, plate tectonics, and the changing environment: Corvallis, Oregon State Univ. Press, 500 p.

Harwood, R. J., 1977, Oil and gas generation by laboratory pyrolysis of kerogen: AAPG Bull., v. 61, p. 2082-2102.

Hays, J. D., J. Imbrie, and N. J. Shackleton, 1976, Variations in the earth's orbit; pacemaker of the ice ages: Science, v. 194, p. 1121-1132.

Heckel, P.H., 1977, Origin of phosphatic black shale facies in Pennsylvanian cyclothems of Mid-Continent North America: AAPG Bull., v. 61, p. 1045-1068.

——— and B. J. Witzke, 1979, Devonian world palaeogeography determined from distribution of carbonates and related lithic palaeoclimatic indicators: Spec. Papers Palaeontology, v. 23, p. 99-123.

Hendricks, T. A., L. S. Gardner, and M. M. Knecktel, 1947, Geology of the western part of the Ouachita Mountains, Oklahoma: U.S. Geol. Survey Oil and Gas Inv., Prelim. Map 66.

Humphreville, R., and R. K. Bambach, 1979, Influence of geography, climate, and ocean circulation on the pattern of generic diversity of brachiopods in the Permian (abs.): Geol. Soc. America Abs. with Programs, v. 11, p. 447.

Hunt, B. G., 1979, The effects of past variations of the earth's rotation rate on climate: Nature, v. 281, p. 188-191.

Irving, E., and T. F. Gaskell, 1962, The paleogeographic latitude of oil fields: Royal Soc. London Geophys. Jour., v. 7, p. 54-64.

——— F. K. North, and R. Couillard, 1974, Oil, climate, and tectonics: Canadian Jour. Earth Sci., v. 11, p. 1-17.

Jenkyns, H. C., 1980, Cretaceous anoxic events; from continents to oceans; Geol. Soc. London Jour., v. 137, p. 171-188.

Kindle, J. C., and J. J. O'Brien, 1974, On upwelling along a zonally-oriented coastline: Jour. Physical Oceanog., v. 4, p. 125-130.

Lancelot, Y., et al, 1977, Initial reports of the Deep-Sea Drilling Project: v. 41, p. 1215-1245.

Leggett, J. K., et al, 1981, Periodicity in the early Paleozoic marine realm: Geol. Soc. London Jour., v. 138, p. 167-176.

Leinen, M., 1979, Biogenic silica accumulation in the central equatorial Pacific and its implications for Cenozoic paleoceanography; summary: Geol. Soc. America Bull., v. 90, p. 801-803.

Leith, H., 1964, Versuch einer kartographischen Darstellung der Produktivitat der Pflanzendecke auf der Erde; Geographisches Taschenbuch, 1964/65: Wiesbaden, Max Steiner Verlag, p. 72-80.

Lisitzin, A. P., 1972, Sedimentation in the world ocean: SEPM Spec. Pub. 17, 218 p.

Lopez-Ramos, E., 1969, Marine Paleozoic rocks of Mexico: AAPG Bull., v. 53, p. 2399-2417.

Lorenz, E. N., 1967, The nature and theory of the general circulation of the atmosphere: World Meteorolog. Organization, 616 p.

Manheim, F., G.T. Rowe, and D. Jipa, 1975, Marine phosphorite formation off Peru: Jour. Sed. Petrology, v. 45, p. 243-256.

Maynard, J. B., 1981, Carbon isotopes as indicators of dispersal patterns in Devonian-Mississippian shales of Appalachian basin: Geology, v. 9, p. 262-265.

McBride, E. F., and A. Thomson, 1970, The Caballos Novaculite, Marathon region, Texas: Geol. Soc. America Spec. Paper 122, 129 p.

McKelvey, V.E., 1967, Phosphate deposits: U.S. Geol. Survey Bull., v. 1252-D, p. D1-D21.

Meyerhoff, A. A., 1970, Continental drift; implications of paleomagnetic studies, meteorology, physical oceanography, and climatology: Jour. Geology, v. 78, p. 1-51.

Milankovitch, M. M., 1941, Canon of insolation and the ice-age problem: Kongl. Serb. Akad., Beograd, 484 p. (English translation by Israel Program for Scientific Translations, for U.S. Dept. of Commerce and National Science Foundation, reference taken from A. L. Berger, 1980).

Mohr, R. E., 1975, Measured periodicities of the Biwabik (Precambrian) stromatolites and their geophysical significance, in G. D. Rosenberg and S. K. Runcorn, eds., Growth rhythms and the history of the earth's rotation: London, John Wiley and Sons, p. 43-56.

Momper, J. A., 1978, Oil migration limitations suggested by geological and geochemical considerations, in Physical and chemical constraints on petroleum migration: AAPG Short Course, p. B1-B60.

Mooers, C. N. K., C. A. Collins, and R. L. Smith, 1976, The dynamic structure of the frontal zone in the coastal upwelling region off Oregon: Jour. Physical Oceanog., v. 6, p. 3-21.

Moore, T. C., Jr., et al, 1980, The reconstruction of sea surface temperatures in the Pacific Ocean of 18,000 b.p.: Marine Micropaleontology, v. 5, p. 215-247.

Nansen, F., 1902, Oceanography of the North Polar basin, Norwegian North Polar expedition, 1893-1896: Scientific Results, v. 3, no. 9, 427 p.

O'Brien, J. J., 1971, A two dimensional model of the wind-driven North Pacific: Investigaciones Pesqueras, v. 35, p. 331-349.

Parrish, J. T., 1981, Atmospheric circulation and upwelling in the Mesozoic and Cenozoic (abs.): Geol. Soc. America Abs. with Programs, v. 13, p. 526.

——— and R. L. Curtis (in press), Atmospheric circulation, upwelling, and organic-rich rocks in the Mesozoic and Cenozoic: Palaeogeography, Palaeoclimatology, Palaeoecology.

——— and R. G. Humphreville, 1981, Upwelling and phosphorites in the Paleozoic (abs.): AAPG Bull., v. 65, p. 969.

——— and A. M. Ziegler, 1980, Climate asymmetry and biogeographic distributions (abs.): AAPG Bull., v. 64, p. 763.

——— ——— 1981, Paleozoic paleogeography and upwelling: Abstracts, Coastal upwelling; its sediment record: Advanced Research Inst. Conf., Sept. 1981, Vilamoura, Portugal.

——— K. S. Hansen, and A. M. Ziegler, 1979, Atmospheric circulation and upwelling in the Paleozoic, with reference to petroleum source beds (abs.): AAPG Bull., v. 63, p. 507-508.

——— A. M. Ziegler, and C. R. Scotese (in press), Rainfall patterns and the distribution of coals and evaporites in the Mesozoic and Cenozoic: Palaeogeography, Palaeoclimatology, Palaeoecology.

Perry, A. H., and J. M. Walker, 1977, The ocean-atmosphere system: London, Longman, 160 p.

Petterssen, S., 1969, Introduction to meteorology, 3d ed.: New York, McGraw-Hill Book Co., 327 p.

Philippi, G. T., 1974, The influence of marine and terrestrial source material on the composition of petroleum: Geochim. et Cosmochim. Acta, v. 38, p. 947-966.

Pinet, P. R., and P. Popenoe, 1980, Cenozoic flow patterns of the Gulf Stream over the Blake Plateau (abs.): Geol. Soc. America Abs. with Programs, v. 12, p. 500.

Pisciotto, K. A., and R. E. Garrison, 1981, Lithofacies and depositional environments of the Monterey Formation, California, in The Monterey Formation and related siliceous rocks of California: SEPM, Pacific Sec. p. 97-122.

Prell, W. L., et al, 1980, Surface circulation of the Indian Ocean during the last glacial maximum, approximately 18,000 yr. b.p.: Quaternary Research, v. 14, p. 309-336.

Ray, E. O., 1971, Petroleum potential of eastern Kentucky, in Future petroleum provinces of the United States—their geology and potential: AAPG Mem. 15, p. 1261-1268.

Richards, F. A., 1960, Some chemical and hydrographic observations along the north coast of South America—I. Cabo Tres Puntas to Curacao, including the Cariaco Trench and the Gulf of Cariaco: Deep-Sea Research, v. 7, p. 163-182.

Riehl, H., 1978, Introduction to the atmosphere: New York, McGraw-Hill Book Co., 410 p.

Robinson, P. L., 1971, A problem of faunal replacement on Permo-Triassic continents: Palaeontology, v. 14, p. 131-153.

Rosen, B. R., 1975, The distribution of reef corals: Rept. Underwater Assoc., v. 1, p. 1-16.

445

Ryther, J. H., et al, 1971, The production and utilization of organic matter in the Peru coastal upwelling current: Investigaciones Pesqueras, v. 35, p. 43-59.

Sancetta, C., 1978, Neogene Pacific microfossils and paleoceanography: Marine Micropaleontology, v. 3, p. 347-376.

Saunders, P.M., 1977, Wind stress on the ocean over the eastern continental shelf of North America: Jour. Physical Oceanog., v. 7, p. 555-571.

Savin, S. M., R. G. Douglas, and F. G. Stehli, 1975, Tertiary marine paleotemperatures: Geol. Soc. America Bull., v. 86, p. 1499-1510.

Schlanger, S. O., and H. C. Jenkyns, 1976, Cretaceous oceanic anoxic events; causes and consequences: Geologic en Mijnbouw, v. 55, p. 179-184.

Schopf, J. W., 1978, Evolution of the earliest cells: Scientific American, v. 239, p. 110-138.

Scotese, C. R., 1979, Phanerozoic continental drift base maps; prepared for R. K. Bambach and C. R. Scotese, Paleogeographic reconstructions—state of the art: Geol. Soc. America Short Course, April, 1979, Blacksburg, Va.

———— et al, 1979, Paleozoic base maps: Jour. Geology, v. 87, p. 217-277.

Scrutton, C. T., 1965, Periodicity in Devonian coral growth: Palaeontology, v. 7, p. 552-558.

———— 1978, Periodic growth features in fossil organisms and the length of the day and month, in P. Brosche and J. Sundermann, eds., Tidal friction and the earth's rotation: Berlin, Springer-Verlag, p. 154-196.

Sharma, G. S., 1978, Upwelling off the southwest coast of India: Indian Jour. Marine Sci., v. 7, p. 209-218.

Sheldon, R. P., 1964, Paleolatitudinal and paleogeographic distribution of phosphorite: U.S. Geol. Survey Prof. Paper 501-C, p. C106-C113.

———— E. K. Maughan, and E. R. Cressman, 1967, Sedimentation of rocks of Leonard (Permian) age in Wyoming and adjacent states, in L. A. Hale, ed., Anatomy of the western phosphate fields; a guide to the geologic occurrence, exploration methods, mining engineering, recovery technology: Intermountain Assoc. Geologists, 15th Annual Field Conf., p. 1-44.

Smith, R. L., 1973, Upwelling, in R. G. Pirie, ed., Oceanography; contemporary readings in ocean sciences: New York, Oxford Univ. Press, p. 126-147.

———— et al, 1971, The circulation in an upwelling ecosystem; the Pisco cruise: Investigaciones Pesqueras, v. 35, p. 9-24.

Stauble, A. J., and G. Milius, 1970, Geology of Groningen gas field, Netherlands, in Geology of giant petroleum fields: AAPG Mem. 14, p. 359-369.

Stehli, F. G., 1970, A test of the earth's magnetic field during Permian time: Jour. Geophys. Research, v. 75, p. 3325-3342.

Summerhayes, C. P., 1981, Organic facies of middle Cretaceous black shales in deep North Atlantic: AAPG Bull., v. 65, p. 2364-2380.

Sundermann, J., and P. Brosche, 1978, Numerical computation of tidal friction for present and ancient oceans, in P. Brosche and J. Sundermann, eds., Tidal friction and the earth's rotation: Berlin, Springer-Verlag, p. 125-143.

Svansson, A., 1975, Interaction between the coastal zone and the open sea: Merentutk. Julk., v. 239, p. 11-28.

Sverdrup, H. U., M. W. Johnson, and R. H. Fleming, 1942, The oceans; their physics, chemistry, and general biology: Englewood Cliffs, N. J., Prentice-Hall, 1087 p.

Swallow, J. C., 1980, The Indian Ocean experiment; introduction: Science, v. 209, p. 588.

Taft, B., 1971, Ocean circulation in monsoon areas, in J. D. Costlow, Jr., ed., Fertility of the sea: New York, Gordon and Breach Pub., v. 2, p. 565-579.

Tissot, B., and D. H. Welte, 1978, Petroleum formation and occurrence: New York, Springer-Verlag, 538 p.

———— G. Deroo, and J.P. Herbin, 1979, Organic matter in Cretaceous sediments of the North Atlantic; contribution to sedimentology and paleoceanography, in M. Talwani, W. Hay, and W. B. F. Ryan, eds., Deep drilling results in the Atlantic Ocean; continental margins and paleoenvironment: Maurice Ewing Ser., v. 3, p. 362-374.

———— et al, 1974, Influence of nature and diagenesis of organic matter in formation of petroleum: AAPG Bull., v. 58, p. 499-506.

———— et al, 1980, Paleoenvironment and petroleum potential of middle Cretaceous black shales in Atlantic basins: AAPG Bull., v. 64, p. 2051-2063.

Trask, P. D., 1932, Origin and environment of source sediments of petroleum: Houston, Gulf Pub. Co., 323 p.

Vail, P. R., R. M. Mitchum, Jr., and S. Thompson, III, 1977, Seismic stratigraphy and global changes of sea level, part 4; global cycles of relative changes of sea level, in Seismic stratigraphy—applications to hydrocarbon exploration: AAPG Mem. 26, p. 83-97,

Van Eysinga, F. W. B., 1975, Geological time table: Amsterdam, Elsevier.

Vassoyevich, N. B., and L. A. Polster, 1966a, Map of oil and gas bearing (sic) of the Kyn deposits of the Volga-Ural region, in A. P. Vinogradov, ed., Atlas of the lithological-paleogeographical maps of the U.S.S.R., v. II: Moscow, Min. Geol., Acad. Sci., U.S.S.R.

———— ———— 1966b, Map of oil and gas bearing (sic) of the Pashiy deposits of the Volga-Ural region, in A. P. Vinogradov, ed., Atlas of the lithological-paleogeographical maps of the U.S.S.R., v. II: Moscow, Min. Geol., Acad. Sci., U.S.S.R.

———— ———— 1967, Map of oil and gas bearing middle Visean deposits of the Volga-Ural region, in A. P. Vinogradov, ed., Atlas of the lithological-paleogeographical maps of the U.S.S.R., v. II: Moscow, Min. Geol., Acad. Sci., U.S.S.R.

Vinogradov, A.P., ed., 1968, Atlas of the lithological-paleogeographical maps of the U.S.S.R., v. I: Moscow, Min. Geol., Acad. Sci. U.S.S.R.

———— ed., 1969, Atlas of the lithological-paleogeographical maps of the U.S.S.R., v. II: Moscow, Min. Geol., Acad. Sci., U.S.S.R.

Wardlaw, B. R., 1980, Middle-Late Permian paleogeography of Idaho, Nevada, Montana, Utah, and Wyoming (abs.): AAPG Bull., v. 64, p. 799.

Wells, J. S., 1971, Forest City basin, in Future petroleum provinces of the United States—their geology and potential: AAPG Mem. 15, p. 1098-1103.

Wells, J. W., 1963, Coral growth and geochronometry: Nature, v. 197, p. 948-950.

Williams, G. E., 1975, Late Precambrian glacial climate and the earth's obliquity: Geological Magazine, v. 112, p. 441-544.

———— 1976, Discussion on Late Precambrian mixtites; glacial and/or non-glacial?: Am. Jour. Sci., v. 276, p. 370-374.

Williams, J. A., 1974, Characterization of oil types in Williston basin: AAPG Bull., v. 58, p. 1243-1252.

Wolfe, J. A., 1978, A paleobotanical interpretation of Tertiary climates in the northern hemisphere: Am. Scientist, v. 66, p. 694-703.

———— 1980, Tertiary climates and floristic relationships at high latitudes in the northern hemisphere: Palaeogeography, Palaeoclimatology, Palaeoecology, v. 30, p. 313-323.

Wooster, W. S., 1969, Equatorial front between Peru and Galapagos: Deep-Sea Research, v. 16 (suppl.), p. 407-419.

———— and J. L. Reid, Jr., 1963, Eastern boundary currents, in M.N. Hill, ed., The sea: v. 2, p. 253-280.

Yentsch, C. S., 1974, The influence of geostrophy on primary production: Tethys, v. 6, p. 111-118.

Ziegler, A. M., and J. T. Parrish, 1981, Paleoclimates and climatically controlled sediments—the stability of global atmospheric circulation patterns (abs.): Geol. Soc. America Abs. with Programs, v. 13, p. 587.

———— J. T. Parrish, and R. G. Humphreville, 1979, Palaeogeography, upwelling, and phosphorites (abs.), in P. J. Cook and J. H. Shergold, eds., Proterozoic and Cambrian phosphorites: Proc, 1st Mtg. Internat. Geol. Correlation Project 156., p. 21.

———— et al, 1977, Silurian continental distributions, paleogeography, climatology, and biogeography: Tectonophysics, v. 40, p. 13-51.

———— et al, 1981, Paleozoic biogeography and climatology, in K. J. Niklas, ed., Paleobotany, paleoecology, and evolution: New York, Praeger Pub., v. 2, p. 231-266.

Zuniga y Rivero, F., et al, 1976, Hydrocarbon potential of Amazon basins of Colombia, Ecuador, and Peru, in Circum-Pacific energy and mineral resources: AAPG Mem. 25, p. 339-348.

446

The American Association of Petroleum Geologists Bulletin
V. 69, No. 3 (March 1985), P. 448-459, 13 Figs.

Numerical Climate Modeling, A Frontier in Petroleum Source Rock Prediction: Results Based on Cretaceous Simulations[1]

ERIC J. BARRON[2]

ABSTRACT

Organic productivity and preservation are primary controls of the distribution of petroleum source beds. Both organic productivity and preservation are in part a function of geographic setting and climate. Therefore, the prediction of paleoclimates is a major frontier in petroleum source rock prediction. A climate model based on the fundamental laws thought to govern atmospheric circulation is applied here to the prediction of upwelling regions associated with very high productivity and bottom conditions that may become anoxic despite a high oxygen supply. The potential and limitations of climate modeling in the prediction of upwelling regions are examined by (1) demonstrating the ability of the model to predict present-day winds and upwelling regions, (2) demonstrating that the model—and the location of upwelling regions—is sensitive to geography (continental positions, sea level, and topography), and (3) investigating whether the model sensitivity to past geographies is realistic, i.e., if predicted upwelling regions for the Cretaceous Period correspond to source rock localities.

The model predictions of ocean and coastal upwelling compare very favorably with present-day observations and with estimates of primary productivity. The location of middle-latitude and high-latitude upwelling locations are sensitive to model-specified geography. The location of low-latitude upwelling is less sensitive to changes in global geography if coastlines are oriented appropriately. Despite some limitations, a comparison of source rock localities with model predictions is promising.

The results of these studies suggest that numerical climate modeling is a potential tool in petroleum source rock prediction.

INTRODUCTION

Organic productivity and preservation are primary controls of the distribution of petroleum source beds in space and during time. The major limiting factor in organic productivity is the nutrient supply within the photic zone (primarily nitrates and phosphates) in the aquatic realm and precipitation in the terrestrial realm (Fogg, 1975). Nutri-

ents are supplied to the photic zone from coastal sources and by remineralization of nutrients below the photic zone. Consequently, the regions of highest aquatic productivity occur in coastal waters and in upwelling regions (divergences), and lowest productivities occur in central oceanic gyres and convergences (e.g., Menzel, 1974). Therefore, productivity can be related ultimately to geographic setting and climate. Geography and atmospheric circulation govern the location of oceanic divergences, convergences, and gyres.

To a large degree, the prediction of regions of high paleoproductivity could be based on predictions of atmospheric circulation patterns for times of different continental configurations during earth history. Regions of high rainfall in the terrestrial realm and coastal upwelling and oceanic divergences in the marine realm could be derived from predictions of atmospheric circulation patterns. With the increased accuracy of continental reconstructions (Smith and Briden, 1977; Barron et al, 1981), predictions of past atmospheric circulations represent one frontier in petroleum source rock prediction.

Typically, less than 1% of the organic matter produced is apparently preserved within the sediment (Dow, 1978; Brawlower and Thierstein, in press). Degradation of organic matter proceeds quickly in oxygen-rich environments. Consequently, the conditions that promote anoxia are perhaps the most important factor in the formation of petroleum source beds (Demaison and Moore, 1980). In fact, high productivity will be a good petroleum source bed predictor only if the oxygen consumption from degradation of organic matter exceeds oxygen supply in a consistent manner. In this regard, organic matter is more likely to exceed the available oxygen in regions of very intense productivity in coastal areas (i.e., coastal upwelling regions). Coastal upwelling zones have been frequently cited as an excellent setting for petroleum source bed genesis (e.g., Parrish, 1982).

Similar to organic productivity, the conditions that promote anoxia and hence the preservation of organic matter are largely dependent on geographic setting and climate. Oxygen is supplied to the aquatic realm from interaction with the aerated surface. The degree of oxygen supply at depth is thus dependent on mixing processes, bottom water formation, and oxygen solubility. In particular, stratification is considered to be of major importance in promoting anoxia (Degens and Stoffers, 1976; Arthur and Natland, 1979; Arthur and Schlanger, 1979; Demaison and Moore, 1980). Stratification is governed by heat (temperature) and moisture (salinity) fluxes at the ocean or lake surface and in the surface winds. Even in regions of relatively low primary productivity, preservation may occur in regions of strong, stable stratification. Oxygen solubility is a function of temperature and salinity. High

[1]Manuscript received, March 15, 1984; accepted, September 10, 1984.
[2]National Center for Atmospheric Research, P.O. Box 3000, Boulder, Colorado 80307.
The National Center for Atmospheric Research is sponsored by the National Science Foundation. Additional support was received from Texaco, ARCO, and the Mobil Research Foundation. The author thanks Michael Arthur, Barry Katz, and John Armentrout for their critical comments, which improved the manuscript.

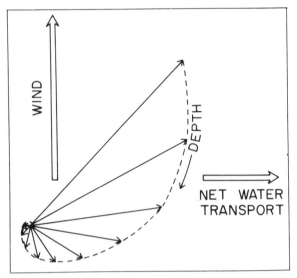

Figure 1—Ekman spiral, illustrating the gradual change in current direction and strength with depth and the 90° relationship of net water transport with wind direction.

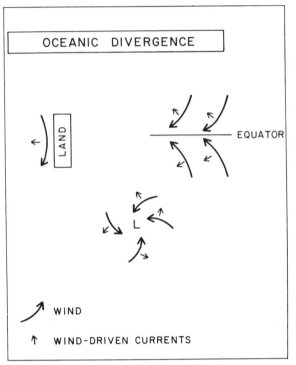

Figure 2—Specific conditions for coastal and open-ocean upwelling.

temperatures and salinity decrease oxygen solubility while stratification of the water column limits oxygen resupply at depth. The prediction of mixing, bottom-water formation, and oxygen solubility, by predicting surface ocean heat and moisture fluxes in the context of oceanic circulation patterns, represents the second frontier in petroleum source rock prediction with respect to climate.

Numerical climate models based on the fundamental physical laws thought to govern atmospheric and oceanic circulation have the potential to be a major new tool in petroleum source rock prediction because they address the controls of both organic productivity and preservation. In particular, the development of atmospheric general circulation models (GCMs) is at a stage where the models can be readily applied to investigate atmospheric circulation characteristics for different paleogeographies (Barron and Washington, 1982). For this reason, and because the controls on organic productivity are apparently less complex than on organic preservation, models are presently more capable of addressing the first frontier in petroleum source rock prediction with respect to climate (productivity), than the second (preservation).

Therefore, the prediction of paleo-upwelling regions, which frequently have very high primary productivity and for which bottom conditions may become anoxic despite a very high oxygen supply, is a logical first goal. Toward this goal, the capability of a GCM to predict present-day upwelling is examined; then the model is applied to a series of mid-Cretaceous geographic sensitivity experiments. The purpose of the sensitivity experiments is to determine to what extent changing geography alters the characteristics of the atmospheric circulation, and hence the locations of paleo-upwelling regions. These problems are investigated through a series of climate model studies based on the mid-Cretaceous to determine the sensitivity of the location of upwelling regions to changes in continental positions, sea level, and topography.

CONDITIONS FOR UPWELLING

Upwelling occurs for any set of conditions resulting in a surface divergence. A surface divergence requires that surface waters be replaced by water from below. For petroleum source rock prediction, persistent, large-scale upwelling that is predominately wind-driven is of primary importance. Parrish (1982) gives a good summary for the variety of upwelling situations in addition to the conditions of primary interest here.

The most important observation relating upwelling to surface winds is that the net transport of the water column beneath the wind is at a right angle to the wind. This is a consequence of the Coriolis force that results from an imbalance between the gravitational force and centrifugal force accompanying the earth's rotation as a particle moves over the earth. Any change in the distance of a body from the axis of rotation will be associated with such a deflecting force. The deflection of a moving particle will occur to the right in the Northern Hemisphere and to the left in the Southern Hemisphere. The deflection of the surface water immediately beneath the wind is relatively small, but, through friction, the water beneath the surface is also set into motion. Consequently, with depth the deflection from the surface-wind direction increases, and the speed of the motion decreases. This spiral of increasing deflection and decreasing motion (Figure 1) is the familiar Ekman spiral. The sum of this spiral gives a mean direction for the water motion, which is at a right angle to the wind direction.

With the observation that the transport of the water column beneath the wind (Ekman transport) is at a right angle to the wind—to the right in the Northern Hemisphere and to the left in the Southern Hemisphere—the specific conditions for persistent, large-scale upwelling can be defined (Figure 2). Therefore, two major situations of interest are coastal and open-ocean upwelling.

Coastal divergence will occur under winds parallel to the coastline. In the Northern Hemisphere the coastline must be to the left of the wind; in the Southern Hemisphere it must be to the right. Typically, coastal upwelling occurs on west-facing coasts within the subtropical high-pressure belt or east-facing coasts within the middle-latitude low-pressure zone, but it also may occur on north- or south-facing coasts associated with appropriate winds. The effects of bottom friction, stratification, local currents, topography, and shoreline configuration are complicating factors in an otherwise simple relationship between wind direction and coastal upwelling.

Open-ocean divergences form in association with the low surface pressure and horizontal convergence of surface winds. Although specific characteristics differ, horizontal convergence of surface winds results in a divergence in the underlying water column. For example, the equatorial region is characterized by converging surface easterly winds from both hemispheres. Remembering that the mass transport of the water column is to the right in the Northern Hemisphere and to the left in the Southern Hemisphere, the wind convergence results in a divergence in the water column. Similar effects occur within the low-pressure belts in each hemisphere that are associated with converging surface winds from opposite directions (polar easterlies and middle-latitude westerlies), or within single low-pressure cells associated with wind convergence (counterclockwise in the Northern Hemisphere and clockwise in the Southern Hemisphere) toward the center of the pressure system.

All of the upwelling conditions described above may result in persistent, large-scale upwelling associated with very high productivity. However, coastal upwelling and open-ocean upwelling associated with epicontinental seas are generally of much greater significance for petroleum source rock prediction. These upwelling regions have a number of characteristics conducive to preserving organic matter in the sediment, including reduced particle settling times, high sedimentation rates, and a smaller oxygen reservoir compared to open-ocean environments today. The major oceanic "anoxic events" (e.g., Schlanger and Jenkyns, 1976) are an exception to this general association of poor preservation in the open-ocean regime.

Importantly, a specific well-documented association between surface winds and upwelling (e.g., Knauss, 1978) can be used as a first-order prediction of paleo-upwelling if the atmospheric winds can be correctly determined.

PREDICTING SURFACE WIND PATTERNS

Predicting surface wind patterns on a globally uniform water planet would be relatively straightforward. Atmospheric circulation would be governed only by planetary rotation and the differential heating from equator to pole.

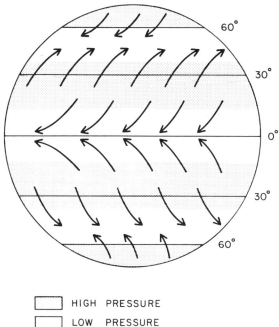

HIGH PRESSURE

LOW PRESSURE

Figure 3—An idealized surface atmospheric circulation, giving major surface-pressure zones and easterly and westerly wind belts.

For such an idealized planet with characteristics similar to the earth, circulation might resemble the schematic representations of the earth's "mean" circulation common in introductory textbooks (Figure 3). The salient features of these schematics are tropical low pressure, subtropical high, middle-latitude low, and polar high pressure associated with tropical easterlies, middle-latitude westerlies, and polar easterlies. A first prediction of a number of paleoclimatic variables for past continental configurations frequently involves moving the continents "beneath" this highly idealized circulation pattern, recognizing only how continents may alter the zonality of this pattern (e.g., Robinson, 1973; Parrish, 1982). Using simple schematic qualitative models is a logical first step toward predicting ancient upwelling regions.

Parrish (1982) predicted the distribution of Paleozoic upwelling regions based on a qualitative model of the present-day mean circulation (Figure 3) and concepts of how continents disrupt the zonal circulation. Surface pressure patterns for an earth-like planet with an idealized continental landmass are illustrated in Figure 4. Because of the different thermal properties of land and sea, the distribution of surface-pressure highs and lows is strongly influenced by the presence of land. For example, in the winter hemisphere the subtropical high is displaced equatorward and is intensified over the continent, while in the summer hemisphere the subtropical high is displaced poleward and is intensified over the ocean.

However, qualitative models cannot address the processes that maintain the circulation, rather they rely solely on analogy with the present day. Therefore, the qualitative

449

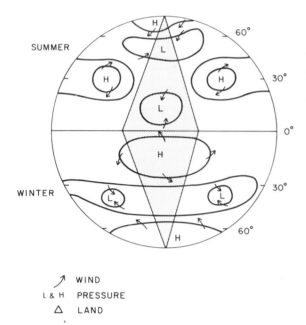

WIND

L & H PRESSURE

△ LAND

Figure 4—An idealized surface atmospheric circulation for summer and winter hemispheres given a hypothetical continent.

models fail to address whether the characteristics of the mean circulation can be altered by the changing continental configuration of the surface of the earth. On a more fundamental level, the energy and angular momentum balance of the earth-atmosphere system must be considered. The manner in which these balance requirements are fulfilled forms the basis of theories on the general circulation of the atmosphere (e.g., Lorenz, 1967). Even without a rigorous consideration of the balance requirements, different geographies appear to have considerable potential to alter how these balances are maintained. For instance, angular momentum is exchanged between the atmosphere and the surface and is transported horizontally by atmospheric motions. Net exchange of angular momentum with the underlying earth above a specific latitude must be balanced by horizontal transport across that latitude. Different frictional torques exerted over land and sea and associated with topography suggest that different continental and topographic distributions must alter the angular momentum exchange between the surface and the atmosphere, and, hence, the atmospheric motions maintaining the balance.

In order to develop the predictive associations between climate and petroleum source rock occurrence, the laws governing atmospheric and oceanic circulation must be considered. In this regard, the physical laws that govern the circulation—expressed in mathematical form as dynamic equations (e.g., Lorenz, 1967)—have been incorporated into numerical climate models. The most fully resolved models, called general circulation models (GCMs), compute a large suite of variables including surface pressure and winds. Numerical climate models have considerable potential to address whether the characteristics of the circulation are altered by changing geography.

MODEL DESCRIPTION

The model used in this study is a three-dimensional, time-dependent, general circulation model of the atmosphere, based on the hydrodynamic and thermodynamic laws applied to the atmosphere. It is a version of the community climate model (CCM) at the National Center for Atmospheric Research (NCAR). This version of the CCM evolved from the spectral climate model described by Bourke et al (1977) and McAvaney et al (1978). The model resolution is 4.5° in latitude and 7.5° in longitude with nine levels in the vertical.

Two principal changes were made in the model from Bourke et al (1977) and McAvaney et al (1978). First, Ramanathan et al (1983) introduced a new radiation-cloudiness formulation that incorporates interactive clouds. Second, a more realistic hydrology treatment has been introduced following Washington and Williamson (1977), in which model-derived precipitation and evaporation rates are used to simulate changes in soil moisture, runoff, and snow cover.

These simulations employ an energy-balance ocean formulation, sometimes referred to as a swamp ocean, which does not include heat storage, transport, or diffusion. Ocean-surface temperatures are predicted based on this energy balance. Sea ice forms and grows if the ocean surface temperature falls below 271.2°K; this formation is accompanied by changes in surface albedo. Energy-balance ocean models are usually restricted to simulations with annual average insolation. The CCM has recently been coupled to mixed-layer and fully resolved ocean models and is capable of simulating the seasonal cycle.

The CCM coupled to an energy-balance ocean satisfactorily simulates most of the features of the observed annual mean temperature structure, and the zonal wind distribution, as well as many other climatic characteristics. Comparisons of the model climatology with present observations can be found in Pitcher et al (1983), Ramanathan et al (1983), and Washington and Meehl (1983).

DETERMINING UPWELLING CONDITIONS FROM MODEL-PREDICTED WINDS

The CCM computes the U (easterly) and V (northerly) components of the wind at nine levels in the atmosphere. Upwelling regions can be located reasonably well by visually examining surface wind vectors, given the simple relationships described earlier. However, an objective determination of divergence or convergence can be computed given the winds.

From the lowermost model level (≈ 70 m above the surface), the surface wind stress is computed from:

$$\tau_x = \rho \, C_D \, V \, \sqrt{V^2 + U^2} \qquad (1a)$$

$$\tau_y = \rho \, C_D \, U \, \sqrt{V^2 + U^2} \qquad (1b)$$

where τ_x is the east-west wind-stress component, τ_y is the north-south component, C_D is a drag coefficient, and ρ is surface air density. If the wind-stress curl is positive, a divergent condition exists. The curl is:

$$(\partial \tau_y / \partial x) - (\partial \tau_x / \partial y). \qquad (2)$$

To determine the mass transport caused by the wind, the Coriolis force must be considered. This force is proportional to the sine of the latitude and results in a greater mass transport at lower latitudes given equal wind velocities and directions.

Remembering that the net transport by the wind is at a right angle to the wind, the convergence or divergence for a fixed boundary (a coastline) can be related to the sign of the τ_x and τ_y components. In order to determine the mass transport caused by the wind, the Coriolis force must also be considered. For example, on a north-south coastline the transport away from the coast is given by:

$$M_x = \tau_y/f \tag{3}$$

where f is the Coriolis parameter.

These formulae represent an objective means of determining open-ocean and coastal upwelling from the winds computed in the CCM, rather than a subjective examination of the pressure or wind fields.

MODEL EXPERIMENTS

The purpose of this research is to perform the preliminary experiments needed to investigate the potential and limitations of climate modeling in petroleum source rock prediction. First, the ability of the climate model to predict present-day surface winds and the location of upwelling zones must be demonstrated. Second, the relative stability of the atmospheric circulation to changes in geography needs to be investigated. This problem must be examined through a series of sensitivity experiments involving modifications of continental positions, sea level, and topography. From these sensitivity experiments, the Parrish (1982) method can be evaluated to determine whether continents can be moved "beneath" a stable atmospheric circulation.

Once the sensitivity of the location of upwelling zones to changes in geography is determined, a specific past geography can be investigated with the goal of petroleum source rock prediction. Ideally, the objective is to demonstrate that the model not only accurately predicts present-day atmospheric circulation but also has the appropriate sensitivity to predict past circulation patterns. Both limitations of the geologic record and climate model experiments serve to guide future research.

Present-day geographic boundary conditions used in the control simulation are illustrated in Figure 5. Note the smoothed outline of these continents given the resolution of the model in latitude (4.5°) and longitude (7.5°). The sensitivity experiments are based on the differences between the mid-Cretaceous and present-day geography, which involves continental positions, global sea level, and topography. The mid-Cretaceous is used here to mean the Albian-Cenomanian with continental positions computed at 100 Ma. The mid-Cretaceous is a large geographic contrast from the present and is a time period of known widespread source rock deposition (e.g., Tissot, 1979).

Three sensitivity experiments, each modifying a single geographic variable, are completed here (Figure 6). First, present-day featureless (no topography) continents are rotated to their mid-Cretaceous positions following Barron et al (1981). In this experiment the total global land

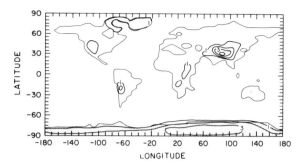

Figure 5—Present-day geographic boundary conditions for a control experiment. Area of permanent ice is shaded, and topography is given in 1-km increments.

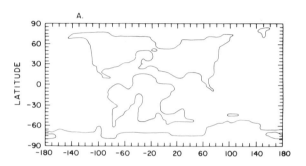

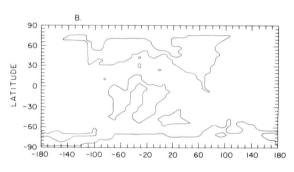

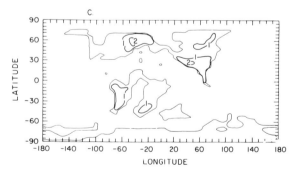

Figure 6—Geographic boundary conditions for three mid-Cretaceous sensitivity experiments. (A) Featureless continents (no topography) with low sea level (present-day areas). (B) Featureless continents with high sea level (Albian-Cenomanian areas). (C) Continents with high sea level and topography given in 1-km increments.

451

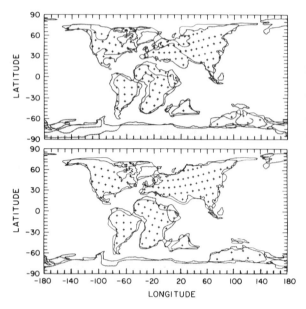

Figure 7—Mid-Cretaceous continental reconstructions including 10° latitude-longitude crosses for reference with model continental outlines for high and low sea level cases superimposed for comparison.

area is the same as for the present-day control. This experiment may be considered a mid-Cretaceous low sea level case. Second, the area of the Cretaceous continents was decreased consistent with the high global sea level. The higher global sea level resulted in approximately a 20% reduction in total land area as measured from the paleogeographic maps of Barron et al (1981) and with the paleogeography of Antarctica similar to that of Tarling (1978). Third, topography was added to the high sea level Cretaceous case. The distribution of topographic highs used in the model simulation is relatively similar to that of Scotese et al (1981) as reported by Parrish and Curtis (1982) for the mid-Cretaceous. In Figure 7, model outlines of high and low sea level Cretaceous continents are superimposed on realistic continental outlines to facilitate comparisons of predictions with modern geographic locations.

Each Cretaceous sensitivity experiment was started from uniform land and sea temperatures. The land-surface albedo was specified as 0.13, except at the interior of Antarctica, which had a somewhat higher surface albedo (more typical of observations for land surfaces with some snow cover). The model computes the atmospheric conditions every 12 hr and, in a sense, is computing the weather as the atmosphere evolves from an initial state. The model reaches quasi-equilibrium after 50 days of simulation time. The results presented here are an average of 100 days of each simulation after the first 50 days. A 100-day average eliminates the day-to-day variability. The prediction of winds therefore better reflects the mean state of the model atmosphere. The present control and the most realistic Cretaceous case were extended to 400 days of simulation. The results do not differ substantially for days 50-150 and 300-400.

Surface winds and inferred upwelling regions predicted for present-day geography in the control simulation are illustrated in Figure 8. Wind vectors are plotted for only 50% of the model grid points. The low-latitude easterlies converging on the equator, the middle-latitude westerlies, and the polar easterlies are clearly evident. From these winds, open-ocean upwelling is computed from equation 2, taking into account the Coriolis parameter. Convergence of winds adjacent to Antarctica are associated with the circum-Antarctic divergence. The North Pacific and Atlantic oceanic divergences are also well simulated. An equatorial upwelling region associated with converging of low-latitude easterlies is predicted to be the strongest open-ocean upwelling area.

A condition for coastal divergence is computed from the coastal winds using equation 3, without considering the Coriolis parameter. The coastal upwelling regions plotted in Figure 8 represent values greater than the mean coastal divergence. In other words, the plotted regions represent the most likely regions of coastal upwelling—wind vectors with the largest alongshore component—rather than the regions with the greatest mass transport, which are latitude dependent through the Coriolis force. Note that regions of coastal divergence are predicted for southwestern Africa, northwestern Africa, western South America, northern South America, the Newfoundland region, northeastern Asia, western Australia, and southern India. Most of the coastal upwelling regions are located in high-pressure belts on the western margins of continents (e.g., southwestern Africa) and low-pressure belts on the eastern margins (e.g., northeastern Asia).

The combined predictions of ocean and coastal upwelling compare favorably with present-day observations and with estimates of primary productivity (e.g., Clapham, 1973) (Figure 9). However, some deficiencies can be noted. First, the predicted upwelling for the tropical region appears to be too strong and too widespread. Second, the observed upwelling off western Mexico is not evident. The predicted wind vectors off Mexico do suggest a coastal divergence; however, the computed value is less than the mean and therefore is not plotted. It is also important to realize that the prediction of upwelling regions is based on a mean annual simulation and upwelling frequently intensifies seasonally. Given these points, the model predictions are remarkably accurate. It appears that the model is capable of predicting present-day surface winds within a reasonable accuracy and, by inference, the regions of high primary productivity. The next step is to determine whether the circulation is sensitive to changes in geography.

The surface pressure distribution for the present-day control and for each of the three Cretaceous sensitivity studies are plotted in Figure 10. Surface winds can be related to the surface pressure through geostrophic arguments, and surface pressure plots give a better visual representation of the nature of surface circulation. Note the familiar pattern of tropical low pressure, subtropical high, middle-latitude low, and polar high pressure in the present-day control simulation. It is also apparent how continentality and topography alter the zonal nature of the

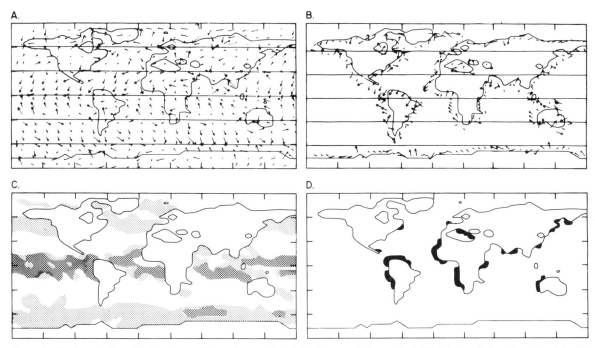

Figure 8—Surface winds and inferred upwelling regions predicted for the present-day control. (A) Surface wind vectors at model grid points (only 50% of the points are plotted). (B) Coastal wind vectors at model grid points. (C) Predicted open-ocean upwelling; regions greater than the mean are darkly shaded. (D) Predicted coastal upwelling; only values of divergence greater than mean values are plotted.

circulation. The position of the subtropical high in relation to the Tibetan Plateau is a good example of these effects.

Most importantly, the surface pressures simulated for each of the Cretaceous geographic sensitivity experiments are substantially different from the present-day and in comparison with each other. For example, the surface pressure pattern is more zonal in the experiments that lack specific topographic features. Although the surface pressure pattern is more zonal than at present for these two cases, the pressure pattern is not what would be expected from the idealized circulation presented in Figure 3. In particular, note that in the Cretaceous low sea level case, the polar high-pressure zone is lacking in the Northern Hemisphere. In contrast, with the exception of variations in the intensity of highs and lows, the low-latitude surface pressure distributions are relatively similar. These results suggest that, except for low latitudes, the circulation patterns are certainly sensitive to geographic variables. The concept of moving the continents "beneath" a stable atmospheric circulation may not be justified. The intensity of surface pressure systems, the zonality of the patterns, and the latitudinal distribution of surface highs and lows are all variable to some extent.

The computed regions of open-ocean divergence for the three Cretaceous sensitivity experiments are given in Figure 11. On a broad scale, the distribution of open-ocean upwelling regions is similar in the sense that each is characterized by equatorial and high-latitude upwelling regions. However, a number of distinctions can be noted. For example, in both high sea level experiments, a substantial

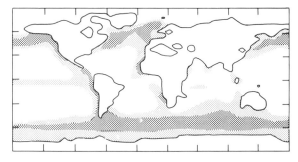

Figure 9—An estimate of present-day primary productivity from Clapham (1973), for comparison with present-day predicted upwelling (Figure 7). Unshaded ocean: < 100 gC/m²/year; light stipple: 100-200 gC/m²/year; dark stipple: > 200 gC/m²/year.

amount of open-ocean upwelling is predicted for epicontinental seas but is not present in the low sea level experiment (e.g., the North American Western Interior seaway of Canada). In the high sea level case, a well-developed low-pressure system exists over the European epicontinental seas with an associated divergence. Similarly, in the high sea level cases, upwelling is predicted near the west coast of Africa and the Angola Basin and in the epicontinental seas bordering the Arctic Basin, but is absent in the low sea level case. This difference between high and low sea level cases reflects both differences in shoreline locations and differences in wind patterns.

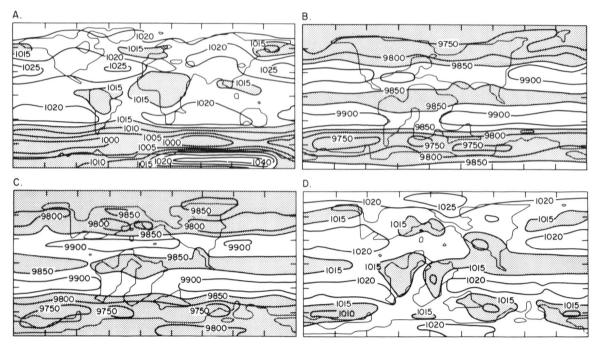

Figure 10—A comparison of predicted surface-pressure distribution for the present-day control and for each of the mid-Cretaceous sensitivity studies.

An examination of the coastal wind vectors (Figure 12) and regions of computed coastal upwelling (Figure 13) predicted for the three sensitivity experiments reveals strong similarities at low latitudes but major differences at higher latitudes. Regardless of the specified geography, the low-latitude easterlies are apparently a stable feature of the circulation. Consequently, the location of coastal upwelling regions is not sensitive to the specified geography as long as the coastline is appropriately oriented. In each of the three Cretaceous sensitivity experiments, the northwestern margin of Africa, the northwestern margin of South America, and part of western South America are characterized by coastal upwelling. The implication is that coastal upwelling conditions at low latitudes may be highly predictable for past geographies.

Two specific examples of major differences in predicted coastal upwelling regions at higher latitudes can be noted in the three sensitivity experiments. The first concerns Tethyan coastal upwelling. For the high sea level cases, the subtropical high-pressure zone is situated over Tethys with an adjacent low-pressure system over the European epicontinental sea. March, January, and July simulations performed by Barron and Washington (1982) show that the positions of these features are in a sense anchored by the presence of a zonal ocean in the subtropics (Tethys). The interesting result is that both of the high sea level cases are characterized by substantial additional coastal upwelling along the Tethyan margins. For the low sea level case, upwelling occurs on a Pacific-facing coast (Mexico) and is almost nonexistent on the south-facing coasts bordering the Tethyan ocean. This result suggests that with sea level changes the location of an upwelling region may not neces-

sarily follow the transgressing or regressing shoreline, but rather may shift in focus entirely because the differences in geography modify the nature of the circulation.

The second example of a major difference in predicted coastal upwelling concerns the continental margins of the Arctic Basin. Unlike the other two sensitivity experiments, the simulation with high sea level and specified Cretaceous topography is characterized by a polar high. For this condition, strong coastal upwelling is predicted for the margins of the Arctic Basin. In fact, much of the coastline has winds conducive to coastal divergence, with about 30% of the coastline having a computed coastal divergence greater than the global mean. The difference between polar high pressure and polar low pressure in the Arctic Basin, which is apparently related to geography, is the difference between a favorable or an unfavorable condition for coastal upwelling given an ice-free Arctic Basin.

DISCUSSION

The results of the present-day model simulation indicate that the model reasonably reproduces observed surface pressure and winds and, by inference, the regions of present-day high productivity. The major deficiencies appear to be too strong and too extensive an equatorial upwelling and the lack of a seasonal cycle in the simulation. The Cretaceous experiments also clearly demonstrate that the model atmosphere is sensitive to changes in geographic boundary conditions, particularly at extratropical latitudes. This result suggests that it is inappropriate to move the continents "beneath" an idealized atmospheric circulation. This last point stresses the importance of

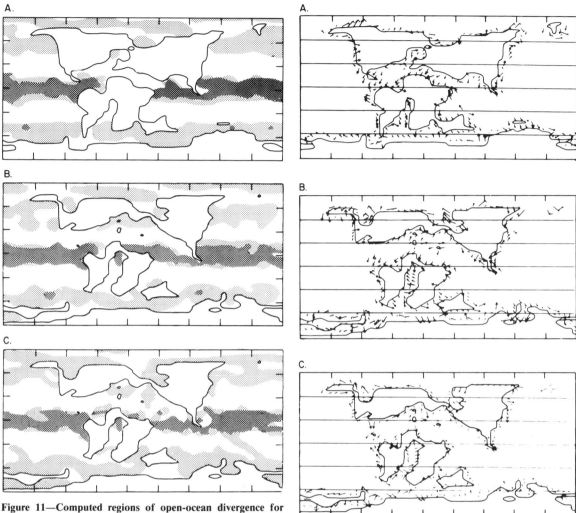

Figure 11—Computed regions of open-ocean divergence for each of the three mid-Cretaceous sensitivity experiments. Dark stipple indicates regions of open-ocean upwelling greater than the mean.

Figure 12—Coastal wind vectors at each model grid point for each of the three mid-Cretaceous sensitivity experiments.

numerical models that consider the laws governing atmospheric and oceanic circulation in developing the predictive associations between climate and petroleum source rock occurrence.

In this regard, a more detailed comparison of the predictions of numerical models and qualitative models is warranted. Parrish and Curtis (1982) applied a qualitative model based on the present zonal circulation (Figure 3) to the prediction of Cenomanian upwelling locations. The Cenomanian geography and topography are similar to the mid-Cretaceous geography and topography used in the present study. In a general sense, the predictions of the qualitative model and the numerical model (Figure 13c) are comparable. In particular, the high-latitude predictions of the two models are quite similar. The low- and middle-latitude predictions differ substantially in many cases, particularly the predictions of the Tethyan margins. Even at low latitudes, important distinctions can be noted.

For example, Parrish and Curtis (1982) predict Cenomanian upwelling off northwestern Africa, but confine the prediction to the region adjacent to Algeria and part of Morocco. The numerical model predicts upwelling that extends equatorward into Mauritania. The numerical modeling result is confirmed both from sedimentologic data and nannoplankton assemblages from the northwestern margin of Africa (e.g., Dean et al, 1978; Tissot et al, 1979; Arthur and Natland, 1979; Roth and Bowdler, 1981).

One major change in the zonal circulation predicted by the numerical simulation with Cretaceous geography is associated with the zonal Tethys ocean, as stated previously. This change in the characteristics of the circulation may explain some of the differences between the predictions of Parrish and Curtis (1982) and the numerical simulations. The close correspondence between predictions of the two models at higher northern latitudes stems directly from the fact that both models are characterized by a polar high. The sensitivity experiments for different Cretaceous sea level and topography indicate that high-latitude surface pressure is quite sensitive to paleogeography. Consequently, with more comparisons of qualitative and quantitative models for different paleogeographies, greater discrepancies will be found at higher latitudes. The utility of the qualitative models is likely to be restricted to low-latitude predictions where trade-wind circulation appears to be relatively stable.

The only major question remaining to establish the utility of numerical climate modeling in petroleum source rock prediction is to demonstrate that the model sensitivity to past geographies is correct, i.e., that the predicted upwelling regions correspond to actual Cretaceous locations. This criterion can only be established through verification of a climate change experiment comparing observations. Verifying the sensitivity of a model is the most difficult problem. Limitations in both the model and the geologic record restrict the ability to verify the model sensitivity.

First, it is conceivable, and perhaps probable, that a seasonal model will have a somewhat different sensitivity to geography than a mean annual model. Thermal contrasts between land and sea are certainly accentuated seasonally, which may alter the pattern of coastal winds. Second, shoreline positions determined from the geologic record are relatively imprecise, and the geographic resolution of the continents is fairly coarse. Coastal divergences are computed for blocklike continental outlines rather than realistic shorelines and do not consider the depth of coastal seas. These factors may also affect the prediction of upwelling locations. Third, and perhaps most importantly, the record of upwelling locations for the Cretaceous is inadequate for verification within reasonable confidence limits for several reasons: (1) not all locations of upwelling during the Cretaceous are preserved as source rocks; (2) many source rocks have no association with upwelling (e.g., anoxic events); (3) many regions have winds conducive to upwelling, but because of local currents, stratification, or variations in bathymetry, a divergence does not occur; and (4) the geologic record of Cretaceous source rocks is not completely sampled. Therefore, verification of the model simulation is problematic.

Despite these limitations, a comparison of Albian-Cenomanian source rock localities and other upwelling indicators with the Cretaceous model predictions appears to be promising. The equatorial upwelling zone and portions of the middle-latitude Southern Hemisphere ocean divergence are associated with siliceous sediments (Ramsay, 1973; Pisciotto, 1981; Miskell, 1983). The Cretaceous high sea level cases are associated with extensive coastal divergence along the Tethyan margin. The Tethyan margin, particularly during the Albian-Cenomanian, is characterized by widespread organic-rich lithologies in continental margin facies (Jenkyns, 1980). Using both sedimentologic data and nannoplankton assemblages (e.g., Dean et al, 1978; Tissot et al, 1979; Arthur and Natland, 1979; Roth and Bowdler, 1981), the northwestern margin of Africa can be interpreted as an upwelling region, as is predicted in the Cretaceous simulations. Similar corre-

A.
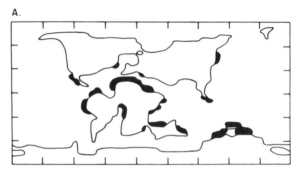

B.
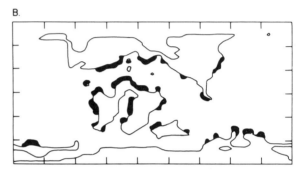

C.
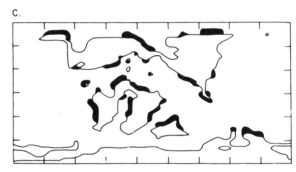

Figure 13—Computed regions of coastal divergence for each of the three mid-Cretaceous sensitivity experiments. Only values of divergence greater than mean values are plotted.

spondence occurs for a source rock locality and upwelling indicators adjacent to the Peruvian Andes and on the western margin of Africa for the Walvis Ridge and the Angola Basin (Jenkyns, 1980; Roth and Bowdler, 1981; Parrish and Curtis, 1982; Miskell, 1983). Mid-Cretaceous organic-rich facies off western Australia from Site 258 (Pisciotto, 1981) to the Carnarvon basin (Jenkyns, 1980) may be associated with predicted open-ocean upwelling in the model simulations. Coastal upwelling predicted in the southern portion of the North American Western Interior seaway and open-ocean divergence in the northern Canadian portion of the seaway may correspond with source rock localities (e.g., Jenkyns, 1980; Parrish and Curtis, 1982). However, many other factors pertain to the deposition of the numerous organic carbon intervals in this region.

Although the comparison of petroleum source rocks and upwelling indicators with model-predicted divergences appears promising, a statement that the predictions accurately represent Cretaceous upwelling regions is premature. Even statistical tests to determine whether the number of source rocks located within predicted upwelling regions exceeds what would be expected from a random distribution are not completely convincing because the samples (source rock discoveries) are not random. The record of mid-Cretaceous source rock localities (e.g., Parrish and Curtis, 1982) is strongly biased toward North America, specifically the Western Interior seaway. Without these data, sample sizes are too small for any statistical confidence. Apparently, source rock predictions based on upwelling determinations from climate models will not be verified confidently without more complete exploration. Enigmatically, such predictions are a key element in the usefulness of climate models for petroleum exploration, but the extent of exploration is a limiting factor in verifying the capabilities of climate models.

Therefore, predictions from climate models for past geographies should not be regarded as fact, but rather as an additional indicator of increased probability of a source rock location in conjunction with the entire suite of more traditional geologic and geophysical data. This increased probability of petroleum source rock location based on climate model predictions combined with a knowledge of other paleo-oceanographic factors can be assessed at several levels. One method might be termed a "bulls-eye approach," in which the highest probabilities are associated with predictions that target the same specific region in each sensitivity experiment. For instance, the most important example stems from predictions for northwestern Africa and South America situated within the stable trade-wind circulation. Perhaps the greatest probabilities can be associated with appropriately oriented coastlines within the stable, low-latitude easterlies. Upwelling also occurs in the Newfoundland region in all of the Cretaceous experiments described here, yet no evidence is found that upwelling occurred in the geologic record. A second approach might be based on specific associations. For example, extensive upwelling predicted for the Tethyan margin is associated with high sea level in the model simulations, as well as in the early South Atlantic Ocean, but not with low sea level. A third approach is to consider only the upwelling predictions for the most realistic Cretaceous simulation. Here again, certain examples are particularly important. The most realistic Cretaceous case is the only simulation that predicts extensive coastal upwelling along the margins of the Arctic Basin.

The results of this study indicate that numerical climate modeling is an important frontier in petroleum source rock prediction. The problems that have been addressed based on the Cretaceous simulations also serve to guide future research and the development of climate models as a tool in petroleum source rock prediction. One major step is to perform a full seasonal simulation that may be readily compared with a whole suite of paleoclimatic variables in addition to known source rock locations. The second step is to broaden the data base used for verification in order to gain greater confidence in the capabilities of the model.

REFERENCES CITED

Arthur, M. A., and J. H. Natland, 1979, Carbonaceous sediments in the North and South Atlantic: the role of salinity in stable stratification of early Cretaceous basins, in M. Talwani, W. Hay, and W. Ryan, eds., Deep drilling results in the Atlantic Ocean; continental margins and paleoenvironments, v. 3: American Geophysical Union, Maurice Ewing Series, p. 375-401.

——— and S. O. Schlanger, 1979, Cretaceous "oceanic anoxic events" as causal factors in development of reef-reservoired giant oil fields: AAPG Bulletin, v. 63, p. 870-855.

Barron, E. J., and W. M. Washington, 1982, Cretaceous climate: a comparison of atmospheric simulations with the geologic record: Palaeogeography, Palaeoclimatology, Palaeoecology, v. 40, p. 103-133.

——— C. G. A. Harrison, J. L. Sloan, and W. W. Hay, 1981, Paleogeography, 180 million years ago to the present: Ecologae Geologicae Helvetiae, v. 74, p. 443-470.

Bourke, W., B. McAvaney, K. Puri, and R. Thurling, 1977, Global modeling of atmospheric flow by spectral methods, in J. Chang, ed., Methods in computational physics, general circulation models of the atmosphere, v. 17: New York, Academic Press, p. 267-324.

Brawlower, T. J., and H. R. Thierstein, in press, Organic carbon and metal accumulation rates in Holocene and mid-Cretaceous sediments: paleooceanographic significance, in J. Brooks and A. Fleet, eds., Marine petroleum source rocks: Oxford, Blackwell Scientific Publications.

Clapham, W. B., 1973, Natural ecosystems: New York, The Macmillan Co., 249 p.

Dean, W. E., J. V. Gardner, L. F. Jansa, P. Cepek, and E. Seibold, 1978, Cyclic sedimentation along the continental margin of northwest Africa, in Y. Lancelot, E. Seibold et al, eds., Abidjan, Ivory Coast, to Malaga, Spain: Initial Reports of the Deep Sea Drilling Project, v. 41, p. 965-986.

Degens, E. T., and P. Stoffers, 1976, Stratified waters as a key to the past: Nature, v. 263, p. 22-27.

Demaison, G. J., and G. T. Moore, 1980, Anoxic environments and oil source bed genesis: AAPG Bulletin, v. 64, p. 1179-1209.

Dow, W. G., 1978, Petroleum source beds on continental slopes and rises: AAPG Bulletin, v. 62, p. 1584-1606.

Fogg, G. E., 1975, Primary productivity, in J. P. Riley and G. Skirrow, eds., Chemical oceanography, v. 2: London, Academic Press, p. 386-455.

Jenkyns, H. C., 1980, Cretaceous anoxic events from continents to oceans: Journal of the Geological Society of London, v. 137, p. 171-188.

Knauss, J. A., 1978, Introduction to physical oceanography: Englewood Cliffs, New Jersey, Prentice Hall, 338 p.

Lorenz, E. N., 1967, The nature and theory of the general circulation of the atmosphere: Geneva, Switzerland, World Meteorological Organization, 161 p.

McAvaney, B. J., W. Bourke, and K. Puri, 1978, A global spectral model for simulation of the general circulation: Journal of the Atmospheric Sciences, v. 35, p. 1557-1582.

Menzel, D. W., 1974, Primary productivity, dissolved and particulate organic matter, and the sites of oxidation of organic matter, in E. D. Goldberg, ed., The sea; v. 5, marine chemistry: New York, John Wiley, p. 659-678.

Miskell, K. J., 1983, Accumulation of opal in deep sea sediments from the mid-Cretaceous to the Miocene: a paleocirculation indicator: Master's thesis, University of Miami, Coral Gables, Florida, 168 p.

Parrish, J. T., 1982, Upwelling and petroleum source beds, with reference to Paleozoic: AAPG Bulletin, v. 66, p. 750-774.

——— and R. L. Curtis, 1982, Atmospheric circulation, upwelling, and organic-rich rocks in the Mesozoic and Cenozoic Eras: Palaeogeography, Palaeoclimatology, Palaeoecology, v. 40, p. 31-66.

Pisciotto, K. A., 1981, Distribution, thermal histories, isotopic compositions, and reflection characteristics of siliceous rocks recovered by the Deep Sea Drilling Project, in The Deep Sea Drilling Project: a decade of progress: SEPM Special Publication 32, p. 129-148.

Pitcher, E. J., R. C. Malone, V. Ramanathan, M. L. Blackmon, K. Puri, and W. Bourke, 1983, January and July simulations with a spectral general circulation model: Journal of the Atmospheric Sciences, v. 40, p. 580-604.

Ramanathan, V., E. J. Pitcher, R. C. Malone, and M. L. Blackmon, 1983, The response of a spectral general circulation model to refinements in radiative processes: Journal of the Atmospheric Sciences, v. 40, p. 605-630.

Ramsay, A. T. S., 1973, A history of organic siliceous sediments in oceans, *in* N. F. Hughes, ed., Organisms and continents through time, v. 12: Special Papers in Palaeontology, p. 199-234.

Robinson, P. L., 1973, Paleoclimatology and continental drift, *in* D. H. Tarling and S. K. Runcorn, eds., Implication of continental drift to the earth sciences, v. 1: London, Academic Press, p. 451-476.

Roth, P. H., and J. L. Bowdler, 1981, Middle Cretaceous calcareous nannoplankton biogeography and oceanography of the Atlantic Ocean, *in* The Deep Sea Drilling Project: a decade of progress: SEPM Special Publication No. 32, p. 517-546.

Schlanger, S. O., and H. C. Jenkyns, 1976, Cretaceous oceanic anoxic events: causes and consequences: Geologie en Mijnbouw, v. 55, p. 179-184.

Scotese, C. R., R. Van der Voo, and W. C. Ross, 1981, Mesozoic and Cenozoic base maps (abs.): AAPG Bulletin, v. 65, p. 989.

Smith, A. G., and J. C. Briden, 1977, Mesozoic and Cenozoic paleocontinental maps: Cambridge, Cambridge University Press, p. 1-63.

Tarling, D. H., 1978, The geological-geophysical framework of ice ages, *in* J. Gribben, ed., Climatic change: Cambridge, Cambridge University Press, p. 3-24.

Tissot, B., 1979, Effects on prolific petroleum source rocks and major coal deposits caused by sea-level changes: Nature, v. 277, p. 463-465.

——— G. Deroo, and J. P. Herbin, 1979, Organic matter in Cretaceous sediments of the North Atlantic, *in* M. Talwani, W. Hay, and W. Ryan, eds., Deep drilling results in the Atlantic Ocean, continental margins and paleoenvironments, v. 3: American Geophysical Union, Maurice Ewing Series, p. 362-374.

Washington, W. M., and G. A. Meehl, 1983, General circulation model experiments on the climatic effects due to a doubling and quadrupling of carbon dioxide concentration: Journal of Geophysical Research, v. 88, p. 6600-6610.

——— and D .L. Williamson, 1977, A description of the NCAR global circulation models, *in* J. Chang, ed., Methods in computational physics, v. 17: New York, Academic Press, p. 111-172.